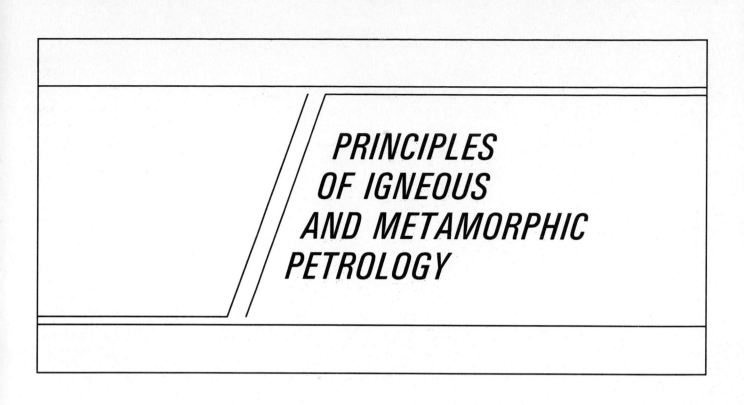

PRINCIPLES OF IGNEOUS AND METAMORPHIC PETROLOGY

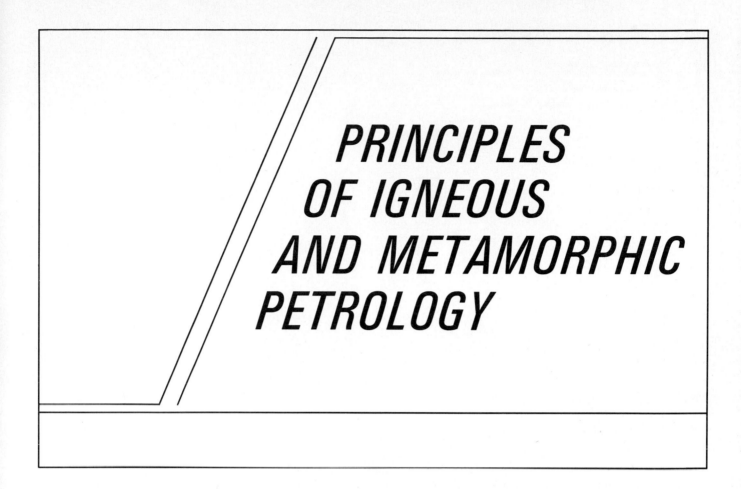

PRINCIPLES OF IGNEOUS AND METAMORPHIC PETROLOGY

Anthony R. Philpotts

Department of Geology and Geophysics
The University of Connecticut

Prentice Hall, Englewood Cliffs, New Jersey 07632

Library of Congress Cataloging-in-Publication Data

Philpotts, Anthony R. (Anthony Robert), (date)
    Principles of igneous and metamorphic petrology / Anthony R. Philpotts.
      p. cm.
    Includes bibliographical references.
    ISBN 0-13-691361-X :
    1. Rocks, Igneous. 2. Rocks, Metamorphic. I. Title.
QE461.P57 1990
552'. 1—dc20   89-48296
                                     CIP

Editorial/production supervision: Debra Wechsler
Cover design: Richard Dombrowski/Donna Wickes
Manufacturing buyer: Paula Massenaro

Figures 11-1, 11-3, 11-4, 11-5, 11-7, 11-8, and 11-9 from H. S. Yoder, ed., *The Evolution of Igneous Rocks: Fiftieth Anniversary Perspectives.* Copyright © 1979 by Princeton University Press. Reprinted with permission of Princeton University Press. Figure 14-15 from R. B. Hargraves, *Physics of Magmatic Processes.* Copyright © 1980 by Princeton University Press. Reprinted with permission of Princeton University Press.

 © 1990 by Prentice Hall
A Division of Simon & Schuster
Englewood Cliffs, New Jersey 07632

Printed in the United States of America

10 9 8 7 6 5 4 3 2

ISBN   0-13-691361-X

Prentice-Hall International (UK) Limited, *London*
Prentice-Hall of Australia Pty. Limited, *Sydney*
Prentice-Hall Canada Inc., *Toronto*
Prentice-Hall Hispanoamericana, S.A., *Mexico*
Prentice-Hall of India Private Limited, *New Delhi*
Prentice-Hall of Japan, Inc., *Tokyo*
Simon & Schuster Asia Pte. Ltd., *Singapore*
Editora Prentice-Hall do Brasil, Ltda., *Rio de Janeiro*

*To my wife, Doreen, who
in the meantime studied
history—all of it.*

CONTENTS

PREFACE

This book is designed to introduce igneous and metamorphic petrology to those who have completed introductory college-level courses in physics, chemistry, and calculus. Its emphasis is on principles and understanding rather than on facts and memorization. With this approach, it is hoped that students will not only gain a sound understanding of petrology but will develop skills that can be applied to the analysis of problems in many other fields of Earth science.

The book is arranged so that it can be used in either a one- or two-semester course. Although it covers both igneous and metamorphic petrology, the contents of chapters allow the text to be used for courses dealing with either of these groups of rocks separately. Such a division, however, is somewhat arbitrary because the principles involved in the formation of these rocks are so similar. Moreover, in the upper mantle, where many petrologic processes have their origins, igneous and metamorphic processes are so interdependent that one cannot be treated without the other. If the book is used for just one of these groups, cross-references will lead the reader to relevant material covered elsewhere in the text.

Today, the basic Earth sciences curriculum at most schools includes at least one freshman-level full-year course in each of physics, chemistry, and calculus. This material forms the basis on which the book is built. Experience indicates, however, that many students have difficulty in applying to the real-world material learned in "pure" mathematics courses. Consequently, most mathematical derivations in the text have been worked through in full. This has also been done because the way in which a problem is formulated and solved is commonly as important as the final equation. By using this approach on tangible geological problems, students can significantly improve their appreciation of the more abstract concepts encountered in mathematics courses; some may even come to enjoy mathematics.

Each chapter ends with a set of problems, which are an integral part of the book. They are not an addendum with which professors can torture students. The importance of these problems to developing an understanding of petrology cannot be overemphasized. Perhaps some measure of their importance is indicated by the fact that the problems took as long to design

and work out as the text took to write. It is also no exaggeration to confess that I have learned more petrology from creating the problems than I have from writing the text. The reader should take note that some important petrologic concepts are discussed only in the problems; this material will be missed if the problems are not done. The text is punctuated at many places by the parenthetic statement (do Prob. x). Do not read beyond such points. Instead, do the problem, for it will provide understanding that is essential to the assimilation of the material to follow. In these cases it was felt that a problem provided the best means of conveying the idea.

Most problems can be done with a pocket calculator, but they are best done using a computer and spreadsheet program. Students with no previous computing experience can learn the use of a spreadsheet in a few hours (one laboratory period). One advantage of the spreadsheet is the speed with which graphs can be created—at the touch of a button—and most problems result in some form of graph. For example, the solution to Prob. 4-9 is a graph showing the cross section of a laccolith (a mushroom-shaped igneous intrusion). Once this problem has been solved using a spreadsheet, the result can be used to investigate how the shape of a laccolith varies with changes in magma pressure, density of overlying rocks, and the depth of intrusion. A new value for any of these variables can be inserted in the spreadsheet, and instantly a graph of the new laccolith results. This is not only fun but is a valuable learning tool. Another advantage in doing problems on the computer is that solutions can be saved on a floppy disk for use in future problems. Solutions to a number of problems early in the book are used again in later problems. For example, the solution to Prob. 2-3, which gives the density of a magma based on its chemical composition, is used numerous times in later chapters.

Throughout the book petrology is introduced mostly in terms of simple models that are amenable to quantitative analysis. These models provide considerable insight into general petrologic processes. This approach differs markedly from that used in most introductory texts, which tend to present the facts about rocks—there are many—and from these to draw qualitative conclusions about processes. In recent years, major advances have been made in petrology through the use of the simple model approach. This in itself is sufficient justification for using this approach in the text. A still more important reason is that it provides a simple means of introducing basic petrologic principles.

To obtain an appreciation of the goals of petrology, it is strongly recommended that after finishing the introduction, readers skim the concluding chapter (Chapter 22). Although some of this material may not be entirely understandable at this stage, it will provide an overview of how the many different topics dealt with in the book relate to the general problem of the origin of rocks.

ACKNOWLEDGMENTS

Probably only those who have written textbooks can fully appreciate the amount of work the preparation of a manuscript involves. The perseverance needed to see this manuscript through to completion came from an interest in petrology that I developed as a student who had the good fortune of being advised by professors with a contagious interest in the subject. A special debt of gratitude is owed professors E. H. Kranck of McGill University, and C. E. Tilley, W. A. Deer, I. D. Muir, S. R. Nockolds, and S. O. Agrell of the University of Cambridge.

The decision to write a petrology book was taken as a result of the annual frustration experienced in searching for a suitable text for the course with which I was involved. The contents of the present book are therefore influenced in no small way by the people with whom I have taught petrology, including Dugald M. Carmichael, A. J. (Mike) Frueh, Norman H. Gray, Ray Joesten, and Randy P. Steinen. It is impossible in the space available to acknowledge in detail how these individuals have influenced this book, but their professional association is gratefully appreciated. I would like specifically to acknowledge Ray Joesten for his elegant derivations of the steady-state geotherm (Eq. 1-14) and the equation describing a reaction surface in terms of T, P, and $X_{CO_2}-X_{H_2O}$. A number of the problems at the end of chapters are also among his favorite exam questions.

Deserving of special recognition are the students who over the years have struggled with the problems given at the end of the chapters. Their questions have helped clarify these problems. Although too numerous to list, a number of students are memorable because of the incisive nature of their questions; these include Pranoti Asher, Chris Doyle, Paul Karabinos, Ingrid Reichenbach, Steve Van Horn, Rebecca Williams (Carmody), and David H. Burkett, who also meticulously edited part of the manuscript.

I would like to thank the following reviewers, not only for their critical comments, but for the encouragement their reviews provided me during the preparation of the manuscript. They are, in alphabetical order: Steven R. Bohlen, U.S. Geological Survey and Stanford University; C. Page Chamberlain, Dartmouth College; Michael L. Cummings, Portland State University; Albert M. Kudo, University of New Mexico; Timothy Lutz, University of Pennsylvania; Jane Selverstone, Harvard University; Robert P. Raiside, Acadia University.

Finally, I would like to thank my family for letting me take root in front of the computer for a couple of years and for keeping me alive in the meantime.

Anthony R. Philpotts
University of Connecticut, Storrs

1 / Introduction

1-1 PETROLOGY AND ITS SCOPE

Petrology is the science dealing with the description, classification, modes of occurrence, and theories of the origins of rocks. Its emphasis is commonly chemical and mineralogical, but it draws heavily on many disciplines, including the basic physical sciences, mathematics, geophysics, structural geology, and geochemistry. Its tools range from the simple hammer and hand lens, to sophisticated devices such as the electron microprobe or the laboratory equipment capable of reproducing conditions deep within the Earth. Its goal is to provide an understanding of the great diversity of rocks found on the surface of the Earth (and other planets), and to provide insight into the nature of those materials within the Earth which are not accessible to direct observation but play such important roles in the Earth's history.

Rocks can be divided into three main groups: igneous, sedimentary, and metamorphic. Those formed from the solidification of molten material are termed *igneous*, whereas those that originate from the deposition of material from water or air are termed *sedimentary*, and those formed from a previously existing rock by some process of change are termed *metamorphic*.

The study of igneous and metamorphic rocks, the subject of this book, is commonly treated separately from the study of sedimentary rocks, mainly because of the different approaches used. Sedimentary rocks are formed by processes which, for the most part, are observable on the surface of the Earth. Careful examination of present-day environments of deposition can, therefore, provide information on the origins of most sedimentary rocks. Igneous and metamorphic rocks, on the other hand, are formed largely by processes operating within the earth and therefore not directly accessible to observation; their origins must, consequently, be deduced through physical–chemical arguments. Also, at the higher temperatures existing within the Earth, reactions proceed more rapidly than on the surface, and thus principles of chemical equilibrium are more applicable to the study of igneous and metamorphic rocks than they are to most sedimentary ones.

Petrologic studies fall into two general categories: the identification and classification of rocks, and the interpretation of these data and the generation of theories on the origin of rocks. The early emphasis in petrology, as in other natural sciences, was on description and classification. There were, nonetheless, many lively discussions concerning the origins of rocks, such as that in the early nineteenth century between Werner and his student von Buch on whether basalt was a sedimentary or volcanic rock. But most of the early work involved the cataloging of the constituents of the Earth's crust. During the latter half of the nineteenth and early part of the twentieth centuries vast amounts of petrologic data were collected, from which came an enormous number of rock names and many different classifications. Despite the surfeit of names, generalizations concerning rock associations and mineral assemblages did emerge, which, in turn, allowed for simplifications in the classifications. Many different rocks that had previously been given separate names could be considered varieties of a single type, and this naturally led to theories explaining the associations. Fortunately, the era of rock naming is over, and modern petrology employs only a small number of rock names.

With recent investigations of the ocean floors, the inventory of rock types available for study in the Earth's crust is almost complete, and today most petrologists are concerned mainly with the genesis of rocks. This change in emphasis has been stimulated by development of experimental techniques that allow us to imitate, in a limited way, rock-forming conditions and processes within the Earth's crust and upper mantle. In fact, with such techniques it is possible to investigate the petrology of parts of the Earth that, while being of great importance as source regions for many igneous rocks, were virtually unavailable for examination and classification by the early petrologists.

Most processes involved in the formation of igneous and metamorphic rocks occur within the Earth and hence are not subject to direct observation (volcanic eruptions are obvious exceptions). Ideas on the origin of these rocks are, therefore, based on interpretations of field observations in the light of experimental and theoretical studies. Nature provides fragmentary evidence of events and processes that have formed rocks, and it is up to the petrologist to assemble this evidence into a coherent story. As new data become available, however, interpretations frequently have to be revised, and this, in itself, makes petrology an active and interesting field for study.

Before beginning a discussion of igneous petrology, it will be useful to review briefly the major structural units of the Earth and the distributions of pressure and temperature within it. This will provide a framework for later discussions of rock-forming processes.

1-2 MAJOR STRUCTURAL UNITS OF THE EARTH

The division of the Earth into three major structural units, *crust*, *mantle*, and *core* (Fig. 1-1), was made early in the history of seismology on the basis of major discontinuities in both compressional (P) and shear (S) wave velocities [see, for example, Press and Siever (1982)]. The discontinuities are best explained as boundaries separating chemically and mineralogically distinct zones. Although their compositions cannot be determined with certainty, reasonable estimates can be made from seismic velocities, the mass and moment of inertia of the Earth, and solar and meteoritic abundances of elements.

The crust contains high concentrations of alkalis, calcium, aluminum, and silicon relative to solar abundances. In continental regions, where the crust is approximately 35 km thick, these elements are so abundant that when combined to form the common minerals quartz and feldspar, the rock approximates the composition of granodiorite. Many light elements and certain heavy ones, such as uranium, thorium, and zirconium, which have difficulty substituting into the structures of common minerals, are also concentrated in this zone. Because some of these elements are radioactive and generate heat, temperature measurements at the Earth's surface can be used to show that this heat-generating granitic zone must be limited to the upper 10 to 20 km of the continental crust (Sec. 1-4 and Prob. 1-4). Increasing seismic velocities with depth in continents indicate the presence of rocks with higher concentrations of iron and magnesium at depth; these are probably basaltic in composition. The oceanic crust, which is only about 11 km thick (including about 5 km of water), is composed largely of basaltic rocks (Fig. 1-2).

The Mohorovičić discontinuity (Moho or M discontinuity), at the base of the crust, marks a sharp increase in seismic velocities which can be accounted for by the disappearance of feldspar. Below this the rock is thought to consist mainly of olivine and pyroxene (and possibly some garnet) and thus is called a peridotite. Samples of this rock are actually brought to the Earth's surface by certain types of explosive volcanic eruption and confirm the seismological deductions as to its composition.

The disappearance of S waves at the mantle–core boundary indicates the presence of liquid below this. P-wave velocities and the density of the core indicate that the liquid is composed largely of iron with some nickel and small amounts of lighter elements such as silicon, oxygen, or sulfur. With increasing depth and pressure this liquid undergoes a phase change to form the solid inner core.

In addition to these traditional structural units, modern seismologists detect many other discontinuities with which they can further subdivide the Earth. Of particular importance to petrology and plate tectonics is a zone extending from depths of approximately 70 to 250 km in which shear-wave velocities markedly decrease; this zone is known as the *low-velocity layer*. Because it has considerably less strength than overlying layers it is also referred to as the *asthenosphere* or weak zone. The more rigid overlying rocks, which include the upper part of the mantle and the crust, constitute the *lithosphere*. The reduced velocities and strengths of rocks in the asthenosphere are thought to result from small degrees of partial melting, possibly as much as 10% in regions of high heat flow. This partial melt can be an important source of magma and a lubricant to ease the tectonic movements of the lithospheric plates.

Between 400 and 1000 km, seismic velocities increase rapidly. Most of this change occurs at two major discontinuities. One at 400 km results from the transformation of olivine to a spinel structure, and the other, at 670 km, marks the conversion of most silicates to a perovskite structure. Below this, velocities increase in a slower and steadier manner down to the mantle–core boundary. This is known as the lower mantle. Little is known about its composition, yet it comprises the largest unit of the Earth.

In later chapters when large numbers are used to describe the depths and pressures under which magmas are generated, it will be easy to lose sight of the fact that we will be dealing with only the extreme outer skin of the Earth. Almost all ter-

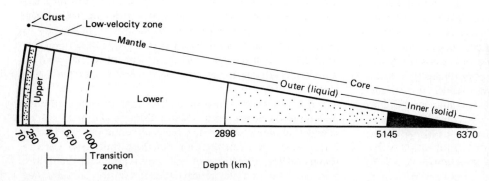

FIGURE 1-1 Major divisions of the Earth based on seismic data. Note that the lithosphere, which is the layer in which almost all rocks that are exposed on the Earth's surface are formed, constitutes a very small fraction of the entire Earth, and that at this scale the thickness of crust is represented by only the outer black line.

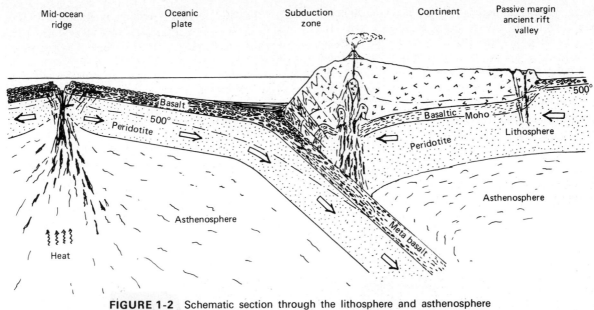

FIGURE 1-2 Schematic section through the lithosphere and asthenosphere showing an oceanic plate moving away from a spreading axis and being subducted beneath a continental plate. Most igneous rocks are formed either at spreading axes or above subduction zones, and most metamorphic rocks are formed in the vicinity of convergent plate boundaries. The range of temperatures and pressures rocks experience are determined largely by plate motions. Note how the depth (pressure) of the 500°C isotherm varies across the section as a function of its position in a plate.

restrial rocks that geologists are able to examine are formed within the upper part of the lithosphere. We, therefore, can examine directly only a minute fraction of the whole Earth. For perspective, then, readers are advised to refer, from time to time, to Figure 1-1.

Despite the small relative volume of the lithosphere, it contains a great variety of rocks and structures. Much of its complexity has been elucidated by modern plate tectonic theories, which show the lithosphere to be a dynamic part of the Earth undergoing constant change. Most igneous and metamorphic rocks owe their origins directly to plate tectonic processes.

Figure 1-2 summarizes the principal components of the lithosphere in terms of their involvement in plate tectonics. The surface of the Earth is divided into 10 major lithospheric plates and several smaller ones. Each moves as a rigid unit with respect to adjoining ones. Where plates move apart, new material must fill the intervening space. Thus, along mid-oceanic ridges, which are divergent plate boundaries, magma rises from the asthenosphere and solidifies to create new lithosphere. Elsewhere material must be consumed if the Earth's circumference is to remain constant. This occurs at convergent plate boundaries where subduction zones form. Plates may also slip past one another along transform faults, as happens on the San Andreas fault.

Lateral and vertical movements of lithospheric plates produce significant lateral variations in the temperature of rocks at equivalent depths in the lithosphere. In Figure 1-2, a line

joining all points at 500°C (the 500°C isotherm) is traced through a typical section of lithosphere. Beneath a stable continental region this isotherm is at a depth of approximately 40 km, a position reflecting a balance between the rate at which heat rises into the lithosphere from the Earth's interior, the rate of heat production from radioactive elements within the lithosphere, and the rate at which heat is conducted to the Earth's surface (Sec. 1-6). At a mid-ocean ridge, the emplacement of large volumes of magma, which have risen from the hotter asthenosphere, moves the isotherm much closer to the Earth's surface. The new, hot lithosphere formed at the ridge cools as it moves away from the spreading axis. The lithosphere consequently thickens, becomes denser, and sinks deeper in the asthenosphere, thus causing oceans to deepen away from ridge axes (Sclater et al. 1975). At subduction zones, the sinking of cool lithospheric plates into the asthenosphere depresses isotherms to considerable depth. Descending plates are eventually heated and become part of the mantle. During heating, descending plates liberate water, which was incorporated during alteration on the ocean floor, and on rising into the overlying asthenosphere, behaves as a flux and causes fusion. Because the melt is less dense than the surrounding rock, it buoyantly rises into the crust, bringing with it heat. This deflects isotherms upward, which, in turn, causes metamorphism and possibly melting of the crustal rocks (Marsh 1979). Well away from the subduction zone, beneath the stable continent, the 500°C isotherm returns to its undisturbed steady-state depth of approximately 40 km.

The variation in the depths of isotherms throughout the lithosphere gives rise to a wide range of possible pressure–temperature conditions. These, in turn, control the processes by which rocks are generated. We therefore turn our attention to the problems of how we determine pressure and temperature within the lithosphere.

1-3   PRESSURE DISTRIBUTION WITHIN THE EARTH

Rocks under high pressure do not have high shear strength, especially over long periods of time. Instead, they flow as if they were extremely viscous liquids. This fact allows us to calculate the pressure for any particular depth in the Earth in the same manner that hydrostatic pressures can be calculated for any depth in water. Indeed, the term *hydrostatic* is often used loosely instead of the correct term *lithostatic* to describe pressures in the Earth resulting from the load of overlying rock. At shallow depths, where open fractures exist or where the rock has high permeability, the pressure on the rock may actually be higher than that on the water in fractures or pores; that is, the lithostatic pressure may be greater than the hydrostatic pressure. On the other hand, exsolving water from crystallizing magma may cause the hydrostatic pressure to exceed the lithostatic pressure and possibly cause explosions.

We will now derive the simple relation between depth and lithostatic pressure in the Earth, assuming that rocks are capable of flowing under high pressures. Although this derivation may appear trivial, it is included here to introduce a particular way of tackling a problem. This same approach will be encountered in later chapters dealing with more complicated problems involving fluid flow, heat transfer, and diffusion.

Consider the forces acting on a small volume of rock at some depth in the Earth. For convenience this volume is given the dimensions dx, dy, and dz (Fig. 1-3), where x and y are perpendicular directions in a horizontal plane, and z is vertical with positive values measured downward (depth). The surrounding rock exerts pressures P_1 to P_6 on each face of the volume. We further stipulate that the pressure at the center of this volume is P, and that there is a vertical pressure gradient, dP/dz.

If the volume does not move horizontally, the forces resulting from the pressures P_3, P_4, P_5, and P_6 must balance one another and need not be considered further. The forces acting in a vertical direction result from the pressures P_1 and P_2 and the acceleration of gravity acting on the mass of the small volume (remember that according to Newton's second law, a force is measured by the mass multiplied by the acceleration that the force produces; $F = ma$). Again, if the volume does not move, the forces must balance; that is, the forces acting in any particular direction must sum to zero. Recalling that the force resulting from a pressure is simply the pressure times the area on which it acts, we can sum the forces acting downward (positive direction of z) on the volume as follows:

$$P_1 \, dx \, dy + (-P_2 \, dx \, dy) + mg = 0$$

where m is the mass of the small volume and g is the acceleration of gravity at the depth considered. Note that the force

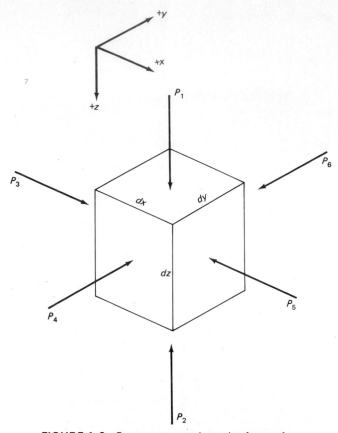

FIGURE 1-3 Pressures exerted on the faces of a small volume (dx dy dz) of rock as a result of the load of the surrounding rocks. Note that the positive direction of z is downward (depth).

resulting from P_2 is negative because it operates in an upward direction.

The mass of the small volume is obviously dependent on its size. We can, however, express mass in terms of the dimensions of the volume and its density (ρ), which allows the force of gravity to be expressed as follows:

$$mg = \rho \, dx \, dy \, dz \, g$$

The two pressures P_1 and P_2 can be related to the pressure P at the center of the volume by the pressure gradient (dP/dz) within the volume. Because the distance from the center of the volume to the top or bottom is $\frac{1}{2}dz$, pressures P_1 and P_2 are given by

$$P_1 = P - \frac{1}{2} dz \left(\frac{dP}{dz}\right) \quad \text{and} \quad P_2 = P + \frac{1}{2} dz \left(\frac{dP}{dz}\right)$$

We can now write the sum of the forces acting downward on the volume as

$$\left[P - \frac{1}{2} dz \left(\frac{dP}{dz}\right)\right] dx \, dy - \left[P + \frac{1}{2} dz \left(\frac{dP}{dz}\right)\right] dx \, dy$$
$$+ \rho \, dx \, dy \, dz \, g = 0$$

which simplifies to

$$\frac{dP}{dz} = \rho g \qquad (1\text{-}1)$$

This indicates that the pressure gradient at any particular depth is simply the product of the density of the material and the acceleration of gravity at that point. To obtain the pressure at that depth we have simply to integrate the expression

$$\int_{P_0}^{P_z} dP = \int_0^z \rho g \, dz$$

from the surface of the Earth to the depth, z, of interest. The pressure at the Earth's surface of 1 atm is so small compared with those at depth that it can be taken as zero. We can then write

$$P_z = \int_0^z \rho g \, dz$$

To integrate the right-hand side of this equation, we must first determine whether ρ and g are constants or functions of depth (z). Most minerals and magmas are rather incompressible. Thus, no serious error is introduced by assuming that their densities remain constant, at least under the pressures encountered in the crust and upper mantle. Large amounts of data do, however, exist on compressibilities (Birch 1966). Therefore, density can be expressed as a function of pressure, if so desired (see Prob. 3-1). We will take it to be constant here.

Variation in the value of q with depth is more complex. At points above the surface of the Earth its value varies inversely as the square of the distance from the center of the Earth. Within the Earth, however, only the underlying mass contributes any net gravitational force; those forces due to the overlying shell of rock sum to zero and hence produce no net gravitational attraction. Therefore, with increasing depth in the Earth, less and less mass remains to cause attraction. But this is partially offset by the increasing density of rocks with depth. The result is that the value of g throughout the crust and upper mantle remains approximately constant, and will be taken as such for our purposes.

Taking the density of material and the acceleration of gravity as constants, we integrate the preceding equation to obtain

$$P = \rho g z \qquad (1\text{-}2)$$

To illustrate this equation, let us determine the pressure at the base of a 35-km-thick granitic crust with an average density of 2800 kg m^{-3}. Substituting these values into Eq. 1-2 and being careful to list units, we obtain

$$P = 2800 \times 9.80 \times 35 \times 10^3 \text{ kg m}^{-3} \times \text{m s}^{-2} \times \text{m}$$

But a kg m s^{-2} is a unit of force, the *newton* (N). The units therefore reduce to N m^{-2}, which is the unit of pressure known as a *pascal* (Pa). The pressure at the base of a 35-km-thick crust is therefore 0.96×10^9 Pa or 0.96 GPa. In cgs units this would

be 9.6 kilobars:

$$1 \, bar = 10^6 \text{ dyn cm}^{-2} = 0.9869 \, atmosphere$$

The pressure at the base of the 35-km-thick crust in this simple calculation is approximately 1 GPa. By taking into account details of the density distribution within the crust, a more accurate determination of pressure can, of course, be obtained (see Prob. 1-1).

1-4 TEMPERATURE GRADIENTS AND HEAT FLOW IN THE LITHOSPHERE

Temperatures in the Earth cannot be determined as easily as pressures. Deep drill holes provide access to the top few kilometers of the crust, but extrapolations from temperatures measured here to lower parts of the crust and the upper mantle are fraught with difficulties. It is not surprising to find that the literature contains numerous, significantly different estimates of the geothermal gradient. Much interest in the internal temperatures of the Earth was piqued in the nineteenth century by attempts, such as that of Lord Kelvin, to calculate the age of the Earth based on cooling rates. A major source of heat, that produced by radioactive decay, was, of course, unknown to these early workers, and thus their calculated ages and temperature gradients were wrong. Today, radiogenic heat is certainly taken into account; nonetheless, markedly different gradients can still be calculated, depending on the assumed distribution of the heat-generating elements (Prob. 1-5). Fortunately, high-pressure experimental investigations of the melting behavior of crustal and upper mantle rocks provide rather tight constraints on the geothermal gradient.

Measurements in deep mines and drill holes indicate that the near-surface geothermal gradient, dT/dz, ranges from 10 to 60°C km^{-1}, with a typical value in nonorogenic regions being near 25°C km^{-1} (Fig. 1-7). If this gradient continued to depth, the temperature would be 625°C at 25 km (0.65 GPa). Experiments indicate that in the presence of water, crustal rocks melt to form granitic magma under these conditions. The transmission of seismic shear waves through this part of the crust indicates that melting at this depth is not a common or widespread phenomenon. Similarly, at a depth of 40 km (1.2 GPa), a geothermal gradient of 25°C km^{-1} would give a temperature of 1000°C, and peridotite, the rock constituting the mantle at this depth, would begin to melt in the presence of excess water. By 52 km (1.6 GPa) the temperature would be 1300°C, which exceeds the beginning of melting point of even dry peridotite. Again, seismic data do not indicate large-scale melting at this depth. Clearly, the near-surface geothermal gradient must decrease with depth. The question, then, is why should the gradient decrease, and can we calculate or predict its change?

A combination of petrological and geophysical observations allows us to place limits on the temperature of several points along the geotherm. Seismic velocities decrease abruptly by about 10% at a depth of 60 to 100 km, the upper boundary of the low-velocity zone. This is probably the contact between solid peridotite above and peridotite that has undergone a small percentage of melting below. A thin film of melt along grain boundaries can greatly reduce the rigidity of a rock and strongly attenuate seismic energy. Experiments show that water-saturated peridotite begins to melt at about 1000°C at a pressure of 2 GPa, which corresponds to a depth of 70 km (Fig. 1-7). The top of the low-velocity zone is the boundary between the lithosphere and the asthenosphere. The base of the low-velocity zone, at 250 km, is marked by a small increase in the rate of increase of velocity with depth. This is interpreted as the depth at which the *P–T* curve for the beginning of melting of mantle peridotite re-crosses the geotherm, so that at depths greater than 250 km, the mantle is solid. This places an upper limit of about 1500°C at 250 km. The *P–T* curve for the melting of iron limits temperature to a minimum of 5000°C at the core–mantle boundary (2898 km). The melting curve for iron crosses the geotherm at 5145 km, the boundary between the liquid outer core and the solid inner core, fixing the temperature at this depth to about 6500°C.

Common experience teaches us that heat flows from regions of high temperature to ones of low temperature, and that it is transferred in several ways depending on the nature of the medium through which it is transmitted. For example, in a vacuum, heat can be transferred only by radiation; in a gas or low-viscosity liquid, it may be transferred by convection, as, for example, in a pot of boiling water; and in rigid, opaque solids it can be transferred only by conduction. These mechanisms are discussed in detail in Chapter 5. Heat transfer through the lithosphere, however, especially in old, stable continental and oceanic regions, is almost entirely by conduction.

Experience also shows that the amount of heat transferred by conduction is proportional to the negative temperature gradient ($-dT/dx$); that is, the greater the temperature decrease in a particular direction x, the greater will be the amount of heat transferred in that direction. This can be expressed mathematically by introducing the quantity known as *heat flux*, J_{Qx}, which is the quantity of thermal energy passing in the x direction through a unit cross-sectional area in a unit of time. We can write this as

$$J_{Qx} = -K\left(\frac{dT}{dx}\right) \qquad (1\text{-}3)$$

where K, the constant of proportionality, is known as the *coefficient of thermal conductivity*. In the case of the Earth, heat flows out because of the geothermal gradient. This gradient (dT/dz) is normally recorded as a positive value with z increasing downward. In the upward direction, however, this gradient is negative, so that a positive quantity of heat is transferred upward and out of the Earth.

Direct measurement of heat fluxes in the field is not practical, but temperature gradients, especially those in the ocean floor and in deep drill holes on continents, can be measured easily, and from these the heat flux can be calculated. If samples of the material in which the gradient is measured are brought back to the laboratory and their coefficient of thermal conductivity measured, the heat flux can be calculated from Eq. 1-3. For example, with a coefficient of thermal conductivity of 2.0 J m⁻¹ s⁻¹ °C⁻¹, a gradient of 25°C km⁻¹ results in a heat flux of 50.0 × 10⁻³ J m⁻² s⁻¹; but a joule per second is

a watt, so the heat flux is 50.0 mW m⁻². By contrast, a geothermal gradient of 60°C km⁻¹, which can be found on mid-ocean ridges, would result in a heat flux of 120.0 mW m⁻², for a similar thermal conductivity. In *cgs* units heat flow is measured in μcal cm⁻² s⁻¹, which is commonly referred to as 1 *heat flow unit* or 1 *HFU*; 1 HFU = 41.84 mW m⁻².

A synthesis of worldwide heat flow data reveals certain regularities in the variations of heat flux in both oceanic and continental regions. Intrusion of basaltic magma along divergent plate boundaries produces steep geothermal gradients and high heat fluxes. Indeed, it is estimated that 60% of all heat lost from the Earth is liberated during the creation of new crust at mid-ocean ridges. Although heat fluxes along the ridges vary considerably, depending on the proximity to local volcanic sources, the mean value in crust younger than 4 Ma (million years) is 250 mW m⁻². Away from ridges, heat flux decreases rapidly until it attains a relatively constant value of 38 mW m⁻² in the oldest ocean floor (200 Ma). The values of heat flux in oceanic regions are found to be very nearly inversely proportional to the square root of the age of the crust. This variation is interpreted simply in terms of the cooling of the lithosphere as it moves away from the oceanic ridges (Oxburgh, 1980). The crust becomes denser as it cools, and isostatic readjustments cause the ocean floor to sink deeper with increasing age (Fig. 1-4).

On continents, heat flux also decreases with increasing age, at least back to about 800 Ma. But the rate of decrease is not what would be expected from conductive cooling of the crust since the time of its formation, as is the case for oceanic crust. The relation is more complex, possibly involving the removal, by erosion, of heat-generating radioactive elements that

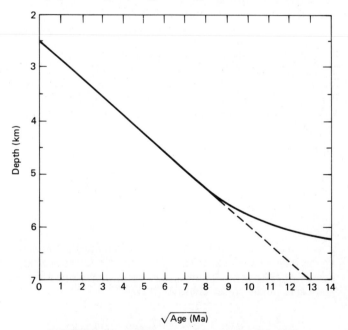

FIGURE 1-4 Variation in the mean depth of the North Pacific and North Atlantic Oceans as a function of the age of the ocean floor. With the exception of the oldest ocean floor (>75/Ma), the depth is proportional to the square root of the age. [After Parsons and Sclater (1977).]

are concentrated in the upper parts of young crust. In young orogenic belts, heat fluxes may be as high as 150 mW m^{-2}, but in crust older than 800 Ma, heat fluxes tend to be about 40 mW m^{-2}, which is similar to the value in ancient oceanic crust.

One of the most important findings to come from the study of heat flow is the relation between the surface heat flux and the concentration of heat-generating radioactive elements (Roy et al. 1968; Lachenbruch and Sass 1977). The heat flux from old eroded plutonic bodies of igneous rock is very nearly linearly related to the local concentration of heat-generating elements, as shown in Figure 1-5. This linear relation can be expressed mathematically as

$$J_Q^0 = J_Q^r + DA_0 \qquad (1\text{-}4)$$

where J_Q^0 is the surface heat flow (depth = 0), J_Q^r the intercept at $A_0 = 0$, A_0 the local radiogenic heat productivity, and D the slope of the line.

The simplest interpretation of this relation is that the slope of the line indicates the thickness of the heat-generating layer. Thus the heat flux at any locality can be interpreted as consisting of two parts; a constant flux of heat, which is given by the intercept J_Q^r, comes from below the layer; to this is added the heat flux generated by radioactive decay in the layer. This type of relation is known as an energy balance or *conservation* equation, for it states that the energy coming out of the top of the layer must equal the energy entering the layer from below plus the energy created within the layer, that is,

$$J_Q\Big|_{z=0}^{\text{out}} = J_Q\Big|_{z=D}^{\text{in}} + \int_{z=0}^{z=D} A_z \, dz \qquad (1\text{-}5)$$

Values of J_Q^r, which are known as the *reduced heat flow*, are characteristic of a given geological province. For example, its value in the eastern United States is 33 mW m^{-2}, in the Basin and Range Province it is 59 mW m^{-2}, and in both the Baltic and Canadian Precambrian Shields it is 22 mW m^{-2}. Values of D, the thickness of the layer containing the radio-

active elements, in these same provinces are 7.5, 9.4, 8.5, and 12.4 km, respectively. The fact that the surface concentration of radiogenic elements cannot extend to depths much greater than 10 km is surprising considering that seismic data indicate that "granitic" rocks at the surface extend to depths of approximately 25 km under most continental areas. Clearly, the radioactive elements must be concentrated near the surface of the Earth. This being so, our model of a layer with a constant concentration, A_0, of radioactive elements throughout its thickness D is unlikely to be valid. The concentration is more likely to decrease downward (see Prob. 1-6).

1-5 HEAT SOURCES IN THE EARTH

Heat flows into the base of the crust from the mantle below in response to temperature gradients that were set up early in the Earth's history. Conversion of kinetic to thermal energy during the Earth's accretion, release of gravitational energy on separation of the core, and generation of heat by decay of short-lived radioactive isotopes, such as ^{26}Al, have all contributed to the thermal energy of the core and mantle. This source accounts for 45 to 75% of surface heat flow through old continental crust and 75 to 90% of surface heat flow through old oceanic crust.

Heat is generated in the crust by the decay of the long-lived radioactive isotopes, ^{235}U, ^{238}U, ^{232}Th, and ^{40}K. Because the concentration of these elements is very low in rocks of the oceanic crust, radiogenic heat contributes no more than 10 to 25% of the heat flux in oceanic regions. In the continental crust, however, radiogenic heat is a major source of thermal energy.

The kinetic energy of particles emitted from the nucleus of radioactive nuclides is dissipated and transformed into heat by collisions with surrounding atoms. Alpha (α) particles ($^4_2$He) emitted by U and Th isotopes have a mass of 4 units, whereas beta (β) particles (electrons emitted from the nucleus) emitted by ^{40}K during radioactive decay to ^{40}Ca have a mass of only 1/1833 unit. ^{40}K also decays to ^{40}Ar by the process of electron capture by the nucleus (see Sec. 21-3), but this does not contribute to heat generation. Because of the difference in mass, a disintegration involving an alpha particle gives off about 7300 times more heat than one involving a beta particle. The concentration of K in crustal rocks, however, is about 10,000 times that of U or Th, so that the contribution to the total radiogenic heat productivity of each of these elements is about the same (see Table 1-1 and Prob. 1-3).

The rate of radiogenic heat production per unit volume of rock, A, is the sum of the products of the decay energy of each of the isotopes present, e_i, and the isotope's concentration in the rock, c_i. This can be written as

$$A = \rho \sum_i e_i c_i \frac{\text{kg}}{\text{m}^3} \frac{\text{J}}{\text{kg s}} \frac{\text{kg}}{10^6 \text{ kg}} \quad \text{or} \quad \frac{\mu\text{W}}{\text{m}^3} \qquad (1\text{-}6)$$

Density appears as a factor in Eq. 1-6 to convert units of power per unit mass to power per unit volume. The concentrations are given in parts per million by weight (ppm = kg/10^6 kg). Values of radiogenic heat productivity for the major rock units of the lithosphere are given in Table 1-1 along with the data

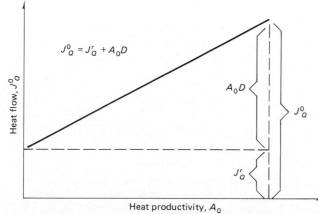

FIGURE 1-5 Heat flow plotted as a function of the radiogenic heat production of surface rocks. [After Roy et al. (1968).] See text for discussion.

TABLE 1-1

Radiogenic heat production[a]

	Concentrations (wt ppm)			Density $(kg\ m^{-3})$	Rate of Radiogenic Heat Production, $A\ (\mu W\ m^{-3})$	Thermal Conductivity, $K\ (W\ m^{-1}\ K^{-1})$
	U	Th	K			
Granite	4.7	20.	36,000	2650	2.95	2.93
Basalt	0.9	2.2	15,000	2800	0.56	2.09
Average crust	1.55	5.75	15,000	2800	1.00	2.51
Peridotite	0.019	0.05	59	3150	0.01	3.35

[a] Decay energies in mW kg^{-1}: U = 9.66×10^{-2}, Th = 2.65×10^{-2}, K = 3.58×10^{-6}.

needed for their computation with Eq. 1-6. Values of A lie in the range 0.008 to 8.0 μW m^{-3}. In the *cgs* system of units, the quantity 1×10^{-13} cal cm^{-3} s^{-1} is known as 1 *heat generation unit* or 1 *HGU*; 1 HGU = 0.4184 μW m^{-3}.

1-6 TEMPERATURES IN THE LITHOSPHERE: THE STEADY-STATE GEOTHERM

We have seen in Sec. 1-4 that the downward extrapolation of near-surface geothermal gradients leads to excessively high temperatures at depth. The gradient must therefore decrease with depth. Several factors contribute to the flattening of the gradient. In continental regions, the large contribution of radiogenic heat to the total heat flux is a major cause for the flattening. Convection of the solid (but plastic) mantle beneath the lithosphere maintains the temperature at or near that of the adiabatic gradient of about 0.6°C km^{-1}. At temperatures above 1200°C, increased transparency of mantle silicate minerals to infrared radiation allows significant heat to be transferred by radiation, thus flattening the gradient in the lower mantle.

Because simple linear extrapolation of the near-surface geothermal gradient leads to erroneous temperatures at depth, it is desirable to develop a more reliable extrapolation that accounts for the change in gradient with depth. Although this is difficult to do for lithosphere in the vicinity of tectonic plate boundaries, it can be done easily for old continental and oceanic areas, where the geothermal gradient has attained a *steady state*; that is, it does not vary significantly with time.

If the geotherm is in a steady state, temperature can vary only as a function of depth; that is,

$$T = T(z) \quad (1\text{-}7)$$

A useful means of approximating a function beyond the region in which its value is known is with a Taylor series expansion (Turcotte and Schubert 1982b). This expansion can be used if $T(z)$ is a function with continuous derivatives of all orders. Physically, this means that the function must vary gradually—there can be no abrupt changes in its value. Experience tells us that this is true of the geotherm. If an abrupt step in the temperature gradient could be produced, heat would transfer from the region of high temperature to that of low temperature until

a smooth gradient was produced. Because we are considering only steady-state geotherms, $T(z)$ will be a function with continuous derivatives of all orders. Given temperature $T_{z=0}$ at the Earth's surface, we expand T downward with a Taylor series in z from $z = z_0$ to $z = z$. Temperature at depth, z, is given by

$$T_z = T_{z_0} + \frac{dT}{dz}(z - z_0) + \frac{1}{2!}\frac{d^2T}{dz^2}(z - z_0)^2$$
$$+ \cdots + \frac{1}{n!}\frac{d^nT}{dz^n}(z - z_0)^n \quad (1\text{-}8)$$

In any complex equation it is worth having a qualitative idea of the meanings of each group of terms. In Eq. 1-8, for example, the first term is the temperature at the Earth's surface. The second group of terms includes the thermal gradient, which is the change of temperature with depth (first derivative). Note that the exclamation mark indicates factorial; that is, $n! = 1 \times 2 \times 3 \times \cdots \times n$. The third group of terms includes the change of gradient with depth (second derivative). The series includes higher-order derivatives, but a good approximation can be obtained with just the first three terms.

To solve Eq. 1-8, we must evaluate the derivatives. From Eq. 1-3 we have

$$\frac{dT}{d(-z)} = -\frac{J_Q^z}{K} \quad \text{or} \quad \frac{dT}{dz} = \frac{J_Q^z}{K} \quad (1\text{-}9)$$

The $(-z)$ is required in the first term to indicate that the gradient used in Eq. 1-3 is in the direction of heat flow, that is, upward $(-z)$. The geothermal gradient, dT/dz, is then a positive quantity. We can then write the second derivative as

$$\frac{d^2T}{dz^2} = \frac{1}{K}\frac{dJ_Q^z}{dz} \quad (1\text{-}10)$$

If we assume that the concentration of heat-producing elements is constant with depth over the interval $z = 0$ to $z = D$, the conservation equation (1-5) becomes

$$J_Q^z = J_Q^D + A(D - z) \quad (1\text{-}11)$$

and

$$\frac{dJ_Q^z}{dz} = -A \quad (1\text{-}12)$$

Thus,

$$\frac{d^2 T}{dz^2} = -\frac{A}{K} \qquad (1\text{-}13)$$

and the Taylor series approximation to temperature becomes

$$T_z = T_{z_0} + \frac{J_Q^{z_0}}{K}(z - z_0) - \frac{A_0}{2K}(z - z_0)^2 \qquad (1\text{-}14)$$

Note that if the temperature at depth z has units of °C, the units of each group of terms on the right-hand side of Eq. 1-14 must also be °C.

The inclusion of successive terms in this series provides ever-closer approximations to the true value of T (Fig. 1-6). The first three terms take into account possible distributions temperature at any depth z with the temperature on the surface of the Earth (T_{z_0}). The first two terms overestimate T, because they linearly extrapolate the near-surface gradient to depth. The first three terms take into account possible distributions of radioactive heat-generating elements, and it may converge closely with the true value of $T(z)$. Incorporation of higher terms in the series would improve the fit still more, but these cannot be evaluated without additional information on the variation of A with z.

Heat flow through oceanic lithosphere approaches a steady state after 180 Ma, with a surface heat flow of 38 ± 4 mW m^{-2} and a reduced heat flow into the base of the crust from the mantle of 25 to 38 mW m^{-2} (Sclater et al. 1980). These limiting values can be used along with other data in Table 1-2 to calculate steady-state geotherms with Eq. 1-14. The results are shown in Figure 1-7. Note that a relatively small difference

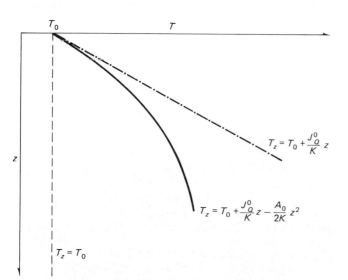

FIGURE 1-6 Plots showing how inclusion of additional terms in the Taylor series expansion of temperature as a function of depth gives closer approximations to the actual geothermal gradient. The dashed line includes only the first term of the series, the dashed–dotted line includes the first and second terms, and the solid line includes the first three terms.

TABLE 1-2

Parameters used in calculating the geothermal gradients in Figure 1-7*[a]

Depth (km)	Ocean Geotherms		Continental Geotherms	
	O_1	O_2	C_1	C_2
0—	$J_Q^0 = 34$	$J_Q^0 = 42$	$J_Q^0 = 46$	$J_Q^0 = 46$
	$A = 0.84$ $K = 2.51$	$A = 0.42$ $K = 2.51$	$A = 2.09$ $K = 2.51$	
10—				
20—	$J_Q^r = 25.1$ $A = 0$ $K = 3.35$	$J_Q^r = 37.7$ $A = 0$ $K = 3.35$	$A = 0.261$ $K = 2.51$	$A = 0.31$ $K = 2.51$
30—				
40—				
			$J_Q^r = 31$ $A = 0$ $K = 3.35$	$J_Q^r = 34$ $A = 0$ $K = 3.35$
50—				

* After Sclater et al (1980).
[a] Units are as follows: J_Q (mW m^{-2}); A (μW m^{-3}); K (W m^{-1} K^{-1}).

in near-surface geothermal gradient (13°C km^{-1} for ocean geotherm 1, O_1, and 17°C km^{-1} for O_2) leads to significantly different temperatures at depth.

Heat flow through stable continental shields approaches a steady state after about 800 Ma, with a surface heat flow of 46 ± 17 mW m^{-2}, and a best estimate of the reduced heat flux from the mantle being 21 to 34 mW m^{-2} (Sclater et al. 1980). Geotherms computed with Eq. 1-14 and data from Table 1-2 are shown in Figure 1-7. The models for continental lithosphere differ in the relative contribution of heat generated in the crust and heat transported from the mantle. Concentration of heat-producing elements in the crust leads to a significant contribution to surface heat flow, thus diminishing the flux from the mantle needed to produce the observed flux at the surface. The net result is lower temperatures at depth than would be produced for a given surface heat flow through a crust depleted in U, Th, and K (see Prob. 1-5).

Two important facts are evident from the calculated geotherms. First, despite differences in surface heat flow, crustal heat productivity, and heat flow into the base of the crust, the range of temperatures at depth beneath old ocean and old continents overlap. Second, temperatures in the oceanic and in the continental crust are too low to produce magmas or metamorphic rocks. The eroded cores of orogenic belts contain rocks that have been formed at temperatures well above the calculated geotherms. For instance, large granite batholiths are formed at temperatures between 700 and 900°C, and metamorphic rocks in the core of these belts commonly exhibit evidence of partial melting, which also requires temperatures in excess of 700°C. Even where metamorphic rocks have not been melted, mineral assemblages provide evidence of high temperatures.

The Al_2SiO_5 polymorphs, andalusite, kyanite, and sillimanite (A, K, and S in Fig. 1-7), which are common metamorphic minerals in pelitic rocks (shales), are particularly useful indicators of metamorphic temperature and pressure. Kyanite

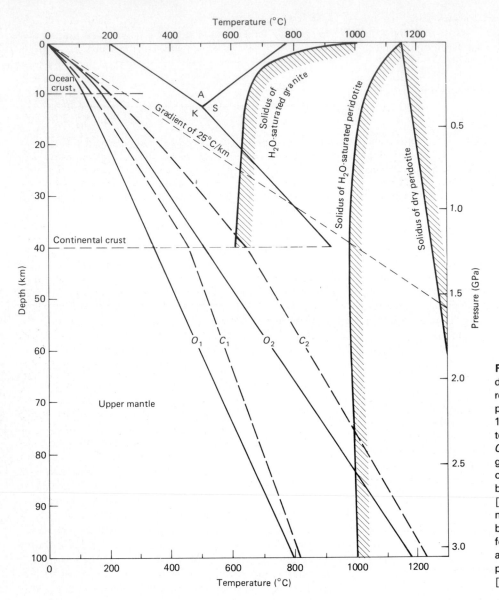

FIGURE 1-7 Possible temperature distributions in a stable lithosphere (far removed from spreading or convergent plate boundaries) calculated using Eq. 1-14 and data in Table 1-2. [After Sclater et al. (1980).] Two oceanic (O_1 and O_2) and two continental (C_1 and C_2) geotherms are shown to cover the range of possible values. Also included are stability fields of the Al_2SiO_5 polymorphs [kyanite, K; andalusite, A; and sillimanite, S—after Holdaway (1971)], the beginning of melting curve (solidus) for water-saturated granite [after Huang and Wyllie (1973)], and the solidus of peridotite both dry and water-saturated [after Kushiro et al. (1968)].

is the stable phase at high pressures and low to moderate temperatures; andalusite is stable at low pressures and moderate temperatures; and sillimanite is the high-temperature phase. All three polymorphs can coexist at a "triple point" at about 500°C and 0.375 GPa (Fig. 1-7). With increasing metamorphic intensity in rocks, these polymorphs are found to change from andalusite to kyanite, or from andalusite to sillimanite, or from kyanite to sillimanite. Examination of Figure 1-7 reveals that the calculated continental geotherms lie entirely in the stability field of kyanite. Indeed, geotherms could pass through the andalusite or sillimanite fields only if the heat flux into the crust from the mantle were substantially increased. We will see that this condition is met above subduction zones (Fig. 1-2) and results from the rise of mantle-derived magmas into the crust. Elevated temperatures also exist in regions of lithospheric extension due to the rise of hot mantle from the asthenosphere. It is these transient effects that are responsible for the formation of rocks. They are consequently of paramount importance

to petrology but cannot be dealt with in our simple steady-state model for geothermal gradients. We will, however, return to their consideration in Chapter 22.

It was suggested above that the top of the low-velocity zone marks the intersection of the geotherm with the pressure–temperature curve for the beginning of melting of mantle peridotite. Such an interpretation is consistent with the calculated geotherms (Fig. 1-7), which intersect the H_2O-saturated beginning of melting curve (solidus) for peridotite between 75 km (C_2) and 150 km (O_1).

Because the geotherms intersect the solidus at a small angle, slight differences in heat flux correspond to large differences in lithospheric thickness. For example, the difference in surface heat flow of 8.4 mW m^{-2} between the two calculated oceanic geotherms leads to a 70-km difference in lithospheric thickness. In general, high surface heat flow causes the geotherm to intersect the peridotite solidus at a shallower depth, whereas low surface heat flow causes it to intersect at greater depth. If worldwide values of surface heat flow are converted into depths of intersection with the peridotite solidus, the lithosphere is found to be up to 300 km thick beneath Precambrian shields

but thins to about 50 km beneath oceanic ridges (Pollack and Chapman 1977). The continents, therefore, resemble ships with very large lithospheric keels descending into the asthenosphere.

PROBLEMS

1-1. Calculate and graph the pressure–depth relation in the outer part of the Earth, where it consists of an upper 25-km-thick granitic layer of density 2.75 Mg m^{-3}, underlain by a 10-km-thick lower basaltic crust of density 3.0 Mg m^{-3}, which in turn is underlain by an upper mantle of density 3.3 Mg m^{-3}; this density continues to a depth of 400 km, where it increases to 3.6 Mg m^{-3}.

1-2. If the continental lithosphere described in Prob. 1-1 is juxtaposed an oceanic one consisting of 5 km of ocean water ($\rho = 1.025$ Mg m^{-3}) and 5 km of basaltic rock ($\rho = 3.0$ Mg m^{-3}), and both sections are underlain by mantle with density of 3.3 Mg m^{-3}, calculate the height of the continent above sea level, assuming isostatic equilibrium.

1-3. Calculate the radiogenic heat productivity of basalt in μW m^{-3} given the following data:

	e_i (mW kg^{-1})	c_i (ppm)
U	9.66×10^{-2}	0.9
Th	2.65×10^{-2}	2.2
K	3.58×10^{-6}	15,000

The density of basalt is 2.8 Mg m^{-3}. Compare individual contributions of U, Th, and K to the total heat productivity.

1-4. If the surface heat flow is 46 mW m^{-2}, compute a steady-state geotherm to 100 km for continental lithosphere having the following properties:

Layer	Thickness (km)	A (μW m^{-3})	K (W m^{-1} K^{-1})
1	10	2.1	2.51
2	30	0.26	2.51
3	60	0.0	3.35

1-5. To examine the effect of radioactive heat generation in the crust, consider a section of crust, 40 km thick, with heat productivity of 2.1 μW m^{-3} and a surface heat flow of 46 mW m^{-2}. K is taken to be 2.51 W m^{-1} K^{-1}. Compute the geotherm within the crust. Comment on the origin of the temperature maximum at a depth of 22 km.

1-6. Compute a steady-state geotherm for a 40-km-thick lithosphere in which the concentration of heat-generating radioactive components decreases exponentially with depth; that is, $A = A_0 e^{-z/D}$, where $A_0 = 2.1$ μW m^{-3}, $D = 40$ km, and the surface heat flow is 46 mW m^{-2}; K is taken to be 2.51 W m^{-1} K^{-1}. Compare the geotherm with that calculated in Prob. 1-4.

1-7. The surface heat flow in the Basin and Range province of the western United States is unusually high—87.86 mW m^{-2}. The heat flow–heat productivity relation for this region is modeled with a 9.4-km-thick crustal layer in which radioactive heat-producing elements are concentrated. For a radiogenic heat productivity for this layer of 2.1 μW m^{-3}, compute the reduced heat flow into the base of the 9.4-km-thick layer from the mantle below. Comment on the value of the reduced heat flow.

1-8. Taking the pressure at the surface of the Earth to be P_0, use a Taylor series to expand pressure (P_z) as a function of depth (z). Determine and discuss the physical significance of the first and second derivatives in this series expansion.

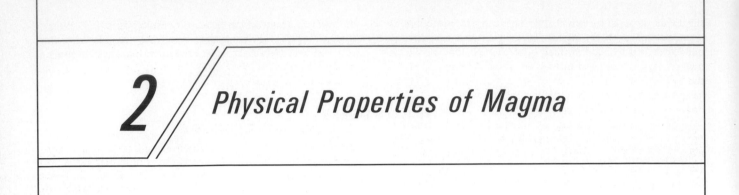

2 / *Physical Properties of Magma*

2-1 INTRODUCTION

Igneous rocks are formed from molten material known as magma, which usually consists of a solution of the Earth's most abundant elements, oxygen and silicon, with smaller amounts of aluminum, calcium, magnesium, iron, sodium, and potassium. Most magmas are therefore silicate melts, but rarer ones contain little or no silica and instead are composed essentially of calcium carbonate, sulfide, or iron oxide. The major elements of common magmas combine, on cooling, to produce the so-called *rock-forming minerals*, quartz, feldspars, feldspathoids, pyroxenes, olivines, and when water is present, amphiboles and mica. In addition, common *accessory minerals*, such as iron-titanium oxides, apatite, zircon, and sulfides, form from minor magmatic constituents that do not readily enter the structures of the major rock-forming minerals.

The history of an igneous rock begins with the formation of a magma at some depth in the Earth. The composition of the magma is determined by the nature of the rock in the source region and by the process of melting. When sufficient melt has formed and coalesced, buoyancy causes it to rise. Further chemical modifications may occur during this period of transport. The magma may rise to the surface and extrude as lava or, if it contains volatiles, may explode onto the surface to be widely distributed as volcanic ash. Most magma, however, solidifies within the earth where slow cooling and crystallization allows for further modification of the initial composition. During this final period, a rock develops its characteristic appearance, which we refer to as *texture*.

A discussion of igneous rocks could begin with a treatment of the initial melting processes in the source region, but this is a theoretical topic and is better left to a later chapter (Chapter 22). Instead, we shall begin with a more tangible subject, the physical properties of magma. It is important to have an accurate conception of the physical nature of magma before considering its emplacement and crystallization. Fortunately, active volcanoes provide opportunities to witness firsthand the eruption of magma onto the Earth's surface. Here, such important properties as magmatic temperature, density, and viscosity can be measured. These will be discussed first. Other physical properties, such as heat capacity, thermal conductivity, and compressibility, can be determined only through sophisticated laboratory measurements. These will be touched upon in later chapters.

2-2 MAGMATIC TEMPERATURES

During periods of volcanic eruption, magmatic temperatures can be measured directly with optical pyrometers and thermocouples. Many eruptions, however, are too violent for this to be done safely. In these cases magmatic temperatures can be determined by heating samples of the erupted rock in the laboratory and measuring their melting points. Such experiments were first done by the Scottish geologist, Sir James Hall, late in the eighteenth century. He withheld publication of his results until 1797 in deference to the views of his distinguished friend James Hutton, who while teaching that nature could be discovered only through observation, considered it rash "to judge of the great operations of the mineral kingdom from having kindled a fire and looked into the bottom of a little crucible." Although such distrust of experimental petrology did not die with Hutton, this approach is accepted today. Modern techniques permit accurate measurements to be made of melting temperatures of rocks under compositionally controlled atmospheres and pressures. Many such determinations have been made on a wide variety of rocks, and most agree well with direct measurements of magmatic temperatures in the field where such measurements can be made.

In discussing magmatic temperatures, it is important to keep in mind that a rock, being a mixture of minerals, does not have a single melting point. Instead, it melts over a temperature range that is commonly several hundred degrees. In addition, this melting range gives only the minimum temperature necessary to have liquid present; magmas could have temperatures well above this. One line of evidence, however, indicates that magmas rarely exceed this temperature range. Magma that cools rapidly by coming in contact with cold rock or by being extruded on the Earth's surface crystallizes to a fine-grained aggregate of minerals or is quenched to a glass. Suspended in this material are almost always larger crystals known as *phenocrysts*. Because these crystals were growing prior to the quenching of the magma, their presence indicates that most magmas have temperatures between the beginning and end of the melting interval. This observation is extremely important, as emphasized by Bowen (1928), for it indicates that magmas do not bring up with them large amounts of excess heat.

Extrusion temperatures for basalt, the type of lava commonly erupted from the Hawaiian volcanoes, commonly range from 1100 to 1200°C. Higher temperatures do occur, but these result from the oxidation of magmatic gases. Most rhyolitic lava, the volcanic equivalent of granite, has temperatures between 800 and 1000°C. Increased pressure causes melting temperatures to rise. But if water is present, the increased pressure allows water to dissolve in the magma, which in turn causes significant lowering of melting temperatures. Basalt and rhyolite can both have their beginning of melting temperatures lowered by several hundred degrees by the addition of water.

Radiation cooling of the surface of lava flows causes solid crusts to form rapidly. The rate of conductive heat transfer through magma and solid crust is, however, so slow that the interior of a flow may remain hot for long periods. For example, in 1959 the eruption of Kilauea Iki produced a lava lake on which a crust developed almost immediately. The crust thickened with time, but three years later the base of the crust, which was at a temperature of 1065°C, had grown to a depth of only 14 m (Fig. 2-1). Cooling of magma at depth in the Earth would be still slower.

Several factors determine the cooling rate of magma, the most important of which are thermal gradients, rate of heat

diffusion through magma and rock, heat capacity of magma and surrounding rocks, and the shape of magmatic bodies. A discussion of the relations between these factors is deferred to Chapter 5 to allow us first to consider the forms and mechanisms of emplacement of igneous bodies.

2-3 MAGMA DENSITIES

Many important petrologic processes are controlled by the density of magma. For example, buoyancy largely determines the rise of magma from a source region. Indeed, magmas that ascend through the lithosphere may have a limited compositional range that is determined by magma density; the lithosphere may act effectively as a density filter. Crystals may float or sink in a magma, and by so doing they can change the composition of the remaining magma (differentiation); this process is clearly determined by contrasts in crystal and liquid densities. Where magma chambers are periodically replenished with surges of fresh magma, as for example beneath oceanic ridges, mixing of magma may occur if magma densities are closely matched, but if densities are different, the magmas remain separate. Magma mixing, which plays an important role in generating certain rock types, some of which are of economic significance, is thus controlled to a large extent by magma densities.

The density of magma can be determined in several ways. First, it can be measured directly at high temperatures. This is technically difficult; so, instead, density measurements are often made at room temperature on glasses formed from the rapid quenching of these liquids. Because glass is a supercooled liquid, the high-temperature density can be calculated from the room-temperature measurements if the coefficient of thermal expansion is known. If V_1 is a volume at T_1, the temperature at which the density measurement is made, the volume, after heating to temperature T, is given by

$$V_T = V_1[1 + \alpha(T - T_1)] \qquad (2\text{-}1)$$

where α, the coefficient of thermal expansion at constant pressure, is defined as

$$\alpha = \frac{1}{V}\left(\frac{\partial V}{\partial T}\right)_P \qquad (2\text{-}2)$$

The *coefficient of thermal expansion* is the relative increase in volume ($\partial V/V$) per unit increase in temperature at constant pressure (Table 2-1). The partial derivative $(\partial V/\partial T)_P$ is used instead of the ordinary derivative to show that this relation is restricted to the change at constant pressure.

Because density is mass per unit volume, Eq. 2-1 can be rewritten as

$$\rho_T = \frac{\rho_1}{1 + \alpha(T - T_1)}$$

where ρ_1 and ρ_T are the densities at temperatures T_1 and T, respectively. Multiplying the top and bottom of the right-hand

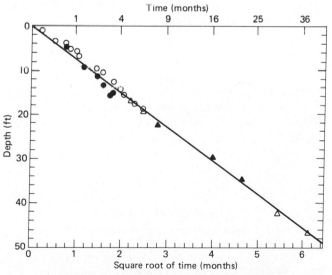

FIGURE 2-1 Thickness of crust in the Alae and Kilauea Iki lava lakes, Hawaii. Solid circles represent the base of the crust, open circles the 1067°C isotherm at Alae Crater, solid triangles the base of the crust, and open triangles the 1065° isotherm at Kilauea Iki Crater. The base of the crust was at approximately 85°C below the initial extrusion temperature. The linear relation between the rate of crustal thickening and the square root of the length of time of cooling is discussed in Chapter 5. [After Peck et al. (1964).]

side of this equation by $[1 - \alpha(T - T_1)]$ gives

$$\rho_T = \frac{\rho_1[1 - \alpha(T - T_1)]}{1 - [\alpha(T - T_1)]^2}$$

Because α is very small (about $10^{-5}\ {}^\circ C^{-1}$), the squared term is negligible, and the denominator is essentially 1. Hence, the density at any temperature T is related to the density at a temperature T_1 by

$$\rho_T = \rho_1[1 - \alpha(T - T_1)] \qquad (2\text{-}3)$$

Another method of determining magma density involves the use of partial molar volumes of the constituent oxides. From the analysis of density measurements of silicate liquids, Bottinga and others (1982) have determined partial molar volumes for most of the common oxide components of magmas (Table 2-1). Hence, a reliable magma density can be calculated from a chemical analysis of the rock.

In any solution, such as a magma, the *partial molar volume* of a component i, denoted by the symbol $\bar{V}_i$, is the change in volume resulting from a change in the number of moles of component $i(n_i)$ at constant temperature, pressure, and numbers of moles of all other components. This can be expressed as

$$\bar{V}_i = \left(\frac{\partial V}{\partial n_i}\right)_{T,P,n_{j \neq i}} \qquad (2\text{-}4)$$

The partial derivative is used to show that this expression is restricted to the volume change at constant temperature, pressure, and number of moles of all constituents except component i. The volume of a magma consisting of many components is given by

$$V = \bar{V}_a n_a + \bar{V}_b n_b + \cdots + \bar{V}_i n_i = \sum_i \bar{V}_i n_i \qquad (2\text{-}5)$$

TABLE 2-1

Partial molar volumes and coefficients of expansion of common oxide components of silicate liquids of 40 to 80 mol % SiO_2 at 1400°C*

Oxide	Molar Volume, V_i ($m^3\ mol^{-1}\ 10^6$)	Coefficient of Expansion, α^a ($^\circ C^{-1}\ 10^5$)
SiO_2	26.75	0.1
TiO_2	22.45	37.1
Al_2O_3	37.80	2.6
Fe_2O_3	44.40	32.1
FeO	13.94	34.7
MgO	12.32	12.2
CaO	16.59	16.7
Na_2O	29.03	25.9
K_2O	46.30	35.9

* From Bottinga and Weill (1970), and Bottinga et al. (1982).
ᵃ Values in this column have been multiplied by 10^5. For example, the coefficient of expansion of SiO_2 is $0.1 \times 10^{-5}\ {}^\circ C^{-1}$.

Thus, the partial molar volume of a component is that quantity which, when multiplied by the number of moles of the component, gives the contribution of the component to the total volume.

Compositions are often more conveniently expressed in terms of mole fractions than as numbers of moles of constituents. The *mole fraction* of a component i, denoted by X_i, is simply $n_i/(n_a + n_b + \cdots + n_i)$. Consequently, the number of moles of each component in Eq. 2-5 can be converted to mole fractions by dividing both sides of the equation by $(n_a + n_b + \cdots + n_i)$, giving

$$\frac{V}{n_a + n_b + \cdots + n_i} = \bar{V}_a X_a + \bar{V}_b X_b + \cdots + \bar{V}_i X_i$$

The volume of magma per total number of moles of all components is referred to as the *molar volume* and is denoted by $\bar{V}$. Hence,

$$\bar{V} = \sum_i \bar{V}_i X_i \qquad (2\text{-}6)$$

The density of a magma is obtained by dividing this molar volume into the molecular weight of the magma, which is given by $\sum_i M_i X_i$, where M_i is the molecular weight of component i. Thus

$$\rho = \frac{\sum_i M_i X_i}{\sum_i V_i X_i} \qquad (2\text{-}7)$$

A final method of obtaining the approximate density of magma is based on the fact that the volume expansion on melting for most rocks is about 10%. Most magmas, therefore, have a density of 90% that of the equivalent solid rock.

The densities of common magmas vary from 2.3 to 3.0 Mg m^{-3}, with the more iron-rich varieties having the higher values. The compressibility of magmas is small; hence pressure differences do not cause large density changes. The density of a melt formed from a Kilauean basalt, for example, increases approximately 133 kg m^{-3} per 1 GPa increase in pressure (Kushiro, 1980). Thus, in rising from a depth of 35 km, the density of this basaltic magma would decrease from 2.743 Mg m^{-3} to 2.610 Mg m^{-3}. If increased pressures result in higher dissolved water contents, the resulting change in composition results in significantly lower magma densities (Bottinga and Weill 1972).

2-4 MAGMA VISCOSITIES

Some lavas are able to flow rapidly over great distances, whereas others barely move, even when erupted on steep slopes. Viscosity is the physical property that describes this resistance to flow. It is extremely important in determining the rates of emplacement of magma and the shapes of igneous bodies, and in determining whether sinking or floating crystals will separate fast enough from their parent magma to change its bulk composition.

Like any fluid, a magma deforms continuously under the action of a shear stress. When a fluid is at rest, there can be no shear stresses. All fluids will react to a shear stress by flowing, but some do this more slowly than others. The property of a fluid that defines the *rate* at which deformation takes place when a shear stress is applied is *viscosity*:

$$\text{viscosity} = \frac{\text{shear stress}}{\text{rate of shear strain}}$$

If a large shear stress applied to a liquid results in slow deformation, the liquid is said to have a high viscosity. Tar and honey are familiar examples of highly viscous liquids, whereas water and gasoline have low viscosities.

As the magnitude of a shear stress applied to a liquid increases, the rate of strain also increases. For many liquids a linear relation exists between the applied shear stress and the strain rate (Fig. 2-2). These liquids are referred to as being *Newtonian*. From the definition of viscosity, the slope in Figure 2-2 is clearly the viscosity. *Non-Newtonian* liquids exhibit a nonlinear relation between applied shear stress and strain rate; their viscosities (slopes) vary with shear stress. Most magmas are Newtonian liquids, but with large contents of suspended crystals or gas bubbles they may show non-Newtonian behavior of the type exhibited by Bingham liquids (Shaw, 1969). These liquids resist deformation until some minimum shear stress (*yield strength*) is applied, after which strain rates increase linearly with shear stress. Thus, at low shear stresses they appear to have infinite viscosity, but above the yield strength they behave like Newtonian liquids. Such behavior has profound effects on velocity profiles across bodies of flowing magma (Sec. 3-7) and on the ability of crystals to move buoyantly through magma (Sec. 13-3).

Viscosity is better appreciated by considering the following example. A shear stress τ (tau) is applied to the liquid between two parallel plates (Fig. 2-3) by moving the upper one

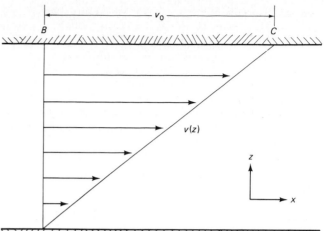

FIGURE 2-3 Laminar flow of liquid between two parallel plates resulting from the upper plate moving in the x direction with a velocity v_0 relative to the stationary lower plate. Note that the positive direction of z is upward (height).

steadily at a velocity v_0 relative to the lower one. Liquid in contact with the upper plate is dragged along at a velocity v_0, whereas that in contact with the lower plate remains stationary. The intervening liquid can be thought of as consisting of a series of thin parallel plates or lamellae, each of which slides over the one beneath. Such movement is referred to as *laminar flow*, in contrast to *turbulent flow*, where both the direction and rate of movement of the liquid experience wide fluctuations. If the length of the arrows in the figure indicates the velocity in the x direction at various heights along the z axis, a marker line AB would move to the position AC after a unit interval of time. The strain rate must therefore be dv/dz, and the viscosity, denoted by η (eta), is given by

$$\eta = \frac{\tau}{dv/dz}\,\frac{N}{m^2}\frac{s}{m}\,m = \frac{N}{m^2}\,s = \text{Pa s} \qquad (2\text{-}8)$$

In SI units, where shear stress is measured in newtons m^{-2}, the velocity in $m\ s^{-1}$, and distance in meters, the units of viscosity are Pa s. The cgs unit of viscosity is the *poise* (P), which is a dyne s cm^{-2} (1 P = 0.1 Pa s). In many fluid mechanics calculations, the viscosity is divided by the density of the fluid. This ratio is referred to as the kinematic viscosity and is defined as $v = \eta/\rho$ with units of $m^2\ s^{-1}$.

It is now possible to make a quantitative statement about the viscosity of magmas. Measurements of viscosity have been made in the laboratory at both atmospheric and higher pressures on a variety of liquids formed by melting common igneous rocks. In addition, viscosities have been calculated from the actual measured rates of flow of lava issuing from active volcanoes. Common magmas cover a wide range of viscosities. At temperatures of 1200°C and atmospheric pressure, basaltic magmas (gabbroic) have viscosities in the range 10 to 100 Pa s (Shaw 1969). Andesitic magmas (dioritic) under the same conditions average about 10^3 Pa s, and most rhyolitic ones

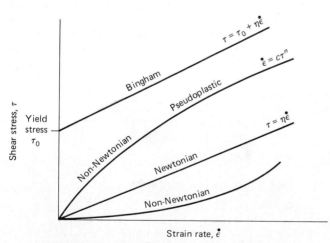

FIGURE 2-2 Shear stress versus rate of shear strain for Newtonian and non-Newtonian liquids. The slope of these lines gives the viscosity of the liquid, which for a Newtonian one remains constant regardless of shear stress. The viscosity of non-Newtonian liquids varies with the applied shear stress.

(granitic) have viscosities above 10^5 Pa s (Clark 1966, p. 299) and can be as high as 10^8 Pa s at 800°C, which is a more normal magmatic temperature for rhyolite. By comparison, ASE 30 motor oil has a viscosity of 0.3 Pa s and that of silicone putty (Silly Putty) is approximately 10^4 Pa s, but Silly Putty's viscosity varies considerably depending on the shear stress (its behavior is represented by the lower curve in Fig. 2-2).

The viscosity of a magma is determined largely by its chemical composition, with the framework building components, SiO_2, $KAlO_2$, and $NaAlO_2$ contributing most to high viscosities. The increase in viscosity from basalt through andesite to rhyolites is attributable to this chemical variation, with basalts containing approximately 50 wt % SiO_2, andesites 55 wt %, and rhyolites 73 wt %. This chemical control makes it possible to calculate reasonable viscosities for magmas of known composition, in much the same way that densities were calculated from partial molar volumes (Bottinga and Weill 1972).

The range of viscosities of magmas becomes even greater when account is taken of the lower temperatures at which rhyolites crystallize compared with basalt. With lower temperatures come marked increases in viscosity. Magmas of rhyolitic composition, which have temperatures several hundred degrees lower than those of basalt, consequently have very much higher viscosities than have basalts. This single fact is largely responsible for the morphological differences between valcanoes and other igneous bodies formed from these different magmas.

The dramatic change in viscosity of magma with temperature generally obeys an Arrhenius relation of the form

$$\eta = \eta_0 e^{E/RT} \qquad (2\text{-}9)$$

where η_0 is a constant, E the activation energy, R the gas constant, and T the *absolute* temperature. Such a relation, which describes the temperature dependency of many reaction rates, can be thought of as consisting of two parts; the preexponential factor, η_0, expresses the frequency of a particular event involved in the reaction (in this case, perhaps the formation of a hole or the breaking of a bond in the liquid structure), and the exponential term gives the fraction of these events that have sufficient energy to permit the reaction to proceed.

Converting Eq. 2-9 to logarithmic form gives

$$\ln \eta = \ln \eta_0 + \frac{E}{RT}$$

which is the equation of a straight line. A plot of $\ln \eta$ versus $1/T$ for viscosities determined at several temperatures yields the activation energy from the slope of the line and the frequency factor from the intercept (see Prob. 2-5). Once the values of η_0 and E are known, the viscosity can be calculated for any temperature.

Pressure has a very much smaller effect on magma viscosity than has temperature. The viscosity of basaltic magma at 1400°C, for example, decreases from 3.5 Pa s at atmospheric pressure to 1.5 Pa s at 2.0 GPa (Kushiro 1980). Indirectly, however, pressure can modify viscosities more dramatically. Higher pressure permits more water to be dissolved in magmas, which causes a lowering of viscosity. For example, at 800°C granitic magma containing 1 wt % H_2O has a viscosity of 10^8 Pa s, whereas with 10 wt % H_2O the viscosity decreases to 10^4 Pa s (Shaw 1965).

Unless one is familiar with the numerical values for the viscosity of liquids, numbers such as 10^2 or 10^8 Pa s have little significance. They can, however, be made more meaningful by considering the effect that viscosity has on the flow rate of lava. Let us consider a lava flow (Fig. 2-4) of thickness h composed of an incompressible Newtonian magma with viscosity η and density ρ that flows with a steady rate down an unconfined plane surface with a slope of θ degrees. For convenience we shall choose a set of orthogonal axes, x, y, and z, such that x parallels the flow direction and z is normal to the plane surface. The lava flows in a laminar manner and for the moment we will neglect any effects due to cooling of the lava. Gravitational attraction causes the lava to flow, and if no other forces are present, the lava accelerates to greater and greater velocities. The viscous force, of course, prevents this from happening.

To analyze this situation we consider the forces acting on a very small volume of lava, $dx\,dy\,dz$. The gravitational force acting in the direction of flow is simply the mass times the acceleration, which is given by $\rho\,dx\,dy\,dz\,g \sin\theta$, where g is the acceleration of gravity. However, the shear stress in the liquid

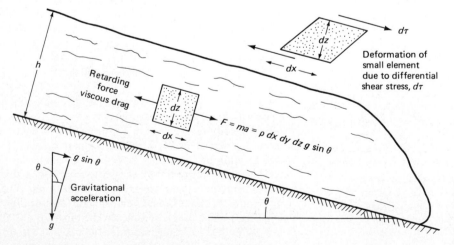

FIGURE 2-4 Forces acting on a small element ($dx\,dy\,dz$) of lava within a flow of thickness h, descending a plane surface with a slope of θ degrees.

retards the flow. Because the shear stress varies with height in the flow, there is a difference in shear stress between the top and bottom of the volume, which we will designate $d\tau$. Because this stress (force/unit area) acts on an area $dx\,dy$, the force retarding the flow is $d\tau\,dx\,dy$. Because the flow rate of the lava is steady, there is no acceleration. Consequently, the sum of the viscous and gravitational forces acting in the plane of flow must be zero, that is,

$$d\tau\,dx\,dy + \rho\,dx\,dy\,dz\,g\sin\theta = 0$$

Dividing both sides of the equation by dx and dy and rearranging gives

$$d\tau = -\rho g \sin\theta\,dz$$

which, upon integration, becomes

$$\tau = -\rho g \sin\theta\,z + \text{constant}$$

The constant of integration equals $\rho g \sin\theta\,h$, because $\tau = 0$ on the upper surface of the lava flow where $z = h$. Hence the shear stress at any point in the flow is given by

$$\tau = \rho g \sin\theta\,(h - z)$$

Because the lava is a Newtonian liquid in laminar flow, $\eta\,dv/dz$ can be substituted for τ (Eq. 2-8), giving

$$\frac{dv}{dz} = \frac{\rho g \sin\theta}{\eta}\,(h - z) \qquad (2\text{-}10)$$

This gives the velocity profile through a lava flow under the specified conditions. Integration of this equation gives

$$v = \frac{\rho g \sin\theta}{\eta}\left(hz - \frac{z^2}{2}\right) + \text{constant}$$

This integration constant is zero, because $v = 0$ where $z = 0$. Hence, the velocity at any height z within the flow is given by

$$v = \frac{\rho g \sin\theta}{\eta}\left(hz - \frac{z^2}{2}\right) \qquad (2\text{-}11)$$

It is now possible, therefore, to use the viscosities of magmas to calculate flow rates for some common lavas. This is left for the reader to do in Probs. 2-7 to 2-9.

When integrating Eq. 2-10, the viscosity was taken to be a constant. But, in reality, the magma would cool along the lower and upper surfaces of the flow, and this would cause a significant increase in the viscosity, which, in turn, would decrease flow velocities. To improve the calculated velocities, it would be necessary, before integrating Eq. 2-10, to express η as a function of height within the flow. This can be done, because a reasonable estimate of the temperature gradient within the flow can be made (see Chapter 5), and if the activation energy in Eq. 2-9 is known, the viscosity can be expressed as a function of z (see Prob. 2-9).

PROBLEMS

2-1. Two lime-silica liquids at 1700°C, one with 30 and the other with 60 mole % CaO, have densities of 2466 and 2665 kg m^{-3}, respectively. Determine the partial molar volumes of SiO_2 and CaO in these liquids at this temperature. (*Note:* Your answers may differ slightly from the values determined from Table 2-1, because the table is based on average values determined from many calculations, such as those done in this problem.)

2-2. The molar volumes of SiO_2 and CaO at 1200°C are 26.744×10^{-6} and $16.05 \times 10^{-6}\,\text{m}^3\,\text{mol}^{-1}$, respectively, whereas at 1600°C they are 26.755×10^{-6} and $17.15 \times 10^{-6}\,\text{m}^3\,\text{mol}^{-1}$, respectively. Calculate the coefficients of expansion, α, for SiO_2 and CaO over this temperature range and give an explanation, in terms of the possible structures of the melts, for the large difference between these two values. (Reminder: $\int dV/V = \ln V + \text{constant}$.)

2-3. Using the partial molar volumes in Table 2-1, calculate the densities of typical rhyolitic (granitic) and basaltic (gabbroic) magmas at 1200°C having the following compositions:

| | Weight % Oxides | |
Oxide	Granite	Basalt
SiO_2	74.0	48.7
TiO_2	0.2	2.8
Al_2O_3	13.5	15.0
FeO	2.0	12.0
MgO	0.3	7.5
CaO	1.2	10.5
Na_2O	3.4	2.5
K_2O	5.4	1.0
Total	100.0	100.0

(Note that the compositions are expressed in weight percent of the oxides, and these must first be converted to mole fractions. The iron in these analyses is expressed as FeO for simplicity. Natural magmas contain both ferrous and ferric iron, the proportions depending on the degree of oxidation of the magma.)

2-4. A sample of granite with density 2670 kg m^{-3} is fused and quenched to a glass that has a density of 2450 kg m^{-3} at 25°C. The coefficient of expansion, α, for the glass is 2×10^{-5} °C^{-1}. Determine the density of the magma at 1025°C, and compare this with the value obtained by simply taking 90% of the density of the rock.

2-5. The viscosity of a particular lava is found by experiment to vary with temperature in the following manner:

T (°C)	1100	1200	1300	1400
η (Pa s)	697	100	22.6	5

By plotting ln η versus $1/T$, determine the value of the activation energy and the preexponential factor in the Arrhenius expression for the temperature dependency of viscosity. (Temperature must be expressed in kelvin.)

2-6. From the equation for the velocity distribution through a lava flow that is in steady-state laminar flow, derive an expression for the maximum velocity and show that this occurs at the upper surface of the flow.

2-7. Plot a graph of the velocity profile through a 2-m-thick basaltic lava flow with a viscosity of 300 Pa s and density 2750 kg m^{-3} descending a 2° slope. What is the shape of this profile?

2-8. Calculate the maximum velocity within the flow given in Prob. 2-7. Do the same for a similar flow of rhyolite with a viscosity of 10^6 Pa s and density 2500 kg m^{-3}, and for a lunar basalt from the Sea of Tranquility having a viscosity of 9 Pa s and a density of 3000 kg m^{-3}. Express your own maximum rate of walking (approximately 4 mi h^{-1}) in the same units of velocity. Why would the flow velocities be too high? (Note that the acceleration of gravity on the Moon is only 1.62 m s^{-2}.)

2-9. Calculate the velocity profile through the same lava flow as given in Prob. 2-7, but because of cooling toward the upper and lower surfaces, the viscosity increases away from the center of the flow. The viscosity at any height z, measured in meters, above the base of the flow is given by $\eta = \eta_c + f(z - h/2)^2$, where η_c, the viscosity at the center of the flow, is 300 Pa s, f is a constant with the value of 10^4 N s m^{-4}, and h is the thickness of the flow.

Useful integrals:

$$\int \frac{dz}{a + bz + cz^2} = \frac{2}{(4ac - b^2)^{1/2}} \tan^{-1} \frac{2cz + b}{(4ac - b^2)^{1/2}}$$

$$+ \text{constant} \qquad \text{when } 4ac - b^2 > c$$

$$\int \frac{z\, dz}{a + bz + cz^2} = \frac{1}{2c} \ln(a + bz + cz^2) - \frac{b}{2c} \int \frac{dz}{a + bz + cz^2} + \text{constant}$$

3 / Intrusion of Magma

3-1 INTRODUCTION

Most field evidence indicates that igneous rocks have formed from upward-moving bodies of either magma, mixtures of magma and crystals, magma and gas bubbles, or even solid rock. Rates of movement vary through many orders of magnitude, with bodies of granite requiring possibly thousands of years to move a few meters, while highly gaseous magmas can break through the Earth's surface at supersonic speeds. In considering the forces responsible for this movement, it is important to bear in mind that magmas, like other fluids, flow only in response to pressure gradients. Perhaps the most obvious cause for such gradients is the loading of magma by overlying denser rocks, but pressures can also result from volume increases on melting in the source region, liberation of gas from vapor-supersaturated magma, and from tectonic forces. Because there is considerable debate over the relative importance of these, we consider some of the arguments for and against each mechanism before discussing rates of intrusion.

3-2 BUOYANT RISE OF MAGMA

Within a body of magma surrounded by denser rock, there is an excess pressure gradient (i.e., in excess of the pressure gradient due to the density of the magma—we can refer to this as the magmastatic pressure gradient to distinguish it from the surrounding lithostatic pressure gradient) which, from Eq. 1-1, is $\Delta\rho\, g$, where $\Delta\rho$ is the density difference between magma and surrounding rock and g is the acceleration of gravity. Magmas are invariably less dense than their solid equivalent ($\rho_{magma} = 0.9 \times \rho_{rock}$), and thus total melting of a volume of rock results in a buoyant force on the magma. If melting is only partial, this force is likely to be greater, for the unfused refractory residue is generally more dense than the initial rock.

To illustrate this process, consider the buoyant rise of magma through a continental crust from a source in the upper mantle (Fig. 3-1). The crust is taken to consist of an upper 25 km with density 2.75 Mg m^{-3} and a lower 10 km with density 3.0 Mg m^{-3}. The magma, with a density of 2.9 Mg m^{-3}, is formed at a depth S kilometers beneath the surface, in the upper mantle, which has a density of 3.3 Mg m^{-3}.

Let us first consider magma that is able to rise buoyantly just to the surface of the Earth along an open tensional fracture, perhaps formed by tectonic plate movements. Once the magma reaches the surface, the buoyant force must be balanced by the downward gravitational force acting on the mass of the column of magma. Under these conditions, the pressure at the base of the magma column is equal to that at the base of the adjoining rock column; thus,

$$2750g \times 25 \times 10^3 + 3000g \times 10 \times 10^3 + 3300g(S - 35) \times 10^3$$
$$= \int_{z=0}^{z=S} \rho_{magma} g\, dz$$

where z is the depth from the surface and S the depth in meters from the Earth's surface to the source. Making the reasonable assumption that the magma is incompressible and g remains constant over the depth considered, we obtain $S = 41.9$ km.

This calculated depth is slightly above the top of the low-velocity zone, which is considered by many to be the source for basaltic magmas. If this same magma were generated at higher levels, it would be unable to rise buoyantly to the surface (Fig. 3-1, No. 2), and, in the absence of other forces, would never form volcanic rocks. Instead, it would form *plutonic* ones, that is, ones that crystallize at depth. A dense magma might also be restricted to plutonic rather than volcanic occurrences. The crust may even act as a *density filter* to magma with densities greater than 2.75 Mg m^{-3} (Stolper and Walker 1980). Finally, if the initial magma were to form a substantial volcano on reaching the Earth's surface, the source would have to be correspondingly deeper (Fig. 3-1, No. 3, and Prob. 3-2).

In the example above, the magma was taken to be incompressible. Thus, its density is unaffected by changes in depth. If, on the other hand, the magma is compressible (see Sec. 2-3), its density will be a function of depth, a fact that must be taken into account when integrating the right-hand side of the preceding equation. This is done by considering the *isothermal coefficient of compressibility* of the magma, which is denoted by β. This coefficient gives the relative decrease in volume $(-\partial V/V)_T$

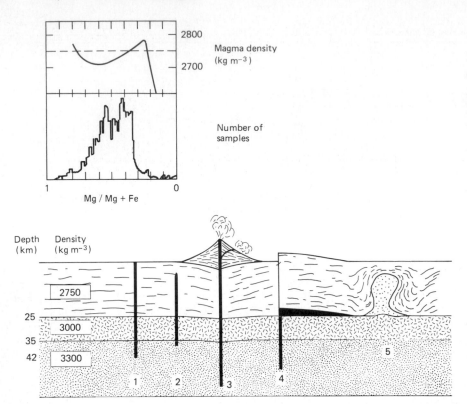

FIGURE 3-1 A section of continential crust and upper mantle showing the depths to the source region from which magma with a density of 2.9 Mg m^{-3} must buoyantly rise along tension fractures to be able to just reach the surface of the Earth (1), intrude to some level within the crust (2), build a volcano of dense rocks on the surface (3), or intrude along the base of the upper crust raising the lighter rocks above it (4). Being less dense than most crustal rocks, granitic magma (5) is able to rise through and forcefully displace the overlying rocks. Also shown is the variation in density of melts derived by partial crystallization of basaltic magmas as a function of the moleculer ratio of Mg/(Mg + FeO) in the melt; the abundance of the continental basaltic rocks drops off sharply where the density of the melts go above the average density of the upper crust (2.75 Mg m^{-3}), suggesting that the crust acts as a density filter for dense magmas. [After Stolper and Walker (1980).]

per unit increase in pressure at constant temperature. This is expressed mathematically as

$$\beta = -\frac{1}{V}\left(\frac{\partial V}{\partial P}\right)_T \qquad (3\text{-}1)$$

Integration of this expression between the limits of the surface of the Earth, where the pressure is P_0 and the volume V_0, and depth z, where the pressure is P_z and the volume V_z, yields

$$\beta(P_z - P_0) = -\ln\frac{V_z}{V_0} \qquad (3\text{-}2)$$

Because P_0 is negligible compared with P_z, Eq. 3-2 can be rewritten as

$$\frac{V_z}{V_0} = \exp(-\beta P_z)$$

and since $V_z/V_0 = \rho_0/\rho_z$, where ρ_0 and ρ_z are the densities of the magma at the surface and depth z, respectively, we obtain

$$\rho_z = \rho_0 \exp(\beta P_z) \qquad (3\text{-}3)$$

But the pressure at depth z, P_z, is obtained by integration of Eq. 1-1,

$$\int dP = \int \rho_z g\, dz$$

If the expression for ρ_z in Eq. 3-3 is substituted in the right-hand side of this equation, we obtain, after rearrangement,

$$\int \exp(-\beta P)\, dP = \rho_0 g \int dz$$

which, upon integration between the limits of the surface of the Earth, where $P = 1$ atm, and a depth S, where $P = P_S$, yields

$$\exp(-\beta P_S) = \exp(-\beta) - \beta\rho_0 gS$$

Thus, the pressure at a depth S in a magma with a coefficient of compressibility of β and density at the Earth's surface of ρ_0 is given by

$$P_S = -\frac{1}{\beta}\ln(e^{-\beta} - \beta\rho_0 gS)$$

Since β is very small, $e^{-\beta}$ is nearly 1, and thus

$$P_S = -\frac{1}{\beta}\ln(1 - \beta\rho_0 gS) \qquad (3\text{-}4)$$

This pressure can then be equated with the pressure at the base of the adjoining rock column, thus allowing us to solve for S, the depth to the source region (see Prob. 3-1).

The mechanism of buoyant rise works well for granitic magmas, which have lower densities than almost all crustal rocks (Fig. 3-1, No. 5) and for magmas rich in volatiles, such as kimberlites and carbonatites. Most basaltic magmas, in contrast, are either denser than upper crustal rocks or neutrally buoyant in the crust. Hence, in the example treated above, the magma, rather than rising to the Earth's surface, could have formed a sheet at the base of the upper crust (Fig. 3-1, No. 4). The overlying rocks would then have floated on the magma, and once the sheet was thick enough, the pressure at the base of the magma column would have, again, balanced the load pressure, and intrusion would have stopped. This mechanism of intrusion is actually more efficient than the first because of the resulting density distribution. Thus, unless fractures are readily available in which magma can rise, basaltic magma would appear more likely to intrude along the base of the crust (see Prob. 3-3). In fact, operation of this process over an extended period could well have formed the lower crust, which, because of its density, is thought to be basaltic in composition. Many high-level bodies of basaltic rock do actually occupy tensional fractures (dike swarms, see Sec. 4-6), in contrast to many bodies of granite which are quite irregular in shape and have clearly forced their way into place (Fig. 3-1).

3-3 VOLUME EXPANSION ON MELTING

The difficulty of buoyantly emplacing large bodies of dense magma high into the Earth's crust has led to consideration of other possible mechanisms of intrusion. One of the most important involves pressures developed in the region of magma generation by increases in volume on melting (Roberts 1970). Most silicates expand 4 to 15% on melting, with 10% being a common value for rocks (see Sec. 2-3). The formation of the first small amounts of melt in a source region, consequently, results in a volume expansion. But the strength of surrounding rock opposes this expansion and causes pressures on the liquid to rise. This, in turn, raises the melting point, and no further melting occurs until the temperature increases. When this does occur, the formation of new liquid results in further pressure increases. Of course, if the process were slow, the surrounding rock would yield by flow, and the excess pressure would be relieved. At faster rates, however, the pressure within the liquid builds until the tensile strength of the surrounding rock is exceeded, at which point brittle failure occurs. The liquid expands into the fracture and, by so doing, decreases its own pressure. This, in turn, allows for more melting and expansion, which causes further fracturing and intrusion, the process continuing as long as the source region remains at a sufficiently high temperature.

Two questions are immediately raised by this process. First, how high a pressure could be generated within an embryonic volume of liquid, and second, how fast would melting have to occur to produce these pressures? The answer to the first question is not simple, because rocks under high confining pressures behave as strong elastic solids when exposed to short-duration stresses, but as rather weak, highly viscous liquids when exposed to long-duration stresses. The amount of pressure that might develop within a volume of rock during melting would, therefore, be rate dependent. If the mechanism is to be

capable of providing pressures for intrusion over extended periods, rocks must have sufficiently high long-term strength.

Energy released from deep-focus earthquakes and the magnitude of regional gravity anomalies both indicate that the long-term strength of rocks in the upper mantle is about 0.05 GPa. Anything greater than this results in fracture or flow. Because the excess pressure that can be sustained within a spherical volume of liquid by surrounding solid rock under lithostatic pressure is two-thirds of the tensile strength of the rock (Roberts 1970), values greater than 0.03 GPa in the liquid would be sufficient to cause brittle failure of the surroundings.

Magmas are rather incompressible liquids; hence, only small amounts of partial melting are necessary to cause significant increases in pressure. For example, with a 10% volume increase on melting, only 0.24% of a rock need melt to produce the excess pressure of 0.03 GPa necessary to cause brittle failure.

Excess pressures would appear likely to accompany the formation of melts. Although these may be small, they would continue to exist as long as melting took place. Higher pressure could occur with more rapid, large-scale melting. In fact, if a region of partially melted rock under a certain excess pressure was suddenly fractured and the pressure released, large-scale melting could occur with considerable volume expansion and intrusion. This would augment the buoyant rise of the magma and would allow for intrusion to higher levels.

Although this mechanism may provide additional force for the rise of dense magmas, the fact remains that once the magma reaches the upper crust where most rocks are less dense, it is more efficient for the magma to spread laterally than to continue rising. The controlling factor may therefore still be the nature of the existing fractures at the time of intrusion, although the added pressure from melting may help exploit some of fractures.

One final point concerning this mechanism is that the source of the pressure is in the region of melting. Should the body of magma become separated from the source, the magma would rapidly lose its excess pressure. In contrast, the buoyant force is due only to the density difference between the magma and surrounding rock, and therefore is not dependent on connection with the source region. But even with buoyant rise, if magma is to ascend through less dense rock, it must also remain connected to deep-seated magma to maintain the necessary pressure for intrusion. We can conclude that whichever mechanism is operative, dense magmas intruded to high levels in the crust must remain connected to deep roots.

3-4 VESICULATION

Prior to eruption, magma may contain considerable quantities of gas held in solution by the confining pressure, but once pressure is lowered, the gas separates and forms bubbles (vesicles), just as it does when the cap is removed from a bottle of carbonated beverage. The vesicular magma is much more buoyant than the original and can rise rapidly to greater heights.

Experimental studies indicate that the maximum amount of gas that can be held in solution by a magma depends on the composition of the magma and, more important, on the confining pressure (see Chapter 11). For example, at atmospheric pressure the amount of water that can be dissolved in rhyolitic and basaltic magmas at their typical magmatic temperatures is negligible. At 0.5 GPa, however, rhyolite can contain 10 wt % H_2O and basalt 8 wt % H_2O. This, of course, does not mean that natural magmas must contain this amount under these condition; all degrees of saturation are possible, depending on the availability of water. But with the pressure decrease that accompanies the rise of magma toward the Earth's surface, even undersaturated magmas must eventually become saturated.

Only a small degree of supersaturation (<1 MPa) is necessary to cause bubbles to nucleate (Fig. 3-2). These bubbles grow as the pressure decreases on the rising magma, or new bubbles nucleate if diffusion rates are too low to transport gas to existing bubbles. During bubble growth the bulk density of the magma (liquid plus gas) decreases, making the magma more buoyant. If the density of the liquid fraction is ρ_l and that of the gas fraction is ρ_g, and the weight fraction of gas is f, the bulk density of the magma, ρ_b, is given by

$$\frac{1}{\rho_b} = \frac{f}{\rho_g} + \frac{1-f}{\rho_l} \qquad (3\text{-}5)$$

The approximate density of the gas can be determined from the ideal gas law ($PV = nRT$)

$$\rho_g = \frac{PM}{RT} \qquad (3\text{-}6)$$

where P is the pressure, M the molecular weight of the gas, R the gas constant, and T the absolute temperature (see Prob. 3-4).

While exsolution of gas is taking place, the velocity of the magma increases (Fig. 3-3), in part because of increased buoyancy, but also because of the expansion of bubbles resulting from both decompression of the gas and diffusion of gas out of the melt into the bubbles. Once the bubbles have grown to the stage that they constitute approximately 75% of the volume of the rising magma, continued growth becomes difficult because the viscous silicate melt that must be displaced has only the thin tortuous sheets of liquid between the bubbles through which to move. Continued exsolution of gas therefore causes pressure within the bubbles to rise above the ambient pressure. Toward the free upper surface of the magma column this excess pressure causes the larger bubbles to burst. This disrupts the magma and breaks it into frothy particles that are ejected into the atmosphere (Fig. 3-2). The term *pyroclastic* (broken by fire) ejecta is used for this material.

Up to the time of disruption, magma consists of a continuous silicate liquid enclosing bubbles of gas. After disruption, the gas forms the continuous phase, with globules of silicate

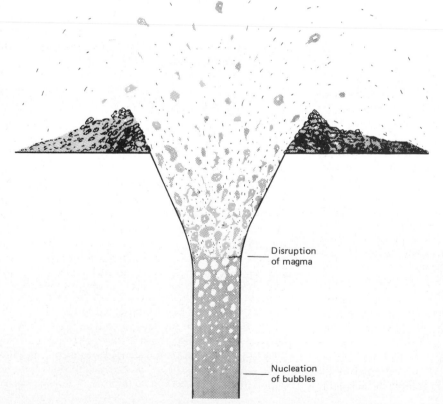

Disruption
of magma

Nucleation
of bubbles

FIGURE 3-2 Magma rising in a volcanic feeder pipe will, at some depth depending on the initial volatile content, become supersaturated with volatiles (mainly H_2O and CO_2) and gas bubbles will nucleate. As the magma continues to rise, the bubbles grow larger both through continued exsolution of gas from the magma and through expansion of gas due to decompression. The bubbles eventually burst and disrupt the magma into a mass of frothy particles that may be ejected from the vent at high velocities (see Fig. 3-3).

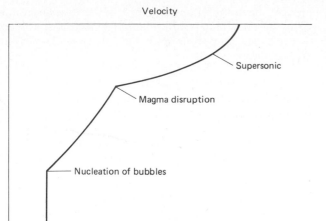

FIGURE 3-3 Typical acceleration of vesiculating magma as it rises in a vent. The velocity below the depth at which gas bubbles nucleate (Fig. 3-2) remains constant as long as the conduit diameter is constant. The growth and expansion of gas bubbles causes magma to accelerate as it approaches the surface. Once the magma is disrupted by the gas bubbles the velocity increases substantially (possibly becoming supersonic) because the low-viscosity gas now forms the continuous medium. [After Wilson and Head (1981).]

melt suspended in the gas. The viscosity of the rising column of magma consequently decreases dramatically following disruption, and the flow velocities can increase to supersonic speeds in the vent (Fig. 3-3). Of course, the gas is what actually moves with this velocity, but it carries, in suspension, the smaller pyroclastic particles (<2 cm). Larger particles will have lower velocities depending on their terminal velocity of fall in the rising stream of gas.

The violence of the disruption is determined in part by the viscosity of the original magma. The low viscosity of basaltic magmas allows gas bubbles to coalesce and rise to the top of a column of basaltic magma during periods of quiescence or slow flow and form large bubbles that burst and throw out, for short distances, relatively large blobs or *bombs* of *scoriaceous* (vesicular) basalt. The magma beneath, meanwhile, has degassed itself and will erupt as a peaceful flow during the next period of activity. This alternation between periods of splatter and quiet flow is known as *Strombolian* activity. In contrast, the high viscosity of rhyolitic magmas impedes the separation of gas bubbles, so that during eruption the bursting of large numbers of evenly spaced small bubbles brings about a more thorough disruption of the magma into many small pieces of *pumice*, which can be ejected violently high into the Earth's atmosphere—the *Plinian* type of eruption.

The initial volatile content of a magma also plays an important role in determining the degree to which disruption will affect the magma. In general, basaltic magmas have much lower volatile contents than do rhyolitic ones, and thus basalts reach the disruption stage only at very shallow depths (see Prob. 3-4). For example, typical Hawaiian basaltic magmas are estimated to contain about 0.45 wt % H_2O, in which case nucleation of gas bubbles does not even begin until the magma is

within 300 m of the Earth's surface. Even so, this amount of gas is sufficient to disrupt the magma at the surface and produce "fountains of fire" over 500 m high. Nonetheless, the general lack of vesicles in dikes beneath eroded Hawaiian volcanoes and other terranes indicates that vesiculation cannot play a role in bringing basaltic magmas up from depth. Rhyolitic magmas, with their higher concentration of volatiles, may be affected by this process, at least at shallow crustal depths.

3-5 TECTONIC PRESSURES ON MAGMA

Our familiarity with squeezing toothpaste from a tube makes the concept of igneous intrusion by tectonic squeezing of a magma chamber a simple one to appreciate. As a result, it is readily invoked, despite the common lack of any supporting evidence. In the case of a squeezed toothpaste tube, flow is caused by a strong pressure gradient resulting from the tube being open at one end. If magma does not have easy access to some lower-pressure region, the magma chamber would simply be deformed by tectonic forces, and there would be no actual intrusion of magma. In the deeper crust and upper mantle, where most magmas originate, solid rocks react to long-term stresses as if they were extremely viscous liquids; under such conditions, there seems little likelihood of tectonic forces causing intrusion. But if movements were sufficiently rapid to cause rupturing of the crust, or if some other mechanism allowed magma to extend itself over a considerable vertical distance, tectonic forces could cause magma to flow toward the surface.

3-6 INTRUSION RATES OF NEWTONIAN MAGMA IN LAMINAR FLOW

The rate at which magma flows depends on the pressure gradient, magma viscosity, the shape of the conduit, and if rocks are to be displaced by the intruding magma, the rate of deformation of the enclosing rocks. Many steeply dipping regional dikes of basaltic composition have intruded tensional fractures along which there may have been minimal displacement of the intruded rocks by the magma, many large bodies of granitic magma have forcefully displaced the country rocks. Rates of intrusion of basaltic dikes are determined largely by the viscosity of the magma and the conduit shape, whereas the rate of deformation of surrounding rocks may determine the rate of intrusion of granite. Because rocks deform much more slowly than liquids, rates of intrusion involving forceful displacement of country rocks are very low, ranging perhaps from meters to millimeters per year. Indeed, many bodies of granitic magma may even continue to rise slowly after they have completely solidified, as long as they are surrounded by denser rock (Ramberg 1981). These slow rates contrast markedly with the rates at which low-viscosity basaltic magma can rise in fissures or volcanic pipes, unimpeded by any necessity to force aside country rocks.

Because volcanoes erupt enormous volumes of lava in very short periods of time through small feeder pipes, magma velocities must be very high. For example, on August 21, 1963, over 7×10^5 m³ of lava was extruded in a period of 12 hours from the Alae pit crater, on the east rift zone of Kilauea, Hawaii; the estimated maximum rate of extrusion was 11.5×10^4 m³ h⁻¹. The largest eruption in historic times, the 1783 fissure eruption at Lakagigar (Laki) in Iceland, is estimated to have poured out 18 km³ h⁻¹ of basalt from a few main fountains along a 10-km-long fissure. Volcanic feeder pipes and dikes that have been exposed by erosion may have diameters or widths of many hundreds of meters, but most are much smaller, and in many cases narrow with increasing depth. It is necessary to conclude, therefore, that magma velocities within feeder conduits can be rather high. Icelandic feeder dikes, for example, average only about 4 m in width, and the estimated average velocity within the dike that fed the 1783 Lakagigar eruption is calculated to be 6 cm s⁻¹, but at times it probably surged to 10 times this value (Thorarinsson 1968).

Reasonable values of flow velocity in volcanic feeders can be calculated if the physical properties of the magma and the dimensions of the pipe are known. To illustrate this, we will consider steady-state laminar flow in a vertical, cylindrical pipe of radius r_0 (Fig. 3-4). The magma has a velocity v in the z direction, and we make the reasonable assumption that it is an incompressible, Newtonian liquid with density ρ_m and viscosity η.

We start by considering the forces acting in the z direction on a small cylindrical volume of magma with radius r and length dz. The buoyant force on the volume is the difference between the pressure on the base (P_z) and the top (P_{z+dz}) of the volume times the cross-sectional area, that is, $-dP\,\pi r^2$. Gravitational attraction on the volume gives a negative force in the z direction of $-\rho_m \pi r^2\,dz\,g$. The shear stress acting on the outer surface of the cylinder, due to the upward movement of magma within the cylinder, gives rise to a force of $\tau \pi 2r\,dz$. It is important to note that because the flow is laminar, the shear stress, τ, and the

pressure, P_z, are functions only of r and z, respectively. The steady-state flow of the magma means that there is no acceleration, and the sum of the forces in the z direction must be zero. We can, therefore, write

$$-dP\,\pi r^2 - \rho_m\,\pi r^2\,dz\,g + \tau \pi 2r\,dz = 0$$

Substituting $\eta\,dv/dr$ for τ (Eq. 2-8) and rearranging gives

$$\frac{dv}{dr} = \frac{1}{2\eta}\left(\frac{dP}{dz} + \rho_m g\right)r$$

which, upon integration, gives

$$v = \frac{1}{2\eta}\left(\frac{dP}{dz} + \rho_m g\right)\frac{r^2}{2} + c$$

The integration constant can be evaluated, because $v = 0$ where $r = r_0$; hence,

$$v = -\frac{1}{4\eta}\left(\frac{dP}{dz} + \rho_m g\right)(r_0^2 - r^2) \tag{3-7}$$

If the magma is rising solely due to buoyancy, the ratio dP/dz is simply the pressure gradient induced in the magma by the density of the intruded rocks. If the country rocks have a density ρ_c, the pressure gradient, dP/dz, will simply be $-\rho_c g$. Hence, the terms in the first set of parentheses in Eq. 3-7 reduce to $(g\rho_m - g\rho_c)$. From this it follows that if the densities of the magma and the intruded rock are the same, the flow velocity will be zero; if the density of the magma is less than that of the surrounding rock, upward flow will occur.

The group of terms in the first set of parentheses of Eq. 3-7 can also be considered to represent the gradient of pressure within the magma that is in excess of the magmastatic pressure gradient ($\rho_m g$), that is, the part of the gradient that can cause flow. If we indicate the gradient of excess pressure as dP_{ex}/dz, Eq. 3-7 can be reduced to

$$v = -\frac{dP_{ex}}{dz}\frac{1}{4\eta}(r_0^2 - r^2) \tag{3-8}$$

Equation 3-8 indicates that the velocity profile across the volcanic pipe is parabolic, with the central part moving the most rapidly. In studying volcanic processes, however, knowledge of the average velocity, $\bar{v}$, is commonly more useful than the maximum velocity. This can be obtained by considering the volume of material that passes any cross-sectional area of the pipe in a unit time. This quantity is known as the *flux* (J) and is given by *Hagen–Poiseuille law*,

$$J = \int_0^{r_0} v 2\pi r\,dr = \int_0^{r_0} -\frac{dP_{ex}}{dz}\frac{1}{2\eta}(r r_0^2 - r^3)\pi\,dr = -\frac{\pi r_0^4}{8\eta}\frac{dP_{ex}}{dz} \tag{3-9}$$

The average velocity is then

$$\bar{v} = \frac{J}{\text{area of pipe}} = \frac{J}{\pi r_0^2} = -\frac{r_0^2}{8\eta}\frac{dP_{ex}}{dz} \tag{3-10}$$

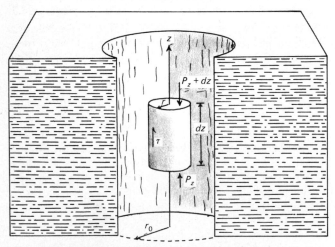

FIGURE 3-4 A small cylindrical volume of magma with radius r and length dz within a volcanic feeder pipe of radius r_0. The steady-state laminar flow of magma within the cylinder results in a shear stress (τ) on the surface of the cylinder. The positive direction of z in this case is upward. See text for discussion.

which is one-half of the maximum velocity at the center of the pipe. Calculations of actual velocities are left to the reader to do in Probs. 3-5 to 3-9.

In this analysis of the flow of magma in a volcanic feeder pipe, the flow has been assumed to be laminar. Experiments indicate that when the value of $2r\rho\bar{v}/\eta$ exceeds 2300 (terms as defined above), a small initial disturbance grows and leads to turbulence, whereas below this value it is damped out, and laminar flow prevails. This group of terms, known as the *Reynolds number*, and symbolized by Re, is a dimensionless number (see below) that gives a measure of the balance between inertial and viscous forces acting on the fluid.

$$\text{Re} = \frac{2r\rho\bar{v}}{\eta} \qquad \text{m} \ \frac{\text{kg}}{\text{m}^3} \ \frac{\text{m}}{\text{s}} \ \frac{\text{m s}}{\text{kg}} \qquad (3\text{-}11)$$

The value of 2300 is said to be the *critical Reynolds number* for flow in a cylindrical conduit. Problem 3-5 illustrates how this number can be used to determine when turbulent flow might be present in a magmatic pipe. If turbulence does occur, the analysis of feeder flow above would be invalid. But with the high viscosities of magmas, especially those rich in silica, laminar flow is likely to be more common than turbulent flow. Turbulent flow might, however, be expected for large radius conduits up which very fluid basalt rises (Shaw and Swanson 1970), and komatiitic magmas with their extremely low viscosities (< 0.3 Pa s) almost certainly would flow turbulently (Huppert and Sparks 1985). Turbulent flow will be considered in Sec. 3-8.

In the previous treatment, magma in a volcanic feeder pipe was taken to be in a steady state of laminar flow. In reality, the pipe must have a beginning, possibly where it taps a magma chamber, and at this point the previous treatment is no longer valid. Magma entering the pipe does not immediately have a well-developed parabolic velocity profile. In fact, the magma will have to travel some distance along the pipe before attaining the steady-state velocity profile previously calculated. This distance, known as the *entrance length*, is of great importance in a wide range of fluid mechanics problems, including the flow of human blood through the aorta.

The entrance length can be appreciated by referring to Figure 3-5, which illustrates a volcanic feeder pipe extending from a magma chamber. At the instant magma enters the pipe, the velocity is constant across the pipe because there has not yet been any viscous drag along the walls. But as soon as magma begins traveling up the pipe, a boundary layer develops, due to the viscous drag on the pipe walls. The velocity of magma near the walls decreases, but because the average velocity must remain constant along the pipe (assuming constant radius), the magma in the center must simultaneously increase its velocity. This process continues until the boundary layer completely fills the pipe. Then flow is said to be fully developed, and, from there on, the velocity profile remains constant.

The entrance length, symbolized by L_e, is given by

$$L_e = 0.1150(\text{Re})r \qquad (3\text{-}12)$$

where r is the radius of the pipe. Because the Reynolds number already includes the radius of the pipe, the entrance length

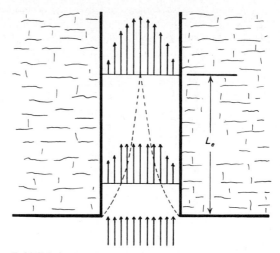

FIGURE 3-5 Magma on first entering a feeder pipe from a magma chamber has the same velocity at all points across the pipe, indicated here by the lengths of the arrows. As the magma rises in the pipe, a boundary layer develops in which velocities are decreased by the viscous drag on the walls. Simultaneously, magma in the central part speeds up in order to maintain a constant flux of magma along the pipe. After a distance known as the entrance length, L_e, the boundary layer completely fills the pipe, and from there on the velocity profile is parabolic.

varies as the square of the radius. The entrance length for a large-diameter pipe could therefore be considerable, and a parabolic velocity profile might not be achieved in the distance available (Prob. 3-5d). This could be of importance in calculations regarding velocities of intrusion and to theories regarding the migration of crystals away from the walls of magma conduits (Chapter 10).

Expressions for velocity profiles across dikes can be derived in much the same way as those for circular pipes. Their derivation will not be given here, but the interested reader should do Prob. 3-10 or refer to standard texts in fluid mechanics for further information. The velocity at any point within a vertical dike, of width W (Fig. 3-6), in which there is steady-state laminar flow, is given by

$$v = -\frac{1}{2\eta}\left(\rho_m g + \frac{dP}{dz}\right)(Wx - x^2) \qquad (3\text{-}13)$$

The similarity to pipe flow should be evident (see Eq. 3-5). The velocity profile is again parabolic. The average velocity in a vertical dike is given by

$$\bar{v} = -\frac{1}{12\eta}\left(\rho_m g + \frac{dP}{dz}\right)W^2 \qquad (3\text{-}14)$$

Equation 3-13 can be generalized for a sheetlike body with any angle of dip θ by multiplying the ρg term by $\sin\theta$. Thus the term in the first set of parentheses becomes $(\rho g + dP/dz)$ for a vertical dike and (dP/dz) for a horizontal sill. The reference direction, of course, would have to be changed so that in each case z is still in the direction of flow.

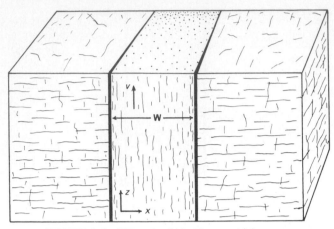

FIGURE 3-6 Dike of width W up which magma flows with a velocity v in the vertical $+z$ direction.

As with flow in pipes, too rapid flow of magma in a dike leads to turbulence, which may, in turn, affect such processes as crystal–liquid separation and cooling. The Reynolds number for flow in a dike is defined as

$$\text{Re} = \frac{\rho W \bar{v}}{\eta} \qquad (3\text{-}15)$$

where v is the average flow velocity (see Prob. 3-16). Again, the critical value marking the onset of turbulence is 2300.

A Reynolds number can also be formulated for a lava flow, which is referred to as open channel flow. This type of flow experiences viscous drag only from the base and sides—the top of the flow is a free surface. This is taken into account by defining the Reynolds number as

$$\text{Re} = \frac{4 r_h \rho \bar{v}}{\eta} \qquad (3\text{-}16)$$

where r_h, the *hydraulic radius*, is the cross-sectional area of the channel divided by the wetted perimeter. For a lava flow that spreads over a great area, the thickness is very small compared with the width. The hydraulic radius thus reduces to the thickness of the flow, a dimension that is easily measured, even in poorly exposed terranes. When the Reynolds number is less than 500, flow is laminar in an open channel, and when greater than 2000, it is turbulent. Between these values is a transition, with the type of flow depending on the detailed nature of the channel bed (Shaw and Swanson 1970). Perhaps the most important effect of the change from laminar to turbulent flow in a lava is the dramatically increased cooling rate. But because of the high viscosities of most lavas, turbulence is a rarity, except in very thick basaltic flows, komatiitic flows, and highly gas-charged ash flows (see Chapters 4 and 5). Where flows do move in a turbulent manner, their average velocity is given by

$$\bar{v} = \left(\frac{2g \sin \theta \, h}{k} \right)^{1/2} \qquad (3\text{-}17)$$

where θ is the surface slope, h the thickness of the flow, and k is a friction coefficient, which has values between about 0.01 and 0.06, depending on the roughness of the surface.

Before leaving the discussion of laminar flow rates through conduits, it is worth remarking on one important aspect that will be encountered again in later chapters. The flux of magma passing along a pipe is given by Eq. 3-9, which is known as the Hagen–Poiseuille law. It consists of two parts: the first group of terms expresses the ease with which the conduit allows material to pass, whereas the second part gives the gradient that provides the driving force for movement. All transport phenomena, whether involving fluid, heat, electricity, or diffusing ions, obey similar laws. Here are two for comparison with Eq. 3-9, which will be encountered in later chapters:

Fluid flow: $\quad J_x = -\dfrac{\pi r^4}{8\eta} \dfrac{\partial P}{\partial x} \quad$ Hagen–Poiseuille law $\quad$ (3-9)

Heat flow: $\quad J_x = -K_T \dfrac{\partial T}{\partial x} \quad$ Fourier's law $\qquad$ (5-3)

Diffusion: $\quad J_x = -D \dfrac{\partial c}{\partial x} \quad$ Fick's law $\qquad$ (18-7)

In each of these, J_x is the flux or amount of the quantity transported per square meter per second in the x direction. The proportionality constants are K_T, the thermal conductivity, and D, the diffusion coefficient. The gradients are expressed as partial derivatives of pressure, temperature, and concentration with respect to x, because only the gradients in the x direction cause transport in that direction. Similar partial derivatives can be written for the gradients in the y and z directions when the flux in those directions is considered.

3-7 FLOW RATE OF A BINGHAM MAGMA

In the derivations of flow rates through conduits, magma was assumed to be a Newtonian liquid. This was valid as long as the magma was largely liquid, but when it contains a significant number of crystals or gas bubbles it can behave as a Bingham liquid; that is, it has a yield strength, which must be overcome before flow can occur. This type of behavior is expected if magmas continue to flow as they crystallize. Because this late-stage flow may affect the appearance of the final rock, it is important for us to examine how flow of Bingham and Newtonian liquids differ. We will do this for the case of a dike.

The shear stress, τ, at any point in a vertical dike with the dimensions and reference directions shown in Figure 3-6 is given by

$$\tau = -\left(\rho_m g + \frac{dP}{dz} \right)\left(\frac{W}{2} - x \right) \qquad (3\text{-}18)$$

From Figure 2-2 we can substitute $\tau_0 + \eta \, dv/dx$ for τ (τ_0 is the yield strength) as long as $\tau > \tau_0$ and $x < W/2$. This gives, on rearranging and integrating, an expression for the flow velocity of a Bingham liquid in a dike:

$$v = -\frac{1}{\eta}\left[\left(\rho_m g + \frac{dP}{dz}\right)\frac{Wx - x^2}{2} - \tau_0\left(\frac{W}{2} - \left|\frac{W}{2} - x\right|\right)\right] \quad (3\text{-}19)$$

which is valid as long as $\tau > \tau_0$. The shear stress in a flowing dike decreases from a maximum at the margins to zero at the center. At some distance in from the margins, the shear stress must drop below the yield strength of the Bingham liquid, and from there through to an equivalent point on the other side of the dike, no additional flow is possible. The result is a central sheet that moves with constant velocity (Fig. 3-7).

The same type of Bingham flow will occur in pipelike conduits and lava flows, which will have a central or upper part, respectively, that flows with constant velocity. The fraction of the total flow represented by the plug depends on the yield strength of the magma; if it is high, most of the magma may flow as a single plug (see Prob. 3-13). The yield strengths of magmas are poorly known, but they can vary over many orders of magnitude. For example, values as high as about 10^6 N m^{-2} have been estimated for dacite flows on the surface of the Earth.

The change from Newtonian to Bingham behavior has a number of important consequences. Most obvious is that flow velocities are diminished because the yield strength of a magma provides an additional resistance to flow. Processes that depend on the shearing of liquid against a contact, such as alignment of phenocrysts or stretching of gas bubbles parallel to contacts, and movements of crystals away from conduit walls (flowage differentiation; Sec. 13-3) will not operate within the region of plug flow. Finally, the sinking of dense crystals through a less dense Bingham liquid will not take place if the crystals do not exert sufficient force to overcome the yield strength of the liquid (see Sec. 13-3).

3-8 INTRUSION RATES OF TURBULENT MAGMA

The discussions above have dealt only with laminar flow. In passing, however, it was mentioned that the very high Reynolds numbers for low-viscosity komatiites and rhyolitic ash flows and thick basaltic flows and dikes indicate that these magmas may have flowed turbulently. These are important types of magma, and we should therefore have some means of evaluating the rate at which they rise from their source or are erupted on the surface. Derivations of the equations of flow of turbulent liquids are not as simple as those for laminar flow, but the interested reader can find them in any of the standard texts on fluid mechanics. Application of these equations, however, is simple.

If magma is assumed to rise buoyantly in a vertical dike of width W, the average turbulent flow velocity, $\bar{v}$, is given by

$$\bar{v} = \left(\frac{g\,\Delta\rho\,W}{k\rho_m}\right)^{1/2} \quad (3\text{-}20)$$

where $\Delta\rho$ is the difference between the densities of country rock and magma, ρ_m is the density of the magma, and k is a friction

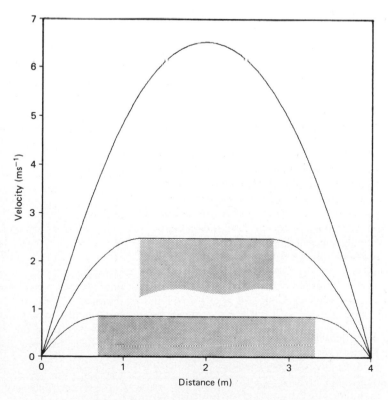

FIGURE 3-7 Laminar flow velocities in a 4-m-wide basaltic dike, in which the magma with a density of 2.6 Mg m^{-3} is buoyantly emplaced in response to a density contrast of 100 kg m^{-3}. The upper curve is for Newtonian behavior with a viscosity of 300 Pa s. The lower two curves are for Bingham behavior, where the viscosity is still 300 Pa s but the yield strength of the magma is 750 Pa (middle) and 1250 Pa (lower). When the magma behaves as a Bingham liquid, its velocity is diminished and a central portion (shaded) flows as a plug, the width of which increases as the yield strength of the magma increases. The maximum shear stress developed at the margin of this dike under its specified conditions is 1960 Pa. Should the yield strength of the magma exceed this value, no flow would be possible.

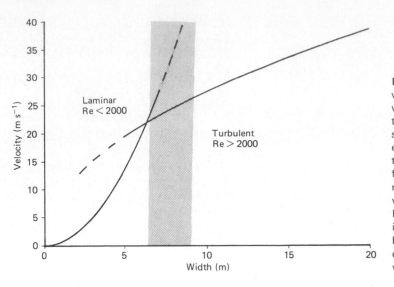

FIGURE 3-8 Variation in average flow velocity in basaltic dikes as a function of width. The basaltic magma is assumed to have a viscosity of 300 Pa s and a density of 2.6 Mg m^{-3}, and to be buoyantly emplaced in response to a density contrast of 200 kg m^{-3}. Because of laminar flow, velocities in narrow dikes increase rapidly with increasing dike width. When widths are greater than about 8 m, the Reynolds number exceeds 2000 (assuming a friction factor of 0.01), and flow becomes turbulent. Velocities in wider dikes therefore do not increase as rapidly with increasing dike width.

coefficient that has values between about 0.01 and 0.06, depending on the roughness of the wall (Huppert and Sparks 1985). This relation is valid as long as the Reynolds number, which is defined as $\rho \bar{v} W/\eta$, is greater than about 2000. The vertical flux of turbulent magma per unit length of dike, J, is simply the product of the average velocity and the width,

$$J = \left(\frac{g\,\Delta\rho}{k\rho_m} \right)^{1/2} W^{3/2} \qquad (3\text{-}21)$$

The most significant difference between the expressions for the average velocity of turbulent and laminar flow of magma in a dike is the way in which dike width affects velocity; in turbulent flow the velocity is proportional to the square root of the width, whereas in laminar flow it is proportional to the width squared. This is brought out in Figure 3-8, where the average velocity in vertical dikes of basaltic magma having $\eta = 300$ Pa s, $\rho_m = 2.6$ Mg m^{-3}, and ρ_c (country rock) $= 2.8$ Mg m^{-3} is plotted as a function of dike width. In dikes with a width up to about 8 m, Reynolds numbers are <2000 and flow is laminar; in wider dikes, flow is turbulent and then velocities increase only slightly with increasing width (Prob. 3-16).

One of the most important consequences of turbulent flow is that temperatures are maintained approximately constant throughout the magma. Turbulence transfers heat rapidly to margins by convection, where it can heat country rocks or be radiated into space in the case of a lava flow. In laminar flow, heat is transferred by conduction, which is a slow process. The heat effects on the surroundings are consequently less pronounced, and cooling rates are very much slower. This point is discussed in Chapter 5.

3-9 DIAPIRIC INTRUSION OF MAGMA

The discussion so far in this chapter has dealt with the passage of relatively fluid magmas through rigid rock. In these cases, igneous injection is rapid. Because magmas (and rocks) have extremely low thermal conductivities, rates of heat loss are typically orders of magnitude less than rates of intrusion (except

for turbulent flows). Consequently, emplacement of magma is completed with little or no loss of heat, that is, adiabatically. Some magmatic flow, however, is so slow that heat losses can be substantial during intrusion. This is particularly true for bodies of magma that rise as *diapirs*. Here, rates of intrusion are determined largely by the high viscosity of the wall rock that must be forced aside by the magma. Bodies of granitic magma rising through the crust or bodies of partially molten peridotite rising through the asthenosphere or lower lithosphere provide important examples of this type of magmatic emplacement.

Diapiric intrusion occurs when a density inversion causes material to buoyantly rise into overlying denser material. Numerous examples of this phenomenon are found in nature; the common thunder cloud, which results from cold dense air overlying warm less dense air, is perhaps the most familiar. Sedimentary beds of salt (a low-density mineral) commonly form diapirs that intrude overlying beds of denser sedimentary rocks. Because magmas are less dense than equivalent solid rock, melting within the Earth typically sets up the conditions for diapiric rise. The book by Ramberg (1981) provides an excellent treatment of the subject; it is well illustrated with many scale models of geologically interesting examples of diapirs.

The sequence of photographs in Figure 3-9 illustrates the development of diapirs, in this case between oil and honey. Figure 3-10 shows a similar-appearing set of diapirs developed between magmas of nepheline monzonite and alkali basalt composition. Both figures illustrate an important feature of diapirs; that is, diapirs form in groups that have characteristic spacings or wavelengths. Regardless of the number of times the oil and honey model of Figure 3-9 is operated, the spacing between the diapirs remains the same, as long as the temperature and quantities of the two fluids are kept constant. In situations where the thickness of the upper layer is great enough so that the upper surface does not interfere with the waves that develop on the interface between the two fluids, Selig (1965) has shown that the spacing between diapirs, λ, is given by

$$\lambda = 2.92 h \left(\frac{\eta_t}{\eta_b} \right)^{1/3} \qquad (3\text{-}22)$$

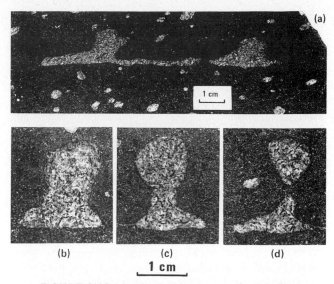

FIGURE 3-10 Small diapiric domes of nepheline monzonite that rose from a thin layer of felsic magma in the upper part of a mafic alkaline sill at Ste. Dorothée, Quebec. Serial sections through the small dome on the right in (a) are shown in photographs (b) to (d). The sheet of felsic liquid, which is interpreted to have separated from the mafic magma as an immiscible liquid, was less dense than the overlying magma; it was therefore gravitationally unstable and tried to rise as diapirs toward the top of the sill. On a larger scale, granite batholiths may have the same general form. [After Philpotts (1972); published with permission of Elsevier Science Publishers.]

FIGURE 3-9 Progressive stages in the development of buoyantly rising domes of oil into overlying denser honey. The wavelength of the disturbance on the oil–honey interface, which determines the eventual spacing of the domes, is determined by the viscosity contrast between the two liquids and the thickness of the oil layer. Diapiric domes of salt, batholiths of granitic magma, and bodies of andesitic magma beneath island arcs presumably rise in a similar manner.

where h is the thickness of the buoyant layer and η_t and η_b are viscosities of the top and bottom layers, respectively. Thus, if the viscosity of the lower layer is decreased while the other parameters remain constant, the spacing between diapirs would increase. This is intuitively what one would expect, because the more fluid the lower layer, the greater is its lateral mobility.

The simple relation between viscosities, source-layer thickness, and diapir spacing can be used to determine any one of these properties if values of the others are known or can be estimated. For example, Philpotts (1972) used the spacing between the diapirs (about 6 cm) developed on the thin layer (about 0.25 cm) of nepheline monzonite in an alkali basalt sill (Fig. 3-10) to determine the ratio of viscosity of the basalt to that of the syenite to be about 600. At a very different scale, Marsh and Carmichael (1974) determined that the thickness of the source layer of andesitic magmas along the top of the Benioff

zone must be about 20 m to account for the typical 70-km spacing between volcanoes along many island arcs.

Despite the success of the diapiric model in explaining the regular spacing of such features as salt domes, serious problems arise when it is applied to igneous bodies that have risen any distance through the lithosphere (Marsh 1982; Spera 1980). Diapiric bodies of magma must remain hot enough to be molten if they are to continue rising through the progressively cooler lithosphere. The high viscosity of the lithosphere (about 10^{20} Pa s) causes ascent rates to be slow, which, in turn, may allow sufficient heat to be lost to wall rocks to cause the magma to solidify. Marsh and Kantha (1978) calculate that for a 6-km spherical diapir of magma to rise to the surface of the Earth without solidifying, its ascent rate would have to be greater than 3 m a^{-1}. If magmatic diapirs are to ascent at these rates, they must do so by reducing the viscosity of the lithosphere through which they travel. This is most easily done through partial melting of the wall rocks (Marsh 1982). But this in turn causes some of the magma to crystallize, thus reducing the amount of liquid that will reach the surface of the Earth. Repeated passage of diapirs through the same part of the lithosphere could raise wall rock temperatures and by so doing increase the percentage of liquid in each successive diapir. Clearly, the rise of a magmatic diapir is a complicated problem involving heat transfer, fluid convection, and partial melting. The topic will not be pursued further here; the interested reader is referred to papers by Marsh (1982) and Spera (1980).

3-10   FACTORS IN SOURCE AFFECTING THE SUPPLY OF MAGMA

The rates at which magmas move are determined by numerous factors, some of which are dependent on the magma itself, such as density, viscosity, and yield strength, whereas others are determined by the environment, such as gravitational acceleration, lithostatic pressure, and conduit shape and roughness. We have also seen that although many magmas flow in a laminar fashion, ones with low viscosities may flow turbulently, especially when in wide conduits or thick flows. For each of the various types of flow, steady-state velocities can be determined. It would be a mistake, however, to conclude that magmas must always flow with these velocities. All magmatic activity is transient; it has a beginning and an end. There will be periods during which flow will be either increasing to the steady-state condition or diminishing toward the end of the activity. These periods of flux may, in fact, constitute the largest fraction of any particular igneous episode. It is important therefore to consider what might affect the rates of flow during these stages of activity.

Once steady-state flow has been reached during an episode of magmatic activity it can be expected to continue as long as magma is available and the driving pressure remains the same. Should either of these factors decrease, magmatic flow would decrease or even stop. Because the driving pressure on a magma is most likely the result of buoyancy, any factors affecting the density of magma could cause fluctuations in flow rate. Apart from this, fluctuations are likely to result from changes in the availability of magma.

Bodies of magma are formed by the partial melting of mantle or crustal rocks followed by the segregation and separation of the liquid from its refractory residue. Magma containing a large fraction of refractory crystals does not have the bulk physical properties (low density and viscosity) necessary for it to ascend rapidly through conduits. When rocks partially melt, the first-formed liquid occupies spaces between grains of different minerals, but because of wetting properties, the liquid spreads along grain edge intersections. The rock therefore mops up the liquid as it is formed, in much the same way that a sponge absorbs water. Magma can flow only slowly between grains at this stage because the channels are narrow and tortuous. Furthermore, surface tension forces are significant at this scale. The most likely type of movement at this stage is the diapiric rise of an entire mass of partially melted rock. Only when magma is segregated from its refractory residue does viscosity become low enough for magma to rise rapidly through conduits.

A greater degree of melting is necessary before liquid is able to segregate and migrate upward. This degree of melting, which is analogous to the quantity of water a sponge can absorb before becoming saturated, is between 5 and 22%, depending on the surface tension between liquid and crystals. Once sufficient melt is produced to wet grain edges, segregation of liquid can take place. But rapid ascent through conduits may still not be possible because this early segregated liquid is likely to carry high percentages of crystals and thus behave as a Bingham liquid. Rapid flow would not occur until the fraction of liquid increased to the point where buoyant forces could overcome the yield strength of the liquid (see Prob. 3-13). Once this happens, the ascent of magma is likely to be relatively rapid compared with that during the segregation stage. Thus magmatic activity is likely to be episodic and flow rates highly variable.

PROBLEMS

3-1. In the example worked out on page 19, magma of density 2.9 Mg m^{-3} that was able to rise buoyantly just to the surface of the Earth through a mantle of density 3.3 Mg m^{-3}, 10 km of lower crust with density 3.0 Mg m^{-3}, and 25 km of upper crust with density 2.75 Mg m^{-3} was shown to have originated at a depth of 41.9 km beneath the surface. In this example the magma was taken to be incompressible, but if the coefficient of compressibility, β, is known to be 5.0×10^{-11} Pa^{-1}, what would the depth to source be if the surrounding rocks are again taken to be incompressible and g to remain constant over the depth considered?

(Before doing any problem, it is worth anticipating the answer. For example, in this case, will the depth to the source be greater or less than that in the example where the magma was taken as being incompressible? In working a problem, equations are often derived that can be solved only by trial and error. By substituting estimated values for the unknown and noting the resulting error, a correct value can soon be obtained.)

3-2. (a) Epp (1984) states that Mauna Loa towers 8192 m above the floor of the Pacific Ocean, but only 4169 m of this is above sea level. He estimates that about 150 m of ocean-floor sediment covers the oceanic crust, which is 6.5 km thick in this region. The densities of the various materials, in Mg m^{-3}, are: seawater 1.0, sediments 2.0, crust 2.84, mantle 3.1, and magma 2.9. Calculate the depth of origin below sea level of an incompressible magma that rises buoyantly to the summit of Mauna Loa.

(b) Magma erupting onto the floor of the caldera of Kilauea, to the southeast of Mauna Loa, rises only 1240 m above sea level. Assuming that densities and thickness of materials beneath this volcano are similar to those beneath Mauna Loa, what difference in the depth to source region might be expected between these two volcanoes? Does this calculated depth for the source of magma agree with evidence from seismic disturbances associated with Hawaiian eruptions? (See page 36.)

3-3. Assume that magma is able to rise buoyantly through the lithosphere just to the surface of the Earth, and that the lithosphere is stratified with respect to density as indicated in the adjoining table. Plot a graph showing the excess pressure, ΔP, in a column of magma as a function of depth for **(a)** a komatiitic magma with density 2.8 Mg m^{-3}, **(b)** a basaltic magma with density 2.7 Mg m^{-3}, **(c)** an andesitic magma with density 2.6 Mg m^{-3}, **(d)** a rhyolitic magma with density 2.4 Mg m^{-3}. The density of each magma is taken to remain constant with depth. On the same graph, plot a line marking the maximum tensile strength of the rocks, which can be taken to be about 10 MPa. At what depth might plutonic bodies of each of these magmas form?

Depth (km)	Density ($Mg\ m^{-3}$)	ΔP (Pa) − o +
0		
	2.3	
3		
	2.75	
25		
	3.0	
35		
	3.3	

3-4. The maximum amount of water soluble in a basaltic melt at normal magmatic temperatures (about 1200°C) increases with pressure according to the approximate relation

$$\text{wt \% } H_2O \text{ in basaltic magma} = 6.8 \times 10^{-6} P^{0.7}$$

where P is measured in pascal. If the pressure on a rising column of water saturated basaltic magma is equal to the lithostatic pressure imposed by the intruded rock, which has a density of 2.8 Mg m^{-3} to a depth of 5 km, the pressure will be 0.1372 GPa and the water content of the magma will be 3.38 wt %. Below this depth, 5 wt % water is held in solution in the basalt, but above this, exsolution of the gas must occur. If the magma is taken to contain the saturation amount of H_2O throughout its rise to the surface, construct a graph showing the *weight* fraction of exsolved gas as a function of depth, assuming that the temperature remains constant at 1200°C, the density of the silicate fraction of the magma remains constant at 2.7 Mg m^{-3}, and the gas obeys the ideal gas law. Also mark the depth at which the *volume* fraction of the gas reaches 0.75 and disruption of the magma takes place.

Repeat the calculation for rhyolitic magma, where the solubility of water in a saturated magma is given by

$$\text{wt \% } H_2O \text{ in rhyolite magma} = 0.411 \times 10^{-3} P^{0.5}$$

Assume that the rising magma, which has a density of 2.4 Mg m^{-3}, becomes saturated at a depth of 5 km and that the magma temperature remains constant at 800°C. What conclusion might you draw concerning the nature of the eruptions in each case?

3-5. During the 1959 eruption of Kilauea Iki, a fountain of lava rose from the vent to a height of 450 m, and on one occasion, to 580 m. The ponding of this lava within the crater allowed for accurate monitoring of the volumes of erupted material in much the same way as a measuring cylinder is used to determine volumes of liquid in the laboratory. The flux of lava was estimated to have varied from 100 to 300 m^3 s^{-1}. Although there were periods of more violent eruption brought about by escaping gas, the steady fountaining is thought to have simply resulted from the velocity of ejection of magma from the orifice.

(a) Using the height of 450 m and neglecting air resistance, calculate the maximum velocity of ejection of lava from the orifice. Assuming a parabolic velocity profile across the orifice (i.e., fully developed flow), determine the average velocity of ejection.

(b) Assuming a cylindrical feeder pipe, use the average velocity from part (a), along with the rate of ejection of 100 m^3 s^{-1}, to determine the diameter of the pipe.

(c) If the magma had a density of 2.75 Mg m^{-3} and viscosity of 300 Pa s, would you expect turbulent or laminar flow in a feeder pipe with the diameter calculated in part (b)?

(d) If the diameter of the feeder pipe calculated in part (b) remains constant down to the magma chamber, how far would the magma have to rise before fully developed flow was achieved in the pipe? Was the assumption of parabolic velocity profile in part (a) justified?

3-6. If the diameter of the Kilauea Iki feeder pipe was 1.65 m, what would the (a) maximum and (b) average velocities have been for the magma if it rose in response to a pressure gradient induced solely by the loading of solid basalt ($\rho = 3.0$ Mg m^{-3}) on the magma ($\rho = 2.75$ Mg m^{-3})? The viscosity of the magma is 300 Pa s.

3-7. If the pressure gradient causing the eruption of 100 m^3 s^{-1} of lava from Kilauea Iki in 1959 was due solely to the loading of solid basalt on the magma, what diameter would you expect the feeder pipe to have? The viscosity and density are the same as in Prob. 3-6.

3-8. The 1959 eruption of Kilauea Iki developed a deep lava lake which eventually covered the vent. During periods of quiescence, the lava from the lake drained back down the vent at the incredible rate of 425 m^3 s^{-1}. Assuming that the lava attained a constant rate of back flow, and the pressure on the base of the column was the same as that on the top (1 atm, or 10^5 Pa), the only driving force would have been that of gravity. Calculate the diameter of the vent using a magma density of 2.75 Mg m^{-3} and viscosity of 300 Pa s.

3-9. Discuss differences in the calculated diameters of the Kilauea Iki feeder pipe in Probs. 3-5, 3-7, and 3-8. Indicate which method of calculation is likely to give the most accurate value. How does the power to which the radius is raised in the various equations affect the sensitivity of the results?

3-10. Derive Eq. 3-13, which expresses the velocity of magma rise at any point across a vertical dike that is in steady-state laminar flow. Begin by considering the forces acting on a small magma volume, $dx\ dy\ dz$.

3-11. Starting with Eq. 3-13, derive an expression for the average laminar flow velocity in a dike.

3-12. Prove mathematically that the maximum velocity in a dike does, indeed, occur at its center.

3-13. What fraction of a 10-m-wide dike of dacite that behaves as a Bingham liquid would flow as a plug if the magma has a yield strength of 10^4 N m^{-2} and the magma is buoyantly emplaced in response to a density contrast between magma and country rocks of 300 kg m^{-3}?

3-14. Starting with Eq. 2-11, which expresses the laminar flow velocity as a function of height within a lava flow, derive an expression for the average laminar flow velocity of a lava.

3-15. Why are each of the transport laws (Hagen–Poissuille, Fourier's, and Fick's) preceded by a negative sign?

3-16. Jurassic basalts in eastern North America, associated with the initial opening of the Atlantic, are similar to those still being erupted today at the mid-Atlantic ridge in Iceland. In both areas erosion has exposed dikes that were feeders to fissure eruptions.

In Iceland these dikes average 4 m in width, but in eastern North America many are as much as 60 m wide. Assuming that magma rose buoyantly in both areas and that the average density of the intruded lithosphere was 3.0 Mg m^{-3} in Iceland and 2.9 Mg m^{-3} in eastern North America, determine whether the flow was laminar or turbulent in these two areas, and calculate the average flow velocities in the respective dikes. The basaltic magma in both areas can be taken to have had a density of 2.6 Mg m^{-3} and a viscosity of 300 Pa s, and the friction factor in the dikes was 0.01.

3-17. Volcanoes along the Aleutian Arc are spaced about 70 km apart, whereas those in the adjoining Kamchatka belt are only 30 km apart. Assuming the viscosity of the lithosphere in the two regions to be similar, speculate on what might account for the different spacings.

4 / *Forms of Igneous Bodies*

4-1 INTRODUCTION

Bodies of igneous rock are referred to as *extrusive* when formed on the Earth's surface, and *intrusive* when formed within the Earth. Intrusive bodies are further subdivided into *plutonic*—large intrusions formed at moderate to great depth, and *hypabyssal*—small intrusions formed near the Earth's surface. Hypabyssal rocks cool rapidly, and most bear more resem-

blance to volcanic rocks than to coarser-grained plutonic ones. Extrusive bodies of igneous rock are relatively well understood, for many of them have actually been studied during their formation (MacDonald 1972; Williams and McBirney 1979; Decker and Decker 1981). The same cannot be said for plutonic bodies, where even their form may be uncertain and their processes of formation often conjectural. We shall, therefore, start with a discussion of the extrusive bodies.

Extrusive Bodies

4-2 FLOOD BASALTS

At many times in the Earth's history, basalt has poured out from long fissures and multiple vents in such large quantities as literally to flood vast areas of the Earth's surface. Rates of extrusion were sufficiently great and viscosities low enough to allow the basalt to spread out as almost horizontal sheets rather than form volcanoes. Later erosion of these areas has commonly led to the development of plateaus, from whence the name *plateau basalt* is derived. They are also termed *flood basalts* in reference to their mode of emplacement.

Flood basalts are the most voluminous type of extrusive rock. The Cretaceous Deccan trap of western India covers an area of at least 500,000 km² and averages 600 m in thickness. In the western United States, the Miocene Columbia River basalts in the state of Washington cover an area of 200,000 km² and in places are more than 1500 m thick (Fig. 4-1). To the southeast, in Idaho, the Snake River plain is underlain by 50,000 km² of basalt. In other areas, erosional remnants, such as the 4500-m-thick Precambrian basalts of the Keweenawan peninsula on Lake Superior, are all that attest to the former existence of these great sheets. Perhaps the most extensive of all were those of the Tertiary Brito-Arctic, or Thulean province (Fig. 4-2) now exposed as erosional remnants in Northern Ireland, Scotland, Iceland, and Greenland. In addition to these major fields there are many smaller areas of flood basalt, such as those of the various Triassic–Jurassic basins of eastern North

America (see Fig. 4-26). Fortunately for man (but not for petrologists), no flood basalts have erupted during historic times. The 1783 fissure eruption of Lakagigar, Iceland, is the most voluminous eruption that has been witnessed, but its volume of 12 km³ erupted over six months is small by comparison with those in the geologic record (Thorarinsson 1968).

A special but extremely important form of flood basalt is that which issues from the fissures along the mid-ocean ridges. Mid-ocean ridge basalt, *MORB* for short, is the most abundant rock type on the surface of the Earth; most ocean floors are covered by it. Knowing the length of the ocean ridges and their spreading rates, it is simple to calculate that an average of 3 km² of new oceanic crust is produced each year. On the ocean floor this new crust consists of basaltic flows, beneath which are sheeted dike complexes, and below that are larger bodies of gabbroic rocks. In total, 6.7×10^{13} kg a^{-1} of new igneous rock is formed along the ridges. The ocean floors, however, are destroyed as rapidly as they are formed, and as a result, MORB is not preserved in the geological record in proportion to its abundance at any instant in time. Only rarely in subduction zones do some ocean floor rocks get incorporated and preserved with continental rocks. These produce the ophiolites, which are discussed in Sec. 14-2.

Flood basalts owe their great thicknesses mostly to the accumulation of large numbers of thin flows, but some thick flows may also be present. Individual flows typically range in thickness from meters to tens of meters and more rarely to as much as 100 m. They can extend laterally for many kilometers

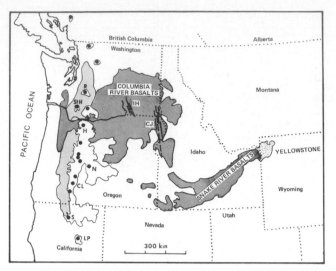

FIGURE 4-1 Volcanic rocks of the northwestern United States. Lavas from the older Cascade strato volcanoes are shown in light stippled pattern, whereas the lavas from the younger High Cascades (to the east of the older lavas) are outlined in clear with some of the main strato volcanoes being marked with letters: Rainier (R), St. Helens (St.H), Hood (H), Newberry (N), Crater Lake (CL), Shasta (S), and Lassen Peak (LP). Flood basalts of the Columbia and Snake Rivers are shown in heavy stipple. The Ice Harbor (IH) and Chief Joseph (CJ) dike swarms were feeders to the Columbia River basalts. The rhyolitic rocks of Yellowstone presumably reflect the present position of a hotspot whose trace has produced the linear belt of Snake River Basalts as the North American plate has moved westward. [Sources of data: Waters (1955) and Smith and Christiansen (1980).]

FIGURE 4-2 Tertiary flood basalts, County Antrim, Northern Ireland. Repeated eruptions of fluid basalt have resulted in the formation of laterally extensive horizontal sheets of lava. During cooling, each sheet developed three distinct zones characterized by different types of jointing (see Fig. 4-3). These zones are responsible for the prominent layering seen here in the cliff face.

and in some cases, for more than 100 km. Commonly, the lateral extent of a flow away from its source is approximately proportional to the third power of its thickness near the source. Volumes of individual flows average several tens of cubic kilometers but can be as much as hundreds of cubic kilometers.

For sheets of lava to spread over great areas, they must do so rapidly before cooling raises their viscosity. It is not surprising, therefore, in light of the calculated flow rates of various types of lava in the problems of Chapter 2, that basalts are the only ones to form these plateaus. Calculations show that flows only tens of meters thick could spread for hundreds of kilometers before cooling made them too viscous to flow (Shaw and Swanson 1970; Danes 1972). In fact, the common lack of crystal alignment in these basalts suggests that they spread rapidly, probably in turbulent flow, and come to rest prior to the onset of significant cooling and crystallization. During final cooling, shrinkage of these sheets commonly results in the development of fractures normal to the cooling surfaces. Because these are often extremely regular, breaking the basalt into five- or six-sided columns, they are known as *columnar joints* (Figs. 4-2 and 4-3). Many thick flows can be divided into three parts on the basis of the style of columnar jointing. A lower zone of regular columns, known as the *colonnade*, extends up into a zone of smaller curving and radiating columns, known as the *entablature*; this in turn is overlain by a zone at the top of the flow which has more widely spaced joints and is amygdaloidal (Fig. 4-3b).

If buoyancy is the force responsible for the ascent of these basaltic magmas, the Earth's surface should subside as the basalts accumulate in thickness. This is borne out by field observations, with many flood basalts occupying downfaulted areas. For example, the lowest members of the Columbia River basalts are now below sea level, but at no time was the basin invaded by the sea; eruptions must therefore have kept pace with subsidence. This type of faulting is commonly associated with extension of the Earth's crust, which at the same time produces the fractures up which the magma rises. In deeply dissected basalt plateaus, large numbers of parallel dikes are exposed, some of which can be shown to have been feeders to the flows. The total thickness of these dikes indicates that considerable crustal extension occurred during the igneous episode. For example, in parts of western Scotland and in Iceland the dikes account for crustal extensions of about 5%. This extension, of course, is associated with the opening of the North Atlantic.

4-3   CENTRAL VOLCANOES

Central volcanoes include all those structures formed by volcanic processes associated with a main central vent or magma chamber (Fig. 4-4). They are classified on the basis of their form and type of volcanic activity (MacDonald 1972). The forms are determined largely by the type of activity, which in turn is determined almost entirely by the composition of magma involved. Basalt is very fluid and flows rapidly, even on gentle slopes, to form what are known as *shield volcanoes*. Magmas with higher silica contents are more viscous and cannot flow as far from central vents. In addition, such magmas are more explosive and the resulting volcanoes, which are composed of both lavas and

(a)   (b)

FIGURE 4-3 Columnar jointing in flood basalts of the Giant's Causeway, County Antrim, Northern Ireland. (a) Individual columns in the lower part of the flow are up to 0.5 m in diameter and are cut by transverse joints that may be concave or convex upward (water in the former). (b) Extremely regular columns extending upward from the base of the flow form the so-called colonnade. Above this is a central zone of smaller, curving, and splaying columns known as the entablature; the top of this zone is cuspate. The top zone has more regular fractures and becomes vesicular or amygdaloidal toward the upper surface of the flow. The lower colonnade of an overlying flow is just visible at the top of the photograph.

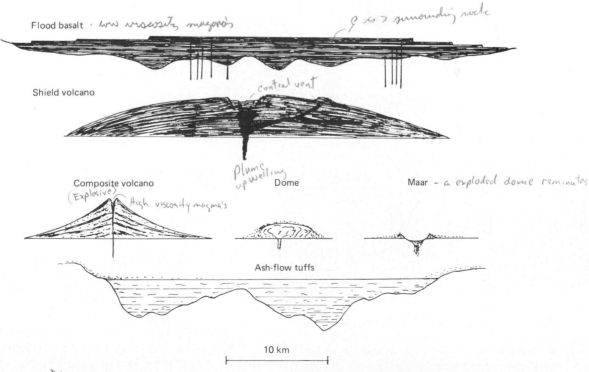

Flood basalt - *low viscosity magma's*   *ρ is > surrounding rock*

Shield volcano   *central vent*

Plume upwelling

Composite volcano (Explosive)   *High viscosity magma's*   Dome   Maar - *a exploded dome reminates*

Ash-flow tuffs

10 km

FIGURE 4-4 Forms of the major volcanic structures and their approximate sizes. Basaltic rocks are colored dark and rhyolitic ones light.

fragmental material, are referred to as *composite*. Silica-rich magmas, such as rhyolitic ones, are so viscous that they do not generally flow far from their vent; instead, they develop large blisters or *domes* above the vent. Explosive activity is so common with these that the bulk of eruptive material may be in the form of extensive fragmental deposits (*tephra*).

Shield volcanoes are built almost entirely from basaltic flows which emanate from summit craters or from fissures on the flanks of the volcanoes. The low-viscosity basalt moves rapidly down slopes and accumulates on the outer parts of the shield. For example, some of the Hawaiian flows have traveled more than 50 km and yet may be only 5 m thick. Slopes rarely exceed 10° and are commonly convex upward, giving rise to a shieldlike form in cross section. But the symmetry of many of these volcanoes is destroyed by the extrusion of large amounts of material from fissures on the flanks of the volcano. Also, the summits of many of these volcanoes have been downfaulted along concentric fractures to produce *calderas*. These form when the roof of a near-surface magma chamber collapses due to magma migrating either into the flanks of the volcano or up along the faults bounding the caldera. Subsequent eruptions are commonly localized along the faults bounding the caldera, and lava lakes are often formed within the caldera itself.

The Hawaiian volcanoes are excellent examples of shield volcanoes. Mauna Loa forms a giant shield, rising over 8 km above the ocean floor (see Prob. 3-2), with only the upper 4.2 km exposed above sea level (Fig. 4-5). A caldera, more than 200 m deep and 15 km in circumference, crowns its summit. During recent years careful measurements with seismographs and tiltmeters on the island of Hawaii have revealed much about the behavior of such volcanoes, in particular of Kilauea, a volcano

situated on the eastern flank of Mauna Loa (Richter et al 1970; Ryan et al 1981; MacDonald 1972).

The conduit through which magma ascends beneath Kilauea has been mapped by locating seismic hypocenters that are associated with the fracturing of rock produced by excess pressures in the buoyantly rising magma (Ryan et al 1981). The seismicity defines a southward-dipping zone that extends to a depth of 60 km. The zone narrows and becomes better defined toward the top of the mantle. At a depth of 14 km, it is elliptical in plan, with maximum and minimum dimensions of 3.8 and 2.1 km, respectively (Fig. 4-5). This zone is not filled with liquid but must, instead, consist of a large percentage of solid rock that is capable of being fractured by the magma pressure. The elliptical shape of the conduit, which has its long dimension striking N67°E, reflects stress inhomogeneities in the upper mantle resulting from plate tectonic motions. On rising through the oceanic crust, the cross-sectional area of the conduit decreases slightly, and between 8.8 and 6.5 km the major dimension of the elliptical section rotates in a clockwise direction (in map view), becoming essentially parallel with the strike of the East Rift Zone of Kilauea (S41°E). From a depth of 6.5 to 5.7 km, seismic activity is largely absent, indicating that the ratio of magma to rock in this zone is sufficiently large to allow for easy access of ascending magma. An offshoot from the main conduit at this level feeds a conduit beneath the East Rift Zone. The aseismic zone is interpreted to be the base of a magma chamber. The top of the chamber, which is at a depth of about 1.5 km, is a seismicly active zone, presumably as a result of intense diking activity. The filling of the magma chamber inflates the volcano by as much as several meters, with subsidence occurring only when there is an eruption or magma moves laterally along the conduit into the East Rift Zone. The flow of magma into fractures in the roof of the chamber develops stresses at the tip of the cracks, which causes further fracturing. The alternating fracturing and injection produces a pulsating seismic wave known as *harmonic tremor*. These waves cease abruptly once magma breaks through to the surface of the Earth.

Despite an apparent rather constant annual flux of about 0.1 km³ of magma into the base of Kilauea, the exact timings, locations, and chemical compositions of eruptions are highly variable. This is to be expected, however, in view of the rather complex plumbing system beneath the volcano. New batches of magma may rise directly to the surface, they may mix with earlier batches that have cooled and changed composition, or they may move laterally into the East Rift conduit to be extruded at a later date. Complexities in older shield volcanoes indicate that Kilauea is a perfectly normal example of this type of volcano.

Although the timing and location of Kilauean eruptions is variable, investigations by the U.S. Geological Survey (Klein 1984) have made it possible to forecast eruptions of this volcano at the 99.9% confidence level. This is done by monitoring surface tilt, rate of tilting, seismicity, and 14-day Earth tidal cycles. Of these parameters, the tilt, which is measured with a simple water tube, gives the earliest warnings and is effective up to 30 days prior to an eruption. The frequency of earthquakes, on the other hand, is effective only within 10 days of an eruption. The apparent constant flux of liquid into the magma chamber results

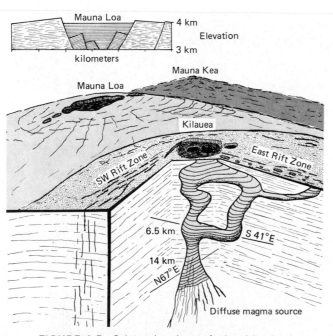

FIGURE 4-5 Schematic view of the island of Hawaii, with the magma conduit beneath Kilauea being exposed in the cutaway. [Drawn from sections presented by Ryan et al. (1981).]

in the volcano being on the verge of eruption at all times. Consequently, minor disturbances, such as a maximum in the 14-day Earth tide cycle, can be sufficient to trigger eruption.

Lava flows on shield volcanoes are generally of two types, pahoehoe and aa (Fig. 4-6). *Pahoehoe* flows have smooth surfaces, which may be wrinkled into ropy-looking masses by the continued flow of hot, fluid lava beneath the partially solidified crust. The hot lava issues forth as tongues along the leading edge of the flow. Cracks formed in the crust are rapidly healed by lava welling up and pouring out onto the surface (Fig. 4-7). The fluid cores of pahoehoe flows can drain completely if the supply of fresh lava is cut off, leaving behind empty *lava tubes* which can be many meters in diameter and of considerable length (Fig. 4-8). In contrast, the surface of an *aa* flow is broken into a mass of clinkery fragments, each of which has an extremely rough surface, from which project many small spines of glass. This rubble is carried along on the surface of the flow until it reaches the leading edge, whereupon it tumbles down the steep front face and is overridden by the flow. This conveyor-type action results in rubbly material occurring in the upper and lower parts of aa flows.

The decrease in confining pressure on a magma, resulting from its upward intrusion and eventual extrusion on the Earth's surface, causes dissolved gases to come out of solution and form bubbles, which are known as *vesicles* (see Sec. 3-4). These may later be filled with such minerals as zeolites, chlorite, calcite, or forms of silica, which are deposited from late magmatic solutions or circulating groundwater. These filled vesicles are called *amygdales* because of their resemblance to almonds, the Greek word for which is *amygdalos*. The diminutive term *amygdule* is also used, although commonly with little regard to absolute size. Rocks containing amygdales are said to be *amygdaloidal*. The

minerals of amygdales usually reflect the composition of the host rock. For example, those in nepheline-bearing basalts commonly contain zeolites, whereas those in more silica-rich basalts may contain chalcedony or quartz.

Both pahoehoe and aa flows contain abundant vesicles or amygdales. These are mostly concentrated in the upper parts of flows, where the buoyantly rising bubbles encounter cooler, more viscous lava. The central part of flows generally contains far fewer vesicles and in thick flows may be completely devoid of them. This is due to the rapid rise of bubbles through basaltic magma. For example, a 5-mm-diameter bubble rises 1.3 m h^{-1} in a lava of density 2.75 Mg m^{-3} and viscosity of 100 Pa s. Vesicles may be present in the base of a flow, where they are trapped by rapid cooling. Here they may have the form of vertically elongated tubes several millimeters in diameter and many centimeters long; these are called *pipe vesicles*. Where these are filled with later minerals, they are named *pipe amygdales*. They frequently curve at their upper end, indicating the final direction of movement of the lava (Fig. 4-9). Similar curvature is seen on cylinders of highly vesicular lava that intrude up into a flow from the base. These structures, which are called *vesicle cylinders*, are commonly several centimeters in diameter and may extend upward for several meters. They are particularly abundant where flows have crossed wet ground.

Compositionally, pahoehoe and aa flows are very similar. In fact, a pahoehoe flow may change into an aa one as it flows away from its source. One striking difference between them is in the form of their vesicles (MacDonald 1972). In pahoehoe, these are mostly spherical, indicating that the bubbles were growing or expanding at the time of solidification of the lava. But vesicles in aa have highly irregular shapes, which indicate that growth of the bubbles had ended prior to solidification.

FIGURE 4-6 The smooth ropy surface of a pahoehoe flow, which covers an older aa flow with typical rough clinkery surface, Kilauea, Hawaii.

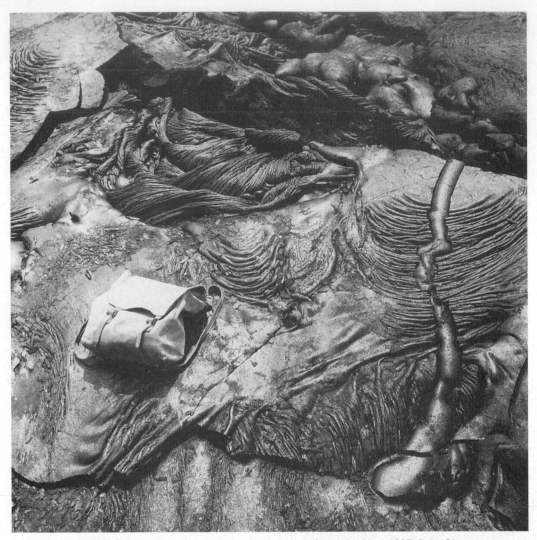

FIGURE 4-7 Freshly extruded pahoehoe lava from the May 1987 Pu'u O'o eruption on Kilauea's east rift zone, Hawaii. A variety of ropy structures indicate variable temperatures and rates of extrusion. Note that lava that rose to fill a crack in the brittle crust must have frozen rapidly by radiation before it could spill onto the surface of the flow.

This implies that pahoehoe lavas have had less time to degas and are derived more directly from their source. This is also supported by the higher temperatures of pahoehoe flows. The rate of flow, however, is an additional factor, for ropy lava, if moved too rapidly, can develop a clinkery surface.

Lava that comes in contact with water is rapidly quenched to a glass that may fragment into sand-size particles. Rocks formed from this material are called *hyaloclastites*. They differ from volcanic ash in that the fragments are equant and generally lack the cuspate shape of the glass shards formed from the lava trapped between expanding bubbles of gas (Fig. 4-20) that are formed when lava is disrupted during eruption from a vent.

When basaltic flows come in contact with water, they may, instead, divide or bud into large sacks of lava which rapidly develop glassy selvages that insulate the molten material

within (Fig. 4-10). These sacks are able to move by rolling down slopes, and since they are largely molten and deformable, their lower surfaces become molded to the irregularities of the bottom on which they come to rest, while their upper surface takes on a smooth convex-upward shape. Such bodies are referred to as *pillows*. When pillows are deposited one on top of the other, their lower side molds itself onto the convex upper surfaces of the lower pillows to produce a tail. This asymmetry to pillows is extremely useful in studying folded volcanic rocks, as it provides a means of determining the original upward direction (top determination) of pillowed lava flows (Fig. 4-10).

Pillows may vary in size and shape from buns to mattresses, and show a variety of internal structures, including radial fractures, amygdales that may be concentrically arranged or radially elongated, or large empty cavities which are formed by shrinkage of the lava on crystallization or leaking of lava

FIGURE 4-8   Lava draining from the interior of a lava flow can leave a lava tube, the sides of which commonly have benches marking periods during which the draining lava remained at a particular level for some considerable time. Stalactite-like dribbles of lava (lavacicles) may hang from the roof. Valentine Cave, Lava Beds National Monument, California.

FIGURE 4-10   Vertically dipping Precambrian pillow lavas, Ungava, northern Quebec. The convex upper surface of the pillows and the tail on their lower side indicate that the original top of these lavas was to the right.

FIGURE 4-9   Meter-thick basalt flow with amygdaloidal lower and upper zones and a massive core, County Antrim, Northern Ireland. Amygdales at the base form titled pipes, indicating that the lava flowed from right to left.

from the pillow. Because of the molding prior to solidification, pillows generally fit together rather closely. They may, however, be surrounded by hyaloclastite, fragments of pillow selvage, sedimentary material, or later cavity-filling minerals, such as chalcedony, quartz, calcite, and zeolites.

The presence of pillows usually indicates a marine environment, especially where there are great thicknesses of them. But there are many examples of freshwater pillows, such as those in the Talcott basalt, which erupted into a lake occupying a downfaulted Mesozoic basin in Connecticut, or those formed in the lake that filled the Tertiary caldera on the island of Mull, Scotland. Many orogenic belts contain thick units of pillowed basalt which were deposited on the ocean floor along with typical geosynclinal sediments. Investigations of present deep ocean floors and oceanic ridges indicate that pillows are the most common form of volcanic material in these regions.

A number of minor rock types containing more silica than basalt can occur on shield volcanoes. These lavas are thought to be derived from basalt by various processes of differentiation operating beneath the volcano. Because of their higher silica contents, and consequently higher viscosities, they may form small, steep-sided cones on the main shield. Similarly shaped cones may also be formed from cinders ejected during periods of more violent disruptive eruption, especially where groundwater or seawater gains access to an active vent and is rapidly converted to steam. The amount of this fragmental material is small in most shield volcanoes.

Composite volcanoes are built of both lava flows and fragmental material ejected from vents during periods of explosive activity. These materials are interlayered and, because of their markedly different resistance to weathering, produce very stratified-looking rocks. Consequently, the name *strato volcano* is also used for this type of volcano. They have the typical form that most people associate with volcanoes, that is, concave-upward slopes ranging from 10 to 36° with a single crater at the summit (Fig. 4-11). Mount Fuji in Japan is perhaps the best known of these, but there are many others, especially in orogenic regions, for example, Mayon in the Philippines, Vesuvius in Italy, and Shasta, Rainier, and Hood in the Cascade Range of the northwestern United States (Fig. 4-1). These volcanoes owe their symmetrical shapes to eruptions from a single summit vent. Although this is common, composite volcanoes can erupt from several vents, or the feeder pipe may shift with time so that later cones may be offset from the earlier center of activity. Mount Shastina on the western flank of Shasta is an example of such a younger vent. In still others, a volcanic cone may be situated within the remains of an earlier one that has been largely destroyed by violent explosive activity. This is the case, for example, with the Kanaga volcano in the Aleutians, shown in Figure 4-11.

The increase in slope toward the summit of composite volcanoes may be gradual, but more often there is a relatively constant gentle slope on the lower parts of the volcano followed by rapid steepening near the summit. This morphology is the direct result of the greater accumulation of erupted material near the summit, especially of tephra, which consists of large, heavy fragments near the vent, grading to finer particles farther down the slopes. The outer flanks consist mainly of lava, which commonly forms *blocky* flows that are somewhat akin to aa, but the blocks are more regular in shape and lack the extremely rough surfaces of the clinkery aa material (Fig. 4-12). Lesser amounts of aa and pahoehoe may also occur. In addition, the lower slopes may be covered with mudflows of loose tephra washed down from above by heavy rainfalls. Indeed, if it were not for the interlayers of lava in these volcanoes, the steep slopes would soon be reduced by erosion. These mudflows, which are named *lahars* to distinguish them from ones not associated with volcanism, can carry in suspension blocks of material measuring up to many meters in diameter. Lahars pose a serious environmental hazard, especially in areas of heavy rainfall. For example, mudflows were responsible for the complete destruction of the Colombian town of Armero and its more than 20,000 inhabitants when the Nevado del Ruiz volcano erupted in November 1985. Torrential rain and melting snow caused volcanic ash laid down by prehistoric eruptions to become saturated with water; the seismic disturbance produced by the eruption of Nevado del Ruiz was then sufficient to trigger the mud slides.

In general, the rocks of composite volcanoes are more siliceous than those of shield volcanoes. Andesites and dacites (see Chapter 6) are the two most common rock types, but smaller amounts of basalt and rhyolite may occur. The magmas from which most of these rocks form are more viscous than basalt, and the lavas therefore cannot flow as far. Moreover, the higher viscosities prevent gas from escaping easily from the magma. Consequently, most bubbles of gas formed during ascent remain in the magma until the time of eruption. The decrease in pressure accompanying ascent brings about expansion

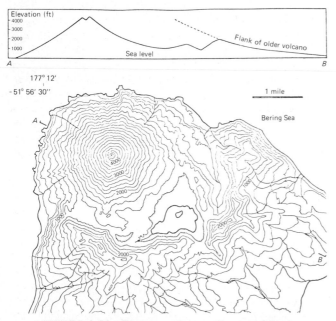

FIGURE 4-11 Kanaga volcano, Kanaga Island, Aleutians. The symmetrical shape is typical of strato volcanoes, with eruptions emanating from a single central vent. Note that the slope of the volcano steepens toward the summit. The arcuate valley surrounding the eastern, southeastern, and southern sides of the volcano marks the rim of an ancient crater formed when a previous volcano at this site was destroyed by explosion.

FIGURE 4-12 Blocky andesitic lava flow that erupted from the cone in the upper left of the photograph, McKenzie Pass, High Cascades, Oregon.

of the gas and eventual disruption of the magma, which results in the periodic explosive activity characteristic of this type of volcano. For example, a bubble of gas in magma extruded onto the surface from a depth of 3 km expands approximately 1000 times in volume (see Probs. 3-4 and 4-4). Of course, gas may escape from magma that remains stationary for a sufficient length of time, or within a volcanic pipe, gas may rise and concentrate in the upper part, leaving behind relatively gas-free magma. This results in periods of explosive activity alternating with periods of lava eruption.

The type of volcanic activity in which ash is ejected high into the atmosphere is known as *Plinian* (Fig. 4-14), after the Roman historian Pliny the Younger, who described in considerable detail the A.D. 79 eruption of Vesuvius. The ash from this eruption not only buried the city of Pompeii but was indirectly responsible for the death of Pliny the Elder, who died of an apparent heart attack while escaping from the eruption with his nephew, Pliny the Younger.

The particles ejected from a volcano during explosive activity vary considerably in size and composition. Many consist of molten material, but others can be of igneous rocks formed during earlier periods of activity, while still others may be of foreign rocks ripped from the walls of the feeder pipe. All of this ejected material is collectively known as *tephra*, which when consolidated forms *pyroclastic rocks* (broken by fire). Ejected bodies composed of molten material may take on streamlined shapes while passing through air; these are known as *volcanic bombs* if their diameters are greater than 64 mm (Fig. 4-13). Smaller particles are called *lapilli*, and the fine material (<2 mm) is referred to as *ash*. Thin filaments of glass known as *Pele's hair* (Fig. 4-13) may also form when droplets of lava separate from their source. Rocks formed from the ac-

cumulation of larger fragments are known as *agglomerates*, whereas those formed from ash are known as *tuff*.

The proportion of lava to pyroclastic rocks in composite volcanoes varies, depending, to a large extent, on the composition of magmas involved. It is not uncommon for the magmas to become more siliceous toward the final stages of the life of a volcano, with rhyolite being the last rock type to be erupted. This change is accompanied by increased explosive activity, and the highly viscous magma, rather than producing flows, may simply rise in the summit crater to form a *dome* (Fig. 4-14).

The type of volcanic structures formed when only highly viscous magmas are erupted are quite different from those already discussed. High viscosities prevent lava flowing rapidly away from a vent; thus as long as there are no violent explosions, lava accumulates over the vent as a *blister* or *dome* (Figs. 4-4 and 4-15). These generally have the form of a flattened hemisphere, but eruption from a fissure, rather than from a vent or eruption on a sloping surface, may result in the formation of an elongated body. The dome grows by magma being intruded into its base; hence the term *endogenous* dome is used to indicate that it was built from within. As a dome is inflated by the addition of new magma, its cooling surface cracks and develops a talus breccia around the body. If pressure within a dome is sufficiently great, the surface may burst and highly viscous lava extrude and flow down the steep sides of the dome. Such a flow, for example, erupted from just below the summit of the Puy de Dome in Auvergne, France. If a dome is built largely through the extrusion of lava, it is said to be *exogenous*. These, however, are less common than endogenous ones. Commonly, the pressures within a dome force up *spines* of solid rock that may be up to hundreds of meters in diameter and hundreds of meters in height. These become highly fractured on cooling and are soon eroded away. A spectacular spine grew to a height of 350 m in nine months during the 1902 eruption of Mount Pelée, in Martinique.

Domes vary considerably in size from a few tens of meters to several kilometers in diameter. For example, the Puy de Sarcoui in Auvergne is 400 m in diameter and 150 m high, whereas Lassen Peak in California is almost 2 km in diameter and 600 m high. Most domes are composed of rhyolite (obsidian), dacite, phonolite, or trachyte (see Chapter 6 for definitions).

When large quantities of viscous magma are erupted, especially on sloping surfaces, slowly moving lava flows can develop. These are thicker than most flows associated with the types of volcanism discussed previously, with thicknesses of several hundred meters not being uncommon. It will be recalled from Eq. 2-11 that the maximum velocity in a laminar flow is proportional to the square of the thickness and inversely proportional to the viscosity. Hence, with highly viscous lavas it is only the exceedingly thick flow that will travel a significant distance before cooling renders it too viscous to move. The surface of these flows is deformed into ridges, which are elongated transverse to the direction of flow. Although this wrinkling resembles the ropy structure of pahoehoe flows when viewed from a great distance (Fig. 4-16), the wavelength of the ridges is measured in meters to tens of meters and is not visible at the scale of the outcrop. Instead, at this scale the surface of the flow

FIGURE 4-13 Molten material ejected from volcanoes may take on various forms. Large bodies (>64 mm), known as bombs (a), commonly have tails with prominent lines and groves formed when the particle separated from its source. The bomb shown here is broken, exposing concentric layers of vesicles within. Smaller particles (<64 mm, >2 mm), known as lapilli (b), are commonly teardrop-like in shape. Fine strands of basaltic glass (c) formed during the separation of lapilli and bombs is known as Pele's hair, after the Hawaiian goddess of fire.

FIGURE 4-14 Plume of volcanic ash $3\frac{1}{2}$ km high venting from Mount St. Helens on June 9, 1982. Visible through the gaping hole blown in the northern side of the mountain by the catastrophic eruption of May 18, 1980 is a 200-m-high dome of dacite. (Photograph by James Zollweg.)

is extremely irregular and consists of broken blocks of obsidian, large crevasse-like gashes, and spines and slabs thrust up from beneath (Fig. 4-17).

Highly viscous flows have much more internal structure than thinner, less viscous ones (Fig. 4-17). An excellent example of such structure is provided by a rhyolite flow in Nevada that was studied in considerable detail by Christiansen and Lipman (1966). This flow, which is up to 250 m thick, can be traced for 11 km from its leading edge back to a feeder vent. The upper and lower parts of the flow are breccia, consisting of glassy and highly vesicular fragments (pumice) formed from the fracturing of the chilled crust by the continued movement of lava

within the flow. The upper breccia is commonly cut by protrusions of viscous lava thrust up from the main central part of the flow, which consists of a strongly foliated rhyolite. The foliation is marked by textural variations resulting from different proportions of glass, spherulites, and microlites (see Chapter 12). When formed, the foliation is vertical and parallel to the walls of the feeder pipe, but it becomes highly distorted during flow on the surface. In the lower part of the lava flow it parallels the base of the flow, but toward the top it becomes much steeper, with the dip direction being opposite to the flow direction. It also is contorted into recumbent folds that are overturned in the direction of flow.

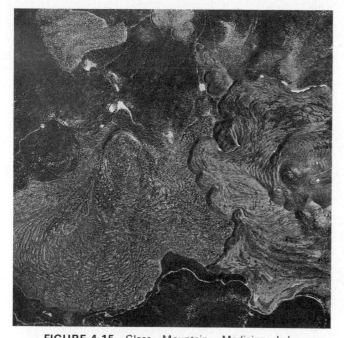

FIGURE 4-15 Glass Mountain, Medicine Lake Highlands, California. A number of thick, viscous rhyolite flows emanate from this obsidian dome (right center). Glass Mountain and several smaller domes are localized along a northwesterly trending fracture. An earlier dome and associated flow occur in the center and lower left of the photograph. The apparent ropy surface of the flows is produced by transverse ridges that have a wavelength of many meters. (Photo courtesy of U.S. Forest Service.)

FIGURE 4-16 The Big Obsidian Flow, Newberry Volcano, central Oregon (N in Fig. 4-1). The obsidian flow erupted from the margin of the caldera at the summit of the Newberry shield volcano, the largest volcano in the conterminous United States. A small rhyolite dome has developed over the vent, and the surface of the 40-m-thick rhyolite flow has been wrinkled into large transverse waves as the lava descended the slope onto the floor of the caldera. Note the many small cinder cones in the background scattered on the broad flanks of the shield volcano. (Photograph courtesy of Oregon Department of Geology and Mineral Industries.)

4-4 PYROCLASTIC DEPOSITS AND CALDERAS

The explosive activity associated with highly viscous magmas produces a variety of pyroclastic deposits that can be classified into three groups on the basis of the way in which fragments are transported from the source to their final site of deposition (Sheridan 1979; Fisher and Schminke 1984).

Air-fall deposits are formed from particles that follow a simple ballistic trajectory, the direction and length of which is determined by the eruption velocity from the vent, the wind velocity, and the gravitational terminal velocity of the particle. The deposits, which are well sorted, blanket the surface evenly with little regard for topography. They form plumes, which thin and become finer grained downwind from the source.

Surge deposits are formed from clouds of ash that move out horizontally from a vent at hurricane velocities following a volcanic explosion. Similar rapidly moving clouds are well documented in association with underground nuclear explosions. The passage of the cloud is short lived and thus only a fraction of the total pyroclastic deposit associated with an eruption is normally formed in this way. Bedding forms such as dunes and antidunes testify to the high velocities of emplacement, as does deposition in the lee of topographic obstructions. These clouds have a high gas-to-solid ratio.

Ash-flow deposits are formed from suspensions of particles that are carried along by hot rising gases. These suspensions have relatively high densities (about 1000 kg m^{-3}) but very low viscosities. Ash flows have much lower gas-to-solid ratios than surges. Consequently, they move like fluid liquids, flowing down valleys at high velocities. Indeed, their velocities can be high enough to allow them to "jump" topographic barriers. Thick ash flows pond in topographic depressions; their thickness is therefore variable and determined largely by the underlying topography. They are poorly sorted and can contain particles ranging in size from ash to boulders.

Explosive activity, such as that which destroyed Krakatoa in 1883, Monte Somma (Vesuvius) in A.D. 79, Mount Mazama (now Crater Lake, Oregon) 6600 years ago, or Mount St. Helen's (Washington State) in 1980 throws volcanic ash high into the atmosphere, where winds may transport it for great distance before depositing it as an *air-fall tuff*. Particles may consist of older rocks torn from the walls of the vent by the explosion, or of magma disrupted by expanding gas bubbles. On traveling through the atmosphere, the latter particles cool rapidly, both through radiation and adiabatic expansion of the gas. They are quenched to glass shards with sharp, cuspate boundaries that mark the outline of the previously surrounding gas bubbles (Fig. 4-20). Although several days may be required for the farthest-traveled particles to settle out, deposition of air-fall ash from a single eruption can be considered an essentially synchronous geological event over a wide area. Extensive

(a)

FIGURE 4-17 Surface features of the Little Glass Mountain obsidian flow, Medicine Lake Highlands, California. (a) Broken blocks of obsidian form the leading edge of the flow. (b) Three-meter-high spine of obsidian thrust upward by pressure within the flow. (c) Large crevasse-like fracture, formed where the obsidian is domed up by the rise of low-density pumice that underlies the obsidian. (d) The fracture surface exposes flow layering, which is common in rhyolite flows.

(b)

(c)

(d)

blankets of volcanic ash therefore provide the geologist with an ideal means of correlating geological events in widely separate areas, as long as individual ash falls can be distinguished from one another. The chemical composition of glass shards and the nature of phenocryst assemblages have both proved useful in "fingerprinting" tuffs. Dating by means of tuffs or tephra is known as *tephrochronology* (Thorarinsson 1981).

Ash flows were first recognized during the devastating 1902 eruption of Mount Pelée, which destroyed the city of St. Pierre in Martinique. The French geologist Lacroix, who was an eyewitness, described *nuées ardentes* (glowing clouds) descending the flanks of the volcano at high speeds. Studies of ancient volcanic regions have since revealed that ash-flow tuffs are one of the most common modes of occurrence of rhyolite in the geologic record (Chapin and Elston 1979). Moreover, many ancient ash-flow tuffs are orders of magnitude larger than those seen to erupt in historic times. The mode of emplacement of these large ash flows is still not completely understood.

Before continuing with our discussion of ash-flow tuffs, it is necessary to consider one remaining major volcanic structure, which unlike those already mentioned, is not a constructional feature; these are the large calderas associated with rhyolitic magmatism. Calderas were mentioned in connection with shield volcanoes, but these are much smaller than those associated with rhyolites. For example, one of the largest shield calderas is that on Mauna Loa, with an area of about 9.2 km², but this is small compared with a caldera such as that at Valles, New Mexico, which is about 430 km², and it is completely dwarfed by the one at Yellowstone, which is about 2500 km² (Fig. 4-18). These large calderas are not formed by the collapse of the crest of a volcano, but result from the foundering of large crustal blocks along ring fractures into large, near-surface magma chambers. Volcanic activity that is associated with these structures emanates mostly from the ring fracture.

Collapse of these large calderas cannot take place without the displacement of large volumes of magma, and it is this magma that has so often erupted in the form of ash flows. The volume of erupted ash flow might be expected to correlate directly with the volume displaced by the subsiding caldera block. Unfortunately, the amount of subsidence in a caldera is often difficult to determine. Smith (1979), however, has shown that the volume of an erupted ash flow correlates strongly with the area of the caldera, which is a much more easily determined dimension than its volume. As seen in Figure 4-19, this correlation holds true for calderas ranging in size over several orders of magnitude. The correlation of the erupted volume

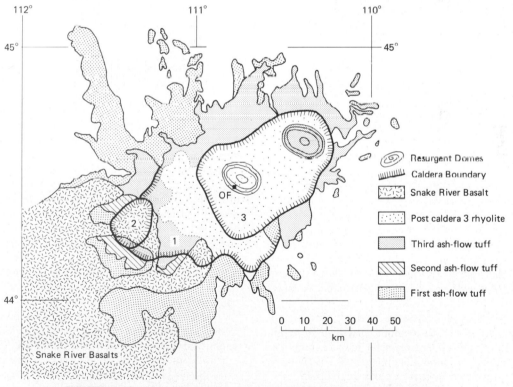

FIGURE 4-18 The Yellowstone volcanic center, Wyoming, was formed mainly during three successive episodes of caldera collapse and concomitant eruption of ash flows. The first caldera, which is the largest, erupted 2500 km³ of ash-flow tuff in a single eruption. Note that many of the ash-flow tuffs extend out from the caldera in long tongues as a result of the avalanches of hot ash filling in existing river valleys. Within the third caldera, two resurgent domes have formed, with the Old Faithful (OF) geyser on the western flank of the western dome being driven by the thermal energy from the underlying magma. [Simplified from Christiansen (1979).]

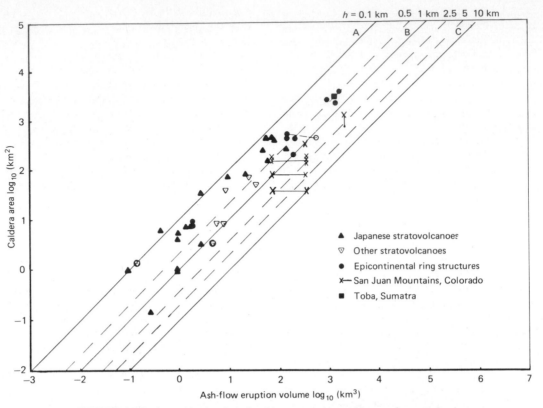

FIGURE 4-19 Logarithmic plot of caldera area versus volume of erupted ash flow. The diagonal lines give depth of drawdown in the magma chamber, assuming vertical walls and a flat roof. Most calderas subside about 0.5 km regardless of the area of the caldera. [After Smith (1979); reprinted by permission of the Geological Society of America.]

with the area of subsidence implies that most calderas sink approximately the same amount during an eruption, which from Figure 4-19 is about 0.5 km. No great significance can be placed on this number because there is considerable uncertainty in the values for the volume of erupted material, especially around ancient calderas, where erosion may have removed ash-flow tuffs or later eruptions may have concealed them. Despite this, it is fair to conclude that the relatively constant amount of subsidence, regardless of size of caldera, indicates that the larger the caldera, the greater is the diameter-to-depth ratio of the subsided block; that is, the subsided block in a large caldera tends to be slablike, whereas that in a small one is more pistonlike.

The fact that many different-sized calderas have undergone similar amounts of subsidence implies that absolute depth of subsidence rather than caldera diameter may be more important in determining the amount of down-drop. This, in turn, suggests that pressure, and its effect on the solubility of volatile constituents in a magma, may be key factors in determining the amount of subsidence.

By far the largest fraction of magma extruded during caldron subsidence is in the form of ash flows. Ash flows form only from magma that initially contains sufficient volatiles so that on decompression the magma is completely disrupted into a suspension of magma particles in gas. Until this disruption occurs, the rate of caldron subsidence is determined by the rate

at which highly viscous magma is able to ascend ring fractures. With the onset of magma disruption the viscosity of rhyolitic magma decreases by at least five orders of magnitude, thus allowing for rapid extrusion of ash flows and rapid subsidence of the caldera (see Prob. 4-5). Once started, vesiculation and disruption of the magma will work downward until depths are reached where pressures are great enough to prevent gas exsolution from disrupting the magma. This, then, may be the depth at which significant cauldron subsidence ceases. The actual depth would depend on the composition of the magma and its initial volatile content. In Sec. 13-5 we will see that the tops of rhyolite magma chambers tend to be enriched in volatiles and hence are ripe for ash-flow eruptions.

Enormous volumes of magma have been erupted in the form of ash flows from calderas. Some of the world's most voluminous ash-flow tuffs occur in Yellowstone National Park, in northwestern Wyoming, where there were three major eruptions (Smith and Christiansen 1980). During the first, which occurred 2.0 million years ago, a caldera over 75 km in diameter subsided and expelled more than 2500 km^3 of magma in a single ash flow (Fig. 4-18). This volume is equivalent to spreading a layer of magma 200 m thick over an area equivalent to that of the state of Connecticut or the country of Wales. The next eruption, which occurred 0.8 million years later, was much smaller, forming only 280 km^3 of ash flow and a 25-km-diameter caldera at the western end of the first caldera. Magmatic activity then moved eastward, and 0.6 million years later another caldera, which measures 45 × 65 km, formed at the eastern end of the first caldera, and 1000 km^3 of magma was erupted, partly concealing the first caldera. Postcaldera eruptions of rhyolite lava flows (not ash flows) have flooded much of the floor of the caldera, and flood basalts from the Snake River Plain have inundated parts of the ring structure. Since the last ash-flow eruption 0.6 million years ago, two *resurgent domes* have grown within the youngest crater, indicating that magma is still present at shallow depths. Such domes commonly form during the later stages of the life of a caldron system (Smith and Bailey 1968). The Old Faithful geyser, located on the western flank of the western dome, and hot springs throughout the area are constant reminders of the proximity of the magma.

We are now in a position to return to our discussion of ash-flow tuffs. By far the greatest volumes of these have been erupted from large calderas rather than from central volcanoes. But from either source they flow rapidly, even on gentle slopes, eventually ponding in topographic depressions. They commonly fill in old stream valleys, as can be seen from many of the tongues of ash-flow tuffs emanating from the Yellowstone calderas (Fig. 4-18).

Considerable controversy surrounds the mechanism by which ash flows move. Clearly, gas is the medium in which the particles are suspended, but does this gas emanate from the particles of magma themselves, or is it air simply entrained in an avalanche of hot ash? In a viscous magma, gas can only be exsolved slowly. Thus, even after disruption of magma by expanding bubbles, gas will continue to exsolve from the particles of magma, and this may be sufficient to keep the particles apart and in constant agitation. Suspensions with which we are familiar that have formed in a similar way, but at much lower temperatures, are common aerosol sprays, which engineers refer to as being *fluidized*. Reynolds (1954) was the first to draw attention to the importance of fluidization in petrologic processes, but for a recent treatment the reader should refer to a paper by Wilson (1984). He showed, by experiment, that it is difficult to fluidize all particles in a pyroclastic flow because of the wide range in grain size. Also, the fluidization process tends to sort particles, transporting the fine pumice fragments to the top. Pyroclastic flows, however, exhibit little sorting. Some geologists believe that insufficient gas would be liberated from a magma during eruption to keep an ash flow fluidized (McTaggart 1960). Instead, they think that fluidization results from the heating of air that becomes entrapped beneath the leading edge of an advancing flow. Both processes undoubtedly play roles, but the exsolution of gas from the particles of magma would appear to be the more important, at least where ash flows have spread over almost level surfaces for thousands of square kilometers.

A fluidized suspension of lava particles, being relatively dense and yet having extremely low viscosity, is capable of flowing at great velocities down the gentlest slopes, and traveling great distances. The ash flow that descended Mount Pelée on May 8, 1902 traveled a distance of 6 km at an average velocity of 20 m s^{-1} and had velocities as great as 150 m s^{-1} on the steep slopes of the mountain.

In contrast to the blanketing deposits of air-fall tuff from Plinian-type eruptions, the distribution of ash-flow tuffs is controlled largely by topography, with flows descending valleys, thinning over topographic rises, and ponding in depressions. Where there have been many such flows, the filling in of topography results in the formation of flat areas or plateaus, similar to those formed by flood basalts. Above a glowing avalanche there is a cloud of ash, which being less dense than the fluidized mass beneath, is not controlled so much by topography and may spread out and blanket large areas with ash. It was such a cloud that actually destroyed the city of St. Pierre, not the ash flow itself, which was diverted by a river valley and entered the sea 3 km north of the city. Unfortunately, an upper stretch of this valley heads directly toward St. Pierre, and it was from here that the cloud above the ash flow was hurled straight at the city.

The temperature of an ash flow is initially that of magma in the vent, but as flow occurs, heat is lost through conduction, convection, and radiation, but also through the internal processes of gas expansion. Boyd (1961), who has analyzed the thermal behavior of ash flows, has shown that if magma is saturated with water prior to eruption, expansion of the gas resulting from the ascent from the source to the surface is approximately 20°C/wt % H$_2$O in the initial magma. Thus, magma originally containing 10 wt % H$_2$O would have its temperature lowered by 200°C due only to adiabatic expansion of gas (see Sec. 7-5). Counteracting the cooling is heat liberated by the exsolution of gas (about 11°C/wt % H$_2$O exsolved) and by crystallization (about 0.4 J kg^{-1}), and by the dissipation of viscous forces which in a rapidly moving flow may be actually great enough to cause heating (Shaw and Swanson 1970).

Although the precise thermal history of an ash flow may be difficult to ascertain, field evidence clearly indicates that when these flows come to rest they are hot enough to weld their constituent particles together. These welded ash-flow tuffs are commonly referred to as *ignimbrites*. Sheets of this rock may be tens of meters thick and display prominent columnar jointing. During the welding processes, particles become flattened parallel to the sheet (Fig. 4-20) giving the rock a prominent foliation, which is accentuated by the different colors of the various particles of glass, devitrified glass, and crystalline material. This general appearance is described as *eutaxitic*. Pumice fragments are generally light colored, but dense obsidians are dark, and when flattened among pumice fragments produce a striking flamelike, or *fiamme*, texture. Glass shards become flattened and are wrapped around phenocrysts and lithic fragments. The particles, which vary from sand size to several tens of centimeters, show little evidence of sorting, unlike those deposited from air falls. This is to be expected, since ash flows must move in a turbulent manner, at least near their source.

Ash-flow tuffs are extremely common in the geologic record. One of the first to be recognized was that which occupies the Valley of Ten Thousand Smokes, Alaska. It was formed in 1912 by a great eruption in the vicinity of the Katmai volcano, but it was not until several years later that geologists examined the deposit and concluded that it had been formed by an ash flow similar to those seen on Mount Pelée. The great explosion that produced Crater Lake, Oregon, in addition to forming ash-fall tuffs over large parts of the northwestern United States, erupted 33 km^3 of ash flows. The Bishop tuff, which has a volume of 600 km^3, erupted 0.7 million years ago from the Long

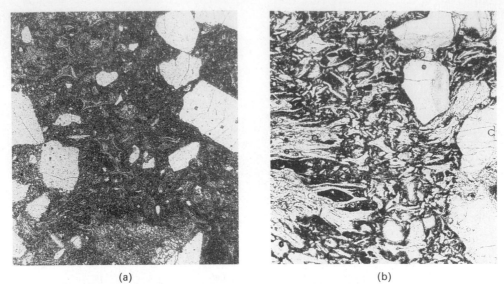

(a)   (b)

FIGURE 4-20   (a) Ash-fall deposits typically consist of glass shards and pumice fragments that show no signs of deformations; Los Alamos, New Mexico. (b) Similar fragments in ash-flow deposits become flattened, welded together, and wrap around phenocrysts; Bishop tuff, California. Both fields of view are 4 mm wide.

Valley Caldera in eastern California. In some areas, successive eruptions have produced large volumes of ash-flow tuffs, as, for example, in the central part of North Island, New Zealand, where they have a total volume in excess 7000 km³. Even more extensive deposits formed during the Miocene in Nevada and covered an area of approximately 100,000 km². Ash-flow tuffs are therefore the only type of volcanic product with volumes approaching those of the great flood basalts.

Intrusive Bodies

4-5 GENERAL STATEMENT

The shapes of intrusive igneous bodies are more difficult to determine than those of volcanic ones because the Earth's erosion surface provides only a limited three-dimensional view of these bodies. Geophysical surveys, including gravity, magnetic, heat flow, and seismic, can impose constraints on our interpretation of the shape of bodies at depth, as long as the physical properties of the igneous rock and country rock are sufficiently different. Structures in rocks surrounding intrusive bodies can also provide information on the form of an intrusion. Finally, from maps of intrusions that are exposed to different depths by erosion, a generalized picture of the form of intrusive bodies has emerged. Only a relatively small number of common intrusive forms exist (Fig. 4-21).

Intrusive bodies are divided into two classes; *discordant bodies* cross-cut the structure of the intruded rocks, whereas *concordant bodies* parallel it (Fig. 4-21). Although purely descriptive, this division has some genetic implications. Near the surface of the Earth, where rocks react to stress by fracturing, discordant bodies are common. At depth, where fractures are less abundant due to the ability of rocks under high confining pressures to relieve stresses by flowing, other features, such as compositional layering and schistosity, can provide planes of weakness along which magma intrudes to form concordant bodies.

Unlike volcanic bodies, intrusive ones must displace rocks to make room for themselves. The way in which this occurs largely controls the form taken by the intrusion. There is considerable uncertainty, however, as to precisely how this takes place. Historically, arguments concerning this point have figured prominently in discussions concerning the origin of granite and was one of the reasons some geologists believed that granite was not of igneous origin at all, but was formed by solid-state replacement of other rocks. Today, the room problem remains one of the most perplexing in igneous petrology.

Magma may simply rise into tensional fractures that result from extension of the Earth's crust. Open fractures, however, extend only to shallow depths. Moreover, the width of bodies formed in this way are likely to be small and limited by the rates of tectonic plate motion.

If magma does not have open fractures to enter, it must displace the rock it is to intrude. This can be done by forcing country rock aside. Near the surface of the Earth this may simply involve the lifting of the overlying rocks. At depth, however, solid rock must be pushed aside, which is likely to occur only if the country rock is rendered plastic by heat from the intrusion. But magmas do not typically contain excess heat; thus, this mechanism of making room for magma is severely limited.

Magma may also provide space for intrusion by having blocks fall from the roof of the magma chamber and sink to lower parts, thus effecting an exchange of solid material from above with molten material from below. This process is known

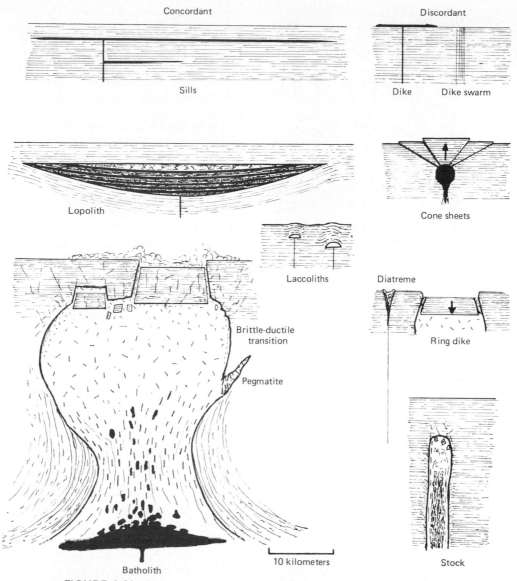

Concordant

Sills

Discordant

Dike   Dike swarm

Lopolith

Cone sheets

Laccoliths

Diatreme

Brittle-ductile
transition

Ring dike

Pegmatite

Stock

Batholith

10 kilometers

FIGURE 4-21 *Forms of major intrusive bodies and their approximate sizes. Rocks formed from basaltic magma are dark colored, whereas those formed from rhyolitic magma are light colored. Bodies on the left, in general, are concordant; that is, they parallel the structure of the intruded rocks. Bodies on the right are discordant and cut across the structure of the intruded rocks.*

as *stoping* because of its similarity to the common mining practice of removing blocks of ore from the roof of underground workings or *stopes*. Evidence of stoping is found in the margins of many igneous bodies, in particular those formed at high levels in the Earth's crust. Zones of igneous rock containing fragments of the country rock rim many intrusions and form what are known as *igneous breccias* (Fig. 4-44). For stoping to be effective, the country rocks must be denser than the magma and be fractured or capable of being fractured. This second requirement limits stoping to the upper part of the crust, for in the lower part, rocks are able to flow slowly under high confining pressures, and hence fractures are not likely to exist except during periods of rather rapid tectonic movement. The rate at which magma can intrude by stoping is determined by the availability of fractured rock and the sinking rate of fragments in

the magma; this, in turn, depends on the viscosity of the magma, the size of the fragments, and the density contrast between fragments and magma (see Sec. 13-3).

4-6 DIKES AND SILLS

Dikes and sills are sheetlike bodies that have great lateral extent relative to their thickness, and their opposing contacts are approximately parallel. Dikes cross-cut the structure of the intruded rocks and thus are discordant; sills parallel the intruded structure and thus are concordant (Fig. 4-22). These bodies range from millimeter-thick films to massive sheets hundreds of meters thick and tens of kilometers long. In general, sills are thicker than dikes, but dikes commonly occur in such large

FIGURE 4-22 Dikes and sills intrusive into gently dipping Ordovician limestone surrounding the Cretaceous Monteregian intrusion at Montreal, Quebec. The two narrow, dark dikes that slope upward to the left and are cut by the wider vertical dike on the right are both offset by the vertical dike; the narrow dikes must therefore be older. The upper narrow dike is offset also by the upper sill (center of photograph), but the lower narrow dike cuts the lower sill. The lower sill must therefore be one of the earliest intrusive rocks in this exposure. The wider vertical dike on the left is composite; a narrow felsic dike is intruded within it. What can be concluded from the intersection of this composite dike with the two sills? (See Prob. 4-11.)

numbers that their cumulative thickness can be immense. The Palisades Sill, which forms the escarpment on the west bank of the Hudson River opposite New York City, is 300 m thick and can be traced along strike for 80 km. The Whin Sill in northern England is only 75 m thick, but it is laterally extensive; it forms a 125-km-long escarpment which provided the Romans with a substantial foundation on which to build Hadrian's Wall. In South Africa, the Karoo Sills are very extensive, with individual sheets having areas of as much as 12,000 km². The widths of dikes are commonly in the range of meters to tens of meters with ones such as the 150-m-wide feeder to the large Muskox intrusion in Arctic Canada (Fig. 14-15) being rare. The so-called *Great Dyke of Zimbabwe* (Rhodesia), which is 5 km wide and 500 km long, is not actually a dike but is, instead, a long, narrow, downfaulted block of igneous rock formed in a previously overlying horizontal sheet-like body. Dikes in Iceland average only 3 to 5 m in width, yet

in one 53-km stretch of coastline G. P. L. Walker measured approximately 1000 dikes with a total thickness of about 3 km. Dikes may extend laterally for great distances, with lengths in excess of 100 km being common for those with widths of a few tens of meters.

Open fractures exist only near the Earth's surface, so at depth dikes and sills must generate their own pathways. This they do in one of three ways (Fig. 4-23) depending on the properties of the intruded rocks (Pollard 1973; Shaw 1980). In each case the direction of propagation is in a plane normal to the direction of minimum principal stress. In the upper part of the lithosphere where rocks are brittle, excess magma pressures are able to fracture rock. Magma flowing into a fracture concentrates great stress on the tip of the fracture, even when excess pressures are small. The rock is therefore wedged apart by the intruding magma (Fig. 4-23a) in much the same way that fractures can be opened in rock by pressurized water. Indeed, the term *hydraulic fracturing* is commonly used for this type of dike propagation. Magma is far more viscous than water, so that the tip of the dike lags behind the tip of the fracture (Fig. 4-24). At greater depth in the Earth where confining pressures are higher, an intrusive sheet may propagate by forcing aside rock along conjugate brittle shear planes, which are oriented at about 30° to the plane of the sheet (Fig. 4-23b). This bulldozer-like action produces a more blunt-ended sheet than does the wedging action. Also, magma may intrude along the shear planes, and fault-bounded blocks of country rock may become incorporated in the sheet. At still greater depths, the mechanism of propagation is essentially the same except that the shear planes become ductile faults, which make an angle of about 45° with the plane of the sheet (Fig. 4-23c). If the intruded rocks become completely plastic, a diapiric dome rather than a sheet-like body would form.

In the zone where sheet propagation is by "hydraulic fracture," the width W of a sheet and its horizontal length L are related to the physical properties of the intruded rock by (Fig. 4-25)

$$\frac{W}{L} = \frac{2.25 P_{ex}}{\rho_{host} V_p^2} \qquad (4\text{-}1)$$

where P_{ex} is the excess pressure of the magma (see Eq. 3-8), ρ_{host} the density of the host rock, and V_p the compressional seismic wave velocity in the host rock. With increasing depth, V_p increases and thus the ratio W/L must decrease. Therefore, we can expect dikes to be longer and narrower at depth and shorter and wider near the surface. Typical values inserted in Eq. 4-1 indicate that near the surface of the Earth W/L has values of 10^{-2} to 10^{-3}, but at depth it has values of 10^{-3} to 10^{-4}. A 10-m-wide dike near the surface of the Earth, for example, might have a length of about 3 km, whereas at depth a 10-m-wide dike would be about 30 km long.

The pattern formed by the 250-km-long Mesozoic Higganum dike set which crosses Connecticut and Massachusetts may well be explained in terms of the width-to-length ratio changing as the dike propagated toward the surface, as implied in Fig. 4-26. The dike consists of short individual segments that are about 4 km long and up to 60 m wide. These dimensions give a typical near-surface value for W/L of about 10^{-2}. The

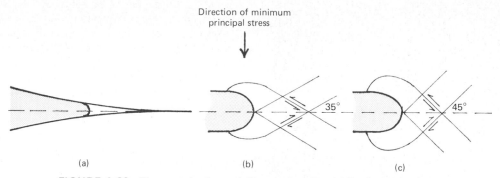

Direction of minimum
principal stress

(a)

(b)

35°

(c)

45°

FIGURE 4-23 Three mechanisms of dike propagation. (a) In the brittle upper crust, excess magma pressure can fracture rock; in such cases, magma lags behind the tip of the fracture. (b) At greater depth where confining pressures are greater, magma forces aside the intruded rock along conjugate brittle shear planes. (c) At still greater depth, deformation takes place on ductile shear planes. In each of these cases, the direction of propagation is in a plane perpendicular to the direction of minimum principal stress. [After Pollard (1973).]

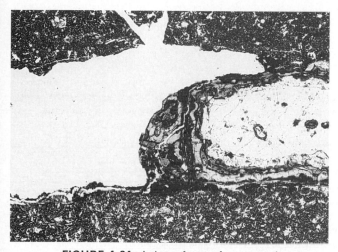

FIGURE 4-24 Lobate front of magma that was quenched to a glass as it advanced along a fracture in the margin of the Mesozoic Fairhaven diabase dike, Connecticut. The dark rim on the lobe is due to alteration and devitrification of the glass. Width of fracture 5 mm.

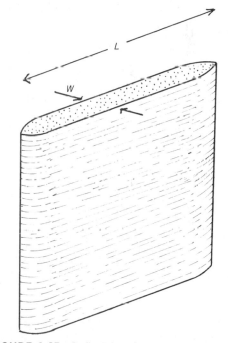

FIGURE 4-25 Definitions of length and width of a vertical dike.

short segments, in turn, form longer groupings extending over distances of about 10 km. These longer groups may reflect the presence of 10-km-long dike segments at depth, which would be expected from Eq. 4-1. The longer groups may, in turn, branch from a main fracture in the lower lithosphere, which controls the overall trend of the dike set. Two other dikes to the west of the Higganum dike exhibit similar segmentation. In each of these dikes, segments are slightly offset from the next, and the segments tend to overlap. Where they overlap they thin, so that their combined thickness remains constant. The strikes of the overlapping sections tend to veer toward the other dike, and if the dikes actually meet, they do so at right angles. Dike sets along spreading axes on the ocean floor can exhibit this same geometrical arrangement of dike segments (Lonsdale 1985). Indeed, this pattern is common in dike sets at all scales and is referred to as *enechelon* (Pollard et al. 1975).

The width of an intrusive sheet, in particular of a feeder dike, can be modified by magmatic erosion. For example, country rocks adjacent to the contact can be melted or blocks ripped off and transported away with the magma. Because flow velocities in dikes are proportional to the width of a dike (Eqs. 3-14 and 3-20), this erosion is more likely to occur where a dike is already widest. Consequently, wider parts of dikes tend to become wider. This widening and resultant concentration of flow at a few points is well illustrated by fissure eruptions, which are fed by dikes. The 1959 eruption of Kilauea Iki began with magma erupting along the entire length of an 800-m-long fissure, but within hours the flow became concentrated at a

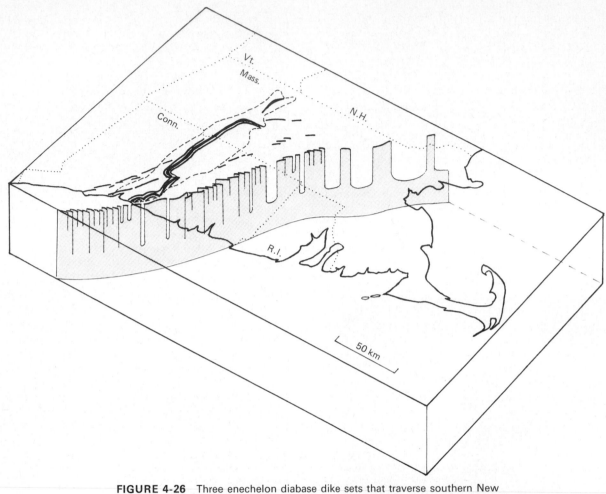

FIGURE 4-26 Three enechelon diabase dike sets that traverse southern New England in a northeasterly direction were the feeders to the three flood basalt units that are preserved as eastward-dipping flows in the downfaulted Mesozoic Hartford basin of Connecticut and Massachusettes (Philpotts and Martello, 1986). The easternmost of these dike sets, the Higganum, fed the earliest basalt. The enechelon segments of this dike can be seen to form two distinct groupings; the shortest segments are about 4 km long, whereas the larger groupings are about 10 km long.

single vent (Richter et al. 1970). Similarly, the huge fissure eruption in 1783 at Lakagigar in Iceland began with lava pouring from a 25-km-long fissure, but soon the flow became concentrated at a few nodes. In addition, the degree of contact metamorphism developed where dikes are thicker is commonly far more intense than can be accounted for simply by the increased thickness of the dike; here the amount of heat liberated by the greater flux of magma is the controlling factor.

There are limits to how thin an intrusive sheet can be. Thin sheets may travel so slowly that heat losses are rapid enough to cause the magma to solidify before it has traveled far. In addition, the yield strength of a Bingham magma may be great enough to halt flow in a thin sheet. The thin dike shown in Fig. 4-24 clearly was frozen to a glass as it flowed along a fracture. If magma is to continue flowing, intrusion must be rapid enough to maintain the magma in a liquid or partially liquid state.

The minimum conduit dimensions necessary for maintenance of magmatic flow have been calculated by Wilson and Head (1981). Cooling rates and magma yield strengths both are important factors in determining these dimensions. Unfortunately, little is known about the yield strengths of magmas. For magmas with a small yield strength, the cooling rate will largely determine the minimum thickness of a sheet. In such cases, the mass flux of magma per unit length of fissure J/L that must pass through a sheet of thickness W to maintain flow is

$$\frac{J}{L} = 1200\eta \left[\left(1 + \frac{1.39 \times 10^{-4} g W^3 (\Delta\rho)\rho_m}{\eta^2} \right)^{1/2} - 1 \right] \quad \text{kg m}^{-1}\,\text{s}^{-1}$$

$$(4\text{-}2)$$

where $\Delta\rho$ is the density difference between host rock and magma, ρ_m the magma density, and η the magma viscosity.

If the flux necessary to maintain flow is determined, the velocity and driving pressure can be calculated. The driving pressure in most cases is determined by the lithostatic pressure on the magma; it therefore has a definite limit, which, in turn, will control the minimum width of the dike (see Prob. 4-2).

Although dikes and sills are not restricted by definition to having a particular attitude, wide dikes generally dip steeply, and most thick sills are relatively flat lying, provided that they have not later been folded. Also, although dikes and sills can be formed from many different rock types, most thick ones are formed from basaltic magma which crystallizes to a slightly coarser grain size than in lava flows; this rock is called *diabase* (*dolerite* in British usage).

The grain size and textures of dikes and sills show marked variations across their widths. In the upper part of the Earth's crust, where country rock temperatures are relatively low, hot bodies of magma cool rapidly along their margins to produce fine-grained rocks. The grain size increases in a systematic manner toward the center of these bodies (see Chapter 5, Eq. 5-21), and in the case of diabase sheets, dikes consistently become coarser grained than sills of comparable thickness formed in the same environment. Deeper-seated bodies cool less rapidly along their margins and hence show less grain-size variation. Dikes and sills commonly contain phenocrysts, which are particularly evident in chilled margins, where they may be aligned by the flow of magma parallel to the contacts. If the early crystallizing minerals are denser than the magma, they may sink and accumulate in the lower parts of sills.

A dike or sill may be the site of more than one injection of magma. Renewed surges of magma prior to the complete solidification of a sheet result in the intrusion of material along the center of the body. If solidification is complete, fractures formed by the cooling and shrinkage of the first sheet can lo-calize the emplacement of later magmas. In the later case, sharp internal contacts are formed, and with sufficient time for cooling between injections, fine-grained chilled margins can form on the younger rocks. These bodies are described as *multiple* where successive injections are of essentially similar compositions, and as *composite* where they are of different compositions. Composite dikes and sills can contain rocks as different as diabase and granophyre (granite). Such composite dikes are common, for example, around many of the Tertiary igneous complexes in northwestern Scotland and northern Ireland. In these composite sheetlike intrusions the granophyre is usually younger than the diabase and typically occurs in the central part of the sheet.

Structures associated with the emplacement of dikes and sills generally indicate dilation. In fact, with many small sheets, especially those with irregular outlines, opposing contacts can be seen to fit back together in considerable detail (Fig. 4-22). There are many bodies, however, such as the 120-m-wide Medford dike in Massachusetts or the giant (up to 500-m-wide) Gardar dikes of southern Greenland (Bridgwater and Coe, 1970), whose emplacement can be shown to have involved no dilation; these are thought to have been emplaced by stoping. Evidence for dilation consists of structures in the country rock which, upon intersecting a sheet at an oblique angle, are offset across the body, as illustrated in Figure 4-27a. Stoping produces no offset of the country rock structures (Fig. 4-27b), nor do dikes formed by replacement.

An interesting type of dilation occurs in those sills formed by the buoyant rise of the overlying rocks. In these the dilation must be in a vertical direction, regardless of the dip of the sill. Also, if the sill climbs from one stratigraphic level to another, the dike connecting the two parts should also have the same vertical thickness (Fig. 4-27c).

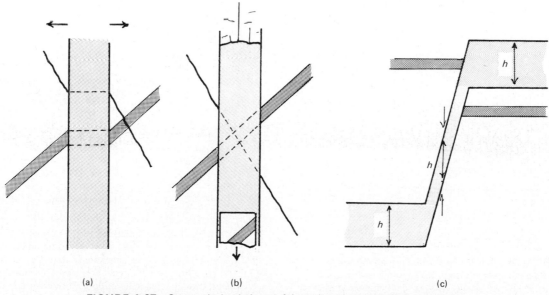

(a)   (b)   (c)

FIGURE 4-27 Geometrical relations of intrusive sheets to the surrounding rocks. (a) Simple dilation and emplacement of magma. (b) No dilation; dike rock is emplaced by replacement or stoping. (c) Emplacement of a dike sill sequence by buoyant forces results in the vertical height *h* being everywhere the same.

In many areas steeply dipping subparallel dikes occur in vast numbers, forming what are known as *dike swarms*. Almost all of these are composed of diabase and are associated with outpourings of flood basalts. They have been emplaced in zones of crustal extension. Because of their great numbers, their cumulative thickness represents considerable crustal strain. Many swarms are formed near divergent plate boundaries, with the strike of the dikes being perpendicular to the direction of plate motion. For example, the dikes that traverse southern New England (Fig. 4-26) formed during the early Jurassic parallel to the initial spreading axis of the early central Atlantic Ocean. Similar dikes have continued to form in the vicinity of the mid-Atlantic Ridge, as evidenced by the intense dike swarms in Iceland. Early spreading of the North Atlantic is recorded in an extensive dike swarm crossing northern England and the west coast of Scotland. Ancient directions of plate motion are presumably recorded by the many different dike swarms found in Precambrian shields.

Steeply dipping dikes can also occur in *radial* patterns, commonly with a large intrusion or volcanic neck located at their focus (Fig. 4-28). Although the distribution suggests that the dikes emanated from the central body, the rocks in the dikes are commonly of a very different composition and age from those of the central body. Johnson (1970), who has made a detailed analysis of the stresses associated with intrusions, has shown that magma pressures within a vertical pipelike body will not produce radial fractures. Instead, they will have the opposite effect and cause compression tangential to the chamber walls. However, the association of radial dikes about an intrusion is so common as to warrant an explanation. In some cases the central body appears to have been localized by a pre-existing intersection of dikes, rather than having been the cause of the radial dikes. In others, radial fractures, formed by doming of the roof of an actively rising magma chamber, have localized dikes that were then cut by the main body as it continued to rise. In still other cases, the presence of a body of magma has simply disturbed a regional stress pattern so as to cause tensional fractures to converge on the magma chamber and hence produce radial dikes.

4-7 RING DIKES, CONE SHEETS, AND CALDRON SUBSIDENCE

When a magma chamber intrudes close to the surface of the Earth, as happens in many volcanic regions, large cylindrically shaped crustal blocks may collapse into the chamber, producing a caldera. Magma forced up along the steeply dipping bounding fracture of the block may issue forth on the surface as volcanic material; that which solidifies beneath the surface forms an annular sheet known as a *ring dike*. These only rarely form a complete ring (Fig. 4-29); most consist of a series of arcuate segments whose curvature nonetheless clearly defines a ring structure.

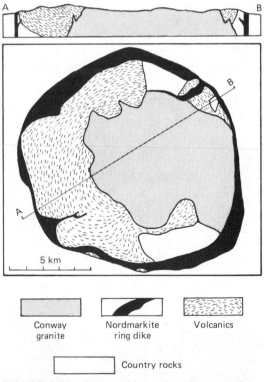

| Conway granite | Nordmarkite ring dike | Volcanics |

| Country rocks |

FIGURE 4-29 Ossipee Mountains ring complex, New Hampshire. The border of the complex is marked by a continuous ring dike of nordmarkite (quartz syenite). Associated but earlier volcanic rocks subsided within the complex. They were intruded by the ring dike and the still younger Conway granite. Note that the granite is restricted to the confines of the ring structure. These Cretaceous age igneous rocks intrude Precambrian and Paleozoic rocks. [Simplified from Kingsley (1931).]

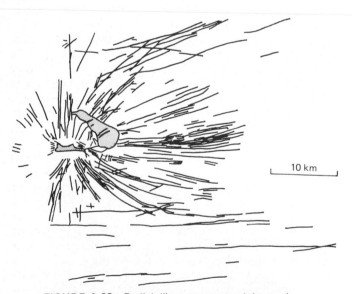

FIGURE 4-28 Radial dike swarm around the stocks of the Spanish Peaks, Colorado. An older east-west-trending dike set is composed of lamprophyres, whereas the younger radial set is mainly andesitic. By contrast, the stocks are composed of syenodiorite (west) and granite and granodiorite (east). [After Knopf (1936) and Johnson (1970).]

Ring dikes were first recognized in association with Tertiary volcanic centers of the west coast of Scotland, where they form steeply dipping arcuate bodies up to several hundred meters in width and many kilometers in length (Anderson 1936). At some of these localities, for example on the Ardnamurchan peninsula (Fig. 4-30), these dikes are concentric about a center that moved laterally with time, producing intersecting ring structures of clearly different ages.

Another type of circular structure, also first recorded from the Tertiary igneous complexes of Scotland, has the form of a cone-shaped dike with its downward pointing apex coinciding with the same centers as defined by ring dikes (Figs. 4-30 and 4-31). These are aptly named *cone sheets*. They are much narrower than ring dikes, rarely exceeding several meters in width, but they commonly occur in such great numbers that their aggregate thickness can be many hundreds of meters. They dip toward the center of the structure at approximately 45°; however, this angle tends to be steeper for sheets closer to the center. On Ardnamurchan (Fig. 4-30), where sets of intersecting cone sheets attest to the lateral movement of the center of igneous activity, downward projection of the sheets indicates a convergence at a depth of approximately 5 km, which is thought to be the top of the magma chamber from which the cone sheets came (Fig. 4-31).

Despite their close association in space and time, ring dikes and cone sheets appear strikingly different. Ring dikes are commonly composed of coarse-grained rocks exhibiting textures characteristic of slow cooling. Thus, they resemble plutonic bodies of rock. Cone sheets, on the other hand, are hypabyssal bodies and are typically composed of fine- to medium-grained rocks. Many cone sheets have chilled contacts against other cone sheets, indicating that they were formed in

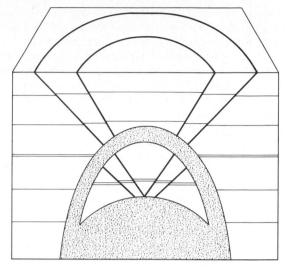

FIGURE 4-31 Block diagram of cone sheets and a ring dike. Cone sheets result from the filling of conical fractures that develop above a magma chamber in response to increased magma pressure. Stratigraphic markers lying above the cone sheets are lifted by the emplacement of magma. A ring dike forms when a decrease in pressure in the magma chamber removes support from the roof, which then collapses along steeply outward-dipping fractures; magma then wells up along the fractures as the central block subsides. Stratigraphic markers within the central block are lowered by the emplacement of the ring dike.

a series of successive injections. Multiple ring dikes, formed from successive injections, are also common, but slower cooling and lack of chilled contacts makes the intrusive sequence more difficult to determine. When there are many dikes the structure is referred to as a *ring complex*.

Ring dikes are formed from a wide variety of rock types, but granitic and syenitic ones (felsic) are the most common. In contrast, most cone sheets are composed of some type of basaltic rock. These differences probably arise from the contrasting forms and modes of intrusion of these bodies. Obviously, a block could sink more easily in a less dense felsic magma than it could in a dense basaltic one. Indeed, many crustal rocks would not sink at all in some basaltic magmas—they would float. Thus, felsic magmas are the most likely ones to form ring dikes, but their high viscosities would necessitate the dikes being moderately wide or the magma to have vesiculated and disrupted in order to maintain the minimum flow rate necessary for intrusion (see Eqs. 3-11 and 4-2 and Prob. 4-5). On the other hand, it is probable that only basaltic magmas, with relatively low viscosities, can flow fast enough in the thin cone sheets to intrude any great distance (see Probs. 4-6 and 4-7).

Although the ring complexes of the Tertiary volcanic districts of Scotland are associated with a period of igneous activity that was predominantly basaltic in composition (first the flood basalts, followed by the basalts of the central volcanoes), the majority of ring complexes discovered in other parts of the world are associated with magmas of granitic and syenitic compositions, especially ones in which sodic amphibole and pyroxene have crystallized due to the high contents of alkalies

FIGURE 4-30 Air photograph of the Tertiary Ardnamurchan ring complex, west coast of Scotland. Igneous activity about three successive centers produced intersecting sets of ring dikes and cone sheets. Only the last two centers are clearly evident at this scale. The youngest center is marked by the circular outcrop pattern of ring dikes. To the southwest of this center, a slightly earlier center is marked by arcuate cone sheets. Width of field is 20 km. (Reproduced with the permission of the Controller of Her Majesty's Stationery Office © Crown copyright.)

(see Chapter 6). For example, a 1300-km-long belt of Jurassic-age ring complexes that extends in a northerly direction across Nigeria and Niger consists largely of granitic ring dikes, some of which are up to 60 km in diameter (Bowden and Turner 1974). Important economic deposits of tin and niobium are associated with these complexes. A similar belt of granitic ring complexes, which is of Jurassic to Cretaceous age, occurs in the northeastern United States, extending 400 km in a northerly direction through the White Mountains of New Hampshire. The Ossipee ring dike (Fig. 4-29) is one of the most southerly of these complexes. Syenitic ring complexes of late Paleozoic age are common in several parts of Scandinavia, including those of the Oslo district, Norway, and the huge Khibina and Lovozera complexes of the Kola Peninsula, Soviet Union, which contain important deposits of phosphorus, niobium, titanium, zirconium, and rare earths. In contrast to the ring complexes of the Scottish Tertiary, these more felsic ones generally lack cone sheets, probably because of the types of magma involved.

The formation of ring dikes and cone sheets has been analyzed by Anderson (1936), who has shown that a set of conical fractures can develop above a magma chamber if the magma pressure rises above the shear strength of the overlying rock. With a decrease in magma pressure, steeply outward-dipping tensional fractures develop along which magma rises to form the ring dikes (Fig. 4-31). Anderson's solution of the stress distribution in the rocks overlying a magma chamber was based on rather special assumptions regarding the shape of the magma chamber (parabolic) and stress distributions along its walls. More realistic assumptions have been used in subsequent analyses (Roberts 1970), but the general fracture pattern is still essentially that suggested by Anderson. The correlation of cone sheets with fractures developed during periods of pressure buildup within a magma chamber requires that the rock above the cone sheet be lifted by the intrusion of magma. This, in fact, is what is found, and in areas where there are many cone sheets, the central block can be many hundreds of meters above its initial position (Fig. 4-31).

The tension fractures in Anderson's analysis dip steeply outward, allowing the central block to sink easily. Although ring dikes are found with such dips, many dip vertically (Fig. 4-29) or even steeply inward, making it necessary for the subsiding block to adjust its shape, perhaps by faulting or folding, or for the ring dike to intrude by piecemeal stoping, that is, by having small blocks fall from the walls of the dike and sink through the magma. Subsidence of the central block can be demonstrated with most ring dikes, even when deeply eroded. The most common evidence is the presence of volcanic crater filling material found within the ring dike at levels below its original position (see, for example, the cross section of the Ossipee ring dike, Fig. 4-29).

The amount of subsidence accompanying the emplacement of ring dikes can be great—hundreds or thousands of meters is not uncommon. The first area in which such large displacements were shown to exist was at Glen Coe, Scotland, where the name *caldron subsidence* was used to describe this phenomenon (Clough et al. 1909). Part of the subsidence can take place along fractures that do not reach the surface of the Earth, in which case a cylindrical block may sink and force magma up along the bounding fracture into the space above

the block (Fig. 4-31). The mechanism can be compared with the movement of a piston in a cylinder, and the farther the block sinks, the larger will be the pluglike body of igneous rock that will be formed above. Many pluglike bodies within ring complexes are formed in this way, as, for example, the Tertiary Mourne granites of Northern Ireland. The body of Conway granite within the Ossipee Mountain complex may also have formed this way, because the granite is completely confined by the ring dike (Fig. 4-29).

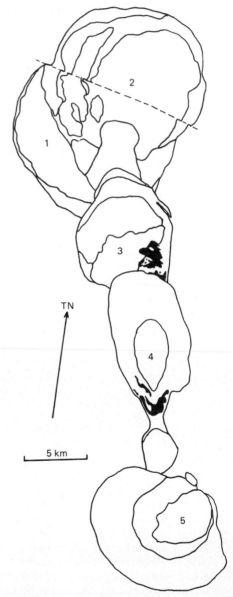

FIGURE 4-32 Sara-Fier complex, northern Nigeria, consists of five ring complexes. With the exception of the second ring complex, the center of igneous activity moved progressively southward with time. Such a linear belt of intrusions could result from the northward migration of a plate over a hot spot. Each of the complexes is composed mostly of granites, with only minor amounts of mafic rocks (black) being exposed in complexes three and four. [Simplified from Turner (1963).]

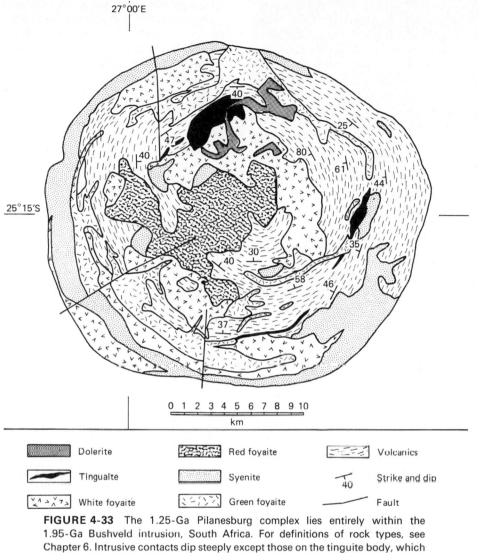

Dolerite   Red foyaite   Volcanics

Tingualte   Syenite   Strike and dip
40

White foyaite   Green foyaite   Fault

FIGURE 4-33 The 1.25-Ga Pilanesburg complex lies entirely within the 1.95-Ga Bushveld intrusion, South Africa. For definitions of rock types, see Chapter 6. Intrusive contacts dip steeply except those on the tinguite body, which is a cone sheet. The legend is not in chronological order (see Prob. 4-8). [Simplified from Mathias (1974).]

Many caldron subsidences that have had long histories during which the underlying magma chambers have changed their composition and position consist of complicated patterns of ring dikes of different rock types (Smith and Bailey 1968). When subsidence occurs more than once, the sinking of the central block commonly results in the formation of successive ring dikes that become progressively younger toward the center, and if subsidence is great enough, the central block may sink so deep as to be no longer exposed. However, there are many exceptions, and ring dikes can be found within, and peripheral to, preexisting ones. Figure 4-32 shows five ring complexes from the Sara-Fier district of Nigeria (Turner, 1963), each of which shows younger ring dikes toward the center, and only in the oldest complex (1) are there volcanic rocks exposed that belong to the original subsiding block. The main center of igneous activity beneath a caldron subsidence may, with time, move laterally far enough to produce a new set of ring fractures, as at Sara-Fier (Fig. 4-32) or Ardnamurchan (Fig. 4-30), but in

many cases the movement is restricted to the confines of the original ring structure, with different parts of the cylindrical block sinking at different times (Fig. 4-18). The Pilanesberg complex of South Africa (Mathias 1974) provides an example of a cauldron subsidence that had a long history of successive intrusive events (Fig. 4-33), and it is left to the reader, in Prob. 4-8, to work out its history.

4-8 DIATREME BRECCIA PIPES

A *diatreme* is a pipelike body of breccia composed of fragments that are derived mainly from the neighboring country rocks but also include rocks from stratigraphically higher and lower levels; some fragments may even have a mantle origin (Fig. 4-34). The matrix of the breccia consists of finely cominuted fragments, but it may also be of igneous material that typically has low silica but high magnesium and potassium contents

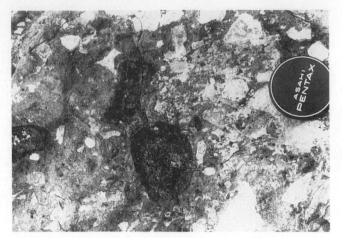

FIGURE 4-34 Diatreme breccia from Ile Bizard at the western end of the Cretaceous Monteregian igneous province, Quebec. The rock consists of an assortment of fragments of lower Paleozoic sedimentary rocks, Precambrian basement rocks, and dark mantle-derived nodules.

(alkaline magma; see Chapter 6). One particularly important type of diatreme is that which contains diamonds, and because of their discovery at Kimberley, South Africa, are known as kimberlite pipes. Much of what we know about diatremes has come from the mining of these pipes. This weight of evidence may bias our ideas on the genesis of diatremes, which may have more than one mode of origin. A review of diatremes, especially those with kimberlite affinities, is given by Mitchell (1986).

Diatremes are believed to be the feeders to small, shallow volcanic craters known as *maars* (Fig. 4-4). These are explosion craters that are surrounded by a low rim of coarse fragmental material that grades outward into a thin blanket of air-fall ash. The volume of ejected material is approximately equal to that of the crater, and the ejecta are composed of the rocks in which the crater develops. Rarely is there any new, or juvenile, volcanic material erupted, nor are lava flows produced. In short, the entire eruptive event appears to be driven by the explosion of gas. These craters are similar in form to those produced by meteorite impact, which also is an explosive event. Indeed, there have been numerous arguments over the terrestrial versus extraterrestrial origin of some craters, and it was not until the discovery of diagnostic high-pressure phases which could have formed only through meteorite impact that the question was resolved.

A subvolcanic rock type that may be related to diatreme breccias is known as *tuffisite*. As the name suggests, this rock resembles a tuff, containing lithic fragments, particles of pumice, broken phenocrysts, shards, and flattened fiamme, but it is of intrusive origin. In contrast with diatremes, tuffisites may form rather irregular bodies with branching veins of breccia invading the country rock. Bodies of breccia associated with porphyry copper deposits may be related to tuffisites. The evidence for involvement of juvenile magma in most tuffisites is clear, whereas in many diatremes it is lacking. Furthermore, tuffisites contain highly siliceous magma in contrast with the silica-poor ones of diatremes.

Diatremes are carrot-shaped bodies with nearly circular cross sections. Contacts invariably dip inward at 80 to 85°, and

thus they can have only limited vertical extent. The diameter near the surface is at most about 700 m, decreasing to about 100 m at a depth of about 2000 m. Below this is a root zone in which the shape of the pipe becomes irregular, commonly branching or spreading into one or several intersecting dikes (Fig. 4-35). The amount of igneous matrix increases through this zone until massive igneous rocks are encountered in the dikes. The contact of a diatreme with surrounding rocks is sharp, with little evidence of subsidiary veining or deformation. The type of country rock also is unimportant in determining the shape of a diatreme, as is clear where a pipe has cut straight through a sequence of layered rocks of highly variable hardness [shale and flood basalt, for example; Williams (1932)]. Country rocks are not domed up by a diatreme, nor do they exhibit significant slumping into the pipe. The emplacement of a diatreme resembles a cleanly drilled hole connecting the top of a dike or intersecting dike set with the surface of the Earth, the drilling products being left to fill the hole.

Fragments in diatremes are fairly angular, but some rounding is evident. In some tuffisites, such as that in West Cork, Ireland (Coe 1966), fragments are not only rounded but some are polished or etched on one side, implying that the orientation of the fragment remained fixed long enough for the surface to be sandblasted by the passage of gas and fine particles. Fragments range from meter-sized blocks down to submicron particles with little or no evidence of sorting. Although thermal metamorphism is lacking, fragments commonly have a hydrothermally altered rim, suggesting that at least steam had passed through the breccia.

One of the most striking features of the breccia is the thorough mixing that fragments have undergone. Fragments of mantle-derived rocks can be found juxtaposed fragments of crustal rocks (Fig. 4-34). This mixing cannot be attributed solely to upward turbulent transport, because the same breccia may contain fragments of country rock that are derived from levels

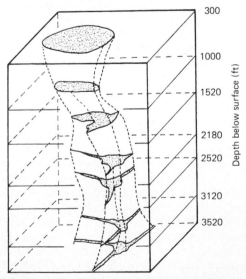

FIGURE 4-35 Cross sections through the diatreme pipe of the Kimberley diamond mine, South Africa. At depth, the carrot-shaped body grades downward into a number of intersecting dikes. [Drawn from data given by Williams, (1932).]

stratigraphically well above the existing level of exposure of the breccia. This is well illustrated by the Cretaceous diatreme breccia at the 1967 World's Fair site in Montreal, Quebec. At its present level of exposure this diatreme cuts Ordovician shales, limestones, and sandstones, rocks that form a large fraction of the fragments, along with a smaller admixture of underlying Precambrian granites and gneisses. But the breccia also contains some blocks of Devonian sandstone which today are exposed nowhere else in the Montreal area. Indeed, it is necessary to travel several hundred kilometers south into New York State to find these rocks. During the Cretaceous, however, these rocks must have covered the Montreal area. Based on stratigraphic thicknesses, these Devonian blocks must have descended over 1700 m in the diatreme breccia.

Despite the thorough mixing of fragments in diatremes, there may be concentric or vertical zonation in the abundance of fragments from different stratigraphic levels. McGetchin and Ullrich (1973), for example, showed that the abundance of mantle-derived fragments increased upward in the ejecta blanket surrounding a maar on an island off the west coast of Alaska; that is, the abundance of deep-seated fragments increased with time as deeper parts of the diatreme were tapped. Based on the distribution of these fragments, McGetchin and Ullrich calculated eruption velocities of 500 m s^{-1}, that is, supersonic (velocity of sound in air about 330 m s^{-1})

Although diatremes are clearly formed by gaseous explosions, there is considerable debate over their ultimate origin. What, for example, is the source of gas? How are fragments transported and mixed? Is the presence of magma essential, and if so, need it be of a particular composition? We will ex-

amine each of these questions briefly. For a more complete discussion, the reader is referred to Chapter 4 in Mitchell (1986).

There are two schools of thought on the origin of the gas. One claims that it is primarily of mantle origin (Wyllie 1980), rising either as a separate phase or in solution in a magma that exsolves the gas on approaching the surface. The other claims that the gas is derived from groundwater or surface water that comes in contact with and is heated by magma (Lorenz 1975; Sheridan and Wohletz 1983).

Long-lived accumulations of vapor of sufficient volume to form a diatreme cannot exist in the mantle, because buoyancy would cause it to diffuse upward. Considerable quantities of vapor could, however, be exsolved from an ascending and crystallizing mantle diapir of magma. Sufficient pressure might be generated in this way to propagate a fracture to the Earth's surface and generate a diatreme; successive eruptions might then tap deeper and deeper sources. However, where igneous rocks are exposed in the lower parts of diatremes, they are neither vesicular nor disrupted into shards. They do contain biotite or amphibole, which indicates that the magma did at least contain some dissolved volatiles.

The difficulty in deriving sufficient gas from the mantle led to the idea that surface waters may play a role. Explosive volcanic activity resulting from groundwater or surface water leaking into a magmatic system is well documented from many volcanic regions; it is known as *hydrovolcanism* or *phreato-magmatism* (Sheridan and Wohletz 1983). Deep craters filled with pyroclastic debris and slumped blocks can form in this way as shown in Figure 4-36. Diatremes are claimed to form preferentially in regions where there are good aquifers, such as

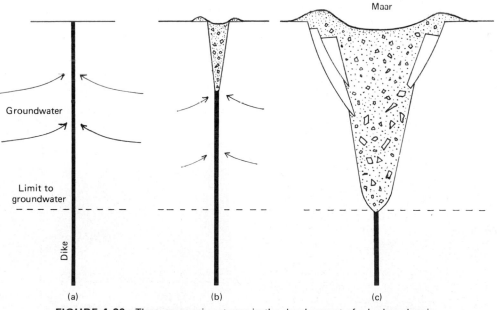

FIGURE 4-36 Three successive stages in the development of a hydrovolcanic breccia pipe. (a) Groundwater or surface water gains access to a conduit filled with magma. (b) Conversion of groundwater to steam causes disruption of the magma and brecciation of the surrounding rock; brecciation commences at the surface and works downward. (c) Brecciation of the rock continues to depths where groundwater is still capable of gaining rapid access to the conduit. A shallow crater, known as a maar, forms above the pipe, and it is surrounded by a blanket of ejecta derived largely from the country rocks.

permeable sedimentary rocks. Even when formed in regions where such rocks are absent, diatremes tend to be located at the intersection of major fractures that could serve as conduits for groundwater. According to the hydrovolcanic hypothesis, diatremes are formed when groundwater that has been transformed to steam by coming in contact with a magma rises along a fracture and explodes through to the surface of the Earth. The brecciation would start at the surface and work downward eventually reaching the magma. Indeed, most hydrovolcanoes have craters that contain fragments that in large part are juvenile, although blocks of country rock can slump in from the walls. Most diatremes, in contrast, consist largely of fragments of country rocks. The hydrovolcanic hypothesis does not by itself explain the presence of mantle-derived fragments in some diatremes.

Regardless of source, fluxes of gas have undoubtedly been involved in the formation of diatremes. The passage of gas through particles brings to mind the fluidization process which was discussed in connection with ash flows (Sec. 4-4). Following the publication by Reynolds (1954), fluidization gained instant popularity as a hitherto unrecognized geologic process. One of its most convincing applications was in explaining the origin of diatreme breccias. Modeling experiments successfully reproduced many of the features seen in diatremes (Woolsey et al. 1975).

As the velocity of gas passing through a body of close-packed particles is increased, a stage is reached where the gas pressure between the particles becomes high enough to lift and slightly separate the particles. At this stage the system becomes *fluidized*, and the material changes from being dense and rather rigid to less dense (about 20% less) and extremely fluid. It is this dramatically increased mobility that offers an explanation for the thorough mixing of fragments in diatremes. It is important to emphasize that the flow of gas through a fluidized system is not great enough to expel material from the conduit but simply to raise the particles into a fluidized state. This is precisely what is found in most diatremes—the breccia, while thoroughly mixed, is still essentially in its correct stratigraphic position; rarely does a diatreme clear its throat and replace the breccia with magma. Another important feature of fluidization is that it must start at the top and work down because it involves the lifting of particles. Diatremes, then, if formed by fluidization, would have to start at the Earth's surface and work downward along some fracture up which gas was escaping. The gas could be coming from either a magmatic source or from groundwater heated by a magma.

Fluidization, however, is not as simple to apply to diatremes as was first thought. Industrial fluidization involves the movement of particles of a limited size range. Diatremes contain fragments that vary in size over six orders of magnitude. Experiments using material of variable grain size (Wilson 1984) indicate that fluidization in such material is complicated. First, there is no single gas velocity at which fluidization of the entire system occurs. The gas flux necessary to lift micron-sized particles is very much less than that necessary to lift centimeter-sized particles. If gas velocities are increased so as to lift the large particles, large bubbles of gas start ascending through the system; also, finer particles are transported upward, resulting in a sorting of the particles. Sorting is noticeably lacking in most diatremes. Also, the constant agitation in a fluidized system would be expected to produce more rounding of the fragments than seen in most diatreme breccias.

Finally, there is the question of associated magmatic rocks. Although igneous rocks are not found with all diatremes, they are found with so many that they probably occur in the root zones of all diatremes and would be exposed if erosion went deep enough. More important, most diatremes are associated with a small class of igneous rocks that are referred to as alkaline (see Chapter 6) and particularly to alkaline ones that are very poor in silica, such as kimberlite. These have associated mantle-derived fragments and diamonds which must come from depths of at least 150 km (Sec. 14-7). This association indicates that hydrovolcanism, although possibly a contributory factor to brecciation, cannot be the ultimate cause of diatreme activity—it is a phenomenon peculiar to alkaline magmatism. Yet the lack of vesiculation or magma disruption in the root zone of diatremes indicates that if the gas required for brecciation is to have exsolved from the alkaline magma, it must have done so at greater depth so that any signs of vapor-phase separation would be obliterated.

4-9 LACCOLITHS

A *laccolith* is a concordant intrusive body with a flat base and domed roof—that is, a mushroom-shaped body (Figs. 4-21 and 4-37). They form at shallow depths, mostly within 3 km of the Earth's surface and in relatively flat-lying stratified rocks. The emplacement of magma domes the overlying rocks upward. By contrast, stratification above and below a sill are essentially parallel. Laccoliths are thought to be fed by a central conduit,

FIGURE 4-37 Maverick Mountain, west Texas, is a laccolith of alkali gabbro intrusive into limestone; its topographic expression corresponds almost exactly with its igneous form. A small cinder cone in the foreground is younger than the laccolith. An old erosion surface (pediment) can be seen surrounding the laccolith. This has been cut into by a younger erosion surface on which the cinder cone has developed.

but exposures rarely allow this to be determined for certain. Dikes and lateral feeders may also occur. They are commonly formed from magmas with intermediate silica contents, but examples ranging in composition from basaltic to rhyolitic are known. Because of their thickness, laccoliths contain coarser-grained rocks than do sills.

Laccoliths were first described by Gilbert in 1877 from the Henry Mountains of Utah. His account of the fieldwork, description of the bodies, and insight into their mode of formation makes for some of the most fascinating reading in all of geology [see Yochelson (1980)]. Gilbert recognized that both sills and laccoliths occur in the Henry Mountains, but sills always had areal extents of less than 1 km², whereas laccoliths were greater than 1 km². From this he concluded that the sills were the forerunners of the laccoliths but that they required a minimum area before they exerted sufficient force to dome upward the overlying rocks. He also drew a correlation between the size of a laccolith and its depth of burial at the time of formation, the larger ones forming at greater depth.

Despite the frontier conditions under which Gilbert's field work was done, his analysis of the formation of a laccolith has a distinctly modern flavor, and it emphasizes some of the important forces that must be involved. For the purpose of the analysis, he treated a laccolith as a pistonlike body that lifted the overlying rock along a ring fracture, as shown in Figure 4-38a. He reasoned that the force lifting the roof came from the pressure in the magma, P_m, which acted on the area of the roof of the laccolith, πr^2; that is,

$$\text{lifting force} = \pi r^2 P_m$$

This force would be opposed by the body force resulting from the mass of the overlying rocks. If the overlying rocks have a thickness T and density ρ_c, then

$$\text{weight of roof rocks} = \pi r^2 T \rho_c g$$

But $T\rho_c g$ is the lithostatic pressure at depth T. We can therefore write

$$\text{weight of roof rocks} = \pi r^2 P_l$$

where P_l is the lithostatic pressure. In lifting the overlying rocks, magma also has to overcome the frictional resistance along the ring fracture resulting from the shear strength τ. This stress acts on the surface area of the ring fracture and gives a resisting shear force of

$$\text{shear force} = 2\pi r T \tau$$

Gilbert argued that for a laccolith to be able to lift the overlying rocks, the lifting force of the magma has to be equal to or greater than the sum of the forces resulting from the mass and shear strength of the overlying rocks; that is,

$$\pi r^2 P_m \geq \pi r^2 P_l + 2\pi r T \tau$$

which on rearranging gives

$$(P_m - P_l)r \geq 2T\tau \tag{4-3}$$

This relation indicated to Gilbert that a minimal radius was necessary for a laccolith to overcome the shear strength of the overlying rock. It also accounted for the radius of laccoliths increasing with increasing depth of formation.

We cannot expect Eq. 4-3 to give a precise description of the parameters governing laccolith formation. It does not, for example, take account of the force necessary to bend the overlying rocks or the pressure necessary to intrude the magma between the two contacts, regardless of the amount of lifting. A more sophisticated analysis of the forces involved in the formation of these laccoliths has since been done by Johnson (1970)

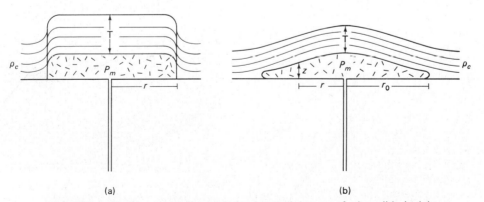

(a)   (b)

FIGURE 4-38 Two possible models for the emplacement of a laccolith. In (a) the overlying rocks are lifted largely as a block that moves on a cylindrical ring fracture; the dimensions are those used in Gilbert's (1877) analysis of these structures. In (b), the overlying rocks are domed upward by the magma as soon as it extends laterally far enough to supply sufficient leverage on the overlying rocks; the dimensions are those used in Pollard and Johnson's (1973) analysis of these structures. See text for discussion.

and Pollard (1973), who have also made a careful study of the actual structures in the country rocks at the periphery of these laccoliths.

Johnson and Pollard, like Gilbert, found that sills in the Henry Mountains range in area up to 1 km², whereas laccoliths are greater than this; sills are from 0.5 to 10 m thick and laccoliths from 10 to 200 m. Sills do not extend out as thin wedges but terminate abruptly, their blunt ends being surrounded by deformed and sheared country rock as shown in Figure 4-23b. The periphery of laccoliths is also blunt, with any one of the configurations shown in Figure 4-39 being common. In plan view the periphery may be smooth but can consist of fingerlike extensions, the spacing of which can be explained on the basis of instabilities similar to those for the regular spacing of diapirs (Pollard et al. 1975). The diorite, of which the Henry Mountain laccoliths are composed, is porphyritic and may have behaved as a Bingham liquid with a yield strength of about 100 Pa. This may have prevented the magma spreading too far laterally.

The important point of these field relations is that laccoliths are not large sills but distinct bodies formed by a very different type of displacement of the overlying rocks. Sills expand until they have an area of about 1 km², at which stage continued lateral spreading becomes more difficult than lifting of the roof. Johnson and Pollard's analysis of these structures, combined with experimental modeling, indicate that the intrusion of a sill acts like a lever on the overlying rocks. When a sill is small the mechanical advantage is also small, because the distance from the point of magma influx to the leading edge of the sill is short. The magma therefore continues to flow laterally as the overlying rocks are simply lifted upward. As the sill expands the leverage increases, until at a critical stage of growth, the rocks at the leading edge and above can be bent rather than simply being lifted, and a laccolith is born.

The extent to which a sill will expand before reaching the critical stage at which a laccolith develops depends not only on the depth of emplacement but on the physical properties of the magma and the country rocks. A fluid basaltic magma might extend laterally a greater distance than the porphyritic diorite of the Henry Mountains, and a more viscous granitic one might extend less. Also, the more competent the overlying rocks are, the more a sill will have to expand before being able to bend and dome the roof rocks.

Pollard and Johnson (1973) give the following equation relating the height z that the roof of a laccolith will be lifted

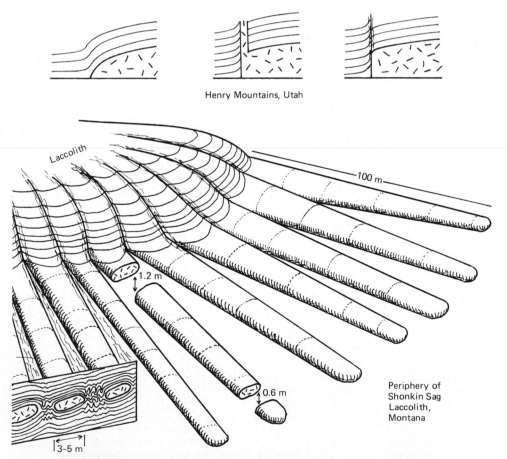

Henry Mountains, Utah

Laccolith

100 m

1.2 m

0.6 m

3–5 m

Periphery of Shonkin Sag Laccolith, Montana

FIGURE 4-39 Three different terminations found at the peripheries of laccoliths in the Henry Mountains, Utah (Johnson and Pollard 1973), and the fingered periphery of the Shonkin Sag laccolith in the Highwood Mountains, Montana. [Drawn from data presented by Pollard et al. (1975).]

as a function of distance from its center r, where r_0 is the radius of the body (see Fig. 4-38b):

$$z = \frac{3(P_m - \rho_c g T)}{16BT^3}(r_0^2 - r^2)^2 \qquad (4\text{-}4)$$

where B is the elastic modulus, which is defined as $E/(1 - v^2)$, E being Young's modulous and v Poisson's ratio [see Johnson (1970) for further explanation].

It is valuable to inspect Eq. 4-4 qualitatively to gain an appreciation of the factors controlling the shape of a laccolith. The first terms in parentheses give the difference between the pressure of the magma, which tries to lift the overlying block, and the load pressure of the overlying rock, which tries to make the roof sink. The difference, then, is the resultant upward pressure on the roof rocks. The terms in the second set of parentheses define the length of the lever arm; the longer it is, the greater will be the height that the roof can be lifted. Note that this factor, being raised to the fourth power, is extremely important. The denominator expresses the rigidity and thickness of the overlying rocks. Clearly, as either of these factors increase, the height to which the roof can be domed must decrease. Note that the thickness T of the overlying rocks is raised to the third power and thus plays an important role in determining the thickness of the laccolith (see Prob. 4-9). The equation, of course, cannot describe completely the shape of a laccolith because as shown in Figure 4-39, considerable faulting and peripheral vertical dike intrusion accompanies the simple flexing of the roof rocks.

In summary, laccoliths are restricted to shallow depths because room for the magma is provided by the doming and bending of the roof rocks. Their size and form is controlled largely by the stratification and physical properties of the surrounding rocks.

4-10 LOPOLITHS AND LAYERED INTRUSIONS

Lopoliths are large conformable, saucer-shaped intrusions (Figs. 4-21, 14-14, and 14-15). Their form superficially resembles that of an inverted laccolith, but these two intrusive types are quite unrelated. Laccoliths are never more than a few kilometers in diameter, whereas lopoliths may be over a hundred kilometers in diameter. Laccoliths form near the surface of the Earth, where they make room for themselves by doming upward the surface rocks. Lopoliths, on the other hand, are so large that they affect the entire crust, and room for the intrusion is made as much by bowing down the floor as it is by lifting overlying rocks. Laccoliths can be formed from a variety of different magmas, but lopoliths are invariably formed from basaltic ones. The feeder conduit to a lopolith is commonly hidden by the intrusion itself, but large regional basaltic dikes have been found to connect with the base of some bodies (Fig. 14-15). Roof rocks, where present, may include relatively unmetamorphosed flat-lying sedimentary rocks or even volcanic rocks formed from the same magma that produced the lopolith.

Because of their enormous size, lopoliths cool very slowly, allowing a variety of features to develop that are not present in small intrusions. Among these, the most striking is prominent flat or gently dipping layering. The layers, which can range in thickness from millimeters to meters (Fig. 4-40) and extend laterally for hundreds of kilometers, may result from variations in the proportions or compositions of minerals, or in the orientation of crystals. Some layers can have concentrations of minerals that are great enough to be of economic value. One of the principal sources of Cr, for example, is from layers rich in chromite. So striking is the layered appearance of rocks in lopoliths that these intrusions are commonly referred to as *layered intrusions* or layered basic (mafic) intrusions, the term *basic* meaning that the rocks were formed from basaltic magma (see Chapter 6). Layering does occur in some other types of intrusive bodies but never on the same extensive scale as in lopoliths.

The largest known lopolith is the 2.0-Ga-old Bushveld complex of South Africa, which has an area of 66,000 km² (Fig. 14-14). Over 8 km of layered rocks is exposed between its floor and roof. The complex is lobate in plan, with rocks dipping at shallow angles toward the center of each lobe. The body probably developed by the coalescence of four lopoliths, which may have been initiated by meteorite impact (Sec. 14-9). Another well-exposed lopolith is the 1.2-Ga-old Muskox intrusion in the Northwest Territories of Canada. It forms an elongate canoe-shaped intrusion with rocks dipping toward its axis at about 30° (Fig. 14-15). Because this body plunges northward, both the floor and roof are exposed. A 150-m-wide, 60-km-long dike feeds into the base of the intrusion; from here the body widens gradually as it extends 50 km northward, reaching a maximum width of 11 km at the point where it disappears beneath the roof rocks. A magnetic anomaly indicates that the intrusion continues northward for at least another 30 km beneath the cover rocks. Within the exposed section, a thickness

FIGURE 4-40 Millimeter- to centimeter-scale layering in gabbroic rocks of the Kiglapait lopolith, Labrador. Graded layering is brought out by concentrations of dark-weathering olivine at the base of layers passing up through pyroxene-rich zones to light-colored plagioclase-rich rock at the top of layers.

of 1.8 km of layered rocks constitutes the bulk of the intrusion. The so-called "Great Dyke of Zimbabwe" (Rhodesia), a 530-km-long by 5.6-km-wide body of layered rocks, may have formed in a Muskox-like body that was later downdropped along two faults that parallel the trend of the original body. Many lopoliths are only partially preserved. The 3.2-Ga-old Stillwater complex in Montana, for example, is only the lower two-thirds of a former lopolith, the upper part having been eroded away or uncomformably overlain by Paleozoic rocks. The remaining 8 km of layered rocks, however, attests to the former existence of a gigantic body. Perhaps one of the most unusual lopoliths is that at Sudbury, Ontario. This 1.7-Ga-old lopolith was emplaced in what is interpreted to have been a huge meteorite crater measuring about 40 km in diameter (Sec. 14-9).

It is not by design that all of the examples of large lopoliths cited here are of Precambrian age. In the Phanerozoic, only relatively few, small lopoliths have formed. This is one of a number of petrologic facts which indicate that the principle of *uniformitarianism* (the present is the key to the past) may not be completely applicable to all processes throughout geologic time. The restriction of large lopoliths to the Precambrian may indicate that the thermal character of the Earth has changed with time, a point to which we return in Sec. 14-8.

4-11 STOCKS

A *stock* is defined as a plutonic body having an areal extent of less than 100 km², whereas a *batholith* is one whose area is greater than this. The division is arbitrary and thus at first may appear to have little significance. For instance, could a stock be the top of a batholith that had not yet been eroded deeply enough to expose its full extend? If the difference between these bodies depended on the whims of erosion, the definitions would have no more significance than to indicate qualitatively the present area of exposure; this could be done more simply with a statement of the present number of square kilometers exposed. There is, however, a distinct group of intrusions that fit the definition of a stock and whose areal extent does not vary significantly with depth of erosion. This is not to say that the term *stock* has not been used to describe bodies that are the upper tips of a batholith. But we ignore these here and take them up in Sec. 4-12.

Stocks, as defined above, have two distinct modes of occurrence. One is as bodies of diorite, granodiorite, or granite (see Chapter 6) intruded beneath large calderas. Here they clearly form an integral part of caldron subsidence, providing the source of magma for eruptions of ash-flow tuffs and a medium into which the surface blocks can sink. The other mode of occurrence involves rocks that are referred to as alkaline (see Chapter 6) and which range in composition from peridotite, through gabbro, to nepheline syenite and even granite. These stocks are typically associated with rift valleys on continents.

In Sec. 4-4, calderas were described as cylindrical blocks that subside along steeply dipping ring fractures into bodies of magma, the subsidence typically being accompanied by the eruption of ash flows. Numerous studies of old, deeply dissected

calderas fully substantiate the deep structures surmised from the surface exposures. Furthermore, these studies indicate that most calderas are the sites of multiple collapses, and that the magma chamber into which the surface block sinks makes room for itself by displacing downward the preexisting rocks along the same, or closely related, ring fractures. These relations, which were first worked out by Clough et al. (1909) for the caldron subsidence of Glen Coe, Scotland, have since been found with many near-surface magma chambers.

The ring fracture of a caldera extends downward and becomes the wall of the underlying magma chamber. Its steep dip and cylindrical shape produces a pluglike body or stock (Fig. 4-41a). The roof of the chamber is the base of the subsided block of the caldera. The floor is formed by rocks that are forced downward into a deeper magma chamber, the floor rocks moving in a manner similar to that of a piston in a cylinder. No room problem exists with this type of emplacement; the floor rocks and deeper magma simply exchange places. This type of emplacement is possible even if no overlying caldera is formed as shown in Figure 4-41b. All movement of magma and country rocks is probably in response to buoyant forces that are developed locally; that is, the sinking block must be denser than the ascending magma. While the magma cools and crystallizes, a light fraction may accumulate near the top of the chamber, where it may bow up the roof to form a *resurgent dome* on the floor of the caldera, as happened in Yellowstone (Fig. 4-18). After complete solidification, sections of the intrusive body may sink (it would now be denser) as new magma wells up around the piston to form a new chamber beneath the roof rocks. The subsiding floor rocks may therefore bear testimony to repeated intrusions and subsidences.

A stock that forms in this way has certain characteristics. It is circular to oval in plan and has vertical or steeply dipping walls and flat or gently sloping roof and floor. The transition from roof to wall takes place over a distance of meters. Contacts with the wall rocks are strongly discordant. The wall rocks, however, show little sign of deformation and certainly are not bent upward. Some stocks may be surrounded by additional ring fractures along which sheared rocks or dikes may be pre-

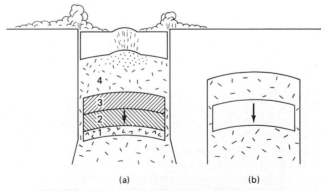

(a)   (b)

FIGURE 4-41 Development of a stock by the sinking of the roof rocks along a ring fracture up which magma rises as the block descends. On the surface, a stock may have an associated caldera (a), but it need not break through to the surface at all (b).

sent. These intrusions are generally composite or multiple, with successive fillings of the magma chamber forming a roughly horizontal stratigraphy as the floor subsides. The rocks are restricted in composition to ones that form from magmas with densities less than those of the intruded rocks. The magmas crystallize to massive rocks except near the feeder ring, where a strong foliation or shearing may be developed.

Stocks composed of alkaline rocks differ from those described above in several important ways, the difference being attributable mainly to the modes of emplacement. First, the magmas involved in alkaline stocks may be denser or lighter than the intruded rock, so buoyancy, at least at the local level, is not important. Many of the rocks are derived from magmas which have come directly from the mantle and which isotopic studies (Sec. 21-3) reveal have not had a long residency time in the crust. These intrusions rarely show evidence of an obvious floor, either through direct exposure or through geophysical means, indicating that they are of considerable vertical extent. They are strongly discordant. Unlike the stocks associated with caldron subsidence, however, these bodies have contacts with

surrounding rocks that are marked by considerable deformation, which may involve plastic flow of the country rocks or intense stoping. The rocks within these stocks commonly exhibit prominent layering or foliation, which dips steeply and is parallel to the contacts. Disruption of the layering attests to vigorous movement of magma during solidification. Multiple injections of strongly contrasting magmas (gabbro-syenite) are also common.

The Cretaceous Monteregian intrusions of southern Quebec provide excellent examples of most of the features characteristic of this type of stock (Philpotts 1974). These intrusions are located at the intersection of two rift systems, the St. Lawrence and Ottawa valleys, which have been active since the late Proterozoic (Fig. 4-42a). A wide range of alkaline rocks, including peridotite, carbonatite, gabbro, nepheline syenite, and granite, form stocks that are circular to elliptical in plan and have vertical contacts (Fig. 4-42b). The smallest stock, Mount Johnson, is nearly circular in plan, whereas some of the larger ones are lobate and consist of stocks within stocks, the relative ages being indicated by truncation of layering. Almost all of the

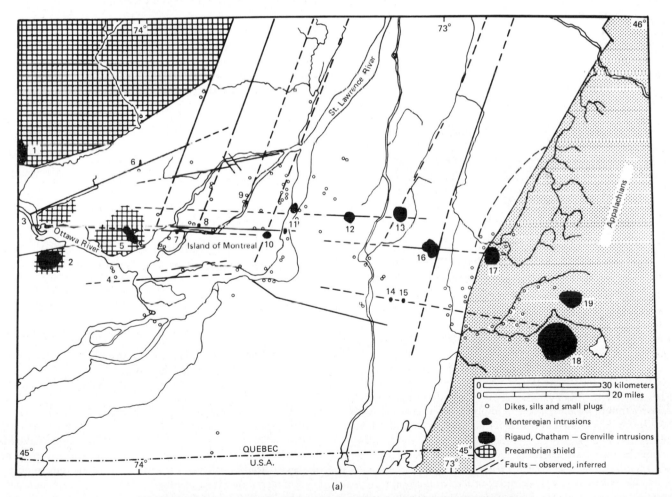

(a)

FIGURE 4-42   (a) The Cretaceous Monteregian intrusions of southern Quebec form a linear belt of stocks extending eastward from the Ottawa graben across the St. Lawrence lowlands into the Appalachian Mountains. The main intrusions are Oka (5), Royal (10), Bruno (12), St. Hilaire (13), Johnson (15), Rougemont (16), Yamaska (17), Brome (18), and Shefford (19).

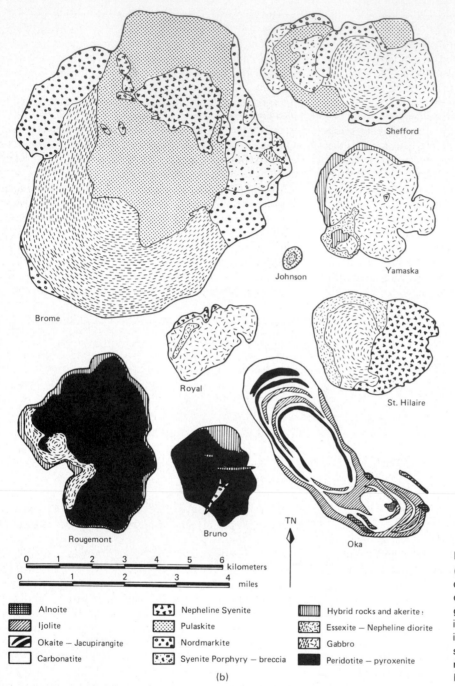

Brome

Shefford

Johnson

Yamaska

Royal

St. Hilaire

Rougemont

Bruno

TN

Oka

0 1 2 3 4 5 6
kilometers
0   1   2   3   4
miles

Alnoite	Nepheline Syenite	Hybrid rocks and akerite
Ijolite	Pulaskite	Essexite – Nepheline diorite
Okaite – Jacupirangite	Nordmarkite	Gabbro
Carbonatite	Syenite Porphyry – breccia	Peridotite – pyroxenite

(b)

FIGURE 4-42   (Continued)
(b) Each of the intrusions forms a vertically elongated cylindrical body; larger ones consist of several intrusions which give the bodies a cuspate outline. Most intrusions exhibit compositional layering that parallels the contacts and dips steeply inward. [After Philpotts (1970); reprinted by permission of Canadian Mineralogist.]

stocks exhibit concentric, steeply dipping layering which was built up by deposition or growth of minerals on the walls of the intrusions. This sequence of solidification is clearly evident where channels have been cut in the earlier deposited layers and then filled in by later layers (Fig. 4-43).

The stocks in the western half of the Monteregian province were emplaced in flat-lying Ordovician sandstone, limestone, and shale. These sedimentary rocks provide ideal markers with which to monitor the deformation associated with magma emplacement. No diapiric doming is associated with any of these stocks. Instead, the sedimentary rocks remain flat lying

to within a few meters to tens of meters of the contact. Here they begin to dip in toward the intrusion, with the dip increasing toward and becoming vertical at the contact (Fig. 4-45). Toward the contact the rocks show increasing signs of plastic deformation and indeed of melting, with refractory layers remaining as brittle material that is segmented by the flow of the surrounding more fusible rock (Fig. 4-44). The contact zone with the igneous rocks of the stock is therefore marked by a zone of breccia with an igneous matrix formed from the partial fusion of the country rocks. Such a liquid is referred to as *rheomorphic* to distinguish it from the magma that rose from depth and brought

FIGURE 4-43 Vertically dipping layering in essexite (nepheline gabbro), Mount Johnson, Quebec. The contact of this almost perfectly cylindrical intrusion (see Fig. 4-42b) is to the left of the outcrop shown here. The layering parallels the contact. Flow of magma eroded a channel into the layers of crystals which were deposited on the wall of the intrusion. Where the younger layers enter the channel they thicken, with the result that the irregularity in the layering caused by the channel is rapidly eliminated following the deposition of several new layers. [After Philpotts (1968); reprinted by permission of Canadian Journal of Earth Sciences.]

FIGURE 4-44 Rheomorphic breccia at the contact of the Brome Mountain Monteregian intrusion, Quebec (see Fig. 4-42). Heat from the intrusion of an alkali gabbro magma fused country rocks that have low melting temperatures; the more refractory country rocks, which include highly aluminous rocks (thin dark bed), quartzite (white fragment beneath hammer), and early mafic dikes (dark fragments), were broken and pulled apart by the flow of the newly melted rocks. Rocks in the lower right are vertically dipping metamorphosed country rocks which project into the zone of rheomorphism. Evidence of selective partial melting and flow can be seen by comparing the two layers on the left side of this projection of country rock.

with it the heat necessary to fuse the country rocks. It will be recalled from Sec. 3-9 that this zone of melted country rock is precisely what is required in the magmatic ascent model of Marsh (1982).

The stocks in the eastern half of the Montergian province, which are generally larger than those in the western part, were emplaced in the folded Appalachian rocks, so the effects of intrusion on the geometry of the surrounding rocks is not as easily determined. Zones of rheomorphic breccia are, nonetheless, present, and where the country rocks dip steeply and strike into the intrusions at a high angle the details of the contact reveal selective melting of the country rocks (Fig. 4-45). For example, layers of refractory quartzite or slate (which is metamorphosed to corundum-bearing rock at the contact) project into the zones of rheomorphic breccia for several meters before being broken off by the viscous flow of the breccia.

Toward the inside of the zone of rheomorphic breccia, fragments of the refractory country rocks become rounded by dissolution in the rheomorphic liquid and eventually disappear. This results in a zone of rather streaky, inhomogeneous rock of granodioritic composition, which is a few tens of meters wide. This granodioritic rock, formed from a mixture of mantle-derived magma and melted crustal rocks, is markedly different

in composition from those in the main part of the intrusions, which have inherited their chemical characteristics only from a mantle source.

A room problem exists with the emplacement of this type of intrusion. Although the rheomorphic breccia zone indicates that magma was making room for itself by partially melting and stoping country rock, it is unclear where the melted country rocks have gone. Mixing with the main body of magma is possible, but the relatively uncontaminated nature of most of the igneous rocks in these stocks indicates that this amount would have to have been small. The melted country rocks would have had a much higher silica content than most of the magmas rising from depth; thus, they would have had a lower density. The rheomorphic magma may, therefore, have been able to rise buoyantly along the margins of the intrusions and collect at the top of the stock. If this happened, the evidence has, unfortunately, been eroded away.

Although stocks formed in association with caldron subsidence and with alkaline rocks have similar outcrop patterns, details of the contacts and internal structures indicate two very different modes of emplacement. The caldron-subsidence stock involves passive upwelling of buoyant magma while a cylindrical block of dense rock sinks. The alkaline stock, on the other hand, involves a vigorous thermal event, with heat being brought from considerable depth to high levels in the crust,

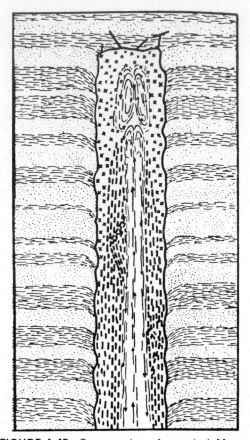

FIGURE 4-45 Cross section of a typical Monteregian stock intruded into flat-lying Paleozoic sedimentary rocks. Near the stock, country rocks dip inward and become partially melted at the contact, which is marked by a zone of rheomorphic breccia (Fig. 4-44). Within the intrusion, compositional layering generally parallels the walls of the body. Active convection of the magma (rising in the core and descending along the walls) is indicated by discordances in the layering (Fig. 4-43). This convection was responsible for renewing the heat that was lost to the surroundings by conduction and in melting of country rocks. A separate convecting layer of less dense, felsic magma may have existed in the upper part of the body. [After Philpotts (1968); reprinted by permission of Canadian Journal of Earth Sciences.]

where it plays a significant role in making room for the magma. In closing, it should be mentioned that although we have referred to alkaline magmas forming these hot, thermally active stocks, they can also form stocks through the caldron subsidence process, as long as densities are appropriate.

4-12   BATHOLITHS

A *batholith* is a plutonic igneous body that has an areal extent greater than 100 km² and extends to great depth. In some definitions batholiths are described as having no floor; although this is not possible, it emphasizes the fact that they extend so

deep that their roots may well be hidden. The name *batholith* itself comes from the Greek word *bathos*, meaning deep. Batholiths are composed of relatively low-density rocks, ranging in composition from quartz diorite to granite, with granodiorite being the most abundant (see Chapter 6). They are never composed of gabbro or ferromagnesian-rich rocks, but some Proterozoic batholiths are composed of anorthosite, a rock consisting essentially of plagioclase (Sec. 14-8) and also having a low density relative to surrounding rocks. This restriction in the composition of rocks to ones of low density is clear evidence that batholiths are emplaced by the buoyant rise of magma through the crust. The low density of the rocks also makes it possible to use gravity surveys to determine the extent of batholiths at depth. Most batholiths are restricted to orogenic belts, where they may be syn- or postorogenic in age.

Many batholiths are extremely large. The Coast Range batholith, which extends from Alaska through the entire length of British Columbia, is 600 km long and 100 to 200 km wide. The Coastal batholith of Peru is 1100 km long and 50 km wide (Fig. 4-49). These batholiths are synorogenic and are elongated parallel to the mountain chain. Postorogenic batholiths are smaller and less elongated, such as the White Mountain batholith in the northern Appalachians of eastern North America, which is elliptical in plan, measuring 45 km by 30 km. Large batholiths are not single intrusive bodies but consist of many smaller batholiths and stocks, each of which may contain distinct rock types. For example, nine different episodes of igneous activity are recognized in the Coast Range batholith in British Columbia, and eight have been found in the Sierra Nevada batholith in California. The vertical extent of batholiths is difficult to determine, but seismic and gravity surveys indicate that they have relatively flat floors at depths of 6 to 10 km but may have roots extending as deep as 15 km. The geophysical evidence thus indicates that batholiths have a form similar to that of a molar tooth (Lynn et al. 1981; Oliver 1977; Kearey 1978; Simmons 1964). These depth determinations are in agreement with heat flow measurements. It will be recalled from Sec. 1-4 and Prob. 1-5 that a layer of granite could not extend to depths greater than about 10 km without generating a heat flux far in excess of what is observed.

Because of their great vertical extent, batholiths can be found in many different depth environments, where they exhibit quite different characteristics. Buddington (1959), for example, drew attention to the difference between batholiths emplaced in the *epizone* (shallow), *mesozone* (intermediate), and *catazone* (deep). Although sharp divisions between these zones cannot be drawn, batholiths emplaced near the surface are quite distinct from those emplaced at depth. The main differences result from changes with depth in the behavior of the crust to long-term stresses. In Chapter 1 we saw that the crust can be divided into an upper brittle zone and a lower ductile zone. Some batholiths are sufficiently large to span both zones, and smaller ones may certainly have passed through both during ascent. Batholith emplacement at great depth involves ductile movements in the surrounding rocks, whereas that at shallow levels involves brittle failure of the country rocks. Because batholiths are so large, their ascent through the crust causes a regional raising of isotherms, which in turn causes the brittle–ductile transition to occur at shallower depths than normal.

Batholiths emplaced at high levels in the Earth's crust are markedly discordant. Exposures of their roofs are numerous, as are roof pendants and xenoliths. *Roof pendants* are isolated bodies of country rocks within batholiths whose point of attachment to the roof has been removed by erosion. The orientation of structures in roof pendants is consistent with that in country rocks outside the batholith. *Xenoliths* (foreign rock), on the other hand, are unattached blocks of country rock, and they may have any orientation. The abundance of xenoliths near contacts indicates that the surrounding rocks were brittle and that stoping was a major means of magma emplacement. Fine-grained chilled margins are found on the earliest phases of intrusion, but little chilling is seen with later intrusive phases. The rocks are typically unfoliated, and phenocrysts have shapes similar to those found in volcanic rocks. Small gas cavities lined with euhedral crystals (*miarolitic cavities*) are common, but *pegmatites*, veins of extremely coarse-grained granitic rock, are generally lacking.

Batholiths emplaced at great depth in the crust are generally concordant. Contacts with country rocks are steep, and chilled margins and xenoliths are completely lacking. Large sheets or *screens* of country rocks may occur in the batholith parallel to the contacts. Evidence of small-scale replacement of country rocks at the contact is common. The igneous rocks tend to be foliated and recrystallized. Pegmatites and associated veins of a fine- to medium-grained sugary-appearing rock known as *aplite* are common, but miarolitic cavities are absent.

Batholiths exhibit vertical and lateral compositional variations. These variations may be abrupt, as for example between bodies formed during different episodes, or gradational within a single body. In general, the proportion of quartz diorite is greater at depth than it is near the surface, and successive intrusions become more siliceous with time. Within a single intrusive body, rocks tend to be more siliceous and hydrous toward the top. Compositional zoning in thick ash-flow tuffs shows that magmas in near-surface chambers can be strongly zoned in composition while still largely liquid (see Sec. 13-5). Although rocks of gabbroic composition never form batholiths, inclusions of basaltic composition, ranging in size from centimeters to meters, are extremely common in all rocks found in batholiths but especially in the more mafic ones which typically occur in the lower parts of batholiths. These inclusion are podlike in form, and in high-level intrusions they resemble basaltic pillows in granite (Fig. 4-46). In deeper bodies they form greatly elongated ellipsoidal patches flattened parallel to the foliation of the surrounding igneous rock.

Nowhere is a complete section through a large batholith exposed. Our concept of one must therefore be pieced together with evidence drawn from various sources, including field studies, geophysical surveys, modeling experiments, and theoretical considerations. The following description of the genesis and form of a batholith is therefore purely interpretive but is probably close to the truth.

The first step in the development of a batholith is the formation of an enormous volume of relatively siliceous magma. We are not in a position at this stage in the book to discuss this aspect in any detail, but a few general statements can be made. The volume of magma involved indicates that large quantities of heat must be developed in the crust. This may

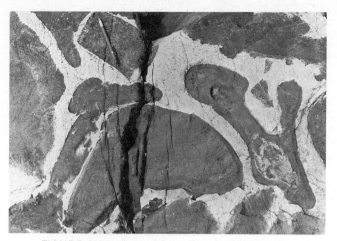

FIGURE 4-46 Pillowlike bodies of basalt formed when mafic magma was injected into still molten granitic magma. Rapid quenching of the basalt, viscosity contrasts between the magmas, and even liquid immiscibility may all have played a role in preventing these contrasting magmas from mixing. Note that the body of basalt on which the pencil rests is angular and must be of a different origin from the globular bodies. Note also that the globular bodies have a marginal zone (slightly lighter) formed from the quenching of the basalt by the lower-temperature granitic magma. The angular basalt fragment lacks this marginal zone. Mount Desert Island, Maine.

occur either by introducing heat from below or developing it in situ from radioactive decay. Some batholiths appear to have formed largely through melting of crustal rocks, whereas others have involved a significant contribution from the mantle. It is possible that basaltic magma ascending from the mantle spreads laterally on reaching the top of the lower crust, where it encounters rocks of lower density than itself. Here it raises temperatures sufficiently to bring about partial melting of crustal rocks. Some mixing of the newly formed melt with the basaltic magma may occur, but density differences tend to keep the liquids stratified with the denser basaltic magma remaining beneath. Some of the basaltic inclusions so common in batholithic rocks are probably incorporated at this stage. The role of the basalt in this model is simply as a transferer of heat from the mantle into the crust, but some chemical transfer also takes place. Heating of the lower crust to temperatures where partial fusion can occur will take place through radioactive heating if the crust is greatly thickened during orogenesis (see Prob. 1-5).

Regardless of the ultimate source of heat causing fusion of the crust, the melt, once formed, is less dense than the overlying rocks, and thus it is gravitationally unstable. Diapirs consequently develop on the upper surface of the zone of melt with a regular spacing determined according to the relation in Eq. 3-22. A batholith begins life as a diapir ascending through the ductile part of the crust. It is probably spherical or tear-shaped at this stage and would look very much like the early rising diapirs of oil shown in Figure 3-9. Batholiths are not likely to be preserved in this stage of development, because they will ascend through the ductile zone before stopping. In the deep part of the crust the only record of the passage of an ascending batholith would be a zone of steeply dipping, strongly foliated, and lineated rocks left in the region where the batholith had passed. Just as narrow tails of oil are left behind the ascending diapirs of oil (Fig. 3-9), a vertically elongated body of granitic rocks might be left behind in this deep zone. Such bodies of granite are found, for example, in the deep crustal rocks exposed in the Grenville province of the Canadian Shield.

The diapirs rise through the ductile part of the crust until their ascent is slowed by encounter with the brittle crust, whereupon they spread laterally in much the same way that diapirs of oil do on reaching the top of the model tank (Fig. 3-9). It is at this stage of development that batholiths are commonly observed in deeply eroded sections through the crust. These batholiths form series of regularly spaced conformable domes, such as those in the Precambrian of Zimbabwe (Fig. 4-47). Lateral spreading of the diapirs compresses and displaces downward the intervening country rocks, just as the spreading diapirs of oil displace most of the honey from between them (Fig. 3-9). Zones or screens of strongly flattened, steeply dipping country rocks are common between closely spaced batholiths and are evidence of the lateral spreading of these bodies. Such a zone is clearly evident between two juxtaposed batholiths in the satellite photograph of part of Queensland, Australia (Fig. 4-48). Lateral spreading may produce an almost continuous belt of magma, with only narrow screens of country rock and root zones marking the original diapirs.

FIGURE 4-48 Landsat image of granite batholiths in the Pilbara district of western Australia. The diapiric rise and lateral spreading of the batholiths has compressed the intruded rocks into tightly downfolded belts between the batholiths. Diabase dikes can be seen crossing the batholiths as narrow, dark, northeasterly trending lines. The dikes are much younger than the batholiths and formed at a time when the crust was behaving brittlely. Width of region is approximately 138 km. (Photo courtesy of NASA.)

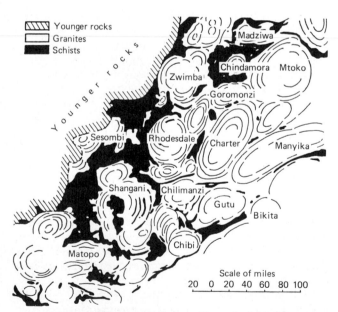

FIGURE 4-47 Regularly spaced granite batholiths in the Archean terrane of Zimbabwe. The spacing was probably determined by the viscosity contrast between the granite magma and the country rocks in the same way that the diapirs were formed in the oil and honey model shown in Figure 3-9. [After Macgregor (1951); reprinted by permission of the Geological Society of South Africa.]

Batholith ascent above the ductile zone must involve largely the rupturing of the overlying rocks. The distance a body rises through this zone will depend on its initial content of thermal energy and the state of stress in the upper crustal rocks. Heat loss by conduction increases as the temperature of the surrounding rocks decreases, as it would at progressively higher levels in the crust. Only large bodies of magma are likely to have sufficient heat for some fraction to remain liquid long enough to intrude far. One exception would be where heat is replenished by the intrusion of new, hot basaltic magma into the base of a batholith. The common occurrence of basaltic inclusions in batholithic rocks and ash-flow tuffs suggests that this indeed may be a common means of reenergizing a magma chamber. The final ascent toward the surface must be controlled by the ease with which country rocks respond to stoping and cauldron subsidence.

Much is known about the upper parts of batholiths because of the excellent exposures provided in deeply dissected mountain chains. For example, the cross section through the Coastal batholith of Peru (Fig. 4-49) is based on exposures covering more than 4.5 km of relief (Myers 1975). This batholith exhibits features that are typical of bodies emplaced at this level in the crust. It consists of numerous plutons that were emplaced by repeated caldron subsidence. The resulting stocks and ring dikes produce a complicated pattern of intrusive rela-

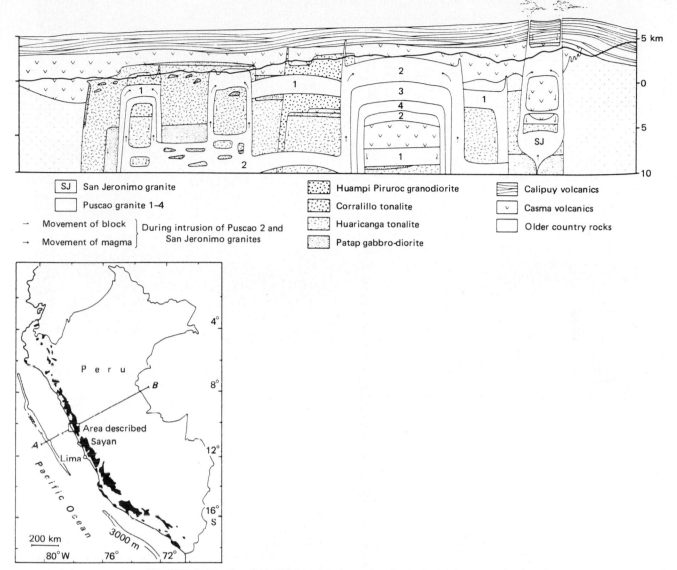

SJ — San Jeronimo granite

Puscao granite 1–4

→ Movement of block
→ Movement of magma

During intrusion of Puscao 2 and San Jeronimo granites

Huampi Piruroc granodiorite

Corralillo tonalite

Huaricanga tonalite

Patap gabbro-diorite

Calipuy volcanics

Casma volcanics

Older country rocks

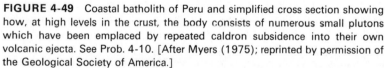

FIGURE 4-49 Coastal batholith of Peru and simplified cross section showing how, at high levels in the crust, the body consists of numerous small plutons which have been emplaced by repeated caldron subsidence into their own volcanic ejecta. See Prob. 4-10. [After Myers (1975); reprinted by permission of the Geological Society of America.]

tions. The plutons even intrude associated volcanic rocks, some of which have dropped considerable distances during caldron subsidence (see Prob. 4-10).

Batholiths probably do not conform to a single static shape. Their enormous size and content of buoyant material (even when solidified) may well cause their form to change continually as they are unroofed and exposed by erosion. Nonetheless, the form outlined here in this genetic description, which can be thought of as resembling a molar tooth with roots extending down into a ductile zone and a cap that is marked by various brittle intrusive relations, is one that many batholiths closely approach. To stress the importance of the relation between process and morphology, we examine a body of igneous rock that was formed by many of the same processes outlined

in the model, but instead of being tens of kilometers across is only 1 cm across (Fig. 4-50).

A flat-lying, meter-thick basaltic sill of Cretaceous age near Montreal, Quebec, carries globules (ocelli) of fine-grained felsic rock which is interpreted to have formed as an immiscible liquid in the basalt (see Sec. 13-6). The origin of this felsic liquid need not concern us at this point; it is simply necessary to recognize that two liquids were present and that the felsic one was less dense and so rose and collected to form a sheet approximately 5 cm below the upper contact of the sill. It did not rise all the way to the top, because the cooled basaltic liquid near the contact was too viscous to allow it to pass. Before the rocks solidified, a second injection of fresh, hot basaltic liquid in the center of the sill raised the temperature of the earlier material

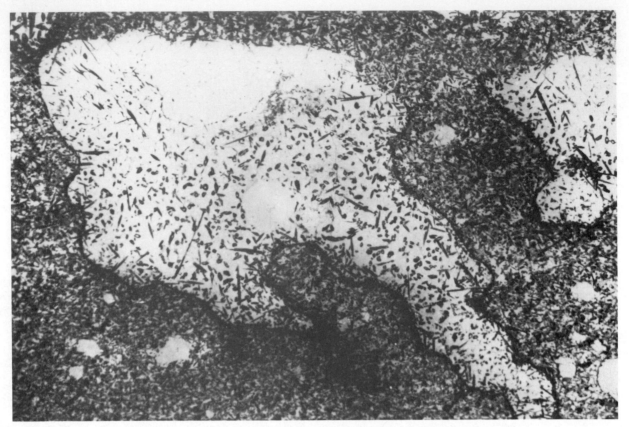

FIGURE 4-50 Diapiric body of nepheline-analcite syenite that has buoyantly risen through a mafic Cretaceous sill near Montreal Quebec. Although the top of the diapir is only 1 cm across, it has been formed by processes similar to those active during the formation of a batholith, and as a result it exhibits many of the physical and chemical attributes of a batholith. See text for discussion.

sufficiently to lower its viscosity. The felsic layer, which was already gravitationally unstable, was then able to develop diapirs that rose through the denser overlying basalt toward the upper contact, eventually coming to rest where they once again encountered highly viscous, possibly even brittle material near the upper contact. One of these diapirs is shown in Figure 4-50.

Let us examine the genesis and form of this diapir, keeping in mind the possible analogy with a batholith. First, a pulse of fresh basaltic magma supplies heat that initiates the rise of the diapir. The heat may cause fusion of minerals that have already solidified; it most certainly will lower the viscosities of both the felsic and, more important, basic liquids. Basalt intruded into the base of the Earth's crust will similarly, not only cause fusion of crustal rocks but will reduce the viscosity of the ductile zone. Although not shown in Figure 4-50, a regularly spaced set of diapirs rises from the felsic layer (see Fig. 3-10). Each diapir rises until it reaches cooler and significantly more viscous basalt near the upper contact of the sill, whereupon it accumulates and flattens to form a body having the shape of a molar tooth, with a root or tail marking the ascent route of the diapir through the underlying basalt. Because the magma in the sill continues to move slightly as the diapir rises, the structure becomes tilted. Batholithic diapirs might also be tilted by tectonic movements at convergent plate boundaries. The upper surface of the diapir is quite flat where it abuts against the overlying more rigid basaltic rock. The felsic material stays molten to temperatures about 200°C lower than those at which the basaltic rock solidifies. Although not shown in Figure 4-50, elsewhere in this sill, fractures have allowed the felsic liquid to escape from the diapirs and actually vein the overlying brittle rock, thus mimicking the brittle behavior in the upper zones of batholiths. Within the diapir there is slight fractionation of the felsic liquid so that more ferromagnesian minerals are present in the root zone and more felsic ones toward the top. One small patch of extremely felsic and hydrous material (contains analcite) has accumulated at the very top of the diapir. Similar compositional variations are found in batholiths, and accumulation of buoyant, highly siliceous, hydrous magma at the top of a batholith is one of the possible causes for the development of resurgent domes.

Large batholiths, and the small diapir in Figure 4-50, have formed as a result of buoyant magma rising slowly through a denser but increasingly more viscous medium following an input of thermal energy. They both effect an upward transfer of low-melting fractions, and in the case of batholiths this process has undoubtedly played a major role in the development of the Earth's crust throughout geologic time.

PROBLEMS

4-1. In historic times, the 1783 fissure eruption of Lakagigar, Iceland, most closely resembles a flood basalt eruption, but it is small by comparison with flood basalts in the geologic record. Between June 9 and July 29 of that year, 9.5 km³ of magma was erupted from a 10-km-long fissure estimated to have been about 4 m wide at depth. What was the average flow velocity in the feeder dike, and was the flow turbulent or laminar? Use typical values for the physical properties of a basaltic magma.

4-2. The average width of basaltic dikes in Iceland is 4 m.
   (a) If the magma that rises in these dikes has a density of 2.7 Mg m⁻³ and a viscosity of 300 Pa s, and they intrude rocks with a density of 2.97 Mg m⁻³, what would the average flow velocity need to be to maintain flow?
   (b) Would this flow be laminar or turbulent?
   (c) Would the density difference be sufficient to provide a buoyant force great enough to produce the required average flow velocity?

4-3. In light of your answers to Probs. 4-1 and 4-2, what might you conclude about the 1783 Lakagigar eruption, which was clearly able to maintain flow for almost two months?

4-4. If essentially all the CO_2 in an ascending andesitic magma has exsolved by the time the pressure has dropped to 50 MPa, how much will this gas expand on rising to the surface (pressure at surface = 0.101 MPa) if its temperature is held constant by the thermal mass of the surrounding lava and it behaves as an ideal gas? If the gas at 50 MPa constituted 5 vol % of the ascending magma, what effect would the gas expansion have on the magma on rising to the surface?

4-5. The Loch Ba granophyric (rhyolitic) ring dike on the island of Mull, Scotland, has an outside diameter of approximately 5.6 km and an inside diameter of 5.5 km.
   (a) If intrusion of the ring dike resulted solely from the subsidence of the central block, what would the ratio of the average rate of intrusion to the rate of subsidence have been?
   (b) The radius of curvature of the ring dike is sufficiently great that at any point the dike can be treated as a planar sheet. The granophyric magma had a density of 2.4 Mg m⁻³ and a viscosity of 10⁷ Pa s. The subsided block consists largely of basaltic rocks with a density of 3.0 Mg m⁻³. Assuming laminar flow, calculate the average intrusion velocity of the magma relative to the surrounding rocks and the rate of subsidence of the central block. Check to see if the assumption of laminar flow is valid.
   (c) If growth of gas bubbles in the ring dike magma disrupted the magma, the viscosity of the magma–gas mixture was 0.1 Pa s, and its density was 1.8 Mg m⁻³, what would have been the average rate of intrusion of the magma and the rate of subsidence of the block?
   (d) In light of your answers to parts (b) and (c), what role is magma disruption likely to play in the emplacement of rhyolitic ring dikes?

4-6. **(a)** For a cone sheet having a width W, a dip of θ degrees, and an initial radius at the magma reservoir of r_0 (see diagram), derive an expression for the average magma velocity $\bar{v}$ at any height z above the source in terms of the initial average velocity $\bar{v}_0$.
   (b) From the equation derived, what might you conclude about the possible dip on cone sheets and the initial radius r_0?

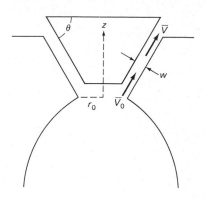

4-7. **(a)** In Prob. 4-2 we saw that an average flow velocity of 10.21 m s⁻¹ was necessary to maintain flow in a 4-m-wide basaltic dike in which the magma, with a density of 2.7 Mg m⁻³ and viscosity of 300 Pa s, intrudes basaltic rocks with a density of 2.97 Mg m⁻³. If a 4-m-wide cone sheet forms with a dip of 30° and an initial radius (r_0) at the top of the magma chamber of 100 m, calculate how high (z) the magma will rise in the cone sheet before its velocity becomes too low to maintain flow. The pressure gradient imposed on the magma at the point of entry into the cone sheet is 3000 Pa m⁻¹. Make use of the relation derived in Prob. 4-6.
   (b) Repeat the problem for cone sheets that dip at 60° and 20°.
   (c) What would be the consequences of the raised block being tilted by the injection of the cone sheet so that the cone sheet would be thicker on one side and narrower on the other? What effect would this have on the distance the magma would rise? What might the resulting outcrop pattern of the cone sheet be on the surface of the Earth?

4-8. By studying the map of the Pilanesberg complex, South Africa (Fig. 4-33), deduce the sequence of intrusive events that led to the formation of this body.

4-9. **(a)** Using Eq. 4-4, draw cross sections through two laccoliths (on a single graph), both having a radius of 2 km, one of which is intruded at a depth of 200 m and the other at 500 m into sandstone having a density of 2.7 Mg m⁻³, Young's modulous of 3 × 10¹⁰ Pa, and a Poisson's ratio of 0.1. The pressure on the magma is due solely to buoyancy resulting from a 10-km-long feeder column of magma with a density of 2.4 Mg m⁻³ intruding rocks with an average density of 2.7 Mg m⁻³.
   (b) Repeat the calculation for a laccolith intruded at a depth of 200 m but having a radius of only 1 km.
   (c) Repeat part (a) for the case where intrusion is into limestone that has a Young's modulus of 6 × 10¹⁰ Pa, a Poisson's ratio of 0.2, and a density of 2.7 Mg m⁻³.
   (d) In light of the results of your calculations, discuss the relative importance of the factors controlling the shape of a laccolith.

4-10. Work out the chronological sequence of igneous rocks shown in the cross section through the Coastal batholith of Peru in Figure 4-49.

4-11. Draw a sketch of Figure 4-22 and work out the chronology of the intrusive events seen in this outcrop. The dark igneous rocks are of various gabbroic compositions, and the light-colored ones are of syenite.

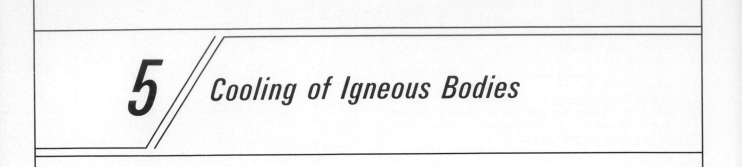

5 / *Cooling of Igneous Bodies*

5-1 INTRODUCTION

The rate of heat loss from magmatic bodies plays an important role in many important petrologic processes, such as nucleation and growth of crystals, settling of crystals, magmatic convection, rates of intrusion, and chemical reactions between magmas and phenocrysts or country rocks. The transfer of heat from an igneous body into its surroundings also causes contact metamorphism. On a still larger scale, cooling of newly formed lithosphere as it moves away from oceanic ridges controls the depth to which plates sink into the asthenosphere (and consequently, ocean depths). Eventually, when plates are subducted, the rate at which heat flows back into the descending slab determines the nature of the resulting metamorphic rocks (see Sec. 1-2).

Heat transfer is a large subject involving considerable mathematics. Only a brief introduction can be given here. Many of the mathematical derivations are complicated, but their results can be applied quite simply to interesting geological problems.

Heat, which is thermal energy, can be thought of as flowing from high- to low-temperature regions. This transfer can take place in three different ways: (1) by *conduction*, in which the thermal energy is transmitted through a substance by the transference of kinetic energy from one atom or molecule to another; (2) by *convection*, in which heat is transferred by the actual bodily movement of hotter material to cooler parts of the system; (3) by *radiation*, where the energy is transferred directly from one place to another by electromagnetic radiation. In many geological situations, conduction is the most important mechanism of heat transfer, but in large bodies of magma or large sections of the mantle, convection can play an important role. Heat loss through radiation is the least important mechanism in the cooling of plutonic masses because of the relative opacity of most rocks and magmas to electromagnetic radiation at the temperatures encountered in the Earth's crust. At higher temperatures, however, such as those deep in the mantle, radiation may play a significant role in the transfer of heat. Extremely felsic magmas, which on rapid cooling form relatively clear glasses, might radiate considerable heat to the margins of an intrusion, giving rise to more rapid cooling than would otherwise be expected from conduction alone. Radiation, of course, plays an important role in the cooling of the surface of lava flows, and if the lava is in turbulent motion, it can be the most important mechanism for cooling the entire flow (Shaw and Swanson 1970). In this chapter we discuss cooling of igneous bodies by conduction and radiation, leaving the topic of convection to Chapter 13.

The literature dealing with heat conduction is extensive. One of the most readable treatises is by Ingersoll, Zobel, and Ingersoll (1954), who develop much of the general theory and give many applications, including geological ones. The most comprehensive treatment of the subject, however, is the standard text *Conduction of Heat in Solids* by Carslaw and Jaeger (1959). But for many geologists the chapter by Jaeger (1968) on the *Cooling and Solidification of Igneous Rocks*, in Volume 2 of *Basalts*, edited by Hess and Poldevaart (1968), is more digestible. Finally, an excellent treatment of many heat transfer problems in geology is given by Turcotte and Schubert (1982b).

5-2 GENERAL THEORY OF HEAT CONDUCTION

If two parallel plates that differ in temperature by ΔT are separated by a distance Δx (Fig. 5-1), the amount of heat Q transferred from one to the other per area A of the plates is proportional directly to the temperature difference and the length of time t during which heat flows, and inversely to the distance between the plates. This can be expressed as

$$Q = -KA \frac{\Delta T \, t}{\Delta x} \qquad (5\text{-}1)$$

The constant of proportionality K is known as the *thermal conductivity*. If the plates are moved closer together, the limiting value of $\Delta T/\Delta x$ is $\partial T/\partial x$, which is the *thermal gradient* at that point. The partial derivative $\partial T/\partial x$ is used here to emphasize that in the general case the temperature gradients in the other two directions ($\partial T/\partial y$, $\partial T/\partial z$) could be considered. In this spe-

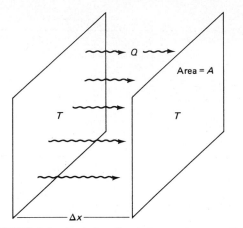

FIGURE 5-1 Heat transfer between two parallel plates.

cific example, however, where heat is transferred only in the x direction, the ordinary derivative can be used.

If the thermal gradient is taken in the direction of heat flow, it will be negative. The rate of heat flow, which is a positive number, is given by differentiating Eq. 5-1,

$$\frac{dQ}{dt} = -KA\frac{dT}{dx} \qquad (5\text{-}2)$$

This rate of heat flow across a unit area is known as the *heat flux* and is given by

$$J_x = \frac{dQ}{dt\,A} = -K\frac{dT}{dx} \qquad (5\text{-}3)$$

This is known as *Fourier's law*, which is similar in its formulation to the expression for fluid flow derived in Chapter 3 (Eq. 3-9). Values of heat flux from the Earth have already been given in Chapter 1. Typical values range from $60\ \text{mW m}^{-2}$ ($1\ \text{W} = 1\ \text{J s}^{-1}$) over stable continents to $335\ \text{mW m}^{-2}$ over oceanic ridges (Prob. 5-1).

Although thermal conductivity tells us the rate at which heat can be transferred into a substance, it does not indicate the rate at which the temperature of the material will change as a result of this influx of heat. And yet for many petrologic problems the temperature, not the quantity of heat transferred, is the critical factor. For example, temperature determines when a magma will crystallize or a metamorphic reaction will take place, not the amount of heat that flows. The temperature change can be determined only by knowing the heat capacity of the substance into which the heat passes.

Heat capacity C_p is the amount of heat necessary to raise the temperature of a unit mass of a substance 1 degree at constant pressure. If the density of the material is ρ, the amount of heat necessary to raise the temperature of a unit volume of the substance by 1 degree is $C_p\rho$. Therefore, this quantity of heat must flow into the unit volume for its temperature to be raised 1 degree. The rate of heat flow is given by the thermal conductivity K, which is the quantity of heat that flows in unit

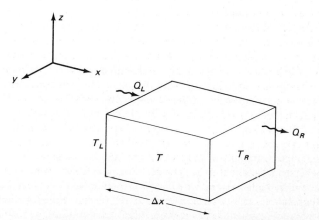

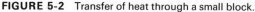

FIGURE 5-2 Transfer of heat through a small block.

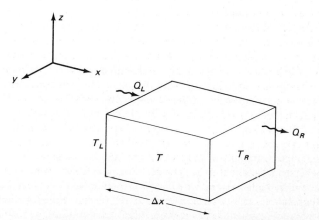

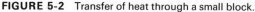

The difference between the rates of heat flow into the left end and out of the right end of the small volume is the amount of heat gained per unit time by the material in the volume:

$$\frac{dQ}{dt} = \frac{1}{2} KA \, \Delta x \left(\frac{d^2 T}{dx^2}\right) - \left(-\frac{1}{2} KA \, \Delta x \left(\frac{d^2 T}{dx^2}\right)\right)$$

$$= KA \, \Delta x \left(\frac{d^2 T}{dx^2}\right) \qquad (5\text{-}8)$$

The rate of heating of the volume can, on the other hand, be expressed in terms of the heat capacity. Multiplication of the mass of the volume ($\rho A \, \Delta x$) by the heat capacity (C_p) gives the amount of heat necessary to raise the temperature of the volume by 1 degree. Multiplication of this amount ($C_p \rho A \, \Delta x$) by the rate of temperature rise (dT/dt) gives the rate of heating of the volume, which can then be equated to the expression for this quantity in Eq. 5-8. Therefore,

$$C_p \rho A \, \Delta x \left(\frac{dT}{dt}\right) = KA \, \Delta x \left(\frac{d^2 T}{dx^2}\right) \qquad (5\text{-}9)$$

and

$$\frac{dT}{dt} = \frac{K}{C_p \rho} \left(\frac{d^2 T}{dx^2}\right) = k \left(\frac{d^2 T}{dx^2}\right) \qquad (5\text{-}10)$$

In the general case where heat, instead of flowing in the x direction only, has y and z components as well, the same general relation holds, but the derivatives of the thermal gradients in each of the mutually perpendicular x, y, and z directions must be summed. For the general case, then,

$$\frac{dT}{dt} = k \left(\frac{\partial^2 T}{\partial x^2} + \frac{\partial^2 T}{\partial y^2} + \frac{\partial^2 T}{\partial z^2}\right) \qquad (5\text{-}11)$$

This is known as *Fourier's equation*, which states that the change in temperature of a substance with time is given by the product of the thermal diffusivity and the derivative of the thermal gradient with respect to distance. The solution to any problem involving heat flow must satisfy this equation.

The significance of Eq. 5-11 is illustrated in Figure 5-3. The temperature distribution near an igneous contact is shown at some time following intrusion. Below this is the first derivative of this temperature distribution; note that a maximum occurs at the contact. The second derivative, at the bottom of the figure, has a maximum at A and a minimum at B. According to Fourier's equation, the rate of temperature change at any point is determined by the second derivative. Hence, at the particular time represented in Figure 5-3, the country rock at point A is heating the most rapidly, and the magma at point B is cooling the most rapidly. As the temperature gradient changes with time, so do the positions of these maxima. Because the second derivative is zero at the contact, the temperature there remains constant and, as will be shown later, remains constant as long as magma and country rock at some distance from the contact still have their initial temperatures.

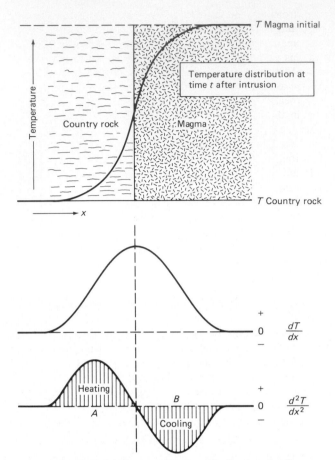

FIGURE 5-3 The temperature distribution in the vicinity of an igneous contact some time after intrusion is shown at the top of the diagram. Beneath this are the first and second derivatives of temperature with respect to distance. According to Fourier's equation (Eq. 5-11), the maximum rate of cooling occurs where the second derivative is most negative, and the maximum rate of heating occurs where it is most positive.

5-3 HEAT CONDUCTION ACROSS A PLANE CONTACT

We will now consider a simple problem, that of magma cooling in the vicinity of a plane contact, such as would occur near the margin of a dike or sill, or perhaps at the margin of a batholith, as long as the contact was not significantly curved. We must first set up the initial and boundary conditions. For simplicity in the calculations, the temperature of the country rock is taken to be initially 0°C, and after intrusion this same temperature occurs at greater and greater distances from the contact. The fact that rocks may initially be at some temperature other than zero is taken into account simply by adjusting the temperature scale up or down as the case may be. The magma is emplaced with a temperature T_0 (adjusted, as previously discussed) and begins to lose heat through the contact to the country rock. The igneous body is, however, of sufficient size that magma with the initial intrusion temperature still remains at some distance from the contact. This distance, of course, increases with time, and eventually no magma with the initial temperature will remain. At this point the boundary conditions are changed, and this poses a different problem, which will be considered later. Mathematically, these boundary conditions can be expressed as follows: If the distance from the contact is denoted by x, negative values being in the country rock and positive ones in the igneous body, then for $x > 0$, T/T_0 tends to the value 1, and for $x < 0$ it tends to 0.

To find the temperature T, at distance x, at time t after intrusion, a solution to the Fourier conduction equation must be found that satisfies the conditions of the problem. This involves the use of Fourier's series and integrals. For the reader with knowledge of this area of mathematics, the solutions presented in the previously cited references are worth consulting. The solution to this problem and to many others concerning heat conduction involves a rather frightening-looking integral known as an *error function* (erf):

$$\text{erf}(y) = \frac{2}{\sqrt{\pi}} \int_0^y e^{-u^2} \, du \qquad (5\text{-}12)$$

Fortunately, values of the error function are given in most mathematical tables, and for convenience a few are given in Table 5-1 [note that $\text{erf}(-y) = -\text{erf}(y)$].

A derivation of the solution to the problem posed here is given by Carslaw and Jaeger (1959, p. 58). The solution, which is based on the thermal diffusivity k of magma and country rock being the same (a valid assumption), is given by

$$\frac{T}{T_0} = \frac{1}{2} + \frac{1}{2} \, \text{erf}\!\left(\frac{x}{2\sqrt{kt}}\right) \qquad (5\text{-}13)$$

The reader should verify that this solution does indeed satisfy the initial and boundary conditions of the problem. For ex-

ample, when $x \gg 0$, the error function tends to the value 1, and hence T/T_0 tends to 1. What does the value of T/T_0 tend to when $x \ll 0$?

This solution reveals an important fact about the maximum possible temperature in the country rock near an intrusion. At the contact, $x = 0$, and therefore $T = 0.5T_0$. That is, the temperature at the contact is half the temperature of the magma (suitably adjusted so that the initial country rock temperature is zero) immediately after intrusion, and it remains at this value as long as the boundary conditions are the same. The country rock cannot be heated to a higher temperature no matter how large the body of magma unless there is an additional supply of heat, as could happen with a fresh surge of hot magma or with the emplacement of another intrusion nearby.

Equation 5-13 also allows us to follow the progress of an isotherm which may correspond to a particular event, such as the temperature of complete solidification of magma. If this temperature is T_1, then T_1/T_0 is a constant and the error function remains constant. Thus, $x = \text{constant} \sqrt{kt}$. That is, the distance of the isotherm from the contact is proportional to $\sqrt{t}$. This explains the linear relation discussed in Chapter 2 between the thickness of the crust on the Alae and Kilauea Iki lava lakes and the square root of the cooling time (Fig. 2-1).

To illustrate Eq. 5-13, we will determine the time necessary for magma 1 m from the contact of a large dike to cool to 900°C following intrusion at 1200°C into country rock at 100°C. The thermal diffusivity of the magma and country rock is 10^{-6} m^2 s^{-1}. Adjusting the temperature scale so that the country rock is at zero gives $T_0 = 1100$°C and $T = 800$°C, and therefore $T/T_0 = 0.727$. This gives a value for the error function of 0.454, which from Table 5-1 indicates that $x/(2\sqrt{kt}) = 0.425$. When $x = 1$ m, the time to cool, t, is 16 days. For comparison, magma 2 m from the contact would take 62 days to cool to the same temperature, and at 4 m it would require 258 days.

In these calculations no account is taken of the latent heat of crystallization of magma. When magma crystallizes, heat is liberated, which causes magma to cool significantly more slowly than predicted by Eq. 5-13 (Prob. 5-5). This heat is not simple to incorporate into the calculations. By adjusting the initial temperature of the magma upward, however, an amount of heat equivalent to that liberated by crystallization can be introduced. Thus, substituting $(T_0 + L/C_p)$ for T_0, where L is the latent heat of crystallization, a closer approach to the rate of cooling is obtained, at least in the country rock. This approximation, unfortunately, gives temperatures in the igneous body that are too high.

There is one relatively simple situation in which heat of crystallization can be taken into account. This involves magma that on cooling near a plane contact, solidifies at a single definite temperature. In this case, a solidification surface migrates inward as cooling proceeds (Fig. 5-4). Most magmas, in fact, solidify over a temperature range (Sec. 2-2), but this is commonly sufficiently small (Chapter 10) that the approximation of magma solidifying at one temperature does not introduce large errors.

The conditions are essentially the same as in the previous case, that is, a plane contact between magma and country

---

TABLE 5-1

The error function: $\text{erf}(y) = \dfrac{2}{\sqrt{\pi}} \displaystyle\int_0^y e^{-u^2} \, du$

y	$\text{erf}(y)$	y	$\text{erf}(y)$	y	$\text{erf}(y)$
0.00	0.000	0.80	0.742	1.60	0.976
0.10	0.112	0.90	0.797	1.70	0.984
0.20	0.223	1.00	0.843	1.80	0.989
0.30	0.329	1.10	0.880	1.90	0.993
0.40	0.428	1.20	0.910	2.00	0.995
0.50	0.521	1.30	0.934	2.20	0.998
0.60	0.604	1.40	0.952	2.40	0.9993
0.70	0.678	1.50	0.966	2.50	0.9996

For computer calculations the error function can be approximated by

$$\text{erf}(y) = 1 - (a_1 t + a_2 t^2 + a_3 t^3) \exp(-(y^2))$$

where $t = 1/(1 + 0.47047y)$, $a_1 = 0.34802$, $a_2 = -0.09587$, $a_3 = 0.74785$.

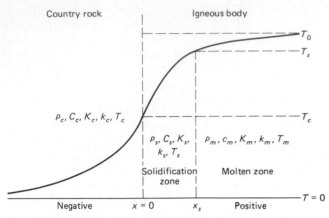

FIGURE 5-4 Temperature distribution near an igneous contact where magma solidifies at a specific temperature T_s, resulting in a solidification zone that progresses into the magma body with time. See text for definitions of terms and discussion.

rock, with distance x from the contact being positive in the magma and negative in the country rock. The initial temperatures of magma and country rock are T_0 and zero, respectively, and these continue to exist but at progressively greater distances from the contact as cooling proceeds. To these conditions must, now, be added a zone of solidified magma that extends from the contact ($x = 0$) a distance x_s, where x_s is a function only of time. Solidification of the magma occurs once the magma temperature has decreased to T_s. The physical properties—density, heat capacity, thermal conductivity and diffusivity, and temperature—of the materials constituting the three different zones—magma, solid igneous rock, and country rock—are designated with subscripts m, s, and c, respectively, as indicated in Figure 5-4.

The temperature gradient through the three zones following some period of cooling is illustrated in Figure 5-4. Magma with the initial temperature, T_0, must first cool to T_s before solidification occurs. This signifies that the magma is initially superheated by an amount $T_0 - T_s$, a condition that is not common. Indeed, as indicated in Sec. 2-2, the almost universal presence of phenocrysts in rapidly quenched magmas indicates that most magmas are not superheated. In the case considered here, therefore, we can make the conditions more realistic by setting $T_s = T_0$, and at the same time simplify the problem.

The initial and boundary conditions set forth in the preceding problem hold true for this case as well. The moving surface of solidification, however, provides additional conditions. There must be continuity of temperature across the solid–liquid interface. Thus, at $x = x_s$, the temperatures of the solid and magma must be equal; that is, $T_s = T_m$. Furthermore, as the interface advances, heat is generated by the liquid–solid phase change. The heat liberated by the crystallization of a certain volume of magma is given by the product of the latent heat and density of the magma, $L \cdot \rho_m$. If the solid–liquid interface advances a distance dx_s, a quantity of heat $L\rho_m \, dx_s$ per unit area is liberated, and this must be removed by conduction. A further condition, then, is that the heat flux leaving the inter-

face must equal the sum of the heat flux entering from the magma and the heat generated by the phase change; that is,

$$K_s\left(\frac{dT_s}{dx}\right) = K_m\left(\frac{dT_m}{dx}\right) + L\rho_m\left(\frac{dx_s}{dt}\right) \qquad (5\text{-}14)$$

Note that each group of terms in this equation has units of W m^{-2}. If the solidification temperature T_s and initial magma temperature T_0 are equal (no superheat), there will be no temperature gradient in the magma; that is, $dT_m/dx = 0$. The heat flux leaving the interface is, therefore, equal simply to the heat flux generated by the phase change.

The solutions to this problem are given in Carslaw and Jaeger (1959). First, the position of the solid–liquid interface at time t is given by

$$x_s = 2\lambda\sqrt{k_s t} \qquad (5\text{-}15)$$

where the dimensionless constant λ is obtained from the relation

$$\lambda[\sigma + \text{erf}(\lambda)]e^{\lambda^2} = \frac{C_s T_s}{L\sqrt{\pi}}$$

by iterative numerical calculations (trial and error). In this equation, $\sigma = K_s k_c^{1/2}/K_c k_s^{1/2}$, with values of σ varying from 0.5 to 3.0 depending on the rock types in contact. Equation 5-15 reveals that, just as in the preceding case, where no latent heat of crystallization was considered, the distance of a particular isotherm (in this case the solidification boundary) from the contact is proportional to $\sqrt{t}$ (see Fig. 2-1).

The temperature at the contact, $T_{x=0}$, is given by

$$T_{x=0} = \frac{\sigma T_s}{\sigma + \text{erf}(\lambda)} \qquad (5\text{-}16)$$

If the thermal properties of both the igneous and country rocks are identical (that is, $\sigma = 1$) and because the error function must have a value between zero and unity, the contact temperature is a constant and has a value between T_s and $0.5T_s$. The latter case corresponds to the latent heat of crystallization being zero, and the result is identical with that obtained in the preceding problem, where latent heat was ignored.

Within the solidified region, the temperature is given relative to the contact temperature by the expression

$$T_{x=0}\left[1 + \frac{1}{\sigma}\text{erf}\left(\frac{x}{2\sqrt{k_s t}}\right)\right] \qquad (5\text{-}17)$$

and in the country rock by

$$T_{x=0}\left[1 + \text{erf}\left(\frac{x}{2\sqrt{k_c t}}\right)\right] \qquad (5\text{-}18)$$

The reader should again verify that these solutions do satisfy the initial and boundary conditions.

The cooling and solidification of a lava flow from its upper free surface can be treated in a similar manner (Fig. 5-5).

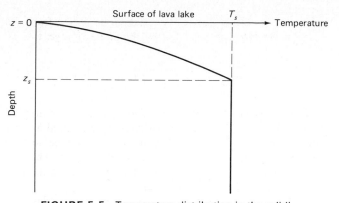

FIGURE 5-5 Temperature distribution in the solidified crust of a lava flow or lake.

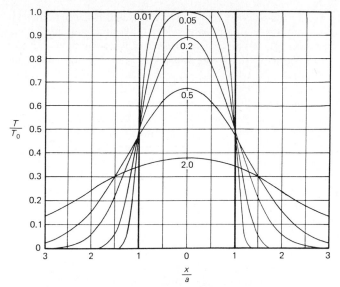

FIGURE 5-6 Temperature distribution across a sheetlike intrusion of thickness $2a$ at various times after intrusion. Times are given on the curves by the dimensionless term kt/a^2, and temperatures are expressed as fractions of the initial magma temperature T_0. [After Carslaw and Jaeger (1959).]

Depth in the lava is measured as positive values of z, with the surface at $z = 0$. The surface temperature is adjusted to be zero, and the initial temperature of the lava is taken to be, as in the preceding problem, its solidification temperature T_s (no superheat). This temperature must also be adjusted by the same amount as the surface temperature. The thermal properties of the liquid and solid lava are taken to be identical. The temperature in the solid zone is then given by

$$T = \frac{T_s \operatorname{erf}(z/2\sqrt{kt})}{\operatorname{erf}\lambda} \tag{5-19}$$

The value of the constant λ can be obtained by iterative calculations of the relation

$$\lambda e^{\lambda^2} \operatorname{erf}(\lambda) = \frac{C_s T_s}{L\sqrt{\pi}}$$

where L is the latent heat of crystallization and C_s is the heat capacity of the solid lava. Equation 5-19 again demonstrates that the thickening of the solidification crust is proportional to $\sqrt{t}$ (Prob. 5-6).

The boundary conditions in the preceding problems require that there always be magma at its initial intrusion temperature at some distance from the contact. No matter how large the intrusion, this condition eventually breaks down and the center of the body begins to cool. At this point we have different boundary conditions and a new problem. Carslaw and Jaeger (1959) give the following solution for a sheetlike intrusion of thickness $2a$, in which distance, x, is measured from the center of the body:

$$\frac{T}{T_0} = \frac{1}{2}\left[\operatorname{erf}\left(\frac{a-x}{2\sqrt{kt}}\right) + \operatorname{erf}\left(\frac{a+x}{2\sqrt{kt}}\right)\right] \tag{5-20}$$

A graphical representation of this equation is given in Figure 5-6.

The early period of cooling in a sheet is identical to that developed in the first problem, with the temperature of the contact remaining at one-half the initial adjusted magma tempera-

ture. As soon as the center of the sheet begins to cool, however, the temperature at the contact begins to fall. The time at which the maximum temperature is attained at any particular point increases with distance from the contact. Therefore, if minerals in contact metamorphic aureoles are formed at the time the maximum temperature is attained at any plane, those in the outer zones would form last. No account is taken of latent heat of crystallization in this solution. Of course, until the solidification surfaces advancing in from both contacts meet in the center, the solutions presented in Eqs. 5-16 to 5-18 can be used. Beyond this time, however, no simple solution involving latent heats of crystallization exists.

Winkler (1949) used Eq. 5-20 to obtain a mean cooling velocity that could then be related to the grain size variation at the margins of sheetlike bodies. This mean velocity is the temperature range through which the mineral crystallizes, divided by the time required to cool this amount. If crystallization begins at $s_1 T_0$ and ends at $s_2 T_0$, where s_1 and s_2 are simply fractions that give the appropriate temperatures in terms of T_0, the mean cooling velocity is given by

$$\frac{s_1 T_0 - s_2 T_0}{t} = \frac{4kT_0(s_1 - s_2)}{x^2\{[\operatorname{inverf}(2s_2 - 1)]^{-2} - [\operatorname{inverf}(2s_1 - 1)]^{-2}\}} \tag{5-21}$$

where x is the distance from the contact and $\operatorname{inverf}(z)$ is the inverse error function, such that if $y = \operatorname{inverf}(z)$, then $\operatorname{erf}(y) = z$. If we assume that grain size is inversely proportional to the mean cooling velocity, then, from Eq. 5-21, the grain size of the rock should be proportional to the square of the distance from the contact (see Chapter 12 for further discussion).

Because most rocks have very similar thermal diffusivities, grain sizes on opposite sides of dikes and sills may be expected to show similar variations, even where the country rocks differ from one side to the other. This is not true for the cooling of a lava flow where the upper surface, in contact with air or water, cools much more rapidly than the lower one, which is in contact with rock or soil. In fact, the upper surface can be considered, for practical purposes, as instantly attaining some very low temperature, such as that of the air or water, and remaining at that temperature throughout the cooling of the flow. Equation 5-19 will then describe the rate at which cooling takes place from the upper surface. Cooling from the lower surface occurs more slowly by heat flowing into the underlying rocks; Eq. 5-17 describes this process. The temperature distribution within the flow becomes skewed as cooling proceeds, with the temperature maximum moving from its initial central position toward the base of the flow. From Figure 5-6 it is evident that if the rate of cooling of magma controls crystal grain size, the coarsest rock in a lava flow would be expected to occur below the center, whereas in a sill it would occur at the center, as long as there were no movement of the crystals after they had formed.

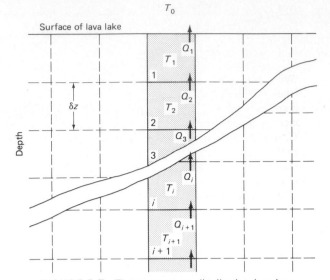

FIGURE 5-7 The temperature distribution in a lava lake can be calculated numerically by first dividing the body into many small cells and then determining the change in temperature of each cell as a function of time. See text for discussion.

5-4   NUMERICAL ANALYSIS

Many heat flow problems are too complicated to have exact, analytical solutions. Nonetheless, useful answers can be obtained by numerical methods, which, in general, involve dividing the region of interest into a number of small volumes or cells. The heat flow through any cell can be analyzed in the same manner as was done earlier in this chapter when we derived Fourier's equation. In numerical analysis, however, instead of allowing the cell to shrink to an infinitesimal volume, as was done in deriving the differential equation, it is maintained at some small but finite volume. The overall thermal behavior of the region can then be determined by summing the effect of all the small cells. As the cells become smaller and more numerous, the results, of course, approach those of the exact solution of the differential equation. In most cases, however, sufficiently accurate results can be obtained with only modest numbers of cells. Because this work is invariably done on a computer, the number of cells used is limited only by the computational time.

Consider, for example, the cooling of an extensive lava lake whose upper surface is kept at ambient air temperature. This problem is considered by Shaw, Hamilton, and Peck (1977), whose paper should be referred to for details. Because of the lateral extent of the lava lake, heat can be considered to flow only in the vertical, z, direction. The loss of heat around the sides of the lake will not be dealt with here. The lava lake is divided into numerous small cells, such as those illustrated in Figure 5-7. Each cell, which has a cross-sectional area A and vertical dimension δz, is numbered consecutively from the surface downward. Although each cell is shown as being of the same dimensions, this is not a requirement, and indeed computations in many problems can be reduced by allowing the cells to expand in less critical regions. The temperature of each

cell, T_i, is taken to be at its center. Physical properties such as thermal conductivity K_i, thermal diffusivity k_i, heat capacity C_i, density ρ_i, and volume V_i can be specified for each cell. The heat flowing from one cell to the next is designated Q_i, and if heat were to flow in directions other than vertical, as occurs around the margins of the lake, there would be heat fluxes across the other faces of the cells.

According to Fourier's law (Eq. 5-1), the quantity of heat passing from the $(i + 1)$th cell into the ith cell (see Fig. 5-7) is

$$Q_{i+1} = -\frac{[K(T_i - T_{i+1})A \, \delta t]}{\delta z} \tag{5-22}$$

where δt is the time involved. Should adjoining cells have markedly different thermal conductivities, K would be the average of K_{i+1} and K_i.

To conserve thermal energy, the change in heat content of any cell, δQ_i, must be equal to the heat flux in, Q_{i+1}, minus the heat flux out, Q_i, plus any sources or sinks of heat, Q_i^*, within the cell. Sources, for example, could be latent heats of crystallization, heat generated by radioactive decay, and heat generated by the dissipation of viscous forces, if the magma were moving rapidly. Heat sinks, on the other hand, could include heat of evaporation of rainwater that enters cracks in the crust of the lava lake, or endothermic reactions in a contact metamorphic aureole. The change in heat content of a cell, then, is

$$\delta Q_i = (Q_{i+1} - Q_i) + Q_i^* \tag{5-23}$$

The change in heat content will, of course, result in a change in cell temperature. The heat capacity per cell is given by $C_i\rho_i V_i$, which when divided into the change in heat content,

δQ_i, gives the change in temperature, δT_i, of the cell. Thus,

$$\delta T_i = \frac{\delta Q_i}{C_i \rho_i V_i} = \frac{Q_{i+1} - Q_i + Q_i^*}{C_i \rho_i V_i} \qquad (5\text{-}24)$$

We are now able to apply Eqs. 5-22 and 5-24 to the problem of the cooling lava lake. The initial temperature of the lava, and thus of each cell, is taken to be 1200°C. Loss of heat through the upper surface to the air, which is at a constant temperature of 25°C, brings about cooling. Loss of heat would also occur through the base of the lava lake. Although this can be analyzed in precisely the same manner, only the cooling at the upper surface is treated here.

The air can be considered to occupy a cell immediately above the surface of the lava. Following eruption, the difference in temperature between this cell and the underlying one results in a quantity of heat, Q_i, flowing upward out of cell number 1 (Fig. 5-7). According to Eq. 5-22, this quantity is given by

$$Q_1 = -\frac{K(25 - 1200)A \, \delta t}{\delta z}$$

The value of Q_1 for any particular cross-sectional area A can be calculated once values of δt and δz have been selected.

Values of δt and δz are chosen so as to provide an adequate description of the cooling history without using excessive amounts of computational time. If too large a time interval is chosen, however, meaningless results can be obtained. It has been shown (Shaw et al. 1977) that provided the condition

$$\frac{K_i \delta t}{C_i \rho_i (\delta z)^2} \leq 0.25 \qquad (5\text{-}25)$$

is satisfied, an acceptable or stable result will be obtained. Note that in this equation, $K_i / C_i \rho_i$ is the thermal diffusivity, k_i.

In the problem of the cooling lava lake, where thermal diffusivities of both molten and solidified lava are taken to be 10^{-6} m^2 s^{-1}, values of δz of 0.2 m and δt of 1 h (3600 s) satisfy the condition expressed in Eq. 5-25 $[k_i \, \delta t/(\delta z)^2 = 0.1]$. The quantity of heat crossing the upper surface of the lava in the first hour is, therefore, given by

$$Q_1 = -\frac{K(25 - 1200)A3600}{0.2} \, \text{J}$$

The quantities of heat transferred between each of the other cells would be calculated in a similar way, but at least during the first time interval, the lack of temperature differences between any of the other cells results in no heat being transferred across their boundaries. For example,

$$Q_2 = -\frac{K(1200 - 1200)A3600}{0.2} = 0 \, \text{J}$$

Once heat fluxes across all cell boundaries are determined and account taken of any sources or sinks of heat, temperature changes in each cell can be calculated from Eq. 5-24. In this example no sources or sinks of heat will be included.

Although their involvement is simple, there is insufficient space to properly introduce them here; the interested reader is referred to the paper by Shaw, Hamilton, and Peck (1977). In the example considered here, Eq. 5-24 reduces to

$$\delta T_i = \frac{k \, \delta t}{(\delta z)^2} (T_{i+1} - 2T_i + T_{i-1})$$

which for cell number 1 gives a temperature lowering of 106 degrees, from 1200°C to 1094°C. Because no heat is transferred between the other cells during this first interval, their temperatures remain unchanged at 1200°C. The new temperatures for each cell provide the input data necessary to calculate the heat fluxes during the next time interval, which in turn provide the new temperatures for each cell as cooling proceeds. The results for the uppermost five cells of the lava lake for the first 5 h are presented in Figure 5-8 (Prob. 5-9).

Calculations of this sort are tedious, and clearly a computer is necessary to generate sufficient data for results to be significant. Nonetheless, the few results presented in Figure 5-8 are worth examining. The temperature gradients derived by the numerical calculation (solid lines) have similar forms to those derived from the exact solution of the differential equation (dashed lines; Prob. 5-8). With the exception of the first cell, temperatures calculated by the two methods are in reasonable agreement. Those in the first cell are considerably higher than they should be. This results from taking the air temperature as if it were for a cell immediately above the surface of the lava. Because temperatures are taken as being at the center of cells, the air temperature of 25°C is consequently displaced 10 cm ($\frac{1}{2}\delta z$) above its actual position. In this region of steep temperature gradients, this small shift causes a serious error in calculated temperatures, at least during the initial stages of cooling. At later stages, the error is drastically reduced. If necessary, of course, the error can be eliminated entirely by decreasing the size of the cells near the upper surface.

With sufficiently small cells and intervals of time, a numerical analysis can approach as closely as necessary an exact solution. The value of the numerical method, however, lies in the ease with which complexities can be incorporated. Heats of crystallization can be introduced in cells in a manner that accounts for the fraction of crystals formed at any particular temperature. Heats of metamorphic reactions in a contact aureole can be inserted at appropriate temperatures and locations. Furthermore, cooling studies need not be restricted to simple geometrical igneous bodies, for any irregular shape can be described by a three-dimensional configuration of small cells.

Another widely used numerical solution to heat conduction problems is the implicit finite-difference technique of Crank and Nicolson (1947). This solution can be used not only for simple heat conduction problems but also for problems involving heat source and sinks and movement of the mass under consideration. Discussion of the latter complexities is left to Sec. 20-4; here we consider only the numerical solution of Fourier's equation (Eq. 5-11) for the conduction of heat in only the z direction, which can be written as

$$\frac{dT}{dt} = k \left(\frac{d^2 T}{dz^2} \right) \qquad (5\text{-}26)$$

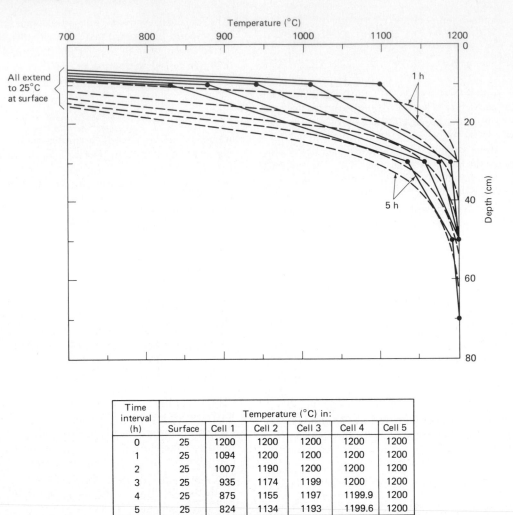

Time interval (h)	Temperature (°C) in:					
	Surface	Cell 1	Cell 2	Cell 3	Cell 4	Cell 5
0	25	1200	1200	1200	1200	1200
1	25	1094	1200	1200	1200	1200
2	25	1007	1190	1200	1200	1200
3	25	935	1174	1199	1200	1200
4	25	875	1155	1197	1199.9	1200
5	25	824	1134	1193	1199.6	1200

FIGURE 5-8 Temperatures near the surface of a lava lake at hourly intervals during the first 5 h of cooling. Solid lines are calculated using the numerical method involving Eqs. 5-22 and 5-24, whereas the dashed lines are calculated using the exact solution of Eq. 5-13.

Temperature in Eq. 5-26 is both a spatial (z) and temporal (t) function. We can represent these two variables on a grid (Fig. 5-9), with the number of grid increments of δz and δt being designated by m and n, respectively. To evaluate Eq. 5-26 numerically, the temperature T at any point m, n must be expressed in terms of the finite differences in δz and δt of the grid.

Given a temperature at any point in the grid, such as $T_{m,n}$ (Fig. 5-9), it is possible to express the temperature at any adjoining point $T_{m+1, n}$ by a Taylor series:

$$T_{m+1,n} = T_{m,n} + \frac{dT(\delta z)}{dz} + \frac{1}{2!}\frac{d^2 T(\delta z)^2}{dz^2} + \frac{1}{3!}\frac{d^3 T(\delta z)^3}{dz^3} + \cdots \tag{5-27}$$

Because points $T_{m+1, n}$ and $T_{m,n}$ are close, higher-order terms need not be included. We can also write the temperature of a point on the other side of m, n as

$$T_{m-1, n} = T_{m,n} + \frac{dT(-\delta z)}{dz} + \frac{1}{2!}\frac{d^2 T(-\delta z)^2}{dz^2} + \frac{1}{3!}\frac{d^3 T(-\delta z)^3}{dz^3} + \cdots \tag{5-28}$$

Adding Eqs. 5-27 and 5-28 together and rearranging gives

$$T_{m+1, n} + T_{m-1, n} - 2T_{m,n} = \left(\frac{d^2 T}{dz^2}\right)(\delta z)^2$$

or

$$\frac{d^2 T}{dz^2} = \frac{T_{m+1, n} + T_{m-1, n} - 2T_{m,n}}{(\delta z)^2} \tag{5-29}$$

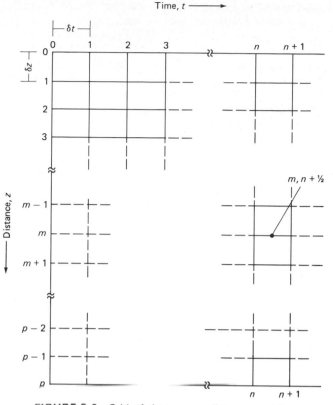

Time, $t \longrightarrow$

FIGURE 5-9 Grid of time versus distance used in deriving the Crank–Nicolson (1947) finite-difference technique of numerical analysis of heat conduction problems. See text for discussion.

Equation 5-29 expresses the second derivative in Fourier's equation in terms of the temperatures at several points in the grid and the finite difference δz between them.

Next we express the first term in Fourier's equation as a finite-difference ratio by considering the temperatures at points m, n and $m, n+1$ on either side of a point $m, n + \frac{1}{2}$ (Fig. 5-9):

$$T_{m,n} = T_{m,n+1/2} + \tfrac{1}{2}(-\delta t)\left(\frac{dT}{dt}\right) \qquad (5\text{-}30)$$

and

$$T_{m,n+1} = T_{m,n+1/2} + \tfrac{1}{2}(\delta t)\left(\frac{dT}{dt}\right) \qquad (5\text{-}31)$$

Subtracting Eq. 5-30 from Eq. 5-31 and rearranging gives

$$\frac{dT}{dt} = \frac{T_{m,n+1} - T_{m,n}}{\delta t} \qquad (5\text{-}32)$$

The second derivative of T with respect to z at this same point, $m, n + \frac{1}{2}$, can be obtained using Eq. 5-29 from the mean of the values determined for points $(m-1, n) - (m, n) - (m+1, n)$

and $(m-1, n+1) - (m, n+1) - (m+1, n+1)$, that is,

$$\frac{d^2 T_{m,n+1/2}}{dz^2} = \frac{\begin{array}{c} T_{m+1,n} + T_{m-1,n} - 2T_{m,n} + T_{m+1,n+1} \\ + T_{m-1,n+1} - 2T_{m,n+1} \end{array}}{2(\delta z)^2} \qquad (5\text{-}33)$$

Substituting Eqs. 5-33 and 5-32 into Eq. 5-26 gives

$$\frac{T_{m,n+1} - T_{m,n}}{\delta t} = \frac{k}{2(\delta z)^2}(T_{m+1,n} + T_{m-1,n} - 2T_{m,n} + T_{m+1,n+1} + T_{m-1,n+1} - 2T_{m,n+1})$$

which on rearranging gives

$$T_{m,n+1}\left[1 + \frac{k\,\delta t}{(\delta z)^2}\right] = \frac{k\,\delta t}{2(\delta z)^2}(T_{m+1,n} + T_{m-1,n} + T_{m+1,n+1} + T_{m-1,n+1}) \cdots + T_{m,n}\left[1 - \frac{k\,\delta t}{(\delta z)^2}\right] \qquad (5\text{-}34)$$

Equation 5-34 can be simplified further if we select values of δt and δz so that $k\,\partial t/(\delta z)^2 = 1$. Equation 5-34 then becomes

$$T_{m,n+1} = \tfrac{1}{4}(T_{m+1,n} + T_{m-1,n} + T_{m+1,n+1} + T_{m-1,n+1}) \qquad (5\text{-}35)$$

Equation 5-35 expresses the temperature at a point in terms of four temperatures at points that differ from it by one finite difference in space and one finite difference in time; this temperature can be thought of as the mean of the four other temperatures.

To use Eq. 5-35 it is necessary to specify the initial and boundary conditions. For example, what was the initial temperature distribution, and what were the temperature gradients across the boundaries at any time t? These conditions would be given as follows:

$$T = f(z) \qquad \text{at } t = 0 \text{ for } 0 < z < 1$$

$$\frac{dT}{dz} = H_0(T) \qquad \text{at } z = 0 \text{ for } t \geq 0$$

$$\frac{dT}{dz} = H_p(T) \qquad \text{at } z = p \text{ for } t \geq 0$$

That is, at $t = 0$ there is a temperature distribution which is a function of z; at any time there is a temperature gradient across the boundary at $z = 0$ which is a function H_0 of T; and at the other boundary, where $z = p$, there is a gradient that is a function H_p of T. Once these conditions are specified, Eq. 5-35 can be solved by trial and error, as will become clear in the example to follow.

Before using Eq. 5-35, however, it is necessary to examine how we can evaluate the temperatures along the top and bottom of the grid in Figure 5-9, that is, for values of $z = 0$ and $z = p$. Clearly, Eq. 5-35 cannot be used because two of the necessary temperatures would fall outside the grid and are thus undefined except in terms of the functions H_0 and H_p that define the boundary conditions. Crank and Nicolson (1947) express these temperatures as follows:

$$T_{0,n+1} = \tfrac{1}{2}(T_{1,n} + T_{1,n+1}) + \tfrac{1}{8}H_0(T_{0,n} + T_{0,n+1}) \qquad (5\text{-}36)$$

and

$$T_{p,n+1} = \tfrac{1}{2}(T_{p-1,n} + T_{p-1,n+1}) + \tfrac{1}{8}H_p(T_{p,n} + T_{p,n+1}) \quad (5\text{-}37)$$

Of course, if the boundary temperatures are constant, these values can simply be entered into the grid.

Let us now apply Eqs. 5-35 to 5-37 to the problem considered in the previous numerical analysis, that of the cooling of a laterally extensive basaltic lava lake. The initial and boundary conditions are easily defined. At $t = 0$, the temperature at all depths in the lake is 1200°C, except on the surface ($z = 0$), where it is instantly 25°C, and it remains at this temperature while the lava at depth continues to cool. During the period of interest, lava with a temperature of 1200°C continues to exist at depth in the lake. We will ignore the heat of crystallization.

The first step in the calculation is to select values of δt and δz so that $k\,\delta t/(\delta z)^2 = 1$. We can, for example, choose a time interval of half an hour (1800 s), in which case $\delta z = 0.0424$ m if $k = 10^{-6}$ m^2 s^{-1}. We can then construct a grid, as shown in Table 5-2, using these values. Because of the iterative type of calculation this table is best constructed using a spreadsheet on a computer. The first column in the table gives the depth in meters in the lava lake. The next column gives the initial temperatures, which are 1200°C except at $z = 0$, where it is 25°C. The temperature at $z = 0$ remains at 25°C at all times; this value can therefore be entered in the first row of the table for all times. The remainder of the columns show the temperatures in the lava lake following successive half-hour intervals. These temperatures are calculated using Eq. 5-35, starting with

the first time increment. If the calculations are done manually, a reasonable estimate is made of the temperatures that are thought might exist after one half-hour period. These estimates are then tested to see if they satisfy Eq. 5-35. If they do not, they are adjusted until they do. If a computer and spreadsheet are used, Eq. 5-35 is entered in each cell of the table except for cells at the greatest depth, where Eq. 5-37 is entered. Because the equations make use of temperatures in cells that are yet to be calculated, the calculation is iterative and continues as long

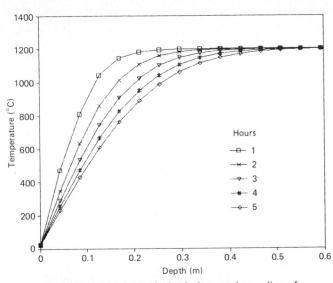

FIGURE 5-10 Numerical solutions to the cooling of a lava lake using the method of Crank and Nicolson (1947). The five curves indicate the temperature distributions at hourly intervals during the first 5 h of cooling of the lake. Calculated temperatures differ by no more than a few degrees from the exact solution of the differential equation. See text for discussion.

TABLE 5-2

Cooling rate of a lava lake

Depth (m)	\multicolumn{11}{c}{Temperature (°C) — Time of Cooling after Formation of Lava Lake (h)}										
	0	0.5	1	1.5	2	2.5	3	3.5	4	4.5	5
0.000	25	25	25	25	25	25	25	25	25	25	25
0.042	1200	570	473	393	349	316	292	273	257	244	233
0.085	1200	1031	810	711	634	580	537	503	475	451	431
0.127	1200	1155	1043	936	859	796	745	703	667	637	610
0.170	1200	1188	1144	1077	1013	956	907	864	827	793	764
0.212	1200	1197	1181	1149	1107	1064	1023	986	951	919	890
0.255	1200	1199	1194	1180	1158	1130	1100	1071	1042	1014	988
0.297	1200	1200	1198	1193	1182	1166	1147	1126	1104	1082	1061
0.339	1200	1200	1199	1197	1193	1185	1174	1160	1145	1129	1112
0.382	1200	1200	1200	1199	1197	1193	1188	1179	1170	1159	1147
0.424	1200	1200	1200	1200	1199	1197	1194	1190	1184	1177	1169
0.467	1200	1200	1200	1200	1200	1199	1198	1195	1192	1188	1183
0.509	1200	1200	1200	1200	1200	1200	1199	1198	1196	1194	1191
0.552	1200	1200	1200	1200	1200	1200	1200	1199	1198	1197	1195
0.594	1200	1200	1200	1200	1200	1200	1200	1200	1199	1199	1198
0.636	1200	1200	1200	1200	1200	1200	1200	1200	1200	1199	1199

as the calculated temperatures are significantly different from the previously calculated ones.

A plot of temperature distributions in the lava lake at hourly intervals during the first 5 h is shown in Figure 5-10. At all times, calculated temperatures agree within a few degrees of those obtained with the exact solution of the differential equation (Prob. 5-8). The Crank–Nicolson numerical method therefore provides a reliable means of solving heat flow problems. It can, for example, be used to solve any of the problems introduced in the earlier part of this chapter (Probs. 5-10 and 5-11). In Chapter 20 we see how it is used to include heat sources and mass movement of the body of rock being investigated.

5-5 COOLING BY RADIATION

Cooling by radiation from a surface is very much more rapid than heat conduction through a body. As a result, many lava flows rapidly cool and form a crust on their upper surface, while beneath, lava may remain molten and continue flowing for considerable time. The radiant heat emitted in time t from a perfectly radiating surface of area A, commonly referred to as *blackbody radiation*, is given by the *Stefan–Boltzmann law*,

$$Q = \sigma A T^4 t \tag{5-38}$$

where σ, the Stefan–Boltzmann constant, has a value of 5.6696×10^{-8} W m^{-2} K^{-4} and T is the absolute temperature. For all natural substances, which are not perfect radiators, the law is expressed as

$$Q = \epsilon \sigma A T^4 t \tag{5-39}$$

where ϵ, the emissivity, must be measured for the particular material at the temperature of interest. Few data are available on values of ϵ for molten lavas, but an approximate value of 0.5 can be used (Shaw and Swanson 1970) without introducing serious discrepancy, because the quantity of radiated heat is far more dependent on temperature.

Many basalts are extruded at approximately 1200°C (1473 K); hence, they radiate heat at a rate of 133 kW m^{-2}. It is instructive to consider what thermal gradient would be necessary in the lava to supply, by conduction, this amount of heat to the surface for radiation. This can be determined from Eq. 5-3 by assuming the reasonable value of 2 W m^{-1} °C^{-1} for the thermal conductivity, in which case $dT/dx = J/K = 133/2 = 665$°C cm^{-1}. Clearly, as soon as cooling takes place to a depth of 2 cm, the thermal gradient would not be great enough to cause sufficient heat to be supplied through conduction to keep up with the radiant cooling of the surface. For this reason the crust on a lava lake or flow may be cool enough to walk on despite the presence of molten material a few centimeters beneath.

In Chapter 4 ash flows and perhaps some flood basalts were said to travel turbulently, in which case heat lost from their upper surfaces by radiation is rapidly replaced by convection of hot material from within the flow onto the surface. The turbulence tends to keep the flow at the same temperature throughout and ensures that radiation from the upper surface

is the main process of cooling (Boyd 1961; Shaw and Swanson 1970; Danes 1972).

The cooling rate of a flow which is so turbulent that its temperature is essentially constant throughout, as might be the case for a rapidly moving ash flow, can be evaluated by expressing the Stefan–Boltzmann law in differential form,

$$-\frac{dQ}{dt} = \epsilon \sigma A T^4 \tag{5-40}$$

The negative sign is placed before the derivative to indicate that the heat is being given up by the flow. As heat is lost, the temperature falls by an amount determined by the heat capacity of the cooling material. Because $dQ/dT = C_p \rho V$, where V is the volume of the material and ρ its density, $C_p \rho V \, dT$ can be substituted for dQ in Eq. 5-40, yielding on rearranging,

$$-\frac{dT}{dt} = \frac{\epsilon \sigma A T^4}{C_p \rho V} \tag{5-41}$$

Equation 5-41 gives the rate of change of temperature with time. Integration of this equation yields the total temperature change in a given interval of time. To do this, we multiply by dt and divide by T^4 to separate variables,

$$-\frac{dT}{T^4} = \frac{\epsilon \sigma A}{C_p \rho V} \, dt$$

If the emissivity and heat capacity remain constant over the temperature range of interest, we can write

$$-\int_{T_0}^{T} \frac{dT}{T^4} = \frac{\epsilon \sigma A}{C_p \rho V} \int_0^t dt$$

which upon integration gives

$$\frac{1}{3T^3} - \frac{1}{3T_0^3} = \frac{\epsilon \sigma A t}{C_p \rho V}$$

or

$$T = \left(\frac{3 \epsilon \sigma A t}{C_p \rho V} + \frac{1}{T_0^3} \right)^{-1/3} \tag{5-42}$$

Equation 5-42 gives the temperature to which cooling takes place in t seconds from an initial temperature T_0. If for purposes of illustration, we take the initial temperature of 1 g of basalt to be 1200°C (1473 K), the emissivity to be 0.5, and the heat capacity to be 0.8 kJ kg^{-1} K^{-1}, the temperature after 1 h (3600 s) would decrease to 300 K or 27°C due to radiation from a surface of 1 cm^2.

The cooling rate of a flow, as determined from Eq. 5-42, is probably too high, because turbulence, unless extreme, is unlikely to bring hot material from within the flow at a sufficient rate to replenish the heat radiated from the surface. Also, the equation does not take into account any other sources of heat within the flow, such as latent heat of crystallization or conversion of gravitational energy through the dissipation of

TABLE 6-6
Classification of lamprophyres after Streckeisen (1979)

Felsic Constituent		Predominant Mafic Mineral			
Feldspar	Foid	biotite diopside augite ± olivine	hornblende diopside augite ± olivine	amphibole- (barkevikite kaersutite) titanaugite olivine biotite	melilite biotite ± titanaugite ± olivine ± calcite
or > pl	—	Minette	Vogesite		
pl > or	—	Kersantite	Spessartite		
or > pl	fsp > foid				
pl > or	fsp > foid			Sannaite	
—	foid			Camptonite	
—	—			Monchiquite	Polzenite
					Alnöite

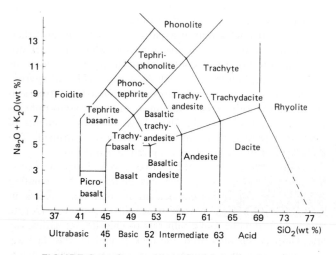

FIGURE 6-4 Compositional fields of volcanic rocks in terms of total alkalis and silica. [After Le Bas et al. (1986); published by permission of Oxford University Press.]

altered, and their classification by this scheme may be erroneous. Rocks falling in the trachybasalt field can be further classifed as hawaiite if $(Na_2O - 2) > K_2O$ and as potassic trachybasalt if $(Na_2O - 2) < K_2O$. Similarly, the field of basaltic trachyandesites can be divided into mugearite (Na) and shoshonite (K), and the field of trachyandesite into benmoreite (Na) and latite (K). Classification by this scheme is almost totally consistent with that based on the QAPF diagram (Fig. 6-1).

6-7 THE IRVINE–BARAGAR CLASSIFICATION OF VOLCANIC ROCKS

In recent years the classification proposed by Irvine and Baragar (1971) has gained wide acceptance. It sets up divisions between different rock types based solely on common usage; that is, in practice most geologists associate a particular rock name with a certain compositional range. The scheme also incorporates the well-established fact that volcanic rocks fall into a number of distinct genetic series, which can be distinguished by simple chemical parameters. These series have the added significance that they can be correlated with distinct tectonic environments.

Volcanic rocks are classified by Irvine and Baragar into three main groups (Fig. 6-5), the *subalkaline*, the *alkaline*, and the *peralkaline* (alkali-rich). Most rocks belong to the first two groups, which are each subdivided into two subgroups. Assigning a rock to any one of these groups is based on simple chemical parameters or normative compositions. Before this is done, however, the chemical effects of alteration must be taken into account, if possible. Many volcanic rocks become oxidized, hydrated, or carbonated by hydrothermal activity during burial or during later metamorphism. These chemical changes can seriously affect the normative composition of a rock, which may, in turn, affect its classification. For example, conversion of ferrous iron to ferric during alteration results in smaller amounts of iron silicates being calculated in the norm; this then produces a norm that appears more saturated in silica than was the original rock. This type of alteration, however, can be corrected for, because in many unaltered volcanic rocks there is a strong positive correlation between the TiO_2 and Fe_2O_3 contents. The primary wt % Fe_2O_3 in many volcanic rocks is given approximately by (wt % TiO_2 + 1.5). H_2O and CO_2 are subtracted from the analysis and the total recalculated to 100%. Norm calculations are carried out according to the CIPW rules, but Irvine and Baragar chose to recalculate the normative minerals to molecular rather than weight percentages. Thus, instead of multiplying the mole proportions by the weight factors given in Table 6-2, the mole proportions are simply recalculated to 100%. Finally, in expressing feldspar compositions, nepheline is recast as albite. Thus, the normative anorthite content is given by $100 \times An/(An + Ab + \frac{5}{3} Ne)$. Analyses of typical samples of each of the main rock types in Irvine and Baragar's classification are given in Table 6-7.

Division into the three main groups is based on the alkali content of the rocks. Rocks in which the molecular amounts of $(Na_2O + K_2O) > Al_2O_3$ fall into the peralkaline group.

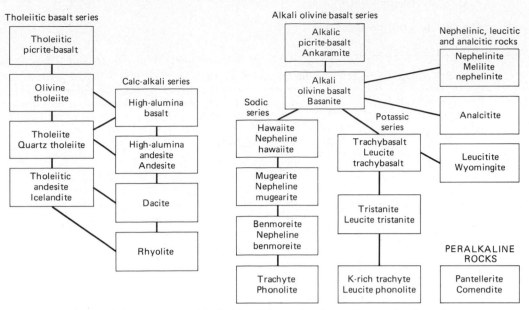

FIGURE 6-5 General classification scheme for the common volcanic rocks. Lines joining boxes link commonly associated rocks. The small print within the boxes refers to variants of the main rock. [After Irvine and Baragar (1971); published by permission of Canadian Journal of Earth Sciences.]

TABLE 6-7

Typical analyses of the rocks listed in Figure 4-5*

	Subalkaline Rocks											Peralkaline Rocks	
	Tholeiitic Basalt Series					Calc-alkali Series							
	Tholeiitic Picrite	Ol Tholeiite	Tholeiite	Tholeiitic Andesite	Icelandite	High-Al Basalt	High-Al Andesite	Andesite	Dacite	Rhyolite		Pantellerite	Comendite
SiO_2	46.4	49.2	53.8	58.3	61.8	49.1	58.6	60.0	69.7	73.2		69.8	75.2
TiO_2	2.0	2.3	2.0	1.7	1.3	1.5	0.8	1.0	0.4	0.2		0.4	0.1
Al_2O_3	8.5	13.3	13.9	13.8	15.4	17.7	17.4	16.0	15.2	14.0		7.4	12.0
Fe_2O_3	2.5	1.3	2.6	3.4	2.3	2.8	3.2	1.9	1.1	0.6		2.4	0.9
FeO	9.8	9.7	9.3	6.5	5.8	7.2	3.5	6.2	1.9	1.7		6.1	1.2
MnO	0.2	0.2	0.2	0.2	0.2	0.1	0.1	0.2	0.0	0.0		0.3	0.1
MgO	20.8	10.4	4.1	2.3	1.8	6.9	3.3	3.9	0.9	0.4		0.1	0.0
CaO	7.4	10.9	7.9	5.6	5.0	9.9	6.3	5.9	2.7	1.3		0.4	0.3
Na_2O	1.6	2.2	3.0	3.9	4.4	2.9	3.8	3.9	4.5	3.9		6.7	4.8
K_2O	0.3	0.5	1.5	1.9	1.6	0.7	2.0	0.9	3.0	4.1		4.3	4.7
P_2O_5	0.2	0.2	0.4	0.5	0.4	0.3	0.2	0.2	0.1	0.0		0.2	0.1

	Alkaline Rocks														
	Alkali Olivine Basalt Series											Nephelinites etc.			
	Alkalic Picrite	Ankaramite	K-poor Alk basalt	K-rich Alk basalt	Trachybasalt	Hawaiite	Mugearite	Tristanite	Benmoreite	Trachyte	Phonolite	Nephelinite	Analcitite	Leucitite	Wyomingite
SiO_2	46.6	44.1	45.4	42.4	46.5	47.9	49.7	55.8	55.6	60.7	60.6	39.7	49.0	46.2	54.1
TiO_2	1.8	2.7	3.0	4.1	3.1	3.4	2.1	1.8	0.9	0.5	0.0	2.8	0.7	1.2	2.3
Al_2O_3	8.2	12.1	14.7	14.1	16.7	15.9	17.0	19.0	16.4	20.5	18.3	11.4	13.0	14.4	9.9
Fe_2O_3	1.2	3.2	4.1	5.8	4.1	4.9	3.4	2.6	3.1	2.3	2.7	5.3	4.9	4.1	3.1
FeO	9.8	9.6	9.2	8.5	7.3	7.6	9.0	3.1	4.9	0.4	1.2	8.2	4.5	4.4	1.5
MnO	0.1	0.2	0.2	0.2	0.2	0.2	0.3	0.1	0.2	0.2	0.2	0.2	0.1	0.0	0.1
MgO	19.6	13.0	7.8	6.7	4.6	4.8	2.8	2.0	1.1	0.2	0.1	12.1	8.3	7.0	7.0
CaO	9.4	11.5	10.5	11.9	9.4	8.0	5.5	4.5	2.9	1.4	0.8	12.8	11.5	13.2	4.7
Na_2O	1.6	1.9	3.0	2.8	3.8	4.2	5.8	5.2	6.1	6.2	8.9	3.8	3.9	1.6	1.4
K_2O	1.2	0.7	1.0	2.0	3.1	1.5	1.9	4.1	3.5	6.7	5.1	1.2	3.0	6.4	11.4
P_2O_5	0.3	0.3	0.4	0.6	0.9	0.7	0.5	0.4	0.7	0.0	0.0	0.9	1.1	0.4	1.8

* From Irvine and Baragar (1971).

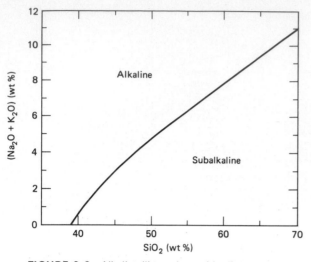

FIGURE 6-6 Alkalis-silica plot with line separating fields of alkaline and subalkaline rocks. [After Irvine and Baragar (1971).]

These rocks typically contain aegerine or a sodic amphibole. The alkali content that separates the subalkaline from the alkaline groups varies with the silica content of the rock (Fig. 6-6). The equation for the boundary between these groups is given by

$$SiO_2 = -3.3539 \times 10^{-4} \times A^6 + 1.2030 \times 10^{-2} \times A^5$$
$$- 1.5188 \times 10^{-1} \times A^4 + 8.6096 \times 10^{-1} \times A^3$$
$$- 2.1111 \times A^2 + 3.9492 \times A + 39.0$$

where $A = (Na_2O + K_2O)$. These two groups can also be distinguished in a plot of the normative contents of olivine–nephelin–quartz (Fig. 6-7). To plot a rock in this diagram the normative minerals are recast as follows: $Ne' = Ne + \frac{3}{5} Ab$, $Q' = Q + \frac{2}{5} Ab + \frac{1}{4} Opx$, and $Ol' = Ol + \frac{3}{4} Opx$. The subalkaline rocks plot on the quartz side of the boundary line, whereas the alkaline ones plot on the nepheline side of it.

The subalkaline rocks are divided into the *calc-alkali* and *tholeiitic* series on the basis of their iron contents in the AFM plot (Fig. 6-8), where $A = Na_2O + K_2O$, $F = FeO + 0.8998 \times Fe_2O_3$, and $M = MgO$ (all in wt %). This plot distinguishes intermediate members of these series very well, but at the mafic and felsic ends there is considerable overlap. Calc-alkali basalts and andesites, however, contain 16 to 20% Al_2O_3, which is considerably more than occurs in tholeiitic basalts and andesites, which contain from 12 to 16%. At the extreme felsic end there is no satisfactory way of distinguishing calc-alkali and tholeiitic members; thus all granitic rocks are assigned to the calc-alkali series. The alkaline rocks are divided into the *alkali olivine basalt* series and the *nephelinitic–leucitic–analcitic* series. Rocks of the latter series typically contain less than 45% SiO_2, have normative color indices greater than 50, and may contain normative leucite.

The naming of rocks within the various subgroups is based on normative plagioclase composition and on normative color index. In the various subalkaline series, the rocks range from *basalt* through *andesite* and *dacite* to *rhyolite* with decreasing normative anorthite content and decreasing normative color index (Fig. 6-9). Two series of rock names are used for the alkaline rocks, depending on whether they are sodic or potassic. This division is made on the basis of the normative feldspar composition (Fig. 6-10). With decreasing normative anorthite content, the sodic series passes from *alkali basalt* through *hawaiite*, *mugearite*, and *benmoreite* to *trachyte* (Fig. 6-11a), whereas the potassic series passes from *alkali basalt* through *trachybasalt* and *tristanite* to *trachyte* (Fig. 6-11b). At

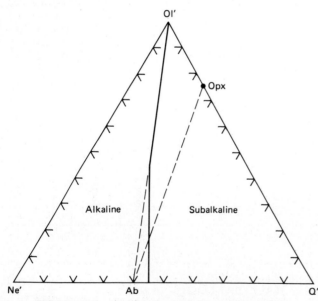

FIGURE 6-7 Ol'–Ne'–Q' projection with line separating fields of alkaline and subalkaline rocks. Plot in percent cation equivalents; see text for explanation. [After Irvine and Baragar (1971).]

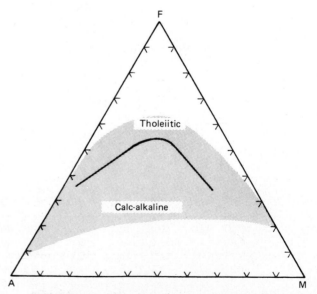

FIGURE 6-8 AFM plot showing line separating fields of tholeiitic and calc-alkaline rocks as proposed by Irvine and Baragar (1971). $A = Na_2O + K_2O$; $F = FeO + 0.8998Fe_2O_3$; $M = MgO$, all in weight percent.

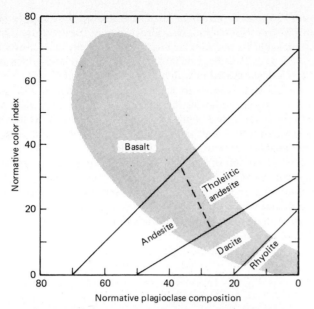

FIGURE 6-9 Irvine and Baragar's (1971) sub-division of the subalkaline rocks in a plot of normative color index versus normative plagioclase composition. Plot in percent cation equivalents. Normative color index $-$ ol $+$ opx $+$ cpx $+$ mt $+$ il $+$ hm. Normative plagioclase composition $= 100$ An$/($An $+$ Ab $+ \frac{5}{3}$Ne$)$; this converts nepheline into albite.

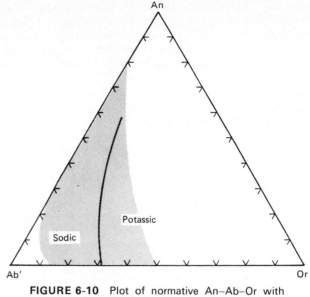

FIGURE 6-10 Plot of normative An–Ab–Or with dividing line separating sodic and potassic alkaline rocks. Ab$' = $Ab $+ \frac{5}{3}$Ne. Plot in percent cation equiv-alents. [After Irvine and Baragar (1971).]

the mafic end of all of these series, basalts containing more than 25% normative olivine are named *picrites*; these rocks contain abundant phenocrystic olivine. *Ankaramites*, which belong to the alkaline group, contain an abundance of augite phenocrysts

which causes the norm to have more than 20% clinopyroxene. Basalts containing more than 5% normative nepheline are named *basanite* if they contain modal nepheline, or *basanitoid* if nepheline is not visible. Finally, nepheline-bearing trachyte is known as *phonolite*.

Although this classification is descriptive, it has consid-erable genetic significance. The calc-alkali series is character-istic of orogenic belts and gives rise to the volcanic rocks of

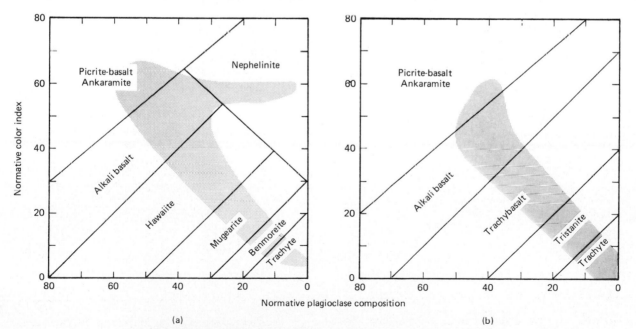

FIGURE 6-11 Plots of normative color index versus normative plagioclase composition (defined as in Fig. 6-9) for (a) sodic alkaline rocks and (b) potassic alkaline rocks. [After Irvine and Baragar (1971).]

island arcs. These rocks are clearly related to subduction zones. The tholeiitic rocks are prominently developed in zones of crustal extension where they commonly develop thick sequences of flood basalts. They constitute the major rock type along oceanic ridges (MORB = mid-ocean ridge basalt) and on many large oceanic islands, such as Hawaii. Many alkali olivine basalts and associated rocks occur in areas of continental rifting and in regions overlying deeply subducted plates; they also occur at intraplate hot spots in both oceanic and continental regions.

6-8 IGNEOUS ROCK NAMES

Despite the relatively small number of rock names proposed in the IUGS and Irvine and Baragar classifications, the petrologic literature contains hundreds of rock names. Although most of these are no longer used, it is necessary at least to be able to find their definitions in order to read the literature. The list in Table 6-5 includes the most commonly encountered rock names. They have been defined, if possible, by relating them to the IUGS classification with a pair of numbers. The first number, 1, 2, and 3, refers to Figures 6-1, 6-2, and 6-3, and the second number indicates the field in which the rock plots in the particular figure. Thus a jotunite (1:9) is a hypersthene monzodiorite and plots in Figure 6-1, field 9. Some rock names are based on textural features, in which case these are briefly stated. The object of this tabulation is not to present a catalog of names to be memorized but simply to provide a convenient list to which reference can be made quickly. The important rock names, including for example those in the IUGS classification, have been placed in bold print, and their definitions should be learned.

PROBLEMS

6-1. Calculate the CIPW norms of the following rocks listed in Table 6-7: dacite, tholeiitic picrite, comendite, hawaiite, phonolite, leucitite.

6-2. Using the CIPW norms from Prob. 6-1, determine the names that should be given to these rocks according to the IUGS classification set forth in Figure 6-1.

6-3. Recalculate the CIPW norms in Prob. 6-1 to molecular norms; then, using the criteria of Irvine and Baragar (1971), determine whether these rocks were properly named in Table 6-7.

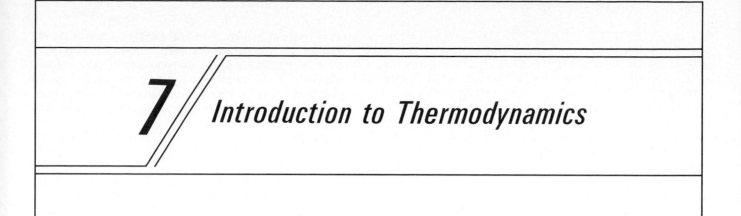

7 / Introduction to Thermodynamics

7-1 INTRODUCTION

Thermodynamics, the study of energy, is one of the most important subjects in all of science. Historically, it evolved from the desire to understand the efficiency of machines, in particular of steam engines. Much of its terminology, therefore, centers around heat and work, especially work associated with expanding gas. Thermodynamics, however, deals with the transfer of other forms of energy, such as that associated with chemical reactions. Although heat and mechanical work done by expanding gas are important in geology, for example in the cooling of a magma or the explosion of a volcano, it is in the study of chemical energies that thermodynamics is of greatest value to petrology. It is particularly useful in the study of processes that take place within the Earth, where they cannot be observed directly. The increased availability in recent years of thermodynamic data for the common minerals has resulted in a rapid growth in the application of thermodynamics to petrologic problems.

The general applicability of thermodynamics stems from the fundamental nature of the principles on which it is based, namely simple observations on the behavior of energy. For example, although energy can be converted from one form to another (kinetic to potential, chemical to thermal, etc.), it can never be destroyed. Furthermore, experience tells us that heat flows from hot to cold bodies, and never the reverse. The first observation, which concerns the conservation of energy, is embodied in the *first law of thermodynamics*, whereas the second one, which deals with the natural direction of processes, leads to the *second law of thermodynamics*. These laws can be expressed in simple mathematical forms, which can then be combined and manipulated to give useful functions from which the equilibrium conditions for a process or reaction can be calculated. In this way it is possible, for example, to determine melting points of minerals, compositions of minerals crystallizing from magma, temperatures and pressures of metamorphic reactions, relative stabilities of minerals with respect to chemical weathering, and compositions of ore-forming solutions.

Little more than a descriptive treatment of petrology could be given if thermodynamics were to be omitted. However, an entire book would be required to fully develop all thermodynamic relations encountered in the petrologic literature. In this and the following two chapters, only some of the more important fundamental concepts are covered. Standard physical chemistry texts will provide the reader with a more extensive coverage of the topic [e.g., Castellan (1971) and Denbigh (1955)]. A number of excellent introductory texts to thermodynamics and its applications to geology are also available (Kern and Weisbrod 1967; Wood and Fraser 1976; Fraser 1977).

7-2 ENERGY IN THE FORM OF HEAT AND WORK

When discussing the energy of processes, it is important to specify the extent of the material being considered. This is done by using the term *system* to designate that part of space under consideration. A system may have real boundaries, such as the walls of a magma chamber, or imaginary ones, as did the small control volume used in Chapter 2 to derive the rate of flow of magma. The system is chosen to suit the particular problem. Systems are *isolated* if they have no interaction with the surroundings, *closed* if they exchange only heat, and *open* if they exchange both heat and material. Truly isolated systems are difficult to find, but their concept plays an important role in derivations of certain theoretical relations. Many geological systems can be considered closed, as for example, a small rapidly cooling dike. A large batholith, on the other hand, might exchange considerable amounts of water and other mobile constituents with its surroundings while cooling and would be considered an open system.

Energy can be expressed in the form of either heat or work, ignoring for the moment energy tied up with chemical reaction. *Heat* is the quantity of energy that flows across the boundary of a system in response to a temperature gradient. *Work* is the quantity of energy that crosses the boundary of a system and is converted entirely into mechanical work in the surroundings, such as the lifting of a weight (Fig. 7-1); a geological example would be the explosive removal of the top of Mount St. Helens. By convention, energy put into a system in

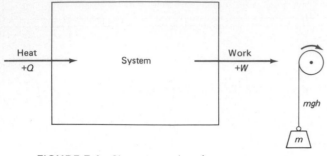

FIGURE 7-1 Sign convention for energy transferred into or out of a thermodynamic system in the form of heat or work.

the form of heat, Q, is positive, whereas that in the form of work, W, is negative. Positive work, then, is done on the surroundings. This convention is inherited from the early days of thermodynamics when there was interest in how much work a machine could do on its surroundings. Some texts [Kern and Weisbrod (1967), for example] have changed the sign convention for work, so that, like heat, it is positive when done on the system. Because the designation of sign is purely arbitrary, the sign convention does not affect thermodynamic conclusions. Care should be taken, however, in reading thermodynamic texts to ascertain the sign convention used.

The type of work most commonly encountered in petrologic processes is that known as *work of expansion*. For example, when a rock melts at some depth in the Earth, the approximate 10% expansion involved with the phase change results in work being done as the volume expands against the opposing pressure (P_{op}) of the surrounding rock. This work of expansion is given by

$$W_{exp} = (\text{force}) \times (\text{distance}) = (P_{op} \times \text{area}) \times (\text{distance})$$
$$= P_{op} \, \Delta V \qquad (7\text{-}1)$$

where ΔV is the volume change. Other types of work include electrical and magnetic, but these are not normally involved in petrologic processes (do Probs. 7-1 and 7-2).

To appreciate work of expansion, consider a gas bubble with volume V_1, pressure P_1, and temperature T in a magma

that is suddenly erupted onto the Earth's surface where the pressure is P_2. The bubble expands to V_2 against the opposing pressure P_2, but its temperature is kept constant by the thermal buffering of the surrounding hot magma. This change can be represented by the simple mechanical analog shown in Figure 7-2. The work of expansion done by this bubble on the surrounding magma is given by

$$W_{exp} = P_2(V_2 - V_1)$$

If the gas behaves ideally ($PV = nRT$), this isothermal expansion can be represented by the P versus V plot in Figure 7-2, in which the shaded area represents the amount of work done.

If the magma had stopped at some intermediate depth where the bubble could have expanded against an intermediate pressure P_i and then erupted onto the surface to complete its expansion, the work done in this two-stage decompression would be

$$W_{exp} = P_i(V_i - V_1) + P_2(V_2 - V_i)$$

This amount of work (Fig. 7-3a) is greater than that done by the single-stage expansion. If the bubble had expanded in three stages, the work would have been still greater (Fig. 7-3b). Clearly, the maximum amount of work that could be obtained from this expansion results from an infinite number of infinitesimal steps, in which case the work done would be

$$W_{exp}^{max} = \int_1^2 P_{op} \, dV$$

where P_{op} is the opposing pressure at any stage of expansion. If the gas is taken to be ideal, P_{op} can be replaced by nRT/V, in which case

$$W_{exp}^{max} = \int_1^2 \frac{nRT}{V} \, dV = nRT \int_1^2 \frac{dV}{V} = nRT \ln \frac{V_2}{V_1} \qquad (7\text{-}2)$$

Supposing that the gas bubble were now to be compressed back to its original state, work would have to be done on the system. This could be done in a single stage of compression by suddenly increasing the pressure to P_1, in which

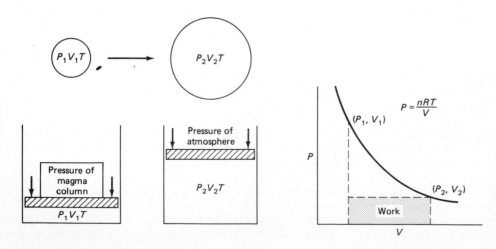

FIGURE 7-2 Mechanical analog of a gas bubble expanding isothermally against a pressure of 1 atm (10^5 Pa), and a graphical representation of the amount of work done during expansion ($P \, \triangle V$) assuming that the gas behaves ideally ($PV = nRT$).

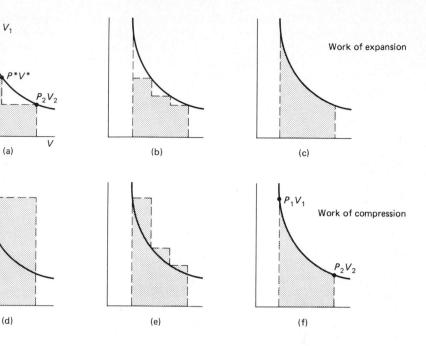

FIGURE 7-3 Work of expansion (shaded area) produced when decompression of an ideal gas takes place in (a) two stages, (b) three stages, (c) an infinite number of infinitesimal steps, and the amount of work necessary to return this gas to its original state by isothermally compressing it in (d) one stage, (e) three stages, and (f) an infinite number of infinitesimal steps.

case the amount of work done (Fig. 7-3d) would be given by

$$W_{comp} = P_1(V_1 - V_2)$$

If the bubble were compressed in a series of stages (Fig. 7-3e), the amount of work required would obviously be less, and the minimum amount of work (Fig. 7-3f) would be done when the pressure at each stage was increased infinitesimally, giving

$$W_{comp}^{min} = \int_2^1 P_{op}\, dV = nRT \int_2^1 \frac{dV}{V} = nRT \ln \frac{V_1}{V_2}$$

$$= -nRT \ln \frac{V_2}{V_1} \qquad (7\text{-}3)$$

We have followed a bubble through an isothermal cycle of expansion and compression. During the first half of the cycle the bubble did work on the surroundings, but during the second part, work was done on the bubble to restore it to its initial state. From Figure 7-3, the amount of work produced during expansion in a finite number of steps is clearly less than the amount of work that has to be performed on the system to restore it to its former state. Therefore, processes such as this, carried out in a finite number of steps, always result in work having to be done on the system. On the other hand, if the cycle of decompression and compression could be carried out in an infinite number of infinitesimal steps, the work of expansion (Eq. 7-2) would equal exactly the work required to compress the gas (Eq. 7-3); that is, the areas under the PV curves ($\int P\, dV$) would be equal. Because the compression part of the cycle carried out in this manner is the exact reverse of the decompression part of the cycle, processes of this type are described as *reversible*, whereas processes carried out in a finite number of steps are *irreversible*. All natural processes are irreversible; reversible processes are unnatural because they would require an infinite amount of time to take place. The concept of a re-

versible reaction, however, plays an important role in determining conditions of equilibrium, as will be seen later in this chapter.

7-3   FIRST LAW OF THERMODYNAMICS

In the isothermal cyclic process described above, irreversible expansion or compression of the bubble results in turbulence in the gas which can be equated with heat. For the temperature to remain constant, this heat has to be liberated into the surroundings. Hence, all natural cycles (natural = irreversible), in addition to requiring work to be done on the system, produce heat in the surroundings. Experience tells us that these two quantities of energy are always equal. This fact was first clearly enunciated by the German physicist Mayer in 1842 and eight years later was quantified in Joule's classic experiment on the mechanical equivalent of heat. Observations such as these lead to the formulation of the *first law of thermodynamics*, which states that *for any cyclical process (reversible or irreversible), the work produced in the surroundings is equal to the heat removed from the surroundings.*

Mathematically, the first law can be expressed as

$$\oint dW = \oint dQ$$

or

$$\oint (dQ - dW) = 0 \qquad (7\text{-}4)$$

where dQ and dW are the differentials of heat and work involved in the cyclic process, and the symbol $\oint$ indicates a cyclic integral, that is, the sum of all the small dQ's and dW's around the cycle.

We saw from considering the expansion and compression of the bubble that $\oint dW$ has no definite value unless the path followed around the cycle is completely specified (that is, number of stages, pressures, etc.). The same is true of $\oint dQ$. Equation 7-4, however, reveals that although $\oint dQ$ and $\oint dW$ by themselves do not have definite values, $\oint(dQ - dW)$ does have a definite value, which for a cyclic process is zero. This means that although the quantities of work and heat taken separately can have various values depending on the path followed by the process, the value of their combination $(dQ - dW)$ is independent of the path; its value is determined only by the initial and final states of the system. Because of this behavior $dQ - dW$ is said to be a *state property*; that is, a property which is dependent only on the state of the system and not on the path followed. For convenience, this state property is given the name *internal energy*, E, and is defined as

$$dE \equiv dQ - dW \qquad (7\text{-}5)$$

Note that the internal energy is not an independent quantity that can be equated with heat and work; it is simply the sum of the energy put into the system in the form of heat (dQ) and work $(-dW)$. Note also that it is defined in differential form; only changes in the internal energy can be measured, not absolute amounts. Integration of Eq. 7-5 results in a constant of integration that cannot be evaluated. This lack of knowledge of the absolute value of internal energy is of no importance because all thermodynamic calculations eventually deal only with differences, and thus the constant of integration disappears. For example, the change in internal energy associated with a reaction going from state 1 to state 2 is given by

$$\int_1^2 dE = \int_1^2 dQ - \int_1^2 dW$$

or

$$E_2 - E_1 = \Delta E = Q - W \qquad (7\text{-}6)$$

So far, we have used, for illustrative purposes, the purely physical process of the expansion and compression of a gas bubble. Equation 7-5, however, is equally applicable to any process in which there is a change in energy, whether it is physical or chemical. The process could, for example, involve a change of state from solid to liquid or from one mineral polymorph to another, or a chemical reaction between several minerals. Each of these processes will involve the transfer of energy in the form of both heat and work and must obey Eq. 7-5.

Many reactions of interest to petrologists involve work of expansion $(P\,\Delta V)$ at constant pressure—for example, the melting of a rock at a specific depth in the Earth. In such a case, Eq. 7-6 can be expressed as

$$E_2 - E_1 = Q_P - P(V_2 - V_1)$$

where Q_P is the heat involved with the reaction at constant pressure. On rearranging, we obtain

$$Q_P = (E_2 + PV_2) - (E_1 + PV_1) \qquad (7\text{-}7)$$

which shows that the heat involved with a reaction taking place at constant pressure is the difference between two groups of

terms which describe the energy, pressure, and volume of the final and initial states of the system. We have already seen that the internal energy is a state property. But PV is also a state property; for example, in the case of an ideal gas it would be equal to nRT. Consequently, $E + PV$ must also be a state property. It is therefore given a special name, *enthalpy* (H), which is defined as

$$H \equiv E + PV \qquad (7\text{-}8)$$

Equation 7-7 now becomes

$$Q_P = H_2 - H_1 = \Delta H \qquad (7\text{-}9)$$

The enthalpy change in a reaction is, therefore, the heat withdrawn from the surroundings at constant pressure.

If heat is given out during a reaction, it is said to be *exothermic* and ΔH is negative. An *endothermic* reaction is one that takes heat from the surroundings and hence ΔH is positive. For example, if forsterite were to react with quartz at 298 K (25°C) to form enstatite, 7.3 kJ mol^{-1} of forsterite would be liberated, and hence $\Delta H_{298} = -7.3$ kJ mol^{-1}. In contrast, the reaction from low quartz (α) to high quartz (β) at 848 K (575°C) is endothermic, and $\Delta H_{848} = +1.2$ kJ mol^{-1}.

7-4 STANDARD HEATS OF FORMATION

Because enthalpy involves the internal energy (Eq. 7-8), absolute values cannot be known. This, however, is not a problem because thermodynamic calculations deal only with changes in enthalpy. For example, in the reaction above where forsterite reacts with quartz to form enstatite, the absolute values of the enthalpies of the minerals cannot be known, but the enthalpy change ($\Delta H = -7.3$ kJ mol^{-1}) accompanying the reaction can be determined and used to calculate the conditions under which this reaction will occur.

Although absolute values of enthalpy cannot be known, it is convenient to think of substances as having such values. By convention, then, we assign an arbitrary "absolute" value of zero to the enthalpy of each of the elements in their standard stable form at 298.15 K (25°C) and a pressure of 10^5 Pa (1 bar). This is represented by $H^\circ_{298,\text{element}} \equiv 0$, where the superscript $^\circ$ indicates 10^5 Pa pressure. With this arbitrary base level it is then possible to define the enthalpy of a mineral in terms of the enthalpy change accompanying the formation of that mineral from the elements at 298 K and 10^5 Pa pressure. An enthalpy defined in this way is referred to as the *standard heat* (enthalpy) *of formation* of the mineral, $H^\circ_{f,298}$. For example, the reaction to form quartz would be

$$\text{Si}_{\text{crystal}} + \text{O}_{2\,\text{gas}} \xrightarrow{\Delta H^\circ_{f,298,Q}} \text{SiO}_{2\,\text{crystal}}$$

and the standard heat of formation would be given by

$$\Delta H^\circ_{f,298,Q} = H^\circ_{298,Q} - (H^\circ_{298,\text{Si}} + H^\circ_{298,O_2})$$

But the terms in parentheses have values of zero, since they refer to the elements in their stable states at 298 K and 10^5 Pa pressure. The enthalpy of quartz under these conditions is therefore equal to the enthalpy change of the reaction, which is $-910,648$ J mol^{-1}. Note that in this reaction the stable form of silicon under these conditions is a crystalline metal, whereas the stable form of oxygen is the diatomic gas (O_2). Standard heats of formation of minerals are given in Table 7-1.

Although the heat of formation of quartz in the previous reaction was given for a temperature of 298 K, this reaction is not likely to proceed rapidly at this temperature—if it did, pocket calculators and computers would have a very short life expectancy. High temperatures are required before silicon metal will react rapidly with oxygen. Therefore, if we were interested in measuring the heat involved with this reaction, it would be necessary to carry out the experiment at high temperatures. How, then, would we determine, from the high-temperature experiments, the enthalpy change associated with the reaction at 298 K?

To answer this we make use of the fact that enthalpy is a state property, and therefore its value is independent of the path followed by the reaction. For instance, instead of trying to react oxygen with silicon at 298 K, we can heat these materials to a high temperature, 1800 K for example. Quartz forms rapidly at this temperature. Once the reaction is complete, the quartz can be cooled to 298 K. Thus, the temperature of the starting materials and end product will both be 298 K, even though the reaction took place at 1800 K. The enthalpy change between starting materials and end product at 298 K will be the same whether the reaction proceeded directly at 298 K or followed the high-temperature path. These two possible ways of carrying out this reaction can be illustrated as follows:

$$(1800 \text{ K}) \quad \text{Si}_{\text{crystal}} \quad + \quad \text{O}_{2 \text{ gas}} \xrightarrow{\Delta H^\circ_{f,1800,Q}} \text{SiO}_{2 \text{ Q}}$$

$$(H^\circ_{1800} - H^\circ_{298})_{\text{Si}} \quad (H^\circ_{1800} - H^\circ_{298})_{\text{O}_2} \quad -(H^\circ_{1800} - H^\circ_{298})_{\text{Q}}$$

$$(298 \text{ K}) \quad \text{Si}_{\text{crystal}} \quad + \quad \text{O}_{2 \text{ gas}} \xrightarrow{\Delta H^\circ_{f,298,Q}} \text{SiO}_{2 \text{ Q}}$$

Because enthalpy is a state property, its integral around a cyclic process must be zero. Thus, if we were to heat Si and O_2 from 298 K to 1800 K, react them together to form quartz, cool the quartz to 298 K, and then break down the quartz to form Si and O_2 again, the sum of the enthalpy changes of all of the steps around this cycle would be zero; that is,

$$(H^\circ_{1800} - H^\circ_{298})_{\text{Si}} + (H^\circ_{1800} - H^\circ_{298})_{\text{O}_2} + H^\circ_{f,1800,Q}$$
$$+ [-(H^\circ_{1800} - H^\circ_{298})_{\text{Q}}] + [-(\Delta H^\circ_{f,298,Q})] = 0$$

By rearranging this we obtain the standard heat of formation of quartz at 298 K,

$$H^\circ_{f,298,Q} = H^\circ_{f,1800,Q} - (H^\circ_{1800} - H^\circ_{298})_{\text{Q}}$$
$$+ [(H^\circ_{1800} - H^\circ_{298})_{\text{Si}} + (H^\circ_{1800} - H^\circ_{298})_{\text{O}_2}] \quad (7\text{-}10)$$

Most of the enthalpy data presented in Table 7-1 have been collected at temperatures other than 298 K and have had to be corrected to this standard temperature. Moreover, many

of the reactions have followed more complicated paths than ones simply requiring heating and cooling. Some have involved dissolving the elements and minerals in acid or high-temperature metallic melts. The heats of solution in these solvents have then be used to calculate the $\Delta H^\circ_{f,298}$. Again, because enthalpy is a state function, the actual path taken in the reaction does not affect the enthalpy change of the overall reaction, which depends only on the initial and final states.

The enthalpy data presented in Table 7-1 are normally used to calculate the enthalpy of formation of a mineral at a particular temperature of interest. For example, we might wish to know the enthalpy of formation of quartz at 1800 K and 10^5 Pa pressure. This can be determined by rearranging Eq. 7-10 and inserting the value for the heat of formation of quartz at 298 K obtained from thermodynamic tables. This gives

$$H^\circ_{f,1800,Q} = \Delta H^\circ_{f,298,Q} + (H^\circ_{1800} - H^\circ_{298})_{\text{Q}}$$
$$- [(H^\circ_{1800} - H^\circ_{298})_{\text{Si}} + (H^\circ_{1800} - H^\circ_{298})_{\text{O}_2}] \quad (7\text{-}11)$$

High-temperature heats of formation can also be read directly from thermodynamic tables, but care must be exercised in doing this. Enthalpy data are commonly presented in two different ways. One lists values calculated according to Eq. 7-11 (Robie et al. 1978). The other also uses Eq. 7-11, but the enthalpy terms for the elements (those in brackets) are dropped (Helgeson et al. 1978); enthalpies calculated this way are referred to as *apparent enthalpies of formation* from the elements. A simple example will illustrate the justification for dropping these terms. To determine the enthalpy change associated with the transformation from low to high quartz at 848 K, we need know only the difference in the heats of formation from the elements of these two forms of quartz at this temperature. Values obtained from Eq. 7-11 for the two polymorphs will have identical terms for the elements (those in brackets) because the same elements are involved in both minerals. These terms, therefore, cancel when we take the difference in the enthalpies of formation of the two polymorphs. The same argument applies to other more complicated reactions, because the same elements are present on both sides of the reaction.

To calculate a high-temperature heat of formation from Eq. 7-11, the change in enthalpy associated with changes in temperature [e.g., $(H^\circ_{1800} - H^\circ_{298})_{\text{Q}}$ in Eq. 7-11] must be evaluated. This can be determined from the heat capacity of the mineral at constant pressure (C_p), because

$$\int_{H^\circ_{298}}^{H^\circ_T} dH = \int_{298}^{T} C_p \, dT \quad (7\text{-}12)$$

Heat capacities, however, vary with temperature; thus C_p must be expressed as a function of T before Eq. 7-12 can be integrated. Variations in the heat capacity of most minerals can be fitted to an expression of the form

$$C_p = a + bT - \frac{c}{T^2} \quad (7\text{-}13)$$

Values of these coefficients are given in Table 7-1 for the common minerals. Substitution of Eq. 7-13 into Eq. 7-12 gives, on

TABLE 7-1

Thermodynamic data for common minerals at 298.15 K and 10^5 Pa (1 bar)

Name	Formula	Formula Weight (kg)	Volume ($m^3\ mol^{-1} \times 10^3$)	$S°$ ($J\ mol^{-1}\ K^{-1}$)	$\Delta H°_f$ ($kJ\ mol^{-1}$)	$\Delta G°_f$ ($kJ\ mol^{-1}$)	a ($J\ mol^{-1}\ K^{-1}$)	$b \times 10^3$ ($J\ mol^{-1}\ K^{-2}$)	$c \times 10^{-5}$ ($JK\ mol^{-1}$)	Temperature Range (K)	$\Delta V°$ ($m^3\ mol^{-1} \times 10^3$)	$\Delta S°$ ($J\ mol^{-1}\ K^{-1}$)	$\Delta H°$ ($J\ mol^{-1}$)
Akermanite	$Ca_2MgSi_2O_7$	.27264	.09281	209.33	−3878.304	−3681.092	251.42	47.70	47.70				
Albite	$NaAlSi_3O_8$	.262224	.10025	207.15	−3931.621	−3708.313	258.15	58.16	62.80	298–473			
							342.59	14.87	209.84	473–1200			
Analcite	$NaAlSi_2O_6 \cdot H_2O$	.220155	.0971	234.30	−3306.168	−3088.202	223.80	101.00	37.15				
Andalusite	Al_2SiO_5	.162046	.05153	92.88	−2576.783	−2429.176	172.84	26.33	51.85				
Andradite	$Ca_3Fe_2Si_3O_{12}$	.505184	.13185	293.42	−5778.125	−5428.652	475.02	65.42	129.24				
Annite	$KFe_3(AlSi_3O_{10})(OH)_2$	.51189	.15432	398.32	−5155.504	−4799.701	445.30	124.56	80.79				
Anorthite	$CaAl_2Si_2O_8$	.27821	.10079	205.43	−4216.518	−3992.783	264.89	61.90	64.60				
Anthophyllite	$Mg_7Si_8O_{22}(OH)_2$	.780872	.2644	538.06	−12086.526	−11361.359	755.97	253.44	160.93	298–903			
							826.52	174.11	55.82	903–1258			
							834.80	174.11	55.82	1258–1800			
Antigorite	$Mg_{48}Si_{34}O_{85}(OH)_{62}$	4.536299	1.74913	3603.93	−71424.608	−66140.756	5139.83	2149.57	1199.47	298–848			
							5166.53	2097.19	1176.88	848–1000			
Aragonite	$CaCO_3$	.100089	.03415	90.21	−1208.017	−1129.157	84.22	42.84	13.97				
Brucite	$Mg(OH)_2$	.058327	.02463	63.14	−926.296	−835.319	101.13	16.79	25.56				
Calcite	$CaCO_3$	.100089	.036934	92.68	−1208.222	−1130.098	104.52	21.92	25.94				
Carbon dioxide	CO_2	.04401	24.465	213.69	−393.522	−394.392	44.22	8.79	8.62				
Chalcedony	SiO_2	.060085	.022688	41.34	−909.108	−854.691	46.94	34.31	11.30				
Chrysotile	$Mg_3Si_2O_5(OH)_4$	.277134	.1085	221.33	−4364.427	−4037.020	317.23	132.21	73.55				
Clinochlore 14-A	$Mg_5Al(AlSi_3O_{10})(OH)_8$	.555832	.20711	465.26	−8857.377	−8207.765	696.64	176.15	156.77				
Clinochlore 7-A	$Mg_5Al(AlSi_3O_{10})(OH)_8$	.555832	.2115	445.60	−8841.616	−8188.511	681.24	211.79	171.04	298–848			
							694.59	185.60	159.75	848–900			
Clinozoisite	$Ca_2Al_3Si_3O_{12}(OH)$	.622882	.1362	295.56	−6879.421	−6483.861	444.00	105.50	113.57				
Coesite	SiO_2	.060085	.020641	40.38	−906.313	−851.616	46.02	34.31	11.30	298–848			
α													
β							59.37	8.12	.00	848–2000			
Cordierite	$Mg_2Al_3(AlSi_5O_{18})$	.584969	.23322	407.23	−9134.505	−8624.391	601.78	107.95	161.50				
Cordierite (hydr.)	$Mg_2Al_3(AlSi_5O_{18}) \cdot H_2O$	.602984	.24122	466.22	−9437.748	−8875.728	649.48	107.95	161.50				
Corundum	Al_2O_3	.101961	.025575	50.96	−1661.655	−1568.264	115.02	11.80	35.06				
Cristobalite	SiO_2	.060085	.02738	50.05	−902.384	−850.565	72.76	1.30	41.38				
Diopside	$CaMg(SiO_3)_2$	.21656	.06609	143.09	−3203.262	−3029.216	220.92	32.80	65.86				
Dolomite	$CaMg(CO_3)_2$	.184411	.064365	155.18	−2329.865	−2167.228	173.87	100.22	41.35				
Enstatite	$MgSiO_3$												
Clino		.100396	.031276	67.78	−1546.766	−1459.923	102.72	19.83	26.28	298–903	.00002	.770	695
Ortho							120.35	.00	.00	903–1258	.00109	1.297	1632
Proto							122.42	.00	.00	1258–1800			
Epidote	$Ca_2FeAl_2Si_3O_{12}(OH)$	.651747	.1392	314.97	−6461.903	−6072.432	492.13	53.62	133.32				
Fayalite	Fe_2SiO_4	.203778	.04639	148.32	−1481.634	−1381.695	152.76	39.16	28.03				
Ferrosilite	$FeSiO_3$												
Clino		.131931	.032952	94.56	−1195.055	−1117.797	110.83	21.21	23.22	298–413	.000056	.377	155
Ortho							87.86	37.66	.00	413–1400			
Forsterite	Mg_2SiO_4	.140708	.04379	95.19	−2175.680	−2056.704	149.83	27.36	35.65				
Gehlenite	$Ca_2Al_2SiO_7$	.274205	.09024	201.25	−3981.766	−3780.612	266.69	33.47	63.26				
Graphite	C	.012011	.0052982	5.74	0	0	16.86	4.77	8.54				
Grossularite	$Ca_3Al_2Si_3O_{12}$	.450454	.1253	254.68	−6624.933	−6263.310	435.21	71.18	114.30				

Mineral	Formula												
Hedenbergite	CaFe(SiO$_3$)$_2$	.248106	.06827	170.29	-2838.827	-2674.488	229.33	34.18	62.80				
Hematite	Fe$_2$O$_3$	.159692	.030274	87.61	-827.260	-745.401							
α							98.28	77.82	14.85	298-950	.703		669
β							150.62	.00	.00	950-1050			
γ							132.67	7.36	.00	1050-1800			
Hydrogen	H$_2$	.002016	24.465	130.57	0	0	27.28	3.26	-.50				
Hydrogen sulfide	H$_2$S	.0340799	24.465	205.69	-20.627	-33.539	32.68	12.38	1.92				
Jadeite	NaAl(SiO$_3$)$_2$	.20214	.0604	133.47	-3021.033	-2842.798	201.50	47.78	49.66				
K-feldspar	KAlSi$_3$O$_8$	.278337	.10887	213.93	-3971.403	-3746.245	320.57	18.04	125.29				
Kalsilite	KAlSiO$_4$	.158167	.05989	133.26	-2131.363	-2015.642							
α							123.14	72.63	22.26	298-810	.828		669
β							177.82	.00	.00	810-1800			
Kaolinite	Al$_2$Si$_2$O$_5$(OH)$_4$	.258161	.09952	203.05	-4109.613	-3789.089	304.47	122.17	90.04				
Kyanite	Al$_2$SiO$_5$	.162046	.04409	83.68	-2581.057	-2430.720	173.19	28.52	53.90				
Laumontite	CaAl$_2$Si$_4$O$_{12}$·4H$_2$O	.470441	.20755	485.76	-7233.661	-6682.028	515.47	186.06	68.74				
Lawsonite	CaAl$_2$Si$_2$O$_7$(OH)$_2$·H$_2$O	.3142	.10132	233.47	-4846.428	-4492.060	342.25	97.74	68.03				
Magnesite	MgCO$_3$	.084321	.028018	65.69	-1111.396	-1027.833	82.55	52.46	19.87				
Magnetite	Fe$_3$O$_4$	.231539	.044524	145.73	-1118.174	-1014.930							
α							91.55	201.67	.00	298-900	.000		0
β							200.83	.00	.00	900-1800			
Margarite	CaAl$_2$(Al$_2$Si$_2$O$_{10}$)(OH)$_2$	.398187	.1294	266.94	-6217.520	-5834.044	428.86	68.41	117.36	298-848			
							415.51	94.60	128.66	848-1000			
Merwinite	Ca$_3$Mg(SiO$_4$)$_2$	.328719	.1044	253.13	-4566.602	-4339.586	305.31	50.04	60.42				
Methane	CH$_4$	.016043	24.465	186.15	-74.810	-50.739	23.64	47.86	1.92				
Microcline (max.)	KAlSi$_3$O$_8$	.278337	.108741	213.93	-3971.403	-3746.245	267.06	53.97	71.34				
Monticellite	CaMgSiO$_4$	.156476	.05147	110.46	-2262.707	-2145.677	154.05	22.34	33.47				
Muscovite	KAl$_2$(AlSi$_3$O$_{10}$)(OH)$_2$	.398313	.14071	287.86	-5972.275	-5591.083	408.19	110.37	106.44				
Nepheline	NaAlSiO$_4$	.145217	.05516	124.35	-2093.008	-1978.496	150.24	27.02	30.66				
Paragonite	NaAl$_2$(AlSi$_3$O$_{10}$)(OH)$_2$	.382201	.13253	277.82	-5928.573	-5548.034	407.65	102.51	110.62				
Pargasite	NaCa$_2$Mg$_4$Al(Al$_2$Si$_6$O$_{22}$)(OH)$_2$	.835858	.2735	669.44	-12623.356	-11912.551	861.07	174.31	210.08				
Periclase	MgO	.040311	.011248	26.94	-601.659	-569.384	42.59	7.28	6.19				
Phlogopite	KMg$_3$(AlSi$_3$O$_{10}$)(OH)$_2$	.417286	.14966	318.40	-6226.072	-5841.646	420.95	120.42	89.96				
Prehnite	Ca$_2$Al(AlSi$_3$O$_{10}$)(OH)$_2$	.412389	.14033	271.96	-6201.060	-5818.007	383.25	158.24	82.01				
Pyrophyllite	Al$_2$Si$_4$O$_{10}$(OH)$_2$	.360316	.1266	239.32	-5628.790	-5255.091	423.30	79.66	48.12				
Pyroxene Ca-Al	CaAl$_2$SiO$_6$	.218125	.0635	146.44	-3280.310	-3105.729	332.34	164.07	72.31				
Quartz	SiO$_2$	.060085	.022688	41.34	-910.648	-856.239							
α							46.94	34.31	62.34	298-848	1.431	.000372	1213
β							60.29	8.12	11.30	848-2000			
Sanidine (high)	KAlSi$_3$O$_8$	.277337	.109008	228.15	-3960.315	-3739.400	267.06	53.97	71.34				
Sepiolite	Mg$_4$Si$_6$O$_{15}$(OH)$_2$(OH$_2$)$_2$·(OH)$_4$	.647861	.2856	613.37	-10116.912	-9251.627	660.74	436.39	78.16				
Silica (amorphous)	SiO$_2$·nH$_2$O	.060085	.029	60.00	-897.753	-848.900	24.81	197.48	94.98				
Silica glass	SiO$_2$	.060085	.02727	47.40	-903.200	-850.559	*	*	*				
Sillimanite	Al$_2$SiO$_5$	.162046	.0499	96.78	-2573.574	-2427.101	167.46	30.92	48.84				
Spinel	MgAl$_2$O$_4$	.142273	.03971	80.63	-2238.008	-2163.153	153.86	26.84	40.62				
Steam	H$_2$O	.0180153	24.465	188.72	-241.818	-228.589	30.54	10.29	.00				
Sulfur	S$_2$	.064128	24.465	228.07	128.365	79.329	36.48	.67	3.77				
Talc	Mg$_3$Si$_4$O$_{10}$(OH)$_2$	.379289	.13625	260.83	-5903.289	-5523.667	345.10	174.10	55.81				
Tremolite	Ca$_2$Mg$_5$Si$_8$O$_{22}$(OH)$_2$	.81241	.27292	548.90	-12319.656	-11592.546	787.52	239.72	187.54				
Wairakite	CaAl$_2$Si$_4$O$_{12}$·2H$_2$O	.434411	.18687	439.74	-6608.850	-6182.496	420.07	186.06	68.74				
Wollastonite	CaSiO$_3$	.116164	.03993	82.01	-1630.965	-1545.758	111.46	15.06	27.28				
Zoisite	Ca$_2$Al$_3$Si$_3$O$_{12}$(OH)	.622882	.1359	295.98	-6879.044	-6483.606	444.00	105.50	113.57				

(Modified from Helgeson et al. 1978. Reprinted by permission of American Journal of Science. Data for silica glass from Robie et al. 1978)

* C_p^0 = 74.639 − (7.2594 × 10^{-3}T) + (5.5704 × 10^{-6}T^2) − (3.1140 × 10^{6}T^{-2}) (valid from 298 to 1500 K).

integration,

$$H_T^\circ - H_{298}^\circ = a(T - 298) + \frac{b}{2}(T^2 - 298^2)$$

$$+ c\left(\frac{1}{T} - \frac{1}{298}\right) \quad (7\text{-}14)$$

Calculation of the high-temperature enthalpy of formation of a mineral is therefore a simple matter using Eqs. 7-11 and 7-14 and the data in Table 7-1 (Prob. 7-3). Because the calculations are tedious, they are best carried out by computer; they can be handled easily on the simple spreadsheet.

The actual reactions involved in forming minerals from the elements are of little interest in themselves, as most do not occur in nature. Elemental silicon, for example, is never found reacting with oxygen to form quartz. The enthalpies of these reactions, however, can be used to calculate the enthalpies of reaction (ΔH_r°) between other minerals, and herein lies the value of the standard heats of formation. To illustrate this, consider the petrologically important reaction of olivine with quartz to form orthopyroxene:

(Forsterite)	(Quartz)		(Clinoenstatite)
Mg_2SiO_4	$+ \quad SiO_2$	$\xrightarrow{\Delta H_{r,298}^\circ}$	$2\,MgSiO_3$
$\uparrow$	$\uparrow$		$\downarrow$
-2175.7	-910.6		-1546.8
$2Mg + Si + 2O_2$	$Si + O_2$		$2(Mg + Si + \tfrac{3}{2}O_2)$

$H_{f,298}^\circ$ (kJ mol^{-1})

For each of these minerals it is possible to write a reaction for their formation from the elements. The enthalpies of these reactions are obtained directly from Table 7-1. These reactions provide another path between the reactants and products. Clinoenstatite, for example, could be broken down into its constituent elements; these elements could then be recombined to form forsterite and quartz; reaction of forsterite with quartz returns us to clinoenstatite. Because enthalpy is a state property, its integral around this cycle must be zero. We can determine the enthalpy change of the reaction ($\Delta H_{r,298}^\circ$), then, by summing all of these terms as we proceed around the cycle in one direction—clockwise, for example. In doing this care must be taken to keep the signs of the enthalpy changes correct. Table 7-1 indicates that the enthalpy of formation of a mineral from the elements is negative; that is, heat is liberated into the surroundings when elements are combined to form the mineral. If the reaction takes place in the opposite direction, that is, the mineral breaks down into the elements, the enthalpy change must be positive. Note must also be taken of the number of moles of each mineral involved in the reaction. For this cycle we can write

$$\Delta H_{r,298}^\circ - 2(-1546.8) + (-910.6) + (-2175.7) = 0$$

from which it follows that $\Delta H_{r,298}^\circ = -7.3$ kJ mol^{-1} of olivine. The enthalpy change of this reaction at higher temperatures can be calculated using the enthalpies of formation of the minerals at the higher temperatures (Prob. 7-4).

7-5 SECOND LAW OF THERMODYNAMICS

Determination of the enthalpy change accompanying a reaction is the first step to understanding the conditions under which a reaction will take place. There remains the important question of the direction of the reaction. Will mineral A change into B, or will B change into A? We know from experience that many everyday processes have a definite direction to them. When cream is stirred in coffee, mixing occurs; if the direction of stirring is reversed, the coffee and cream do not unmix. It is therefore a matter of experience that the process of stirring results in mixing; the opposite is never observed. This implies that some fundamental principle governs the direction of the process. If this principle can be determined, it could be used to indicate the directions of reactions with which we do not have everyday experience, such as those occurring in the Earth. This principle is embodied in the second law of thermodynamics, and it involves an obscure property of material known as entropy (S).

In discussing the first law of thermodynamics it was emphasized that the heat involved with a change from one state to another has no definite value unless the path is specified. This can be illustrated by considering different ways in which a gas can be expanded from one state to another. Imagine that this is done reversibly, although we know in reality such a process would take infinite time. Consider first the isothermal expansion of the gas from an initial state A to a final state B (Fig. 7-4). During this expansion a quantity of heat, Q_4, is absorbed from the surroundings in order to keep the temperature constant. If the temperature is not maintained constant during

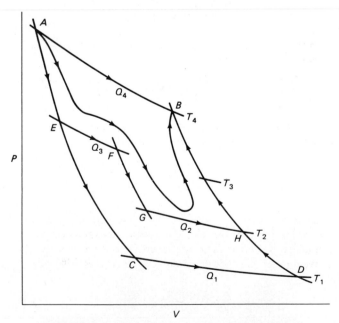

FIGURE 7-4 Various paths that can be followed by a gas changing its pressure and volume from state A to B. The curves labeled T_1 to T_4 are isothermal lines arranged in order of increasing temperature, whereas the steeper curves are adiabatic lines.

the expansion, many other paths can be followed between A and B. For example, the gas could expand in an insulated container where no heat would be absorbed from the surroundings, and as a result its temperature would fall, for example to T_1 (C in Fig. 7-4). Such a change, in which no heat is transferred in or out of the system, is said to be *adiabatic*. From point C the gas could expand isothermally to point D with the absorption of a quantity of heat Q_1. Adiabatic compression would then take the gas to point B. The expansion of the gas from A to B could also involve several isothermal steps, such as the path $AEFGHB$, in which two quantities of heat, Q_3 and Q_2, are absorbed along the isothermal lines T_3 and T_2, respectively. Even when the path does not follow adiabatic or isothermal lines, as along the irregular curve in Figure 7-4, the path can be treated as a large number of infinitesimal adiabatic and isothermal steps, with each of the latter involving the absorption of a quantity of heat, dQ, so that the total heat absorbed between A and B is $\int_A^B dQ_{rev}$. The subscript $_{rev}$ indicates that the process takes place reversibly.

Each one of the paths between A and B in Fig. 7-4 involves a different quantity of heat; that is,

$$Q_4 \neq Q_1 \neq (Q_3 + Q_2) \neq \int_A^B dQ_{rev}$$

Because the amount of heat involved in the change from A to B depends on the path followed, heat cannot be a state property. If, however, the various quantities of heat are divided by the absolute temperatures at which the heat absorption takes place, a function is created that is independent of the path followed. This function must, therefore, be a state function. Thus,

$$\frac{Q_4}{T_4} = \frac{Q_1}{T_1} = \frac{Q_3}{T_3} + \frac{Q_2}{T_2} = \frac{dQ_{rev}}{T}$$

This state property is given the name *entropy* and symbol S. It is defined by the equation

$$dS \equiv \frac{dQ_{rev}}{T} \qquad (7\text{-}15)$$

As with any state property, the total change in entropy accompanying a reversible cycle is zero; that is,

$$\oint dS = \oint \frac{dQ_{rev}}{T} = 0$$

A reversible cycle, however, is not a natural one. In an irreversible or natural cycle the amount of heat generated in the surroundings is greater than in a reversible one (review the sign convention in Fig. 7-1). Consequently, we can write

$$\oint \frac{dQ_{irrev}}{T} < \oint \frac{dQ_{rev}}{T} = \oint dS = 0 \qquad (7\text{-}16)$$

or in general form

$$dS \geq \frac{dQ}{T} \qquad (7\text{-}17)$$

where the equality sign applies to the reversible case and the inequality sign to the irreversible one. For a real reaction to take place (irreversible) dQ/T must be less than dS. Also, for a process taking place in an isolated system dQ_{irrev} must be zero, and therefore $dS > 0$. *That is, for a real reaction to occur in an isolated system entropy must increase. The reaction will continue until equilibrium is attained, at which point dS becomes zero and the entropy is a maximum.* This statement is but one of many different ways of expressing the *second law of thermodynamics*.

It is important to emphasize that in Eq. 7-15, entropy is defined using the heat involved in a *reversible* reaction. The fact that a reversible reaction is not possible does not invalidate the definition. Entropy is a state function, and therefore its change in value depends only on the initial and final states and not on the path of the reaction or whether it was carried out reversibly or irreversibly. All natural reactions are irreversible, and this simply means that $dQ_{irrev}/T < dS$.

We will see later how entropy can be measured. But first we will investigate the physical significance of entropy.

7-6 ENTROPY

Thermodynamic terms such as pressure, work, and heat are familiar from everyday experiences, but entropy, despite its importance, is not. But our expectations that stirring cream in coffee will cause mixing, or that oxygen in the air is unlikely to suddenly all move to one end of a room, or that heat will flow from high to low temperatures are based on processes in which entropy strives for a maximum. This suggests that entropy is a measure of the degree of randomness in a system. This relation was first formalized by the Austrian physicist Boltzmann, who showed that entropy can be defined in terms of the number of possible arrangements of the particles constituting a system. Entropy, so defined, is given by

$$S = k \ln \Omega \qquad (7\text{-}18)$$

where k is the Boltzmann constant (gas constant per molecule, $R/N_0 = 1.3806 \times 10^{-23}$ J K^{-1}) and Ω is the number of possible arrangements. From this relation, entropy is clearly related to the amount of disorder or randomness in a system.

To illustrate this relation, we will consider the entropy change accompanying the transformation from low- to high-temperature albite. Albite is a framework silicate with four asymmetric tetrahedral sites, three occupied by silicon and one by aluminum. In the low-temperature form, aluminum enters one specific site, but at high temperature it may be in any of the four sites. Consequently, the high-temperature form has a greater capacity for randomness and should therefore have the higher entropy.

In low-temperature albite the atoms can be arranged in only one way. Admittedly, the silicon atoms could be switched around in the silicon sites, but silicon atoms are indistinguishable, and thus this would not result in distinguishable arrangements. The entropy due to occupancy of the tetrahedral sites in low albite is therefore

$$S_{low\ Ab} = k \ln 1 = 0$$

At high temperature the aluminum can enter any of the tetrahedral sites. But 1 mol of albite ($NaAlSi_3O_8$) contains N_0 (Avogadro's number = 6.022×10^{23}) atoms of aluminum and $3N_0$ atoms of silicon that must be distributed over $4N_0$ tetrahedral sites. The number of possible ways of arranging these is

$$\Omega = \frac{(4N_0)!}{(N_0)!\,(3N_0)!}$$

Hence,

$$S_{\text{high Ab}} = k[\ln(4N_0)! - \ln(N_0)! - \ln(3N_0)!]$$

Because N is very large we can use Stirling's approximation, that is,

$$\ln N! = N \ln N - N$$

The entropy of the high-temperature form is then

$$S_{\text{high Ab}} = kN_0[4 \ln 4 - 3 \ln 3]$$

but $kN_0 = R$, the gas constant (8.31443 J K^{-1} mol^{-1}), so that

$$S_{\text{high Ab}} = 18.70 \text{ J mol}^{-1}\text{ K}^{-1}$$

Therefore, the entropy change due to the disordering of the aluminum and silicon in the tetrahedral sites is $S_{\text{high Ab}} - S_{\text{low Ab}}$, which is 18.70 J mol^{-1} K^{-1}.

It should be emphasized that this calculated entropy change is a maximum because no account is taken of any crystal chemical restrictions on the possible groupings of ions. Also, this calculation pertains only to the change in the configuration of the aluminum and silicon in the tetrahedral sites. The albite structure may have other sources of randomness which contribute to the absolute entropy of this mineral (see Prob. 7-5).

7-7   THIRD LAW OF THERMODYNAMICS AND THE MEASUREMENT OF ENTROPY

The third law of thermodynamics states that *the entropy of a pure, perfectly crystalline substance is zero at the absolute zero of temperature*. The entropy of such a substance at temperature T is then

$$S_T = \int_0^T \frac{dQ_{\text{rev}}}{T} = \int_0^T \frac{C_p}{T}\,dT \qquad (7\text{-}19)$$

where C_p, the heat capacity at constant pressure, is a readily measured physical property. Entropy is normally determined by graphically evaluating the integral in Eq. 7-19. This is done by plotting C_p/T versus T and measuring the area under the curve (Fig. 7-5).

The terms *pure* and *perfectly crystalline* in the third law are very important ones. Substances such as glass or intermediate composition plagioclase would still have entropy at absolute zero because of their randomness in structure (configurational entropy). For any substance the entropy can be

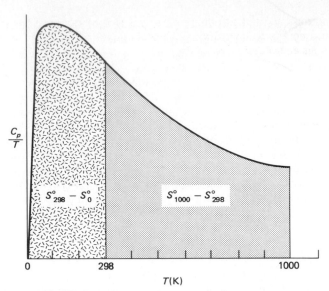

FIGURE 7-5 The entropy of a substance is given by the area under the curve in a plot of C_p/T versus T. Entropies of substances at 298 K and 10^5 Pa (1 bar) pressure (lightly shaded area) are listed in Table 7-1. Higher-temperature entropies can be obtained by adding the area under the higher-temperature part of the curve (darkly shaded area).

thought of as consisting of two parts, one thermal and the other configurational; that is,

$$S = S_{\text{thermal}} + S_{\text{configurational}}$$

The S_{thermal} becomes zero at the absolute zero of temperature for all substances, but the $S_{\text{configurational}}$ becomes zero at this temperature only for pure, perfectly crystalline substances. This need not concern us, however, for the entropy data at 298 K in Table 7-1 take this into account.

The entropy of a substance at high temperature and 10^5 Pa pressure can be calculated from the data in Table 7-1 through the following relation:

$$S_T^\circ = \int_{298}^T \frac{C_p}{T}\,dT + S_{298}^\circ \qquad (7\text{-}20)$$

The heat capacities at high temperatures, however, are given by Eq. 7-13, with the coefficients being listed in Table 7-1. Integration of Eq. 7-20, therefore, gives

$$S_T^\circ = \left[a \ln T + bT + \frac{c}{2T^2} \right]_{298}^T + S_{298}^\circ \qquad (7\text{-}21)$$

7-8   GIBBS EQUATION: THERMODYNAMIC POTENTIALS

The first law of thermodynamics gives the relations between the various forms of energy, whereas the second law gives the sense of direction for reactions by introducing the concept of entropy, which is given absolute values by the third law. These can now be combined into a general relation governing reactions and equilibrium.

Rearranging Eq. 7-5 gives

$$-dE - dW + dQ = 0$$

From the second law (Eq. 7-17) $T\,dS \geq dQ$; hence,

$$-dE - dW + T\,dS \geq 0$$

The work can be expressed as work of expansion ($P\,dV$) plus any other form of work (dU). Thus,

$$-dE - P\,dV - dU + T\,dS \geq 0 \qquad (7\text{-}22)$$

This is a general relation indicating that at equilibrium, which is equivalent to the reversible situation, the left-hand side of the equation must be zero, and for a spontaneous reaction it must be greater than zero.

Let us consider a reaction that takes place under constant pressure and constant temperature, the most common conditions encountered in petrologic problems. In this case, $P\,dV$ and $T\,dS$ can be written as $d(PV)$ and $d(TS)$. Equation 7-22 therefore becomes

$$-dE - d(PV) + d(TS) \geq dU$$

or

$$-d(E + PV - TS) \geq dU \qquad (7\text{-}23)$$

The combination of terms ($E + PV - TS$) is a state variable and is given the name *Gibbs free energy* (G); that is,

$$G \equiv E + PV - TS = H - TS \qquad (7\text{-}24)$$

Equation 7-23 becomes

$$-dG \geq dU$$

and in the case where only work of expansion is done, which is the most common geological situation,

$$-dG \geq 0 \qquad (7\text{-}25)$$

Thus, for a spontaneous reaction (irreversible) to occur at constant P and T, $-dG$ must be positive; that is, the free energy must decrease. The reaction will proceed until equilibrium is attained (reversible), at which point $-dG = 0$, and the free energy is a minimum.

Although various forms of energy are transferred during a reaction proceeding at constant P and T, it is the Gibbs free energy that controls the feasibility and direction of the reaction. This is illustrated graphically in Figure 7-6. The change in internal energy (ΔE) accompanying a reaction taking place at a particular pressure and temperature consists of the work done on the system ($-P\,dV$) and the enthalpy change (ΔH). The enthalpy change can be further subdivided into thermal energy

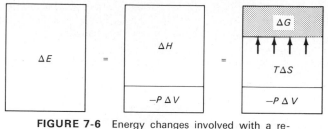
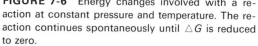

FIGURE 7-6 Energy changes involved with a reaction at constant pressure and temperature. The reaction continues spontaneously until $\triangle G$ is reduced to zero.

due to the entropy change ($T\,\Delta S$) and the change in free energy (ΔG). As the reaction proceeds, entropy is always increasing and striving for a maximum consonant with the state of the system; hence, $T\,\Delta S$ is also increasing and does so at the expense of ΔG. The free-energy change is therefore that part of the heat removed from the surroundings that can be used to increase the randomness of the system. As the reaction proceeds and the system becomes more random, this fraction of the enthalpy change decreases and becomes zero at equilibrium. At this point it is clear from Figure 7-6 that $T\,\Delta S = \Delta H$. This also follows directly from Eq. 7-24, which, for an infinitesimal change, would be

$$dG - dE + P\,dV + V\,dP - T\,dS \quad S\,dT \qquad (7\text{-}26)$$

But, if P and T are constant,

$$dG = dE + P\,dV - T\,dS$$

which, from Eq. 7-8, gives

$$dG = dH - T\,dS$$

which, in turn, for a finite change, becomes

$$\Delta G = \Delta H - T\,\Delta S \qquad (7\text{-}27)$$

At equilibrium $\Delta G = 0$; hence,

$$T\,\Delta S = \Delta H$$

or

$$T_{\text{equil}} = \frac{\Delta H}{\Delta S} \qquad (7\text{-}28)$$

Equation 7-28 gives a simple means of determining the equilibrium temperature for a reaction. Consider, for example, the reaction of cristobalite melting at 10^5 Pa pressure to form silica liquid. Data for this reaction at high temperature indicate that ΔH is $+8071$ J mol^{-1} and ΔS is $+4.05$ J mol^{-1} K^{-1}. Therefore, the equilibrium temperature or melting point would be $T_{\text{equil}} = \Delta H/\Delta S = 1992$ K or 1719°C.

The equilibrium temperature for any reaction, such as that for the melting of cristobalite, is the temperature at which $\Delta G = 0$. This is shown in Figure 7-7 as the point of intersection

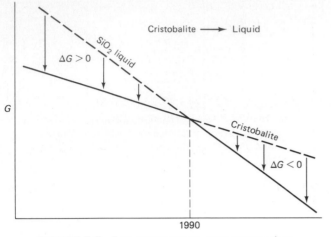

FIGURE 7-7 Free energy versus temperature plots at constant pressure for cristobalite and silica liquid. The intersection of the two curves is the melting point of cristobalite.

of the free-energy curves for cristobalite and silica liquid (glass at low temperature). At higher temperatures, ΔG is negative; hence, the reaction proceeds with the melting of cristobalite. Below this temperature, ΔG is positive, so the reaction can proceed only in the opposite direction, causing cristobalite to crystallize. The most stable form is always the one with the lowest free energy. Of course, thermodynamics indicates only what the equilibrium state should be, but kinetic factors may prevent this from being achieved. Volcanic glasses, for example, could lower their free energy by crystallizing, but the kinetics of this process are slow at low temperatures.

For many petrologic problems we need to calculate the ΔG of reaction under conditions different from those for which the data in Table 7-1 are applicable. To do this it is necessary to know how free energy changes with temperature and pressure. From Eq. 7-22 it is clear that for a reversible reaction involving only work of expansion, dE is equivalent to $T \, dS - P \, dV$, which can be substituted for dE in Eq. 7-26, giving

$$dG = T \, dS - P \, dV + P \, dV + V \, dP - T \, dS - S \, dT$$

which reduces to

$$dG = -S \, dT + V \, dP \qquad (7\text{-}29)$$

In a closed system, that is, one in which no matter is transferred in or out, G is a function only of temperature and pressure $[G = f(T, P)]$. We can express the total change in G (dG) resulting from a change in T and P as the sum of the change due to T and the change due to P. This is known as a *total differential* of the function, and it is represented as follows:

$$dG = \left(\frac{\partial G}{\partial T}\right)_P dT + \left(\frac{\partial G}{\partial P}\right)_T dP \qquad (7\text{-}30)$$

The terms in parentheses are known as *partial derivatives*, as they denote the variation in G with respect to only one of the two variables, while the other variable, shown as a subscript outside the parentheses, is held constant. Comparison of Eqs. 7-30 and 7-29 reveals that

$$\left(\frac{\partial G}{\partial T}\right)_P = -S \qquad (7\text{-}31)$$

and

$$\left(\frac{\partial G}{\partial P}\right)_T = \bar{V} \qquad (7\text{-}32)$$

The bar over the $\bar{V}$ signifies molar volume. This is introduced because values of G are typically given as molar quantities. Because all substances have positive entropy, free energy always decreases with increasing temperature at constant pressure, and because liquids have higher entropies than corresponding solids, their free energy decreases more rapidly than that of solids (Fig. 7-7). Molar volumes are also always positive, therefore increasing pressure at constant temperature causes the free energy to rise.

Similar relations can be derived for the free-energy change of a reaction. The change in the G of a reaction with temperature is

$$\left(\frac{\partial \Delta G}{\partial T}\right)_P = -\Delta S \qquad (7\text{-}33)$$

and the change with pressure is

$$\left(\frac{\partial \Delta G}{\partial P}\right)_T = \Delta V \qquad (7\text{-}34)$$

In these cases, the change in the ΔG of the reaction is determined by the entropy change and volume change of the reaction. Both of these terms can be either positive or negative, so generalizations about the variation of the ΔG of a reaction cannot be made. However, in the case of reactions that evolve a gas, ΔS and ΔV will both be positive, and therefore the ΔG of reaction will decrease with increasing temperature and increase with increasing pressure.

Other useful relations can be derived from Eq. 7-30 simply by utilizing two properties of a total differential, that is the cross-derivative rule and the cyclic rule. We have seen by comparing Eqs. 7-29 and 7-30 that

$$\left(\frac{\partial G}{\partial T}\right)_P = -S \qquad \text{and} \qquad \left(\frac{\partial G}{\partial P}\right)_T = V$$

If we take derivatives of these expressions with respect to the variable held constant, we obtain

$$\left[\frac{\partial}{\partial P}\left(\frac{\partial G}{\partial T}\right)_P\right]_T = -\left(\frac{\partial S}{\partial P}\right)_T \qquad \text{and} \qquad \left[\frac{\partial}{\partial T}\left(\frac{\partial G}{\partial P}\right)_T\right]_P = \left(\frac{\partial V}{\partial T}\right)_P$$

But from the *cross-derivative rule* for a total derivative that is exact,

$$\left[\frac{\partial}{\partial P}\left(\frac{\partial G}{\partial T}\right)_P\right]_T = \left[\frac{\partial}{\partial T}\left(\frac{\partial G}{\partial P}\right)_T\right]_P \qquad (7\text{-}35)$$

Therefore,

$$-\left(\frac{\partial S}{\partial P}\right)_T = \left(\frac{\partial V}{\partial T}\right)_P$$

But from Eq. 2-2, $(\partial V/\partial T)_P = \alpha V$, where α is the isobaric coefficient of thermal expansion. Therefore,

$$\left(\frac{\partial S}{\partial P}\right)_T = -\alpha V \qquad (7\text{-}36)$$

This allows us to evaluate the variation in entropy with pressure from two easily measured physical properties, α and V.

Similarly, by differentiating Eq. 7-32, we obtain

$$\left(\frac{\partial^2 G}{\partial P^2}\right)_T = \left(\frac{\partial V}{\partial P}\right)_T$$

but from Eq. 3-1, $(\partial V/\partial P)_T = -\beta V$, where β is the isothermal coefficient of compressibility. Therefore,

$$\left(\frac{\partial^2 G}{\partial P^2}\right)_T = -\beta V \qquad (7\text{-}37)$$

And by differentiating Eq. 7-31 we obtain

$$\left(\frac{\partial^2 G}{\partial T^2}\right)_P = -\left(\frac{\partial S}{\partial T}\right)_P$$

but $(\partial S/\partial T)_P = C_p/T$. Thus,

$$\left(\frac{\partial^2 G}{\partial T^2}\right)_P = -\frac{C_p}{T} \qquad (7\text{-}38)$$

The *cyclic rule* for total differentials states that if any three variables, x, y, and z, for example, are connected by a functional relation, the three partial derivatives satisfy the following relation:

$$\left(\frac{\partial x}{\partial y}\right)_z\left(\frac{\partial y}{\partial z}\right)_x\left(\frac{\partial z}{\partial x}\right)_y = -1 \qquad (7\text{-}39)$$

This rule is easily remembered by writing the three variables in any order in a row, and then repeating them below so that none of the vertical columns match. These vertical pairs give the partial derivatives, with the subscripted variable, which is held constant, being the third variable. For example,

$$\begin{matrix} zyx \\ xzy \end{matrix} \quad \text{becomes} \quad \left(\frac{\partial z}{\partial x}\right)_y\left(\frac{\partial y}{\partial z}\right)_x\left(\frac{\partial x}{\partial y}\right)_z = -1$$

If the cyclic rule is applied to the three variables P, T, and V, we obtain

$$\left(\frac{\partial P}{\partial T}\right)_V\left(\frac{\partial T}{\partial V}\right)_P\left(\frac{\partial V}{\partial P}\right)_T = -1 \qquad (7\text{-}40)$$

But from Eq. 2-2, $(\partial V/\partial T)_P = \alpha V$, and from Eq. 3-1, $(\partial V/\partial P)_T = -\beta V$, which, when substituted into Eq. 7-40, give

$$\left(\frac{\partial P}{\partial T}\right)_V\left(\frac{1}{\alpha V}\right)(-\beta V) = -1$$

Hence,

$$\left(\frac{\partial P}{\partial T}\right)_V = \frac{\alpha}{\beta} \qquad (7\text{-}41)$$

Thermodynamic relations, then, can be manipulated with the cyclic and cross-derivative rules into useful forms for specific applications.

7-9 FREE ENERGY OF FORMATION AT ANY TEMPERATURE AND PRESSURE

Finally, we will derive an expression for the free energy of formation of a phase at any temperature and pressure. We define the molar free energy of formation of a phase from the elements at 298.15 K and 10^5 Pa (1 bar) as

$$\text{free energy of formation} = \Delta H^\circ_{f,298} - 298\,\Delta S^\circ_{298} \qquad (7\text{-}42)$$

The ΔS in this expression refers to the difference in entropies of the phase and its constituent elements in their standard states $(S_{\text{phase}} - \sum S_{\text{elements}})$. In Sec. 7-4 it is shown that if the ΔH°_f is used to calculate the ΔH of a reaction between phases, the terms for the enthalpy of the elements on the reactant and product sides of a reaction cancel. The same is true for the entropies of the elements. The amount of calculation can thus be decreased by simply ignoring the enthalpies and entropies of the elements.

A free energy of formation that ignores the enthalpies and entropies of the elements is known as an *apparent free energy of formation*; it is given by

$$\Delta \bar{G}^\circ_{f,298} = \Delta H^\circ_{f,298} - 298.15 \times S^\circ_{298} \qquad (7\text{-}43)$$

where S°_{298} is the entropy of the phase. The apparent free energy of formation of a phase will be less than that given in Eq. 7-42 by an amount $298 \times \sum S_{\text{elements}}$. For example, the $\Delta \bar{G}^\circ_{f,298}$ of corundum calculated from Eq. 7-43 and values of $\Delta H^\circ_{f,298}$ and S°_{298} from Table 7-1 is $-1,676,849$ J mol^{-1}. But the free energy of formation of corundum given in column 7 of Table 7-1, which is calculated according to Eq. 7-42, is $-1,568,264$ J mol^{-1}. The difference between these two values (108,585 J mol^{-1}) is simply

$298 \times (\frac{3}{2}S_{O_2} + 2 \times S_{Al})$. Both free energies of formation are equally usable, but when it comes to determining values at higher pressures and temperatures, needless calculations are created by carrying the terms for the elements. We will therefore use the *apparent* free energy of formation. Readers should take care to ascertain which type of free energy of formation is being used when they consult other texts.

To calculate the apparent free energy of formation of a phase at temperature T (K) and pressure P (Pa), we first determine its value at standard conditions from Eq. 7-43. If the temperature is to be raised, appropriate terms must be added for changes in enthalpy, dH (Eq. 7-12), and entropy, dS (Eq. 7-20). If the pressure is to be changed, the free energy must also be adjusted according to Eq. 7-32. Combining all these terms into one expression gives

$$\Delta \bar{G}_{f,T,P} = \left(\Delta H^{\circ}_{f,298} + \int_{298}^{T} C_p \, dT \right)$$
$$- T \left(S^{\circ}_{298} + \int_{298}^{T} \left(\frac{C_p}{T} \right) dT \right) + \int_{10^5}^{P} V \, dP \quad (7\text{-}44)$$

Substituting the polynomial expression for C_p (Eq. 7-13) and integrating, we obtain

$$\Delta \bar{G}_{f,T,P} = \Delta H^{\circ}_{f,298} - TS^{\circ}_{298} + a(T - 298) + \frac{b}{2}(T^2 - 298^2)$$
$$+ c\left(\frac{1}{T} - \frac{1}{298} \right) - T\left[a \ln\left(\frac{T}{298} \right) + b(T - 298) \right.$$
$$+ \left. \frac{c}{2}\left(\frac{1}{T^2} - \frac{1}{298^2} \right) \right] + \int_{10^5}^{P} V \, dP \quad (7\text{-}45)$$

If a computer has not yet been used in solving problems in this text, Eq. 7-45 will rapidly convince the reader of its advantages. The last term in this equation has not been integrated. If pressure remains constant, the term becomes zero. Because the compressibilities of minerals and magmas are extremely small, V can be considered a constant, so when pressure does vary, this term becomes $V(P - 10^5)$. But for a gas, V is certainly not a constant, and the variation of V with P must be known before we can integrate this term. This problem is dealt with in Sec. 8-3.

PROBLEMS

7-1. If the molar volume of a peridotite is 5×10^{-5} m³, and its volume increase on totally melting is 10%, how much work is done when 1 mol of peridotite melts at a depth where the pressure is 2 GPa? Note that the system being considered is the 1 mol of rock. Be certain to get the sign convention correct (see Fig. 7-1).

7-2. If the molar volume of granitic magma is 7×10^{-5} m³, and on crystallizing it decreases by 10%, compare the work done by magma crystallizing near the top of a batholith, where the pressure is 0.05 GPa, with magma crystallizing near the base of the batholith, where the pressure is 0.5 GPa. Be careful of the sign convention.

7-3. From data in Table 7-1, and using Eq. 7-14, determine the enthalpy of formation of kyanite and andalusite at 466 K and 10^5 Pa (1 bar). If kyanite were to change into andalusite under these conditions, what would be the enthalpy of reaction? Is the reaction exothermic or endothermic? (Be careful of the sign convention; write the reaction kyanite → andalusite, then ΔH_r is the final enthalpy minus the initial.)

7-4. Using the ΔH°_f at 298 K and heat capacity data in Table 7-1, calculate the enthalpy of reaction at 10^5 Pa and 1500 K for the reaction

$$\text{forsterite} + \text{quartz} \longrightarrow 2 \text{ enstatite}$$

Be certain to take into account transformations that occur in both quartz and enstatite; each transformation has an enthalpy change.

7-5. In dolomite, calcium has two different possible sites to occupy at high temperatures, but at low temperatures it preferentially enters one of these sites, and magnesium occupies the other. Calculate the configurational entropy associated with the complete disordering of dolomite.

7-6. If entropy is a function of temperature and pressure, that is, $S = S(T, P)$, **(a)** write the total differential of $S(T, P)$, and **(b)**

show that the total differential of $S(T, P)$ is given by

$$dS = \left(\frac{C_p}{T} \right) dT - \alpha V \, dP$$

where α is the coefficient of thermal expansion (Eq. 2-2). [*Hint:* Use derivatives of $G(T, P)$.]

7-7. Convection within the mantle or within a magma chamber causes the thermal gradient to approach the adiabatic gradient ($dQ = 0$ and $dS = 0$). Using the expression for dS in Prob. 7-6b, along with the cyclical rule for partial derivatives, show that the adiabatic gradient is given by

$$\left(\frac{\partial T}{\partial P} \right)_S = \frac{T \bar{V} \alpha}{C_p}$$

7-8. The granitic magma in Prob. 7-2, which has a molar volume of 7×10^{-5} m³, has a molecular weight of 0.168 kg ($\rho = 2.4$ Mg m⁻³), a coefficient of thermal expansion, α, of 2×10^{-5} K⁻¹, a heat capacity, C_p, of 0.8 kJ kg⁻¹ K⁻¹, and a temperature of 900°C (1173 K).

  (a) If magmatic convection has established an adiabatic temperature gradient within the body, calculate the value of $(\partial T/\partial P)_S$ from the relation in Prob. 7-7. (Note 1 J = 1 Pa m³.)

  (b) Using the relation $dP = \rho g \, dz$ (Eq. 1-1), calculate the adiabatic temperature gradient, $(\partial T/\partial z)_S$, in this convecting body of magma, and compare this value with a typical geothermal gradient in the upper continental crust.

7-9. From the data in Table 7-1 and using Eq. 7-21, calculate the entropies of kyanite and andalusite at 466 K and 10^5 Pa. What entropy change accompanies the change of kyanite to andalusite under these conditions?

7-10. Using the results for ΔH° and ΔS° from Prob. 7-3 and 7-9, respectively, calculate the free-energy change, ΔG_r (Eq. 7-27), for the reaction of kyanite to andalusite at 466 K and 10^5 Pa. From the value of the ΔG_r°, what can you conclude about this reaction under these conditions of temperature and pressure?

7-11. For the reaction kyanite → andalusite at 466 K and 10^5 Pa calculated in Prob. 7-10, determine the change in the ΔG_r if the pressure is increased to 10^8 Pa at 466 K. Which of the minerals will be more stable under these new conditions? (Recall that $1\ J = 1\ Pa\ m^3$.)

7-12. For the reaction kyanite → andalusite calculated in Prob. 7-11, calculate how much the temperature would have to be increased at a pressure of 10^8 Pa in order to return the value of ΔG_r to zero, that is, to reestablish equilibrium. Assume that the entropy change for the reaction remains constant and is equal to the value calculated in Prob. 7-9.

7-13. Using Eq. 7-45, calculate the ΔG_f of calcite and of aragonite at 800 K and 10^5 Pa. Which phase is more stable under these conditions?

7-14. Using Eq. 7-45, calculate the ΔG_f of calcite and of aragonite at 298.15 K and 0.5 GPa. Which phase is more stable under these conditions?

8 / Free Energy and Phase Equilibrium

8-1 INTRODUCTION

In Chapter 7 our experiences of the behavior of energy in nature were formalized into laws dealing with the conservation of energy, the natural direction of processes, and the absolute zero of temperature. Once given mathematical expression, these laws were used to relate important thermodynamic functions to easily measured properties. The thermodynamic functions were, in turn, used to determine the direction of reactions and the conditions necessary for equilibrium. In this chapter we deal with the practical aspects of applying these thermodynamic principles to determining the equilibrium relations between minerals.

We have seen that energy can take various forms—mechanical, thermal, chemical, electrical. In each case, the energy can be expressed as the product of two terms, one *intensive* and the other *extensive*. Intensive ones can be measured at any point in a system and are independent of the extent of the system; they can be thought of as environmental factors. They include pressure, temperature, and chemical and electrical potentials. The extensive ones, on the other hand, must be integrated over the entire system. These include volume, entropy, number of moles, and electrical charge. Changes in energy can then be expressed as follows:

Mechanical	Thermal	Chemical	Electrical
$P\,dV$	$T\,dS$	$\mu_i\,dn_i$	$\mathscr{E}\,dQ$

where μ_i is the chemical potential of substance i (Chapter 9), n_i the number of moles of i, $\mathscr{E}$ the electrical potential, Q the electrical charge, and all other terms are as defined in Chapter 7.

Writing energies as the products of intensive and extensive variables provides insight into the meaning of the more abstract thermodynamic terms. Just as electrical charge is the capacity of a system for electrical energy, so entropy is the capacity of a system for thermal energy. Similarly, while electrical potential determines the level to which the electrical capacity

is exploited, temperature determines how much the thermal capacity will be exploited. Similarly, volume is the capacity of a system for mechanical energy, and pressure is the mechanical potential that exploits that volume. The capacity for chemical energy is the number of moles of substance present in the system, and the chemical potential is the environmental factor that exploits that capacity.

Another important reason for distinguishing these two types of variable is that at equilibrium, any intensive variable must have the same value throughout a system; that is, no gradients in the potentials of mechanical (P), thermal (T), chemical (μ_i), and electrical ($\mathscr{E}$) energy can exist. Although perhaps self-evident, this principle is so important that it is commonly referred to as the *zeroth law* of thermodynamics.

In determining the equilibrium between minerals, we constantly have to evaluate how variations in the intensive and extensive variables affect equilibrium. One particularly important question is how many of these variables can be changed simultaneously without disturbing equilibrium. If a system can be described in terms of n variables, and r equations relate these variables, then $(n - r)$ variables can be changed independently while maintaining equilibrium. The number $(n - r)$ is called the *variance* or number of *degrees of freedom* of the system.

We can illustrate the significance of variance by examining a simple system consisting of 1 mol of ideal gas. Because only 1 mol is considered, the number of moles is not a variable. The system, then, can be described in terms of the two intensive variables P and T and the one extensive variable, V. But these three variables are related by the ideal gas law ($PV = RT$), so that the variance of the system must be 2 $(3 - 1)$; that is, any two variables can be changed independently, but the value of the third is determined by the equation relating the variables. We will have occasion to refer often to the variance of systems. The numbers of variables and relations between variables may be considerably greater than those considered in this simple example, but the same principle will be involved (see Sec. 10-5).

8-2 FREE-ENERGY SURFACE IN *G–T–P* SPACE

In Sec. 7-8 we saw that in a closed system, that is, one in which no matter is transferred in or out, the change in the Gibbs free energy, dG, is a function only of T and P; it can therefore be expressed as a total differential in terms of T and P (Eq. 7-30). If we consider a system consisting only of a phase α, the free energy of the system at any T and P could be obtained by integrating Eq. 7-30; that is,

$$\int_{\bar{G}_0^\alpha}^{\bar{G}_{T,P}^\alpha} dG = \int_{T_0}^{T} \left(\frac{\partial \bar{G}^\alpha}{\partial T}\right)_P dT + \int_{P_0}^{P} \left(\frac{\partial \bar{G}^\alpha}{\partial P}\right)_T dP \qquad (8\text{-}1)$$

where $T_0 = 298$ K, $P_0 = 10^5$ Pa, and $\bar{G}_0^\alpha =$ free energy per mole at 298 K and 10^5 Pa. From Eqs. 7-31 and 7-32, we know that $(\partial \bar{G}^\alpha / \partial T)_P$ is $-\bar{S}^\alpha$ and $(\partial \bar{G}^\alpha / \partial P)_T$ is $\bar{V}^\alpha$, respectively. If we assume that S and V remain constant over the range of conditions considered, integration of Eq. 8-1 gives

$$\bar{G}_{T,P}^\alpha = \bar{G}_{T_0,P_0}^\alpha - \bar{S}^\alpha(T - T_0) + \bar{V}^\alpha(P - P_0) \qquad (8\text{-}2)$$

This relation is the equation of a surface in $G–T–P$ space, which has an intercept on the G axis of G_{T_0,P_0}^α (Fig. 8-1). If P is held constant, the slope on the surface is $-\bar{S}^\alpha$; if T is held constant, the slope is $\bar{V}^\alpha$. In reality, S and V would be functions of T and P, so that slopes would change with T and P, and the surface would be curved. For simplicity of illustration, however, S and V have been taken as constants.

If the system contains a second phase β, a polymorph of phase α, for example, another surface in $G–T–P$ space can be constructed (Fig. 8-2). Phase β will most likely have different values of S and V from those of phase α. Thus, the free-energy surfaces of the two phases will not be parallel and will intersect along a line where $\bar{G}^\alpha = \bar{G}^\beta$. But if $\bar{G}^\alpha = \bar{G}^\beta$ and phase α changes into phase β, the free energy would not change; that is, $\Delta \bar{G}_r = 0$. According to Eq. 7-25, this condition describes equilibrium. The line of intersection of the two surfaces therefore gives the values of T and P under which phases α and β are in equilibrium with each other; that is, they can coexist.

We know from Eq. 7-25 that a spontaneous reaction is one in which dG is negative; that is, the reaction proceeds in a direction that minimizes G. On the low-temperature/low-pressure side of the line of intersection of the two surfaces in the $G–T–P$ plot of Figure 8-2, the free energy of the system can be lowered if phase β changes into phase α. In this region, then, phase α is stable relative to phase β. At pressures and temperatures above the line of intersection of the two surfaces, the β phase has the lower free energy and is therefore stable relative to the α phase. It is important to emphasize that these stabilities must be given relative to one another and not as absolute stabilities. For example, although the α phase has a lower free energy than the β phase at low temperature and is therefore the more stable of the two under these conditions, a third phase γ might exist that could have a lower free energy than either α or β. We would still be able to say that α was more stable than β under these conditions, but both phases would be unstable relative to the γ phase.

Let us determine the variance along the line $\bar{G}^\alpha = \bar{G}^\beta$ ($\Delta G = 0$) in Figure 8-2. The system has three variables, G, T, and P, which are related through Eq. 8-2 for phase α. A similar but independent equation can be written for phase β. Thus, the three variables are related through two independent equations,

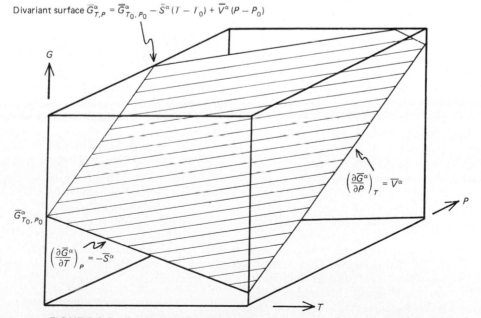

Divariant surface $\bar{G}_{T,P}^\alpha = \bar{G}_{T_0,P_0}^\alpha - \bar{S}^\alpha(T - T_0) + \bar{V}^\alpha(P - P_0)$

G

$\left(\dfrac{\partial \bar{G}^\alpha}{\partial P}\right)_T = \bar{V}^\alpha$

$\bar{G}_{T_0,P_0}^\alpha$

$\left(\dfrac{\partial \bar{G}^\alpha}{\partial T}\right)_P = -\bar{S}^\alpha$

P

T

FIGURE 8-1 Free-energy surface of phase α plotted in $G–T–P$ space. The surface is ruled with lines of constant free energy.

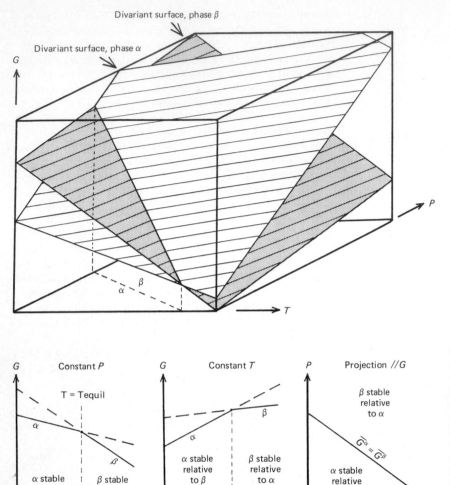

FIGURE 8-2 Intersecting free-energy surfaces of phase α and phase β (shaded). Both surfaces are ruled with lines of constant G. The line of intersection of the two surfaces, where the free energies of the two phases are equal, indicates the pressure and temperature conditions under which the two phases are in equilibrium with each other. Sections through the G–T–P plot at constant T and constant P are shown below, along with a projection of the line of intersection of the two surfaces on the P–T plane.

and the variance must be $3 - 2 = 1$. The assemblage of $\alpha + \beta$ is therefore said to be *univariant*. Any one of the three variables can be changed, but the values of the other two are then fixed by the equations. If only one of the phases is present, the three variables are related by only one equation, and the variance of the system would be 2. This *divariant* surface would be the surface for the α or the β phase in the G–T–P plot.

The three-dimensional plot of G–T–P space is commonly reduced to two dimensions by taking sections through it at constant P or constant T, or projecting the data parallel to the G axis onto the P–T plane (Fig. 8-2). In a section at constant P or T, the two surfaces for phases α and β appear as lines that intersect at a point where the univariant line $\bar{G}^\alpha = \bar{G}^\beta$ pierces the plane of the section. The variance of the assemblage $\alpha + \beta$ therefore becomes zero (*invariant*) under these conditions. This follows from the fact that the number of variables is reduced to two by holding one of them constant while two equations still relate the variables. With only two variables and two equations, there can be no variance; that is, only one set of conditions will permit phases α and β to coexist. Should a

rock be found containing these phases, we would know precisely the conditions under which it formed.

Despite the simplicity of the sections at constant T and P, the most useful diagram is the P–T projection. In this plot, the variance of the system is not decreased by the projection (nothing is held constant). Thus the univariant line $\bar{G}^\alpha = \bar{G}^\beta$ remains a univariant line. Even the divariant surfaces for phases α and β can be contoured with lines of constant G, but in most diagrams only the univariant line is shown. It is important when examining P–T diagrams illustrating phase equilibria to realize that they are projections from G–T–P space.

To plot univariant lines in P–T projections, we must know how to determine their slopes. The difference in free energy (G) between the two surfaces for the α and β phases in G–T–P space is a function only of T and P in a closed system. According to the cyclic rule for total differentials (Eq. 7-39), we can therefore write

$$\left(\frac{\partial \Delta G}{\partial T}\right)_P \left(\frac{\partial P}{\partial \Delta G}\right)_T \left(\frac{\partial T}{\partial P}\right)_{\Delta G} = -1$$

from which it follows that

$$\left(\frac{\partial P}{\partial T}\right)_{\Delta G} = -\frac{(\partial \, \Delta G/\partial T)_P}{(\partial \, \Delta G/\partial P)_T}$$

But according to Eqs. 7-33 and 7-34, $(\partial \, \Delta G/\partial T)_P = -\Delta S$ and $(\partial \cdot \Delta G/\partial P)_T = \Delta V$. Thus

$$\left(\frac{\partial P}{\partial T}\right)_{\Delta G} = \frac{\Delta S_r}{\Delta V_r} \qquad (8\text{-}3)$$

This is known as the *Clapeyron equation*. It shows that the slope of a univariant reaction ($\alpha \rightarrow \beta$) in a *P–T* projection is given by the entropy change ($S^\beta - S^\alpha$) divided by the volume change ($V^\beta - V^\alpha$) of the reaction. The univariant lines can have positive or negative slopes depending on the signs of ΔS and ΔV. A metamorphic reaction in which a gas is liberated will invariably have positive values of ΔS and ΔV, and thus its slope in a *P–T* diagram will be positive. Similarly, most melting is accompanied by positive values of ΔS and ΔV (ice is an exception), and thus most melting reactions also have positive slopes in *P–T* diagrams.

Standard units for ΔS and ΔV in Eq. 8-3 give units of J m^{-3} K^{-1} for $\partial P/\partial T$. But a joule is the amount of work done when a force of 1 newton causes a displacement of 1 meter (1 J = 1 N m). Thus the units of $\partial P/\partial T$ become N m^{-2} K^{-1} or Pa K^{-1}.

For purposes of illustration the phases α and β have been taken to be polymorphs, such as α and β quartz. But the general principles outlined here are equally applicable when α and β represent assemblages of phases constituting the reactants and products of a reaction. They can be crystalline minerals, magmas, or gases. The entropy and volume changes in these cases are the total changes in these properties determined for the balanced chemical reactions.

8-3 PLOTTING UNIVARIANT LINES IN *P–T* DIAGRAMS

Although the Clapeyron equation (Eq. 8-3) gives the slope of a univariant line in a *P–T* diagram, it does not locate the line with respect to absolute values of pressure and temperature. To determine this, we must know the equilibrium temperature of the reaction at some specified pressure. We might, for example, determine it at 10^5 Pa using the standard data in Table 7-1 and Eq. 7-28. Such a calculated temperature, however, might be far removed from the particular temperature (and pressure) range of interest. The known starting point might also be derived from experimental work carried out at high pressures and temperatures. In either case, it is necessary to be able to extend the standard thermodynamic data in Table 7-1 to conditions of high pressure and temperature.

We will illustrate the manipulation of thermodynamic data by considering a much-studied reaction, that of the polymorphic transformation of kyanite to andalusite. At low pres-

sures and temperatures this reaction is too slow to be studied in the laboratory. Even at high pressures and temperatures the reaction is extremely sluggish, but careful experiments have shown that the two polymorphs are in equilibrium at 800°C and 0.72 GPa (Richardson et al. 1969). Is this result in agreement with the thermodynamic data in Table 7-1, and if so, what slope should the reaction have in a *P–T* diagram?

Using Eqs. 7-14, 7-21, and 7-27 and data from Table 7-1, the free energies of formation ($\Delta \bar{G}^\circ_{f,1073}$) of kyanite and andalusite at 800°C (1073 K) and 10^5 Pa are determined. With these we calculate the ΔG_r of the transformation of kyanite to andalusite as follows:

$$\Delta G^\circ_{r,1073} = \Delta G^{\text{And}}_{f,1073} - \Delta G^{\text{Ky}}_{f,1073} = -5344 \text{ J mol}^{-1}$$

Because ΔG_r is negative, the reaction kyanite $\rightarrow$ andalusite proceeds spontaneously at 800°C and 10^5 Pa. But the value of ΔG_r can be changed by increasing the pressure. If it is increased by 5344 J mol^{-1}, ΔG_r will become zero and the reaction will be at equilibrium. The change in ΔG_r with pressure is given by Eq. 7-34, which in its integrated form gives

$$\Delta G^P_{r,1073} = \Delta G^\circ_{r,1073} + \Delta V_r(P - 10^5)$$

From Table 7-1 we obtain $\Delta V^\circ_{r,298}$. Molar volumes of the minerals will increase slightly with increasing temperature, but because the coefficients of thermal expansion of most minerals are small and similar, the value of ΔV_r remains almost constant. The effect of the coefficient of compressibility is even less important. The value of $\Delta V^\circ_{r,298}$ can therefore be used at high pressures and temperatures. For the kyanite to andalusite reaction it is $+7.44 \times 10^{-6}$ m^3 mol^{-1}.

If we are to reestablish equilibrium by raising the pressure at 800°C, $\Delta G^P_{r,1073}$ will have to become zero (the condition for equilibrium). In that case,

$$\Delta G^\circ_{r,1073} = -\Delta V_r(P - 10^5)$$

from which it follows that P must be 0.718 GPa, a pressure in good agreement with the experimental data.

The slope of the univariant line at 800°C and 0.72 GPa is given by $\Delta S^P_r/\Delta V^P_r$, according to the Clapeyron equation (Eq. 8-3). As stated above, the volume change of the reaction at 298 K and 10^5 Pa can be used at high pressures and temperatures without introducing serious error. The value of ΔS°_r can be calculated from the data in Table 7-1 for 800°C and 10^5 Pa. According to Eq. 7-36, $(\partial S/\partial P)_T = -\alpha V$. But we are dealing with the difference in volumes (ΔV_r) between *solids* which have very small and similar values of α. We can therefore use the $\Delta S^\circ_{r,1073}$ for the high-pressure value (8.13 J mol^{-1} K^{-1}). The resulting slope of the univariant line at 800°C and 0.72 GPa is

$$\left(\frac{\partial P}{\partial T}\right)_{\Delta G} = \frac{8.13}{7.44 \times 10^{-6}} = +1.1 \times 10^6 \text{ Pa K}^{-1}$$

which again is in good agreement with the experimentally determined slope. The reason that the thermodynamic calculation agrees so well with the experimental results is that the initial thermochemical data used for Table 7-1 were adjusted to conform with as many high-quality experimental studies as possible [see Helgeson et al. (1978)].

Univariant curves for reactions involving a gas are not as simply calculated as those involving condensed phases (solid or liquid), because ΔV_r and ΔS_r are strongly dependent on pressure and temperature. The free energy of a gas at high pressures can, however, be calculated easily from Eq. 7-32, which upon integration gives

$$\bar{G}_T^P = \bar{G}_T^\circ + \int_0^P \bar{V}\,dP \qquad (8\text{-}4)$$

The free energy of the gas at temperature T and 10^5 Pa ($\bar{G}_T^\circ$) can be determined directly from the data in Table 7-1. Integration of the $\bar{V}\,dP$ term requires that we know the functional relation between V and P.

If the gas behaves ideally ($V = RT/P$), Eq. 8-4 reduces to

$$\bar{G}_T^P = \bar{G}_T^\circ + RT \ln \frac{P}{10^5} \qquad (8\text{-}5)$$

where pressure is measured in pascal. If the pressure is measured in bars (1 bar = 10^5 Pa), Eq. 8-5 becomes

$$\bar{G}_T^P = \bar{G}_T^\circ + RT \ln P \qquad \text{bars} \qquad (8\text{-}6)$$

Thus, pressure is a measure of the free energy of an ideal gas.

Under the high pressures encountered in the Earth, most gases are not likely to behave ideally, so they will not obey Eq. 8-6. But the functional form of this equation is so convenient that it is preserved for real gases by inventing a function, f, known as the *fugacity*, such that

$$\bar{G}_T^P = \bar{G}_T^\circ + RT \ln f \qquad (8\text{-}7)$$

The fugacity therefore measures the free energy of a real gas in the same way that pressure measures the free energy of an ideal gas. The fugacity of a gas is related to its pressure by the *fugacity coefficient*, γ, where

$$f = P\gamma \qquad (8\text{-}8)$$

For an ideal gas $\gamma = 1$ and $f = P$. Tables of fugacities and fugacity coefficients for high pressures and temperatures are given for H_2O by Burnham et al. (1969) and for CO_2 by Mel'nik (1972).

For a reaction involving condensed phases and a gas, the free energy change can be obtained by integrating Eq. 8-4:

$$\Delta G_{r,T}^P = \Delta G_{r,T}^\circ + \Delta V_{r,T,c}(P - 1) + RT \ln f \qquad (8\text{-}9)$$

where $\Delta V_{r,T,c}$ refers only to the volume change in the condensed phases; normally, the value of ΔV_{298}° can be used. The free-energy change resulting from the volume change of the gas is

accounted for in the ($RT \ln f$) term, where the fugacity is given for the specified P and T. If we are interested in knowing at what pressure this reaction is at equilibrium, we set $\Delta G_{r,T}^P$ to zero, and Eq. 8-9 becomes

$$\Delta G_{r,T}^\circ = -\Delta V_{r,T,c}(P - 1) - RT \ln f \qquad (8\text{-}10)$$

which can be solved for P by trial and error (Prob. 8-8). Note that P and f have units of bars (1 bar = 10^5 Pa = 10^5 J m^{-3}).

8-4 SCHREINEMAKERS RULES FOR INTERSECTING SURFACES IN *G–T–P* SPACE

So far we have considered the location and orientation of only a single univariant reaction ($\alpha \to \beta$) in G–T–P space and its projection onto the P–T plane. The minerals involved in this reaction, however, might, under certain conditions of pressure and temperature, be involved in other reactions. More than one univariant line may, therefore, occur in the pressure and temperature range of interest. Because each line would most likely have a different slope, the lines will intersect at a point in the P–T diagram, and under these conditions of pressure and temperature the assemblage of minerals represented by the two lines will have zero degrees of freedom. The possible ways in which univariant lines can intersect at an invariant point follow certain simple geometrical rules. These were first worked out by the Dutch physical chemist Schreinemakers and are thus known as *Schreinemakers rules* (Zen 1966).

A univariant line describing a reaction, such as the one illustrated in Figure 8-2, divides the P–T plane in two (Fig. 8-3). To the left of the line, phase α is more stable than phase β, because phase α's free energy is lower than that of phase β. To the right of the line, phase β is more stable than phase α, because under these conditions phase β has the lower free en-

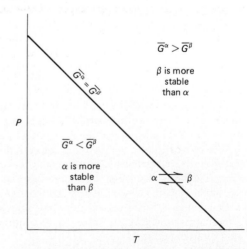

FIGURE 8-3 Relative stabilities of phases α and β on either side of the line along which the two phases are in equilibrium because their free energies are the same.

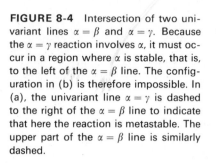

FIGURE 8-4 Intersection of two uni-variant lines $\alpha = \beta$ and $\alpha = \gamma$. Because the $\alpha = \gamma$ reaction involves α, it must oc-cur in a region where α is stable, that is, to the left of the $\alpha = \beta$ line. The config-uration in (b) is therefore impossible. In (a), the univariant line $\alpha = \gamma$ is dashed to the right of the $\alpha = \beta$ line to indicate that here the reaction is metastable. The upper part of the $\alpha = \beta$ line is similarly dashed.

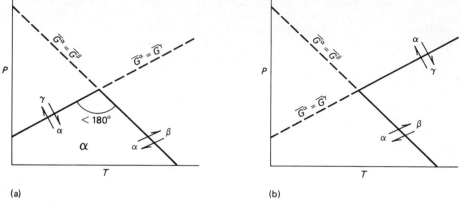

(a)

(b)

ergy. This condition would be equally true if α and β repre-sented assemblages of minerals rather than single minerals. Also, note the deliberate use of the phrase "more stable than ..." to indicate that we are discussing relative, not absolute, stabilities.

If a second univariant line were present, as would happen if α, for example, were able to transform into a third poly-morph γ, the P–T plane would again be divided by this line, with α being more stable on one side of the line and γ being more stable on the other. Let us assume that this second reac-tion has a positive slope on the P–T diagram. The univariant line for the $\alpha \rightarrow \gamma$ reaction will intersect the univariant line for the $\alpha \rightarrow \beta$ reaction as shown in Figure 8-4a and b. However, for the reaction $\alpha \rightarrow \gamma$ to be stable, the free energies of α and γ must both be lower than that of β. This is satisfied only by the relations shown in Figure 8-4a. To the right of the $\alpha - \beta$ univariant line, β is more stable than α. So a reaction involving α on the right side of this line is said to be *metastable*. That is, the reaction can take place, but it does not involve the most stable phases under these conditions.

The distinction between metastable, stable, and unstable can be illustrated by likening a chemical reaction to the simple mechanical situation of a rectangular block resting on a surface (Fig. 8-5). When the block has its long axis parallel to the hori-zontal surface, the block's center of gravity is as low as it can possibly be. The potential energy of the block is therefore at a minimum, and this condition is described as *stable*. If, on the other hand, the block were resting with its long axis in a ver-

tical position, the center of gravity would be higher than in the stable configuration. The potential energy would therefore be higher. By knocking the block over, the potential energy would be reduced. But energy is required to topple the block, because as the block tilts sideways, its center of gravity is raised slightly because of the pivoting action about the lower edge. The amount of energy necessary to raise the block during the pivoting is known as the *activation energy*; this amount of energy must be expended to topple the block. Once the center of gravity passes over the pivot point, both the activation energy introduced during the tilting and the initial excess potential energy are transformed to kinetic energy as the block falls. In this state the block is *unstable*. Had the activation energy not been intro-duced, however, the block could have remained in its vertical position indefinitely, even though in this position the block is not in its lowest-energy state. Under these conditions, the block is said to be *metastable*.

The univariant line $\alpha = \gamma$ in Figure 8-4 is metastable to the right of the univariant line $\alpha = \beta$. This part of the line is therefore dashed. It will be noted that the line for the $\alpha - \beta$ reaction is dashed above the line for the $\alpha = \gamma$ reaction because, under these conditions, the $\alpha = \beta$ reaction must also be meta-stable. It follows, then, that *the stable end of a univariant line must lie in the field in which the free energies of both the reac-tants and products are lower than the free energies of any other related phases in the system.*

At the pressure and temperature represented by the point of intersection of the two univariant curves in Figure 8-4, the

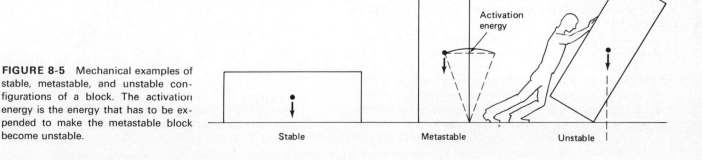

FIGURE 8-5 Mechanical examples of stable, metastable, and unstable con-figurations of a block. The activation energy is the energy that has to be ex-pended to make the metastable block become unstable.

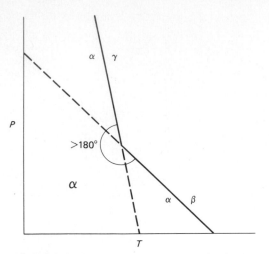

FIGURE 8-6 Impossible arrangement of univariant lines, because both reactions make the phase α unstable in the other reaction. The sector in which phase α is stable must subtend an angle of less than 180° at the invariant point.

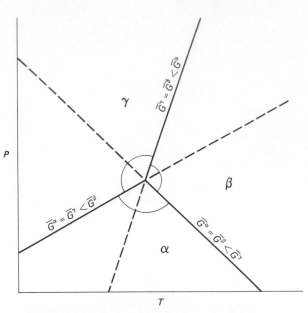

FIGURE 8-7 Thermodynamically permissible arrangement of univariant lines separating the divariant fields of the phases α, β, and γ.

system is invariant; that is, phases α, β, and γ can coexist only at this specific set of *P–T* conditions. Along the lines extending from this point of intersection the system is univariant, with phases α and β or α and γ coexisting. Under conditions of pressure and temperature represented by the field between the univariant lines, the system is divariant, and only one phase exists (α in Fig. 8-4a). Because the stable end of a univariant line must be in a field where its reactants and products are more stable than any other phase (or phases), *the angle between univariant lines bounding a divariant field at an invariant point must be less than 180°.* The arrangement of univariant lines in Figure 8-6, for example, would be impossible because the α phase in both reactions would be made metastable by the other reaction.

Because the two univariant lines that intersect to give the invariant point represent coexising α − β and α − γ, a third univariant line representing the coexistence of γ and β must also pass through the invariant point. Because this reaction involves γ, it must lie in the field where γ is stable; that is, within 180° in a clockwise direction from the stable end of the line α = γ. But it also involves β and must therefore occur within 180° in a counterclockwise direction from the stable end of the line α = β. These conditions are met in only one sector (Fig. 8-7). Note that the resulting divariant fields for β and γ both subtend angles of less than 180° around the invariant point.

The set of univariant lines around the invariant point is now complete. Each line, while dividing the *P–T* plane into fields of different relative stabilities of the phases involved, is itself divided into stable and metastable parts by the other reactions. The various parts of the *P–T* diagram can now be labeled in terms of the relative free energies of each of the phases (Fig. 8-8).

The relations between the free energies and the stability of the phases are most clearly seen in isothermal *G–P* plots (Fig. 8-9). At temperatures below the invariant temperature (T_{IP}), the trace of the free-energy surface of the β phase is above the point of intersection of the free-energy surfaces of the α and

γ phases (Fig. 8-9b). Reactions with the β phase (1 and 2 in Fig. 8-9) are therefore metastable. At temperatures above the invariant point, the trace of the free-energy surface of the β phase lies below the intersection of the free-energy surfaces of the α and γ phases (5 in Fig. 8-9c), making the α → γ reaction metastable at these temperatures. The intersections of the α and β surfaces and the β and γ surfaces both become stable under these conditions. At the temperature of the invariant point (Fig. 8-9d), the free-energy surfaces for the three phases intersect at a point. Only under these conditions do all three phases have the same free energies; that is, they are equally stable (see Prob. 8-1).

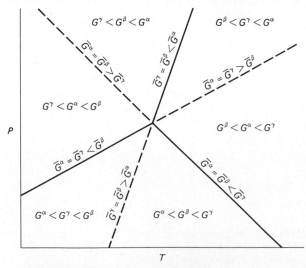

FIGURE 8-8 Relative free energies of phases α, β, and γ in various parts of the *P–T* projection.

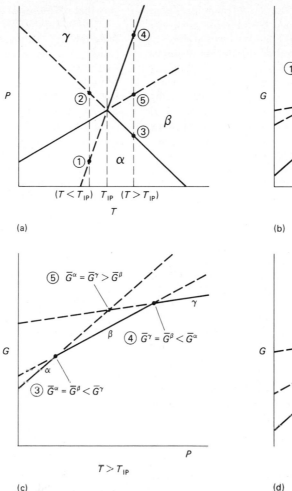

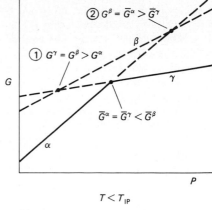

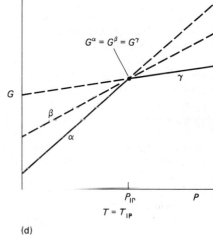

FIGURE 8-9   Isothermal sections through $G–T–P$ space showing the relative positions of the $\alpha = \beta$, $\alpha = \gamma$, and $\beta = \gamma$ univariant lines at a temperature below the invariant point (b), above the invariant point (c), and at the invariant point (d). The relative positions of these sections are shown in the $P–T$ plot (a).

8-5   SCHREINEMAKERS RULES APPLIED TO MULTICOMPONENT SYSTEMS

In the example above we considered a system consisting of three phases, α, β, and γ. The principles determining the arrangement of univariant lines about an invariant point, however, apply equally to systems containing more phases, the only difference being that as extra phases are added, the number of univariant lines increases. This number depends simply on how many independent reactions can be written between the minerals involved.

If we designate the number of *phases* (minerals, magmas, fluids) present by ϕ, and the *minimum* number of *components* necessary to describe the formulas of the phases by c, the number of degrees of freedom (variance) of the system is given by

$$f = c - \phi + T + P + n \tag{8-11}$$

where n is the number of environmentally controlled (intensive) variables in addition to temperature (T) and pressure (P); the fugacity of water might, for example, play such a role. If temperature and pressure are the only intensive variables, Eq. 8-11

is commonly written as

$$f = c + 2 - \phi \tag{8-12}$$

where it is understood that the 2 refers to temperature and pressure. This relation, which is known as *Gibbs phase rule*, is developed more fully in Sec. 10-5.

If a system consists of $(c + 2)$ phases, the variance will be zero. In a $P–T$ diagram, $(c + 2)$ univariant lines radiate from the invariant point, each line involving $(c + 1)$ phases. These univariant lines divide $P–T$ space into $(c + 2)$ divariant fields. One or more assemblages of c phases occurs in each field; one of these assemblages is unique to each field. Thus in the example worked out in Figure 8-7, the formulas of the phases α, β, and γ (polymorphs of silica, for example) can all be expressed in terms of one component (SiO_2). At the invariant point, three $(c + 2)$ phases (α, β, and γ) coexist. If any one of the phases is absent, the system is univariant. For example, if γ is absent, the system consists of α and β, and the $P–T$ conditions are given by the univariant line $\alpha = \beta$. It is convenient to describe each of the univariant lines by the phase that is absent. According to the symbolism used by Schreinemakers, this absent phase is placed between parentheses. Thus, the $\alpha = \beta$ reaction is designated by (γ). Similarly, the univariant lines for the $\alpha = \gamma$ and $\beta = \gamma$ reactions are designated by (β) and (α), respectively.

One reason for using this method of labeling reactions is that it leads to a simple mnemonic scheme for working out the sequence of univariant lines around an invariant point without having to resort to basic principles each time. We start by writing each reaction as a simple equation. The equation is then rewritten, placing each phase in parentheses and replacing the equals sign with the phase absent from that reaction. Thus, for the one-component system with phases α, β, and γ, we write the three reactions as

$$\alpha = \beta \qquad \alpha = \gamma \qquad \beta = \gamma$$
$$(\alpha)|(\gamma)|(\beta) \qquad (\alpha)|(\beta)|(\gamma) \qquad (\beta)|(\alpha)|(\gamma)$$

The phases in parentheses (phase absent) now refer to reactions. From the mnemonic for the first equation, we read that the stable part of the (α) reaction $(\beta = \gamma)$ lies on one side of the univariant line (γ) and the (β) reaction lies on the other. Thus, either of the arrangements of lines in Figure 8-10a or b would satisfy this requirement. According to the mnemonic from the second equation, the (α) reaction must also lie on the opposite side of the (β) line from the (γ) reaction. Thus, if we chose the option in Figure 8-10a, this condition can be satisfied only by making the high-pressure end of the (γ) reaction metastable (Fig. 8-10c). If the option in Figure 8-10b is chosen, the only

way that reactions (α) and (γ) can be on opposite sides of reaction (β) is for the low-pressure end of the (γ) reaction to be metastable. The univariant lines are now all in their correct sectors, and the information from the third reaction, that is, that the (β) and (γ) univariant lines must be on opposite sides of the (α) reaction, serves simply to confirm that no mistakes were made in orienting the lines with the information from the first two equations.

This mnemonic scheme leads to two possible arrangements of univariant lines that are reverse images of each other. Both are thermodynamically consistent, and without additional information, selection of the correct one might be difficult. Of course, if thermodynamic data for the phases are available, the decision is simple. But even when this is lacking, knowledge of the field occurrence of the phases can provide the answer. If, for example, the mineral β were known to occur only in high-temperature rocks, the scheme in Figure 8-10c would be the correct one. Similarly, if one of the phases is a liquid or a gas, it must appear on the high-temperature side of a reaction with a solid.

The actual slopes of the univariant lines cannot be determined from the mnemonic scheme; this requires information on the ΔS and ΔV of the reactions or experimental data. Commonly, however, the slope of at least one of the reactions is known, and then the scheme provides some restrictions on the possible slopes of the remaining univariant lines. Two reactions must, of course, be known to determine the position of an invariant point in $P–T$ space.

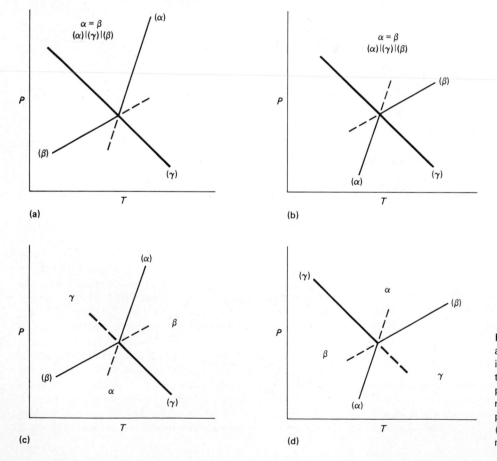

(a)

(b)

(c)

(d)

FIGURE 8-10 Sequence of steps in arranging univariant lines around an invariant point in a one-component system. The line $\alpha = \beta$, designated by the phase absent (γ), is taken as given. The remaining two lines must plot on opposite sides of (γ) as shown in (a) or (b). Part of the γ reaction then becomes metastable as shown in (c) or (d).

Although the example worked out above is for a one-component system, the mnemonic scheme works equally well for systems containing more components. Consider, for example, the two-component system $MgO-SiO_2$. Common minerals in this system include periclase (MgO), forsterite (Mg_2SiO_4), enstatite ($MgSiO_3$), and quartz (SiO_2). Because the system has two components, it would be invariant if all four minerals were present together ($f = 0 = c + 2 - \phi$). Radiating from this invariant point would be four ($c + 2$) univariant lines, each representing the coexistence of three minerals ($c + 1$).

The compositions of the minerals and the reactions between them can be graphically represented by a simple line that at one end is pure MgO and at the other is pure SiO_2 (Fig.

8-11e). Enstatite ($1MgO + 1SiO_2$), for example, would plot halfway along this line, whereas forsterite ($2MgO + 1SiO_2$) would plot only one-third of the way toward SiO_2. Reactions between the minerals can be determined simply by inspecting the relative positions of the minerals on this compositional line. A reaction product must lie between the reactants. For example, periclase and enstatite could react to form forsterite. The four possible univariant reactions between periclase, forsterite, enstatite, and quartz and their mnemonic schemes are as follows:

1. Periclase   Enstatite   Forsterite

     MgO $+$ $MgSiO_3$ $=$ Mg_2SiO_4

     (Pe)   (En) $|(Q)|$ (Fo)

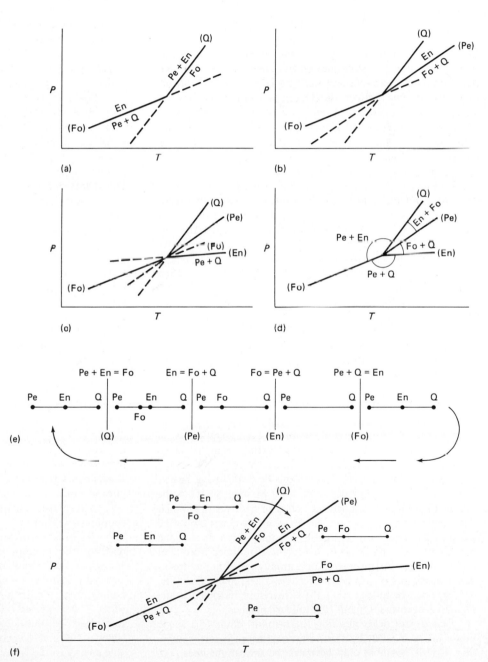

FIGURE 8-11 Sequence of steps involved in arranging the univariant lines around the invariant point involving periclase (Pe), forsterite (Fo), enstatite (En), and quartz (Q) in the two-component system $MgO-SiO_2$. See text for explanation.

2. Forsterite   Quartz   Enstatite

$$Mg_2SiO_4 + SiO_2 = 2MgSiO_3$$

   (Fo)   (Q) |(Pe)| (En)

3. Periclase   Quartz   Enstatite

$$MgO + SiO_2 = MgSiO_3$$

   (Pe)   (Q) |(Fo)| (En)

4. Periclase   Quartz   Forsterite

$$2MgO + SiO_2 = Mg_2SiO_4$$

   (Pe)   (Q) |(En)| (Fo)

Thermodynamic data indicate that all of these reactions have positive slopes in *P–T* space and that quartz and periclase are stable on the high-temperature side of the (Fo) reaction (Fig. 8-11a). According to the mnemonic scheme for the first equation, the (Fo) reaction lies on the opposite side of the (Q) reaction from the univariant lines for (Pe) and (En). Starting with the (Fo) line as given, the (Q) line is drawn with a positive slope; it intersects the (Fo) line to create the invariant point (Fig. 8-11a). Because the (Q) reaction involves enstatite, the reaction is metastable below the (Fo) reaction. Similarly, because the (Fo) reaction involves enstatite and periclase, it must be metastable to the right of the (Q) reaction. The univariant lines for (Pe) and (En) can, according to the first reaction, be placed anywhere on the right side of the (Q) reaction. However, according to the second reaction, the (Fo) and (Q) lines must lie on one side of the (Pe) line and (En) on the other. Also, because the (Pe) reaction involves enstatite, only that part above the (Fo) line can be stable. The (Pe) line can therefore be placed only in the sector shown in Figure 8-11b. The third reaction requires that the (Pe) and (Q) lines lie on one side of the (Fo) line and (En) on the other. This can be satisfied only by placing the (En) line as shown in Figure 8-11c. The diagram is now complete but can be checked by examining the fourth reaction. This indicates that the (Pe) and (Q) lines should lie on the opposite side of the (En) reaction from the (Fo) line, which indeed they do.

The four univariant lines in Figure 8-11d separate four divariant fields, each characterized by assemblages of two (*c*) phases. These fields each subtend angles of less than 180° at the invariant point. Some of these assemblages, such as that of Pe + Q in the lower sector, can be read directly from the phases involved in the reactions. Others, such as that of Pe + Fo, must be deduced. Consider the assemblage Pe + En being heated until it crosses the (Q) reaction to form Fo (Fig. 8-11a). If the initial assemblage contained an excess of En over that which was needed for the reaction, the reaction products would be Fo plus the excess En. Thus, En is stable to the right of the (Q) reaction as long as no Pe is present. Continued heating of the assemblage Fo + En would eventually result in the breakdown of En to Fo + Q when it crosses the (Pe) reaction (Fig. 8-11b). The assemblage En + Fo is therefore restricted to the sector between the (Q) and (Pe) univariant lines.

Divariant fields marking the stability limits of a single phase are also given by this diagram. Enstatite, for example (Fig. 8-11d), is stable only between the univariant lines (Fo)

and (Pe). Here again, the sector in which a single phase is stable cannot exceed 180°.

The reactions given by the univariant lines radiating from the invariant point can be represented graphically as shown in Figure 8-11e. We will start with the assemblage in the upper left of the diagram and proceed in a clockwise direction around the invariant point. The phases stable in the sector bounded by the (Q) and (Fo) univariant lines are Pe + En and En + Q. We see that the latter assemblage is possible by considering what the reaction products would be if a mixture of Pe + Q, containing an excess of Q, was raised above the (Fo) reaction. Clearly, En + Q would remain. The first compositional line in Figure 8-11e therefore has the minerals Pe, En, and Q marked on it. If a rock has a bulk composition that lies between En and Q, it would consist of the minerals En and Q. A rock with a composition between En and Pe would, on the other hand, consist of En and Pe.

On crossing the (Q) univariant line, Pe reacts with En to form Fo. Thus, on the compositional line, the mineral Fo is added. Now rocks having bulk compositions between En and Pe will consist of Pe + Fo or Fo + En, depending on whether they fall to the left or right of the Fo composition.

Continuing in a clockwise direction, the (Pe) reaction is encountered next, and En breaks down to form Fo + Q. The disappearance of En from the compositional line means that rocks of appropriate composition will contain Fo + Q. Note that the assemblage Pe + Fo is still stable in MgO-rich rocks.

Next, Fo breaks down to form Pe + Q on crossing the (En) reaction, leaving only Pe and Q as stable minerals on the composition line. Any composition in this system will therefore consist of some mixture of Pe and Q.

Finally, on crossing the (Fo) reaction, Pe and Q react to form En, and we then have the assemblage with which we started. It is important to note that each one of these steps involves changing only one mineral at a time. If you find that two minerals have been changed in going from one sector to the next, a mistake has been made in positioning one or more of the univariant lines.

We are now in a position to label all of the univariant lines and divariant fields with the appropriate mineral assemblages (Fig. 8-11f) to complete the *P–T* diagram. Such diagrams provide a convenient means of analyzing the significance of mineral assemblages found in rocks. Most rocks will fall in divariant fields in *P–T* diagrams, because they typically have existed over a range of pressures and temperatures. As a result, most rocks will contain *c* phases, which in this example is two. Rocks containing more than *c* phases must lie on univariant lines or at invariant points, or be out of equilibrium.

Univariant and invariant assemblages are not as uncommon as might at first be expected. If a change in intensive variables causes a divariant assemblage to reach a univariant reaction, the enthalpy change of the reaction may be large enough to buffer and control the intensive variables. For example, by analogy, it is not an unusual occurrence to find water boiling in a pot in a kitchen, yet the univariant reaction water → steam is actually invariant under the ambient pressure of 10^5 Pa; that is, you know that the temperature of the water is 100°C if it is boiling. A stove may transmit considerable energy to the water, but this heat is used in vaporization rather than in raising temperature. Thus, the boiling buffers the reaction at 100°C as long as there is water to boil. In a similar way, reactions, such as the partial melting of the upper mantle or the growth of a metamorphic mineral in a contact aureole of an igneous intrusion, can produce univariant or invariant assemblages that will, through their enthalpies of reaction, maintain the intensive variables at the special values of univariant or invariant equilibria, as long as reactants remain.

As the number of components increases, so does the number of univariant lines radiating from an invariant point. The simple mnemonic scheme outlined above, however, can still be used to determine the sequence of univariant lines. Let us consider the system MgO–Al_2O_3–SiO_2. Because it contains three components, an assemblage of five minerals ($c + 2$) in this system will be invariant, and radiating from an invariant point will be five univariant lines. This system contains many minerals, any group of five of which will produce an invariant assemblage. We will consider the invariant point associated with the five minerals andalusite (An), corundum (Co), spinel (Sp), enstatite (En), and cordierite (Cd).

Because the system contains three components, the compositions of the minerals can be expressed conveniently in a triangular plot where each apex of the triangle represents one of the three components (Fig. 8-12). From such a plot, the reactions between the minerals can easily be determined. First, draw a set of lines joining the compositions of each of the minerals, but no two lines may intersect. The set of lines shown in triangle I in Figure 8-12 is one possible way of drawing these lines, but any of the arrangements shown in triangles I to V would be acceptable. These lines divide a triangle into a number of smaller triangles. A rock with a bulk composition plotting within any one of these subtriangles consists of the minerals at the apexes of the subtriangle. For example, a rock composed of 40% SiO_2, 30% MgO, and 30% Al_2O_3 (× in triangle I) would consist of Co + En + Cd.

Next, remove any one line and replace it with another line joining a different pair of minerals. For example, the line joining En and Co can be replaced with one joining Cd and Sp (II in Fig. 8-12). These switching lines constitute a reaction, which in this case is En + Co = Cd + Sp. Because this reaction does not involve andalusite, it is referred to as the andalusite-absent reaction or (An). Note that when this one line is switched, the remainder of the lines do not change; that is, only one line is changed at a time. Continuing in this manner, the remainder of the reactions are easily determined (Fig. 8-12 II to V).

We now write out the balanced reactions and the mnemonic scheme for each of the five univariant lines radiating from the invariant point.

From the ΔS and ΔV of each reaction the slopes of the univariant lines are determined. The ΔS of the reaction also indicates which assemblage is favored at high temperature. For example, with increasing temperature, reactions (An) and (Sp) proceed to the right, but the remainder proceed to the left.

Start by placing the (An) reaction on the P–T diagram with its appropriate slope (it is instructive to draw this yourself and then compare it with Fig. 8-12). From the mnemonic, we know that the (En) and (Co) lines must plot on one side and the (Cd) and (Sp) lines on the other side of the (An) reaction. Two complementary diagrams could be drawn if we did not know that the (An) reaction proceeds to the right with increasing temperature; that is, Cd and Sp are stable together on the high-temperature side of the (An) line. Thus, the (Cd) and (Sp) lines must be placed on the low-temperature side of the (An) line, and (En) and (Co) on the high-temperature side. These lines can be plotted temporarily anywhere on each side of the (An) line. A short segment of the metastable extension of each line should be plotted beyond the invariant point; this will help in positioning the remainder of the univariant lines.

Next, consider the (En) reaction. According to the mnemonic, the (Cd) and (Co) reactions must lie on one side of the (En) line and the (An) and (Sp) reactions on the other. Depending on how these lines were drawn on either side of the (An) line in the first step, they may need to be switched or adjusted to be consistent with the mnemonic for the second reaction. Continuing in this manner, then, the positions of the univariant lines are adjusted until they are consistent with the mnemonic for each reaction. This leads to the unique solution shown in Figure 8-12.

Once the univariant lines have been positioned in their correct sectors the slopes of the lines can be adjusted to agree with the slopes determined from the thermodynamic data. This should not move a univariant line into a new sector unless a mistake has been made in the positioning of the univariant lines or if there is an error in the thermodynamic data. Each of the lines can now be labeled with the appropriate reaction. Care must be taken to get the correct assemblages on each side of each line. In doing this, it is useful to recall that the sector subtended at the invariant point by a phase assemblage cannot exceed 180°.

A final check on the correctness of the scheme of univariant lines can be obtained by constructing the compositional triangles in each divariant field based on the labeling of the univariant lines bounding the fields. Lines can be drawn between any pair of phases that a reaction indicates is stable in the particular sector. In the lower right sector of Figure 8-12, for example, Cd + Sp are stable together and can therefore be joined with a line. But this assemblage could also include Co

Phase Absent	Reaction	Mnemonic	ΔV ($m^3 \times 10^3$)	ΔS ($J\ mol\ K^{-1}$)	dP/dT ($Pa\ K^{-1} \times 10^{-5}$)
(An)	5En + 5Co = Cd + 3Sp	(En)(Co) |(An)| (Cd)(Sp)	0.068	55.42	8.1
(En)	Cd + 5Co = 5An + 2Sp	(Cd)(Co) |(En)| (An)(Sp)	−0.024	−36.37	15.2
(Co)	2Cd + Sp = 5En + 5An	(Cd)(Sp) |(Co)| (En)(An)	−0.092	−91.79	10.0
(Cd)	An + Sp = En + 2Co	(An)(Sp) |(Cd)| (En)(Co)	−0.009	−3.81	4.2
(Sp)	2En + 3An = Co + Cd	(En)(An) |(Sp)| (Co)(Cd)	0.042	43.99	10.5

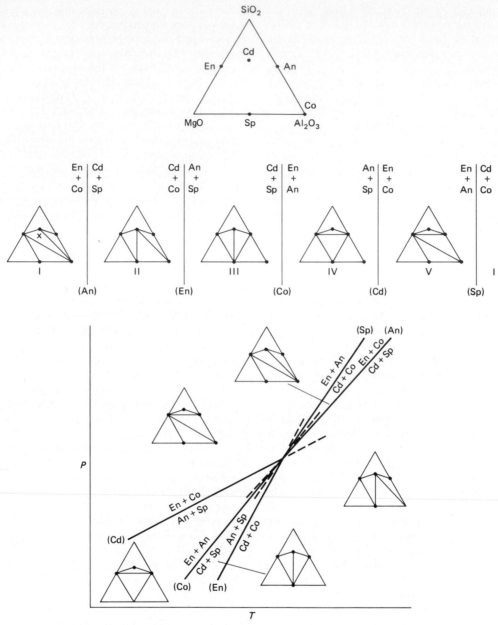

FIGURE 8-12 Construction of a *P–T* diagram for the univariant lines at the invariant point involving enstatite (En), cordierite (Cd), andalusite (An), corundum (Co), and spinel (Sp) in the three-component system MgO–Al$_2$O$_3$–SiO$_2$. See text for explanation.

if Co had been present in excess when the (An) reaction formed the Cd + Sp. Consequently, Cd, Sp, and Co can all be joined with lines. Had En been in excess instead of Co, then Cd + Sp + En would also be stable. These phases, therefore, can also be joined with lines. The resulting compositional diagrams in each sector differ by only one line from the diagram in the adjoining sector. If they do not, univariant lines have been incorrectly positioned or labeled.

8-6 DEGENERATE SYSTEMS

In general, as many univariant lines radiate from an invariant point as there are phases in the system; this will be two more than the number of components. In some systems, however, reactions between certain phases may be described in terms of a smaller number of components than are needed to describe the other reactions being considered. In such cases, the number of univariant lines decreases and the system is said to be *degenerate*.

A system becomes degenerate in two ways. First, two or more phases may have the same composition (polymorphs, for example). Second, in systems consisting of three or more components, three or more phases may be collinear or four or more phases may be coplanar in ternary or quaternary compositional diagrams, respectively. The result is that some reactions will involve fewer than $c + 1$ phases. Those phases taking part in reactions involving fewer than $c + 1$ phases are termed *singular phases*, and the remainder are called *indifferent phases*. Two types of degeneracy occur depending on whether the indifferent phases plot on opposite or the same sides of the line joining the singular phases. This can be illustrated by considering a specific example.

The invariant point involving corundum, periclase, spinel cordierite, and quartz in the system $MgO–Al_2O_3–SiO_2$ is degenerate for two reasons. First quartz, cordierite, and spinel plot on a straight line in the compositional diagram; second, periclase, spinel, and corundum also plot on a straight line (Fig. 8-13). In the first case Q, Cd, and Sp are considered the singular phases, whereas Pe and Co, which plot on opposite sides of the collinear group, are the indifferent phases. In the second case, Pe, Sp, and Co are the singular phases, and the indifferent ones are Q and Cd, which both plot on the same side of the collinear group.

As before, reactions between phases can be determined by inspection of the compositional triangle (Fig. 8-13). However, unlike the previous example, where five different reactions were obtained, only three occur at this invariant point.

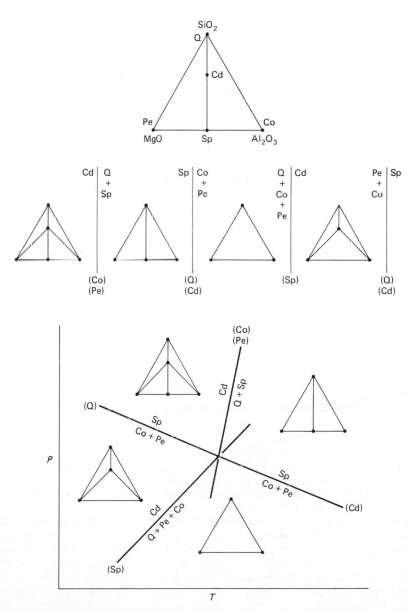

FIGURE 8-13 Construction of a *P–T* diagram for the univariant lines at the invariant point involving quartz (Q), cordierite (Cd), corundum (Co), spinel (Sp), and periclase (Pe) in the three-component system $MgO–Al_2O_3–SiO_2$. See text for explanation.

They are

$$
\begin{aligned}
\text{(Co)(Pe)} \quad &\text{Cd} = 5\text{Q} + 2\text{Sp} \\
\text{(Q)(Cd)} \quad &\text{Sp} = \text{Co} + \text{Pe} \\
\text{(Sp)} \quad &5\text{Q} + 2\text{Pe} + 2\text{Co} = \text{Cd}
\end{aligned}
$$

the mnemonics for which are

$$
\begin{array}{ccc}
\text{(Cd)} & |\text{(Co)(Pe)}| & \text{(Q)(Sp)} \\
\text{(Sp)} & |\text{(Q)(Cd)}| & \text{(Co)(Pe)} \\
\text{(Q)(Pe)(Co)} & |\text{(Sp)}| & \text{(Cd)}
\end{array}
$$

The first reaction does not involve Co or Pe; so it can be referred to as the (Co) or (Pe) line. Similarly, the second reaction is both the (Q) and (Cd) reaction.

    We will start with the (Sp) reaction on the *P–T* diagram. According to the mnemonic, the (Q), (Pe), and (Co) reactions lie on one side and the (Cd) reaction on the other side of the (Sp) line. Because Cd is stable at lower temperatures than Q + Pe + Co, the (Cd) reaction must lie on the high-temperature side of the (Sp) line and (Q), (Pe), and (Co) on the low-temperature side. But (Q) and (Cd) are the same reaction. Therefore, the reaction of Sp = Co + Pe passes straight through the invariant point with no change in slope. On the high-temperature side it is labeled (Cd) and on the low-temperature side, (Q).

    The mnemonic for the second reaction indicates that the (Sp) reaction lies on one side of the (Q) or (Cd) reaction and the (Co) and (Pe) reactions on the other. But the (Co) and (Pe) reactions are the same; thus they must plot as a single line. Also, from the third reaction we see that the (Co) and (Pe) reactions must plot on the same side of the (Sp) reaction as the (Q) reaction. The univariant lines for (Co) and (Pe) therefore plot as a single line, as shown in Figure 8-13.

    We see, then, that if the indifferent phases lie on the same side of the collinear or singular phases, a single univariant line passes straight through the invariant point [(Q) and (Cd) in Fig. 8-13]. This degeneracy does not lead to a decrease in the number of univariant lines, but two of the lines will have iden-tical slopes. If the indifferent phases lie on opposite sides of the singular phases, two univariant lines become one. This there-fore leads to a decrease in the number of univariant lines about the invariant point.

8-7 SUMMARY AND CONCLUSIONS

We have seen that the Gibbs free energy of a phase (or group of phases) varies as a function of temperature ($-S$) and pressure (V). Because different phases are not likely to have identical molar entropies and volumes, free-energy surfaces will not be parallel in *G–T–P* space. Where these surfaces intersect, the free energies of the two will be identical. Thus, if the phases are related by a reaction, the line of intersection of the free energy surfaces defines the pressure and temperature conditions under which the reactants and products are at equilibrium ($\Delta G = 0$).

    The line of intersection of two free-energy surfaces is said to be univariant, because only one of the two intensive vari-ables, *T* or *P*, can be independently varied. If a second uni-variant line is present involving the same components, the system is invariant at the point of intersection of the two lines. Only under this one set of *P–T* conditions are the assemblages of phases from both univariant lines stable together.

    The position and slope of univariant lines in *P–T* space can be calculated from the thermodynamic data for the phases. In addition, Schreinemakers rules provide a simple geometrical means of correctly orienting univariant lines about an invari-ant point. The univariant lines divide *P–T* space into divariant fields; that is, regions in which both *P* and *T* can vary inde-pendently without modifying the mineral assemblage present in that field. In nature, rocks may form under conditions corre-sponding to those of divariant fields, univariant lines, or, under unusual conditions, invariant points. *P–T* diagrams therefore provide an invaluable tool for the analysis of the mineral as-semblages found in rocks, and they will be used extensively throughout the remainder of this book.

PROBLEMS

8-1. For the system illustrated in Figure 8-9, draw three isobaric (constant pressure) *G* versus *T* diagrams for conditions where $P < P_{\text{IP}}$, $P > P_{\text{IP}}$, and $P = P_{\text{IP}}$, where P_{IP} is the pressure at the invariant point. Label all isobaric invariant points and indicate which parts of each diagram are stable and which are metastable.

8-2. The melting of a pure mineral can be written as a simple reaction. For diopside it is

$$\text{diopside}_{\text{crystal}} = \text{diopside}_{\text{liquid}}$$

At 10^5 Pa (1 bar), diopside melts at 1665 K. If the S°_{1665} of crystalline and liquid diopside are 532.2 and 619.6 J mol^{-1} K^{-1}, respectively, and their volumes are, respectively, 0.06609×10^{-3} and 0.07609×10^{-3} m^3 mol^{-1}, calculate the melting point at 2 GPa (20 kbar) using the Clapeyron equation. (Recall that 1 J = 1 Pa m^3.)

8-3. The Al$_2$SiO$_3$ polymorphs andalusite, kyanite, and sillimanite can coexist at an invariant point at 500°C and 0.376 GPa.

    (a) Calculate the slopes of the three univariant lines radiating from the invariant point using values of S°_{298} and V°_{298} from Table 7-1.

    (b) Plot the univariant lines in a *P–T* diagram (200 to 800°C and 0 to 1 GPa).

    (c) Compare your results with the experimentally determined values of Holdaway (1971), which are fitted by the equations

$$K = A \qquad P = 13.3(T - 200) - 0.0026(T - 200)^2$$

$$A = S \qquad P = 14.0(770 - T)$$

$$K = S \qquad P = 20.0(T - 315) + 0.0009(T - 315)^2$$

where P is in bars and T in °C.

(d) Discuss the possible causes for the differences.

8-4. The mineral jadeite, which is known to occur in metamorphic rocks formed at high pressures, can be related to two common low-pressure minerals by the reaction

$$\text{albite} = \text{jadeite} + \text{quartz}$$

(a) Balance the reaction, and using data in Table 7-1, calculate the ΔG of reaction at 298 K and 10^5 Pa. From the sign of $\Delta G^\circ_{r,298}$, comment on the feasibility of the reaction under these conditions.

(b) Repeat part (a), but for 600°C and 10^5 Pa.

(c) Calculate what pressure increases would be needed at 298 and 873 K to bring the reaction to equilibrium.

(d) How does the slope of this univariant line compare with the slope obtained from the Clapeyron equation and data for 298 K and 10^5 Pa?

(e) Compare your results with those obtained experimentally by Holland (1980). Plot Holland's results on your $P–T$ diagram using his empirical equation for the univariant line.

8-5. At high pressure in the two-component system $NaAlSiO_4–SiO_2$, the four minerals nepheline, jadeite, albite, and quartz coexist at an invariant point.

(a) Write balanced reactions for the univariant lines that radiate from this invariant point.

(b) Using the S°_{298} and V°_{298} values from Table 7-1, calculate the slopes of these univariant lines.

(c) Using the mnemonic scheme and Schreinemakers rules, construct a $P–T$ diagram for this set of univariant lines. While not changing the sectors in which the univariant lines plot, adjust the slopes of the lines to agree as closely as possible with the slopes determined from the Clapeyron equation.

8-6. Construct a $P–T$ diagram for the set of univariant lines associated with the invariant point for the assemblage periclase, andalusite, spinel, enstatite, and cordierite in the system $MgO–Al_2O_3–SiO_2$. Use the values of S°_{298} and V°_{298} from Table 7-1 to determine the slopes of the lines. Note that the slopes of some of the lines calculated with the data for 298 K must be wrong, because they are not consistent with the arrangement of univariant lines required by Schreinemakers rules.

8-7. The invariant assemblage quartz, enstatite, sillimanite, cordierite, and sapphirine ($Mg_2Al_4SiO_{10}$) in the system $MgO–Al_2O_3–SiO_2$ is degenerate.

(a) Write balanced reactions for all univariant lines at this invariant point.

(b) Determine which phases are singular and which are indifferent. How many invariant lines do you expect around the invariant point?

(c) Determine the slope of the sapphirine-absent reaction from data in Table 7-1.

(d) Construct a $P–T$ diagram for this set of univariant lines. Label all reactions, and draw triangular diagrams indicating the stable mineral assemblages in each divariant field.

8-8. Muscovite and quartz react to form sillimanite, potassium feldspar, and steam at high temperatures.

(a) From the data in Table 7-1, calculate the ΔG°_r for this reaction at a range of temperatures from 400 to 800 K, and plot the value of ΔG°_r against T. Determine the equilibrium temperature for this reaction at 10^5 Pa. (This is most easily done using a spreadsheet.)

(b) Assuming steam behaves ideally, calculate the equilibrium pressure for this reaction at 600, 650, 700, 750, and 800 K using Eq. 8-10. This equation has to be solved by trial and error; guess a pressure and see if it is correct. Plot a $P–T$ diagram for the univariant line. (Recall that the gas constant $R = 8.3144 \text{ J mol}^{-1} \text{ K}^{-1}$.)

(c) Although steam behaves far from ideally under these conditions, the resulting $P–T$ diagram is typical of all reactions in which a gas phase is liberated (this includes most important metamorphic reactions).

8-9. At high pressure, the reaction considered in Prob. 8-7 will intersect the quartz–coesite polymorphic phase change. Without determining the precise nature of the distribution of univariant lines at this invariant point, make a qualitative prediction of what happens to the slope of this univariant line as it passes through the invariant point. The answer lies in the application of one of Schreinemakers rules.

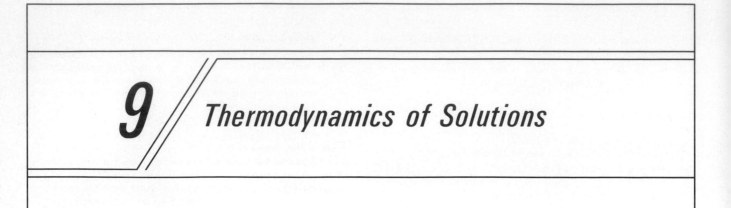

9 / Thermodynamics of Solutions

9-1 INTRODUCTION

So far we have dealt only with systems and minerals of fixed composition. Many systems in nature, however, are open; that is, material can be added or subtracted. In addition, most of the common rock-forming minerals belong to solid solution series in which compositions can vary widely. Magmas and intergranular solutions in metamorphic rocks are other examples of natural materials of variable composition. Changes in composition bring about changes in energy. Thus, when systems strive for equilibrium, compositional adjustments must be made so as to minimize free energies.

Changes in energy resulting from changes in composition are of two types. First, there is the simple addition of material and the energy that it brings with it. For example, when you add 1 gallon of gasoline to your gas tank, you expect a certain amount of work in return, depending on the efficiency of the automobile. Similarly, a surge of new magma into a cooling body of magma introduces a certain amount of heat, which depends on the temperature, heat capacity, and volume of the added magma.

Material added to a system may react with the initial material and produce additional energy. The amount of energy created in this way depends on the nature of the reaction. Consider, for example, two systems, one consisting of oil and the other of sulfuric acid. To each is added an equal volume of water. Because oil and water do not mix, the change in energy in this system depends only on the heat capacity and temperature of the added water. In the other system, however, the water dissolves readily in the sulfuric acid, and, as anyone knows who has performed this experiment, so much heat is generated that the water–sulfuric acid solution may even boil. Clearly, the energy associated with the addition of material to a system depends on the nature in which that material is taken into the system.

Most petrologically important processes involve changes in composition. Magmas, for example, are complex solutions of seven major oxides and a number of minor and trace elements (Sec. 6-1). As a magma begins to crystallize, its composition changes depending on the minerals that form. Its composition can also change through additions of new magma or assimila-tion of xenoliths of country rock. During metamorphism many important reactions evolve gas (H_2O or CO_2), which is added to the fluid phase that migrates along grain boundaries through the rock. The composition of this fluid is continually being modified by the various reactions that take place. In turn, the composition of this fluid in part determines what reactions can occur.

Igneous and metamorphic rock-forming minerals are mostly members of solid solution series; they, therefore, have a wide range of possible compositions. At the time of formation, however, the most favored composition is that which minimizes the free energy of the system. As conditions of temperature and pressure change, so will the equilibrium composition. But many minerals, once formed, lack the ability to change composition because of very low solid diffusion rates. Thus initial compositions may be preserved through a wide range of temperatures and pressures, especially in the cores of large grains. These initial compositions preserve a record of the conditions under which the minerals formed. The problem is how to read that record. Herein lies one of the most important applications of thermodynamics to petrology—the determination of the temperatures (geothermometry) and pressures (geobarometry) of formation of rocks from mineral compositions.

In this chapter we deal with the basic principles governing the free energy of solutions. It introduces the important concept of chemical potential, which is the intensive measure of the chemical contribution of a component to the free energy of a solution. Through chemical potential, the free energy of a solution can be related to the free energies of its pure components, values of which can be obtained from thermodynamic tables. Once we know how the free energy of a solution varies with composition, we can determine how to minimize the free energy and thus determine equilibrium compositions.

9-2 CONSERVATIVE AND NONCONSERVATIVE COMPONENTS OF A SOLUTION

The composition of a solution can be described in many different ways. Consider, for example, how we might describe the composition of a clinopyroxene that is a member of a solid solution whose general formula is

$$Ca^{VIII}(Mg_xFe_{1-x})^{VI}Si_2O_6$$

One mole of this pyroxene could be made from

1 mol Ca, X mol Mg, $1 - X$ mol Fe, 2 mol Si,
$$6 \text{ mol O} = 10 \text{ mol atoms}$$

or

1 mol CaO, X mol MgO, $1 - X$ mol FeO,
$$2 \text{ mol SiO}_2 = 4 \text{ mol oxides}$$

where X is a mole fraction. In the first case 10 mol of atoms is needed to make 1 mol of pyroxene, whereas in the second case only 4 mol of oxides is needed. Clearly, the number of moles of the components defined in these ways is not a conservative measure of composition; that is, the numbers are not conserved.

Molar amounts can provide a conservative measure of composition if the components chosen to describe the system are selected carefully. In the case of the clinopyroxene we could form 1 mol of pyroxene from

X mol $CaMgSi_2O_6$ plus $1 - X$ mol
$$CaFeSi_2O_6 = 1 \text{ mol pyroxene}$$

These components, in addition to being conservative, have the advantage that they correspond to minerals for which thermodynamic data are available in Table 7-1.

The composition of a solution can also be expressed in terms of weight percent or fractions of its components, and regardless of how the components are defined, weight always provides a conservative measure of composition. For this reason it will be found that most of the diagrams in Chapter 10 relating the compositions and temperatures of silicate liquids to silicate minerals are plotted in terms of weight percent.

9-3 FREE ENERGY OF SOLUTIONS

In Sec. 7-8, the free energy in a closed system was seen to be a function only of temperature and pressure (review Eq. 7-30). In an open system, however, the free energy is a function not only of temperature and pressure, but also of composition; that is, $G = G(T, P, n_a, n_b, \ldots, n_i)$, where n_i is the number of moles of component i. Thus, in the example of the pyroxene above, we can write $G = G(T, P, n_{Di}, n_{Hd})$. Any change in G can therefore be expressed in terms of the total differential, which for this pyroxene is

$$dG = \left(\frac{\partial G}{\partial T}\right)_{P, n_{Di}, n_{Hd}} dT + \left(\frac{\partial G}{\partial P}\right)_{T, n_{Di}, n_{Hd}} dP$$
$$+ \left(\frac{\partial G}{\partial n_{Di}}\right)_{T, P, n_{Hd}} dn_{Di} + \left(\frac{\partial G}{\partial n_{Hd}}\right)_{T, P, n_{Di}} dn_{Hd} \quad (9\text{-}1)$$

The partial derivative of G with respect to the number of moles of a particular component is given the name *chemical potential* and is defined as

$$\mu_i = \left(\frac{\partial G}{\partial n_i}\right)_{T, P, n_j \neq i} \quad (9\text{-}2)$$

That is, the chemical potential of component i is the partial molar free energy of i at constant T, P, and number of moles of all other (j) components. Equation 9-1 can therefore be rewritten for the clinopyroxene as

$$dG = -S\,dT + V\,dp + \mu_{Di}\,dn_{Di} + \mu_{Hd}\,dn_{Hd} \quad (9\text{-}3)$$

or for the general case as

$$dG = -S\,dT + V\,dP + \sum_i \mu_i\,dn_i \quad (9\text{-}4)$$

An important property of chemical potential is that when multiplied by the number of moles of the component, it gives the contribution of that component to the total free energy of the solution. Thus the total free energy of a solution is simply the sum of all the $n_i\mu_i$ terms; that is,

$$G = \sum_i n_i\mu_i \quad (9\text{-}5)$$

If the solution were to contain just one component, diopside for example in the case of the pyroxene, Eq. 9-5 would reduce to $G^{Px} = n_{Di}\mu_{Di}$ or $\mu_{Di} = G^{Px}/n_{Di}$, which is $\bar{G}^{Px}$. In this limiting case of the solution becoming the pure end member, the chemical potential becomes the molar free energy of the pure substance.

If Eq. 9-5 is differentiated, we obtain

$$dG = \sum_i n_i\,d\mu_i + \sum_i \mu_i\,dn_i \quad (9\text{-}6)$$

which if combined with Eq. 9-4, yields

$$S\,dT - V\,dP + \sum_i n_i\,d\mu_i = 0 \quad (9\text{-}7)$$

This is known as the *Gibbs–Duhem equation*. Note that in this equation, the potentials or intensive variables all appear as derivatives, whereas in Eq. 9-4 only T and P appear as such.

In the special case where temperature and pressure are constant, the free energy of a solution can be changed only through changes in composition, and the Gibbs–Duhem equation reduces to

$$\sum_i n_i\,d\mu_i = 0 \quad (9\text{-}8)$$

This shows that the chemical potentials of the various components must vary in a related way. For the example of the pyroxene we can write

$$n_{Di}\,d\mu_{Di} + n_{Hd}\,d\mu_{Hd} = 0 \quad (9\text{-}9)$$

which on rearranging gives

$$d\mu_{Hd} = -\left(\frac{n_{Di}}{n_{Hd}}\right)d\mu_{Di} \qquad (9\text{-}10)$$

Thus, if a variation in composition changes the chemical potential of diopside, the chemical potential of hedenbergite would change according to Eq. 9-10.

Let us consider what happens to the free energy of a system consisting of two crystals, one of augite and the other of pigeonite, if a small quantity of diopside component (dn_{Di}) is transferred from the pigeonite to the augite at constant temperature and pressure ($dT = 0$, $dP = 0$). Because G is an extensive property, its values are additive. Therefore,

$$dG^{Sy} = dG^A + dG^P \le 0$$

where the superscripts Sy, A, and P stand for system, augite, and pigeonite, respectively. The $\le$ sign indicates that if the change is to be spontaneous, dG must be negative, and at equilibrium dG is zero. From Eq. 9-3 we can write

$$dG^A = \mu_{Di}^A \, dn_{Di}^A + \mu_{Hd}^A \, dn_{Hd}^A$$

Since only the number of moles of diopside change, this equation reduces to

$$dG^A = \mu_{Di}^A \, dn_{Di}^A$$

Similarly, we write for the pigeonite crystal

$$dG^P = \mu_{Di}^P \, dn_{Di}^P$$

But $dn_{Di}^A = -dn_{Di}^P$, so that

$$dG^{Sy} = \mu_{Di}^A \, dn_{Di}^A - \mu_{Di}^P \, dn_{Di}^A \le 0$$

or

$$dG^{Sy} = (\mu_{Di}^A - \mu_{Di}^P) \, dn_{Di}^A \le 0 \qquad (9\text{-}11)$$

Thus, for a spontaneous change to occur, dG^{Sy} must be less than zero. This can occur in one of two ways; either $dn_{Di}^A < 0$ or $(\mu_{Di}^A - \mu_{Di}^P) < 0$, which is to say that $\mu_{Di}^P > \mu_{Di}^A$. The first possibility does not satisfy the conditions of the problem, that is, diopside component is transferred from pigeonite to augite ($dn_{Di}^A > 0$). We come to the important conclusion, therefore, that if the diopside component is to be transferred spontaneously from pigeonite to augite, the chemical potential of diopside in pigeonite must be greater than the chemical potential of diopside in augite. In other words, *material diffuses from regions of high chemical potential to regions of low chemical potential.*

It is also evident from Eq. 9-11 that at equilibrium ($dG^{Sy} = 0$), $\mu_{Di}^A = \mu_{Di}^P$; that is, the transfer of diopside from pigeonite to augite proceeds until, at equilibrium, the chemical potentials of diopside in augite and in pigeonite are equal. This is analogous to heat being transferred from regions of high temperature to regions of low temperature until at equilibrium the

temperatures become the same. *At equilibrium, then, potentials such as T, P, and μ_i must be everywhere the same throughout the system.*

9-4 FREE ENERGY OF IDEAL SOLUTIONS

If two end members of a solution are placed together, given time, they will mix and form a homogeneous solution. Obviously, this occurs more rapidly if the end members are gases than if they are solids, but this is a kinetic, not a thermodynamic, distinction. If mixing occurs spontaneously, the free energy of the system is lowered by an amount known as the free energy of mixing, ΔG_{mix}. From Eq. 7-24 we can write

$$\Delta G_{mix} = \Delta H_{mix} - T\Delta S_{mix} \qquad (9\text{-}12)$$

In general, there will be both an enthalpy and entropy of mixing. If, however, the enthalpy of mixing is zero, we have what is known as an *ideal solution*, where

$$\Delta G_{mix} = -T\Delta S_{mix} \qquad (9\text{-}13)$$

To calculate the free energy of mixing of an ideal solution, we must be able to determine the ΔS_{mix} (nonideal solutions will be examined in Sec. 9-5).

For illustrative purposes we will determine the entropy of mixing of pure diopside and hedenbergite to form an intermediate composition pyroxene with the formula $Ca(Mg_xFe_{1-x}) \cdot SiO_2O_6$, which in terms of end members would be Di_xHd_{1-x}. The solution will be assumed to behave ideally. The initial state of the system is represented in Figure 9-1a by pure diopside and hedenbergite separated by a partition. Once the partition is removed, the two pyroxenes mix to form the homogeneous pyroxene of the final state.

The entropy in the initial state of the system is simply the weighted average of the entropies of the two end members; that is,

$$S_{initial} = X_{Di}\bar{S}^{Di} + (1 - X_{Di})\bar{S}^{Hd} \qquad (9\text{-}14)$$

Graphically, this is obtained by drawing a straight line between the entropies of the end members on an S versus mole fraction plot, and then selecting the value on this line corresponding to the given value of X_{Di} (Fig. 9-1b). When the partition is removed, the initial entropies of the two pyroxenes are retained, but to these are added an entropy resulting from the spontaneous mixing of the pyroxenes. The final entropy of the system will be the sum of the initial entropies plus the configurational entropy resulting from mixing. That is,

$$S_{final} = X_{Di}\bar{S}^{Di} + (1 - X_{Di})\bar{S}^{Hd} + S_{config} \qquad (9\text{-}15)$$

If the system eventually contains 1 mol of homogeneous pyroxene N_0 (Avogadro's number) sites are available on which to place N_{Di} ions of Mg^{2+} and N_{Hd} ions of Fe^{2+}. It is important at this point to check the formula of the pyroxene to see that 1 mol will indeed have N_0 sites for the magnesium and iron

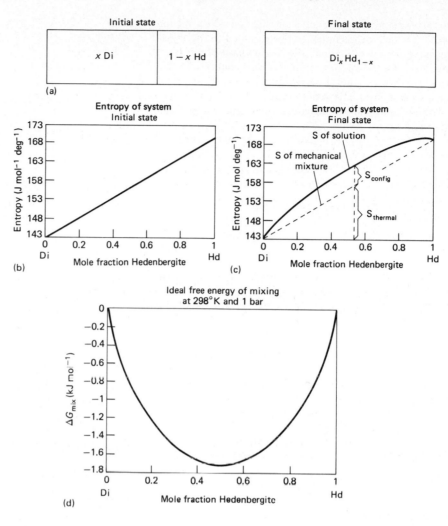

FIGURE 9-1 Mixing of diopside and hedenbergite to form a clinopyroxene with composition $CaMg_xFe_{1-x}Si_2O_6$. (a) In the initial state of the system, diopside and hedenbergite are separated by a partition, which is then removed and the two pyroxenes mix to form a homogeneous solid solution. (b) Initial and (c) final entropies of the system, assuming ideal mixing. (d) Ideal free energy of mixing at 298 K and 1 bar.

(this depends on how the formula is written). The number of possible ways of arranging these ions (Sec. 7-6) is

$$\Omega = \frac{N_0!}{N_{Di}!\, N_{Hd}!} \tag{9-16}$$

According to Eq. 7-18, then, the configurational entropy is

$$S_{config} = k \ln \frac{N_0!}{N_{Di}!\, N_{Hd}!} \tag{9-17}$$

where k is the Boltzmann constant, which is equal to R/N_0. Thus

$$S_{config} = \frac{R}{N_0}\left[\ln N_0 - (\ln N_{Di}! + \ln N_{Hd}!)\right] \tag{9-18}$$

Using Stirling's approximation ($\ln N! = N \ln N - N$ when N is large) and recalling that $N_{Di} + N_{Hd} = N_0$, Eq. 9-18 becomes

$$S_{config} = R\left(\frac{N_0}{N_0}\ln N_0 - \frac{N_{Di}}{N_0}\ln N_{Di} - \frac{N_{Hd}}{N_0}\ln N_{Hd}\right) \tag{9-19}$$

But $N_{Di}/N_0 = X_{Di}$ and $N_{Hd}/N_0 = X_{Hd}$ and $X_{Di} + X_{Hd} = 1$. Therefore,

$$S_{config} = R\left(\frac{N_{Di}+N_{Hd}}{N_0}\ln N_0 - X_{Di}\ln N_{Di} - X_{Hd}\ln N_{Hd}\right)$$

which on rearranging gives

$$S_{config} = -R(X_{Di}\ln N_{Di} - X_{Di}\ln N_0 + X_{Hd}\ln N_{Hd} - X_{Hd}\ln N_0)$$

which simplifies to

$$S_{config} = -R\left(X_{Di}\ln \frac{N_{Di}}{N_0} + X_{Hd}\ln \frac{N_{Hd}}{N_0}\right)$$

or

$$S_{config} = -R(X_{Di}\ln X_{Di} + X_{Hd}\ln X_{Hd})$$

But $X_{Hd} = 1 - X_{Di}$, so

$$S_{config} = -nR[X_{Di}\ln X_{Di} + (1 - X_{Di})\ln(1 - X_{Di})] \tag{9-20}$$

The factor n, the number of equivalent sites on which mixing takes place in each formula unit, has been added to this equation to make it generally applicable. In the case of pyroxene $[Ca(Mg, Fe)Si_2O_6]$, n is 1, but 1 mol of olivine contains $2 \times N_0$ mixing sites $[(Mg, Fe)_2SiO_4]$, so n would be 2; and in garnet $[(Mg, Fe, Ca)_3Al_2Si_3O_{12}]$, n would be 3.

The entropy of any homogeneous pyroxene can be found by combining Eqs. 9-20 and 9-15:

$$S_{final} = X_{Di}\bar{S}^{Di} + (1 - X_{Di})\bar{S}^{Hd} - nR[X_{Di} \ln X_{Di} + (1 - X_{Di}) \ln(1 - X_{Di})] \quad (9\text{-}21)$$

Assuming an ideal solution, the variation in the entropy of clinopyroxene as a function of composition at 298 K and 10^5 Pa is given in Figure 9-1c. The entropy of intermediate pyroxenes is clearly higher than that which would result from a simple admixture of end members. It is higher by the amount contributed by the configurational entropy. That is, the ΔS_{mix} in an ideal solution is the configurational entropy (Eq. 9-20).

If diopside and hedenbergite mix ideally, G_{mix} (Eq. 9-13) is given by

$$\Delta G_{mix} = nRT[X_{Di} \ln X_{Di} + (1 - X_{Di}) \ln(1 - X_{Di})] \quad (9\text{-}22)$$

The graph of this equation is given in Figure 9-1d. Note that ΔG is always negative; that is, mixing takes place spontaneously, and the greatest lowering of free energy occurs where the mole fraction is 0.5 (Prob. 9-1).

We are now in a position to calculate the free energy of formation of intermediate members of an ideal solution, given the free energies of formation of the pure end members. We proceed in the same way as with entropy; a weighted average of the free energies of formation of the end members is added to the ΔG_{mix}. Again, for the example of a clinopyroxene,

$$\Delta\bar{G}_f = X_{Di}\bar{G}_f^{Di} + (1 - X_{Di})\bar{G}_f^{Hd} + nRT[X_{Di} \ln X_{Di} + (1 - X_{Di}) \ln(1 - X_{Di})] \quad (9\text{-}23)$$

The graph of Eq. 9-23 is given in Figure 9-2. The first two terms of the equation give the dashed line; this is equivalent to a simple mechanical mixture (no solution) of end members. Actual free energies are lower than this by the amount contributed by the ΔG_{mix}, which is negative.

Let us consider what significance the slope of the $\Delta\bar{G}_f$ versus X_{Di} line has in Figure 9-2. We can define the slope of this line at any point in terms of the intercepts of the tangent with the G axes. We will define the intercept on the G axis at the diopside end member as μ_{Di} and at the hedenbergite end member as μ_{Hd}. The slope at any mole fraction X_{Di} is then

$$\left(\frac{\partial \Delta G}{\partial X_{Di}}\right)_{T,P} = \frac{\mu_{Di} - \mu_{Hd}}{1} = \mu_{Di} - \mu_{Hd} \quad (9\text{-}24)$$

But the slope can also be obtained by differentiating Eq. 9-23:

$$\left(\frac{\partial \Delta G}{\partial X_{Di}}\right)_{T,P} = (\Delta\bar{G}_f^{Di} + nRT \ln X_{Di}) - (\Delta\bar{G}_f^{Hd} + nRT \ln X_{Hd})$$

$$(9\text{-}25)$$

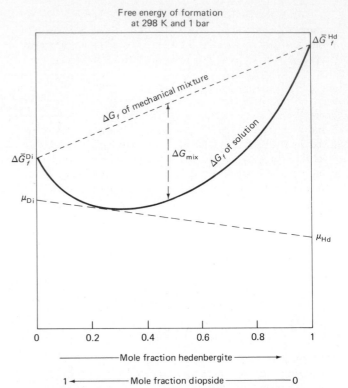

Free energy of formation at 298 K and 1 bar

FIGURE 9-2   Free energy of formation of diopside–hedenbergite solutions at 298 K and 1 bar, assuming ideal mixing. Intercepts on the ΔG axes of the tangent to the free-energy curve give the chemical potentials of diopside and hedenbergite in the solution at the point of tangency.

from which it follows that

$$\mu_{Di} = \Delta\bar{G}_f^{Di} + nRT \ln X_{Di} \quad (9\text{-}26)$$

and

$$\mu_{Hd} = \Delta\bar{G}_f^{Hd} + nRT \ln X_{Hd} \quad (9\text{-}27)$$

The intercepts of the tangent line at $X_{Hd} = 0$ and $X_{Hd} = 1$ are, in fact, the chemical potentials of diopside and hedenbergite, respectively, in the pyroxene at the point of tangency on the free-energy curve. Note that as the point of tangency approaches pure diopside the intercept, μ_{Di}, approaches the free energy of formation of pure diopside; that is, as $X_{Di} \rightarrow 1$, $\mu_{Di} \rightarrow \Delta\bar{G}_f^{Di}$. Similarly, as $X_{Hd} \rightarrow 1$, $\mu_{Hd} \rightarrow \Delta\bar{G}_f^{Hd}$. This same conclusion was previously deduced from Eq. 9-5. Indeed, the molar free energy of formation of a pure substance ($\Delta\bar{G}_f^{Di}$) is commonly replaced by the symbol μ_{Di}^*, the * indicating that this is the chemical potential of diopside in the pure end member at the specified temperature and pressure. Equation 9-26 can therefore be written in a general form as

$$\mu_i = \mu_i^* + nRT \ln X_i \quad (9\text{-}28)$$

Thus as X_i approaches 1, $\ln X_i$ approaches zero and μ_i approaches the chemical potential of the pure end member (see Prob. 9-3).

The relation between the chemical potential and mole fraction of a component in an ideal solution (Eq. 9-28) can also be derived from the functional relation between free energy and the pressure of an ideal gas. From Eq. 8-6 we have

$$\bar{G}_{i,T,P} = \bar{G}_{i,T}^{\circ} + RT \ln P_i \qquad \text{bars} \qquad (9\text{-}29)$$

where $\bar{G}_{i,T}^{\circ}$ is the molar free energy of pure gas i at temperature T and a pressure of 1 bar. Although this equation was introduced to evaluate how changes in the intensive parameter P affect the free energy of an ideal gas, it is equally applicable to situations where pressure is decreased by the gas being diluted with another gas; that is, it applies to partial pressures. For ideal gases, the partial pressure (P_i) is related to the pressure of the pure gas (P_i^*) by *Raoult's law*:

$$P_i = X_i P_i^* \qquad (9\text{-}30)$$

Substituting $X_i P_i^*$ for P_i in Eq. 9-29 gives

$$\bar{G}_{i,T,P} = \bar{G}_{i,T}^{\circ} + RT \ln P_i^* + RT \ln X_i \qquad (9\text{-}31)$$

But according to Eq. 8-6,

$$\bar{G}_{i,T}^{\circ} + RT \ln P_i^* = \bar{G}_{i,T,P}^*$$

where $G_{i,T,P}^*$, is the free energy of the pure gas at pressure P^* and temperature T. Thus Eq. 9-31 becomes

$$\bar{G}_{i,T,P} = \bar{G}_{i,T,P}^* + RT \ln X_i \qquad (9\text{-}32)$$

But the free energy per mole is the chemical potential. Thus the chemical potential of a component i in an ideal gas solution is

$$\mu_i = \mu_i^* + RT \ln X_i \qquad (9\text{-}33)$$

where μ_i^* is the molar free energy of the pure i gas at specified pressure (bars) and temperature. This, then, is identical to Eq. 9-28.

9-5 FREE ENERGY OF NONIDEAL SOLUTIONS

In nonideal solutions ΔH_{mix} is not zero and the relation between chemical potential and mole fraction no longer holds. To retain the simple functional form of Eq. 9-33, a term known as the *activity*, a_i, is created, such that for a nonideal solution,

$$\mu_i = \mu_i^* + RT \ln a_i \qquad (9\text{-}34)$$

The activity can be thought of as the effective mole fraction of a component in a nonideal solution, just as the fugacity is the effective pressure of a nonideal gas (Eq. 8-7). The activity measures the chemical potential of a component in a nonideal solution in the same way that mole fraction measures the chemical potential of a component in an ideal solution.

The activity of a component is related to its mole fraction by an *activity coefficient*, γ_i;

$$a_i = \gamma_i X_i \qquad (9\text{-}35)$$

The activity coefficient is a function of temperature, pressure, and composition; it can be greater than or less than 1, and when equal to 1, $a_i = X_i$, and the solution is ideal. The value of γ_i cannot be determined from thermodynamic properties but must be determined experimentally from measurements of a_i and X_i. Typical relations of a_i and X_i are shown in Figure 9-3. Note that as $X_i \to 1$, $a_i \to 1$, and therefore $\gamma_i \to 1$. At low concentrations, a_i and X_i commonly are linearly related, making it possible to express the activity as

$$a_i = K_i X_i \qquad (9\text{-}36)$$

where K_i is known as the *Henry's law constant*. This formulation of the activity is useful in evaluating the distribution of trace elements between coexisting phases.

The free energy of a two-component (1, 2) nonideal solution can be expressed in a form similar to that for an ideal solution (Eq. 9-23),

$$\Delta \bar{G}_f = X_1 \Delta \bar{G}_f^1 + (1 - X_1) \Delta \bar{G}_f^2 + \Delta \bar{G}_{\text{mix}} \qquad (9\text{-}37)$$

Because the $\Delta \bar{G}_{\text{mix}}$ may contain a ΔH_{mix} term, and ΔS_{mix} may differ from the ideal entropy of mixing, the nonideal $\Delta \bar{G}_{\text{mix}}$ can be greater or less than the ideal $\Delta \bar{G}_{\text{mix}}$. Indeed, under certain conditions of pressure, temperature, and composition, the $\Delta \bar{G}_{\text{mix}}$ in some solutions can become strongly positive. Before considering the variation of free energy with composition in nonideal solutions, it is necessary to consider the stability of a phase as a function of $G(T, P, X)$.

For a binary solution (ideal or nonideal) we can write from Eq. 9-4

$$dG = -S \, dT + V \, dP + \mu_1 \, dn_1 + \mu_2 \, dn_2 \qquad (9\text{-}38)$$

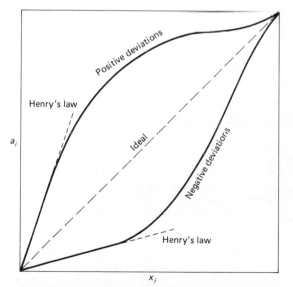

FIGURE 9-3 Relation between activity, a_i, and mole fraction, x_i, of a component i in solution. Both positive and negative deviations from ideal behavior are shown. The linear relation between activity and mole fraction at low concentrations of i is referred to as Henry's law.

To determine if this solution is thermally stable, we take first and second derivatives of G with respect to T, keeping the other variables constant:

$$\frac{\partial G}{\partial T} = -S \quad \text{and} \quad \frac{\partial^2 G}{\partial T^2} = -\frac{\partial S}{\partial T}$$

But $\partial S / \partial T = C_P / T$, and because $C_P > 0$ and $T > 0$, $-\partial S / \partial T < 0$. Therefore, $\partial^2 G / \partial T^2 < 0$. Thus for a phase to be thermally stable, it must not only have a negative slope on a G–T plot ($\partial G / \partial T = -S$), but it must be concave downward ($\partial^2 G / \partial T^2 < 0$) (Fig. 9-4a).

To determine whether a phase has mechanical stability, we take first and second derivatives of Eq. 9-38 with respect to P:

$$\frac{\partial G}{\partial P} = V \quad \text{and} \quad \frac{\partial^2 G}{\partial P^2} = \frac{\partial V}{\partial P} = -V\beta < 0$$

The second derivative is negative because the volume and coefficient of compressibility (β) are both positive. On a G–P plot a mechanically stable phase must have a positive slope (V) that is concave downward (Fig. 9-4b).

Finally, to determine whether a phase is chemically stable, we take first and second derivatives of Eq. 9-38 with respect to the mole fraction of, for example, component 2, recognizing of course that $X_2 = 1 - X_1$:

$$\frac{\partial G}{\partial X_2} = \mu_2 - \mu_1 \quad \text{and} \quad \frac{\partial^2 G}{\partial X_2^2} = \frac{\partial \mu_2}{\partial X_2} > 0$$

According to Eq. 9-34, addition of component 2 to the solution increases the chemical potential of component 2; therefore, $\partial \mu_2 / \partial X_2$ must be positive. On a G–X plot, then, a stable phase must plot as a concave upward line. If it becomes convex upward, the phase will become unstable (Fig. 9-4c). To find out how the chemical potential of one of the components varies with composition in a chemically stable solution, we take first and second derivatives of Eq. 9-28:

$$\frac{\partial(\mu_1 - \mu_1^*)}{\partial X_2} = \frac{-RT}{1 - X_2} < 0 \quad \text{and} \quad \frac{\partial^2(\mu_1 - \mu_1^*)}{\partial X_2^2} = \frac{-RT}{(1 - X_2)^2} < 0$$

Because the first and second derivatives are both negative, a phase must plot as a negatively sloping concave downward line in a $\mu_1 - \mu_1^*$ versus X plot if it is to be stable (Fig. 9-4d). If this condition is not met, the phase becomes unstable and the solution unmixes.

Examination of Figure 9-2 reveals that the variation of free energy with composition in an ideal solution satisfies the requirements for chemical stability. Because the $\Delta \bar{G}_{mix}$ in an ideal solution is always negative, the free energy versus mole fraction plot is always concave upward. The same, however, is not always true for nonideal solutions. The $\Delta \bar{G}_{mix}$ in nonideal solutions can even become positive, which makes the solution chemically unstable. We will now examine how the $\Delta \bar{G}_{mix}$ in nonideal solutions can vary.

9-6 NONIDEAL SOLUTION: THE REGULAR SOLUTION MODEL

The $\Delta \bar{G}_{mix}$ of a nonideal solutions cannot be deduced from thermodynamic properties of the end members of a solution; it must be measured. Once the measured values are obtained, a number of physical models of solutions can be used to account for the variation of $\Delta \bar{G}_{mix}$ with composition. We will examine two of these, the *symmetric regular solution model* and the *asymmetric regular solution model* (Thompson 1967).

In an ideal solution, the $\Delta \bar{G}_{mix}$ results only from the entropy of mixing (Eq. 9-22), which is calculated from the number of possible ways of arranging the components on the available sites in the solution (Eq. 9-16). It was assumed that the occupying of a site by one component did not affect the statistical probability of what component might occupy an adjoining site. If, however, different pairs of molecules have different interaction energies, this assumption would not be valid. The interaction between components in a regular solution is accounted for in the expression for the $\Delta \bar{G}_{mix}$ by adding an excess energy term to the ideal mixing term; that is,

$$\Delta \bar{G}_{mix} = \Delta \bar{G}_{ideal} + \Delta \bar{G}_{excess} \tag{9-39}$$

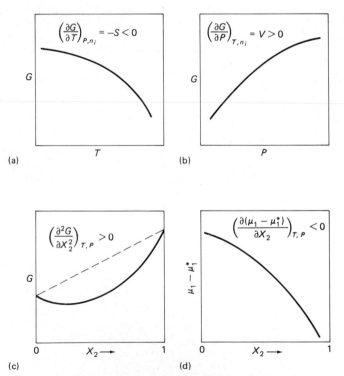

FIGURE 9-4 Necessary conditions for a solution to be stable. (a) The G versus T plot must have a negative slope and be concave downward for thermal stability. (b) The G versus P plot must have a positive slope and be concave downward for mechanical stability. (c) The G versus X plot must be concave upward for chemical stability. (d) The μ-μ^* versus X plot must have a negative slope for chemical stability.

Prior to mixing in a two-component system (A and B) there would be separate groups of A–A and B–B molecules. After mixing, juxtaposed A–B molecules will be present. We let W_{AA} represent the increase in potential energy in bringing two molecules of A from infinity to adjoining sites in the pure A end member, and W_{BB} represents the equivalent increase in potential energy for the B end member. If the increase in potential energy in bringing one molecule of A and one of B from infinity to adjoining sites in the solution is W_{AB}, the change in potential energy resulting from mixing pure A and B would be

$$W = W_{AB} - \frac{W_{AA}}{2} - \frac{W_{BB}}{2} \qquad (9\text{-}40)$$

W, which is known as the *interchange energy*, can be positive or negative. The $\Delta \bar{G}_{\text{excess}}$ is then given by

$$\Delta \bar{G}_{\text{excess}} = nWX_A X_B \qquad (9\text{-}41)$$

The factor n is the number of equivalent sites on which mixing takes place (see Eq. 9-20). The $\Delta \bar{G}_{\text{mix}}$ of a regular binary solution is therefore

$$\Delta \bar{G}_{\text{mix}} = \underbrace{nRT(X_A \ln X_A + X_B \ln X_B)}_{\text{ideal}} + \underbrace{nWX_A X_B}_{\text{excess}} \qquad (9\text{-}42)$$

Note that as $X_A \to 1$, $X_B \to 0$, and therefore $\Delta \bar{G}_{\text{excess}} \to 0$. The value of $\Delta \bar{G}_{\text{excess}}$ is a maximum at a mole fraction of 0.5, and it decreases symmetrically above and below this value. For this reason this particular model is known as the *symmetric regular solution model*.

The ideal term in Eq. 9-42 is always negative and will always dominate the expression for $\Delta \bar{G}_{\text{mix}}$ at high temperatures. The excess term, on the other hand, can be positive or negative. It can, therefore, increase or decrease the $\Delta \bar{G}_{\text{mix}}$ over that which would occur from ideal mixing alone. At low temperatures, the excess term dominates the expression for $\Delta \bar{G}_{\text{mix}}$. These relations are illustrated in Figure 9-5. The ideal component of $\Delta \bar{G}_{\text{mix}}$ is plotted separately for temperatures ranging from 500 to 1300 K (Fig. 9-5a) so that it can be compared with the actual $\Delta \bar{G}_{\text{mix}}$ that would result from an interchange energy of $+15$ kJ mol^{-1}. At temperatures in excess of 900 K, the $\Delta \bar{G}_{\text{mix}}$ versus X plot for this regular solution is concave upward, but the values of

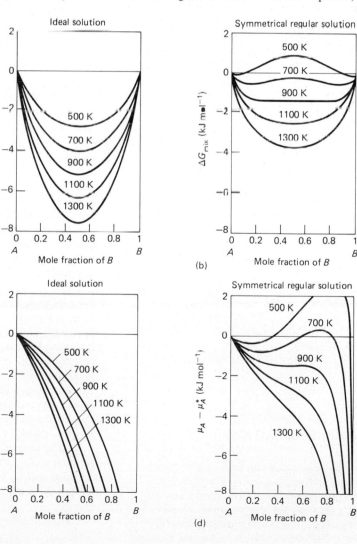

FIGURE 9-5 (a) The ideal component of the free energy of mixing of a binary solution at temperatures ranging from 500 to 1300 K. (b) Actual free energies of mixing (ideal + excess) of a symmetric regular solution having an interchange energy of 15 kJ mol^{-1} at temperatures ranging from 500 to 1300 K. Below 900 K, part of the ΔG curve is concave downward, and hence unstable. (c) Variation in the ideal component of the chemical potential of component A as a function of X_B. (d) Actual variation (ideal + excess) in chemical potential of component A at the same temperatures as in (b).

$\Delta \bar{G}_{mix}$ are far less negative than they would be if the solution behaved ideally. Below 900 K, the excess term causes part of the $\Delta \bar{G}_{mix}$ versus X plot to become concave downward, and over this compositional range the solution is unstable and must unmix; that is, exsolution will occur.

The variation in the chemical potential of one component $(\mu_A - \mu_A^*)$ as a function of composition in a symmetrical regular solution is given by

$$\mu_A - \mu_A^* = \underbrace{nRT \ln(1 - X_B)}_{\text{ideal}} + \underbrace{nWX_B^2}_{\text{excess}} \qquad (9\text{-}43)$$

This also consists of ideal and excess terms. A graph of the ideal part of this equation is plotted in Figure 9-5c. At all temperatures, the plots have negative slopes and are concave downward. But when the excess term is added, the slopes are greatly diminished, and at temperatures below 900 K, part of each curve has a positive slope. Solutions having compositions in the range where the slopes are positive are unstable.

Because of the form of the expression for the $\Delta \bar{G}_{mix}$ of the symmetrical regular solution model, instabilities will first occur with falling temperature at a mole fraction of 0.5. With falling temperature the region of instability expands, but it does so symmetrically about the mole fraction of 0.5 (Fig. 9-5b). In some solutions, however, deviations, from ideality are greater near one end member than they are near the other. For these solutions the *asymmetric regular solution model* (Margules' formulation) is used. Here a separate value of W is used for each of the components, and then the $\Delta \bar{G}_{mix}$ is given by

$$\Delta \bar{G}_{mix} = nRT(X_A \ln X_A + X_B \ln X_B) + nX_A X_B(X_B W_A + X_A W_B) \qquad (9\text{-}44)$$

or

$$\Delta \bar{G}_{mix} = nRT[(1 - X_B) \ln(1 - X_B) + X_B \ln X_B] + n(1 - X_B)X_B[X_B W_A + (1 - X_B)W_B] \qquad (9\text{-}45)$$

With the asymmetric regular solution model, variation in the chemical potential of one component as a function of composition is given by

$$\mu_A - \mu_A^* = nRT \ln(1 - X_B) + nX_B^2[W_A + 2(W_B - W_A)(1 - X_B)] \qquad (9\text{-}46)$$

Note that if $W_A = W_B$, then Eq. 9-46 reduces to Eq. 9-43.

9-7   UNMIXING OF NONIDEAL SOLUTIONS: EXSOLUTION

Examples of unmixing of nonideal solutions can be found in a number of important petrologic systems. For example, alkali feldspars (perthite), plagioclase (peristerite), pyroxenes, amphiboles, spinels, and numerous sulfides all form solutions that may unmix at low temperatures. The unmixing produces intergrowths of the two stable phases, forming what are known as exsolution textures. These textures commonly exhibit striking geometrical patterns which result from exsolving phases orienting themselves crystallographically so as to minimize the stress caused by having one phase in another (Fig. 9-6a). Unmixing also occurs in magmas to form silicate liquids of strikingly different compositions (Fig. 9-6b).

Using the symmetric regular solution model as an example of a nonideal solution, unmixing is seen to occur only if the interchange energy, W, is positive. If W is positive, below some critical temperature, T_c, the solution has a higher free energy than it would if it unmixed. Below this temperature the solution is unstable, which, as shown in Sec. 9-5, occurs when $d\mu_A/dX_B$ becomes positive. The critical temperature can, therefore, be determined by differentiating Eq. 9-43 and setting the derivative equal to zero:

$$\frac{d\mu_A}{dX_B} = -\frac{nRT_c}{1 - X_B} + 2nWX_B = 0 \qquad (9\text{-}47)$$

Because this critical temperature occurs at a mole fraction of 0.5 in the symmetric regular solution model, it follows that

$$T_c = \frac{2WX_B(1 - X_B)}{R} = \frac{W}{2R} \qquad (9\text{-}48)$$

The critical temperature in the example worked out in Figure 9-5, where $W = 15$ kJ mol^{-1}, would therefore be $(15,000/2 \times 8.3144)$ 902 K. Note that in Figure 9-5d, the plot for $\mu - \mu_1^*$ versus X_B at 900 K has a short segment in which the slope is positive; that is, the solution is unstable.

At any given temperature below T_c, the $\Delta \bar{G}_f$ versus X plot for a nonideal binary solution is of the general form shown in Figure 9-7a. In this diagram, the $\Delta \bar{G}_f$ of the two pure end members have been assigned equal values, to simplify the illustration. The following argument, however, is just as valid if the two end members have different values.

From Figure 9-7a we see that a solution having a mole fraction of $0.5A$ and $0.5B$ has a higher free energy (h) than if this bulk composition were composed of two solutions, one richer in A and the other richer in B. Indeed, from inspection of Figure 9-7a, the free energy of the system is seen to be minimized if the solution unmixes into the two compositions indicated by α and β. If the solution unmixes to these two compositions, the free energy of the system would be lowered to the point l, which is simply the sum of $\Delta \bar{G}_f^\alpha + \Delta \bar{G}_f^\beta$. It will be noted that a tangent to the ΔG versus X curve at point α is also tangent at point β, indicating that $\mu_A^\alpha = \mu_A^\beta$ and $\mu_B^\alpha = \mu_B^\beta$. At the particular temperature and pressure represented in Figure 9-7a, solutions having compositions of α and β are the equilibrium phases that should form (given appropriate kinetics) for bulk compositions between α and β.

The precise compositions of α and β at any given temperature and pressure can be determined by differentiating Eq. 9-42 and setting the derivative equal to zero (minimum) (see Prob. 9-7). From this it follows that

$$T_{\text{binode}} = \frac{W(1 - 2X_B^\alpha)}{R \ln[(1 - X_B^\alpha)/X_B^\alpha]} \qquad (9\text{-}49)$$

 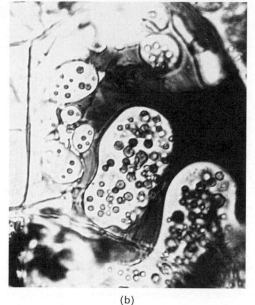

FIGURE 9-6   Exsolution textures developed in solid (a) and liquid (b) solutions. (a) Two directions of augite exsolution lamellae in a crystal of orthopyroxene. (b) Exsolution of an iron-rich silicate liquid (dark) from an iron-poor silicate liquid (clear) in the residual liquid in a basalt from Mauna Loa, Hawaii.

The term *binode* indicates that at this temperature, the two nodes on the free-energy curve indicate that two phases coexist and have lower free energies than intermediate compositions. Also note that because we are treating a symmetric regular solution, $X_B^\alpha = X_A^\beta$.

The compositions α and β in Figure 9-7a are for one particular temperature. We know from Figure 9-5b that if the temperature is lowered, the compositional difference between α and β will increase. The locus of points tracing the binodes in a T–X plot is known as a *solvus* (Fig. 9-8). A solvus in a binary T–X

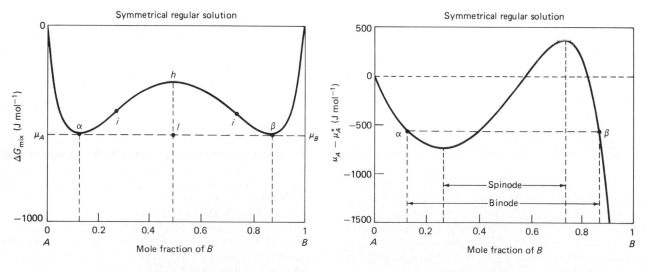

(a)

(b)

FIGURE 9-7   (a) Variation in the ΔG_{mix} of a symmetric regular solution at 700 K in which the interchange energy is 15 kJ mol^{-1}. Points α and β define the compositions of coexisting phases on the solvus at this temperature—α and β both have the same chemical potentials of A and B. The inflection points, i, on the free-energy curve mark the compositions of the spinode at this temperature. (b) Variations in the chemical potential of A ($\mu_A - \mu_A^*$) as a function of X_B. Compositions α and β mark the binode, and the points where the slope on the curve is zero mark the spinode.

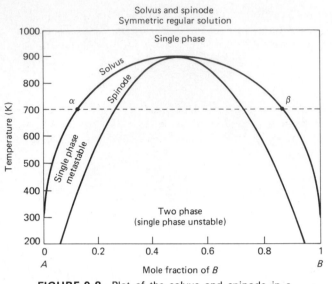

FIGURE 9-8 Plot of the solvus and spinode in a binary (*A–B*) symmetric regular solution model having an interchange energy of 15 kJ mol⁻¹. Compositions α and β at 700 K are the same points as in Figure 9-7. Between the solvus and spinode, single-phase solutions are metastable and, given time, should exsolve. Below the spinode, all single-phase solutions are unstable and must unmix.

diagram is a line at temperatures above which a homogeneous solution is stable but below which two separate phases have a lower free energy and hence are more stable than a single phase. At any temperature and bulk composition below the solvus, the equilibrium compositions of coexisting phases are those indicated by the compositions on the solvus on opposite sides of the two-phase region (α and β, for example). Clearly, these compositions change with temperature; if the compositions are preserved, for example by rapid eruption and quenching of magma, they can serve as an indicator of temperature. For example, a pair of alkali feldspars having compositions of 0.45 and $0.55 X_{Ab}$ must have formed at a higher temperature than a pair having compositions of 0.35 and $0.65 X_{Ab}$.

We can also examine the unmixing of nonideal solutions in terms of the variation of the chemical potential of a component as a function of composition. In Figure 9-7b, the variation in the chemical potential of *A* ($\mu_A - \mu_A^*$) is plotted versus the mole fraction of the other component, X_B. We know that once the slope becomes positive in such a plot, the solution is no longer stable. The two phases that form by unmixing must have the same chemical potential of a component in both. In addition, because of the symmetric regular solution model, if the mole fraction of component *B* in one phase is X_B^α, the mole fraction of *B* in the other phase must be $1 - X_B^\alpha$; that is, $X_B^\alpha = X_A^\beta$. These conditions are satisfied by only one pair of compositions for α and β at this temperature, and this then defines the solvus, or binode, at this temperature.

Despite the thermodynamic arguments presented above, when solutions cool below a solvus, exsolution occurs only if there is time to nucleate the appropriate disparate phases and

to diffuse the necessary components to these phases for growth. In many silicates, especially at lower temperatures, diffusion rates are so low that solutions formed above the solvus may remain homogeneous, but metastable, at temperatures below the solvus. But over a certain composition range, even rapidly cooled high-temperature solutions are not able to remain homogeneous.

Consider a composition on the ΔG versus X curve in Figure 9-7a that falls between the two inflection points (*i*). Any fluctuation in composition, no matter how small, in a solution in this range would lower the free energy of the system, because the curve is concave downward; that is, a line joining any two compositions on the free-energy curve in this interval falls below the curve. Beyond the inflection points, however, small fluctuations of composition cause the free energy of the system to rise, because the free-energy curve is concave upward; in this range it is necessary to nucleate phases of very different composition to bring about a lowering of free energy—phases on either side of the solvus. Between the inflection points, phase separation cannot be prevented, but because it involves only small fluctuations in composition, the separation typically takes place on a submicroscopic scale; hence x-ray diffraction or electron microscopy is normally required to detect this type of exsolution. The locus of points marking the inflection points in a *T*–*X* diagram is known as a *spinode* (Fig. 9-8).

To determine the composition of the spinode at any temperature and pressure, we need to find where the inflection points are on the ΔG versus X curve. This is done by setting the second derivative of Eq. 9-42 equal to zero (Prob. 9-8). From this we obtain

$$T_{\text{spinode}} = \frac{2W X_B (1 - X_B)}{R} \qquad (9\text{-}50)$$

In Eq. 9-48, the critical temperature on the solvus, T_c, was shown to be $W/2R$. Therefore, $W/R = 2T_c$, which when substituted into Eq. 9-50 gives

$$T_{\text{spinode}} = 4 T_c X_B (1 - X_B) \qquad (9\text{-}51)$$

The inflection points on the ΔG versus X plot correspond to the two points on the $\mu_A - \mu_A^*$ versus X plot having zero slope (Fig. 9-7b). Thus T_{spinode} could equally well have been determined by setting the first derivative of Eq. 9-43 equal to zero (Prob. 9-6c).

Figure 9-8 shows both the binode, or solvus, and the spinode. With slow cooling, exsolution will strive to achieve the equilibrium compositions given by the solvus. Because of the kinetics of diffusion, especially in solid solutions, metastable high-temperature solutions are commonly preserved to low temperatures. Within the spinode, however, even metastable solutions become unstable and break down into small domains (mostly submicroscopic) of solutions having compositions given by the spinode.

Although the model presented here for the development of the spinode is satisfactory for symmetric regular *liquid* solutions, it tends to give somewhat too high temperatures for spinodes in *solid* solutions. Compositional fluctuations induce strains in crystals that must be taken into account in considering the energetics of spinodal exsolution. However, this is beyond the scope of this book.

9-8 EQUILIBRIUM CONSTANT OF A REACTION

Many reactions involve phases that are members of solutions. Commonly, such reactions do not involve the elimination of a phase but simply the adjusting of compositions of the solutions so as to minimize free energies. Consider, for example, an olivine crystal reacting with silica in a magma to form an orthopyroxene crystal. Each of these phases is a member of a solution, and as the reaction proceeds, the compositions, and thus the chemical potentials of the components in these solutions, change. The silica content of the magma decreases as the olivine reacts to form orthopyroxene; this in turn decreases the chemical potential of silica in the magma. The reaction does not proceed until the magma contains no silica, but only until the chemical potential of silica in the magma is decreased to a level where it is in equilibrium with the olivine.

Let us consider the simple reaction that might occur between olivine and orthopyroxene in a peridotite during metamorphism. Both minerals belong to solid solution series in which iron can be exchanged for magnesium. Equilibrium between these minerals requires that the iron and magnesium be distributed in such a way as to minimize the free energy of the system. This equilibrium can be achieved through a simple exchange reaction:

(forsterite) (ferrosilite)　　(fayalite) (enstatite)

$$Mg_2SiO_4 + 2FeSiO_3 \rightleftharpoons Fe_2SiO_4 + 2MgSiO_3 \quad (9\text{-}52)$$

This reaction, of course, does not proceed all the way to the right or to the left. Instead, at equilibrium, the olivine and the orthopyroxene both have specific Mg/Fe ratios. The question is: What determines these ratios?

As this reaction proceeds toward equilibrium, the reactants are not pure forsterite, ferrosilite, fayalite, and enstatite. Instead, the forsterite, for example, is in solution in an olivine and the ferrosilite is in solution in an orthopyroxene. Thus, if we are to write an expression for the ΔG of this reaction as it approaches equilibrium, it must be done in terms of the free energies of the reactants in their respective solutions; that is,

$$\Delta G_r = n_{Fa}^{Ol}\mu_{Fa}^{Ol} + n_{En}^{Opx}\mu_{En}^{Opx} - n_{Fo}^{Ol}\mu_{Fo}^{Ol} - n_{Fs}^{Opx}\mu_{Fs}^{Opx}$$

For the reaction to balance, the number of moles of fayalite must equal the number of moles of forsterite, and the numbers of moles of ferrosilite and enstatite must be twice that of fayalite. If we let there be 1 mol of fayalite, and allow the reaction to proceed to equilibrium ($\Delta G = 0$), we can write

$$\mu_{Fa}^{Ol} + 2\mu_{En}^{Opx} - \mu_{Fo}^{Ol} - 2\mu_{Fs}^{Opx} = 0 \qquad (9\text{-}53)$$

If we assume that both olivine and orthopyroxene form ideal solid solutions, each of the chemical potential terms can be replaced with the expression from Eq. 9-28; for example,

$$\mu_{Fa}^{Ol} = \mu_{Fa}^{*Fa} + RT \ln X_{Fa}^{Ol}$$

where μ_{Fa}^{*Fa} is the chemical potential or molar free energy of pure fayalite at the specified temperature and pressure. Making this type of substitution for each term in Eq. 9-53 and rearranging yields

$$\underbrace{\mu_{Fa}^{*Fa} + 2\mu_{En}^{*En} - \mu_{Fo}^{*Fo} - 2\mu_{Fs}^{*Fs}}_{\Delta G_r} + \underbrace{RT \ln \frac{(X_{Fa}^{Ol})(X_{En}^{Opx})^2}{(X_{Fo}^{Ol})(X_{Fs}^{Opx})^2}}_{K} = 0 \qquad (9\text{-}54)$$

The first group of terms is simply the free-energy change of reaction between the pure end members as written in Eq. 9-52. We can therefore write

$$-\frac{\Delta G_r}{RT} = \ln \frac{(X_{Fa}^{Ol})(X_{En}^{Opx})^2}{(X_{Fo}^{Ol})(X_{Fa}^{Opx})^2} \qquad (9\text{-}55)$$

The group of terms in the right-hand term is the familiar *equilibrium constant* of a reaction, K, which is the product of the molecular concentrations of the resultants of a reaction divided by the product of the molecular concentrations of the reactants; if any coefficient in the reaction is not 1, the concentration of that substance is raised to the power of that coefficient. Thus, in this reaction, the X_{En} and X_{Fs} in the orthopyroxene are raised to the power 2. Equation 9-55 then becomes

$$\ln K = -\frac{\Delta G_r}{RT} \qquad (9\text{-}56)$$

Equation 9-56 indicates that for a reaction where the reactants and products are in solution in phases, the equilibrium constant, K, is a function of the ΔG of the reaction between the pure end members. If the solutions behave ideally, the equilibrium constant is a direct measure of the compositions of the phases. If the solutions behave nonideally, the equilibrium constant is expressed in terms of activities; compositions can then be determined only after the activity coefficients are known.

Because the ΔG_r in Eq. 9-56 is likely to change with changes in temperature, and to a lesser extent, pressure, the equilibrium constant will also change. This means that the compositions of the minerals involved in the reaction may provide a measure of the temperatures and pressures at the time of reaction. A reaction that provides a sensitive measure of temperature can be used as a *geothermometer*, whereas one that is sensitive to changes in pressure can be used as a *geobarometer*.

To analyze how the equilibrium constant varies with temperature and pressure, we write Eq. 9-54 for a general case involving i different chemical potentials

$$R \ln K = -\frac{\sum v_i \mu_i^*}{T} \qquad (9\text{-}57)$$

where v_i is the coefficient of the ith component in the reaction ($v = 1$ for fayalite and forsterite and 2 for enstatite and ferrosilite in Eq. 9-54). The μ_i^* are functions of T and P; we therefore can express them in differential form by their total differentials:

$$R\, d \ln K = -\sum v_i \left[\left(\frac{\partial \mu_i^*/T}{\partial T} \right)_P dT + \left(\frac{\partial \mu_i^*/T}{\partial P} \right)_T dP \right]$$

$$= -\sum v_i \left[\left(\frac{\partial \mu_i^*/T}{\partial T} \right)_P dT + \frac{1}{T} \left(\frac{\partial \mu_i^*}{\partial P} \right)_T dP \right] \quad (9\text{-}58)$$

The group of terms in the first set of parentheses is the same as $(\partial \bar{G}_i/T/\partial T)_P$, which, from the rules of differentiation of a product, is

$$\frac{1}{T} \left(\frac{\partial \bar{G}_i}{\partial T} \right)_P - \frac{1}{T^2} \bar{G}_i$$

But $(\partial \bar{G}_i/\partial T)_P = -\bar{S}$ (Eq. 7-31), so that these terms can be expressed as

$$-\frac{\bar{S}_i}{T} - \frac{\bar{G}_i}{T^2} = -\frac{T\bar{S}_i + \bar{G}_i}{T^2} = -\frac{\bar{H}_i}{T^2}$$

When multiplied by v_i and summed over the i different pure end members, this becomes $-\Delta H_r/T^2$. The group of terms in the second set of parentheses, $(\partial \mu_i^*/\partial P)_T$, can, according to Eq. 7-32, be replaced by $\bar{V}_i$, which when summed over the i different pure end members becomes ΔV_r. Equation 9-58 can then be rewritten as

$$R\, d \ln K = \left(\frac{\Delta H_r}{T^2} \right)_P dT - \left(\frac{\Delta V_r}{T} \right)_T dP \quad (9\text{-}59)$$

Equation 9-59 allows us to determine how the equilibrium constant of a reaction varies as a function of temperature and pressure. If pressure is held constant ($dP = 0$), Eq. 9-59 becomes

$$\left(\frac{\partial \ln K}{\partial T} \right)_P = \frac{\Delta H_r}{RT^2} \quad (9\text{-}60)$$

This is known as the *van't Hoff* equation or *Gibbs–Helmholtz* equation. It indicates that if a reaction is endothermic ($\Delta H_r > 0$),

ln K, and thus K, increases with increasing temperature; that is, the reaction, as written, proceeds further to the right (increased products). If the reaction is exothermic ($\Delta H_r < 0$), K decreases with increasing temperature, which is to say that the reaction proceeds more to the left, thus increasing the concentrations of the reactants.

It also follows from Eq. 9-60 that a good geothermometer should have a large value of ΔH_r. In the example given in Eq. 9-52 involving olivine and orthopyroxene, ΔH_r is small, and thus the exchange of iron and magnesium between these minerals is rather insensitive to temperature changes. By contrast, the exchange of iron and magnesium between coexisting orthopyroxene and augite is sensitive to temperature changes and therefore makes a good geothermometer [see, for example, Saxena (1973)].

If the temperature is held constant ($dT = 0$) in Eq. 9-59, we have

$$\left(\frac{\partial \ln K}{\partial P} \right)_T = -\frac{\Delta V_r}{RT} \quad (9\text{-}61)$$

Most petrologic reactions involving condensed phases (solids and liquids) have very small values of ΔV_r. Consequently, changes in the equilibrium constant for these reactions are insensitive to changes in pressure. Very few reactions are suitable for geobarometry. Most reactions involving gases have large ΔV_r, but the gas phase commonly escapes during the reaction or we may not know its composition. In either case we are unable to evaluate the equilibrium constant.

The equilibrium constant provides a simple means of studying reactions in petrology. Even when a lack of thermodynamic data makes quantitative evaluation of the constant impossible, it is still feasible to use the constant in a qualitative way to order observations on the compositions of minerals. For example, many useful studies have been made of the variation in composition of metamorphic minerals with increasing temperature of metamorphism, even though no thermodynamic data were available at the time of these studies. Knowledge of the how the equilibrium constant is related to fundamental thermodynamic properties, however, allowed these compositional variations to be analyzed properly. We will have many occasions in the remainder of this book to refer to the equilibrium constant.

PROBLEMS

9-1. Prove that the maximum lowering of free energy resulting from mixing at constant temperature and pressure in an ideal binary solution occurs for a mole fraction of 0.5.

9-2. Forsterite and fayalite form the olivine solid solution series, which can be considered ideal. For 298 K and 10^5 Pa, calculate the free energy of 3 mol of forsterite and 2 mol of fayalite both before and after mixing. (Take care to consider the number of equivalent sites on which mixing occurs in the olivine formula.)

9-3. Plot a graph of $\mu_i - \mu_i^*$ versus X_i for a component i in an ideal solution at 298 and 1000 K.

9-4. At high temperatures, the alkali feldspars (Ab–Or) form a complete solid solution series. The solution, however, is far from ideal, and at low temperatures it unmixes into Na-rich and K-rich feldspars (perthite). The critical temperature on the solvus is given by Waldbaum and Thompson (1969) as T_c (K) = 921.23 + 13.4607P_c (kbar). Assuming a symmetric regular solution model, calculate the interchange energy, W, for this solid solution series at 1 bar.

9-5. Using an interchange energy of 15.5 kJ mol^{-1} in a symmetric regular solution model of alkali feldspars, calculate **(a)** the expected compositions of coexisting alkali feldspars on the solvus at 700 K and 1 bar, and **(b)** the compositions of coexisting feldspars on the spinode at 700 K, ignoring any strain energies in the crystal.

9-6. The thermodynamic properties of the alkali feldspar solid solution series are better accounted for by an asymmetric rather than symmetric regular solution model (Waldbaum and Thompson 1969). The two interchange energies, W_{Ab} and W_{Or}, are found to be functions of T and P:

$$W_{Ab} = 26{,}471 + 0.3870P - 19.381T \qquad \text{J mol}^{-1}$$

$$W_{Or} = 32{,}099 + 0.4690P - 16.136T \qquad \text{J mol}^{-1}$$

where P is measured in bars (10^5 Pa) and T is in kelvin.

(a) Plot ΔG_{mix} for the alkali feldspars as a function of mole fraction of orthoclase, X_{Or}, at 700 K and 1 bar. From the graph, determine the compositions of the coexisting alkali feldspars on the solvus, and compare these compositions with those obtained in Prob. 9-5a.

(b) From the plot in part (a), determine μ_{Ab} and μ_{Or} in the two feldspars coexisting on the solvus at 700 K and 1 bar relative to their values in the pure end members.

(c) From a plot of $\mu_{Ab} - \mu_{Ab}^*$ versus X_{Or}, determine the composition of the coexisting feldspars on the spinode at 700 K and 1 bar.

9-7. Derive Eq. 9-49 from Eq. 9-42.

9-8. The spinode is determined by the inflection points on the ΔG versus X plot. From Eq. 9-42, determine the positions of the inflection points, and develop the expression in Eq. 9-50 relating temperature, mole fraction, and interchange energy.

9-9. Many rocks contain coexisting plagioclase and alkali feldspar. Equilibrium between these two minerals involves the exchange of the albite component between the plagioclase and alkali feldspar structures; that is, $Ab^{Plag} \rightleftharpoons Ab^{Alk}$. Assuming ideal mixing in both feldspar types at high temperature, show through fundamental thermodynamic relations how the equilibrium constant for this exchange reaction could be used as a geothermometer.

10 / Phase Equilibria in Igneous Systems

10-1 INTRODUCTION

When first encountered, the great variety of igneous rocks and the large number of names and textural terms used to describe them can be bewildering. Diagrams showing the relations between mineralogical parameters, such as the composition of feldspar, abundance of ferromagnesian minerals, and so on, or the variation between major chemical constituents provide a means of attaching names to rocks, but they do not, in themselves, provide an explanation for the compositions and frequencies of rock types. These diagrams can be likened to the periodic table of the elements; they provide a means of classifying, but the underlying important factors justifying such a classification have to be sought elsewhere. Thermodynamics provides the rationale for rock compositions.

Ideally, when a magma crystallizes, an assemblage of minerals is formed that provides the minimum free energy possible for that particular bulk composition under the existing conditions. A close approach to this ideal situation is common in igneous rocks, particularly plutonic ones, and even when equilibrium is not achieved, mineral assemblages can still be understood in terms of reactions that are striving to bring about this minimization of free energy. Thermodynamics, therefore, provides a simple explanation for the mineralogical composition of igneous rocks, and as will be seen later, it can also be used to account for the composition of magmas.

10-2 TWO-COMPONENT SYSTEMS

Magmas are simply high-temperature solutions of the Earth's most abundant elements. The thermodynamic properties of solutions (Chapter 9) should therefore provide a basis for analyzing the stability and composition of magmas. This task may, at first, appear formidable, because the typical magma contains 10 major oxides and many minor elements. But most processes involved in the crystallization of magma can be described in terms of only two or three components, thus greatly simplifying the task. We shall first consider two-component systems.

In Prob. 8-2 you were asked to calculate the melting point of diopside from the entropies and enthalpies of the solid and liquid forms of this mineral. The solution, which involves determining the temperature at which the free energy of the liquid and solid are equal, is illustrated in Figure 10-1. In a plot of free energy versus temperature, both crystalline and liquid diopside have negative slopes $[(\partial G/\partial T)_P = -S]$, but that of the liquid is steeper because it has the larger entropy. Consequently, at high temperatures the liquid has the lower free energy and is stable, whereas at low temperatures the solid has the lower free energy and is stable. The two curves intersect at the melting point (T_m) of diopside.

At some temperature T below the melting point, the free energy of liquid diopside is higher than that of solid crystalline diopside. Although rapid cooling to this temperature might temporarily preserve liquid diopside in a metastable, supercooled (glassy) state, crystallization brings about a lowering of free energy by an amount ΔG. Under equilibrium conditions, therefore, liquid diopside can never be stable at temperatures below the melting point, unless some way exists of lowering the free energy of the liquid to the value of the solid. Solutions provide a means of doing this.

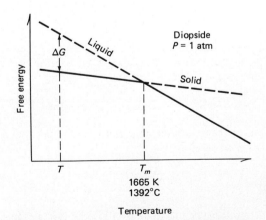

FIGURE 10-1 Free energy versus temperature plot at constant pressure for solid and liquid diopside.

If we assume, for the moment, that magma containing diopside in solution behaves ideally, Eq. 9-28 indicates that the chemical potential of diopside in that magma is given by

$$\mu_{Di}^L = \mu_{Di}^{*L} + RT \ln X_{Di}^L \qquad (10\text{-}1)$$

where μ_{Di}^{*L} is the chemical potential of diopside in a pure diopside liquid at the specified temperature and pressure. Note that if the solution becomes pure diopside, the mole fraction of diopside in the liquid (X_{Di}^L) becomes 1, its logarithm becomes zero, and $\mu_{Di}^L = \mu_{Di}^{*L}$. In any other solution, however, X is a fraction, its logarithm is negative, and $\mu_{Di}^L < \mu_{Di}^{*L}$. As long as the solution remains ideal, the lowering of chemical potential is independent of the substance causing the dilution. For instance, molten diopside could be diluted with molten olivine or molten anorthite, and the chemical potential of diopside would depend only on the mole fraction to which diopside was diluted.

At temperature T below the melting point of pure diopside (Fig. 10-1) it should be possible to dilute molten diopside so that its chemical potential is ΔG lower than that of pure molten diopside at this temperature. The chemical potential of molten diopside would then be the same as that of pure crystalline diopside (μ_{Di}^S) at this temperature, and equilibrium would be established. Note that we are considering only the case where the crystalline phase is pure; that is, crystals of diopside exhibit no solid solution with the component that dissolves in the magma. Many important rock-forming minerals, of course, do form solid solutions; these will be considered in Sec. 10-9. The equilibrium between pure crystalline diopside and molten diopside in solution can be expressed as

$$\mu_{Di}^S = \mu_{Di}^L = \mu_{Di}^{*L} + RT \ln X_{Di}^L \qquad (10\text{-}2)$$

This indicates that although pure, molten diopside can stably coexist with crystalline diopside at only one temperature (the melting point of pure diopside at a specified pressure), diopside-bearing solutions can be in equilibrium with diopside over a range of temperatures, as long as there is appropriate dilution of diopside in the melt.

To determine the temperature range over which this solid–liquid equilibrium can exist, we first rearrange Eq. 10-2 as follows:

$$RT \ln X_{Di}^L = \mu_{Di}^S - \mu_{Di}^{*L} \qquad (10\text{-}3)$$

But $\mu_{Di}^S - \mu_{Di}^{*L}$ is simply the negative of the free-energy change per mole accompanying the fusion of pure diopside at this temperature; that is, ΔG in Figure 10-1. Hence we can write

$$RT \ln X_{Di}^L = -\Delta G_m = -(\Delta H_m - T \Delta S_m) \qquad (10\text{-}4)$$

where the subscript m stands for melting. On rearranging and simplifying, we obtain

$$\ln X_{Di}^L = -\frac{\Delta H_m}{RT} + \frac{\Delta S_m}{R} \qquad (10\text{-}5)$$

At the melting point of pure diopside the ΔG of melting is, of course, zero, and thus $\Delta S_m = \Delta H_m/T_m$, where T_m is the melting point of pure diopside. For many reactions, such as the melting of a substance, the value of ΔS varies little with temperature. Thus, over a limited temperature range, $\Delta H_m/T_m$ can be substituted for ΔS_m without introducing serious error, and Eq. 10-5 becomes

$$\ln X_{Di}^L = \frac{\Delta H_m}{R}\left(\frac{1}{T_m} - \frac{1}{T}\right) \qquad (10\text{-}6)$$

This is known as the *cryoscopic equation*, because it describes the freezing-point (melting-point) depression of a substance brought about by its dilution in a solution.

A plot of the cryoscopic equation is given for diopside in Figure 10-2a. The melting point of pure diopside is 1392°C (1665 K), and its latent heat of fusion is 142.6 kJ mol^{-1} (Navrotsky et al. 1980). Dilution of diopside to a mole fraction of 0.75 lowers its melting point to 1347°C. By diluting the diopside more, still lower temperatures are achieved. The resulting line in this plot is referred to as the *liquidus* of diopside. At temperatures and compositions above the liquidus of a particular phase, that phase must be entirely in solution (molten). At the liquidus, however, the solution is saturated in the phase, and precipitation (crystallization) can occur with further cooling.

The latent heat of fusion of a phase determines the slope of the liquidus of that phase. For example, diopside ($\Delta H_m = 142.6$ kJ mol^{-1}) has its melting point lowered 45°C by dilution to a mole fraction of 0.75. Cristobalite, on the other hand, with a latent heat of fusion of only 8.2 kJ mol^{-1}, has its melting point lowered 737°C by this same dilution. It should be noted, however, that these minerals have very different molecular weights (Di = 0.2166 kg; Cr = 0.0601 kg). Had equivalent weight fractions been considered, the difference in melting point depression would have been very small.

So far, the material causing the dilution of the diopside in the melt has not entered into consideration, except for the stipulation that it cannot form a solid solution with diopside. As the melting point of diopside is lowered, the concentration of the solvent increases, and a stage is reached where the solvent becomes so concentrated that it too starts to precipitate (crystallize).

Let us consider, for illustrative purposes, diluting diopside with molten anorthite. Other minerals could be chosen, but this combination forms an important two-component (diopside–anorthite) solution that approximates the compositions of basalts. The lowering of the melting point of diopside, then, requires increasing the concentration of anorthite in the melt. This is feasible only if anorthite is capable of forming melts at these temperatures. Pure anorthite, however, can form a liquid only above 1557°C, but its melting point can be lowered, according to the cryoscopic equation, by diluting the anorthite in the liquid. In our example, this can be done by dissolving diopside in the melt. The latent heat of fusion of anorthite is poorly known, but lies between 125 and 140 kJ mol^{-1} (Navrotsky et al. 1980). The liquidus for anorthite in Figure 10-2b is obtained using a value of 136 kJ mol^{-1}.

The abscissas in Figure 10-2a and b are, in fact, the same. Because we are dealing with a two-component solution, a decrease in the mole fraction of diopside must be accompanied by a corresponding increase in the mole fraction of anorthite;

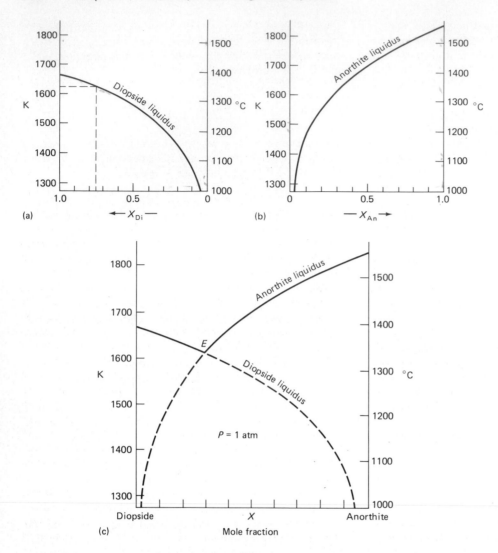

FIGURE 10-2 (a) Diopside liquidus plotted as a function of mole fraction of diopside in melt. (b) Similar plot for liquidus of anorthite. (c) Calculated phase diagram for system diopside–anorthite at atmospheric pressure.

that is, $X_{Di}^L = 1 - X_{An}^L$. We can, therefore, combine these diagrams to form what is known as a two-component or *binary-phase diagram* (Fig. 10-2c).

In this diagram, the lowering of the melting point of diopside is seen to be accompanied by enrichment of the melt in anorthite. Indeed, when the liquidus of diopside has decreased to 1337°C, the melt has become so enriched in this component that the liquidus of anorthite is encountered. Any further attempt to descend the diopside liquidus would be accompanied by crystallization of anorthite. This would prevent the liquid becoming more enriched in this component and thus prevent any further lowering of the melting point of diopside. These arguments apply equally to the lowering of the melting point of anorthite due to dilution by diopside. Our conclusion, therefore, must be that under equilibrium conditions and at the pressure specified for the diagram, the temperature of the liquid cannot be lowered beyond the point of intersection of the liquidus lines for the phases involved. This point of intersection is known as the *eutectic* and is marked with an *E* in Figure 10-2c. Below the eutectic, the liquidus of a phase is metastable and can be followed only if the other phase fails to crystallize due to kinetic factors. For this reason the liquidus of diopside and anorthite are dashed below the eutectic.

In deriving the cryoscopic equation, the solution was taken to behave ideally. This, of course, may not always be valid. Indeed, many petrologically important systems exhibit strong deviations from ideality. The cryoscopic equation, however, can still be used if we substitute activity (a_i) for mole fraction (X_i). It will be recalled from Sec. 9-5 that the activity can be thought of as the effective mole fraction of a component in solution. Activity is related to the mole fraction by the activity coefficient ($\gamma_i X_i = a_i$). For a nonideal solution the cryoscopic equation can be written as

$$\ln a_{Di}^L = \ln \gamma_i X_i^L = \frac{\Delta H_m}{R}\left(\frac{1}{T_m} - \frac{1}{T}\right) \tag{10-7}$$

Figure 10-4 presents the experimentally determined phase diagram for the system diopside–anorthite. Compositions in this diagram are plotted in terms of weight percent rather than mole fractions. Once this difference is accounted for (Prob. 10-1), comparison with Figure 10-2c reveals differences which indicate that diopside–anorthite solutions are not strictly ideal. Differences between the calculated and experimental data can be used to evaluate the activity coefficients of diopside and anorthite in these melts (see Prob. 10-2).

10-3 LEVER RULE

Before continuing with the discussion of phase diagrams, it is necessary to describe how compositions are read from such diagrams. In Figure 10-2c, compositions are plotted along the abscissa, with diopside at one end and anorthite at the other. Any point in the diagram, regardless of the temperature, represents a composition that can be read directly from the abscissa. Other compositional information, such as the proportion of liquid to crystals, can also be read from these diagrams, if they are plotted in weight rather than mole fractions. Weight is a conservative measure of composition and is therefore used in most phase diagrams.

If a line AB (Fig. 10-3) represents compositions of mixtures of two components A and B, any point f along this line represents specific weight fractions of A and of B. Clearly, if f were placed at the midpoint of the line, it would indicate equal amounts of A and B. If, on the other hand, f were closer to A, as shown in Figure 10-3, it would represent a composition richer in A, and conversely, if it were closer to B, it would be richer in B. The actual fractions of A and B in any composition f are determined from the relative lengths of the lines fB and fA, respectively. The fraction of A is given by the ratio fB/AB, and the fraction of B, by the ratio fA/AB. Thus the longer the line fB, the greater the amount of A present, and the longer the line fA, the greater the amount of B present. Because of the analogy of this relation to the moments of a lever about a fulcrum (Fig. 10-3), this way of reading compositions is commonly referred to as the *lever rule*. For example, for a lever we could state that the weight A is to the length L_2 as the weight B is to the length L_1. By analogy, then, the weight of A in composition f is to the line fB as the weight of B in f is to the line fA.

It should be noted that the absolute length of a line in such a diagram is not important, for it is the ratio of lengths that conveys the compositional information. Consider the point f on the line AB; the ratio fA/AB is 0.25, and thus the composition represented by point f is composed of 25% B and 75% A. But this same composition f could be expressed in terms of A and C, the latter being some compound with a composition between A and B. In this case the ratio fA/AC is 0.5, and therefore, composition f contains 50% C and 50% A. In both cases the absolute length of line fA remains the same, but the percentage of A is different.

Before continuing with the discussion of phase diagrams, the lever rule must be fully appreciated, and you should become proficient at graphically manipulating compositions. It is strongly recommended that Prob. 10-3 be done before proceeding to the next section.

10-4 SIMPLE BINARY SYSTEMS WITH NO SOLID SOLUTION

Let us return to a consideration of the binary-phase diagram for the system diopside–anorthite (Fig. 10-4). Along the liquidus of anorthite, melts are saturated in anorthite, and cooling brings about crystallization of this mineral. Similarly, the diopside liquidus is the saturation curve for diopside. At the eutectic, where the two liquidus lines intersect, both anorthite and diopside crystallize simultaneously upon withdrawal of heat. The heat removed comes from the crystallization of diopside and anorthite (latent heat of crystallization $= -\Delta H_m$) and does not cause a lowering of the eutectic temperature. Indeed, as will be shown later, the temperature of a crystallizing liquid at a eutectic, at a specified pressure, remains constant until all of the liquid has crystallized. The eutectic liquid, then, is converted into crystals of anorthite and diopside, both of which must, if there is equilibrium, be at the same temperature as the liquid. Through the point E in Figure 10-4, we can, therefore, draw a horizontal line (constant temperature) that joins the composition of the eutectic liquid to the two phases crystallizing from it. This line is known as the *solidus*.

The liquidus and solidus divide the phase diagram into a number of regions. Above the liquidus the system is entirely liquid, and below the solidus it is completely solid. Between these, however, both liquid and crystals coexist. The eutectic in the diopside–anorthite system divides this region in two; on one side liquid coexists with anorthite and on the other it coexists with diopside. These various regions consist of either one or two phases, and are labeled according to those present. In one-phase regions, such as that of the liquid in this diagram, any bulk composition forms a single phase with a composition identical to that of the bulk composition. In two-phase regions, on the other hand, a bulk composition consists of two phases, the compositions of which are indicated by the ends of the horizontal (isothermal) lines drawn in that region. Thus, within the region marked liquid plus anorthite (L + An), any bulk composition consists of liquid and anorthite, with the composition of the liquid being given by the liquidus of anorthite at the temperature of interest. Below the solidus, bulk compositions consist of crystals of diopside and anorthite, the compositions of which are represented by the two sides of the diagram.

Let us consider the cooling and crystallization of a liquid consisting of 75% anorthite and 25% diopside (x in Fig. 10-4). First, we construct a vertical line to mark the bulk composition. Such a line is known as an *isopleth*, a line of constant composition. Once constructed, the crystallization sequence can be read directly from the phase assemblages encountered by the isopleth as it descends through each phase region. This, of course, assumes that the bulk composition remains unchanged

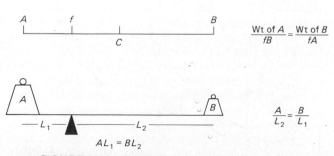

$$\frac{\text{Wt of } A}{fB} = \frac{\text{Wt of } B}{fA}$$

$$\frac{A}{L_2} = \frac{B}{L_1}$$

$$AL_1 = BL_2$$

FIGURE 10-3 Graphical representations of the abundance of components in a binary mixture is analogous to the position of a fulcrum on a lever to the weights at each end of the lever. See text for discussion.

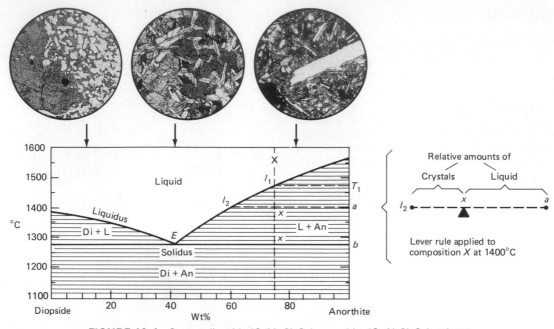

FIGURE 10-4   System diopside ($CaMgSi_2O_6$)–anorthite ($CaAl_2Si_2O_8$) at 1 atm pressure. Slight deviations from a binary system for diopside-rich compositions have been ignored [after Weill et al. (1980)]. Photomicrographs illustrate typical textures in gabbroic rocks resulting from crystallization in different regions of this system.

during the cooling; that is, nothing is added or subtracted from the system.

On cooling, the liquid first encounters the liquidus of anorthite at l_1 and temperature T_1. Further cooling, under equilibrium conditions, brings about crystallization of anorthite, which in turn enriches the melt in diopside. But for the liquid to remain in equilibrium with crystals of anorthite, the enrichment in diopside must be accompanied by cooling at a rate that keeps the liquid on the anorthite liquidus.

When the temperature has fallen to 1400°C, the composition of the liquid will have changed to l_2. It is instructive to consider what fraction of the original liquid will have crystallized by this temperature. We know that the bulk composition of the system has remained unchanged (the isopleth), and that at 1400°C liquid with composition l_2 coexists with anorthite crystals. The lever rule indicates that the fraction of liquid in this system is xa/l_2a and that the fraction of anorthite crystals is xl_2/l_2a. The ratio of liquid to crystals at this temperature is, therefore, xa/xl_2. This proportion is readily apparent from inspection of the phase diagram.

With falling temperature, the proportion of liquid to crystals decreases as the liquid changes its composition along the liquidus, eventually reaching a ratio of xb/xE at the eutectic. At this point, however, the onset of diopside crystallization prevents the composition of the liquid from changing further. Because the eutectic liquid crystallizes both diopside and anorthite and yet does not change its composition, these phases must crystallize in precisely the same proportions as they are present in the melt, that is, 58% diopside and 42% anorthite. Once all the eutectic liquid has crystallized, the solids, which

then consist of 75% anorthite and 25% diopside, are free to cool.

In the treatment above, equilibrium was maintained between crystals and liquid at all times, and thus cooling of the melt gave rise to what is referred to as *equilibrium crystallization*. If, on the other hand, crystals are removed from or otherwise prevented from being in equilibrium with the melt from which they form, *fractional crystallization* results. This could happen, for example, if crystals were much denser than the liquid; they could then separate by sinking. Such fractionation effectively changes the bulk composition, depleting it in the phase that is separating. Of course, equilibrium and complete fractional crystallization are extremes, and in nature we can expect to find all possible gradations between these.

Fractional crystallization need not take place continuously but can occur in stages. For example, melt with a composition of 75% anorthite and 25% diopside could cool, under equilibrium conditions, to 1400°C, at which temperature the melt would have composition l_2. At this stage, all the crystals could be removed, leaving only liquid l_2. This would provide a new starting composition through which could be constructed another isopleth, which is closer to the eutectic than the original. Regardless of whether fractionation is stepwise or continuous, its effect is always to produce liquids that are closer in composition to the eutectic.

The phase assemblages encountered on heating a solid mixture of 75% anorthite and 25% diopside until completely molten are the reverse of those encountered on cooling a liquid of this composition. Solids heated to the eutectic temperature begin to melt with the formation of a eutectic liquid. Only liquid of this composition can exist at this low temperature. Because diopside and anorthite both contribute to the liquid, melting occurs only where these different grains are in contact. (An example of this grain boundary melting is shown in Figure 10-5 between quartz and alkali feldspar.) The solids are consumed in the same proportions as they are present in the eutectic (58% Di, 42% An). Because the starting composition is rich in anorthite, there comes a time when the last grain of diopside melts, and the system then consists of eutectic liquid and anorthite crystals. The temperature has remained constant at the eutectic temperature throughout this melting episode, but with the disappearance of diopside, it is again free to rise. Continued heating causes the remaining anorthite to dissolve, changing the melt's composition along the liquidus. When the melt reaches composition l_1, all anorthite has dissolved, and there can be no further enrichment in this component. The melt, consequently, leaves the liquidus and rises along the isopleth with constant composition.

Let us now examine this phase diagram in terms of what it can tell us about the origin of igneous rocks. First, and perhaps most important, although separate minerals have relatively high melting points, mixtures of minerals melt at considerably lower temperatures, especially when present in eutectic pro-

portions. It is not surprising, then, to find that in the largely solid outer part of the Earth, most igneous rocks have compositions close to eutectics. They are the products of an Earth whose limited thermal energy is sufficient to melt only the lowest-melting fractions. Basalts, for example, have compositions close to the diopside–anorthite eutectic, and granites, as we will see later, have compositions close to the eutectic between quartz and alkali feldspar (Fig. 10-5). Note that regardless of the proportion of anorthite to diopside in a region undergoing partial melting, the first-formed melt must be of eutectic composition. Even if there is sufficient heat to produce liquids of noneutectic composition, these will, on cooling, descend the liquidus toward the eutectic, and fractionation can again produce eutectic liquids. The first important conclusion from phase diagrams, then, is that many common igneous rocks have eutectic compositions.

Any liquid with a composition on the anorthite side of the eutectic, such as the one discussed previously, first crystallizes anorthite. Indeed, if the composition is far from the eutectic, anorthite crystallizes over a considerable temperature interval before diopside appears at the eutectic. The early-formed anorthite crystals are consequently larger than later-formed crystals, and a *porphyritic texture* results, with *phenocrysts* of

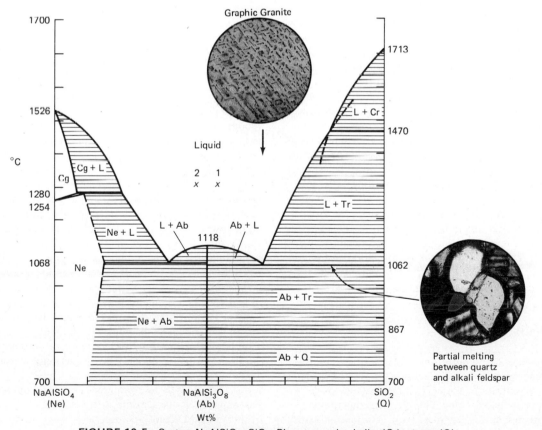

FIGURE 10-5   System $NaAlSiO_4$–SiO_2. Phases are cristobalite (Cr), quartz (Q), tridymite (Tr), albite (Ab), nepheline (Ne), and carnegieite (Cg) [after Greig and Barth (1938) and Tuttle and Bowen (1958)]. Photomicrographs of graphic granite and of experimentally produced partial melt between grains of quartz and feldspar; the melt is quenched to an isotropic glass.

anorthite (Fig. 10-4). Phenocrysts are formed from the phase whose liquidus is first encountered by the cooling magma. If volcanic eruption of a melt on the anorthite liquidus results in rapid quenching, the liquid may never crystallize diopside, and the phenocrysts may instead be surrounded by glass. The texture is then said to be *vitrophyric*. Had the melt first encountered the diopside liquidus, phenocrysts of this mineral would have formed (Fig. 10-4). Regardless of which phase forms the phenocryst, the surrounding liquid, whether quenched to a glass or crystallized to an aggregate of minerals, is always enriched in the eutectic composition.

An interesting situation develops at the eutectic, for here both minerals crystallize together. Unlike phenocrysts, which are free to develop their own crystal faces, minerals forming at the eutectic must accommodate each other's growth habits. This commonly results in characteristic textures known as *eutectic intergrowths*. Plagioclase and pyroxene, for example, form what is referred to as an *ophitic texture* (Fig. 10-4), with plagioclase laths embedded in large single crystals of pyroxene. A particularly striking texture involving quartz and alkali feldspar is known as *graphic granite* (Fig. 10-5), because the grains of quartz, which are embedded in large single crystals of feldspar, have shapes that resemble the characters of cuneiform (wedge-shaped) writing.

In Chapter 2, the common occurrence of phenocrysts in igneous rocks was used as evidence that most magmas are not superheated (Bowen 1928). We are now in a position to attach specific numbers to this statement. The presence of phenocrysts indicates that a magma is on the liquidus and not at some higher temperature. In the system diopside–anorthite, then, the presence of phenocrysts would indicate that the temperatures could not be more than 280°C above the eutectic. We might, therefore, estimate that basaltic magmas have temperatures near 1300°C. Such an estimate would be slightly high (actual temperatures are 1100 to 1200°C) because natural pyroxenes contain some iron and plagioclase contains considerable sodium, both of which have lower melting temperatures. The estimate is, nonetheless, of the correct order of magnitude.

Although discussion of more complex systems will allow us to account for other features of igneous rocks, the explanations provided by the simple binary system for the compositions of common igneous rocks, some of their main textures, and their temperatures of formation are of paramount importance. Later phase diagrams should be examined with these same points in mind.

10-5 PHASE RULE

In our discussion of crystallization in the diopside–anorthite system, reference was made to the fact that temperature remains constant at the eutectic while crystallization takes place. Although heat is continually being removed, the temperature does not change because heat is generated by the crystallization process. A more familiar example of this phenomenon is provided by water that has been cooled to the point that ice begins to form; as long as both water and ice are present and the pressure is 1 atm, the temperature remains at 0°C. Another restriction encountered in the diopside–anorthite system was the fixed

composition of the eutectic liquid. In fact, equilibrium between phases imposes numerous restrictions on variations in temperature, pressure, and composition. These relations can be quantified by the *phase rule*, which was discussed briefly in Sec. 8-5.

In Sec. 8-1 we saw that variables can be classified as extensive or intensive. At equilibrium, the intensive ones, T, P, and μ_i, must be constant throughout the system. In any phase, however, changes in these variables are related by the Gibbs–Duhem equation (Eq. 9-7). For example, in the two-component system diopside–anorthite, we can write for the liquid phase

$$S^L \, dT - V^L \, dP + n_{Di}^L \, d\mu_{Di} + n_{An}^L \, d\mu_{An} = 0 \qquad (10\text{-}8)$$

If no other phase is present, we have only this one equation and four unknowns. As soon as any three variables are specified, the fourth is determined by the equation. We say, therefore, that there are three degrees of freedom, or that the variance of the system is 3. As long as only one phase is present, whether it be liquid or some other phase, there is only one Gibbs–Duhem equation and four unknowns. Thus a single phase always gives a variance of 3 in a two-component system, and thus pressure, temperature, and composition can all be varied independently.

If two phases are present in a two-component system, for example liquid and diopside, Gibbs–Duhem equations can be written for both phases:

$$S^L \, dT - V^L \, dP + n_{Di}^L \, d\mu_{Di} + n_{An}^L \, d\mu_{An} = 0 \qquad (10\text{-}8)$$

$$S_{Di}^{DiS} \, dT - V_{Di}^{DiS} \, dP + n_{Di}^{DiS} \, d\mu_{Di} + n_{An}^{DiS} \, d\mu_{An} = 0 \qquad (10\text{-}9)$$

where the superscript S signifies solid—crystalline diopside in this case. Note that the last term in Eq. 10-9 is zero, because the diopside contains no anorthite; that is, there is no solid solution between these two minerals. We now have two equations and four unknowns. The variance of the system is therefore 2. This means that only two of the three possible variables are independently variable. For example, if the pressure is chosen to be 1 atm and the composition of the melt to be 60 wt % anorthite (l_2 in Fig. 10-4), the temperature is fixed at 1400°C, the liquidus for these particular conditions. A special but commonly encountered condition is that of constant pressure (isobaric). For example, the phase diagram in Figure 10-4 represents the phase relations in the system diopside–anorthite at a pressure of 1 atm. If two phases are present in a two-component system, the variance of 2 is reduced to 1 by keeping the pressure constant. Such a variance is described as isobarically univariant (one degree of freedom) and appears as a line in an isobaric phase diagram. The liquidus of anorthite in Figure 10-4 is one such line. Here temperature and composition are not independently variable, for as soon as one is selected, the other is determined.

If three phases are present in a two-component system, as happens at the eutectic in the system diopside–anorthite, three Gibbs–Duhem equations can be written:

$$S^L \, dT - V^L \, dP + n_{Di}^L \, d\mu_{Di} + n_{An}^L \, d\mu_{An} = 0 \qquad (10\text{-}8)$$

$$S_{Di}^{DiS} \, dT - V_{Di}^{DiS} \, dP + n_{Di}^{DiS} \, d\mu_{Di} + n_{An}^{DiS} \, d\mu_{An} = 0 \qquad (10\text{-}9)$$

$$S_{An}^{AnS} \, dT - V_{An}^{AnS} \, dP + n_{Di}^{AnS} \, d\mu_{Di} + n_{An}^{AnS} \, d\mu_{An} = 0 \qquad (10\text{-}10)$$

and yet there are still only four unknowns. The variance is consequently reduced to 1 (univariant), and at constant pressure this becomes isobarically invariant. In Figure 10-4, therefore, the eutectic is represented by a point (invariant). If pressure is thought of as a variable plotted perpendicular to the page in Figure 10-4, the eutectic is then a univariant line that pierces the 1-atm phase diagram at the isobaric invariant eutectic. There is only one temperature and composition that the eutectic can possibly have at this pressure.

Clearly, if four phases are present in a two-component system, as would happen if the eutectic assemblage in the system diopside–anorthite were joined by a gas composed of these components, the four Gibbs–Duhem equations would uniquely define all the variables. Such a phase assemblage can therefore exist at only one set of conditions of temperature, of pressure, and of compositions of liquid and gas. Such an assemblage would be truly invariant.

The relations above can be generalized for a system composed of c components and ϕ phases. The Gibbs–Duhem equation for each phase will contain terms for each of the possible intensive variables, T, P, and μ_i. There will be $c + 2$ of these terms (one for each component + T + P). There will be ϕ equations (one for each phase), and thus the number of independent variables will be $c + 2 - \phi$. The number of degrees of freedom, or variance, which is commonly designated by the letter f, is given by

$$f = c + 2 - \phi \tag{10-11}$$

This is known as *Gibbs phase rule*.

It will be noted that zero is the minimum possible value for the variance, and in that case $c + 2 = \phi$. In other words, there can never be more than $c + 2$ phases, if equilibrium exists. Rocks can therefore be expected to contain relatively small numbers of minerals. When they do not, a lack of equilibrium can be suspected, and textural evidence for incomplete reactions can be sought. This point is discussed further in later sections.

10-6 BINARY SYSTEMS WITH BINARY COMPOUNDS

In many binary systems, the end components can react together to form intermediate compounds. For example, in the system nepheline–quartz (Fig. 10-5), albite forms, and at high pressure, jadeite is also present. Both of these minerals can be described in terms of the two components $NaAlSiO_4$ (nepheline) and SiO_2 (quartz), and are therefore referred to as *binary compounds* in this system.

Binary compounds are divided into two types based on the way in which they melt. If they form a liquid of their own composition, they are said to melt *congruently*. This is the type of melting already encountered with diopside and anorthite. Another large group of petrologically important minerals, however, break down on melting to form a solid and a liquid, both of which have different compositions from the mineral from which they form. These are said to melt *incongruently*,

because of the compositional mismatch between the mineral and the liquid. We discuss first the simpler case of congruent melting.

At atmospheric pressure in the system nepheline–quartz, albite is the only stable binary compound. It melts congruently at 1118°C and effectively divides the binary system in two. Indeed, we could draw two separate binary diagrams, nepheline–albite and albite–quartz, and their treatment would be exactly the same as that for the system diopside–anorthite. Certain important petrological points can be made, however, if we keep them in a single diagram.

First, the presence of a congruently melting binary compound results in two separate eutectics, in this case between albite and tridymite and albite and nepheline. These eutectics correspond closely in composition to the two common rock types: granite and nepheline syenite, respectively.

The melting point of albite produces a *thermal maximum* on the liquidus in the central part of this diagram. This maximum has a profound effect on liquid fractionation paths in the system. This can be illustrated by considering the crystallization of two liquids, 1 and 2, which have very similar compositions in terms of the binary components, but which fall on either side of the composition of albite. Cooling of liquid 1 results in the crystallization first of albite, which enriches the remaining liquid in silica. Eventually, albite and tridymite crystallize together at the "granite" eutectic. Liquid 2 also crystallizes albite first, but because of the slight difference in composition, its residual liquid becomes enriched in nepheline and eventually crystallizes at the "nepheline syenite" eutectic. Thus, from two very similar composition melts, very different final liquids are produced.

Consideration of the crystallization paths in the vicinity of this thermal maximum should convince the reader that there is no way of changing a liquid from one side of the maximum to the other by fractional crystallization. Melts that start with compositions on the silica-rich side must eventually crystallize tridymite, and those on the other side must eventually crystallize nepheline. This division, then, justifies the classifying of igneous rocks into oversaturated (containing quartz) and undersaturated (containing feldspathoids), as discussed in Chapter 6.

Some igneous bodies do, however, contain over- and undersaturated rocks. In light of the phase diagram, we can see what a serious problem this poses and why it has attracted the attention of so many petrologists. As will be seen later, many ingenious schemes have been designed to cross the thermal barrier.

Both quartz and nepheline have high-temperature polymorphs, the stability fields of which can be shown in the phase diagram. The silica polymorphs show no solid solution with albite, and thus the boundaries between their stability fields can be represented by horizontal lines. The liquidus, on intersecting the boundary between the fields of cristobalite and tridymite, should, theoretically, show an inflection as a consequence of the different latent heats of fusion of these two phases (recall that the slope of the liquidus is a function of the latent heat of fusion). The differences are, however, very small, but application of Schreinemakers rules (Sec. 8-4) indicates that the liquidus of the high-temperature polymorph must be steeper than that of the low-temperature one, as shown in exaggerated form in Figure 10-5. The polymorphism of nepheline is complicated by solid solution with albite. Its discussion is, therefore, deferred to Sec. 10-10.

Incongruent melting is not as familiar to most of us as congruent melting, yet many of the rock-forming minerals exhibit this phenomenon. The process is easily understood if examined in terms of free energy. This will be done for the incongruent melting of enstatite to form olivine and liquid. As was done with diopside, free energies of solid and liquid enstatite can be plotted as functions of temperature (Fig. 10-6). Where the two curves intersect, solid and liquid enstatite must be in equilibrium ($\Delta G = 0$). This is the congruent melting point of enstatite (T_{cm} in Fig. 10-6). The free-energy plot of a compositionally equivalent mixture of olivine and liquid, however, has a steeper slope than that of solid enstatite, and over the temperature interval T_{im}–T_L has lower values of free energy than either solid or liquid enstatite. Over this interval, solid or liquid enstatite can lower its free energy by changing into a mixture of forsterite and liquid. The congruent melting point of enstatite is therefore metastable. At low temperatures enstatite has the lowest free energy and is thus the stable phase for this composition. At temperature T_{im} a mixture of forsterite and liquid becomes the stable assemblage, and this, then, marks the incongruent melting point of enstatite (T_{im}). With rising temperature, the composition and proportion of liquid coexisting with forsterite continuously changes, so a family of G versus T curves could have been plotted, but only two have been shown for clarity. With rising temperature these curves become steeper and the amount of forsterite decreases, becoming zero at temperature T_L. The liquid then has the composition of enstatite. Only above this temperature does the free energy of

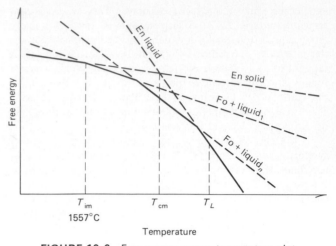

FIGURE 10-6 Free energy versus temperature plot at constant pressure of solid and liquid enstatite and two compositionally equivalent mixtures of forsterite and liquid. Because these olivine–liquid mixtures have a lower free energy than that of enstatite at its congruent melting point, enstatite melts incongruently to forsterite + a more siliceous liquid.

liquid enstatite become low enough for melt of this composition to be stable.

Let us now examine the binary phase diagram Mg$_2$SiO$_4$–SiO$_2$, which contains the binary incongruent-melting compound enstatite. Only the silica-poor side of the system is shown in Figure 10-7, in order to focus on the incongruent melting.

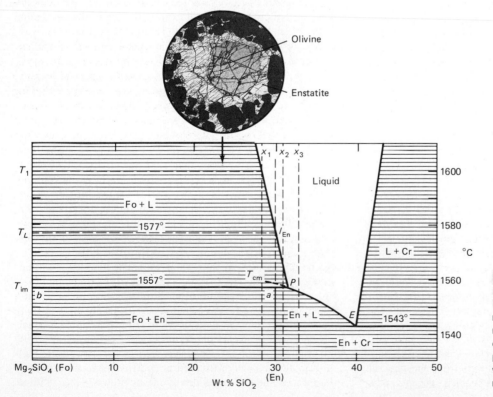

FIGURE 10-7 Part of the system Mg$_2$SiO$_4$–SiO$_2$ at 1 atm pressure [after Bowen and Andersen (1914)]. Photomicrograph of orthopyroxene reaction rims around olivine. Grain boundaries of olivine prior to reaction are indicated by presence of chromite grains (opaque), which accumulated with olivine from the magma. Stillwater complex, Montana.

A eutectic (*E*) between enstatite and cristobalite has a temperature of 1543°C and a composition of 38% silica. From this point, the enstatite liquidus rises to what would be the congruent melting point of pure enstatite (T_{cm}), but before it reaches there, it is intersected (*P*) by the forsterite liquidus at 1557°C. Above this temperature, which is the incongruent melting point of enstatite (T_{im}), the enstatite liquidus is metastable.

Heating of enstatite to 1557°C results in the formation of liquid of composition *P* and crystals of forsterite. The presence of these three phases (En, Fo, Liquid) in a binary system at 1 atm pressure results in an isobaric invariant assemblage ($f = c + 2 - \phi = 1$). The composition of the liquid (*P*) and the temperature must remain constant until enstatite has completely melted, at which point the proportion of liquid to forsterite is given by the ratio of line lengths *ab/aP*. Continued heating causes the forsterite to dissolve in the liquid, which changes composition along the forsterite liquidus. With rising temperature, the proportion of forsterite crystals decreases and becomes zero at temperature T_L, where the liquid has the composition of enstatite (l_{En}). The disappearance of forsterite results in a gain in degree of freedom, and the liquid is then free to leave the liquidus and rise along the isopleth with further heating.

Liquid of composition x_1 in Figure 10-7 will, on cooling, intersect the forsterite liquidus at 1600°C. Continued cooling under equilibrium conditions brings about crystallization of forsterite and enrichment of the melt in silica. When the liquid reaches point *P* (1557°C), it begins to react with the crystals of forsterite to form enstatite. Because the system now has three phases, it is isobarically invariant, and the liquid remains at *P* until the enstatite-forming reaction has gone to completion. Reaction points in phase diagrams, such as *P*, are known as *peritectics*. The reaction at *P* consumes both liquid and forsterite, and depending on the bulk composition, one or the other of these phases can be consumed first, thus terminating the reaction. In the unique case of the composition being precisely that of enstatite, both liquid and forsterite are consumed simultaneously to form enstatite. The system is then totally solid. For composition x_1, however, the isopleth falls on the silica-poor side of enstatite, and cooling below the peritectic must involve only forsterite and enstatite. At the peritectic, therefore, all liquid must be consumed in the reaction, and only solids forsterite and enstatite remain. These minerals would show clear textural evidence of reaction, with enstatite rimming and eating into the olivine grains, as shown in the inset of Figure 10-7.

Liquid of composition x_2 would initially follow a similar sequence of crystallization steps to that of liquid x_1, but the position of its isopleth indicates that at the peritectic, forsterite is the first phase to be totally consumed by the reaction. The system, then, would consist only of liquid and enstatite, and thus cooling could resume, with the melt descending the enstatite liquidus to the eutectic with cristobalite. Liquid of composition x_3, which lies on the silica-rich side of the peritectic, does not intersect the forsterite liquidus at all on cooling. Instead, enstatite forms initially as a primary phase rather than as a reaction rim on forsterite.

Unlike congruently melting compounds, incongruent ones do not change the number of eutectics present in a system.

Under equilibrium conditions, crystallization must terminate at a peritectic or a eutectic, and from these, different assemblages form (forsterite + enstatite, and enstatite + cristobalite, respectively). With fractional crystallization, however, liquids, regardless of their initial compositions, can eventually reach the eutectic. The peritectic itself actually increases the opportunities for disequilibrium. Reaction rims, produced at the peritectic, commonly act as barriers to slowly diffusing reactants. It is not unusual, therefore, especially where cooling is rapid, to find rocks containing both quartz and magnesium-rich olivine, the olivine being mantled and separated from the quartz by orthorhombic pyroxene.

Liquids of peritectic composition, like those of eutectics, correspond closely to some common igneous rocks. For example, many continental flood basalts, when plotted in terms of the components of Figure 10-7, have compositions near the peritectic. Several explanations can be found for this. Melting of any mixture of enstatite and cristobalite must first produce a liquid of eutectic composition. Mixtures of enstatite and forsterite, on the other hand, do not melt until heated to the peritectic temperature, and then, regardless of the proportions of these minerals, the first-formed melt must be of peritectic composition. Thus, from a source region consisting of forsterite and enstatite, peritectic liquid is bound to be formed. It is interesting to note that this liquid, when crystallized, will produce some tridymite (or quartz), a phase not present in the source. Should liquids be formed with compositions above the peritectic, as might happen in a region of high heat flow, cooling to the peritectic will bring about the reaction that causes the system to become isobarically invariant. Because liquid of peritectic composition continues to exist for whatever time is necessary for the reaction to go to completion (assuming equilibrium), the chances of such liquid being tapped are greater than for liquid that is continually changing its composition along the liquidus.

Although enstatite melts incongruently at low pressures, above about 0.5 GPa it melts congruently. The reason for the change can be appreciated by examining the effect of pressure on the free-energy surfaces that were shown in Figure 10-6 for solid and liquid enstatite, and a chemically equivalent mixture of forsterite and liquid. At atmospheric pressure, the congruent melting point of enstatite is metastable because the equivalent mixture of forsterite and liquid has a lower free energy. With increasing pressure the free-energy surfaces all increase (Eq. 7-32), but not at the same rates (Fig. 10-8). The free energy of forsterite and liquid, in fact, rises so rapidly that above 0.5 GPa, it lies above the intersection of the solid and liquid enstatite surfaces. Above this pressure, therefore, enstatite melts congruently because coexisting solid and liquid enstatite have lower free energy than the forsterite and liquid.

This change in the melting behavior of enstatite significantly affects the phase relations in the system Mg_2SiO_4–SiO_2. The univariant line marking the boundary between the liquidus surfaces of enstatite and forsterite (Fig. 10-8) changes from a peritectic at low pressure to a eutectic at high pressure. A single eutectic at low pressure, therefore, gives rise to two at high pressure, and perhaps more important, a thermal divide splits the system in two at high pressures. Fractional crystallization cannot move liquids from the silica-poor to the silica-rich side

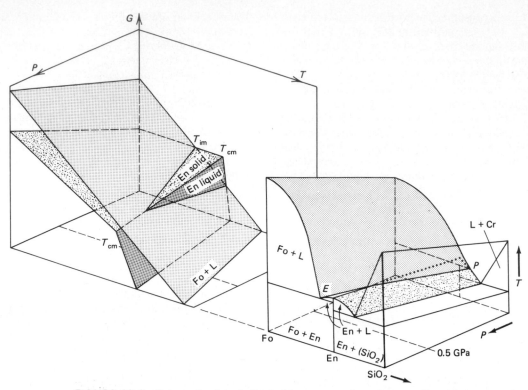

FIGURE 10-8   Schematic plot of effect of pressure on free-energy surface of solid and liquid enstatite, and of chemically equivalent mixture of forsterite and liquid. Above a pressure of approximately 0.5 GPa, enstatite melts congruently, and the lower-pressure peritectic (*P*) changes to a eutectic (*E*). See text for discussion.

at high pressure. Also, melting of forsterite–enstatite mixtures no longer produce an oversaturated liquid as they do at 1 atm, and with increasing pressure the melts become progressively enriched in forsterite as the forsterite–enstatite eutectic moves toward forsterite. The olivine content of tholeiitic basalts may therefore be related to the depth at which the magma is generated (see Sec. 22-6 and Table 22-1).

10-7   BINARY SYSTEMS WITH LIQUID IMMISCIBILITY

There remains only one other phase relation involving liquids in binary systems that can be encountered without introducing solid solution, and that is liquid immiscibility. In developing the cryoscopic equation, the melt was assumed to behave ideally. Deviations from ideality, however, are common, and they can be large. In some systems, the lowering of free energy that should accompany mixing is less than expected, and in still others intermediate composition liquids can have free energies that are actually higher than liquids that are richer in one or the other component. The result is a saddle-shaped free-energy curve (Fig. 9-7a), which permits the coexistence of two compositionally different phases, but both have identical chemical potentials of their components.

Many binary systems with silica exhibit liquid immiscibility. For example, systems involving SiO_2 with MgO, FeO, or CaO exhibit immiscibility, but those with alkalis or alumina do not. We shall look into the reasons for this difference following a discussion of the compositions of immiscible liquids. We shall use the system $FeO–SiO_2$ as an example (Fig. 10-9).

At a pressure of 10^5 Pa, the system $FeO–SiO_2$ has one binary compound, fayalite (Fe_2SiO_4), which melts congruently and produces two eutectics, one with wustite (FeO) and the other with tridymite. The pyroxene ferrosilite ($FeSiO_3$) is stable only at high pressures, but its composition is included in Figure 10-9 for reference. Within the area marked 2L, single liquids unmix into two. This region of immiscibility extends to low temperatures, but only that part above the cristobalite liquidus is stable. Silica-rich liquids can, however, be readily supercooled (excellent glass formers), and thus metastable immiscibility is easily achieved. Immiscible glasses formed in this way are of considerable commercial value.

Let us consider the equilibrium crystallization of three different liquids, each of which encounters the immiscibility field in a different way. Liquid of composition x_1 first encounters the liquidus of cristobalite, and crystallization of this phase enriches the melt in FeO, changing its composition to l_1. At this point it encounters the two-liquid field, and any further enrichment in iron is impossible. But the continued crystallization of cristobalite does generate FeO in the melt, but instead of this

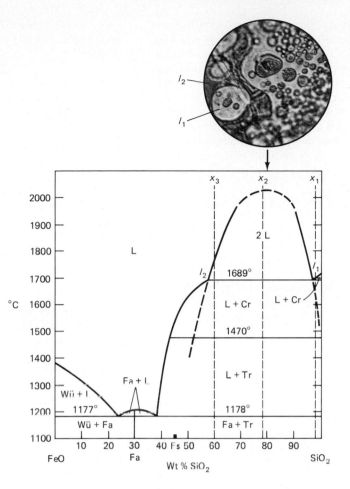

FIGURE 10-9 System FeO–SiO$_2$ at 1 atm pressure. Phases are wüstite (Wü), fayalite (Fa), ferrosilite (Fs), tridymite (Tr), and cristobalite (Cr) [after Bowen and Schairer (1932)]. Photomicrograph of immiscible silicate liquids in residual liquid of basalt, Kilauea, Hawaii.

entering liquid l_1, a second liquid, l_2, forms. We can think of this as a reaction in which l_1 is converted to cristobalite plus l_2. The presence of these three phases in a two-component system produces an isobaric invariant assemblage; the temperature and compositions of liquids l_1 and l_2 cannot change until a phase has been lost. This does not, however, mean that the proportions of the two liquids cannot change; indeed, they must if cristobalite is crystallizing. When liquid l_2 first appears, it forms small iron-rich droplets (brown in thin section) in the silica-rich liquid (clear in thin section), but as crystallization continues, the amount of iron-rich liquid increases. Eventually, all liquid l_1 is converted to cristobalite plus liquid l_2; then cooling resumes as the remaining single liquid descends the cristobalite liquidus to the eutectic with fayalite.

Liquid of composition x_2, on cooling, first encounters the immiscibility field and unmixes into approximately equal amounts of two liquids, one of which becomes progressively richer in silica, and the other richer in iron. The actual temperature at which unmixing first occurs is uncertain due to lack of experimental data. Eventually, the two liquids reach compositions l_1 and l_2, at which temperature cristobalite begins to crystallize. Again, this forms an isobarically invariant assemblage. Eventually, liquid l_1 is consumed and the iron-rich liquid, l_2, cools down the cristobalite liquidus to the eutectic.

Liquid of composition x_3 is relatively iron-rich to begin with, and so, on cooling, it first encounters the two-liquid field

on the iron-rich side. This would bring about the nucleation of droplets of silica-rich liquid in the iron-rich host. Further cooling would bring the two liquids to compositions l_1 and l_2, and then the remainder of the crystallization is the same as in previous cases.

The heat removed from the isobaric invariant assemblage cristobalite + liquid 1 + liquid 2 is the latent heat of crystallization of cristobalite. The nutrients for the growth of this mineral (SiO$_2$) are concentrated in liquid 1, so it is likely that cristobalite will form in this liquid. But recall that equilibrium requires that the chemical potential of a component be the same in all phases. The chemical potential of silica in the iron-rich liquid, l_2, must therefore be the same as in liquid l_1 and the same as in cristobalite. Therefore, cristobalite can form from both liquids, but most will crystallize from the silica-rich one. We will have occasion to return to this point in discussing the evidence for liquid immiscibility in natural rocks.

The immiscibility field in the system FeO–SiO$_2$ extends over approximately the same compositional range as those in the other systems exhibiting immiscibility previously mentioned. One side of the miscibility gap has a composition of almost pure SiO$_2$, whereas the other side has a composition corresponding closely to that of pyroxene. Herein lies the explanation for the immiscibility.

Silica-rich melts have structures similar to those of framework silicate minerals except that long-range order is lacking

(Hess 1980). They consist of networks of linked SiO_4 tetrahedra in which most oxygen form bridging Si—O—Si bonds. Such melts are said to be highly polymerized. Oxides of divalent cations with high ionization potentials also combine with oxygen in melts to form bonds of the type M—O—M, where M represents the cation. These cations, when mixed with silica-rich melt, are unable to bond with the bridging oxygens linking SiO_4 tetrahedra unless they break up the silica network. The following reaction then describes the depolymerization of the silica network brought about by addition of these cations to silica-rich melt:

$$\text{liquid silica} + \text{liquid metal oxide} = \text{silicate melt}$$

$$\text{Si—O—Si} + \quad \text{M—O—M} \quad = 2(\text{Si—O—M})$$

In this reaction the bridging oxygen is changed to a nonbridging oxygen. Examples of nonbridging oxygens can be found in such silicate structures as the pyroxenes, where they bond with the cations between silica chains. In the melt, therefore, the presence of cations with high ionization potentials requires there be silicate ionic groups that can provide nonbridging oxygens. Of course, pyroxene-like chains are only one of the possible groups that can do this; rings and isolated tetrahedra can also provide these oxygens. When immiscible liquids form, a melt splits into a silica-rich fraction composed of a network of linked SiO_4 tetrahedra, and a relatively silica-poor fraction of less polymerized melt in which nonbridging oxygens bond with cations of high ionic potential. The latter melt tends to have a pyroxene composition, as seen by comparing the position of the silica-poor side of the immiscibility field in Figure 10-9 with the composition of ferrosilite.

Systems that do not exhibit immiscibility with silica involve cations with low ionization potentials that are able to accommodate themselves in the continuous linked network of silica-rich melts. Alumina, when present, is able to enter this network, substituting for silicon as long as cations such as K^+, Na^+, or Ca^{2+} are available for charge balance.

The immiscibility in Figure 10-9 occurs at temperatures that are too high to occur in the Earth at low pressure. We will, however, encounter an extension of this binary miscibility gap at much lower temperatures in three-component systems. It is therefore worth considering what petrological principles can be read from this simple diagram, leaving aside for the moment the question of temperature.

Two liquids coexisting in the Earth for any length of time could possibly separate from each other if there is sufficient density contrast and viscosities are low enough. For example, the silica-rich liquid, l_1, could separate from the iron-rich one, l_2, by floating. Immiscibility therefore offers a possible means of fractionating magmas. All three liquids whose crystallization histories were considered pass through the immiscibility field and thus could have been fractionated by this process. Yet each of these liquids eventually crystallizes the same minerals, and nothing in the assemblage itself indicates that immiscibility had been involved earlier. There might, of course, be certain features pertaining to the distribution of cristobalite that could be interpreted as resulting from immiscibility. But short of rapid quenching to form immiscible glassy globules, as happens in many volcanic rocks (Fig. 10-9), little evidence is likely to be preserved of this potentially important means of fractionating magma.

10-8 COMPLEX BINARY SYSTEMS WITH NO SOLID SOLUTION

Most previous sections in this chapter have dealt with binary systems that have each illustrated only one type of phase relation. Many binary systems combine some or all of these relations in a single diagram; consequently, they appear complex. But the combining of these phase relations does not change their behavior, so complex diagrams can be read with no additional problems.

The petrologically important system $KAlSiO_4$–SiO_2 (Fig. 10-10) exhibits a variety of binary phase relations. First, in addition to the polymorphs of the minerals forming the end components, kalsilite (Ks) and orthorhombic $KAlSiO_4$ (OK), and quartz (Q), tridymite (Tr), and cristobalite (Cr), the binary compounds leucite (Lc) and potassium feldspar (Ksp) are present. Leucite melts congruently at 1686°C, but K-feldspar (sanidine) melts incongruently to leucite and liquid at 1150°C, and thus only one additional eutectic is introduced into the system by these binary compounds. The eutectic between orthorhombic $KAlSiO_4$ and leucite at 1615°C is at too high a temperature and too potassic a composition to be of much

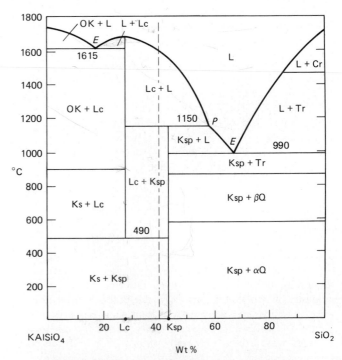

FIGURE 10-10 System $KAlSiO_4$–SiO_2. Phases are kalsilite (Ks), orthorhombic $KAlSiO_4$ (OK), leucite (Lc), potassium feldspar (Ksp), low (αQ)- and high (βQ)-temperature quartz, tridymite (Tr), cristobalite (Cr), and liquid (L). [After Schairer and Bowen (1955) and Scarfe et al. (1966).]

importance in nature. But the one between K-feldspar and tridymite at 990°C is the "granite" eutectic in this potassium-bearing system. We have already met its sodic equivalent in the system albite–quartz (Fig. 10-5).

Let us consider the crystallization of the liquid marked by the isopleth in Figure 10-10. Under equilibrium conditions, leucite starts crystallizing from this liquid at 1600°C, which enriches the melt in silica. Cooling continues until, at 1150°C, the liquid reaches the peritectic, *P*, and reacts with the leucite to form K-feldspar. Because the isopleth immediately below the peritectic is in the field of leucite plus feldspar, all of the liquid must be consumed at the peritectic. Once this has happened, leucite and feldspar cool until, at 490°C, leucite becomes unstable and breaks down to form an intergrowth of kalsilite and K-feldspar, which remains stable to low temperatures.

We have not previously encountered minerals that become unstable on cooling, but such behavior is common. Leucite, although unstable below 490°C, does occur in some volcanic rocks, but this is due to metastability brought on by rapid cooling. With slow cooling, however, leucite does invert to the kalsilite–K-feldspar intergrowth. In doing so it commonly maintains the morphology of the original leucite crystal and is thus referred to as *pseudoleucite*. Although pure potassic pseudoleucites have been found, most in igneous rocks contain considerable sodium in the form of nepheline. These likely involve some reaction or exchange of alkalis with the melt during their formation.

Because our chosen composition eventually crystallizes a feldspathoid rather than quartz, it would be classified as undersaturated (in silica); indeed, any composition on the silica-poor side of K-feldspar would be so classified. During part of its crystallization history, particularly at the peritectic, oversaturated liquids are generated. These liquids, if separated from the leucite crystals, would crystallize tridymite at the "granite" eutectic. Thus from an undersaturated melt an oversaturated one can be generated by fractional crystallization. This behavior is very different from that in the system $NaAlSiO_4$–SiO_2, where congruently melting sodic feldspar produces a thermal barrier, separating the under- and oversaturated rocks. The removal of leucite crystals from undersaturated magmas to produce oversaturated ones has been invoked to explain the association of these disparate rocks. Although the explanation is reasonable in terms of the phase diagram, we shall see later that leucite is restricted, by pressure, to near-surface environments. Moreover, field evidence indicating the separation of leucite crystals from magma has been found in only a few rare instances.

10-9 BINARY SYSTEMS WITH COMPLETE SOLID SOLUTION

So far we have considered systems in which there is no solid solution; that is, each solid phase has a specific composition, and only the liquid has variable composition. But many rock-forming minerals, such as the olivines, pyroxenes, amphiboles, and feldspars, belong to solid solution series. By changing the concentration of a component in a solid the chemical potential of that solid component is changed (Eq. 9-34). Thus the cryo-

scopic equation derived in Sec. 10-2, where the chemical potential of the solid was taken to be fixed at the value of the pure mineral, is not applicable to systems with solid solution. Instead, the liquidus in these systems must be calculated in a manner that takes into account the variable chemical potentials in both the solid and liquid.

Consider equilibrium between a liquid and a solid, both of which form binary solutions. The olivines, ranging in composition from forsterite (Mg_2SiO_4) to fayalite (Fe_2SiO_4), provide a good example of such a system. At any pressure and temperature, olivine liquid will have specific chemical potentials of forsterite (μ_{Fo}^L) and fayalite (μ_{Fa}^L) depending on its composition. Similarly, crystalline olivine will have specific chemical potentials of forsterite (μ_{Fo}^S) and fayalite (μ_{Fa}^S), depending on its composition. If a crystal of olivine is in equilibrium with its melt, we can write

$$\mu_{Fo}^L = \mu_{Fo}^S \qquad \text{and} \qquad \mu_{Fa}^L = \mu_{Fa}^S$$

Using Eq. 9-34, we can write for the chemical potentials of forsterite in liquid and solid, respectively

$$\mu_{Fo}^L = \mu_{Fo}^{*L} + RT \ln a_{Fo}^L \qquad \text{and} \qquad \mu_{Fo}^S = \mu_{Fo}^{*S} + RT \ln a_{Fo}^S$$

where μ_{Fo}^{*L} and μ_{Fo}^{*S} refer to the chemical potentials of pure forsterite liquid and solid, respectively, at the particular pressure and temperature considered. Equating these two equations, we obtain

$$\mu_{Fo}^{*L} + RT \ln a_{Fo}^L = \mu_{Fo}^{*S} + RT \ln a_{Fo}^S$$

which gives, on rearranging,

$$\mu_{Fo}^{*L} - \mu_{Fo}^{*S} = RT \ln \frac{a_{Fo}^S}{a_{Fo}^L} \qquad (10\text{-}12)$$

But $\mu_{Fo}^{*L} - \mu_{Fo}^{*S}$ is the ΔG of fusion of pure forsterite at the specified pressure and temperature. If we assume that the ΔH and ΔS of fusion are constant over the temperature interval of interest, the same approximation for ΔG can be used as in Eq. 10-6, and then Eq. 10-12 can be expressed as

$$\ln \frac{a_{Fo}^L}{a_{Fo}^S} = \frac{\Delta H_m^{Fo}}{R} \left(\frac{1}{T_m^{Fo}} - \frac{1}{T} \right) \qquad (10\text{-}13)$$

where ΔH_m^{Fo} is the heat of fusion of forsterite and T_m^{Fo} is the melting point of pure forsterite. Equation 10-13 is a perfectly general equation that assumes nothing about the behavior of either solution. In the limiting case where the solid phase shows no solid solution, its activity would be 1, and Eq. 10-13 reduces to the previously derived cryoscopic equation (Eq. 10-6). Equation 10-13, then, is the general form of the *cryoscopic equation*.

To construct a phase diagram using this equation, it is necessary to know how the activities of the components are related to their mole fractions in both the solid and liquid solutions (Wood and Fraser 1976). It will be recalled from Eq. 9-20 that a factor *n*, the number of equivalent sites on which mixing takes place in each formula unit of a solid solution, had to be introduced into the expression for the configurational entropy resulting from mixing. This same factor can be used to define activity where there is ideal mixing on n equivalent sites. The activity of an ideally mixed component in a solid solution in which the formula has n equivalent sites on which mixing takes place is given by

$$a_i = (X_i)^n \tag{10-14}$$

If the mineral has two distinct sites (M1 and M2, for example) on which mixing can take place, the activity of an ideal solution is given by

$$a_i = X_i^{M1} \times X_i^{M2} \tag{10-15}$$

where X refers to the mole fraction of the cation (or anion) of the ith component occupying the particular M site. Olivine has two different octahedral sites, but they are so similar that magnesium and iron express little if any preference for one over the other. As a result, the mole fraction of Mg^{2+} in one site is the same as in the other, and the activity of forsterite in crystalline olivine can be expressed as

$$a_{Fo}^s = X_{Mg}^{M1} \times X_{Mg}^{M2} = (X_{Fo}^S)^2 \tag{10-16}$$

Note that we would have obtained this same expression if we had assumed that olivine contained only one equivalent site (Eq. 10-15).

An expression for the activity of forsterite in the melt is more difficult to formulate than that for the solid because less is known about the structure of melts. The simplest model of a silicate melt such as that of olivine is that it consists of cations and silicate anions that maintain, because of charge balance, the same general short-range order that they have in the crystalline state. Mixing is most unlikely to occur between cation and anion sites because of charge balance considerations. The entropy of mixing in such a melt would therefore be the sum of the entropies of mixing on the separate cation and anion sites. We can write the melting of olivine as

$$(Mg, Fe)_2SiO_4 = 2(Mg, Fe)^{2+} + SiO_4^{4+}$$

But because only one type of anion group is present in this system, there can be no contribution to the entropy of mixing from the anions. The entropy of mixing must therefore be due only to the mixing of the cations. The activity of forsterite in the melt can therefore be represented by

$$a_{Fo}^L = (X_{Fo}^L)^2 \tag{10-17}$$

We can now substitute the expressions for the activity of forsterite in the solid (Eq. 10-16) and in the liquid (Eq. 10-17) in Eq. 10-13 to obtain

$$2 \ln \frac{X_{Fo}^L}{X_{Fo}^S} = \frac{\Delta H_m^{Fo}}{R} \left(\frac{1}{T_m^{Fo}} - \frac{1}{T} \right) \tag{10-18}$$

If a value for T in Eq. 10-18 is assumed, a ratio of mole fraction of forsterite in liquid to that in solid can be calculated, but we cannot determine the composition of either uniquely;

there is one too many unknowns. A similar equation to Eq. 10-12 can, however, be written for the fayalite component, from which it follows that

$$2 \ln \frac{X_{Fa}^L}{X_{Fa}^S} = \frac{\Delta H_m^{Fa}}{R} \left(\frac{1}{T_m^{Fa}} - \frac{1}{T} \right) \tag{10-19}$$

where ΔH_m^{Fa} is the heat of fusion of fayalite and T_m^{Fa} is the melting point of pure fayalite. Because there are only two components, $X_{Fa}^S = 1 - X_{Fo}^S$ and $X_{Fa}^L = 1 - X_{Fo}^L$. We can therefore rewrite Eq. 10-19 as

$$2 \ln \frac{1 - X_{Fo}^L}{1 - X_{Fo}^S} = \frac{\Delta H_m^{Fa}}{R} \left(\frac{1}{T_m^{Fa}} - \frac{1}{T} \right) \tag{10-20}$$

Solving Eqs. 10-18 and 10-20 simultaneously, we determine uniquely the compositions of liquid and solid coexisting at any temperature. This gives the liquidus and solidus of olivine (see Prob. 10-5).

The experimentally determined phase diagram for the system Mg_2SiO_4–Fe_2SiO_4 at atmospheric pressure is given in Figure 10-11. Let us consider the equilibrium crystallization of a liquid of composition x in this system. This melt, on cooling, reaches the liquidus at 1675°C (l_1) and crystallizes olivine with a composition of 80% forsterite (s_1). Crystallization of this magnesium-rich olivine enriches the residual melt in fayalite. As the melt descends the liquidus, equilibrium requires that crystalline olivine simultaneously change composition along the solidus. Thus the solid continuously reacts with the liquid to produce more fayalitic olivine. For example, when the liquid reaches l_2, the solid has a composition s_2, and the proportion of liquid to solid is given by s_2a/l_2a. With continued cooling, the melt descends the liquidus until it reaches l_3, at which point the solid has composition s_3, which is identical with the starting bulk composition (see the isopleth). Consequently, the amount of liquid becomes zero at this temperature, and with only one

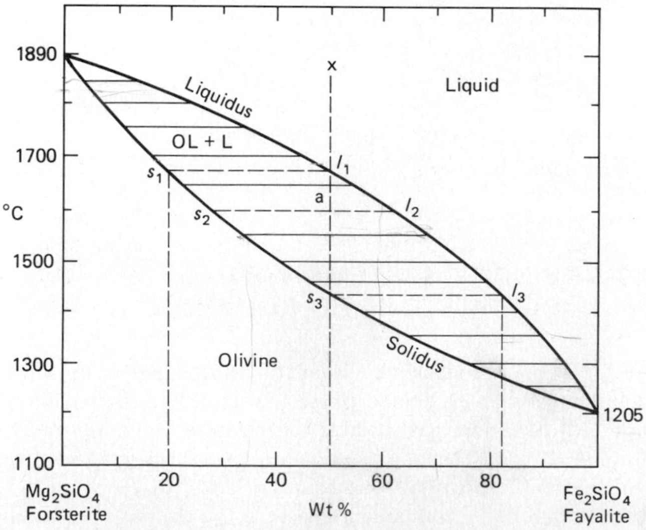

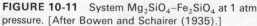

FIGURE 10-11 System Mg_2SiO_4–Fe_2SiO_4 at 1 atm pressure. [After Bowen and Schairer (1935).]

phase remaining (s_3), the system gains a degree of freedom and crystalline olivine is free to leave the solidus and cool.

Because this crystallization process involves only two phases, liquid and solid (solution), it is never isobarically invariant. Thus the reaction that takes place with cooling is univariant and occurs continuously with falling temperature. This is in contrast to the reaction at a peritectic in a binary system, where two solid phases plus liquid produce an isobaric invariant assemblage; in that case there is a discontinuity in the liquidus (the peritectic). These two reaction processes are commonly referred to as *continuous* and *discontinuous*, respectively.

The continuous reaction required to maintain equilibrium during crystallization of a solid solution phase increases chances for fractionation to occur. Mantling of early-formed refractory cores by later crystallizing material through which diffusion can occur only slowly causes residual melts to become increasingly enriched in the low-melting fraction. Zoned crystals are formed with high-temperature cores and low-temperature rims. This type of zoning, which is referred to as *normal*, is common in many minerals, especially in volcanic rocks, where rapid cooling provides insufficient time for equilibration. The ferromagnesian minerals typically become more iron-rich toward their rims, and the feldspars become richer in alkalis.

Whether minerals exhibit zoning or not depends on diffusion rates, which in turn depend to a large extent on the degree of polymerization in the mineral. For example, in a rock containing simultaneously formed olivine, pyroxene, and plagioclase, the olivine might be homogeneous, the pyroxene exhibit zoning only toward the rims, and plagioclase be strongly zoned throughout. The olivine structure, consisting of isolated silica tetrahedra, requires the least disruption to bring about homogenization of the iron and magnesium in the octahedral sites. In pyroxene, on the other hand, silica chains allow for easy diffusion only parallel to their lengths. Finally, the homogenization of plagioclase requires not only the diffusion of sodium and calcium but also of silicon and aluminum. Because this involves disruption of the framework structure, zoning in plagioclase is especially common, even with slow cooling rates.

Zoning is often more common on the rims of crystals than in the cores. Although this can be due to changes in cooling rate, Figure 10-11 provides another explanation. The solidus of any solid solution phase becomes flatter toward the low-temperature end, whereas the liquidus becomes steeper. Consequently, the composition of the solid has to change more during the later stages of crystallization than during the early stages, and this increases the likelihood of zoning.

Melting in systems involving solid solution begins only when the temperature has risen to the solidus for the particular composition. This may be at considerably higher temperature than the minimum solidus temperature for the system as a whole. For example, 1118°C is the lowest temperature at which a melt can exist in the plagioclase system at atmospheric pressure, and this only for a composition of pure albite (Fig. 10-12). Plagioclase of An_{50} composition, on the other hand, would have to be heated to 1280°C before melting would begin. Once formed, though, this liquid could be fractionated down to the pure albite composition at 1118°C. The melting and crystallization processes in systems exhibiting solid solution are, therefore, not necessarily the reverse of each other. This contrasts with

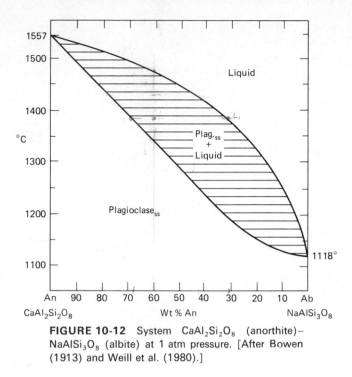

FIGURE 10-12 System $CaAl_2Si_2O_8$ (anorthite)–$NaAlSi_3O_8$ (albite) at 1 atm pressure. [After Bowen (1913) and Weill et al. (1980).]

the behavior in eutectic binary systems, where the initial melt formed on heating and the residual melt formed on cooling always have identical compositions.

10-10 POLYMORPHISM IN BINARY SOLID SOLUTIONS

Minerals belonging to solid solution series can undergo polymorphic transformations with changes in temperature or pressure, but these changes are compositionally dependent. A transformation is thermodynamically identical to that treated in Sec. 10-9, where one solution (solid) transformed (melted) into another solution (liquid). The fact that one solution is a solid and the other is a liquid in no way affects the general applicability of the results. Indeed, the cryoscopic equation describes equally well the compositional dependence of the temperature of a polymorphic transformation; we have simply to change the superscripts from liquid and solid to high- and low-temperature polymorphs, the ΔH_m to the heat of transformation, and T_m to the temperature of polymorphic transformation of the pure end member. The phase diagrams relating to polymorphism are therefore topologically identical to those described above for melting. For example, solid solution can either raise or lower the temperature of a polymorphic transformation. In an analogous manner the melting temperature of fayalite is raised by the addition of forsterite, and the melting temperature of forsterite is lowered by the addition of fayalite.

We have already encountered a polymorphic transformation whose temperature is raised by solid solution. Pure $NaAlSiO_4$, at low temperatures, forms the mineral nepheline, but above 1254°C, at atmospheric pressure, it forms carnegieite (Fig. 10-5). Albite is soluble in both phases, but more so in

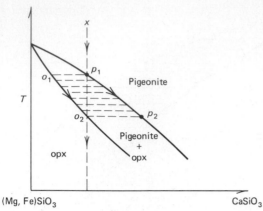

FIGURE 10-13 Idealized phase relations in part of system $(Mg, Fe)SiO_3$–$CaSiO_3$ showing polymorphic transformation of pigeonite to orthopyroxene (opx).

nepheline. Consequently, the transformation temperature rises with increasing albite content. At 1280°C, however, carnegieite and albite melt, preventing any further rise in the transformation temperature.

The change from orthopyroxene to pigeonite with increasing temperature exemplifies a transformation that is lowered by solid solution with $CaSiO_3$ (Fig. 10-13). Cooling or heating through such a polymorphic phase change involves similar types of steps to those encountered in crossing a liquidus and solidus. Cooling down the isopleth in Figure 10-13 from point x results in the appearance of orthopyroxene of composition o_1 from the pigeonite at p_1. Because the orthopyroxene has a different composition from pigeonite, its formation must involve diffusion of calcium through the solid. This is most easily achieved by nucleating numerous units of the low-temperature polymorph throughout the high-temperature phase. With continued cooling pigeonite changes its composition to p_2 and the orthopyroxene to o_2, whereupon, under equilibrium conditions, the transformation would be complete, because the low-temperature polymorph now has the starting composition. Theoretically, the polymorphic transformations in these solid solution phases should extend across the entire binary system, but in this particular example, other phase relations intervene (see Sec. 10-20). In other systems, however, polymorphic transformations can extend across the entire system.

Although polymorphic transformations in end members of solid solution series occur at definite temperatures, intermediate members transform over a temperature range and involve compositional readjustments. Thus the transformation of olivine to a spinel structure at a depth of approximately 400 km in the mantle should not be expected to occur at one depth but should be spread out over a range of depths.

10-11   BINARY SYSTEMS EXHIBITING PARTIAL SOLID SOLUTION

In Section 9-7 we dealt with the thermodynamics of nonideal solutions and why certain minerals belonging to solid solution series unmix into distinct compositional phases on cooling be-

low the solvus. The alkali feldspars, clino- and orthopyroxenes, calcium-rich and calcium-poor amphiboles, ilmenite and hematite, and chalcopyrite and pyrrhotite are but a few of the common mineral pairs that form solid solutions at high temperatures but unmix on cooling. The question to be addressed here is what effect the presence of a solvus has on the liquidus and solidus relations.

The presence of a solvus indicates that mixing in the solids is far from ideal. Even at temperatures well above a solvus, nonideal mixing can make its presence known through deviations in the shape of the liquidus and solidus from those expected in ideal systems. If the deviations from ideality are great enough, the solvus can extend to temperatures that are high enough to intersect the solidus, and thus significantly change the phase relations.

We shall first consider a phase diagram in which a solvus intersects the solidus. At first, this diagram appears very different from that for an ideal solid solution, such as that of the olivine binary. It is, however, closely related, and it is helpful, in studying the more complex system, to keep track of the simple component parts. To help do this, a series of diagrams are presented in which a solvus moves closer to and eventually intersects the solidus (Fig. 10-14). Such stages might develop through increasing pressure. Typically, a solvus is raised to higher temperatures by increased pressure, but so is the solidus. Thus it is necessary, for purposes of illustration, to have the solvus rise more rapidly. Once the solvus intersects the solidus, its upper part becomes metastable, being replaced by the assemblage solids plus liquid. The presence of two solids with liquid at the point of intersection results in isobaric invariance. A peritectic, P, is formed on the liquidus.

The system MnO–FeO provides an example of such a system. The solid solution between these two is interrupted by a solvus, which intersects the solidus at 1430°C. The solids are restricted to either manganese-rich or iron-rich compositions by the solvus. Manganese-rich ones are labeled MnO_{ss} in Figure 10-14d, the subscript ss indicating that it is a solid solution. The iron-rich solid is labeled FeO_{ss}. Beneath the solvus only mixtures of MnO_{ss} and FeO_{ss} phases are stable, the compositions of which are given by points on the solvus at the temperature of interest.

Let us consider the equilibrium crystallization of a liquid of composition x. It reaches the liquidus of MnO_{ss} at l_1 with the formation of crystals of composition s_1. As the liquid cools, the solid continually reacts with it, becoming more iron-rich, until its composition reaches s_2. Simultaneously, the liquid changes composition to P, the peritectic, at which point the MnO_{ss} reacts with the liquid to form FeO_{ss} of composition s_3. The reaction continues under the isobaric invariant conditions until all MnO_{ss} is consumed. Had the isopleth passed to the manganese-rich side of s_3, all liquid would have been consumed at the peritectic instead. (Why?) Because only FeO_{ss} and liquid remain, continued cooling causes the solid to descend the FeO_{ss} solidus. When it reaches s_4, the isopleth composition, the last liquid of composition l_4 crystallizes. The gained degree of freedom allows the FeO_{ss} to leave the solidus and cool until it intersects the solvus at s_5. Continued cooling below this point causes exsolution of MnO_{ss} from the FeO_{ss}, with the compositions of both phases descending opposite sides of the solvus to low temperatures.

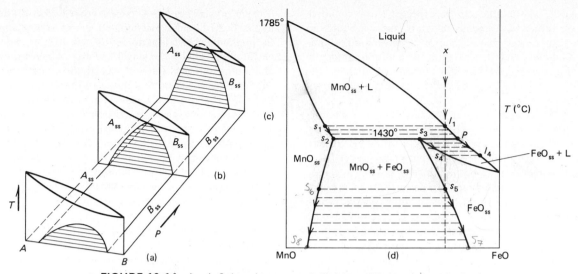

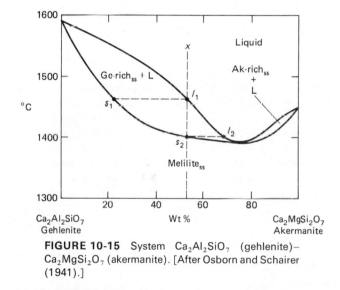

FIGURE 10-14 (a–c) Solvus in system *A–B* rising with increasing pressure until it intersects the solidus. (d) Schematic 1-atm binary-phase diagram for the system MnO–FeO.

This system exhibits both continuous reactions (s_1–s_2 and s_3–s_4) and a discontinuous one (peritectic). Departures from equilibrium can, therefore, produce both zoned and mantled crystals, which on cooling intersect the solvus over a range of temperatures and compositions. This can produce complicated exsolution patterns. Such disequilibrium enriches the residual liquid in the low melting fraction (Prob. 10-8).

10-12 BINARY SYSTEMS WITH LIQUIDUS PASSING THROUGH A MINIMUM

All points on the liquidus in systems such as those considered in Secs. 10-9 to 10-11 fall between the melting points of the pure end members. In systems that deviate markedly from ideality, however, the liquidus more commonly decreases to a minimum between the end members (Fig. 10-15). In such systems the liquid is always closer in composition to the minimum than the coexisting solid solutions, regardless of which side of the minimum it is on, but at the minimum, the liquid and solid have identical compositions. A minimum resembles a eutectic in that the lowest possible temperature at which a liquid can exist occurs within the binary, but it differs from a eutectic in having only two phases present, liquid and one solid (solid solution), instead of liquid and two solids. Two phases are present along the entire length of the liquidus, and the minimum is just the lowest point on the continuously variable isobaric univariant liquidus; there is no discontinuity such as occurs at a eutectic. In still other rarer systems, the liquidus rises to a maximum within the binary system, but no petrologically relevant systems are known to exhibit this type of behavior.

The melilites provide an excellent example of a system exhibiting a liquidus minimum (Fig. 10-15). Gehlenite ($Ca_2Al_2SiO_7$) and akermanite ($Ca_2MgSi_2O_7$), which form a

FIGURE 10-15 System $Ca_2Al_2SiO_7$ (gehlenite)–$Ca_2MgSi_2O_7$ (akermanite). [After Osborn and Schairer (1941).]

complete solid solution series at high temperature, melt at 1590°C and 1454°C, respectively. The liquidus decreases from the end members to a minimum temperature of 1385°C at a composition of 72% akermanite. Liquids, such as that of composition *x*, on cooling to the liquidus, crystallize gehlenite-rich melilite, whereas those on the other side of the minimum crystallize akermanite-rich melilite. Crystals of composition s_1 form first from liquid *x*, but under equilibrium conditions, they continuously react with the liquid, changing composition to s_2 while the liquid cools from l_1 to l_2. At this stage, all liquid crystallizes if equilibrium has been maintained, and the melilite leaves the solidus and continues cooling.

Fractional crystallization in systems such as this enriches residual melts in the minimum composition; thus melilites containing 72% akermanite can be expected to be common. Melilites are rather rare igneous minerals, and occur only in very silica-poor rocks. They do, however, have compositions near the minimum, but they contain, in addition, moderate amounts of iron and a third melilite component, soda melilite ($CaNaAlSi_2O_7$), which cannot be shown in this binary diagram (see Fig. 10-37).

The alkali feldspars provide an example of a much more common solid solution series exhibiting a minimum in the liquidus. This system also has a solvus which can, under certain conditions, intersect the solidus, producing another type of binary liquidus phase relation.

At low pressures potassium feldspar melts incongruently to leucite and silica-rich liquid (Fig. 10-10), neither of which have compositions that can be expressed in terms of the two components $NaAlSi_3O_8$ and $KAlSi_3O_8$ (Fig. 10-16). Additional components are needed to describe all phases that can form in the potassium-rich part of the system, so this part is said to be nonbinary. At high pressures, however, potassium feldspar melts congruently, and the system is binary throughout.

At 0.5 GPa, the solidus is well above the solvus, so crystallization in this system is similar to that in the melilite system (front part of Fig. 10-16a). An alkali feldspar crystallizes which, initially, is richer than the liquid in either the albite or potassium feldspar component, depending on which side of the minimum it falls. Eventually, under equilibrium conditions, a single

feldspar crystallizes whose composition is that of the starting liquid. If the liquid is fractionated, compositions approaching the minimum, which is at approximately 70% *Ab*, will form. On reaching the solvus, the single feldspar unmixes into sodium- and potassium-rich fractions (perthite) which upon further cooling descend on opposite sides of the solvus.

If water is added to this system at 0.5 GPa, it dissolves in the liquid and lowers the chemical potential of alkali feldspar. Water is not soluble in the solid feldspar, nor does it cause hydrous minerals to form. Therefore, water does not affect the chemical potential of the solids, and the result, according to the cryoscopic equation, is a lowering of the alkali feldspar liquidus. The system is now really ternary, but because the added component, H_2O, does not form a crystalline phase, the representation as the simpler binary system is permissible, as long as we remember that the melt is not truly binary but contains water. The pressures in such systems are commonly recorded in the petrologic literature as P_{H_2O}, the subscript indicating the hydrous nature of the system.

As the water content of the alkali feldspar melt increases, so the liquidus temperature progressively decreases. At 0.5 GPa, however, the melt becomes saturated in water when it reaches 10 wt %, and no further lowering of the liquidus temperature can occur. Any additional water at this pressure forms a separate vapor phase. By increasing the pressure, however, more water can be dissolved in the melt and the liquidus temperature

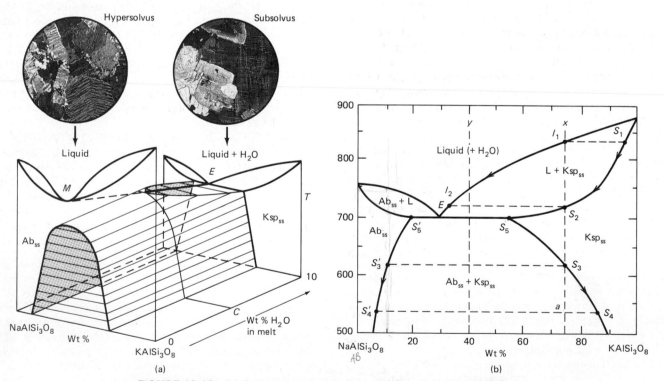

FIGURE 10-16 (a) Schematic representation of effect of addition of H_2O to liquid in system $NaAlSi_3O_8$–$KAlSi_3O_8$ at 0.5 GPa. The minimum (*M*) in the anhydrous system changes to a eutectic (*E*) above a critical concentration (*C*) of H_2O in the melt. Photomicrographs illustrate typical hypersolvus and subsolvus textures. (b) H_2O-saturated phase diagram for alkali feldspars at 0.5 GPa. All phases coexist with an H_2O-rich vapor. [After Morse (1970).]

is lowered still further. Conversely, decreasing the pressure decreases the ability of the melt to contain water and raises the liquidus temperature. Silicate melts, like carbonated beverages, require pressure to keep most dissolved gases in solution. Thus low liquidus temperatures due to dissolved water can occur only under high pressure.

Addition of water to alkali feldspar melt lowers liquidus temperatures, but it does not affect the alkali feldspar solvus, because water does not enter the crystalline feldspar. Consequently, there comes a stage, as water is added to the system (Fig. 10-16a), where the alkali feldspar minimum intersects the solvus. Then, because two solid phases coexist with the melt, a eutectic (or peritectic) relation must exist. The phase relations in the water-saturated alkali feldspar system at 0.5 GPa are depicted in Figure 10-16b—all phases coexist with a water-rich vapor.

Let us consider the equilibrium crystallization of a hydrous melt of composition x in Figure 10-16b. Cooling of the melt to the liquidus at l_1 results in the appearance of crystals with composition s_1, which is rich in the K-feldspar component. Continued cooling causes the liquid to become more sodic as it descends the liquidus to l_2. Simultaneously, the solid continuously reacts with the liquid, changing its composition along the solidus to s_2. Because s_2 falls on the isopleth for composition x, all liquid must have crystallized by this stage.

It is important to bear in mind that the liquid in this system contains water. Thus as melt x cools and crystallizes anhydrous alkali feldspar, water must continuously be transferred from the melt to the vapor phase in order to maintain the concentration of water at the saturation limit (10 wt %) in the ever-decreasing amount of residual liquid. Crystallization of any anhydrous mineral from a water-saturated melt must cause nucleation and growth of bubbles of vapor. As will be seen in Sec. 13-6, this vapor phase plays a pivotal role in the generation of *pegmatites*, rocks with extremely coarse grain size and commonly economically important concentrations of rare elements.

Returning to the cooling of composition x, we see that following the disappearance of the liquid, the solid at s_2 is free to cool. On reaching the solvus at s_3, exsolution lamellae of albite solid solution (s_3') form. Further cooling depletes the host crystal in sodium, changing its composition along the solvus, while the exsolution lamellae become enriched in sodium. At any particular temperature the compositions of the two alkali feldspars are given by points on either side of the solvus, such

as s_4 and s_4' at 530°C. The proportion of exsolved phase to host crystal is, then, s_4a to $s_4'a$.

Although liquid x does not reach the eutectic under equilibrium conditions, a small amount of fractionation would make it do so. Also, if the composition were slightly closer to the eutectic, such as y in Figure 10-16b, the residual liquid would reach the eutectic, regardless of whether there is fractionation. Indeed, many liquids in this system reach the eutectic, at which point two different feldspars crystallize, one a potassium feldspar solid solution of composition s_5 and the other an albite solid solution of composition s_5'. Both of these feldspars develop exsolution lamellae of the other on cooling.

Crystallization in the anhydrous alkali feldspar system gives rise to only a single primary feldspar, whereas in the water-saturated system many liquids crystallize two different alkali feldspars. In the first case, crystallization takes place above the solvus and is said to be *hypersolvus*. In the second, the solidus intersects the solvus, so crystallization is described as being *subsolvus*. We will encounter these terms again in discussing granites, but their significance is clear from the phase relations in Figure 10-16a. Hypersolvus rocks, which are characterized by only one primary feldspar (later unmixing to perthite), form under relatively dry conditions and at relatively high temperatures. In contrast, subsolvus rocks, which are characterized by two different primary alkali feldspars, form under relatively wet conditions and at relatively low temperatures (see insets in Fig. 10-16a).

In examining the intersection of the alkali feldspar solidus with the solvus in Figure 10-16a, the astute reader may have wondered precisely how these two surfaces first make contact. For example, does the minimum coincide precisely with the top of the solvus, in which case a eutectic results; or does the minimum come down on the sodium- or potassium-rich side of the solvus, in which case a peritectic results (Fig. 10-17)? This is not a trivial question, for it implies significant differences in the phase relations.

Because the eutectic in Figure 10-16b lies to the sodium-rich side of the solvus, the minimum may also be on this side, in which case the peritectic relation would be as shown in Figure 10-17a. If the minimum is on the other side of the solvus, the peritectic relation would be like that shown in Figure 10-17b. Reactions at the peritectic result in mantling of one feldspar by the other. Such textures are common in rocks, but are complicated by the solid solution of the third feldspar component, anorthite, which preferentially enters the albite solid solution

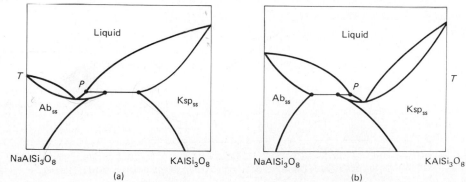

FIGURE 10-17 Possible peritectic relations in the alkali feldspar system.

(a)   (b)

FIGURE 10-18 A series of binary systems grading from complete high-temperature solid solution (left) to no solid solution (right).

and thus raises the liquidus surface of albite. The result is that the peritectic in Figure 10-17b represents the reaction most commonly encountered in rocks, that is, potassium feldspar rimming plagioclase. The position of the minimum is, however, affected by other factors in the melt, such as the activity of silica, and possibly under some conditions the peritectic of Figure 10-17a exists. This could then explain the *rapakivi* texture, where potassium feldspar is rimmed by plagioclase (Prob. 10-10).

Regardless of which peritectic exists, the intersection of the solidus with the solvus increases with increasing water content. As a result, the peritectic moves closer to the solvus and on crossing it, becomes a eutectic, giving the phase relations in Figure 10-16b. The ability of a peritectic to change to a eutectic (or the converse) with changing concentration of a third component is common in systems exhibiting partial solid solution. We will encounter it again in the pyroxene system, where a eutectic between augite and pigeonite changes to a peritectic with increasing iron concentrations.

As the amount of solid solution in a binary system decreases, the intersection of the solidus with the solvus increases. The series of phase diagrams in Figure 10-18 illustrates this change. The first diagram is similar to the "dry" alkali feldspar system. This is followed by one similar to the "wet" alkali feldspar system. In the successive diagrams the amount of solid solution decreases until a diagram similar to the diopside–anorthite system, with essentially no solid solution, is formed. The system with which we began our discussion of binary

phase diagrams is, therefore, an end member of this more general type of diagram.

This completes the discussion of possible phase relations involving the liquidus in binary systems. No new relations exist in systems containing more components, but their geometrical representation is more complex. Consequently, before proceeding to the discussion of ternary systems, readers should ensure they have a firm understanding of the binary-phase relations by doing Probs. 10-1 to 10-10.

10-13 TERNARY SYSTEMS

Although binary systems illustrate all of the types of melting behavior found in multicomponent systems, they fall short of simulating accurately those in magmas. The addition of a third component greatly increases their verisimilitude and forms systems that closely portray the phase relations in common magmas. Graphical representations of systems with the additional component are, however, difficult because four dimensions are necessary to represent all possible variables. Nonetheless, by keeping some of these constant, useful and readily intelligible, diagrams can be constructed. Graphical representations of more complex systems become increasingly difficult to read and are consequently less useful.

Ternary systems can have four independent variables, temperature, pressure, and two of the chemical components; the third component is not independent, for it is specified as

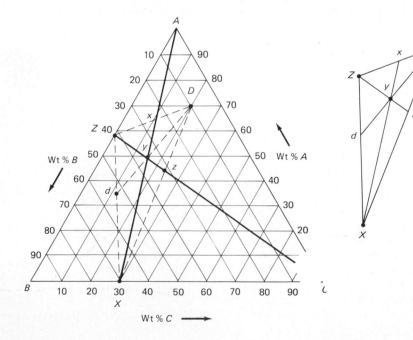

FIGURE 10-19 Graphical representation of ternary compositions. See text for discussion.

soon as the concentrations of the other two components are set. By considering isobaric conditions, as was done in most of the binary systems, the variables are reduced to three, and the system can be represented in three dimensions.

Compositions in ternary systems are normally plotted on triangular graphs, where the apexes represent the three components (Fig. 10-19). A composition plotted at an apex indicates 100% of that component, whereas a point on the opposite side of the triangle represents 0% of that component. Point X in Figure 10-19, for instance, contains no A and is a mixture of B and C in the proportions XC (70%) and XB (30%), respectively. If A is added to this mixture, the ratio of B to C will be unchanged, and compositions, such as Y (50% A, 35% B, 15% C), will form along the line XA. Because Y lies halfway along the line XA, it must, according to the lever rule, contain equal proportions of the compositions at both ends of the line, in this case A and the mixture of B and C. The fraction of A in composition Y is therefore given by YX/XA. The fraction of C in Y would similarly be YZ/ZC, and the amount of B could be determined with a similar line drawn through B and Y. Just as the addition of A to composition X produced Y, subtraction of A from Y can produce X. Addition or subtraction of a component in these diagrams, thus, simply involves a linear change from the starting composition either toward or away from the component, respectively. The amount added or subtracted is determined with the lever rule (Prob. 10-11).

Although compositions are usually plotted in equilateral triangles, this is by no means essential. Indeed, it is often necessary to describe a composition in terms of a different set of components that do not lie at the apexes of an equilateral triangle. For example, composition Y might crystallize minerals having compositions X, Z, and D. The percentage of each mineral in Y can be determined graphically in the same way as before, regardless of whether triangle XZD is equilateral. The fraction of X is xY/xX, of D is dY/dD, and of Z is zY/zZ.

The triangular compositional diagram can now have a third dimension added on which temperature can be represented. The resulting three-dimensional diagram portrays the phase relations at a particular pressure and is therefore described as isobaric. Figure 10-20a is a perspective view of the simplest type of ternary system, consisting of three congruently melting phases A, B, and C. Each face of this triangular prism is a simple binary-phase diagram, each with its own eutectic. From each binary eutectic a line extends into the ternary system, marking compositions of ternary liquids that are in equilibrium with the binary eutectic minerals; this is known as a *cotectic*. The three cotectic lines meet at a ternary eutectic, which is the lowest possible temperature at which a melt can exist in the system at the specified pressure. The liquidus line in the binary diagram becomes a liquidus surface in the ternary system. Because the liquidus surface of each phase is bounded by an intersection with another liquidus surface, the line of intersection is known as a *boundary curve*. In this particular diagram, the boundary curves are cotectics, but other types of boundary curves will be encountered in other diagrams.

Although the perspective representation in Figure 10-20a is easily visualized, quantitative reading of such diagrams is difficult. For example, there is no way of knowing what composition is represented by point X; it could be a mixture of A

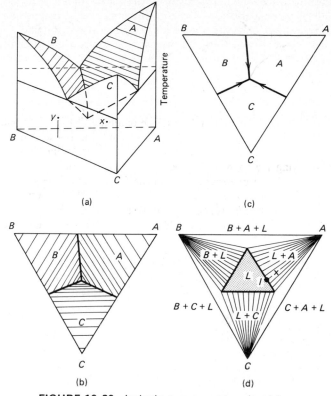

FIGURE 10-20 Isobaric ternary system involving three pure phases, A, B, and C, which melt congruently. In (a), temperature is represented in perspective on the vertical axis. In (b), phase relations are projected onto triangular base, and temperature is indicated by isotherms. In (c), only directions of falling temperature on boundary curves are indicated. (d) Isothermal section through system.

and B, B and C, or any intermediate composition. Composition Y, however, can be read because a vertical line shows not only the temperature of Y (height), but its composition projected onto the triangular base of the diagram. If every composition had to be indicated by a projection line, the diagram would soon become so congested as to be unusable. Instead, the three-dimensional diagram is reduced to two dimensions by projecting temperatures onto the triangular base of the prism (Fig. 10-20b). Compositions can now be read directly from the triangular diagram. Isotherms on the liquidus surface of each phase appear in the projection as lines that can be read in the same way as contours on a topographic map to indicate directions of falling or rising temperature. The boundary curves on the liquidus surface project as lines, and the ternary eutectic as a point. To avoid clutter, the isotherms are often omitted from the diagram, and directions of falling temperature are represented by arrows on the boundary curves (Fig. 10-20c).

The lowering of melting points in ternary systems is described by the same cryoscopic equation (Eq. 10-6) used in binary systems. It will be recalled that this equation is independent of the substance causing the lowering of melting. Thus, in Figure 10-20, the lowering of the liquidus of A is dependent only on the mole fraction of A in the liquid (assuming ideal mixing in the liquid), regardless of whether dilution is effected by *B*, *C*, or a mixture of these. Isotherms will therefore be straight lines of constant mole fraction in ideal systems, and if the molecular weights of the two diluting components (*B* and *C*) are the same, the isotherms will be parallel to the *BC* side of the triangle.

In binary systems, the lowering of the liquidus of one of the phases is limited by the liquid becoming saturated in the other phase used to cause the dilution. This produces the binary eutectic. In ternary systems, however, the liquidus can be lowered still further by diluting the liquid with the third component. Eventually, of course, no further lowering of temperature is possible when the liquid becomes saturated in both the other phases, and this then produces the ternary eutectic. Clearly, addition of a fourth component would allow further lowering of the melting point until a quaternary eutectic was reached.

A useful way of presenting the phase relations in a ternary system is through a series of isothermal, isobaric sections. One such section is shown in Figure 10-20d for a temperature above the ternary eutectic but below any of the binary eutectics. At this temperature, any composition lying within the central triangular area is liquid. The sides of this triangular area are formed by the intersection of the three liquidus surfaces with the plane of the isothermal section. A liquid having a composition on one of these lines (e.g., the one closest to *A*) would be saturated in crystals of *A*. Compositions lying within the triangular area bounded by the liquidus of *A* and *A* itself consist of crystals of *A* and a liquid. The liquid must have a composition on the liquidus. This composition can be determined by drawing a straight line from *A* through the bulk composition to the point of intersection with the liquidus. For example, a bulk composition of *x* in Figure 10-20d at the temperature and pressure of this diagram would consist of crystals of *A* and liquid of composition *l*, and by the lever rule the fraction of crystals would be *xl/Al*. Because the line *Al* ties together the compositions of coexisting phases, in this case crystals of *A* and liquid *l*, it is known as a *tie line*. Note that tie lines also radiate from *B* and *C*, tying the compositions of the respective liquids with their coexisting crystals.

In an isothermal section, cotectic lines become points where the lines pierce the isothermal plane. These points are the corners of the triangle bounding the field of liquid. A liquid having a composition at one of these points would be in equilibrium with two crystalline phases, for example, *A* and *B* for a liquid on the *AB* cotectic. A bulk composition lying anywhere within the triangular area bounded by *A*, *B*, and the liquid on the *AB* cotectic consists of crystals *A*, *B*, and cotectic liquid. Regardless of the bulk composition, the liquid has only one possible composition, that of the point of the cotectic. In terms of the phase rule, a ternary system consisting of three phases (crystals of *A* and *B*, and cotectic liquid, for example) is divariant, but the two variables are used in specifying the pressure and temperature of the isobaric, isothermal section. The resulting invariance means that the liquid cannot change its composition and remain in equilibrium with the two crystalline phases.

Isothermal, isobaric sections through ternary systems can be divided into fields consisting of one, two, or three phases:

for example, liquid (*L*), liquid + crystal (*L + A*), and liquid + two crystalline phases (*L + A + B*). Within the liquid + crystal field, the liquid can vary its composition only along the liquidus surface, and consequently these fields are ruled with tie lines to indicate the permissible compositions of coexisting phases. Within the liquid + two crystalline phase field, the compositions of all phases are fixed and are given by the apexes of the triangular field.

At successively lower-temperature isothermal sections, the field of liquid shrinks, whereas the triangular fields of liquid plus two crystalline phases expand. Eventually, the liquid field shrinks to a point at the ternary eutectic. At temperatures below this, isothermal sections would be of solids only and would simply be the triangle *ABC*.

Isothermal sections are a convenient means of presenting phase data without any possible confusion that might result from more complex projections. For example, it is simple to determine what would happen to a part of the Earth's mantle consisting of the solids *A*, *B*, and *C* in the proportions of composition *x* in Figure 10-20d if it were heated to the temperature of the isotherm in this diagram. All of the *B* and *C* and some of the *A* would melt to form a liquid of composition *l*, leaving only crystals of *A*. If we are also interested in the amount of liquid formed, that too can be read easily from the ratio of *xA/Al*.

10-14 SIMPLE TERNARY SYSTEMS WITH CONGRUENTLY MELTING PHASES WITH NO SOLID SOLUTION

The system albite–nepheline–sodium silicate is a ternary system with three congruently melting minerals. Nepheline exhibits some solid solution toward albite, but for the purpose of this discussion all of the minerals can be considered to have the stoichiometric compositions indicated by the corners of the ternary diagram (Fig. 10-21a). Also, at temperatures above 1280°C, nepheline forms the polymorph, carnegieite, but this does not affect the phase relations in the lower-temperature part of the system.

Consider the slow, equilibrium crystallization of a magma having the composition of *x* in Figure 10-21a. This composition lies within the primary liquidus field of albite. Thus, on cooling, albite will be the first mineral to crystallize. When albite begins to crystallize, the magma must become depleted in albite. Subtraction of an albite component from the melt (to form the crystals) is represented graphically by a line that extends from *x* in a direction directly away from albite. With continued crystallization of albite, the composition of the melt moves away from *x* toward l_1. At l_1 the liquidus of nepheline is encountered; that is, the melt becomes saturated with nepheline, as well as albite.

It is important in tracing the sequence of crystallization in a ternary system to keep track of the composition and amount of the solids that form. During the first stage of crystallization of melt *x*, the composition of the solids formed must plot at the albite corner of the diagram, because this is the only min-

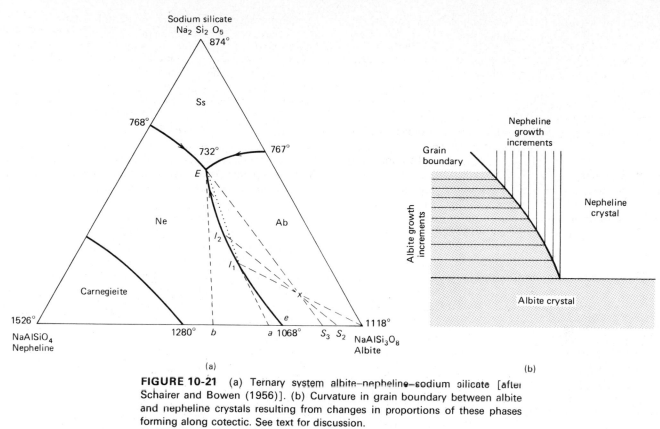

Sodium silicate
Na$_2$Si$_2$O$_5$
874°

(a)
(b)

FIGURE 10-21 (a) Ternary system albite–nepheline–sodium silicate [after Schairer and Bowen (1956)]. (b) Curvature in grain boundary between albite and nepheline crystals resulting from changes in proportions of these phases forming along cotectic. See text for discussion.

eral that crystallizes during this stage. The amount of solid formed is determined with the lever rule; for example, at the instant the melt reaches the cotectic (and before any nepheline has formed) the melt has the composition l_1, and the solids consist only of albite. The liquid l_1 and the solid (albite) must sum to the starting bulk composition, x. The ratio of solid to liquid is therefore given by the ratio of l_1x/xAb, which in this case is 50:50. Eruption and rapid quenching of the magma at this stage of crystallization would result in the formation of a vitrophyric rock containing 50 wt % of albite phenocrysts in a glassy groundmass having the composition l_1.

If, instead of erupting, the magma continues to cool and crystallize slowly at depth, the residual melt will change its composition along the cotectic as albite and nepheline crystallize. Any mixture of solid albite and nepheline has a composition that plots on the line between these minerals. But only one composition on this line will effect the necessary change in liquid composition to keep the melt on the cotectic. For example, melt of composition l_1 must crystallize albite and nepheline in the proportion indicated by point a if it is to move along the cotectic toward l_2. Clearly, the direction of change along the cotectic at any point is given by the tangent to the cotectic at that point. The intersection of this tangent with the line joining the crystallizing minerals must represent the bulk composition of these minerals forming at that instant. Point a indicates that the amount of albite crystallizing from melt l_1 is given by aNe/NeAb and that of nepheline is given by aAb/NeAb.

The tangent to the cotectic allows us to determine the instantaneous rate of crystallization of the cotectic phases, but

the total amount of solid formed during the entire cooling period must also be kept track of. Once the cotectic is reached, the early formed albite has a cotectic mixture of albite and nepheline added to it. The bulk composition of these solids consequently moves away from the albite corner of the ternary diagram along the line toward nepheline. When the melt cools to composition l_2 on the cotectic, the total solid formed has a composition of s_2 on the line nepheline–albite. This is determined by drawing a straight line from l_2 through the bulk starting composition, x, to the line Ne–Ab. This graphical construction simply means that the phases on either end of this line (i.e., the liquid l_2 and the mixture of solids, s_2) must sum to the starting composition, x.

When the cotectic liquid reaches the ternary eutectic, the bulk composition of the solids is represented by s_3. At the eutectic, sodium silicate joins albite and nepheline as crystallizing phases and the system becomes isobarically invariant (four phases in a ternary system). The liquid, consequently, can no longer change its composition. Nonetheless, albite, nepheline, and sodium silicate must crystallize from this liquid, so these minerals must form in the same proportions as present in the ternary eutectic. To the solids at s_3 a eutectic mixture of albite, nepheline, and sodium silicate are added, moving the bulk composition of the solids from s_3 toward the starting composition x. When the solids reach x, the last eutectic liquid disappears and crystallization is complete (Prob. 10-12).

In the discussion above the liquid has been taken to be in equilibrium with the solids at all stages during solidification. It is possible, however, that during crystallization the solids may separate from the liquid, for example, by crystals sinking to the bottom of a magma chamber, where they are no longer in contact with the liquid from which they crystallized. Such crystallization is said to occur *fractionally* and the liquid undergoes *fractionation*. Fractionation is an extremely important process, for it may account for much of the variation in the composition of igneous rocks.

Let us consider the fractional crystallization of magma with a composition of x in the ternary system of Figure 10-21a. When this magma becomes 50% crystallized, the residual liquid has a composition of l_1 and the solids consist only of albite. If all of these crystals are removed, by sinking, for example, only magma of composition l_1 would remain. When this new magma eventually solidifies, it would produce a rock having the composition of l_1, which contains considerably less albite than would have formed from the original magma, x. In this example crystals were removed at only one stage; the removal, however, could be continuous, giving rise to complete fractional crystallization. In either case the result of fractional crystallization, whether partial or complete, is to produce rocks that are enriched in the low-melting eutectic fraction.

The sequence of appearance of phases indicated in the phase diagram would account for the obvious textures of a rock having a composition such as x. Albite, for example, would form phenocrysts, and sodium silicate would appear only in the groundmass. The phase diagram also provides information on the relative rates at which various minerals crystallize. When the melt has the composition l_1, albite and nepheline crystallize in the proportions indicated by point a; that is, approximately twice as much albite would be forming as nepheline at this stage. By the time the cotectic reaches the eutectic, however, the proportions of albite and nepheline crystallizing change to point b, which corresponds to approximately equal proportions of these minerals.

If magma crystallizes slowly, changes in the proportions of crystallizing phases along a cotectic will affect the way in which grains are intergrown. For example, when the liquid has the composition l_1, albite crystals would be expected to grow at twice the rate of nepheline crystals. If two such crystals were juxtaposed as shown in Figure 10-21b, the thicker growth bands on the albite crystal would cause the boundary between these crystals to grow at a steep angle to the albite face. As the residual liquid approached the eutectic, the proportion of nepheline would increase until equal amounts of the two minerals were forming, at which point the angle of the grain boundary would be near 45°. Variation in the curvature of cotectics is therefore one of the factors that determines the shape of grain boundaries.

An important petrologic consequence of a curved cotectic is the possibility of producing monomineralic rocks from the mixing of different magmas on the cotectic. Many large igneous bodies are formed by repeated injections of magma, and many of the rocks are products of mixing of these successive pulses. Let us assume that an initial pulse of magma with composition x has cooled and fractionated to the eutectic composition when a second batch of magma is intruded with a less evolved composition, say magma of composition l_1. Mixing of eutectic liquid with that at l_1 produces liquids that have compositions lying in the primary field of albite. Thus magmas that were initially

crystallizing several phases, precipitate only albite on mixing. In this way layers of monomineralic rock (albite in this case) may be formed in large intrusive bodies.

10-15 TERNARY SYSTEMS WITH CONGRUENTLY MELTING BINARY PHASES

In Sec. 10-6 a congruently melting binary compound was shown to divide a binary system into two simple binaries, each with its own eutectic. Similarly, a congruently melting binary phase divides a ternary system into two simple ternary diagrams, each with its own ternary eutectic. For example, the system diopside–nepheline–silica contains the binary phase albite, and the line joining diopside and albite divides this system in two (Fig. 10-22). On the silica-rich side of this join, three cotectics meet at a ternary eutectic involving diopside, albite, and tridymite. On the silica-poor side of the join, three cotectics meet at another ternary eutectic, involving diopside, albite, and nepheline.

The directions of falling temperature on the boundary curves in Figure 10-22 are, for the most part, self-evident. Of particular interest, however, is the diopside–albite boundary curve, for it is common to both eutectics. To see how temperatures vary along this boundary, consider the crystallization of the two liquids labeled 1 and 2 in Figure 10-22. Both plot in the primary field of diopside; thus they both start by crystallizing diopside, which is then followed by albite when the liquids reach the cotectic. From here on, the liquids follow the cotectic until they reach a eutectic. Because liquid 1 reaches the diopside–albite cotectic to the left of the diopside–albite join, crystalliza-

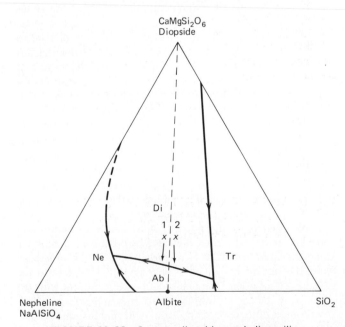

FIGURE 10-22 System diopside–nepheline–silica. Small volume of nonternary olivine omitted from nepheline–diopside side of diagram. [Simplified from Schairer and Yoder (1960).]

tion of these two minerals must move the composition of the liquid away from the join toward the Di–Ab–Ne eutectic. The temperature on this part of the cotectic must therefore decrease toward this eutectic. Liquid 2, on the other hand, reaches the cotectic to the right of the diopside–albite join, and crystallization of these minerals causes the composition of the second liquid to move away from the join toward the Di–Ab–Tr eutectic. On this side of the join, the temperature of the cotectic must decrease toward the Di–Ab–Tr eutectic. If temperatures on the diopside–albite cotectic decrease toward both eutectics, there must be a temperature maximum between the eutectics, and this obviously occurs where the diopside–albite join intersects the cotectic.

The same conclusion can be reached by applying *Alkemade's theorem*, which states that the direction of falling temperature on any boundary curve is always in the direction away from the intersection of the boundary curve with the line joining the two phases that coexist on the boundary curve. This rule holds even when the boundary curve or the join has to be extended to obtain an intersection.

The temperature maximum on the cotectic in Figure 10-22 acts as a thermal barrier that prevents any crystallizing magma from crossing the dashed line joining diopside and albite. This join behaves in much the same way that a topographic divide prevents water from crossing from one drainage basin into another. Any magma with a composition to the left of the diopside–albite join must eventually crystallize at the Di–Ab–Ne eutectic, whereas any magma to the right of the join eventually crystallizes at the Di–Ab–Tr eutectic. The diopside–albite join, therefore, divides rocks that plot in this ternary system into ones that contain nepheline and ones that contain tridymite (quartz). It will be recognized that this division is, in fact, the primary one used in the IUGS classification of igneous rock (Sec. 6-5). Rocks that are classified as alkaline or undersaturated (with respect to silica) plot to the left of the diopside–albite join in Figure 10-22, whereas those that are classified as subalkaline, oversaturated, or tholeiitic plot to the right of this join. This classification can now be seen to have a firm physical chemical justification.

Ternary systems may contain more than one binary compound, but as long as each compound melts congruently, no complications are introduced. Each group of three minerals can be treated as a simple ternary diagram with a single eutectic. Congruently melting ternary compounds can be treated in exactly the same way.

10-16   TERNARY SYSTEMS WITH AN INCONGRUENTLY MELTING BINARY PHASE

In Sec. 10-6 it was shown that if a binary compound melts incongruently in a binary system, an additional isobaric invariant point is introduced, the peritectic or reaction point, but there is still only one eutectic. Similarly, in a ternary system, the presence of an incongruently melting binary compound introduces a ternary peritectic, but there is still only one ternary eutectic.

In the ternary system forsterite–anorthite–silica (Fig. 10-23), the binary compound enstatite melts incongruently (re-

view Fig. 10-7). A boundary curve separating the liquidus fields of forsterite and enstatite extends into the ternary system from the binary peritectic. Along this line, olivine continuously reacts with liquid to produce enstatite. The direction of falling temperature on such boundary curves is commonly marked with a double arrow to indicate a reaction relation and to contrast it with a cotectic boundary curve, which is marked with a single arrow. The olivine–enstatite boundary is not a valley on the liquidus surface (Fig. 10-23a) but is a line of inflection. Above the inflection is the steep liquidus of forsterite, and below it is the gentler-sloping liquidus of enstatite, which continues downward until it intersects the silica liquidus; this produces the cotectic valley that leads down to the ternary eutectic (*E*). Where the forsterite–enstatite boundary line intersects the liquidus surface of anorthite, coexistence of the four phases liquid, forsterite, enstatite, and anorthite produces an isobaric invariant assemblage–the ternary peritectic (*P*). At this peritectic all olivine must be consumed by reaction before the liquid is free to continue cooling down the enstatite–anorthite cotectic to the ternary eutectic.

The directions of falling temperature on the boundary curves in Figure 10-23b can be determined using Alkemade's theorem. Consider first the forsterite–enstatite boundary curve; this line does not intersect the forsterite–enstatite join, so the join must be extended toward silica in order to find the point of intersection and temperature maximum on the boundary curve. This point is the binary peritectic. The temperature on the boundary curve must therefore decrease into the ternary system. The fact that this boundary curve does not intersect the forsterite–enstatite join between the minerals indicates that the boundary curve is a reaction line. Had the point of intersection fallen between the two, it would have indicated that a positive amount of each mineral crystallizes along the boundary, and thus the boundary would be a cotectic. The proportion of olivine and enstatite crystallizing from this liquid would depend on the position of the point of intersection relative to each mineral. The closer it was to enstatite, the smaller would be the proportion of olivine, the amount becoming zero when the intersection occurs at enstatite. When the point of intersection is to the right of enstatite, as it is in Figure 10-23, a negative amount of forsterite is indicated; that is, olivine is being consumed by reaction.

The direction of falling temperature on the enstatite–anorthite boundary curve is determined by tracing the metastable extension of this curve to the point of intersection with the enstatite–anorthite join (*m* in Fig. 10-23b); this point is the temperature maximum on this boundary. The temperature must therefore decrease from *m* toward the ternary eutectic.

In considering crystallization in ternary systems with a peritectic, it is essential to keep track of the composition of both the liquid and the solids. Consider the crystallization of a liquid with composition *A* in Figure 10-23c. First, *A* plots in the subtriangle An–Fo–En; under equilibrium conditions, therefore, it must eventually crystallize to these three minerals. But only at the ternary peritectic, *P*, do these minerals coexist with a melt. Liquid *A* must therefore eventually crystallize at this point. Initially, however, olivine crystallizes from liquid *A*. This depletes the melt in forsterite and changes the composition of the liquid in a straight line away from forsterite until it reaches

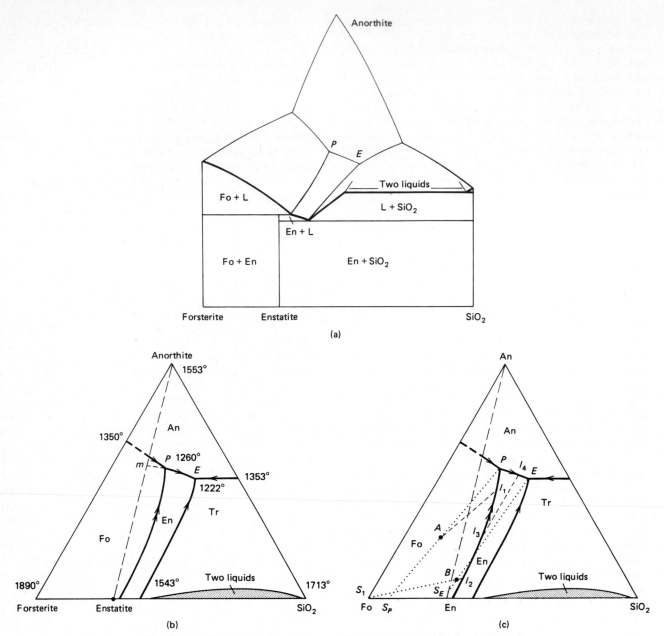

FIGURE 10-23 System forsterite–anorthite–silica at atmospheric pressure. A small field of nonternary spinel is omitted from the anorthite–forsterite side of diagram. (a) Schematic perspective representation. (b) Projection of liquidus relations onto compositional triangle. (c) Construction lines used in following crystallization paths. See text for discussion. [Simplified from Andersen (1915).]

the reaction curve at l_1. Because forsterite is the only mineral to crystallize during this initial stage, the composition of the solids must be represented by s_1. As the liquid follows the reaction curve from l_1 to the ternary peritectic, P, forsterite reacts with the liquid to produce enstatite. The bulk composition of the solids, therefore, moves from s_1 toward enstatite, reaching point s_P when the liquid reaches the peritectic. The composition of the solids at any stage of crystallization is determined by drawing a line from the liquid through the initial bulk composition (A) to the line joining the crystallizing phases. The additional phase and decreased variance at the peritectic means that the liquid can no longer change its composition. It must therefore crystallize forsterite (negative amount), enstatite, and anorthite in the proportions given by point P (any other point would change the liquid's composition). Addition of a mixture of solids of composition P to the already formed solids moves the bulk composition of the solids from s_P toward P. But in doing so, the solids approach and eventually reach composition A, at which point the peritectic liquid is exhausted and a mixture of crystals of forsterite, enstatite, and anorthite are free to cool.

The crystals of forsterite that first form in liquid A are euhedral, but when the liquid reaches the reaction curve they become rounded and rimmed by enstatite. Such reaction textures are common in tholeiitic rocks. For instance, early crystallizing olivine in the Stillwater complex of Montana sank and accumulated along with chromite near the base of the intrusion. The liquid trapped between these crystals had the peritectic composition and therefore reacted with the olivine to produce orthopyroxene. The outline of the original olivine grains is preserved by the position of the chromite grains (see Fig. 13-14).

Let us now consider the crystallization of a liquid with composition B in Figure 10-23c. This liquid also first crystallizes olivine, but because the liquid plots in the subtriangle An–En–SiO$_2$, it must eventually crystallize at the ternary eutectic. Early crystallization of olivine moves the composition of the liquid from B to l_2 on the reaction curve, whereupon enstatite begins to form. As the liquid follows the reaction curve, the solids change their bulk composition from olivine toward enstatite. When the liquid reaches l_3, all forsterite must be consumed by reaction. This can be shown by drawing a straight line from the liquid through the bulk composition, B, to the forsterite–enstatite join; this line passes through enstatite, and thus no forsterite remains. The loss of this phase increases the variance by one, and the liquid is free to leave the isobaric univariant reaction curve. Because enstatite is the only phase crystallizing, the liquid moves directly away from enstatite across the enstatite liquidus to l_4, where it intersects the anorthite liquidus. Note that whereas enstatite originally forms only as a reaction product around forsterite, on leaving the reaction curve it crystallizes directly from the melt and would consequently form euhedral crystals. At l_4, anorthite begins to crystallize, so that the solids are now represented by points on the line En–An. For example, when the liquid first reaches the ternary eutectic, the solids have a composition represented by point s_E. To this composition a eutectic mixture of En–An–Tr is added, which eventually moves the composition of the solids to the initial starting composition, B, at which point the eutectic liquid is exhausted.

Under equilibrium conditions, liquid of composition B would give rise to a rock composed of phenocrysts of enstatite, smaller phenocrysts of anorthite (because it forms later) in a eutectic groundmass of enstatite, anorthite, and tridymite. If the starting liquid had been a little less rich in anorthite, the fractionating liquid would have encountered the tridymite liquidus before that of anorthite, in which case there would have been phenocrysts of tridymite, and anorthite would have been restricted to the eutectic groundmass. Slight variations in the initial composition of the liquid can result in very different sequences of crystallization (Prob. 10-13).

When crystallization involves reactions, disequilibrium commonly results. For example, once olivine is completely rimmed with orthopyroxene, further reaction between liquid and olivine involves solid diffusion through the orthopyroxene. This is an extremely slow process, and except in very large, slowly cooled intrusions, is not likely to go to completion. Thus it would not be surprising in a rock of composition B in Figure 10-23c to find orthopyroxene phenocrysts with cores of olivine. The consequence of disequilibrium is to increase the quantity of residual liquid. Indeed, it is possible for liquids plotting in the subtriangle An–Fo–En, such as liquid A, to fractionate all the way to the ternary eutectic. This happens simply because early crystallizing olivine fails to react with the liquid to produce enstatite. Thus, if a liquid at the peritectic is unable to react with olivine because of kinetic factors, the liquid is free to continue down the cotectic to the ternary eutectic.

Another source of disequilibrium is to have early crystallizing olivine sink and separate from the liquid with which it should react. In this way, a magma of composition A in Figure 10-23c could fractionate all the way to the ternary eutectic. This process is extremely important in producing a diversity of igneous rocks in large intrusions where there is sufficient time for separation of early crystallizing minerals from the magma.

10-17 TERNARY SYSTEMS WITH LIQUID IMMISCIBILITY

In Figure 10-23 a small region of the ternary system Fo–An–SiO$_2$ is labeled "two liquids." Compositions within this field are metastable and should unmix into silica-rich and silica-poor liquids. As indicated in Sec. 10-7, many silicate systems exhibit this immiscibility, but not those involving feldspar. Thus we see in Figure 10-23 that as soon as a small amount of anorthite is added to the system, the two-liquid field disappears. Similar immiscibility fields occur on the silica liquidus at temperatures near 1690°C in many silicate systems at atmospheric pressure. This immiscibility occurs at too high a temperature to be of significance in natural occurrences. In some systems, however, this same immiscibility reappears at less silica-rich compositions and at temperatures well within the range found in crustal rocks. We will examine the best known of these, the system fayalite–leucite–silica (Fig. 10-24).

The system Fa–Lc–SiO$_2$ contains the common high-temperature two-liquid field along the Fa–SiO$_2$ side of the ternary. Addition of only a few percent leucite to these liquids is sufficient to depress the two-liquid field below the silica liquidus.

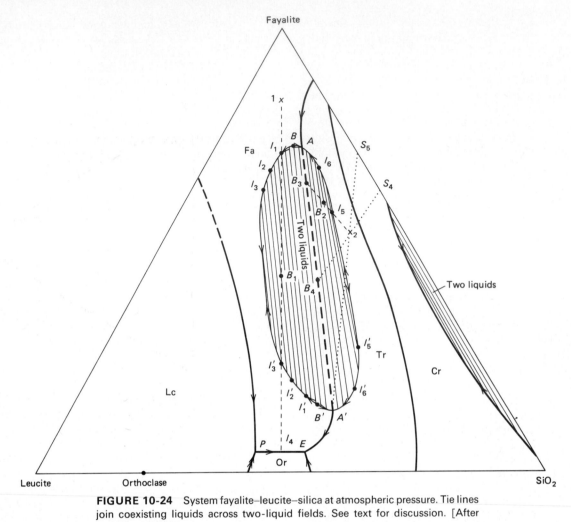

FIGURE 10-24   System fayalite–leucite–silica at atmospheric pressure. Tie lines join coexisting liquids across two-liquid fields. See text for discussion. [After Roedder (1979).]

The two-liquid field does, however, extend metastably below this liquidus and reappears in the central lower-temperature region of the ternary diagram. The region of immiscibility can best be visualized as an anticline plunging downward from the Fa–SiO$_2$ side of the ternary toward the Fa–Lc side. It plunges beneath the liquidus surface of cristobalite on leaving the side of the ternary system but then crops out as an elliptical region in the low-temperature region along the fayalite–tridymite cotectic. The composition of coexisting liquids across these two-liquid fields are given by the tie lines. Note that the fayalite—tridymite coetectic, on reaching the two-liquid field, reappears on the other side of the two-liquid field at the same temperature; that is, both liquids *A* and *A'* coexist with fayalite and tridymite. On the silica side of this cotectic, immiscible liquids coexist with tridymite, whereas on the fayalite side they coexist with fayalite. The irregular sigmoid trace of the fayalite—tridymite cotectic in this system is a result of the same nonideal mixing in the liquids that produces the immiscibility. Note also that the binary-phase orthoclase melts incongruently, as is evident from the fact that its liquidus field does not extend to the composition of orthoclase; instead, the liquidus of leucite extends out over orthoclase.

Let us first consider the crystallization of liquid 1 in Figure 10-24. Because this liquid lies in the primary field of fayalite, crystallization of this mineral will cause the composition of the liquid to move in a straight line away from the fayalite apex of the ternary system until the liquid reaches the liquidus of orthoclase at l_4. In doing this, however, the composition of the liquid would have to traverse the low-temperature two-liquid field, which is impossible if equilibrium is maintained. Therefore, although the straight line away from the fayalite apex gives the variation in the bulk composition of the liquid, the actual compositions of the liquids are given by the margins of the two-liquid field. Thus, as liquid 1 moves away from the fayalite apex, it intersects the two-liquid field at l_1, whereupon droplets of immiscible liquid nucleate in it with composition l_1'. Continued crystallization of fayalite moves the bulk composition of the liquids on the straight line away from fayalite, while the immiscible liquids change along the margin of the two-liquid field. When the bulk composition of the two liquids reaches B_1, for example, the fayalite-rich liquid has changed from l_1 to l_2, and the fayalite-poor one from l_1' to l_2'. The proportions of these two liquids at this stage are given by the lengths of the lines $B_1 l_2'$ and $B_1 l_2$, respectively. As the bulk composition of the liquids moves away from the fayalite apex, the proportion of the fayalite-poor liquid increases. Eventually, the bulk liquid composition reaches l'_3, at which point the last drop of fayalite-rich liquid disappears at l_3. The liquid, now single-phased, proceeds from l'_3 to l_4, whereupon orthoclase begins crystallizing, and the cotectic liquid descends to the ternary eutectic, where fayalite and orthoclase crystallize with tridymite.

It is interesting to note that although during a considerable fraction of the crystallization history of liquid 1 two liquids are present, no textural evidence of this would remain in the final equilibrium crystallization products, which would consist simply of phenocrysts of fayalite in a groundmass of orthoclase, tridymite, and fayalite. Only if immiscible liquids were rapidly quenched would the evidence of immiscibility be preserved. Immiscible liquids can be preserved in residual glasses in some basaltic lavas where cooling is too rapid to permit complete crystallization (see Fig. 13-16). Although textural evidence of immiscibility is unlikely to be preserved in coarsely crystalline rocks formed from a liquid that has traversed a two-liquid field, significant fractionation of the two liquids could occur within the two-liquid field. The fayalite-rich and fayalite-poor liquids have very different densities; therefore, they may separate by floating or sinking in each other. This effect could be augmented by the ability of immiscible globules to coalesce and form larger globules, which could then separate more rapidly.

Let us consider the crystallization of one other liquid, 2, which lies in the primary field of tridymite. Crystallization of tridymite causes the composition of the liquid to move away from the silica apex of the ternary system until it reaches the boundary of the two-liquid field at l_5. At this point an immiscible fayalite-poor liquid nucleates with a composition of l'_5. Continued crystallization of tridymite causes the bulk composition of the two liquids to keep moving directly away from the silica apex. Thus when the bulk composition of the two liquids is B_2, for example, the immiscible liquids have compositions l_6 and l'_6. Eventually, when the bulk composition of the liquids reaches B_3, the immiscible liquids have compositions A and A'. But these compositions lie on the fayalite–tridymite cotectic, so the tridymite and two liquids are joined by fayalite. These four phases create an isobaric invariant assemblage. Only the proportions of the two liquids, not their compositions, change as tridymite and fayalite crystallize. For example, when the bulk composition of the liquids has changed to B_4, there will be equal amounts of the two liquids, and the solids will have a bulk composition given by the point s_4 (this point is obtained by drawing a straight line from the bulk composition of the liquids through the initial bulk composition to the line joining silica and fayalite). Eventually, the bulk composition of the liquids reaches A', at which point the last drop of liquid A disappears and the solids have the composition s_5. Continued crystallization of tridymite and fayalite proceeds as the remaining liquid moves down the cotectic to the ternary eutectic.

In this second example, crystallization causes the liquids to intersect the tridymite–fayalite cotectic in the two-liquid field. Although liquids A and A' are in equilibrium with both tridymite and fayalite under these conditions, crystallization involves largely the formation of fayalite from liquid A. This is evident from the fact that the tangent to the cotectic within

the two-liquid field passes almost through fayalite. We can think of liquid A as being converted into a mixture of crystals of fayalite and liquid A'. As a consequence, crystals of fayalite grow almost entirely in the fayalite liquid—the source of their nutrients. These crystals are in equilibrium with the fayalite-poor liquid, but they do not grow from it, at least not during the two-liquid stage.

10-18 TERNARY SYSTEMS WITH ONE BINARY SOLID SOLUTION WITHOUT A MINIMUM

Many ternary systems contain phases that belong to solid solution series. Systems containing MgO and FeO, for example, will have ferromagnesian minerals that typically exhibit some range of solid solution, and systems containing alkalis and CaO may contain feldspar solid solutions. In this and the following two sections, we consider the phase relations in systems containing solid solutions. We consider first a system where one of the binaries that make up the ternary system exhibits complete solid solution. In following sections we consider solid solutions with a minimum, partial solid solution, and finally, more than one solid solution.

One of the most important ternary systems in petrology is that of diopside–albite–anorthite (Fig. 10-25). One of the binaries that bounds this system is that of the plagioclase feldspars, and the third component, diopside, is one end member of the clinopyroxene solid solution series. This system therefore contains two of the most important minerals in basalts, and liquids in this system can be considered representative of simplified basaltic magmas.

An easy way to visualize the phase relations in the ternary system Di–Ab–An is to start with the binary plagioclase system (see Sec. 10-9 and Fig. 10-12). Addition of diopside to any composition on the plagioclase liquidus lowers the liquidus temperature according to the cryoscopic equation. The plagioclase liquidus–solidus loop in a system containing, for example, 5 wt % diopside looks almost the same as that in the pure plagioclase binary, except that it moves to lower temperatures, especially at the anorthite end. This lowering of temperature of the plagioclase liquidus continues until liquids become saturated in diopside. This occurs along the cotectic marking the boundary between the fields of plagioclase and diopside. The composition of plagioclases coexisting with liquids on this cotectic and the projected composition of the cotectic liquids are shown by the dashed lines on the binary plagioclase system (Fig. 10-25a).

Because the plagioclase liquidus has no minimum, the temperature on the cotectic line in the system Di–Ab–An decreases continuously from the anorthite–diopside eutectic to the diopside–albite eutectic. Consequently, this system has no ternary minimum or eutectic; liquids on the cotectic simply fractionate toward the diopside–albite eutectic. The resulting phase diagram is simple, with just two liquidus surfaces, one for diopside and the other for plagioclase (Fig. 10-25b).

Let us consider the equilibrium crystallization of a liquid having the composition A in Figure 10-25c. This liquid first

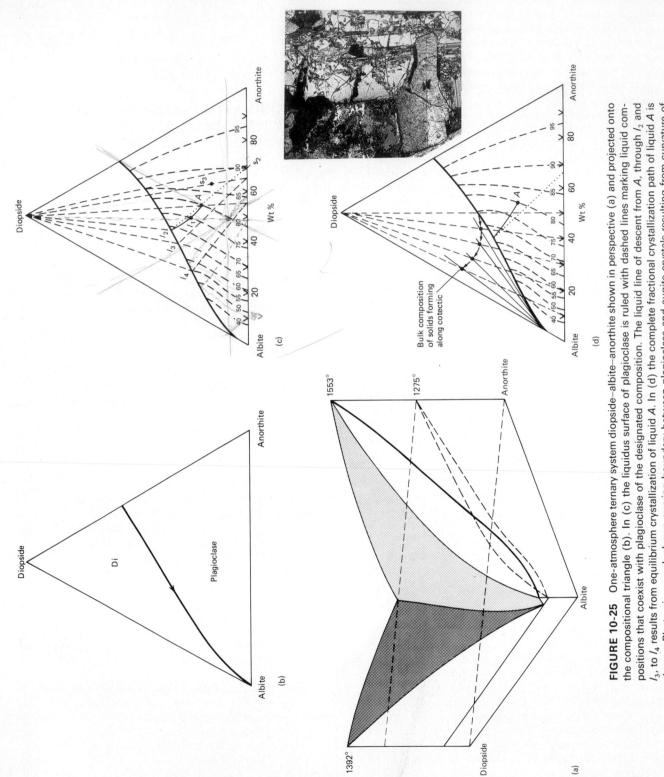

FIGURE 10-25 One-atmosphere ternary system diopside–albite–anorthite shown in perspective (a) and projected onto the compositional triangle (b). In (c) the liquidus surface of plagioclase is ruled with dashed lines marking liquid compositions that coexist with plagioclase of the designated composition. The liquid line of descent from A, through l_2 and l_3, to l_4 results from equilibrium crystallization of liquid A. In (d) the complete fractional crystallization path of liquid A is shown. Photomicrograph shows curving boundary between plagioclase and augite crystals resulting from curvature of cotectic. See text for discussion. [After Bowen (1915a), Barron (1972), and Weill et al. (1980).]

crystallizes plagioclase on cooling. Unfortunately, the composition of the plagioclase that forms initially cannot be read directly from a simple diagram such as that in Figure 10-25b. Although we might suspect that the first-formed plagioclase would be relatively anorthite-rich, its precise composition cannot be determined unless the diagram indicates the compositions of plagioclase that coexist with liquids on the plagioclase liquidus in the ternary system. The simplest way to do this is with lines on the plagioclase liquidus joining liquid compositions that coexist with a particular plagioclase composition (Fig. 10-25c). Such lines indicate that liquid A first crystallizes plagioclase of An_{80} composition.

Crystallization of An_{80} from liquid A would cause the liquid to change its composition in a straight line away from An_{80}. But as the composition changes, the coexisting plagioclase must become more albitic, and this, in turn, causes the path followed by the liquid to curve toward the plagioclase–diopside cotectic. This path can be deduced quite simply from Figure 10-25c. For example, when the plagioclase has a composition of An_{75}, the liquid must lie on the line marking the composition of liquids that can coexist with An_{75}. At the same time, the line joining the liquid and An_{75} must pass through the initial bulk composition, A. The liquid at this stage must therefore have a composition of l_1. Similarly, we can deduce that when the liquid reaches the cotectic at l_2, the coexisting plagioclase will have a composition of An_{69}.

Once the cooling liquid reaches the cotectic, coprecipitation of plagioclase and diopside causes the liquid to descend the cotectic. The three coexisting phases, liquid, plagioclase, and diopside, can be joined by lines to form a *phase triangle*. The first three-phase triangle joins l_2 with diopside and An_{69}. As the liquid descends the cotectic, the phase triangle pivots from its diopside apex as the coexisting plagioclase becomes more sodic. When the liquid reaches l_3, for example, the phase triangle joins plagioclase of An_{65} composition with diopside. The bulk composition of the solids at this stage must lie on the diopside–plagioclase side of this phase triangle. At the same time, a line drawn from the bulk composition of the solids to the liquid at l_3 must pass through the initial bulk composition, A. The solids coexisting with l_3 will therefore have a composition of s_3.

As the liquid continues to descend the cotectic, the three-phase triangle moves to the left, and the diopside–plagioclase side of this triangle moves progressively closer to the initial bulk composition. When the diopside–plagioclase join passes through point A, the last drop of liquid must crystallize. The composition of this final liquid is determined by noting that liquid of composition A must eventually crystallize to a mixture of diopside and plagioclase of composition An_{55} (this is determined by drawing a straight line from diopside through A to the plagioclase binary). The only liquid that can coexist with An_{55} and diopside is l_4. This, then, is the final liquid to crystallize from the starting composition A under equilibrium conditions (Prob. 10-15).

Equilibrium crystallization requires that plagioclase remain in contact with the liquid and that crystals continuously adjust their composition to be in equilibrium with the liquid. Solid diffusion rates in plagioclase are, however, very low, even at magmatic temperatures, as is evidenced by the common oc-

currence of zoned plagioclase crystals in igneous rocks. Equilibrium crystallization is therefore likely to occur in nature only under conditions of very slow cooling. Disequilibrium, or fractional crystallization, is more likely to occur.

Let us again consider the crystallization of liquid A, but this time under conditions of complete fractionation; that is, crystals do not react with the liquid. This occurs, for example, if crystals separate from the liquid as they form, or if crystals become mantled with layers through which diffusion is unable to transfer the reactants in the time available.

Liquid A again first crystallizes plagioclase of An_{80} composition, which causes the liquid to change composition away from An_{80} (Fig. 10-25d). Consequently, the plagioclase becomes more sodic, but because it does not react with the liquid, the path followed by the fractionating liquid is less curved than under equilibrium conditions. In addition, the line joining the liquid and coexisting plagioclase does not pass through the initial bulk composition, but lies to the left of it. This is because a fraction of the initial bulk composition is lost to the early crystallizing plagioclase, which is either removed from the system or is hidden in the cores of zoned crystals. The fractionating liquid reaches the cotectic with diopside at a more sodic composition than it does under equilibrium conditions. The liquid then follows the cotectic, and because the crystals are prevented from reacting, the liquid fractionates all the way to the diopside–albite eutectic. The result is that fractional crystallization leads to a much wider range of liquid compositions than does equilibrium crystallization.

Equilibrium and complete fractional crystallization are two extreme ways of crystallizing a melt. In nature crystallization is likely to follow some intermediate path. Volcanic and hypabyssal rocks tend to form under conditions closer to those of fractional crystallization, whereas plutonic ones form under conditions approaching those of equilibrium crystallization.

Many basaltic rocks form from liquids similar to those on the plagioclase diopside cotectic. The common *ophitic texture*, where plagioclase laths are embedded in large crystals of clinopyroxene, is believed to result from the cotectic crystallization of these minerals. The abundance of plagioclase and clinopyroxene in this cotectic mixture, of course, depends on where the tangent to the cotectic line intersects the plagioclase–diopside (clinopyroxene) join. Along much of the cotectic, the tangent intersects the join at about 40 wt % plagioclase, which is similar to the amount found in typical ophitic intergrowths. Toward the low-temperature end of the cotectic, however, the tangent swings progressively toward pyroxene (Fig. 10-25d): that is, more pyroxene and less plagioclase crystallizes. Evidence of this changing proportion can be seen in the curvature of pyroxene–augite grain boundaries as they grow into patches of residual liquid in basaltic rocks. The inset in Figure 10-25d, for example, shows a zoned crystal of plagioclase attached approximately at right angles to a zoned crystal of augite. The boundary between these two crystals is essentially a graph of the growth rate of one crystal against that of the other. Because the proportion of pyroxene to plagioclase crystallizing from the fractionating liquid increased, the grain boundary is concave toward the plagioclase. Thus, preserved in this grain boundary is a record of the changing fractionation trend of the clinopyroxene–plagioclase cotectic in this natural system.

10-19 TERNARY SYSTEMS WITH ONE BINARY SOLID SOLUTION WITH A MINIMUM

A ternary system in which a binary solid solution exhibits a liquidus minimum has either a ternary minimum or ternary eutectic. The system SiO_2–albite–orthoclase provides an important example of such a system, for it indicates the phase relations in simplified granites. Figure 10-26a is a perspective view of this system at low pressure. To simplify the illustration, orthoclase has been shown as melting congruently, where in fact it melts incongruently to leucite and a silica-rich liquid (Fig. 10-10). At pressures above 0.3 GPa, however, orthoclase does melt congruently, and even at lower pressures the simplification is of no concern as long as orthoclase-rich compositions are avoided.

The liquidus minimum on the feldspar boundary extends into the ternary system as a U-shaped valley on the feldspar liquidus. A *ternary minimum* (M) is created where this valley intersects the liquidus of the silica phase. Tridymite is the stable silica polymorph at the minimum under atmospheric pressure, but with a slight increase in pressure (0.02 GPa) quartz becomes stable. Under most conditions, then, quartz and a single alkali feldspar crystallize together at the ternary minimum. As in the system diopside–albite–anorthite, the precise composition of the feldspar crystallizing cannot be read from the simple phase diagram, unless the feldspar liquidus is contoured with lines joining liquids that coexist with a specific feldspar composition (Fig. 10-26b). With such lines the paths followed by liquids during crystallization are easily determined (Prob. 10-16). In general, these paths trend toward and then down the U-shaped valley to the ternary minimum. Under equilibrium conditions, of course, the line joining a liquid and its coexisting feldspar must, at all times, pass through the initial bulk composition, as was explained in the system diopside–albite–anorthite. Under fractional crystallization, however, a liquid will descend more rapidly to the bottom of the U-shaped valley, and then turn down the valley to the minimum (Fig. 10-26b).

Although the liquidus minimum on the feldspar binary is well above the alkali feldspar solvus, ternary liquids descending the U-shaped valley crystallize feldspars that get progressively closer to the critical temperature on the solvus; they do not, however, reach it, for the ternary minimum at atmospheric pressure is 990°C, whereas the critical temperature on the solvus is 648°C (see Prob. 9-4). Consequently, at low pressures a single feldspar crystallizes from low-temperature liquids in the ternary system. In nature such liquids give rise to granites that contain a single alkali feldspar. Because these crystallize at temperatures above the alkali feldspar solvus, they are known as *hypersolvus* granites.

With increasing pressure the critical temperature on the alkali feldspar solvus rises (see Prob. 9-4), but so do liquidus temperatures. Pressure, by itself then, does not significantly change the phase relations in the SiO_2–albite–orthoclase system. If water is present, however, increased pressure causes the water to dissolve in the melt, which, in turn, lowers the liquidus temperatures because of the decreased chemical potentials of the other components. The water does not, however, change the minerals that crystallize from the melt, so we can still plot

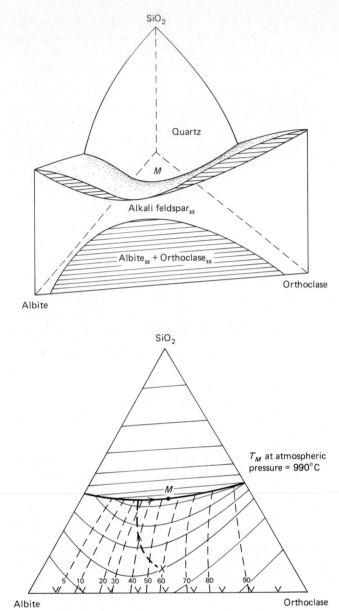

FIGURE 10-26   (a) Perspective representation of ternary system SiO_2–albite–orthoclase at low pressure. Incongruent melting of orthoclase to leucite plus liquid is omitted. The system has a minimum at *M*. (b) Projection of system onto the compositional triangle. Schematic isotherms on liquidus shown with solid lines. Dashed lines join liquid compositions that coexist with a particular composition of alkali feldspar. Heavy dashed line traces fractional crystallization path of liquid *X*. [After Tuttle and Bowen (1958) and Barron (1972).]

the phase relations in a ternary diagram even though the system, which now consists of SiO_2–albite–orthoclase–H_2O, is quaternary. Figure 10-27 shows the water-saturated liquidus-phase relations in this quaternary system at two different pressures.

At low pressure no significant amount of water dissolves in the liquids in the system SiO_2–albite–orthoclase–H_2O, so

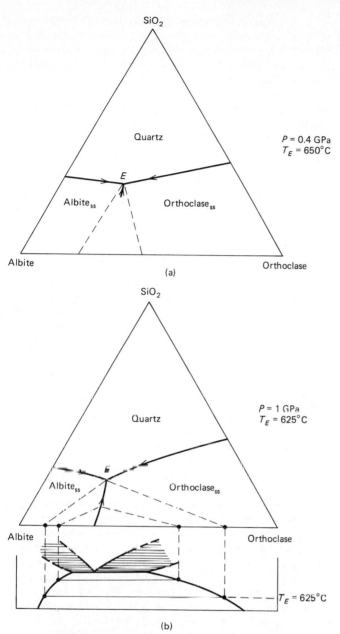

FIGURE 10-27 Ternary projection of water-saturated liquidus relations in the quaternary system SiO_2–Ab–Or–H_2O at 0.4 GPa (a) and 1.0 GPa (b). At these pressures the system has a eutectic (*E*) rather than a minimum, because two alkali feldspars coexist with quartz. Compositions of feldspars coexisting with particular liquids are joined with dashed lines. Positions of these feldspars on the alkali feldspar solvus are shown at the bottom of (b). [After Tuttle and Bowen (1958) and Luth et al. (1964).]

temperature on the alkali feldspar solvus. Two feldspars, consequently, must crystallize from this liquid, which changes the minimum into a ternary eutectic. The compositions of these two feldspars are given by the coexisting alkali feldspars on the solvus at the eutectic temperature.

Extending up from the ternary eutectic in Figure 10-27a are three cotectics, quartz–albite$_{ss}$, quartz–orthoclase$_{ss}$, and albite$_{ss}$–orthoclase$_{ss}$. As the alkali feldspar cotectic extends away from the ternary eutectic, its temperature rises until it reaches the critical temperature for that pressure. At this point, a single alkali feldspar crystallizes from the melt rather than two, and the boundary between albite$_{ss}$ and orthoclase$_{ss}$ ceases to exist; this is a critical point on the boundary curve. Below the critical point, the boundary curve marks the bottom of a V-shaped valley on the liquidus. With rising temperature, the angle of the V increases, and at the critical point the valley becomes U-shaped and continues as such to the minimum on the alkali feldspar side of the system. As the pressure increases, the critical endpoint on the alkali feldspar cotectic gets progressively closer to the alkali feldspar join, and eventually reaches it. At 1 GPa, for example, an alkali feldspar cotectic extends all the way from the ternary eutectic to the feldspar binary (Fig. 10-27b).

The compositions of feldspars crystallizing from liquids on the alkali feldspar cotectic in the water-saturated 1-GPa system can be read directly from the compositions on the solvus at the temperature of the liquid. Two such pairs of compositions are shown for a liquid partway down the cotectic and another at the ternary eutectic in Figure 10-27b. Natural granites that crystallize under water-saturated conditions also crystallize two alkali feldspars, an albite solid solution and an orthoclase solid solution (Fig. 10-27b). Because this crystallization takes place below the solvus, these rocks are known as *subsolvus* granites. Such rocks clearly indicate the presence of water in the magma, which, in turn, indicates that the magma must have been under considerable pressure, and hence at considerable depth, at the time of formation.

All degrees of water saturation are possible in granitic magmas. At high pressure, if the concentration of water in the melt is decreased, liquidus temperatures rise, and the phase relations resemble those at lower pressures, where a critical point exists on the alkali feldspar cotectic. At still lower concentrations, the ternary eutectic is replaced by a minimum and a single feldspar crystallizes. We can conclude, therefore, that hypersolvus granites are dry, whether formed at low or high pressures, and subsolvus granites are wet, which requires at least moderate pressures to keep the water in solution in the magma.

10-20 TERNARY SYSTEMS WITH MORE THAN ONE SOLID SOLUTION SERIES

In this section we consider two important phase diagrams, one involving the feldspars and the other, the pyroxenes. The feldspar diagram involves two solid solution series, the alkali feldspars and the plagioclase feldspars. The pyroxene diagram also

the phase relations are identical to those shown in Figure 10-26 for the "dry" system. With increased pressure, however, the critical temperature on the alkali feldspar solvus increases, whereas the liquidus temperatures decrease, as long as water is present in excess (water-saturated). When the pressure reaches 0.4 GPa, the water-saturated minimum drops below the critical contains two prominent solid solution series, one involving the common augites, and the other, the orthopyroxenes. The pyroxene diagram is made more complicated by the fact that the two solid solution series exhibit considerable solid solution between each other. In addition, a third pyroxene series—pigeonites—can form under certain conditions.

The feldspar ternary system is very much like the diopside–albite–anorthite system in that it is divided in two by a cotectic, which extends down into the ternary system from the anorthite–orthoclase eutectic. Unlike the diopside–plagioclase cotectic, however, the orthoclase$_{ss}$–plagioclase cotectic does not extend all the way to the alkali feldspar side of the ternary system. Instead, it is terminated at a critical end point (c in Fig. 10-28), which is produced by the complete solid solution that occurs between feldspars at liquidus temperatures in the alkali-rich part of this system.

Feldspars near the alkali feldspar side of the ternary feldspar system form a continuous solid solution series at liquidus temperatures. Indeed, at atmospheric pressure the critical temperature on the alkali feldspar solvus is 200°C below the liquidus minimum. Addition of anorthite, however, causes the solvus to rise and expand until, on the anorthite–orthoclase side of the system, the solid solution is very limited. The solvus in the ternary system can be visualized as an anticline plunging from the anorthite–orthoclase side down to the albite–orthoclase side of the ternary system (Fig. 10-28b). The crest of this anticline is the trace of the critical temperature on the ternary feldspar solvus (K–K' in Fig. 10-28b). At K', the critical temperature on the solvus reaches the solidus in the ternary system and produces the critical endpoint on the plagioclase–orthoclase$_{ss}$ boundary (c).

On the higher-temperature part of the boundary curve in the feldspar system, plagioclase and potassium-rich alkali feldspar coprecipitate along a cotectic. The composition of the alkali feldspar and plagioclase coexisting with a liquid such as l in Figure 10-28d would be A and P, respectively. The proportions in which these feldspars crystallize would be given by PX/XA, respectively, X being the point where the tangent to the cotectic at l intersects the join AP. The phase relations along the upper part of this boundary curve can be thought of in terms of a simple pseudobinary diagram with plagioclase$_{ss}$ and alkali feldspar$_{ss}$ crystallizing on either side of the solvus from a cotectic liquid (Fig. 10-28e). The representation is not truly binary because if the solid phases plot in the plane of the diagram, the liquid must plot at more albitic compositions. For this reason the liquidus in Figure 10-28e is dashed.

As liquids descend the cotectic, the point of intersection of the tangent line with the plagioclase–alkali feldspar join shifts toward the alkali feldspar composition until, at point s, it actually passes through the alkali feldspar. Beyond this point the boundary between the fields of plagioclase and alkali feldspar is a reaction curve, with plagioclase reacting with the liquid to produce alkali feldspar, as represented in the pseudobinary diagram in Figure 10-28f. This reaction relation explains the common rimming of plagioclase by alkali feldspar. Below point s, the lowest temperature in the pseudobinary diagram is no longer on the boundary curve, but is at the minimum, which occurs in the alkali feldspar field. This can also be seen from the shape of the isotherms in Figure 10-28a and c.

As liquids descend the boundary curve still further, compositions of coexisting feldspars approach one another until at point K', the critical temperature on the ternary feldspar solvus, they become identical; that is, only one feldspar crystallizes. This point is the critical endpoint on the plagioclase–alkali feldspar boundary (c in Fig. 10-28). Between this point and the alkali feldspar side of the ternary system the phase relations are as shown in the pseudobinary diagram of Figure 10-28g.

In the low-temperature part of the ternary feldspar system, then, there are three different ways in which feldspar can crystallize. Along most of the boundary curve plagioclase and potassium-rich alkali feldspar crystallize together. Between point s and the critical endpoint, however, alkali feldspar forms as a reaction rim around plagioclase. Finally, below the critical endpoint, only a single alkali feldspar forms. All three modes of crystallization are commonly encountered in igneous rocks.

Some igneous rocks exhibit feldspar phase relations that differ from those shown in Figure 10-28. This is because most magmas are more complex systems than the simple ternary one. For example, if water is present and the pressure is sufficient to dissolve the water in the melt, liquidus temperatures can be lowered enough to make the cotectic boundary extend all the way through the system from anorthite–orthoclase eutectic to an albite$_{ss}$–orthoclase$_{ss}$ eutectic. In some rocks, potassium-rich alkali feldspar is rimmed by plagioclase—the *rapakivi* texture—suggesting a reaction relation that is the reverse of the one shown in Figure 10-28f. A discussion of all the possible phase relations in the feldspars is given by Abbott (1978).

Pyroxene phase relations can be discussed most simply in terms of the ternary system $CaSiO_3$–$MgSiO_3$–$FeSiO_3$ (Fig. 10-29a). Most of the common pyroxenes have compositions that plot in the lower half of this system; that is, they plot below the line joining the compositions of diopside ($CaMgSi_2O_6$) and hedenbergite ($CaFeSi_2O_6$). The lower half of this ternary plot is commonly referred to as the *pyroxene quadrilateral*.

Common augite, which is monoclinic (clinopyroxene = Cpx), has compositions that plot between diopside and hedenbergite. Augite that coexists with calcium-poor pyroxene typically contains significantly less than 50 mol% $CaSiO_3$, especially at intermediate Mg/Fe ratios. Pigeonite is also monoclinic but generally contains only about 11 mol % $CaSiO_3$. At high pressure pigeonite can range in composition from the magnesium to the iron end member. In most rocks formed under crustal conditions, however, pigeonite is restricted to intermediate compositions of magnesium and iron. Orthorhombic pyroxenes (Opx), which typically contain between 4 and 5 mol% $CaSiO_3$, can have compositions ranging from the magnesium end member (enstatite) to iron-rich compositions. The iron end member, ferrosilite, is not stable under crustal pressures (< 1.15 GPa); instead, it forms fayalite plus quartz.

Before considering the ternary pyroxene phase relations it is important to understand the enstatite–diopside binary (Fig. 10-29b). Addition of $FeSiO_3$ to this system lowers temperatures, but it does not significantly change relations between the phases. The enstatite–diopside binary can therefore be used as a model for pseudobinary diagrams drawn at any constant $Mg/(Mg + Fe)$ ratio in the ternary system.

At low temperatures enstatite is orthorhombic and shows limited solubility toward diopside. At high temperatures

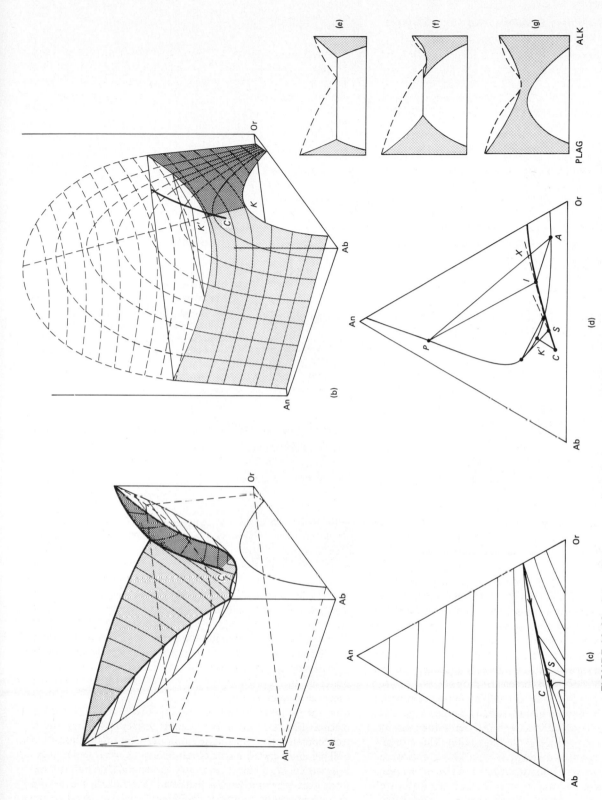

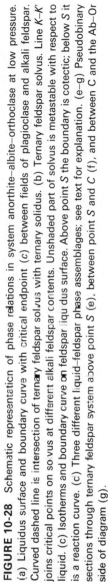

FIGURE 10-28 Schematic representation of phase relations in system anorthite–albite–orthoclase at low pressure. (a) Liquidus surface and boundary curve with critical endpoint (c) between fields of plagioclase and alkali feldspar. Curved dashed line is intersection of ternary feldspar solvus with ternary solidus. (b) Ternary feldspar solvus. Line *K–K′* joins critical points on solvus at different alkali feldspar contents. Unshaded part of solvus is metastable with respect to liquid. (c) Isotherms and boundary curve on feldspar liquidus surface. Above point *S* the boundary is cotectic; below *S* it is a reaction curve. (c) Three different liquid–feldspar phase assemblages; see text for explanation. (e–g) Pseudobinary sections through ternary feldspar system above point *S* (e), between point *S* and *C* (f), and between *C* and the Ab–Or side of diagram (g).

183

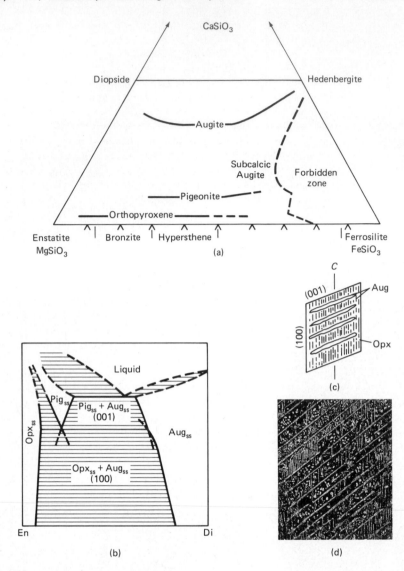

FIGURE 10-29 (a) Compositional range of common pyroxenes in terms of $CaSiO_3$–$MgSiO_3$–$FeSiO_3$. Pyroxenes within "forbidden zone" are unstable with respect to calcium-rich clinopyroxene + fayalitic olivine + quartz. This zone shrinks with increasing pressure, disappearing above 1.15 GPa. (b) Simplified schematic pseudobinary representation of phase relations near the En–Di side of the pyroxene quadrilateral. A high-temperature polymorph of enstatite and a small field of incongruent melting involving olivine have been omitted. Exsolution lamellae involving pigeonite and augite approximately parallel (001), whereas those involving opx and augite approximately parallel (100). (c) Orientation of exsolution lamellae in pigeonite that slowly cools and inverts into opx to form inverted pigeonite. (d) Photomicrograph of inverted pigeonite. Norite, Lake St. John anorthosite massif, Quebec.

enstatite inverts to a monoclinic form, which is able to dissolve considerably more diopside. This high-temperature form is in fact magnesian pigeonite. Other polymorphs of enstatite are known, but they are omitted here for simplicity and because they have little effect on the ternary pyroxene phase relations. In Figure 10-29b the clinopyroxene on the diopside side of the binary is labeled augite$_{ss}$ in order to simplify correlation with the ternary system. It shows limited solubility toward $MgSiO_3$. At low temperature it can coexist with orthoenstatite, and at high temperature, with clinoenstatite (pigeonite). This binary diagram, therefore, has two different solvi, a low-temperature one between orthopyroxene and augite and a high-temperature one between pigeonite and augite. The high-temperature one intersects the solidus to create a eutectic between pigeonite$_{ss}$ and augite$_{ss}$.

Coexisting calcium-rich and calcium-poor pyroxenes, on cooling slowly, develop exsolution lamellae. Pigeonite and augite exsolve from one another on planes that very nearly parallel (001), whereas orthopyroxene and augite exsolve from one another on planes that very nearly parallel (100). Pigeonite that cools slowly first develops fine (001) lamellae of augite. On cooling to the clino- to ortho-inversion temperature (Fig. 10-29b),

the pigeonite must exsolve a significant amount of augite in order to change to the stable orthopyroxene composition. It does this by exsolving augite onto the thin lamellae that form during the initial cooling through the pigeonite–augite solvus. This dramatically thickens these lamellae, which are eventually preserved in the orthopyroxene host that forms by inversion from the pigeonite (Fig. 10-29c and d). More augite exsolves from the orthopyroxene as it continues to cool, but this takes place as thin lamellae nearly parallel to (100). Orthopyroxene that contains the thick exsolution lamellae of augite at an oblique angle to the c axis is referred to as *inverted pigeonite*. It is particularly common in slowly cooled subalkaline gabbroic intrusions. Pigeonite itself is restricted to subalkaline lavas and shallow intrusions, where it cools too rapidly to invert to the orthorhombic form.

The phase relations on the enstatite–diopside binary decrease by about 6°C for each 1 mol % increase in Fe/(Fe + Mg) on extending into the ternary system (Fig. 10-30). For example, the inversion of pigeonite to orthopyroxene + augite, which occurs at just over 1300°C in the iron-free system, occurs at only 1100°C when Mg/(Mg + Fe) is 0.7. Similarly, the critical temperature on the pigeonite–augite solvus decreases from 1460°C

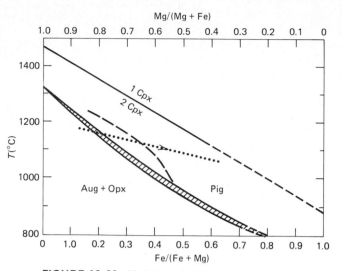

FIGURE 10-30 Variation of critical temperature on augite–pigeonite solvus and minimum stability of pigeonite as a function of mole proportions of Mg/(Mg + Fe) in the Di–En–Hd–Fs system at atmospheric pressure. Curves are dashed at high iron contents to show that pyroxenes within "forbidden zone" are unstable. Long dashed line indicates minimum stability of natural pigeonites containing nonquadrilateral components. Dotted line indicates typical fractionation bath of basaltic magma. [After Lindsley (1983).]

magma crosses the inversion curve. In many magmas it takes place when Mg/(Mg + Fe) is approximately 0.7. The pigeonite formed during the later stages of crystallization, on slow cooling, changes to inverted pigeonite.

The only significant difference between the phase relations in the binary and ternary systems is that coexisting phases in the ternary system may have different values of Mg/(Mg + Fe). These values can be shown clearly in ternary isothermal sections, such as the one illustrated in Figure 10-31 for 1200°C. Here pigeonite is seen to be more iron-rich than coexisting augite and orthopyroxene. Similarly, the cotectic liquid that coexists with augite and pigeonite is considerably more iron-rich than the crystalline phases.

Because magmas contain more components than do liquids in the simple pyroxene quadrilateral, they have lower melting points. The actual liquidus temperatures and liquidus phase relations in the pyroxene quadrilateral are, therefore, not of great importance to the interpretation of pyroxenes in natural rocks. But many of the additional components in magmas do not exhibit significant solid solution in the pyroxenes, and therefore their presence does not substantially change the subsolidus phase relations in the pyroxene quadrilateral. Consequently, the subsolidus relations can be used to interpret pyroxenes that crystallize from complex magmas at temperatures well below the liquidus in the pyroxene quadrilateral. Moreover, compositions of the pyroxenes provide useful geothermometers in subalkaline rocks (Lindsley 1983).

A complete three-dimensional representation of the subsolidus phase relations in the pyroxene quadrilateral would be far too complicated to be of great use. For purposes of geothermometry, however, we need include only the augite–orthopyroxene and augite–pigeonite solvi and the reaction of pigeonite to form orthopyroxene + augite. These phase relations can be illustrated in a series of *isothermal sections* which when superimposed on one another form a *polythermal diagram* (Fig. 10-32).

Let us consider the 1200°C isothermal section at atmospheric pressure (Fig. 10-32a). Coexisting pyroxenes with Mg/(Mg + Fe) > 0.85 consist of augite and orthopyroxene, with compositions that plot on the solvus. Compositions slightly more iron-rich than this consist of augite + orthopyroxene + pigeonite (three-phase triangle in Fig. 10-32). This assemblage represents the reaction of the high-temperature pigeonite to the

to 1300°C over this same interval. Many subalkaline basaltic magmas have magnesium numbers (Mg/(Mg + Fe) above 0.7 and temperatures near 1200°C. Consequently, they start crystallizing below the stability field of pigeonite with the formation of augite + orthopyroxene (Fig. 10-30). With cooling and fractional crystallization these magmas become more iron-rich, following paths such as the one shown in Figure 10-30. The minimum stability temperature of pigeonite, however, decreases more rapidly with iron enrichment than does the temperature of most fractionating magmas. Consequently, pigeonite becomes stable during the later stages of crystallization of these magmas, with the result that the assemblage augite + orthopyroxene is replaced by augite + pigeonite. The precise composition at which this change occurs depends on when the fractionating

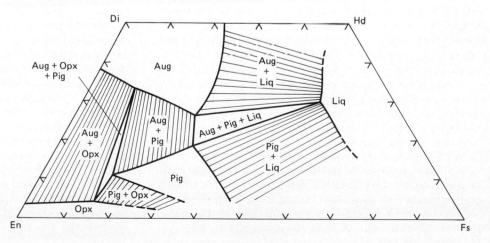

FIGURE 10-31   Isothermal section through system Di–En–Hd–Fs at 1200°C and atmospheric pressure. [After Lindsley (1983).]

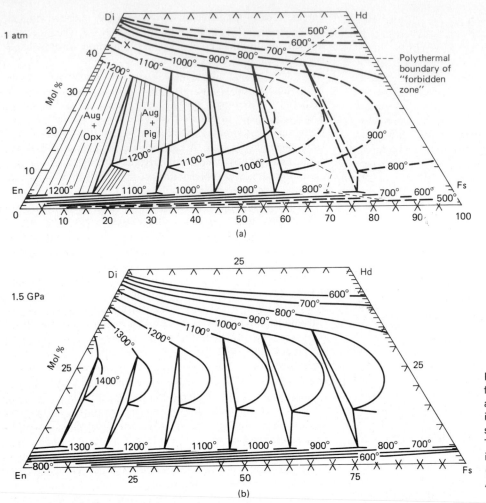

FIGURE 10-32 Polythermal relations for augite + opx, augite + pigeonite, and augite + opx + pigeonite (triangle) in the Di–En–Hd–Fs system at atmospheric pressure (a) and 1.5 GPa (b). Tie lines are drawn only on the 1200°C isothermal section. [From Lindsley (1983); published with permission of American Mineralogist.]

low-temperature assemblage of augite + orthopyroxene. Note that in the isothermal section, the pigeonite is more iron-rich than the coexisting pyroxenes (see also the T versus X representation in Fig. 10-30). Compositions more iron-rich than the three-phase triangle consist of augite and pigeonite, but with increasing iron content the compositions of coexisting augite and pigeonite approach one another until only a single clinopyroxene (*subcalcic augite*) forms when $Mg/(Mg + Fe) <$ 0.6. At still more iron-rich compositions, pyroxenes plotting to the right of the dashed line in Figure 10-32 are not stable with respect to calcium-rich augite + quartz + fayalitic olivine. This part of the pyroxene quadrilateral is referred to as the *forbidden zone*. It becomes smaller with increasing pressure and disappears above 1.15 GPa.

At lower temperatures, the phase relations are similar, except that augite has a higher calcium content and orthopyroxene and pigeonite have lower calcium contents. An analysis of a single pyroxene from any coexisting pair provides an estimate of the temperature at which this pair last equilibrated. Each pair provides two independent estimates of this temperature, and if three pyroxenes are present, three estimates can be made. Orthopyroxene compositions, however, are not very sensitive to temperature changes, as is evident from the close spacing of the isotherms on the orthopyroxene side of the augite–

orthopyroxene solvus. Finally, if only a single type of pyroxene is present, its composition provides only a minimum temperature of crystallization, that is, the temperature of the solvus below which a second pyroxene would form. For example, a single pyroxene with composition X in Figure 10-32a must have formed above 1000°C; otherwise, orthopyroxene would also have been present (Prob. 10-19).

Increasing pressure does not change the pyroxene phase relations much, except for decreasing, and eventually eliminating, the forbidden zone. The augite–orthopyroxene field widens at a given temperature, but the augite–pigeonite field is unaffected. Also, the three-pyroxene triangle, at a given temperature, shifts to more iron-rich compositions with increasing pressure. This has the effect of increasing the calcium content of pigeonites that coexist with augite + orthopyroxene. For example, at low pressure such a pigeonite typically contains about 11 mol % $CaSiO_3$, whereas at 1.5 GPa it contains 16 mol %.

The phase relations in Figure 10-32 are for the pure Di–En–Hd–Fs system. Natural pyroxenes, however, may contain other components, such as Al, Ti, Fe^{3+}, Cr, and Na, which can modify the phase relations. For example, the minimum stability of most natural pigeonites having $Mg/(Mg + Fe) > 0.6$ is approximately 50°C higher than that shown in Figure 10-30 for pure Ca–Mg–Fe pigeonite. This results from the greater solubility of the nonquadrilateral components in augite and orthopyroxene relative to that in pigeonite (Prob. 10-20). For compositions where $Mg/(Mg + Fe) < 0.5$, the lower stability limits of natural and pure pigeonites are identical. How each of the nonquadrilateral components affects the phase relations in the simple system is only partially understood. It is clear, however, that the temperatures in Figure 10-32 are applicable only to pyroxenes in which the quadrilateral components total more than 98%. When nonquadrilateral components exceed 2%, Figure 10-32 can still be used to determine temperatures, if compositions are correctly projected into the pyroxene quadrilateral. The scheme for doing this is given by Lindsley (1983).

10-21 QUATERNARY SYSTEMS

Quaternary systems do not introduce new principles with which to analyze rocks, but they do provide an almost complete representation of the phase relations in most common rocks. With the additional compositional variable, however, graphical representation becomes more complex. Only an introductory treatment of these systems will therefore be given here.

The four components (three independent compositional variables) require three dimensions for graphical representation. A tetrahedral plot is used for this purpose, where each apex represents one of the components (Fig. 10-33). The weight fraction of a component in any composition plotting within the tetrahedron is determined by extending a line from the component through the composition to the opposite face of the tetrahedron; the weight fraction of that component is then given by dividing the length of the line from the tetrahedral face to the composition by the total length of the line (XZ/AZ in Fig. 10-33). The amounts of the other components are determined with similar construction lines drawn from the other components.

Although compositions can be plotted easily in this way, compositions cannot be read from the diagram if only the point is given. Composition Y in Figure 10-33, for example, could be positioned on the front face, ACD, within the tetrahedron, or on the back face, ABD. Other information, such as a construction line or projected point (Y' in Fig. 10-33), must be included so as to define the composition uniquely. Herein lies a limitation to graphical representations in quaternary systems.

Because three dimensions are used for the compositional variables, temperature and pressure cannot also be plotted as variables. Diagrams must therefore represent phase relations at a given pressure (isobaric) and temperature (isothermal). Although such diagrams are used in the study of metamorphic rocks (Chapter 15), it is common when dealing with igneous diagrams to plot only those phase relations involving liquids. By limiting the diagram in this way, temperature can be included as a variable, the temperatures of important points being labeled on the diagram.

Each face of the tetrahedron in a quaternary liquidus diagram is formed by an isobaric ternary diagram. The front right face in Figure 10-34, for example, is the system diopside–forsterite–anorthite. Each of the primary liquidus phase *fields* in such a ternary system extends into the quaternary system as a primary phase *volume*, which is bounded by surfaces that are extensions of the ternary boundary lines into the quaternary system. Three boundary surfaces intersect along an isobaric univariant line, and four intersect at an isobaric invariant point (eutectic or peritectic). In Figure 10-34 the ternary eutectic involving Di–Fo–An extends into the quaternary system as a cotectic along which diopside, forsterite, and plagioclase crystallize together. No quaternary invariant point is reached along this cotectic because the fourth component, albite, forms a solid

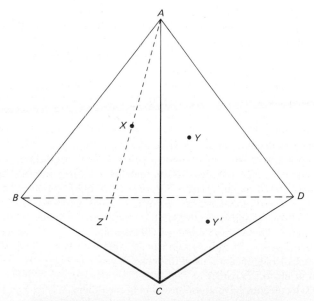

FIGURE 10-33 Perspective tetrahedral representation of quaternary system. See text for discussion.

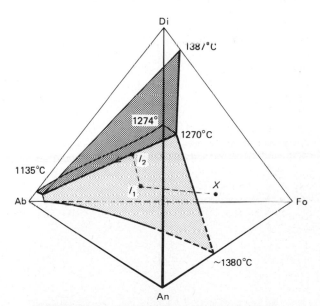

FIGURE 10-34 Simplified perspective representation of system diopside–forsterite–anorthite–albite. A small volume of the nonquaternary phase spinel is omitted from the Fo–An boundary surface (dashed region). See text for discussion. [Simplified from Yoder and Tilley (1962).]

solution with anorthite. The cotectic thus falls continuously in temperature from 1270°C on the Di–Fo–An face to 1135°C, where it passes out of the tetrahedron through the Di–Fo–Ab face. Where this line passes through the face is known as a *piercing point*.

Cooling of a magma with composition X in Figure 10-34 first results in crystallization of forsterite, which drives the magma composition directly away from forsterite until it reaches the boundary surface with plagioclase at l_1, whereupon a calcic plagioclase begins to crystallize. Crystallization of olivine and plagioclase causes the magma to change its composition along the boundary surface from l_1 to l_2. This path is curved slightly because the plagioclase continuously changes its composition as the liquid changes its composition. The degree of curvature depends on how closely crystallization approaches equilibrium. At l_2 the liquid reaches the primary field of diopside, and from there on crystallization of forsterite, plagioclase, and diopside forces the liquid to descend the cotectic. The distance that the liquid travels down the cotectic depends on the degree of fractionation.

In the system Di–Fo–An–Ab the perspective representation in Figure 10-34 is relatively clear, but in more complex systems—those with binary, ternary, or quaternary compounds, for example—phase relations can be far too complex to be portrayed clearly in this way. As a result, the phase relations are commonly presented in the form of a flowchart (Fig. 10-35). Such charts lack any compositional information, but this is not a serious loss because compositions of points in a perspective quaternary diagram cannot be read without additional information in any case. The flowchart does, however, provide a simple means of following the sequence of phases that appear during the crystallization of a magma.

Let us examine the flowchart for the system Di–Fo–An–Ab (Fig. 10-35). In the bounding ternary system Di–Fo–An there are three binary eutectics (Fig. 10-34). In the flowchart these binary eutectics are shown as points where the liquidus lines of single phases come together (Di + Fo, for example).

From each binary eutectic a cotectic leads to a ternary eutectic represented by the triangular point Di + Fo + An. A quaternary cotectic then leaves this point. It would normally intersect two other cotectics to produce a quaternary isobaric invariant point, but this does not happen in this system because of the solid solution in the plagioclase. The quaternary cotectic then passes through the system, with the plagioclase becoming progressively more albitic. Directions of falling temperature on the flowchart are indicated with arrows (double arrows for reaction curves), and the temperatures of special points are given. The piercing point is indicated by a line drawn across the cotectic line.

Because of the simplicity of the Di–Fo–An–Ab system, the flowchart does not provide any advantage over the perspective representation in this case. But with increasing complexity, the flowchart becomes more useful. Consider the quaternary system Di–Fo–An–SiO$_2$ (Fig. 10-36a). This system is complicated by the presence of a binary phase, enstatite, which forms a solid solution with diopside. The phase relations in this system can be seen with difficulty in the perspective representation, but this is about the limit of complexity that can reasonably be shown in perspective. The flowchart of this system (Fig. 10-36b) is certainly easier to read.

The flowchart in Figure 10-36 begins with the isobaric ternary invariant points on the faces of the tetrahedron. Two of these (Fo + An + Di and Fo + Di + En) are peritectics, so that from them, two reaction curves extend into the quaternary system to meet at a quaternary peritectic involving Fo + Di + An + En. The other line that intersects at this peritectic is a cotectic coming from the Fo + Di + An face of the tetrahedron. Because of solid solution between diopside and enstatite at these temperatures, a temperature maximum exists along this cotectic represented by the dashed line in Figure 10-36b. The assemblage at the quaternary peritectic is isobarically invariant. Only after all of the forsterite has reacted out does the liquid start cooling toward the next invariant point, where tridymite begins to crystallize. Two other cotectic lines intersect at this point, one involving Di + En + Tr and the other En + An + Tr. Enstatite is reacted out at this second quaternary isobaric invariant point. At still lower temperatures liquids coexisting with clinopyroxene, anorthite, and tridymite fractionate out of the tetrahedron. In this system, then, the flowchart provides a far clearer indication of the phase relations than does the perspective representations (Prob. 10-21).

Finally, we examine the quaternary system Ca$_2$SiO$_4$–Mg$_2$SiO$_4$–NaAlSiO$_4$–SiO$_2$, which is commonly referred to as the expanded basalt tetrahedron because it contains liquids that approximate the range in composition from oversaturated to strongly undersaturated basaltic magmas. A large number of phases exist in this system. The stable subsolidus tie lines between these phases are given in Figure 10-37a. Clearly, if the liquidus boundaries between all phases were included in the diagram, the representation would be unintelligible. However, the flowchart for this system can be read with ease. The geologically significant part of this chart is given in Figure 10-37b. It is left to the reader to study the diagram. Note that quaternary eutectics have four lines coming into them, whereas peritectics have three lines coming in and one line going out. After familiarizing yourself with the chart, attempt Prob. 10-22.

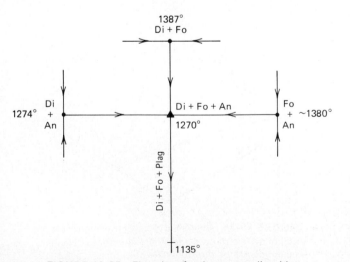

FIGURE 10-35 Flowchart for the system diopside–forsterite–anorthite–albite shown in perspective in Figure 10-34. See text for discussion.

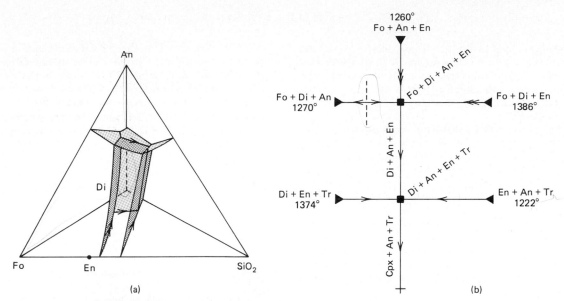

FIGURE 10-36 (a) Schematic perspective representation of the system diopside–forsterite–anorthite–SiO$_2$. Small volume of the nonquaternary phase spinel is omitted from the Fo–An join. (b) Flowchart for same system.

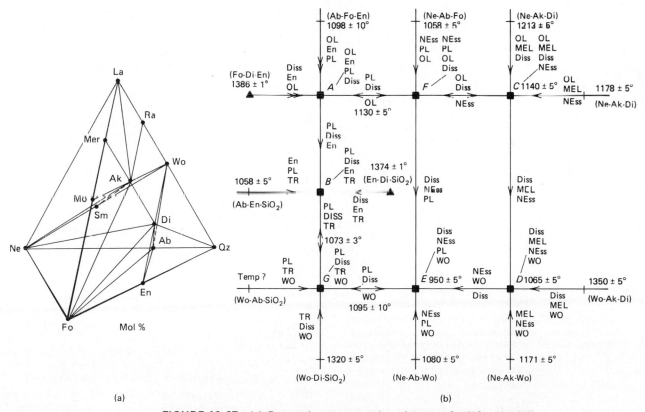

FIGURE 10-37 (a) Perspective representation of system Ca$_2$SiO$_4$–Mg$_2$SiO$_4$–NaAlSiO$_4$–SiO$_2$ with tie lines between coexisting subsolidus phases. Abbreviations are as follows: larnite, La; rankinite, Ra; quartz, Q; wollastonite, Wo; merwinite, Mer; akermanite, Ak; soda melilite, Sm; monticellite, Mo; diopside, Di; albite, Ab; enstatite, En; nepheline, Ne; forsterite, Fo. (b) Flowchart for geologically significant part of this system. Additional abbreviations are olivine, Ol; plagioclase, Pl; melilite, Mel; tridymite, Tr. [After Schairer and Yoder (1964); published with permission of the Geophysical Laboratory, Carnegie Institution of Washington.]

PROBLEMS

10-1. **(a)** Using the ideal cryoscopic equation, calculate the liquidus lines for diopside and anorthite in the binary system diopside–anorthite at atmospheric pressure. The melting points for diopside and anorthite are 1665 and 1830 K, respectively, and their heats of fusion, are 142.6 and 136 kJ mol^{-1}, respectively.

(b) Recalculate mole fractions into weight percent (mol. wt., Di = 0.2166 kg, An = 0.2782 kg), and plot a binary phase diagram for the system diopside–anorthite [$T(°C)$ versus wt %]. What are the temperature and composition of the eutectic?

10-2. **(a)** Compare the calculated phase diagram for the system Di–An from Prob. 10-1 with the experimentally determined one in Figure 10-4. What might explain the differences?

(b) Calculate the activity coefficients (γ_i) for diopside and anorthite at the experimentally determined eutectic by comparing calculated and actual temperatures for this composition. What conclusion can you draw about the ideality of the solutions?

10-3. **(a)** On a line between two components A and B, plot a point representing a magma containing 50 wt % A and 50 wt % B. If on cooling slightly 10 wt % of this magma forms crystals of pure A, what is the composition of the remaining liquid (90% of total)? Solve graphically.

(b) Graphically determine the bulk composition of a mixture of 20% crystals of B and 80% of a liquid with a composition 30% A and 70% B.

10-4. **(a)** List the sequence of appearance and disappearance of phases involved in the equilibrium cooling of a liquid indicated by the isopleth in Figure 10-10. What is the composition of the rock formed from this magma?

(b) If some of the same magma in part (a) were erupted and rapidly quenched when it is 50% solidified, what composition would the glass (supercooled liquid) have in terms of Ks–SiO$_2$, and what would the phenocrysts be?

(c) Some of the magma erupted in part (b) remains at depth and has all of the crystals removed from it when it is 50% solidified. The remaining liquid then crystallizes under equilibrium conditions. List the sequence of crystallization of this separated liquid, and give the composition and name of the rock that it finally forms.

(d) Discuss the range of rock types that can form from the magma in part (a) by the different crystallization paths in parts (a) and (c).

10-5. Using Eqs. 10-18 and 10-20, calculate the liquidus and solidus in the system Mg$_2$SiO$_4$–Fe$_2$SiO$_4$ at atmospheric pressure assuming ideal mixing in both the solid and liquid. The melting points of fayalite and forsterite are 1490 and 2163 K, respectively, and their heats of fusion are 92.174 and 122.256 kJ mol^{-1}, respectively. Convert mole fractions to wt % (mol. wt. Fa = 0.20379 kg, Fo = 0.14069 kg), and compare the calculated phase diagram [$T(°C)$ versus wt %] with the experimentally determined one in Figure 10-11 (plot both in the same diagram). What can you conclude about the ideality of both solid and liquid olivine solutions?

10-6. Thermochemical data from which heats of fusion can be determined are not available for all minerals; forsterite is one such mineral. However, if the effect of pressure on the melting point and the volume change on melting are known, the enthalpy of fusion can be determined from the Clausius–Clapeyron equa-

tion (Eq. 8-3). The melting point of forsterite at atmospheric pressure (10^5 Pa) is 2163 K, and the volume change on melting is 2.703×10^{-6} m^3 mol^{-1}. If the rise in melting point with pressure is 47.7 K GPa^{-1}, calculate the heat of fusion of forsterite at atmospheric pressure. (Recall that 1 J = 1 Pa m^3.)

10-7. In Figure 10-12, if a magma containing 60% anorthite cools and crystallizes under equilibrium conditions until only one-third of the liquid remains, what is the composition of this residual liquid?

10-8. Discuss the textures resulting from disequilibrium crystallization of magma of composition X in Figure 10-14d. Assume that some fraction of every composition that crystallizes is preserved. Pay particular attention to where and when exsolution would occur.

10-9. List the sequence of events in the equilibrium crystallization of a liquid containing 90% Ab in Figure 10-16b.

10-10. Draw sketches contrasting textural differences to be expected in thin sections of feldspars that have crystallized under disequilibrium conditions with the phase relations as shown in Fig. 10-17.

10-11. **(a)** On a sheet of triangular graph paper, plot a magma that contains 25% A, 50% B, and 25% C. This magma assimilates an equal weight of xenoliths that contain 80% crystals of A and 20% crystals of C. What is the resulting bulk composition?

(b) The heat necessary to cause assimilation normally comes from the heat of crystallization of a mineral on the liquidus. If the assimilation in part (a) brings about 20% crystallization of A from the contaminated bulk composition, what is the composition of the final liquid?

10-12. **(a)** Describe the equilibrium crystallization of a liquid containing 40% Ne, 40% Ab, and 20% sodium silicate in the system in Figure 10-21.

(b) What composition does the liquid have following 50% crystallization?

(c) What solids are present following 50% crystallization, and what are their relative abundances?

(d) What is the instantaneous composition of the solids crystallizing from this liquid following 50% crystallization?

(e) What composition do the solids have when this liquid first reaches the eutectic?

10-13. Consider the equilibrium crystallization of a liquid halfway between liquids A and B in Figure 10-23c.

(a) Where will crystallization end, and what will be the composition of the rock formed?

(b) When 50% of the liquid in part (a) has crystallized, what composition do the solids have?

(c) If all of the crystals in part (b) are suddenly removed, trace the equilibrium crystallization of the remaining liquid.

(d) Compare the rock types formed in parts (a) and (c).

10-14. List the sequence of steps involved in the equilibrium crystallization of a magma having a composition of 20% Fa, 20% Lc, and 60% SiO$_2$ in Figure 10-24.

10-15. **(a)** What composition plagioclase will first crystallize from a melt containing 10% Di, 45% Ab, and 45% An in the system Di–Ab–An (Fig. 10-25c)?

(b) Assuming equilibrium crystallization of this liquid, what composition will the plagioclase have when the first diopside crystals form?

(c) Assuming equilibrium crystallization, what is the composition of the final liquid to crystallize from this same starting composition?

10-16. In Figure 10-26b, trace the steps involved in the equilibrium crystallization of a liquid containing 15% Q, 35% Ab, and 50% Or. First determine what the composition of the final feldspar must be.

10-17. List the steps involved in the equilibrium crystallization of a liquid containing 25% Di and 75% En in the pseudobinary diagram of Figure 10-29b. Note the directions in which exsolution lamellae develop and draw a sketch of the resulting texture.

10-18. Would an andesitic lava with a temperature of 1100°C and crystallizing pyroxene with Mg/(Mg + Fe) of 0.6 be likely to have crystals of pigeonite or orthopyroxene?

10-19. A diabase dike that is known to have crystallized near the surface of the Earth contains coexisting orthopyroxene $(Ca_4Mg_{67}Fe_{29})$ and augite $(Ca_{35}Mg_{48}Fe_{17})$. Using the Lindsley pyroxene geothermometer, estimate the temperature at which these pyroxenes crystallized. Would you expect the rock also to contain pigeonite, especially on the rims of the orthopyroxene; if so, what composition would you expect it to have?

10-20. Explain why the greater solubility of nonquadrilateral components in augite and orthopyroxene relative to that in pigeonite increases the stability field of augite + orthopyroxene (see Sec. 10-10).

10-21. Many basaltic rocks contain plagioclase phenocrysts. In terms of Figure 10-36, this means that they plot on the plagioclase (An) saturation surface. For such rocks it is possible to simplify this quaternary system by projecting the phase boundaries that intersect the anorthite surface from the An apex onto the triangular base Di–Fo–SiO$_2$. The result is an isobaric ternary diagram indicating the liquidus phase relations in anorthite-saturated liquids. Using the information given in Figure 10-36, draw this ternary projection and label all boundary lines with appropriate arrows indicating directions of falling temperature and cotectic and reaction relations.

10-22. In Figure 10-37b a temperature maximum is indicated on the olivine–diopside–nepheline boundary line (between *F* and *C*). This temperature maximum occurs where the boundary line pierces the Fo–Di–Ne plane (locate this plane in Fig. 10-37a). List the sequence of steps involved in the fractional crystallization of two magmas that both lie on this boundary line but on opposite sides of the temperature maximum. Plot the approximate paths that these fractionating liquids might follow in the perspective representation of Figure 10-37a. Note that a peritectic must plot outside the tetrahedron formed by the phases that coexist at the peritectic. Moreover, a peritectic must be away from the face of the tetrahedron opposite the phase that is consumed at the peritectic. A eutectic, on the other hand, must plot within the tetrahedron formed by the phases involved. What textural difference would the plagioclase exhibit in the rocks formed from these two liquids?

11-1 INTRODUCTION: COMPOSITION OF VOLCANIC GASES

All magmatic components under specified conditions of temperature, total pressure, and magma composition have a definite vapor pressure, but for most it is so small, even at the low confining pressures on the Earth's surface, that these components form only condensed phases—solids and liquids. A few components, however, can have high vapor pressures, and if their concentrations are high enough, or the confining pressure low enough, they may form a separate gaseous phase. These are the components referred to as the volatile constituents of magmas. Even the normally condensed components can become volatile if pressures are extremely low (approaching a vacuum). Such pressures existed in the early solar nebula during accretion of the planets, and under these conditions differences in the vapor pressures of the common rock-forming minerals played an important role in determining what minerals could condense to form the planets.

The volatile constituents of magmas, although normally present in only small amounts, can profoundly affect liquidus temperatures, phases that crystallize, and magmatic fractionation trends. They play an important role in the generation of magmas, and as such are probably responsible for the shallow depth of the asthenosphere and the Earth's unique style of tectonism.

The true significance of volatile constituents to magmatic processes has become apparent only through experimental studies on silicate systems and natural rocks. Magmas, on cooling and crystallizing, lose most of their volatiles. The final rock may therefore preserve only a poor record of the volatiles that were originally present. Hydrous minerals, fluid inclusions, and vesicles in volcanic and hypabyssal rocks do provide some evidence. In laboratory studies, however, it is possible to control the composition of the volatiles and to evaluate their effect on phase relations. The results, in turn, can be used to interpret the role volatiles may have played in the origin of a rock.

That magmas contain volatiles, and in some cases, large quantities, is obvious from volcanic explosions such as that of Mount St. Helens in 1980. Sampling of the gaseous emanations of active volcanoes has provided a wealth of information on the common volatile constituents associated with near-surface igneous activity. Samples have been obtained from such different environments as strato volcanoes over subduction zones (Mount St. Helens), shield valcanoes over hot spots in oceanic plates (Kilauea), and submarine fumeroles on midocean ridges. Not all gases liberated from volcanoes, however, have their origin in the magma. Indeed, a very large fraction can be recycled groundwater or ocean water. The steep temperature gradients associated with high-level igneous intrusions are ideal for setting up large hydrothermal convection cells. The circulating, hot solutions can drastically alter the composition of the igneous rocks through which they pass. Moreover, many elements, in particular the chalcophile ones, may be redistributed and concentrated by these solutions to form important mineral deposits.

Gaseous volcanic emanations consist largely of the elements hydrogen, oxygen, carbon, and sulfur, which form H_2O, CO_2, CO, SO_2, H_2S, H_2, S and O_2. Other minor constituents include N_2, Ar, HCl, HF, and B (BVSP, 1981). Although the abundance of these gases varies considerably from one volcano to another and even in the same volcano at different times, H_2O and CO_2 are always by far the two most abundant gases, constituting from 30 to 80 and 10 to 40 mol %, respectively, of the vapor phase (Anderson 1975). There is considerable uncertainty, however, as to how much of each of these gases is juvenile (coming to the Earth's surface for the first time) and how much is meteoric (has, at some time, resided in the hydrosphere or atmosphere) and derived from recycled groundwater or contact metamorphism of intruded rocks (CO_2 from limestone, for example). Hydrothermal convection cells are certainly capable of recycling large volumes of water. For example, the entire volume of water in the oceans is estimated to pass through the hydrothermal systems along oceanic ridges once every 8 million years. Over geologic time, however, it is generally agreed that the hydrosphere and atmosphere have been formed and maintained largely by juvenile volcanic gases. For some of the gases (oxygen and helium, for example), isotopes provide a simple means of identifying juvenile gases (sec. 21-4), and direct evidence of juvenile gases is found in fluid inclusions in mantle-derived nodules.

Although the abundance of the various volatiles in volcanic gases is determined in part by elemental abundances, concentrations of individual gas species are also controlled by reactions between the species. For example, carbon dioxide and carbon monoxide are related by the reaction

$$CO_2 \rightleftharpoons CO + \tfrac{1}{2}O_2 \qquad (11\text{-}1)$$

for which an equilibrium constant (Eq. 9-55) can be written as

$$K = \frac{(fCO)(fO_2)^{1/2}}{(fCO_2)} = e^{-\Delta G/RT} \qquad (11\text{-}2)$$

At constant pressure the equilibrium constant is a function of temperature (Eq. 9-60). Also, the gases behave almost ideally at the temperatures and pressures at which they are released from volcanoes. We can therefore rewrite Eq. 11-2 as

$$\frac{pCO}{pCO_2} = pO_2^{-1/2} e^{-\Delta G/RT} \qquad (11\text{-}3)$$

where partial pressures are measured in bars. If we assume for simplicity that these gases are the only ones present, the sum of their partial pressures must equal the total pressure on the gas (1 bar); that is,

$$P_{CO_2} + P_{CO} + P_{O_2} = 1 \text{ bar} \qquad (11\text{-}4)$$

The partial pressure of oxygen, however, is extremely small (about 10^{-10} bar) and can therefore be ignored without introducing serious error. Also, for an ideal gas,

$$p_i = X_i P_{\text{total}} \qquad (11\text{-}5)$$

We can therefore rewrite Eq. 11-3 as

$$\frac{XCO}{XCO_2} = PO_2^{-1/2}\, e^{-\Delta G/RT} = \frac{VCO}{VCO_2} \qquad (11\text{-}6)$$

Equation 11-6 shows that the volume proportions of CO and CO_2 in volcanic emanations are a function of the oxygen fugacity and temperature. As will be shown in Sec. 11-5, the ferromagnesian minerals that crystallize from magmas are capable of maintaining oxygen fugacities at fixed values (buffering) at a given temperature. These minerals will therefore determine the ratio of CO and CO_2 in the magmatic gases. For example, at 1150°C (1423 K) the ΔG for reaction 11-1 is 161 kJ mol^{-1}, and if the oxygen fugacity is buffered at $10^{-11.2}$ bar by the ferromagnesian minerals, the volume ratio of CO to CO_2 would be 0.5; that is, twice as much CO_2 as CO would be present in the volcanic gas.

In passing, it is worth noting that reactions such as Eq. 11-1 are actually used to control oxygen fugacities in experimental laboratory studies. Carefully metered volumes of CO and CO_2 are mixed and allowed to flow into a furnace.

Once the gases reach equilibrium (a few seconds) an oxygen fugacity is established that is determined by the volumes of CO and CO_2 and the temperature of the furnace.

Many other equilibria exist in magmatic gases in addition to the one given in Eq. 11-1 (Heald et al. 1963). Some of these are

$$H_2O \rightleftharpoons H_2 + \tfrac{1}{2}O_2 \qquad (11\text{-}7)$$
$$CO_2 + H_2 \rightleftharpoons CO + H_2O \qquad (11\text{-}8)$$
$$CH_4 \rightleftharpoons C + 2H_2 \qquad (11\text{-}9)$$
$$H_2S + O_2 \rightleftharpoons SO_2 + H_2 \qquad (11\text{-}10)$$

Clearly, the concentrations of all these volatile constituents are highly interrelated and are sensitive to the oxygen fugacity, which in most cases is buffered by the ferromagnesian minerals in the rock. Also, because equilibrium constants change with temperature, the concentrations of the gaseous species are sensitive to temperature. Pressure can also be of importance, as can be illustrated by considering the interaction of Eqs. 11-7 and 11-10. Expressions for the equilibrium constants for these two reactions can be combined to give the following equation:

$$\frac{(fSO_2)}{(fH_2S)} - \frac{K_{11\text{-}10}(fO_2)^{3/2}}{K_{11\text{-}7}(fH_2O)} \qquad (11\text{-}11)$$

At a specific temperature the equilibrium constants are fixed, and fO_2 is fixed by the buffering of ferromagnesian minerals. Thus, as the fugacity of water increases with depth, the fugacity of H_2S will increase relative to that of SO_2. The major sulfur-bearing gaseous species at depth may well be H_2S despite the prevalence of SO_2 in surface samples of volcanic gases. A detailed analysis of the composition of a Kilauean gas sample by Nordlie (1971) provides an excellent example of the difficulties involved in determining the composition of volcanic gases.

In this chapter we focus on the effect of volatiles, in particular water, carbon dioxide, sulfur, and oxygen, on the phase relations between solids and liquids. We consider first the solubility of these volatiles in common magmas, and then study how their presence can alter the phase relations in some of the important systems considered in Chapter 10. Finally, a consideration of fractional crystallization and melting in volatile-bearing systems will reveal the crucial importance of volatiles to the genesis of many igneous rocks.

11-2 SOLUTION OF H₂O IN SILICATE MELTS

Chapter 10 dealt with phase diagrams in which components have such low vapor pressures that even at atmospheric pressure only condensed phases form. In such systems, the lowering of the liquidus of a phase, resulting from the addition of another component, is limited only by the liquid becoming saturated in a new crystalline phase. In volatile-bearing systems, however, addition of a component may lead to the formation of a separate volatile-rich phase. Once this happens, further addition of that component to the system does not lower liquidus temperatures but simply adds to the amount of separate volatile phase. In other words, the liquid becomes saturated in that volatile component. Because the amount of lowering of liquidus temperatures is critical to the formation of magmas in a planet that does not have excess heat, it is important to know what determines the saturation level of a volatile component in a silicate melt.

The simplest way of tackling this problem is to think of the formation of a separate volatile-rich phase in terms of a reaction. For example, if the volatile component is H_2O, the reaction would be written as

hydrous silicate melt = less hydrous silicate melt

$$+ \ H_2O(vapor) \quad (11\text{-}12)$$

Water in a separate vapor phase has a greater volume than when it is dissolved in the melt. Consequently, the volume change for this reaction is positive, being large at low confining pressures but becoming smaller at higher pressures, because of the greater compressibility of the separate vapor phase. Because the left-hand side of this reaction has the smaller volume, increased pressure favors the formation of hydrous melt. In other words, the solubility of water in a melt is a function of pressure. This conclusion is not surprising in view of our everyday experience with carbonated beverages, where releasing the pressure causes carbon dioxide to start coming out of solution.

The solubility of water in melts has been studied in many synthetic systems and in the common magma types (Fig. 11-1). The saturation limit versus pressure relations for all these melts are similar. The solubility at atmospheric pressure is negligible, but with increasing pressure it rises rapidly at first and then at a more moderate, steady rate at pressures above about

0.2 GPa. Mafic melts generally can dissolve less water than felsic ones. For example, at 0.5 GPa and 1100°C, molten basalt can dissolve up to 8.2 wt% H_2O, but molten granitic pegmatite can dissolve up to 11.0 wt%.

To understand more fully the solubility of water in silicate melts, it is necessary to analyze how water actually dissolves in the melt. Does it, for example, remain as molecular water and simply fit into cavities in the structure of the melt, or does it react with the melt? Answers to this question have come from a detailed study of the solubility of water in molten albite (Burnham and Davis 1974; Burnham 1979), which provides a good model for many other silicate melts. The general conclusion is that solution of water involves reaction with the melt.

Burnham's model is based on the reasonable assumption that the structure of a melt at near-liquidus temperatures differs from that of the crystalline solids in lacking only long-range order. The basic silicate structures persist in the melt. Thus an albite melt consists of small units of albite structure which are misoriented with respect to surrounding units. Each unit, however, maintains its structure and can be represented by the formula $NaAlSi_3O_8$.

Being a framework silicate, albite consists of SiO_4 and AlO_4 tetrahedra linked together by shared oxygen anions (bridging oxygens). Albite melt would therefore also be expected to have this structure. One way in which water can enter this melt is by a hydrolysis reaction with a bridging oxygen (Fig. 11-2). This oxygen ion remains attached to one of the tetrahedra and by gaining a hydrogen ion is converted to a hydroxyl ion. The remaining hydroxyl ion completes the tetrahedral coordination around the other tetrahedron. This form of solution can be represented by the reaction

$$H_2O(v) + O^{2-}(m) \rightleftharpoons 2OH^-(m) \quad (11\text{-}13)$$

where (v) represents vapor phase and (m), melt. Because the linkage of the framework structure is broken (depolymerized) by this solution, it dramatically reduces viscosities (Sec. 2-4).

Solution by hydrolysis of bridging oxygen ions is possible in any framework silicate melt. But in albite melts one out of every four tetrahedral groups is an AlO_4 tetrahedron. More-

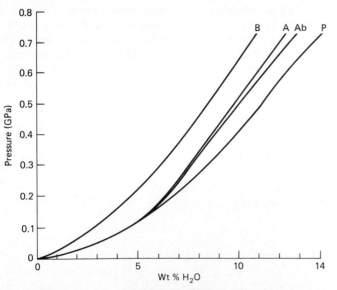

FIGURE 11-1   Solubility of water as a function of pressure in molten basalt (B), andesite (A), albite (Ab), and granitic pegmatite (P) at 1100°C. [After Burnham (1979); published with permission of Princeton University Press.]

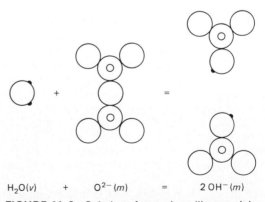

FIGURE 11-2   Solution of water in a silicate melt by hydrolysis of a bridging oxygen of the SiO_4^{4-} tetrahedral framework. Small open circles represent silicon atoms in tetrahedral coordination with oxygen.

over, to maintain local charge balance, a sodium ion must be juxtaposed this tetrahedron. Structures that contain the AlO_4 tetrahedron and charge-balancing cations provide another means of water dissolving in the melt, which involves dissociation of water:

$$H_2O \rightleftharpoons OH^- + H^+ \qquad (11\text{-}14)$$

Again, a bridging oxygen linkage is broken, but instead of the bridging oxygen becoming a hydroxyl ion, it remains an oxygen ion but maintains charge balance by attracting to it the Na^+ that previously was near the AlO_4 tetrahedron. Charge balance on the AlO_4 tetrahedron is maintained by one of its oxygen ions being hydrolyzed by addition of the H^+ released from the dissociation of water. The remaining hydroxyl ion from the dissociation of water is added to complete the other silica tetrahedron. This solution, which is represented graphically in Figure 11-3, can be represented by the reaction

$$H_2O(v) + O^{2-}(m) + Na^+(m) \rightleftharpoons OH^-(m) + ONa^-(m)$$
$$+ H^+(m) \quad (11\text{-}15)$$

As water is added to an albite melt, it first dissolves according to Eq. 11-15 until all of the Na^+ has been exchanged. This occurs when the mole fraction of water in the melt (X_w^m) reaches 0.5 (i.e., $H^+ = Na^+$). Further solution of water then occurs according to Eq. 11-13. Because the first solution mechanism involves the dissociation of water, the relation between the activity and mole fraction of water in the melt has the form

$$a_w^m = k(X_w^m)^2 \qquad (11\text{-}16)$$

where $X_w^m \leq 0.5$. Above this mole fraction, the activity is given by

$$a_w^m = 0.25k \, \exp\left[\left(\frac{6.52 - 2667}{T}\right)(X_w^m - 0.5)\right] \quad (11\text{-}17)$$

Values for $\ln k$ are given in Figure 11-4 for a range of pressures as a function of temperature. Values of k depend largely on pressure and only very slightly on temperature.

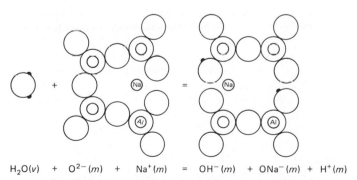

$$H_2O(v) \quad + \quad O^{2-}(m) \quad + \quad Na^+(m) \quad = \quad OH^-(m) \quad + \quad ONa^-(m) \quad + \quad H^+(m)$$

FIGURE 11-3 Solution of water in the albite structure involving dissociation of water and movement of the Na^+ ion to maintain local charge balance. See text for explanation. [After Burnham (1979); published with permission of Princeton University Press.]

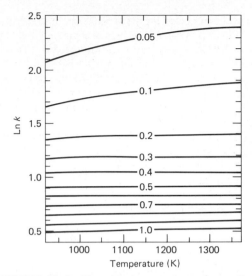

FIGURE 11-4 Plot of $\ln k$ versus temperature. See Eqs. 11-16 and 11-17 for definition of k. [After Burnham (1979); published with permission of Princeton University Press.]

If a separate vapor phase of pure H_2O coexists with an albite melt, the activity of water in the melt must be unity. Thus, by setting $a_w^m = 1.0$, we can calculate the saturation mole fraction of water in the melt at a specified pressure. For example, if the pressure is 0.2 GPa and the temperature is 1100°C (1373 K), $\ln k = 1.4$. According to Eq. 11-16, then, $X_w^m = 0.5$. Thus, under these conditions, an albite melt would become saturated in water when the mole fraction of water reached 0.5. At this same temperature but under a pressure of 0.3 GPa, $\ln k = 1.19$. According to Eq. 11-16, the saturation mole fraction of water would be 0.55. But at this concentration Eq. 11-16 is no longer applicable; we must instead use Eq. 11-17, which gives $X_w^m = 0.54$. Note that because the saturation mole fraction of water in albite melt at 0.2 GPa and 1373 K is 0.5, at pressures below this, solution will occur primarily by the mechanism in Eq. 11-15. Above this pressure, however, solution can also take place by the mechanism of Eq. 11-13.

In Figure 11-1 we saw that different silicate melts dissolve different weight percentages of water under identical pressures and temperatures but that the form of the P versus wt % H_2O plots for each melt are similar. This suggests that a general solution model may be applicable to all of these melts. Of course, some difference is to be expected in such a plot because the molecular weight of basalt is greater than that of granite. Nonetheless, when differences in molecular weights are accounted for, differences in solubility remain. These differences can be eliminated entirely, however, if the mole fraction of water in the melt is calculated in terms of what Burnham (1979) refers to as $NaAlSi_3O_8$-equivalent masses. The chemical analysis of a melt is recast so as to indicate the number of exchangeable cations (alkalis and transition elements), which allow solution to occur by the mechanism of Eq. 11-15, and the number of Si—O—Si bonds in excess of the exchangeable cations, which allow solution to occur by the mechanism of Eq. 11-13. When calculated in this way, the P versus X_w^m curves for basalt, andesite, granitic pegmatite, and albite are identical (Fig. 11-5).

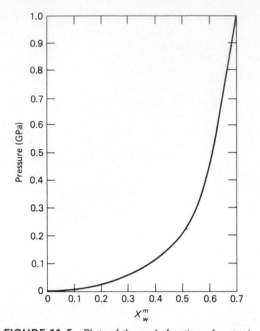

FIGURE 11-5   Plots of the mole fraction of water in the four different melts of Figure 11-1, where the mole fraction is calculated by recasting the rock analyses in terms of albite-equivalent masses. See text for explanation. [After Burnham (1979); published with permission of Princeton University Press.]

The simplicity of this result suggests that the albite–H_2O solution model may be applicable over a wide range of silicate melt compositions. It also implies that the solubility of water involves mainly the aluminosilicate fraction of the melt and is not particularly sensitive to other components in the magma.

11-3   SOLUBILITY OF CO_2 IN SILICATE MELTS

Carbon dioxide, the other major volatile constituent in volcanic gases, behaves quite differently from H_2O in silicate melts. If CO_2 were to dissolve in a silicate melt by combining with O^{2-} [this would be equivalent to the hydrolysis reaction (Eq. 11-13)], the resulting carbonation reaction would be

$$CO_2(v) + O^{2-}(m) \rightleftharpoons CO_3^{2-}(m) \qquad (11\text{-}18)$$

If the O^{2-} in the melt is a bridging oxygen, as for example in molten albite, the carbonate ion formed places a small, highly charged C^{4+} ion near a small, highly charged Si^{4+} or Al^{3+} ion of the tetrahedron. The resulting repulsion means that this reaction is not likely to proceed to the right, at least not at low pressures. Therefore, in framework-dominated melts, the amount of solution of CO_2 by this mechanism is small. Nor is it likely that holes in the structure of these melts will accommodate molecular CO_2. According to Bohlen and others (1982), at pressures greater than 1.5 GPa, molten albite does dissolve some CO_2. Addition of a small amount of H_2O to a framework melt dramatically increases the solubility of CO_2 by

depolymerizing the melt and opening its structure so that it can accept molecular CO_2 (Eggler 1973).

In melts containing nonbridging oxygens (i.e., oxygens not joining SiO_4 or AlO_4 tetrahedra), solution of CO_2 by the carbonation reaction (Eq. 11-18) is feasible. In this case the highly charged C^{4+} ion is not repelled as much by the larger, less-charged ion (Mg^{2+}, Fe^{2+}, Ca^{2+}, for example). Also, the less polymerized melt structure is able to accommodate molecular CO_2 as well. Consequently, as the percentage of nonbridging oxygens increases in the melt, so does the solubility of CO_2. Thus molten diopside (50% bridging oxygens) can dissolve more CO_2 than molten albite (100% bridging oxygens), and molten forsterite (no bridging oxygens) can dissolve still more (Eggler 1973, 1974). The solubility of CO_2, even in the more mafic melts, is, however, far less than the solubility of H_2O. Diopside melt at 2.0 GPa, for example, dissolves only about 5 wt % CO_2, but it can dissolve 17 wt % H_2O.

11-4   SOLUBILITY OF SULFUR IN SILICATE MELTS

Sulfur can be present in the magmatic vapor phase in a number of different species: S_2, SO_2, and H_2S. The relative abundances of these species are determined by the fugacities of S_2, O_2, and H_2, which in turn determines how sulfur dissolves in the melt. If H_2S is abundant, sulfur could dissolve by a hydrolysis reaction with bridging oxygens:

$$H_2S(v) + O^{2-}(m) \rightleftharpoons SH^-(m) + OH^-(m) \qquad (11\text{-}19)$$

If the magma is dry and the oxygen fugacity low, solution of sulfur will take place with the formation of sulfide in the melt:

$$\tfrac{1}{2}S_2(v) + O^{2-}(m) \rightleftharpoons S^{2-}(m) + \tfrac{1}{2}O_2(v) \qquad (11\text{-}20)$$

and at high oxygen fugacities it will dissolve as sulfate:

$$\tfrac{1}{2}S_2(v) + O^{2-}(m) + \tfrac{3}{2}O_2(v) \rightleftharpoons SO_4^{2-}(m) \qquad (11\text{-}21)$$

Solution by the mechanism of Eq. 11-19 seems to be minor because the solubility of sulfur in most melts decreases with increase in number of bridging oxygens. Solution by the mechanism of Eq. 11-20 seems most likely in magmas under the normal range of oxygen fugacities. At oxygen fugacities above those of the nickel–nickel oxide buffer ($fO_2 > 10^{-5}$ bar; see Sec. 11-8), however, Eq. 11-21 will describe the solution.

Fincham and Richardson (1954) showed that the solubility of sulfur in silicate melts can be related to the composition of the vapor phase simply by expressing the sulfur content of the melt in terms of what they referred to as the sulfide and sulfate capacities of the magma, which are defined, respectively, as

$$C_S = \text{wt \% } S(m) \times \left(\frac{fO_2}{fS_2}\right)^{1/2} \qquad (11\text{-}22)$$

and

$$C_{SO_4} = \text{wt \% } S(m) \times (fS_2 \cdot fO_2^3)^{-1/2} \qquad (11\text{-}23)$$

The sulfide capacity is applied when $fO_2 < 10^{-6}$ bar and the sulfate capacity when $fO_2 > 10^{-5}$ bar. At a given temperature and pressure the value C_S or C_{SO_4} is a constant. The logarithm of this constant varies inversely with absolute temperature. Its variation with pressure is unknown.

The explanation for Fincham and Richardson's finding is readily apparent from examining the equilibrium constants for Eqs. 11-20 and 11-21. Consider first the reaction at low oxygen fugacity ($< 10^{-6}$). The equilibrium constant is given by

$$K = \frac{(aS^{2-})(fO_2)^{1/2}}{(aO^{2-})(fS_2)^{1/2}} \tag{11-24}$$

At fixed temperature and pressure, K is a constant. Also, if the composition of the melt is constant other than for sulfur, the activity of O^{2-} will not vary much. Consequently, $(aS^{2-}) \times (fO_2/fS_2)^{1/2}$ must be essentially constant. This then accounts for the functional form of the "sulfide capacity" (C_S). Also, from Eq. 9-56 we know that $\ln K = -\Delta G/RT$, thus accounting for the variation of C_S with temperature. The equilibrium constant for Eq. 11-21 is

$$K = \frac{(aSO_4^{2-})}{(aO^{2-})(fS_2)^{1/2}(fO_2)^{3/2}} \tag{11-25}$$

Using the same arguments as before, $(aSO_4^{2-}) \times (fS_2 \times fO_2^3)^{-1/2}$ must be a constant at a given temperature and composition of the melt. This then accounts for the functional relation of C_{SO_4} in more oxidized melts.

Doyle (1987) has shown that Richardson and Fincham's way of expressing the sulfur contents of melts works so well that the proposed solution mechanisms are probably accurate. It also explains why in magmas with low oxygen fugacities (Eq. 11-22), increasing oxygen fugacity decreases the sulfur content of the melt. By constrast, at high oxygen fugacities (Eq. 11-23), increasing oxygen fugacity increases the sulfur content of the magma. Tholeiitic basalts typically have low oxygen fugacities and thus contain sulfur in solution as sulfide. Many andesites, however, are more oxidized and have sulfur dissolved as sulfate. Because of the difference in the effect of fO_2 on C_S and C_{SO_4}, we can expect that magmas intermediate between basalt and andesite, having fO_2 near that of the nickel–nickel oxide buffer, will have a minimum sulfur-bearing capacity.

Fincham and Richardson (1954) showed that not only is the solubility of sulfur in a silicate melt dependent on the composition of the vapor phase but that the composition of the melt plays a major role, especially its ferrous iron content. Indeed, the solubility of sulfur in silicate melts is so strongly correlated with the FeO content that Eq. 11-20 could be rewritten as

$$\tfrac{1}{2}S_2(v) + FeO(m) \rightleftharpoons FeS(m) + \tfrac{1}{2}O_2(v) \tag{11-26}$$

The equilibrium constant for this reaction can be written as follows:

$$K = \frac{(aFeS)(fO_2)^{1/2}}{(aFeO)(fS_2)^{1/2}} \tag{11-27}$$

from which we see that for a given temperature and sulfur fugacity, increasing the oxygen fugacity decreases the activity of FeS in the melt and consequently the amount of sulfur held in solution. This effect is made more pronounced by the fact that increasing the oxygen fugacity also decreases the activity of FeO in the melt (because of conversion to Fe_2O_3). Lowering the activity of FeO in the melt, for example, by crystallizing Fe-bearing oxides or silicates, also decreases the capacity of the melt to carry sulfur.

In addition to being a volatile constituent in silicate melts, sulfur has the ability to form immiscible sulfide liquids (MacLean 1969; Haughton et al. 1974). These liquids consist largely of iron and sulfur with small amounts of oxygen (they crystallize to pyrrhotite and minor magnetite), but if copper or nickel are present in the magma, these elements partition strongly into the sulfide melt. Large magmatic ore deposits associated with mafic and ultramafic rocks, such as those of the Sudbury district of Ontario (Naldrett 1969), are thought to have formed by the segregation of such sulfide liquids.

The separation of an immiscible sulfide liquid from a magma can be illustrated with the aid of the pseudoternary system FeS–FeO–SiO$_2$ (Fig. 11-6). Not all phases in this system plot within the ternary diagram. Pyrrhotite, for example, contains more sulfur than FeS (troilite), and the oxygen content of wüstite depends on the oxygen fugacity and temperature. For purposes of illustration, however, no serious errors are introduced by considering the system a ternary one that coexists with a vapor phase consisting of sulfur and oxygen. For a complete discussion of this system, see MacLean (1969). Figure 11-6 shows the vapor-saturated phase relations in this system for liquids in equilibrium with solid iron; that is, in addition to the phases shown in Figure 11-6, a vapor phase and solid iron are always present. The composition of the vapor phase varies considerably throughout the system, depending on the phases present. The total pressure is near atmospheric.

The phase diagram is dominated by three two-liquid fields and one three-liquid field. The three liquids involved are a sulfide-rich one with a composition near FeS, a silica-rich one, and an iron-rich silicate one. The two-liquid field involving sulfide liquid and iron-rich silicate liquid is the geologically important part of this diagram. The arcuate boundary of this field reflects the increasing solubility of sulfur in silicate melts as a function of the FeO content. At the highest silica content, this liquid can contain up to about 1 wt % FeS in solution, but as the FeO content increases, the solubility of FeS increases. For example, when the FeO/SiO$_2$ ratio in the melt is the same as that in fayalite, the melt can dissolve up to almost 20 wt % FeS. In still more FeO-rich melts, FeS becomes completely soluble and the immiscibility field disappears.

To illustrate how an immiscible sulfide liquid might form from a magma that initially was undersaturated in sulfide, consider the equilibrium crystallization of liquid X in Figure 11-6. This liquid first crystallizes fayalite ($+$iron), which changes the liquid composition to point a on the cotectic where tridymite begins to crystallize. The liquid then descends the cotectic to the isobaric invariant point b at 1140°C. At b, droplets of immiscible sulfide liquid with composition c nucleate and grow. Further crystallization of fayalite and tridymite decreases the amount of liquid b and increases the amount of liquid c, but

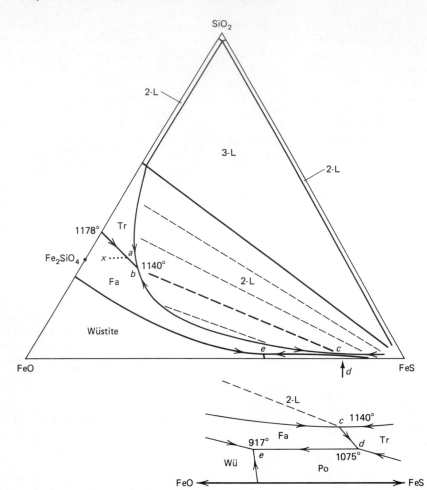

FIGURE 11-6 Liquidus phase relations in the vapor-saturated pseudoternary system FeS–FeO–SiO$_2$. Inset shows details of the low-temperature sulfur-rich part of this system. [After MacLean (1969).]

the compositions and temperatures of these liquids remain unchanged. Eventually, when all of liquid *b* has disappeared, the sulfide liquid at *c* is free to cool again (see enlargement of phase relations in lower part of Fig. 11-6). This it does by following the fayalite–tridymite cotectic from *c* down to the isobaric invariant point *d* at 1075°C. Point *d* is a peritectic where tridymite reacts with the liquid to form fayalite and pyrrhotite. Because the initial magma composition *X* lies in the triangle SiO$_2$–Fa–FeS, all liquid is consumed at this point. If the immiscible sulfide liquid had separated from the silicate fraction during differentiation, the liquid at *d* would have continued cooling to the ternary eutectic *e* at 917°C, where fayalite, pyrrhotite, and wüstite (+iron) would have crystallized.

Although the phase relations in Figure 11-6 are for rather reducing conditions (in equilibrium with iron), they are not very different from those under more oxidizing conditions. The two-liquid field has essentially the same shape, but the tridymite–fayalite cotectic is replaced by a tridymite–magnetite one, and magnetite, tridymite, and pyrrhotite crystallize at the ternary eutectic. Figure 11-6 does not show any of the other important chemical fractionations that occur during the differentiation of a magma, such as the variation in the Mg/Fe ratio. The abundance of these other constituents, however, affects the solubility of sulfur in magma and hence the size of the immiscibility field (Doyle 1987). Most basaltic magmas reach this immiscibility field at some stage in their crystalliza-

tion (inset in Fig. 11-6). If unmixing occurs early in the crystallization history of a magma, the dense sulfide liquid may be able to separate from the silicates and segregate to form an ore body. This early-separating liquid would take with it most of the nickel present in the magma. If unmixing occurs late, the sulfide liquid has greater difficulty separating from the silicates. Also, because nickel enters early-crystallizing ferromagnesian minerals—especially olivine, late-separating sulfide liquids may contain no nickel and therefore be of little economic value.

11-5 EFFECT OF H$_2$O ON MELTING IN SILICATE SYSTEMS

Having examined the solubility of common volatile constituents in magmas, we will now evaluate their effect on liquidus phase relations, beginning with water. The system albite–H$_2$O has been the most carefully studied volatile-bearing silicate system (Burnham and Davis 1974; Burnham 1979), and the effects of water in this system are generally applicable to other silicate melts. Qualitatively, the effect of dissolved water on the liquidus of albite is easy to predict. From the cryoscopic equation (Eq. 10-7) we know that any lowering of the activity of albite in the melt lowers the liquidus temperature. Thus the more water that is dissolved in the melt, the lower will be the liquidus of albite. But the solubility of water in the melt is strongly dependent on pressure. We would expect, therefore, that if water is available to dissolve in the melt, the melting point of albite will decrease with increasing pressure.

To quantify this relation, we must know the solubility of water in albite melt as a function of temperature and pressure. This can be determined easily from Eqs. 11-16 and 11-17 and the data in Figure 11-4 (see Prob. 11-1). If the albite melt is saturated in water and the vapor phase is pure H_2O, the activity of H_2O in the melt must be unity. The vapor phase is not actually pure H_2O, but it is so nearly so that no serious error is introduced by assuming that $a_w^m = 1.0$. The lightly shaded curved surface in Figure 11-7a shows the water saturation surface for albite melt as a function of temperature and pressure. This surface is contoured with lines of constant mole fraction of water in the melt. Hydrous albite melts can exist only on the high-pressure side of this surface. Should a melt cross this surface by having the pressure drop, water would exsolve from the melt and form gas bubbles (vesicles).

For albite to be in equilibrium with a melt, it is necessary for the chemical potential of albite in both phases to be the same; that is,

$$\mu_{Ab}^m = \mu_{Ab}^s \qquad (11\text{-}28)$$

Using Eq. 9-34, we can rewrite Eq. 11-28 as

$$\mu_{Ab}^{m*} + RT \ln a_{Ab}^m = \mu_{Ab}^{s*} + RT \ln a_{Ab}^s \qquad (11\text{-}29)$$

which on rearranging gives

$$0 = \Delta G_{mAb}^* + RT \ln \frac{a_{Ab}^m}{a_{Ab}^s} \qquad (11\text{-}30)$$

where ΔG_{mAb}^* is the free-energy change of melting of pure albite at the given T and P. But according to Eq. 7-34, $\Delta G_{mAb}^* = \Delta G_{mAb}^0 + P\Delta V_{mAb}$, where ΔG_{mAb}^0 is the free-energy change of melting of pure albite at temperature T and 1 bar, and ΔV_{mAb} is the volume change on melting of albite. After dividing through by RT, Eq. 11-30 can be rewritten as

$$0 = \frac{\Delta G_{mAb}^0 + P\Delta V_{mAb}}{RT} + \ln \frac{a_{Ab}^m}{a_{Ab}^s} \qquad (11\text{-}31)$$

The first group of terms in Eq. 11-31 can be evaluated from data in thermodynamic tables. Values of $(\Delta G_{mAb}^0 + P\Delta V_{mAb})/RT$ have been calculated by Burnham (1979) and are given here in Figure 11-8. The activity of albite in the solid is unity as long as we are dealing only with the system albite–H_2O. The activity of albite in the hydrous melt clearly depends on the amount of water dissolved in the melt. From the Gibbs–Duhem equation (Eq. 9-7) for this two-component system,

$$X_{Ab}^m \, d\mu_{Ab}^m + X_w^m \, d\mu_w^m = 0 \qquad (11\text{-}32)$$

For μ_i we can substitute $\mu_i^* + RT \ln a_i$ (Eq. 9-34), giving

$$X_{Ab}^m \, d \ln a_{Ab}^m + X_w^m \, d \ln a_w^m = 0 \qquad (11\text{-}33)$$

But the activity of water in molten albite is given by Eq. 11-16 if the mole fraction of water is less than 0.5. Taking logarithms

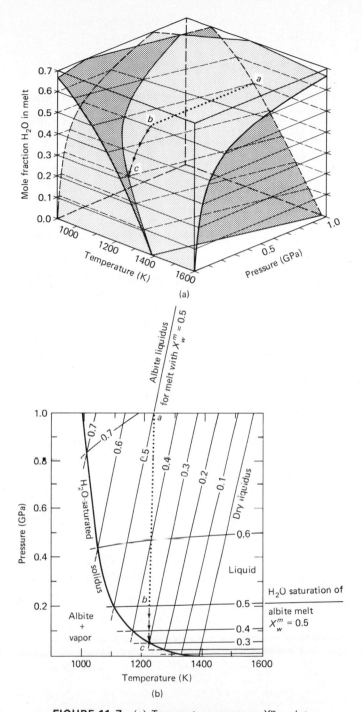

(a)

(b)

FIGURE 11-7 (a) Temperature–pressure–$X_{H_2O}^m$ plot of the water saturation limit of molten albite (lightly shaded surface). This surface is contoured with lines of constant mole fraction of water in the melt. The dark-shaded surface is the liquidus of albite, which is also contoured with lines of constant mole fraction of water in the melt. The line of intersection of these two surfaces is the water-saturated solidus of albite. This line and contours of constant mole fraction of water on the saturation surface and liquidus surface are projected onto the P–T plane in (b). Other points in figure are discussed in the text. [Drawn from data in Burnham (1979); published with permission of Princeton University Press.]

of both sides of Eq. 11-16 and differentiating, we obtain $d \ln a_w^m = 2d \ln X_w^m$, which can be substituted into Eq. 11-33 to give

$$X_{Ab}^m d \ln a_{Ab}^m + 2X_w^m d \ln X_w^m = 0 \qquad (11\text{-}34)$$

But $X_w^m = 1 - X_{Ab}^m$, which on substituting into Eq. 11-34 and simplifying gives

$$\frac{da_{Ab}^m}{a_{Ab}^m} = \frac{2dX_{Ab}^m}{X_{Ab}^m} \qquad (11\text{-}35)$$

On integrating, this gives

$$a_{Ab}^m = (X_{Ab}^m)^2 = (1 - X_w^m)^2 \qquad (11\text{-}36)$$

This simple expression, then, relates the activity of albite in the melt to the mole fraction of water in the melt as long as $X_w^m \leq 0.5$. A similar but more complex expression can be derived from Eq. 11-17 for melts with higher concentrations of water [see Burnham (1979)].

We are now in a position to calculate the albite liquidus for any given pair of values of temperature, pressure, or mole fraction of water in the melt. For example, if we choose a pressure of 2 kbar (0.2 GPa) and a temperature of 1200 K, the value of $(\Delta G_{mAb}^\circ + P \Delta V_{mAb})/RT$ is 0.91 (Fig. 11-8). According to Eq. 11-31, then, the activity of albite in the melt is 0.4, which Eq. 11-36 indicates corresponds to a mole fraction of albite in the melt of 0.63. The mole fraction of water in the melt under

these conditions would then be 0.37 (Prob. 11-4). It is possible, therefore, to calculate the liquidus surface for albite and contour it with lines of constant mole fraction of water in the melt, as shown in Figure 11-7a by the dark-shaded curved surface.

At the base of Figure 11-7a is the liquidus of albite where the mole fraction of water in the melt is zero. This is commonly referred to as the "dry liquidus." It rises from a temperature of 1391 K at atmospheric pressure to just below 1600 K at 1.0 GPa. As the mole fraction of water increases in the melt, this liquidus shifts to progressively lower temperatures. But to do this the melt must be able to dissolve water, and we have already seen that this is strongly dependent on pressure and that melts can exist only on the high-pressure side of the water saturation surface in Figure 11-7a (lightly shaded surface). As a result, only that part of the albite liquidus in Figure 11-7a lying on the high-pressure side of the water saturation surface is stable. The line of intersection of the albite liquidus surface and the water saturation surface is known as the *water-saturated solidus* of albite. Only albite plus vapor is stable below this.

The phase relations shown in perspective in Figure 11-7a are commonly projected onto the $P–T$ plane, as shown in Figure 11-7b. To read information from such a projection, it is necessary to have a clear understanding of the complete $T–P–X_w^m$ diagram. To help illustrate this, consider the path followed by a hydrous magma as it rises to the surface. Be certain to locate the path in both the perspective and projection diagrams.

Let us assume that an albite melt is in equilibrium with a solid albite source at the base of the crust where the pressure is 1 GPa. Enough water is present to produce a mole fraction of water in the melt of 0.5 (point a in Fig. 11-7). The temperature of this melt must be 1230 K. We will assume that the magma rises rapidly toward the surface of the Earth so that its temperature remains essentially constant and its mole fraction remains constant at 0.5. It will therefore follow the dotted line $a–b$ in Figure 11-7. As soon as the magma starts to rise, it leaves the liquidus surface, and any solid albite suspended in it would rapidly dissolve. As the magma rises it approaches the water saturation surface, which for a melt with a mole fraction of water of 0.5, occurs at a pressure of 0.2 GPa (point b in Fig. 11-7). At this point a separate vapor phase forms bubbles in the magma. Note that there is still no crystallization of albite. Further rise of the magma and lowering of pressure requires that the melt descend the water saturation surface in Figure 11-7 (dashed arrowed line, $b–c$). This progressively lowers the mole fraction of water in the melt and increases the volume of vapor phase. During this stage, the initial vapor bubbles grow into large pockets of gas, which are likely to float rapidly toward the upper part of the magma chamber, where they can form pegmatites or hydrothermal solutions. As the rising magma descends the water saturation surface in Figure 11-7, the lowering of the mole fraction of water in the melt brings the magma closer to the albite liquidus. At point c, where $X_w^m = 0.3$ and the pressure is 0.05 GPa, the magma reaches the water-saturated albite solidus and crystallizes to form albite and vapor.

Another useful way of presenting the albite liquidus in the hydrous system is with lines of constant activity of water.

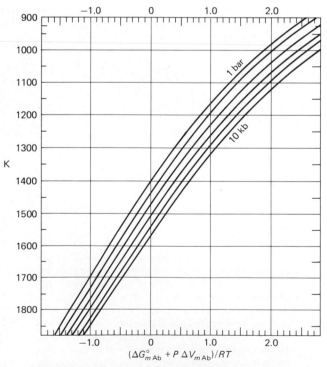

FIGURE 11-8 Free energy of melting of pure albite at P and T expressed in terms of $(\Delta G_{mAb}^0 + P \Delta V_{mAb})/RT$. [After Burnham (1979); published with permission of Princeton University Press.]

Two such lines are already present in Figure 11-7b: that for the dry melt, where the $a_w^m = 0$, and that for the water-saturated solidus, where $a_w^m = 1$. Values of the activity of water between these extremes can be calculated from Eqs. 11-31, 11-36, and 11-16 or 11-17 (see Prob. 11-6). Figure 11-9 shows the same P–T projection of the albite liquidus as in Figure 11-7b, but now it also has lines of isoactivity of water in the melt. Note that these lines do not lie in the planes of constant mole fraction of water in the melt but all radiate from the anhydrous atmospheric melting point of albite.

Isoactivity lines are particularly useful in considering the melting of albite in the presence of a mixed vapor phase, such as H₂O–CO₂. At low to moderate pressures CO₂ is insoluble in albite melt, and thus it behaves essentially as an inert gas that reduces the activity of water in the vapor phase and in the coexisting melt. If a large amount of CO₂ were present, as might be the case if impure limestones were undergoing decarbonation reactions, the activity of water would be low. If the activity of water were only 0.1, melting of albite under a total pressure of 1 GPa would not occur until the temperature exceeded 1420 K. At this same pressure, if the activity of water were 1.0 (no CO₂ in the environment), melting of albite would occur at the much lower temperature of 1000 K.

The melting behavior of albite in the presence of H₂O has been dealt with in some detail here because it provides a good model for the behavior of most hydrous silicate melts. Although the liquidus temperatures of different minerals vary, the effect of dissolved water in lowering liquidus temperatures is similar. Even rocks, which consist of a number of minerals and melt over a range of temperature, show similar lowering of liquidus and solidus temperatures. Figure 11-10 shows the water-saturated solidi—also known as the *beginning of melting curves*—for a number of minerals and common rocks. Most of these solidi have similar negative slopes. The diopside and enstatite curves are steeper than the others, as might be expected in light of the discussion of the way in which water enters silicate melts (Sec. 11-2). At higher pressures than shown in this diagram, each of the solidi passes through a temperature minimum and then begins to rise with further increase in pressure. At these high pressures the vapor phase is so compressed that it contributes very little to the volume change for the reaction from hydrous to less hydrous melt (Eq. 11-12). Instead, the volume decrease on crystallizing begins to dominate and causes melting points to increase with increasing pressure.

One of the most important conclusions to be drawn from the data in Figure 11-10 is that water has a remarkable fluxing ability in lowering the melting points of silicates. For each of the materials in Figure 11-10, solution of a few weight percent water lowers solidus temperatures by several hundred degrees. For example, 5 wt % H₂O lowers the solidi of granitic and basaltic magmas by about 200°C. We can conclude, therefore, that if water is present as a pure vapor phase in a region undergoing melting, the first melt forms on the water-saturated solidus. For granitic melts this is at about 600°C at a pressure of 1 GPa. Such conditions are commonly attained in the highest grades of regional metamorphism. It is not surprising, then, to

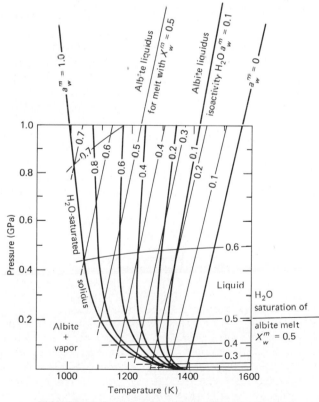

FIGURE 11-9 Projection of the liquidus of albite onto the P–T plane showing lines of constant activity of water. Contours of constant mole fraction of water are also shown on the liquidus and water saturation surfaces (see Figure 11-7). [After Burnham (1979); published with permission of Princeton University Press.]

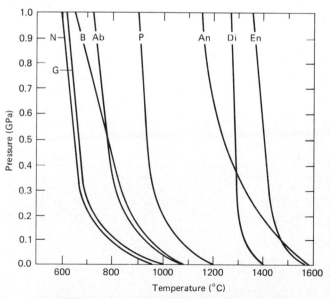

FIGURE 11-10 Beginning of melting curves under water-saturated conditions for nepheline syenite (N), granite (G), tholeiitic basalt (B), albite (Ab), peridotite (P), anorthite (An), diopside (Di), and enstatite (En). (Data from various sources.)

find evidence of partial melting in such rocks (see *migmatites—mixed igneous–metamorphic rocks* in Sec. 22-6). The amount of melt formed at the water-saturated solidus depends on the amount of water available, which may be quite small. Once all of the water in the vapor phase has entered the melt, further melting produces water-undersaturated melts, which in turn require higher temperatures. Also, if the vapor phase is not pure H_2O, melting does not start at the water-saturated solidus but occurs only at the solidus for the particular activity of water (Fig. 11-9).

Figure 11-10, while showing the amount of lowering of solidus temperatures resulting from solution of H_2O in the melts, does not indicate how the compositions of melts vary with temperature. Of course, with single minerals, such as albite, the only variation possible is in the content of water. But when melting involves more than one mineral, as in a rock, the composition of the liquid can change significantly with pressure. This is due to the different solubilities of water in the melts formed from the different minerals. If solubilities are similar, there will be little change in the composition of the melts; if the solubilities are very different, the changes will be great.

Because albite and anorthite are both framework silicates with the same number of exchangeable cations, their ability to dissolve water is similar, and thus they show similar lowering of melting points with increasing water pressure (note that their water-saturated solidi in Fig. 11-10 are parallel). As a consequence, the phase diagram for the plagioclase feldspars under water-saturated conditions at high pressure is almost identical to the anhydrous diagram at atmospheric pressure, except that at high pressure the dissolved water shifts the phase fields to lower temperatures (Fig. 11-11).

By contrast, the water-saturated solidus of diopside decreases less rapidly with increasing water pressure than does that of anorthite, and thus the eutectic between these minerals shifts toward anorthite at high water pressures (Fig. 11-12).

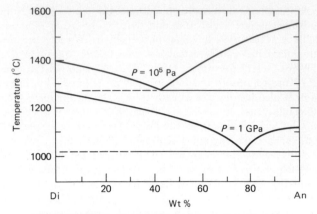

FIGURE 11-12 System diopside–anorthite at atmospheric pressure (10^5 Pa) and dry, and water-saturated at 1.0 GPa. Alumina entering diopside makes diopside-rich parts of this system deviate slightly from binary behavior. [After Yoder (1965).]

Yoder (1965) has shown that this eutectic (actually, cotectic) in the pseudobinary system diopside–anorthite shifts from a composition of 42 wt % anorthite in the anhydrous system at atmospheric pressure to 78% anorthite in the water-saturated system at 1 GPa. Yoder has speculated that the plagioclase-rich composition of this eutectic under these conditions may be responsible for the composition of the rock-type anorthosite.

Similar changes in the extent of liquidus fields can be illustrated from ternary systems, such as that of forsterite–diopside–SiO_2 (Fig. 11-13). In this system, SiO_2 is the only mineral with a framework structure, and thus its melt is capable of dissolving more water than those of forsterite, diopside, or enstatite, each of which shows similar water solubilities. As a result, the phase relations involving SiO_2 are very sensitive to

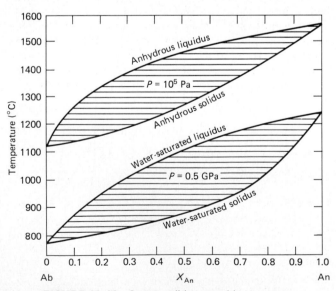

FIGURE 11-11 System albite–anorthite at atmospheric pressure (10^5 Pa) and dry, and water-saturated at 0.5 GPa. [After Yoder et al. (1957).]

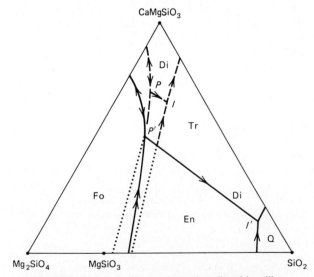

FIGURE 11-13 System forsterite–diopside–silica. Dashed lines show position of phase boundaries on the liquidus in the anhydrous system at atmospheric pressure, and the solid lines are for the water-saturated liquidus at 2 GPa. [After Kushiro (1969).]

water pressure, but those between the ferromagnesian minerals are not. The peritectic between forsterite, diopside, and enstatite in the anhydrous system at atmospheric pressure (P in Fig. 11-13) changes composition only slightly to P' in the water-saturated system at 2 GPa. By contrast, the invariant point I in the anhydrous system at atmospheric pressure, which is a peritectic involving diopside, pigeonite, and tridymite, shifts by almost 60 wt % SiO_2 to I in the water-saturated system at 2 GPa, where it involves diopside, enstatite, and quartz (Kushiro 1969). Kushiro has suggested that the silica-rich composition of this invariant point under hydrous conditions may be responsible for the silica-rich nature of many rocks in the calcalkali series, which are typically more hydrous than those in the tholeiitic series.

Even in systems containing only framework silicates, the composition of phase boundaries can shift if the melts of these phases have different solubilities of water. In the system quartz–albite–orthoclase–H_2O (Figs. 10-26 and 10-27), for example, water can dissolve in an SiO_2 melt only by the mechanism of Eq. 11-13, but melts of albite and orthoclase can, in addition, dissolve water by the mechanism of Eq. 11-15 because of the exchangeable cations. Thus, with increasing pressure the solubility of water in feldspar-rich melts is greater than in silica-rich ones, and consequently, the liquidus field of quartz expands relative to those of the feldspars. Because molten albite and orthoclase have similar solubilities of water, their cotectic does not change composition much with increasing water pressure. Thus, with increasing water pressure the "granite" minimum (or eutectic at high P_{H_2O}) moves away from quartz to more feldspar-rich compositions along the albite–orthoclase cotectic.

Addition of water to anhydrous systems can also bring about changes in the phase relations by stabilizing hydrous minerals, such as amphiboles and micas. Hydrous minerals break down at high temperatures to anhydrous minerals plus a vapor phase. At low to moderate pressures this stability limit is raised by increasing pressures, but at very high pressures the anhydrous phases are again more stable. The stability limit of a hydrous phase can be calculated from Eq. 8-10 (see Prob. 8-8 for the calculation of the stability limit of muscovite + quartz). Regardless of the mineral or group of minerals involved in a dehydration reaction, the stability limit in the presence of a pure H_2O vapor phase plots as a curved line with positive slope in a P–T diagram (Fig. 11-14). The beginning of melting curve of a rock, however, plots as a curved line with negative slope in such a diagram. At some point, therefore, these curves will intersect (A in Fig. 11-14), and above this pressure the hydrous mineral is in equilibrium with a melt.

If the vapor phase in equilibrium with a hydrous mineral is not pure H_2O (a mixture of H_2O and CO_2, for example), the stability limit of the hydrous phase decreases as the activity of water in the vapor decreases (Fig. 11-14). A series of isoactivity of water lines can be drawn on the dehydration reaction surface, which resemble mirror images of the isoactivity lines on the solidus surface. Thus, as the activity of water decreases, the univariant lines for the dehydration reaction rotate in a counterclockwise direction, whereas the lines on the solidus rotate in a clockwise direction. The point of intersection of the solidus with the stability limit of the hydrous mineral therefore moves to progressively higher pressures with decreasing ac-

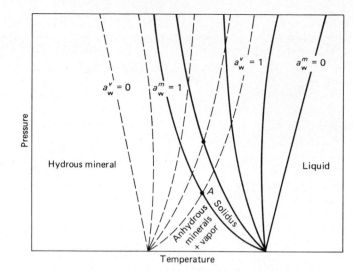

FIGURE 11-14   Plot of the stability limit of a hydrous mineral in the presence of a vapor phase with variable activity of water; dashed lines indicate constant activity of water in the vapor. Solid curves are the vapor saturated beginning of melting of a rock at different activities of water.

tivity of water. Indeed, below a critical activity of water the lines diverge and the hydrous mineral can never be stable at magmatic temperatures.

The stability limits of the hydrous minerals vary with composition and with associated minerals. In general, the stability limits of muscovite, biotite, and hornblende extend well above the solidus of granites, as long as the activity of water is moderately high. In the presence of excess water hornblende is stable up to the liquidus temperatures of some basaltic magmas—camptonite, for example, typically has phenocrysts of hornblende. At pressures above 3 GPa, however, the stability limit of hornblende drops well below the solidus of basalts. Hornblende can therefore play a magmatic role only in the crust or uppermost mantle. Phlogopite, by contrast, is stable to higher temperatures and much higher pressures than is hornblende. Phlogopite may therefore be present in mantle rocks to considerable depth, where it could provide a source for the potassium in alkaline mafic magmas.

11-6 FRACTIONAL CRYSTALLIZATION OF HYDROUS MAGMA

At depth in the Earth, most magmas are probably neither water-saturated nor completely dry. Even when melts form on the water-saturated solidus, the limited supply of volatiles in most regions in the Earth is likely to cause melts to become undersaturated in water as the volume of melt increases. Once magma starts ascending, however, confining pressures decrease, and at some level the magma is likely to become vapor-saturated—most lavas, for example, contain vesicles. The pressure, and consequently the depth, at which vapor saturation

(a)   (b)

FIGURE 11-15   Hale pegmatite, Middletown pegmatite district, Connecticut. (a) Core zone with metersized microcline crystals (sledgehammer for scale) extending into massive quartz; around the core, microcline forms a graphic intergrowth with quartz. (b) Banded aplite from steeply dipping upper contact of pegmatite. Gradational zoning is due to variations in abundance of fine-grained muscovite. Near the hammer handle a zone contains a few larger quartz crystals growing perpendicular to the layering. The contact of the banded aplite with the pegmatite of (a) is abrupt and concordant.

occurs depends on the magma's initial content of volatiles, its saturation limit, its degree of crystallization, and the water content of minerals that crystallize from it (or carbonate content of minerals in CO_2-rich magma).

The degree of solidification of a magma at the time a vapor phase forms is of importance to a number of processes, such as explosive volcanism and pegmatite formation. The formation of pegmatites is of particular interest because of their economic value. Pegmatites provide most of the feldspar used in the ceramics industry, and their large muscovite crystals were used extensively as electrical insulators in the manufacture of electronic tubes. Pegmatites also provide ores of many rare elements, such as Li, Be, Cs, Nb, Ta, W, Zr, U, Th, and the rare earths; they are also the source of many gemstones. Extensive field investigations of pegmatites, many of which are well exposed by mining, has made them some of the best known of all intrusive igneous bodies (Cameron et al. 1949). They have also been investigated experimentally (Jahns and Burnham 1969; London 1987). We will deal with them here to illustrate the factors involved in the fractional crystallization of hydrous magma.

The main evidence for the role of volatiles in the formation of pegmatites is their extremely coarse grain size, but the composition of some of their minerals also indicates the presence of volatiles, such as H_2O (mica), B (tourmaline), CO_2 (calcite), F_2 (topaz, apatite, fluorite), and Cl_2(apatite). Strictly speaking, *pegmatite* is a textural term describing rocks with exceptionally coarse and variable grain size. Most pegmatites are of granitic composition, but nepheline syenite ones are common in alkaline igneous complexes, and even ones approaching gabbroic compositions can be found. Used loosely, without qualifiers, the term is normally understood to indicate a granitic composition. The conclusions we will draw here, however, will apply to pegmatites of any composition.

Crystals with diameters in excess of a meter are not uncommon in pegmatites (Fig. 11-15). As will be shown in Chapter 12, the growth of such large crystals is possible only if few nucleation centers form and nutrients are able to move rapidly through the magma. The depolymerization of silicate melts brought about by high concentrations of water dramatically increases diffusion rates, and in the vapor phase these rates are still higher. Large crystals commonly nucleate on the walls of pegmatites, and as they grow inward they become larger, forming cone-shaped crystals; this is particularly characteristic of alkali feldspar crystals. Not all the material in pegmatites is coarse-grained. Indeed, one particular rock type that characteristically occurs in pegmatites, known as *aplite*, consists of millimeter-sized grains of quartz and feldspar with an allotriomorphic granular texture. The term *sacharoidal* is also used in reference to the sugary appearance of this rock. Aplite may form dikes cutting pegmatites, but most occurs as zones or irregular patches among the coarser crystals (Fig. 11-16). In

FIGURE 11-16 Pegmatite and aplite from Gotta Quarry, Middletown pegmatite district, Connecticut. Note juxtaposition of single coarse alkali feldspar crystal on left with fine-grained aplite.

FIGURE 11-17 Three-meter-thick zoned pegmatite, Lewiston, Maine. Dark crystals are tourmaline; note that tourmaline crystals at top are larger than ones at base. Quartz-rich core is overlain by a zone rich in muscovite (reflecting light) and underlain by a zone of graphic granite.

granitic pegmatites quartz and feldspar are commonly intergrown in graphic intergrowths (Fig. 10-5 and 11-15). In silica-undersaturated alkaline pegmatites similar textures involving albite, nepheline, and sodalite are found.

Most pegmatites form lenticular bodies (Fig. 11-17), but more continuous dike- and sill-like forms are common. They may occur in the upper parts of batholiths and stocks or intrude the overlying rocks. Some occur in metamorphic rocks and have no obvious related magmatic source. Most pegmatites are zoned in both grain size and composition. Numerous patterns are recognized, some of the more common of which are shown in Figure 11-18. Most have quartz cores (Fig. 11-15) and concentrations of potassium-bearing minerals toward the top.

Jahns and Burnham (1969) propose that the development of a separate vapor phase in the magma is responsible for the development of pegmatites. At the temperatures and pressures under which most pegmatites form, this vapor phase is a supercritical fluid with densities similar to those of water at room temperature and pressure. If a vapor phase appears early

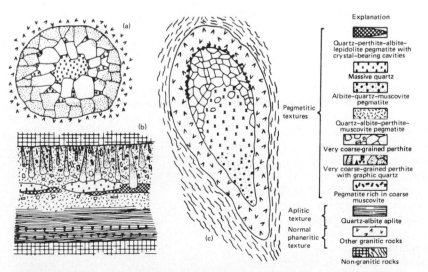

FIGURE 11-18 Typical internal structure of zoned pegmatites. [After Jahns and Burnham (1969); published with permission of the Economic Geology Publishing Co.]

enough in the crystallization history of a magma, it will be able to segregate easily and form large pegmatites. The vapor phase is also capable of altering and replacing early formed minerals to produce assemblages that would otherwise be difficult to explain by simple magmatic crystallization.

Once a separate vapor phase forms, large crystals start growing because of the higher diffusion rates through the vapor. The size of crystals depends, in large part, on how interconnected the vapor phase is. Also, as soon as this phase forms, certain elements, in particular those with large ionic radius and charge, leave the highly polymerized silicate melt in favor of the less polymerized vapor phase. This partitioning is so strong that elements that may be present in trace amounts (ppm) in the magma become highly concentrated in the vapor phase. It is from such solutions, for example, that large beryl and tourmaline crystals grow (Fig. 11-17). Some of these volatile constituents have significant fluxing effects on the magma. Experiments at 0.1 GPa show that the water-saturated solidus of a quartz–albite–orthoclase mixture is lowered by 60°C by addition of 2 wt % B_2O_3 and by 160°C by the addition of 2 wt % F (Manning and Pichavant 1983).

The pressure on the vapor phase may weaken surrounding rocks and cause them to fracture, and by so doing release a pulse of hydrothermal fluid from which minerals will be deposited as the solution cools. The accompanying drop in pressure on the pegmatitic magma brings about rapid crystallization. Consider, for example, a pegmatitic magma on the water-saturated solidus of albite at point *c* in Figure 11-7a. Rupturing of the pegmatite walls and loss of the volatile pressure would drop the magma below the solidus. The suddenness of such an event brings about rapid crystal nucleation and the development of a fine-grained aplite. Repeated losses of pressure could be responsible for the formation of banded aplite (Figure 11-15b).

These, then, are some of the processes involved with the formation of pegmatites. Their effectiveness depends on how early in the crystallization history the magma becomes water-saturated. If saturation occurs early, bubbles are free to rise and segregate to form large pegmatites; if saturation occurs late, restricted movement of the vapor phase may permit only small miarolitic cavities to form. Also, bubbles rising great distances through magma may be able to scavenge more of the extremely minor constituents of the magma than bubbles that travel only a few meters. Pegmatite formation, then, depends on how rapidly a fractionating hydrous magma reaches the water-saturation limit.

Let us assume that a totally liquid magma initially contains a weight fraction of water Wt°. When crystallization begins, the assemblage of minerals that forms has an average weight fraction of water *C*. The weight fraction of water in the residual magma at any stage of crystallization (Wt) can be determined by writing a mass balance equation, in which the weight of water in the residual liquid plus the water that has entered the minerals must equal the amount of water in the initial magma. If the weight fraction of magma that has crystallized is *F*, the mass balance equation can be written as

amount in	+ amount in =	amount in
residual magma	crystals	original magma

$$Wt(1 - F) \quad + \quad C(F) \quad = \quad Wt° \qquad (11\text{-}37)$$

On rearranging, the wt % H_2O in the magma at any stage of crystallization is

$$\text{wt \% } H_2O \text{ in magma} = 100 \times \frac{Wt° - C(F)}{1 - F} \qquad (11\text{-}38)$$

We can use Eq. 11-38 to trace the change in concentration of water in a magma during crystallization and to predict when the magma would become water-saturated. Formation of bubbles of vapor as a result of crystallization is commonly referred to as *resurgent boiling* (also, *second boiling*) to contrast it with the boiling that would occur if the melt were raised to very high temperatures.

If a granitic magma initially contained 1 wt % H_2O and the only crystals that formed during the early stages of solidification were quartz and feldspar, which contain no water ($C = 0$), the concentration of water in the magma would increase along the path shown in Figure 11-19 as crystallization proceeded. If this magma were crystallizing at a depth where the pressure was 0.4 GPa, we know from Figure 11-1 that the water saturation limit would be 10 wt %. This concentration would be reached after 90% of the magma had crystallized. In the residual 10 wt % liquid, then, bubbles of the vapor phase would nucleate and grow as crystallization went to completion. This would occur so late in the crystallization that little segregation of the vapor phase would be possible, and consequently, no large bodies of pegmatite would form.

During crystallization of this magma, compaction of crystals could displace residual liquid into the upper part of the magma chamber, where it would collect and form a new batch of fractionated magma. For purposes of illustration, consider that this filter pressing occurred when 80% of the original magma had crystallized. The new batch of fractionated magma would contain 5 wt % H_2O (Fig. 11-19), and it would reach the 10 wt % H_2O saturation limit when only 50% crystallized. This would be early enough for considerable segregation of the vapor phase to occur, and large pegmatites would result.

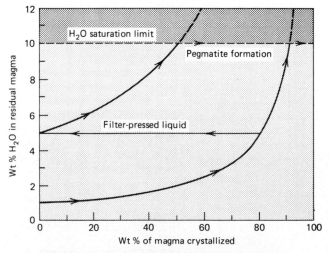

FIGURE 11-19 Plot of the change in concentration of water in a granitic magma as a result of crystallization of anhydrous minerals. The water saturation limit of 10 wt % is for a pressure of 0.4 GPa.

In conclusion, then, large pegmatites form only in magmas that initially are rich enough in water to become saturated before they are more than two-thirds crystallized; at higher degrees of crystallization segregation of the vapor phase becomes increasingly difficult. The initial water content necessary to satisfy this condition depends on the water saturation limit of the magma, which in turn, is a function mainly of pressure. Granitic magma, for example, requires about 4 wt % H_2O at 0.5 GPa but only 1.5 wt % at 0.1 GPa.

11-7 EFFECT OF CO₂ ON MELTING IN SILICATE SYSTEMS

At low to moderate pressures the low solubility of CO_2 in silicate melts makes this gas a relatively inert component. Indeed, in many experimental studies CO_2 is added to hydrous systems as an inert component to lower the activity of H_2O. With increasing pressure (>1.5 GPa), however, CO_2 does become soluble, especially in the less polymerized mafic melts. Consequently, at high pressures, CO_2 does affect liquidus temperatures, and through carbonation reactions, it influences the minerals that crystallize. In no system are these effects more dramatic than in the system $CaO-MgO-SiO_2$ (Eggler 1976; Wyllie and Huang 1976). This system is of particular interest because it models simple peridotites and is therefore of imporance when considering melting in the upper mantle.

The compositions of minerals that form in this simplified peridotite system in the presence of CO_2 are shown in the quaternary plot $CaO-MgO-SiO_2-CO_2$ of Figure 11-20. Apart from the carbonates, calcite, dolomite, and magnesite, all minerals plot on the ternary face $CaO-MgO-SiO_2$. Melts in this system, however, contain variable amounts of CO_2 and thus plot in the tetrahedron. Although the phase relations in the quaternary system can be illustrated more simply when

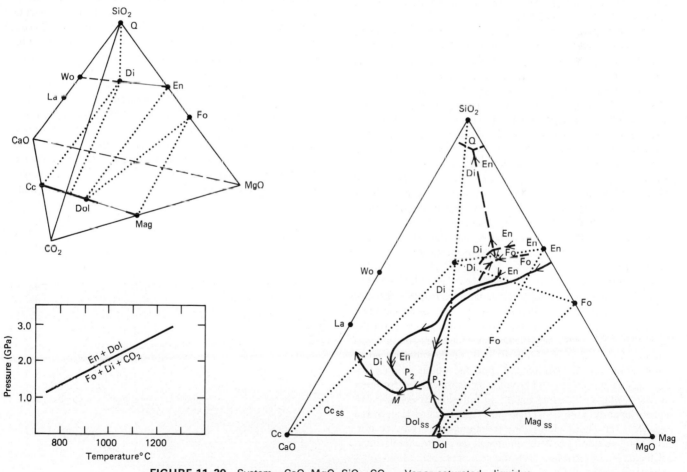

FIGURE 11-20 System $CaO-MgO-SiO_2-CO_2$. Vapor-saturated liquidus phase relations at 3.0 GPa are projected onto the ternary $CaO-MgO-SiO_2$ face from CO_2. Insets show the compositions of phases in the quaternary system and the decarbonation reaction enstatite (En) + dolomite (Dol) = forsterite (Fo) + diopside (Di) + CO_2 [after Eggler (1976)]. Also shown in the ternary projection with short dashes is the position of the Fo–En–Di eutectic in the vapor-free system at 3.0 GPa, and with long dashes, the Fo–En–Di peritectic and the Di–En–Q invariant point at 2.0 GPa [after Kushiro (1969)].

projected onto the $CaO–MgO–SiO_2$ face, it is important to keep in mind that the boundary lines shown in this ternary diagram are projected from within the tetrahedron.

The ternary projection in Figure 11-20 shows the CO_2-saturated liquidus relations in the system $CaO–MgO–SiO_2–CO_2$ at 3.0 GPa. The most significant difference between this diagram and its equivalent one for low pressures (Fig. 11-13) is in the behavior of forsterite and diopside. The assemblage olivine + augite is characteristic of all basaltic rocks formed under crustal conditions regardless of their degree of silica saturation. Above a pressure of 2.9 GPa, however, this assemblage is no longer stable in the presence of a CO_2 vapor phase; instead, it is converted to enstatite + dolomite according to the reaction

(forsterite)   (diopside)   (vapor)

$$Mg_2SiO_4 + CaMgSi_2O_6 + 2CO_2 \rightleftharpoons$$

(enstatite)   (dolomite)

$$Mg_2Si_2O_6 + CaMg(CO_3)_2 \quad (11\text{-}39)$$

As with other carbonation or hydration reactions (Eq. 8-9, Prob. 8-8), the equilibrium curve for this reaction has a positive slope that increases with pressure, and the assemblage with the gas is stable on the high-temperature, low-pressure side. This equilibrium curve is shown in Figure 11-20.

As a result of the carbonation reaction, the cotectic between olivine and clinopyroxene that exists in this system at low pressure (Fig. 11-13) is replaced at 3.0 GPa by two cotectics involving either enstatite + diopside or enstatite + forsterite (Fig. 11-20). The narrow primary field of enstatite which now separates the fields of diopside and forsterite extends across the diopside–forsterite join, which means that extremely silica-undersaturated larnite-normative melts can be in equilibrium with enstatite. Indeed, the enstatite field extends to very low silica concentrations, and at the same time the melts become enriched in carbonate ($<40\%$ CO_2) and hence plot well into the quaternary system.

Let us follow the path of a fractionally crystallizing CO_2-saturated liquid that initially has a composition on the enstatite–forsterite boundary in the ternary subsystem diopside–forsterite–enstatite. Crystallization of enstatite and forsterite drives this liquid down the cotectic between these two minerals, first crossing the diopside–forsterite join and then the diopside–dolomite join. Note that by the time the liquid crosses the diopside–dolomite join, the boundary curve between enstatite and forsterite, when extrapolated, no longer intersects the enstatite–forsterite join. Forsterite must, therefore undergo reaction with the liquid along the low-temperature part of this curve. This reaction line is terminated at an isobaric quaternary peritectic, P_1, where it intersects the field of dolomite solid solution. At this point all forsterite reacts out to leave only enstatite and dolomite solid solution to continue crystallizing down to the next peritectic, P_2, where enstatite is eliminated by the reaction to produce diopside and dolomite solid solution. From this point the residual melt fractionates down to a minimum, M, at 1110°C where diopside and a carbonate solid solution with Ca/Mg ratio of 70:30 crystallize.

The magnitude of the effect of CO_2 on the melting of peridotite compositions at 3 GPa can be seen by comparing the compositions of the liquids at the peritectic P_1 and the eutectic between forsterite, enstatite, and diopside in the volatile-free system under the same pressure (Fig. 11-20). Heating of an olivine-bearing peridotite mantle under CO_2-saturated conditions at 3.0 GPa would produce a melt of peritectic composition P_1 containing 17% SiO_2 and 27% CO_2. Under volatile-free conditions this same mantle would first form a melt at the forsterite–diopside–enstatite eutectic. Under hydrous conditions the first melt would form at the forsterite–enstatite–diopside peritectic (Figs. 11-20 and 11-13). Thus melting of identical mantle peridotite under these three different sets of conditions could generate a critically silica-undersaturated carbonate melt ($+CO_2$ vapor), a silica-saturated melt (no vapor), and a silica-oversaturated melt ($+H_2O$ vapor). A vapor containing both CO_2 and H_2O could produce melts with a wide range of compositions between these extremes, depending on the composition of the vapor. Moreover, the range of magma compositions that could be generated from these melts is still greater if they were to crystallize fractionally. Magma formed at P_1 in Figure 11-20 could fractionate down to a carbonatitic magma at the minimum, and a melt at the olivine–enstatite–diopside peritectic under hydrous conditions could fractionate down to a granitic residue.

Also of particular interest is the suddenness with which the phase relations under the CO_2-saturated conditions change with increasing pressure (Fig. 11-21). At low pressures CO_2 has only a slight effect in lowering solidus temperatures relative to those in the volatile-free system, because of its limited solubility in the melt. Above about 2.5 GPa, however, the solidus begins dropping more rapidly with increasing pressure, and between 2.7 and 2.9 GPa it drops 350°C. This dramatic lowering results

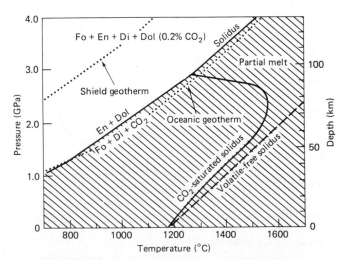

FIGURE 11-21 Schematic representation of the solidus for a peridotite containing 0.2 wt % CO_2. Below 2.9 GPa CO_2 exists as a vapor; above this pressure it is present as dolomite because of the carbonation reaction of Fo + Di + CO_2 = En + Dol. Also shown are geotherms for shield and young oceanic regions. [After Eggler (1976) and Wyllie and Huang (1976).]

from the increased solubility of CO_2 in the melt as the decarbonation reaction is approached. This seems consistent with Burnham's model for silicate melts (Sec. 11-2), where the structure of the melt mimics the structure of the crystalline phases that form from it. One can imagine that as the carbonation reaction is approached, an increasing fraction of the melt would have a carbonate structure.

As the pressure increases from 2.7 to 2.9 GPa and the solidus temperature drops precipitously, the composition of the first-formed melt shifts from silica-saturated to silica-under-saturated. Melts ranging in composition from olivine tholeiite, through alkali olivine basalt, melilitite, kimberlite, to carbonatite can all form from a common source peridotite over a short interval of depth. Also, carbonated melts formed in the deeper zones will, on rising to depths of about 90 km where the pressure drops below 3.0 GPa, start crystallizing and liberating CO_2 vapor, which may, then, be responsible for the explosive emplacement of these magmas.

This experimental work of Eggler (1976) and Wyllie and Huang (1976) confirms work by others that CO_2 cannot exist as a vapor phase in the mantle below depths of about 100 km beneath continents because estimated geotherms lie well below the carbonation reaction. If CO_2 is present, it must be in the form of dolomite. Beneath young oceans, however, the geotherm may lie on the high-temperature side of the carbonation reaction and then CO_2 would exist as a vapor phase. Under such conditions the geotherm would cross the solidus at a depth of approximately 90 km. The small amount of melt that would be formed could well explain the presence of the low-velocity zone in such regions.

11-8 ROLE OF OXYGEN FUGACITY IN PHASE EQUILIBRIA

Oxygen, by volume, is by far the most abundant component in silicate melts. The amount in vapors in equilibrium with these melts, however, is extremely small, and its partial pressure, measured in bars, typically ranges from 10^{-8} to 10^{-11}. Despite these low values, small changes can bring about major changes in mineralogy, which in turn can change the path of magmatic fractionation. These changes can take place even when a separate vapor phase is not present through changes in the fugacity of oxygen. Visualize, for example, a magma with no separate vapor phase in a sealed container which is connected with an external reservoir of oxygen through a membrane that is porous only to oxygen. By changing the pressure and temperature on the reservoir we can change the fugacity of oxygen in the reservoir. Oxygen will then diffuse either into or out of the magma chamber through the semipermeable membrane until the fugacities of oxygen in the magma and the reservoir are the same.

Changes in mineral assemblages resulting from changes in oxygen fugacity are caused largely by changes in the oxidation state of iron (also Mn and possibly Ti). For example, the minerals hematite and magnetite are related by the reaction

$$2Fe_3O_4 + 0.5O_2 \rightleftharpoons 3Fe_2O_3 \qquad (11\text{-}40)$$

Ferromagnesian silicates can also be involved, as for example in

$$3Fe_2SiO_4 + O_2 \rightleftharpoons 2Fe_3O_4 + 3SiO_2 \qquad (11\text{-}41)$$

For any such reaction an equilibrium constant can be written, which, for Eq. 11-41, for example, would be

$$K = \frac{(a_{Mt})^2(a_Q)^3}{(a_{Fa})^3(fO_2)} = \exp\left(\frac{-\Delta G}{RT}\right) \qquad (11\text{-}42)$$

Knowing the value of ΔG for the reaction, K can be calculated as a function of temperature (not very sensitive to P, see Sec. 9-8). If the activities of magnetite, quartz, and fayalite are known, the fugacity of oxygen can be determined for any temperature. Because quartz is always pure SiO_2, the activity of quartz will be unity, as long as the crystalline phase is present. Magnetite and fayalite may also be present in their pure forms. If not, their composition must be known before their activity can be determined. If we assume for the moment that all of these minerals are present as pure phases, the oxygen fugacity in equilibrium with quartz, fayalite, and magnetite (QFM) is given by rearranging Eq. 11-42 to

$$\ln fO_2 = \frac{\Delta G^{QFM}}{RT} \qquad (11\text{-}43)$$

A plot of $-\log_{10} fO_2$ versus T for Eq. 11-43 is given in Figure 11-22.

Equation 11-43 shows that if a rock contains quartz, fayalite, and magnetite, the oxygen fugacity in equilibrium with this assemblage is determined by the temperature. At constant

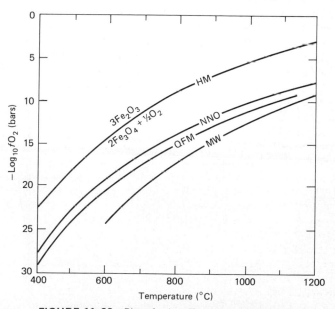

FIGURE 11-22 Plot of $-\log fO_2$ (bars) versus temperature for a number of common oxygen buffering reactions: hematite–magnetite (HM), nickel–nickel oxide (NNO), quartz–fayalite–magnetite (QFM), and magnetite–wüstite (MW). See text for reactions.

temperature, then, the oxygen fugacity is fixed. This assemblage of minerals is therefore said to be an oxygen buffer—the QFM buffer. Different mineral assemblages buffer the oxygen fugacity at different values. In addition to the QFM buffer, Figure 11-22 shows the buffer for coexisting hematite and magnetite (HM), and for the reaction involving magnetite and wüstite (MW)

$$3FeO + 0.5O_2 \rightleftharpoons Fe_3O_4 \qquad (11\text{-}44)$$

and nickel–nickel oxide (NNO)

$$Ni + 0.5O_2 \rightleftharpoons NiO \qquad (11\text{-}45)$$

The buffers QFM, MW, and NNO have been used widely in experimental work to control oxygen fugacities because they cover the range of fugacities normally encountered in the crust and upper mantle. Most igneous rocks, in fact, are formed under fugacities near the QFM buffer (Carmichael and Ghiorso 1986).

Through reactions such as those plotted in Figure 11-22, oxygen fugacities place restrictions on the minerals that can coexist and crystallize from melts. For example, there are no conditions under which hematite and fayalite can crystallize together. Similarly, if the oxygen fugacity in a magma is buffered by some external reservoir at a value equal to that of the NNO buffer, fayalite could not crystallize.

To appreciate the effect that oxygen fugacity can have on crystallizing melts, we will examine the simple binary system $FeO–Fe_2O_3$ (Fig. 11-23). This system contains the phases iron-oxide liquid, iron, wüstite, magnetite, and hematite, all of which can contain variable amounts of oxygen depending on the oxygen fugacity. The system has liquidi for each of the crystalline phases, two eutectics—one between hematite and magnetite, the other between wüstite and iron, a peritectic (P) involving magnetite and wüstite, and solvi between each pair of juxtaposed crystalline phases. In this respect the diagram is

not unlike binary systems already discussed in Chapter 10. It differs, however, in that for any composition of phase or pair of phases it is possible to specify for a given temperature the oxygen fugacity that must exist. The phase diagram can thus be contoured with lines of equal oxygen fugacity (isobars). In fields where two phases coexist, such as hematite and magnetite, the oxygen isobars are horizontal lines, because such assemblages, regardless of the relative abundances of the phases, is an oxygen buffer for a given temperature. In fields of a single phase, such as wüstite or liquid, the isobars are sloping lines because oxygen fugacity here is a function of both temperature and composition.

Because changes in oxygen fugacity can affect phase assemblages just as easily as changes in temperature (and pressure) in systems such as that of $FeO–Fe_2O_3$, the oxygen fugacity must be specified before we can interpret what will happen to a composition with changing temperature. Two extreme cases can be considered; one where the bulk composition of the system remains constant—no oxygen is allowed to enter or leave the system; the other is where the oxygen fugacity is maintained at a constant value—this requires that oxygen be able to enter or leave the system. Conditions ranging between both extremes are found in nature.

Consider first the cooling of a melt containing 50% Fe_2O_3 under conditions of constant bulk composition. At 1600°C the oxygen fugacity in equilibrium with this melt is 10^{-3} bar (Fig. 11-23). On cooling, the melt reaches the magnetite liquidus at 1530°C where the oxygen fugacity is 10^{-4} bar. The melt then descends the liquidus to the peritectic, P, where magnetite reacts with the liquid to produce wüstite. Once all liquid is consumed by the reaction, magnetite and wüstite are free to cool. As the temperature drops, the oxygen fugacity continues to fall, and wüstite changes composition to progressively lower oxygen contents. At 560°C wüstite breaks down to α-iron and magnetite, which then cools to room temperature.

Next consider the cooling of this same melt under conditions of constant oxygen fugacity from a temperature of 1530°C

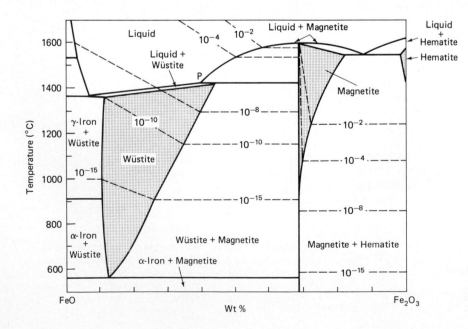

FIGURE 11-23 System $FeO–Fe_2O_3$. Dashed lines are lines of constant oxygen fugacity measured in bars. [After Muan (1958).]

on the magnetite liquidus where the oxygen fugacity is 10^{-4} bar. As heat is removed, magnetite crystallizes, but the melt cannot change its composition because only this one composition of melt on the magnetite liquidus is stable at an oxygen fugacity of 10^{-4} bar. As more magnetite crystallizes from the melt, the bulk composition changes along the 10^{-4} isobar until it eventually coincides with the magnetite composition, at which point no liquid remains. Note that under these conditions the melt never descends to the peritectic, nor does any wüstite form. Once all of the liquid has disappeared, the temperature again begins to fall. To remain on the 10^{-4} isobar, the magnetite becomes richer in oxygen until at 1080°C it reaches the solvus with hematite. The temperature then again remains constant while magnetite is converted to hematite. Once all magnetite is consumed, the hematite is then free to continue cooling. In this second example, then, the bulk composition of the system has to change from its initial value of 50% Fe_2O_3 to its final value of 100% Fe_2O_3 in order to maintain the oxygen fugacity at 10^{-4} bar. This change is effected by oxygen being added to the system from an external source.

Oxygen fugacity is also important in silicate systems involving iron, as for example, in the system $SiO_2–FeO–Fe_2O_3$ (Fig. 11-24). The binary diagram of Figure 11-23 forms one of the bounding systems to this ternary. Addition of SiO_2 to this binary introduces the phases tridymite and fayalite. The points of intersection of the lines of constant oxygen fugacity with the liquidus in the binary system (Fig. 11-23) can be extended across the liquidus surface in the ternary system (Fig. 11-24). Note that fayalite melts incongruently to iron plus liquid. Also,

the peritectic in the binary system $FeO–Fe_2O_3$ involving magnetite and wüstite becomes a cotectic after descending a short distance into the ternary system. It is important to keep in mind that this ternary system is part of the larger system $Si–Fe–O_2$ (inset in Fig. 11-24). When dealing with crystallization at constant oxygen fugacity, oxygen must be added to or removed from the system. For this reason, we must know the location of the oxygen apex in the larger system.

We will again consider crystallization under the two extreme conditions of constant bulk composition and constant oxygen fugacity. Let us first consider the crystallization, under constant bulk composition, of a melt with a composition X that plots in the primary field of fayalite (enlarged inset in Fig. 11-24). Fayalite, on crystallizing, causes the composition of this melt to move directly away from fayalite, thus enriching it in oxygen. This, in turn, causes the oxygen fugacity to increase from 10^{-11} to $10^{-9.4}$ bar by the time the liquid reaches the cotectic with magnetite. Crystallization of magnetite and fayalite drives the liquid down the cotectic to the ternary eutectic with tridymite, where the oxygen fugacity is 10^{-9} bar.

If this same melt is to crystallize at constant oxygen fugacity, it is first necessary to determine in which direction the temperature decreases on the isobars on the liquidus surface of fayalite. This can be done by drawing a line from fayalite to the oxygen apex; this crosses the isobars at temperature maxima. Crystallization of fayalite from melt X under an oxygen fugacity of 10^{-11} bar will, therefore, cause the liquid to proceed along the isobar toward the tridymite field. Once this happens, however, a line drawn from the liquid to fayalite no longer

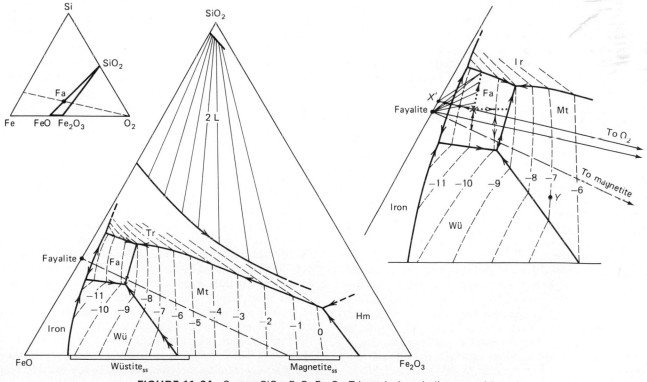

FIGURE 11-24 System $SiO_2–FeO–Fe_2O_3$. Triangular inset indicates position of system in terms of $Si–Fe–O_2$. [After Muan (1955).]

passes through the original bulk composition X. The bulk composition must, therefore, change, either by adding or subtracting oxygen. To determine which, a line is drawn through X and the oxygen apex; points along this line have a constant ratio of silicon to iron but variable oxygen contents. The point of intersection of this construction line with the line joining the liquid composition to fayalite is the new bulk composition.

Continued crystallization of fayalite and cooling of the liquid along the 10^{-11} isobar eventually brings the liquid to the cotectic with tridymite. However, as tridymite and fayalite crystallize, the liquid cannot descend the cotectic, for this would increase its oxygen fugacity. Instead, the bulk composition must continue to lose oxygen as it changes its composition along the line extending from X away from the oxygen apex. When the bulk composition reaches the join between fayalite and tridymite (X'), the liquid is totally crystallized, and the system consists only of fayalite and tridymite.

The sequence of crystallization and the compositions of the fractionating liquids are strikingly different in these two modes of crystallization. In nature, then, very different rock types could result from a single parental magma, depending on whether crystallization occurs under buffered or unbuffered conditions. Although conditions approaching both extremes have been identified in natural occurrences, most magmas follow some intermediate path between these extremes.

Because oxygen fugacity plays such an important role in determining what minerals can form in a rock, it is commonly possible to determine from the mineralogy of a rock the oxygen fugacity that existed at the time the rock formed (or at least last equilibrated). A number of mineral assemblages have been used for this purpose but none with more success than that involving coexisting magnetite and ilmenite (Buddington and Lindsley 1964).

Many rocks contain both an ilmenite–hematite solid solution (rhombohedral phase) and a magnetite–ulvöspinel solid solution (spinel phase). These phases plot in the system FeO–TiO$_2$–O$_2$ (Fig. 11-25), and they can be related through the following oxidation reactions:

(spinel phase)	(rhombohedral phase)	(spinel phase)	
$3Fe_2TiO_4 + 0.5O_2 \rightleftharpoons$	$3FeTiO_3$	$+ Fe_3O_4$	(11-46)
$2Fe_3O_4 + 0.5O_2 \rightleftharpoons$	$3Fe_2O_3$		(11-47)

From these reactions and Figure 11-25 it is clear that at low oxygen fugacities the spinel phase is rich in ulvöspinel and it coexists with an ilmenite-rich rhombohedral phase. As the oxygen fugacity increases (moving toward O$_2$ in Fig. 11-25) the spinel phase becomes richer in magnetite, and at high oxygen fugacities the spinel phase is rich in magnetite and the rhombohedral phase contains considerable hematite. The compositions of these coexisting phases are also dependent on temperature (total pressure has negligible effect). Thus it is possible, in rocks containing both of these phases, to determine the oxygen fugacity and the temperature at the time the minerals last equilibrated.

To apply this oxygen geobarometer and geothermometer it is necessary to know the compositions of both the spinel

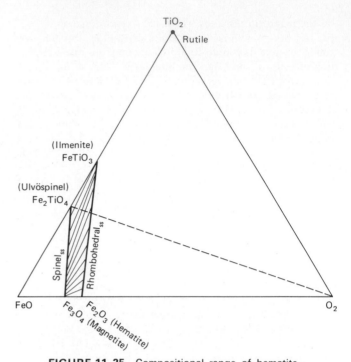

FIGURE 11-25 Compositional range of hematite–ilmenite solid solutions (rhombohedral phase) and magnetite–ulvöspinel solid solution (spinel phase) in terms of TiO$_2$–FeO–O$_2$. Tie lines indicate compositions of typical coexisting rhombohedral and spinel phases. Dashed line indicates addition of oxygen (increasing fO_2) to ulvöspinel.

and rhombohedral phases. These would normally be obtained from electron microprobe analyses. The spinel phase is expressed in terms of mole percent magnetite and ulvöspinel and the rhombohedral phase in terms of mole percent hematite and ilmenite. With these compositions, the oxygen fugacity and temperature can be read directly from the graph in Figure 11-26. Over much of the range of conditions the oxygen fugacity can be determined to within one order of magnitude and the temperature to $\pm 30°C$.

While simple in principle, difficulties can be encountered in applying this geothermometer. Both the spinel and rhombohedral phases develop exsolution lamellae on cooling (ilmenite$_{ss}$–hematite$_{ss}$ and ulvöspinel$_{ss}$–magnetite$_{ss}$). In plutonic rocks where cooling is slow, these exsolution lamellae may be expelled from their host grains, which can make determining the original composition of the phases impossible. Many temperatures determined for plutonic rocks are consequently lower than expected. Postmagmatic oxidation can also change phases. For example, the spinel phase is commonly oxidized on cooling to a mixture of magnetite and ilmenite. The ilmenite forms lamellae in the magnetite and resembles an exsolution texture. But the solid solution between magnetite and ilmenite at high temperatures is negligible. Analyses of magnetite grains containing such lamellae of ilmenite must be recalculated to the original spinel composition (Buddington and Lindsley 1964).

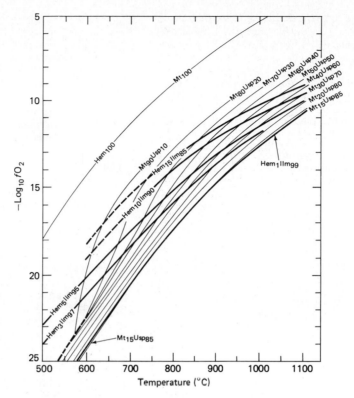

FIGURE 11-26 Projection of the compositions of hematite–ilmenite solid solutions and magnetite–ulvöspinel solid solutions in terms of oxygen fugacity and temperature projected onto the fO_2–T plane. Compositions of phases are in mole percent. Compositions of coexisting rhombohedral and spinel phases uniquely define the oxygen fugacity and temperature at which the phases equilibrated. [After Buddington and Lindsley (1964); published with permission of Oxford University Press.]

PROBLEMS

11-1. From Eq. 11-16 or 11-17 and the data for ln k in Figure 11-4, determine the mole fraction of water in a water-saturated albite melt at 0.5 GPa and 1000°C.

11-2. If a tholeiitic magma that initially is nearly saturated with an immiscible sulfide liquid and has an oxygen fugacity close to that of the QFM buffer rises to a high level in the Earth's crust, where its oxygen fugacity is raised to that of the NNO buffer, describe what might happen to the sulfide content of this magma.

11-3. Trace the sequence of events in the complete fractional crystallization of a liquid in the system SiO_2–FeO–FeS having a composition that plots at the top of the letter F of Fa marking the field of fayalite in Figure 11-6. How does the variation in the sulfide content of the immiscible silicate liquid conform to the general pattern of solubility of sulfur in silicate melts?

11-4. **(a)** Calculate the mole fraction of albite and mole fraction of water in an albite-saturated melt at 1300 K and 1.0 GPa.
(b) Under these conditions of pressure and temperature, what is the maximum mole fraction of water that an albite melt can contain?
(c) Is it possible for the albite liquidus to exist at 1300 K and 1.0 GPa for a hydrous melt?

11-5. Repeat Prob. 11-4 for a temperature 1100 K and pressure of 0.1 GPa.

11-6. If the mole fraction of water in an albite melt on the liquidus is 0.2 and the pressure is 0.4 GPa, what is the temperature of the melt and the activity of water in the melt? Check your answer by inspecting Figure 11-9.

11-7. **(a)** If the vapor phase in a metamorphic rock containing only albite is buffered so that $a_w = 0.4$ and the pressure is 0.8 GPa, at what temperature would the first melt appear, and what mole fraction of water would it contain?
(b) If the magma separates from its source, so that the activity of water is no longer buffered, and the magma rises toward the surface at constant temperature, at what pressure would the melt become water-saturated, and at what pressure would it eventually solidify?

11-8. Create a graph similar to that in Figure 11-19 giving curves for magmas that initially contain 1, 2, 3, 4, 5, 6, 7, and 8 wt % H_2O. Assume that only anhydrous phases crystallize.

11-9. At what degree of solidification would a granitic magma that initially contains 2 wt % H_2O reach a saturation level of 10 wt % H_2O if the crystallizing mineral assemblage consists of 80 wt % quartz and alkali feldspar and 20 wt % muscovite? (The muscovite contains 4.2 wt % H_2O.)

11-10. **(a)** If a magma that initially contains 2 wt % H_2O crystallizes only quartz and alkali feldspar during the first 50% of its crystallization, but then crystallizes a mixture of quartz, alkali feldspar, mica, and hornblende that contains an average of 1 wt % H_2O during the final 50% of its crystallization, at what stage of crystallization would the magma reach a saturation level of 10 wt % H_2O?
(b) If no hydrous phases had crystallized from this magma, at what stage of crystallization would the magma have become water-saturated?

11-11. **(a)** A granitic magma initially containing 2 wt % H_2O first crystallizes phenocrysts of biotite containing 3.5 wt %

H_2O. Following 20% solidification of the magma, quartz and alkali feldspar join the biotite, and the crystallizing phases now contain only 0.1 wt % H_2O. Trace the concentration of H_2O in the magma as a function of the degree of solidification, and indicate at what stage of solidification the magma would reach a saturation level of 10 wt % H_2O.

(b) By comparing the answers to part (a) and Prob. 11-10 discuss the significance of early versus late crystallization of hydrous phases to the development of water-saturated magmas.

11-12. The boundary line between the fields of magnetite and wüstite in Figure 11-24 changes from a reaction to a cotectic as it descends into the ternary system.

(a) What composition does the wüstite have when the reaction relation changes to a cotectic?

(b) From Figure 11-23 determine at what oxygen fugacity wüstite of this composition is in equilibrium with magnetite.

(c) Knowing this oxygen fugacity, determine the point on the magnetite–wüstite boundary in Figure 11-24 where this change from reaction to cotectic occurs.

11-13. Describe the crystallization of a melt of composition Y in the enlarged inset of Figure 11-24 under conditions of **(a)** constant bulk composition, and **(b)** constant oxygen fugacity. **(c)** If a magma of composition Y were to crystallize fayalite in its residual liquid, what might you conclude about its conditions of crystallization?

11-14. If an andesite contains phenocrysts of magnetite and ilmenite that have compositions of 45 mol % ulvospinel and 15 mol % hematite, respectively, what were the oxygen fugacity and temperature at the time of eruption?

12 // *Crystal Growth*

12-1 INTRODUCTION

With the exception of volcanic glasses and a few rare rock types, all igneous and metamorphic rocks are composed of crystalline mineral grains. The compositions of these minerals can be explained simply in terms of thermodynamic phase equilibria, but the growth of these grains is controlled by poorly understood kinetic factors. Yet that important property of a rock known as texture is determined largely by crystal growth. A few experimental studies have shed some light on the processes of crystal growth that operate over the time span typically encountered in cooling lavas [see reviews by Kirkpatrick (1975) and Lofgren (1980)]. Growth rates in plutonic igneous rocks and metamorphic rocks, however, are so slow that they will probably always remain outside the reach of the experimentalist. Only through understanding the principles are we likely, therefore, to gain insight into these slower growth processes (Dowty 1980; Brandeis et al. 1984).

The basic principles of crystal growth are equally applicable to igneous and metamorphic rocks—only the medium in which growth occurs is different. Crystal growth in both types of rock is dealt with in this chapter. Crystal growth can occur only after thermodynamic intensive variables exceed, by some finite amount, the equilibrium conditions for the formation of that phase. This overstepping (undercooling, overheating, supersaturating) provides the energy to form a nucleus on which growth can occur and the driving force to sustain that growth. Once formed, a nucleus provides a sink to which the crystallizing components diffuse. The distance over which diffusion transports material depends on the rates of diffusion and the time available. Material that is too far from a nucleus must form a separate nucleus. The number of nuclei that form depends, therefore, on the ease of nucleation and the time available for material to diffuse and grow onto these nuclei. The interplay between the rate of nucleation and the rate of crystal growth, therefore, determines the eventual grain size of the rock.

Crystal growth is a complex process involving numerous steps, each of which proceeds at its own rate, with the slowest one determining the overall growth rate of the crystal. Different minerals growing together may have different rate-controlling steps, as may different faces on a single crystal. Gradients of temperature, composition, and stress affect these rates. A rock's texture is a complicated record of these different rate processes.

12-2 NUCLEATION

In Chapter 10, phase diagrams were used to show what minerals should form from particular compositions at given pressures and temperatures. It was tacitly assumed that, when a liquid, for example, is cooled below the liquidus, a mineral crystallizes. At the liquidus itself, the ΔG of crystallization is zero, and therefore the rate at which crystals form will also be zero. Even when the temperature drops below the liquidus, experiments indicate that crystals do not form immediately. Indeed, a significant degree of undercooling is found to be necessary before crystallization begins. Conversely, with solid-state reactions that take place with rising temperatures, a certain degree of overheating is necessary before crystals of the reaction products start to form.

The necessity for overstepping the equilibrium condition is related to energies involved in forming the first minute crystals of the new phase, which are known as *nuclei*. Fluctuations in the composition of the reactants momentarily bring about clusters of the components necessary to form the new phase. Most of these clusters are unstable, as will be shown below, and they disappear. Only those reaching a critical size survive and grow. Nuclei may form randomly in the host from which they are growing, in which case nucleation is said to be *homogeneous*, but if it takes place on an already existing surface, such as another crystal or a gas bubble, it is said to be *heterogeneous*. Olivine crystals in a slowly cooling magma, for example, may nucleate homogeneously, but the tourmaline crystals in the pegmatite illustrated in Figure 11-17 clearly nucleated heterogeneously on the roof and floor of the pegmatite. Nucleation of metamorphic minerals is likely to occur heterogeneously.

The reason that small clusters of the new phase have difficulty in surviving is related to the development of an interface between the embryo and the medium in which it is growing.

Any surface between two phases involves an interfacial energy, the magnitude of which depends on the degree of mismatch of structures across the boundary. When a crystal is large the surface energy constitutes only a very small fraction of the total free energy of the crystal and can be ignored. But when the crystal is very small, the surface energy is significant.

Let us consider all of the energies involved in the growth of a new crystal. First, there is the free-energy change involved in forming the new phase, independent of surface energies. This is the free-energy change with which we are already familiar and which can be obtained from tables of thermodynamic data. We will refer to this as the free-energy change of the volume, ΔG_{vol}. Second, there is the free energy associated with the creation of a new interface (ΔG_{intf}). Third, the growth of a crystal may involve the development or release of strain (ΔG_{strain}). A crystal of ice growing in a milk bottle left out on a winter's day is capable of cracking the bottle. Clearly, the crystallization of this ice would require a greater driving force than would ice growing in an unconfined space. In other cases the medium from which a crystal grows may be strained, and the growth of the crystal releases the strain energy. This certainly occurs when a structurally deformed rock recrystallizes.

For a nucleus to grow, the overall free-energy change associated with its formation must be negative. This free-energy change can be expressed as

$$\Delta G_{growth} = \Delta G_{vol} + \Delta G_{intf} + \Delta G_{strain} \qquad (12\text{-}1)$$
$$\quad (-) \qquad (+) \qquad (+ \text{ or } -)$$

The main driving force for growth of the crystal comes from ΔG_{vol}, which becomes increasingly negative the more the equilibrium conditions are overstepped. The free-energy change associated with the creation of a new interface is always positive and therefore acts to prevent crystallization. The strain energy either inhibits or enhances growth, depending on whether it is positive or negative. For ΔG_{growth} of a nucleus to be negative, the equilibrium conditions must be overstepped sufficiently to make $-\Delta G_{vol}$ greater than $\Delta G_{intf} + \Delta G_{strain}$. This can be done, for example, in the case of nucleation of a crystal in a magma by supercooling the magma. In a metamorphic rock it might be necessary to superheat the rock or deform the rock (thus making ΔG_{strain} negative). In both igneous and metamorphic environments, ΔG_{vol} can also be changed by changing the composition of the host so that it becomes supersaturated.

Let us, for purposes of illustration, consider the homogeneous nucleation of a single olivine crystal in a basaltic magma at constant temperature, pressure, and composition of magma. Under these conditions the value of ΔG_{vol} is fixed, and because the crystal is growing in a liquid, ΔG_{strain} is zero. If we assume that the embryonic cluster of olivine is essentially spherical, the free-energy change associated with the formation of an embryo of radius r will be

$$\overset{\text{volume}}{} \qquad \overset{\text{interface}}{}$$
$$\Delta G_{growth} = \tfrac{4}{3}\pi r^3\,\Delta G_{vol} + 4\pi r^2\,\Delta G_{intf} \qquad (12\text{-}2)$$

Recall that ΔG_{vol} is negative and ΔG_{intf} is positive. The functional form of Eq. 12-2 is shown in Figure 12-1. For very small values of r an increase in radius causes an increase in the free

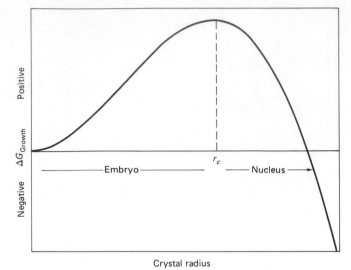

FIGURE 12-1   Relation of free energy of growth of an embryo to its radius. The embryo forms a stable nucleus once it reaches a critical radius, r_c. Increase in radius beyond r_c lowers free energy.

energy of growth; this, therefore, is an unnatural process and the embryo dissolves. Above a critical value of the radius, r_c, increasing r brings about decrease in the free energy of growth, and thus a stable nucleus is formed that will grow. The critical radius is given by $-2\Delta G_{intf}/\Delta G_{vol}$ (see Prob. 12-1).

The rate at which critical nuclei form in a cooling magma depends strongly on the degree of undercooling. As the temperature falls below the liquidus, the increasing negative value of ΔG_{vol} increases the likelihood that nucleation will occur. Falling temperature, however, makes diffusion more difficult and decreases the chances of embryos forming with critical radius. The nucleation rate at a given temperature therefore depends on which of these effects dominates. We know from Eq. 7-33 that $(\partial\,\Delta G/\partial T)_P = -\Delta S$, and $\Delta H_m/T_m$ can be substituted as an approximation for ΔS, where ΔH_m is the enthalpy of melting and T_m the melting point of the pure mineral. Upon integrating Eq. 7-33, we obtain

$$\Delta G_{vol} = \frac{\Delta H_m(T_0 - T)}{T_0} \qquad (12\text{-}3)$$

which indicates that ΔG_{vol} is a linear function of the degree of undercooling ($T_0 - T$). Diffusion, on the other hand, like viscosity (Eq. 2-9), changes exponentially with temperature, as indicated by

$$D = D_0 \exp\!\left(\frac{-Q}{RT}\right) \qquad (12\text{-}4)$$

where D is the diffusion coefficient at temperature T, D_0 a constant, and Q an activation energy. With falling temperature, then, the increased negative value of ΔG_{vol} at first causes the nucleation rate to increase, but with continued lowering of temperature the diffusion term comes to dominate and the nucleation rate passes through a maximum and then decreases

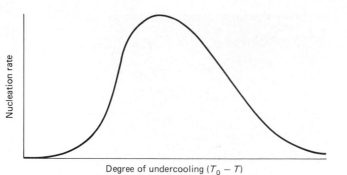

FIGURE 12-2 Variation in rate of nucleation in a melt as a function of degree of undercooling.

to zero (Fig. 12-2). Curves of this type have been experimentally determined, for example, for quartz, alkali feldspar, and plagioclase nucleating in granitic and granodioritic melts. The maximum nucleation rates for these crystals occurred at 200 to 400°C of undercooling (Swanson 1977). Dikes provide a natural occurrence where the degree of undercooling would be expected to increase toward the margins because of increased cooling rates. The number of nuclei would therefore be expected to increase toward the margins until the maximum nucleation rate is reached and then decrease again as higher degrees of supercooling make it more difficult for nuclei to form and the melt cools to a glass. Such a variation is illustrated in Figure 12-11.

Crystal nucleation rates for metamorphic reactions that take place on heating behave quite differently from magmatic ones because, while the rising temperature makes the value of ΔG_{vol} more negative, the rising temperature also increases diffusion rates. Thus the nucleation rate steadily increases with degree of overstepping of the equilibrium conditions. More nuclei of a mineral are therefore likely to form in a contact metamorphic rock near the igneous intrusion where temperatures are high, rather than farther away, where they are low. Because grain size is inversely related to the number of nuclei formed, the grain size would be expected to increase away from the intrusion (other factors being equal) to a distance where the temperature had never risen above that necessary for nucleation. Figure 12-3 illustrates such a variation in the contact metamorphic zone at the margin of a basaltic dike from County Mayo, Ireland. Grossular garnets formed in the limestone country rock by heat liberated from the dike show a steady decrease in number and increase in size away from the contact.

With metamorphic nucleation, strain energies must also be taken into account. The most common case involves rocks that were deformed prior to heating, so that when new minerals grow or the original grains recrystallize, strain energy is released. From Eq. 12-1 it is clear that with slow heating, the most strained parts of a rock will be the regions where ΔG_{growth} will first become negative; that is, nucleation will occur first where strain is greatest. Moreover, the greater the strain in a rock, the greater will be the number of nuclei to form, and consequently, the finer grained it will be.

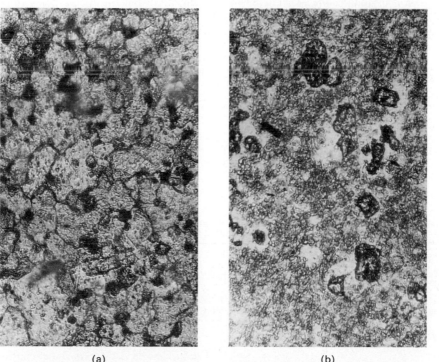

(a)   (b)

FIGURE 12-3 Variation in the abundance and size of grossular garnet crystals in metamorphosed limestone near a diabase dike, County Mayo, Ireland. Near the contact [(a) 0.3 m from contact] garnet crystals (high relief and dark) are small but present in large numbers because of a high nucleation rate. Farther from the contact [(b) 2.8 m from contact] garnets are larger but fewer in numbers. The width of each field is 0.7 mm.

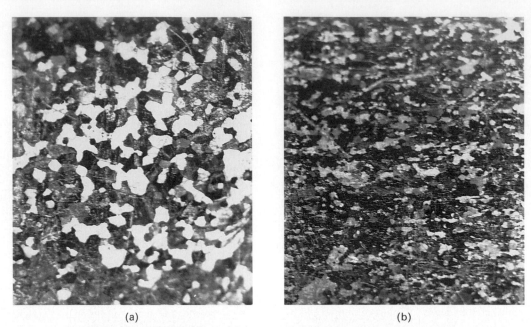

(a)   (b)

FIGURE 12-4   Annealed aluminum metal. Both specimens were first deformed by being passed through a roller, but the spacing between the rollers for specimen 2 was half that for specimen 1. The greater deformation in specimen 2 created more nucleation sites when the samples were recrystallized by heating to high temperature. Specimen 2 is therefore finer grained. The width of each field is 2 cm.

The effect of strain on nucleation can be illustrated with a metallurgical example. Figure 12-4 shows two pieces of aluminum metal that came originally from the same bar. Both were flattened by being passed through a roller, but specimen 2 was flattened to half the thickness of specimen 1 by moving the rollers closer together. Specimen 2 was therefore more strained than specimen 1. Both specimens were then held at high temperature in a furnace where they recrystallized (annealed). Because of the greater strain in specimen 2, more nuclei formed and the annealed metal is much finer grained than the less strained specimen.

The same effects are to be expected in metamorphic rocks, with more deformed rocks recrystallizing to a finer grain size. Indeed, one of the finest-grained metamorphic rocks, known as *mylonite*, is formed in fault zones where strain is at a maximum. The grain size of these rocks was originally interpreted to result from grinding and milling in the fault zone (hence the name). Although some comminution may occur in this way, the fine grain size is now interpreted to be due largely to crystallization at many nucleation sites.

12-3   CRYSTAL GROWTH RATES

Once a stable nucleus has formed it continues to grow and forms a crystal whose size is determined by the concentration of nutrients in the surroundings and the proximity of neighboring nuclei. Four steps are involved in the growth process. First, nutrients must diffuse to the nucleus through the medium in which the crystal is growing; diffusion rates vary consider-ably depending on whether the medium is a gas, fluid, silicate melt, or solid rock. Second, nutrients, on arriving at the nucleus, may have to react and arrange themselves into building units that are acceptable to the crystal. Third, building units must then attach themselves to the crystal surface; this may involve nucleation of new surfaces or the growth of dislocations. Finally, attachment of the building units produces heat of crystallization and an increase in concentration of components not entering the crystal. Both of these must be dissipated before further growth can occur. The rate at which a crystal grows is determined by the slowest of these steps. We will now examine the time dependency of each of these processes, assuming that the temperature and concentration of nutrients in the environment at some distance from the crystal remain constant during the period of interest.

Diffusion-Controlled Growth.   If diffusion controls the rate at which a crystal face grows, it simply means that this is the slowest of the chain of steps involved in the growth of that face. The other steps still occur, but because they take place more rapidly they have to wait for diffusion to bring nutrients into the chain. Diffusion, then, is described as being the rate-determining process.

The distance a face advances, say in the x direction, in a given time depends on the flux of nutrients brought to that face by diffusion. When the crystal first starts to grow the region immediately in contact with the face becomes depleted in nutrients, and a steep concentration gradient develops. With time, however, the gradient becomes shallower (Fig. 12-5), the flux of nutrients brought to the crystal face decreases, and the crystal grows more slowly.

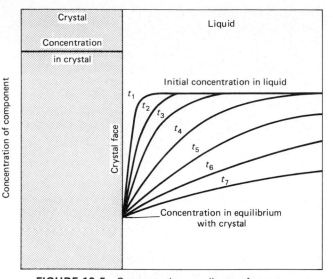

FIGURE 12-5 Concentration gradients of a component at successive times (t_i) in advance of a crystal face whose growth rate is determined by diffusion of nutrients through the surroundings. The concentration of the component immediately in contact with the crystal face is the equilibrium concentration. The concentration in the crystal is shown as being constant, but it may vary during growth (see Fig. 12-15).

The relation between the diffusional flux and the concentration gradient is given by *Fick's law,*

$$J_x = -D \frac{\partial c}{\partial x} \qquad (12\text{-}5)$$

where J_x is the flux of material that diffuses in the x direction in response to a concentration gradient in that direction of $\partial c/\partial x$, and D is the diffusion coefficient. This equation is identical in form to the Hagen-Poiseuille law (Eq. 3-9) for the flux of fluids and to Fourier's law (Eq. 5-3) for the flux of heat—they all are transport laws. The change in concentration gradient with time resulting from diffusion is similar to the change

in thermal gradient with time as a result of the diffusion of heat (compare, for example, Figs. 5-8 and 12-5). Because the flux of material reaching the crystal decreases as the concentration gradient decreases, the growth rate is given by

$$\frac{dx}{dt} = K \left(\frac{D}{t} \right)^{1/2} \qquad (12\text{-}6)$$

where K is a constant. Integration of this equation gives t in terms of x ($t = x^2/4K^2 D$), which, when substituted back into Eq. 12-6, gives

$$\frac{dx}{dt} = \frac{k_D}{x} \qquad (12\text{-}7)$$

where k_D is a constant for the diffusion process that combines all of the other constant terms. This equation expresses the growth rate in terms of a property that we can observe in a rock—grain size. Equation 12-7 indicates that for diffusion-controlled growth the larger a crystal becomes, the slower it grows.

Phase Boundary Reaction-Controlled Growth. In the growth of complex silicates, addition of atoms one at a time may not be possible. For example, local charge balance may require that certain groups of ions be attached simultaneously. If the reaction to form these building units determines the rate of crystal growth, there is no reason for the growth rate to change with time. The growth rate would therefore have the form

$$\frac{dx}{dt} = k_R \qquad (12\text{-}8)$$

where k_R is a constant for the reaction process.

Surface Nucleation-Controlled Growth. Once nutrients are organized into acceptable building units they must attach themselves to the surface of the crystal. This happens in a number of different ways (Fig. 12-6a). For example, on a perfectly planar surface a building unit could attach itself only

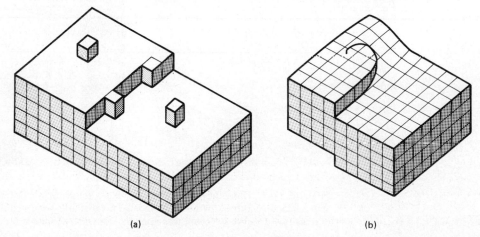

FIGURE 12-6 (a) Possible attachment sites on the surface of a crystal. Sites which provide the largest number of bonds are likely to be occupied first (i.e., corners > edges > faces). (b). The edge produced by a screw dislocation is self-perpetuating, as growth develops a spiral ramp.

(a)

(b)

by starting a new surface layer. Such attachment, however, satisfies only a few bonds and actually increases the proportion of surface area to volume. Steps on the surface present more favorable sites for attachment, and corners are still better. As a result, precipitation on a face is likely to fill in and complete all irregularities before a new surface layer will nucleate. Attachment of the first block of the new layer is a statistical, process—the larger the face, the more chance there is that an attachment will occur somewhere on that face. We can conclude therefore that the growth rate for surface nucleation-controlled growth must be proportional to the surface area of the crystal. If all the faces on the crystal are growing at the same rate, the x direction can be thought of as the radius of the crystal, and then the area of the growing faces would be approximately $4\pi x^2$ (assuming a sphere). The growth rate could then be expressed as

$$\frac{dx}{dt} = k_s x^2 \qquad (12\text{-}9)$$

where k_s is a constant for surface nucleation-controlled growth.

Screw Dislocation-Controlled Growth. Attachment of nutrients to the crystal may also occur on screw dislocations. Dislocations are imperfections in crystal structures, and the type known as a screw dislocation (Fig. 12-6b) is of particular interest in crystal growth because the step it produces on the growing surface is self-perpetuating, with each new building block being added to form a spiral ramp. A num-

ber of these dislocations may exist on a single face. The rate at which these dislocations grow, if they are the rate-determining step, is independent of time and the size of the crystal but will depend only in some manner on the concentration of nutrients. The growth rate determined by this mechanism is therefore a constant

$$\frac{dx}{dt} = k_{\text{SD}} \qquad (12\text{-}10)$$

where k_{SD} is a constant for screw dislocation-controlled growth.

Dissipation of Heat of Crystallization and Impurities-Controlled Growth. When nutrients transfer from a melt onto a crystal face the latent heat of crystallization causes the temperature of the melt on the face to increase. At the same time, components in the melt that do not enter the crystal become concentrated in the melt at the crystal face and, according to the cryoscopic equation (Eq. 10-6), lower the liquidus temperature, which in turn would decrease the degree of supersaturation. Both heat and material must therefore diffuse away from the crystal surface before growth can continue. When this process determines the growth rate of the crystal, any irregularity that protrudes beyond the general surface of the crystal will extend into cooler and more supersaturated melt and will grow more rapidly. A smooth crystal face is therefore unstable. Instead, the crystal develops a dendritic or skeletal form (Fig. 12-7a). Such habit is common in rapidly cooled volcanic rocks such as pillow lavas, but also in silica-rich residual patches of melt in more slowly cooled

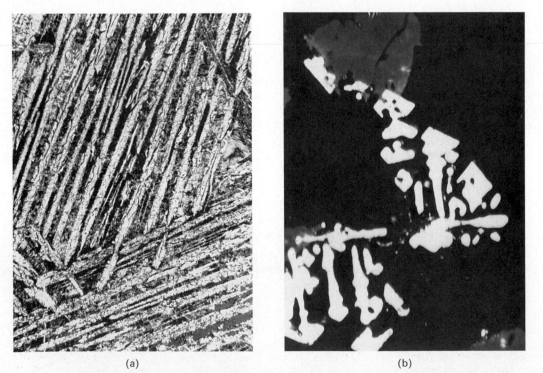

(a)   (b)

FIGURE 12-7   Dendritic crystals of (a) olivine in komatiite from Monroe Township, Ontario (width of field, 1 cm), and (b) magnetite in the silica rich residue of a diabase dike, New Haven, Connecticut. The width of field is 0.2 mm.

rocks where magnetite, ilmenite, and sphene commonly grow with skeletal form (Fig. 12-7b). Dendritic growth of a crystal cannot be described in terms of the simple advance of a crystal face because of the complex branching nature of the crystal-melt boundary. However, the rate of advance of the tips of dendrites would be expected to remain constant, because these parts of the crystals are always growing into melt of constant composition and temperature.

12-4 CRYSTAL MORPHOLOGY DETERMINED BY RATE-DETERMINING GROWTH PROCESSES

Experiments by Lofgren (1974) on plagioclase and by Kirkpatrick (1974) on plagioclase, pyroxenes, and melilite have shown that at small degrees of undercooling of the melt large euhedral crystals are formed that resemble phenocrysts in igneous rocks. At greater degrees of undercooling, crystals tend to grow with skeletal form and are also more *acicular*, that is, their length to breadth ratio increases. At still greater degrees of undercooling, crystals have a branching, *dendritic* form. Finally, at the greatest degrees of undercooling, radiating crystalline fibers form a *spherulitic* texture.

These changes are interpreted to result from a change in the rate-determining step from surface nucleation controlled growth at small degrees of undercooling to growth controlled by dissipation of impurities at high degrees of undercooling. At the lowest temperatures growth rates are slow because of the low diffusion rates. As a result, the layer of liquid on the crystal face that is enriched in those components not entering the crystal is thin, and small wavelength perturbations are large enough to penetrate it. This results in the crystal face advancing as a series of closely spaced fibers. At higher temperatures, and consequently smaller degrees of undercooling, the thickness of the zone enriched in components not entering the crystal increases. Larger perturbations are therefore necessary to penetrate this layer, thus explaining the dendritic morphology and at still higher temperatures the skeletal forms. At the smallest degrees of undercooling, where surface nucleation becomes the rate-determining process, diffusion has sufficient time to dissipate unwanted material, and the crystal faces grow as planar surfaces. Textural variations (Fig. 12-8) similar to those obtained in the experiments can be found in pillows (Bryan 1972).

The cooling rates and degrees of undercooling in the experiments and in the pillows are very large compared with those that must occur in large plutonic bodies. Could different rate-determining steps operate if the degree of undercooling were only a few degrees instead of tens of degrees? While experiments at very small degrees of undercooling are impractical, evidence of the rate-determining steps under these conditions has been obtained from rocks themselves.

Large diabase dikes are sufficiently simple bodies that their cooling histories can be interpreted with some degree of confidence (Chapter 5). Near their centers, magma may take many years to cool through the interval over which crystal nucleation and growth occur. Dikes, therefore, have the potential of providing data for time spans that are much longer than those achievable in laboratory experiments.

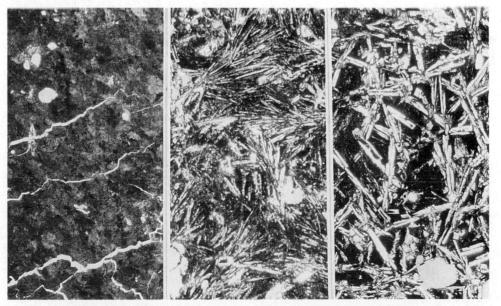

FIGURE 12-8 Textural variation from glassy selvage (left) to core (right) of basaltic pillow from Mid-Atlantic Ridge. Near the margin of the pillow, feldspar crystallized from glass to produce patches of slightly birefringent material. With slower cooling, radiating and branching fibrous plagioclase crystals formed (center). In the center of the pillow, where cooling was slowest, crystals have planar faces, but some skeletal forms are still present. The width of each field is 1.0 mm.

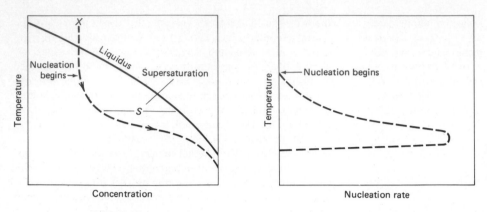

FIGURE 12-9 Temperature–composition diagram showing the path followed by a magma of initial composition X that cools below the liquidus of a mineral and becomes progressively more supersaturated (s) until nucleation occurs. Also shown is the rate of nucleation as a function of undercooling.

In studying the textural variations across wide, flash-injected diabase dikes, Gray (1978) found that the total number of nuclei of a single mineral (actually, the number of crystals) in a unit volume of diabase, N, is related to the distance from the contact, x, by

$$N \propto x^n \tag{12-11}$$

where n is a constant that depends, in part, on the growth-controlling mechanism during nucleation. Thus, from this simple relation it may be possible to determine the growth-controlling mechanism at the much smaller degrees of undercooling that existed in the centers of these dikes.

On cooling below the liquidus of a mineral, a magma becomes supersaturated in that phase, as long as no nuclei form. The degree of supersaturation, s, increases with the amount of undercooling, ΔT(Fig. 12-9). Once nucleation begins, however, the degree of supersaturation decreases until it is just sufficient to drive the growth of the crystals. In relatively slowly cooled melts where nucleation occurs over a small temperature interval, the rate of nucleation, dN/dt, is found to be relatively insensitive to temperature changes and be very nearly proportional to the degree of supersaturation raised to a power m, which is related to the number of molecules forming the critical nucleus (typically, between 2 and 8) (Gray 1978); that is,

$$\frac{dN}{dt} = ks^m \tag{12-12}$$

Thus, as the degree of supersaturation of the magma increases with undercooling, the greater will be the rate of nucleation. But once nuclei form and decrease the supersaturation, the rate of nucleation decreases to zero (Fig. 12-9). The total number of nuclei formed during this interval determines the number of crystals per unit volume of rock that are eventually formed and hence the grain size of the rock. The actual number of nuclei formed can be determined by integrating Eq. 12-12.

The variation in the supersaturation must be known as a function of time before Eq. 12-12 can be integrated. Supersaturation depends on the degree of undercooling, the slope of the liquidus (Fig. 12-9), and the amount of material removed from the melt by crystallization, which in turn depends on the crystal growth mechanism. For a flash-injected diabase dike (no feeder flow), Gray (1978) shows that integration of Eq. 12-12

leads to Eq. 12-11. Figure 12-10 shows plots of Eq. 12-11 for three different growth-controlling mechanisms—dislocation-, diffusion-, and phase boundary reaction-controlled growth—when the value of m in Eq. 12-12 is chosen to be 3.2. The slopes of these three lines corresponds to values of n in Eq. 12-11 of -0.9, -1.5, and -2.2, respectively. If a different value of m had been chosen, the slopes of the lines in Figure 12-10 would have been different for the three rate-determining processes (different n). Unfortunately, the value of m cannot be determined from field data. However, the slopes in Figure 12-10 are the ones that best fit the data for minerals from diabase dikes.

Although the growth-controlling mechanism during nucleation cannot be determined with certainty from the slopes of lines in $\log N$ versus $\log x$ plots, the fact that such plots give straight lines is evidence for the operation of a single rate-determining process. Moreover, if a line has a break in slope, the two segments of the line must reflect different growth-controlling mechanisms. Similarly, if two minerals have different slopes, they must have had different growth-controlling mechanisms. Clinopyroxene, for example, typically has a significantly more negative value of n than does coexisting plagioclase.

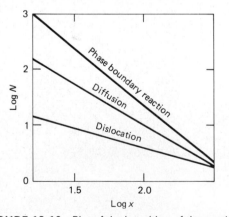

FIGURE 12-10 Plot of the logarithm of the number of crystals of a mineral per unit volume of diabase versus the logarithm of the distance into a dike from the contact. Assuming the value of m in Eq. 12-12 is 3.2, the three lines indicate the variations to be expected with the three growth-controlling mechanisms—phase boundary reaction, diffusion, and dislocation. [After Gray (1978).]

The main textural variation in diabase dikes is determined by the ratio of the number of plagioclase to clinopyroxene crystals per unit volume of rock. If this ratio is less than 1, the texture is *intergranular*; if it is greater than 1, the texture is *ophitic*; and if it equals 1, the texture is *subophitic* (Fig. 12-11). The number of plagioclase crystals in a volume of rock at a given distance from the contact is given by $N_{Plag}x^{nPlag}$. Similarly, the number of clinopyroxene crystals is given by $N_{Cpx}x^{nCpx}$. The ratio of the number of plagioclase to clinopyroxene crystals, then, can be expressed as a function of distance from the contact by

$$\frac{N_{Plag}}{N_{Cpx}} \propto x^{(nPlag - nCpx)} \qquad (12\text{-}13)$$

Because the value of n for clinopyroxene is typically more negative than that for plagioclase, the value of N_{Plag}/N_{Cpx} will increase into the dike. Thus a dike with an intergranular texture ($N_{Plag}/N_{Cpx} < 1$) near the contact will have an ophitic texture toward the center of the dike as long as the dike is sufficiently wide.

One special case where a record of the rate-controlling crystal growth mechanism may be preserved in the rock involves sector-zoned crystals (Hollister 1970; Gray 1971). This type of zoning, which is common in a number of igneous and metamorphic minerals (titanaugite, plagioclase, staurolite, and chloritoid, for example), is visible in thin section because different sectors of a crystal may have different colors, birefringence, or extinction angle. It is normally produced by different faces on the crystal growing with slightly different compositions. As these faces grow they produce pyramid-shaped sectors extending out from the center of the crystal (Fig. 12-12a). In titanaugite, the (100) sector has higher concentrations of Ti, Al, and Fe^{3+} than has the (010) or (001) sectors, and in staurolite the (010) sector has a higher concentration of Ti than has the (001) or (110) sectors.

The boundary between two sectors marks the locus of points of intersection of two crystal faces at various stages of crystal growth; that is, as one crystal face advances some distance x, the other face advances a distance y, and the sector boundary marks the point of intersection of the two faces at each stage of growth. The boundary can be thought of as a graph of the growth rate of one face plotted against the growth rate of the other face. If, for example, the growth rates on the two faces remain constant, the boundary between the two sectors would be planar and in section would be a straight line (Fig. 12-12b). This is not to say that the rates need be the same on both faces; one face might grow faster than the other. If the growth rates changed at different rates during growth, the sector boundary would be curved (Fig. 12-12c). Crystals that exhibit strong zoning, especially of the oscillatory type, preserve this same geometrical information about the shape of the crystal during growth.

Gray (1971) carefully measured the shape of the "*hourglass*" sector zoning in a titanaugite crystal from a 5.5-m-thick

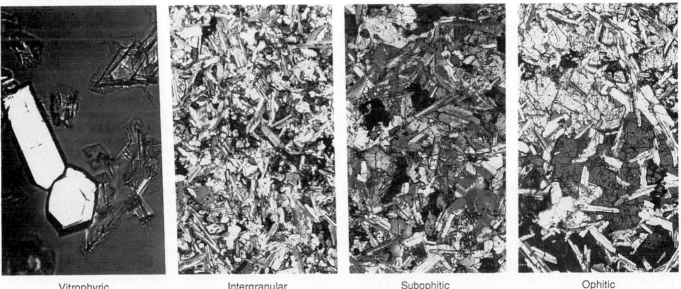

Vitrophyric   Intergranular   Subophitic   Ophitic

FIGURE 12-11 Textural variation from margin (left) to middle (right) of Mesozoic diabase dike, Hartford Basin, Connecticut. At the margin, olivine microphenocrysts were quenched in a glass that contains dendritic crystals of plagioclase and augite; the texture here is vitrophyric. At slower cooling rates an intergranular texture developed with numerous pyroxene crystals occurring in each space between the plagioclase laths. At still slower cooling rates, a subophitic texture developed, with numbers of plagioclase and pyroxene crystals being approximately equal. At the slowest cooling rates, the number of pyroxene crystals is less than the number of plagioclase crystals, and the texture is ophitic. The width of field in photomicrograph of margin is 0.1 mm; in all others it is 2.5 mm.

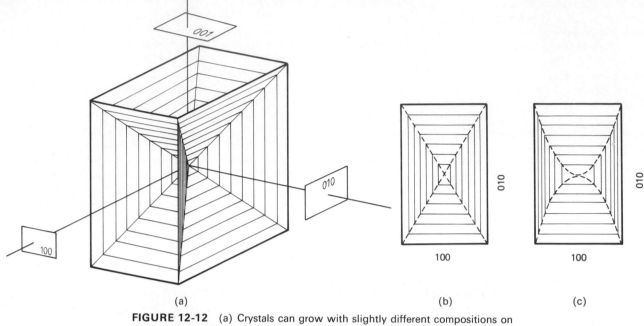

(a)   (b)   (c)

FIGURE 12-12   (a) Crystals can grow with slightly different compositions on crystallographically different faces, giving rise to sector zoning. The shapes of sectors depend on growth rates of adjoining faces. If these rates remain constant, the sector boundary is straight (b); if they change at different rates, the boundary is curved (c).

mafic alkaline sill near Montreal, Quebec. As the name "hourglass" suggests, the boundaries between sectors are curved. Gray found that the (010)–(100) sector boundary is a parabolic cylinder (Fig. 12-12c). This indicates that as the crystal grew, the rate of advance of the (010) face decreased relative to that of the (100) face. Because these changes produce a parabolic sector boundary, they must be systematic. Indeed, if the (010) face grew by a diffusion-controlled mechanism and the (100) face by a phase-boundary reaction mechanism, the resulting sector boundary would be parabolic (Gray 1971).

If the growth of the (010) face is diffusion-controlled, we can obtain an expression for the distance it advances (x_{010}) in time t by integrating Eq. 12-7 to give $(x_{010})^2/2k_D = t$. The constant of integration will be zero if the crystal at time zero is approximated by a point. Similarly, the distance the (100) face advances (x_{100}) in time t can be obtained by integrating Eq. 12-8 to give $x_{100}/k_R = t$. Any point on the sector boundary must represent a simultaneous solution to these two equations; that is, the time in both equations must be the same. Combining these two equations, we obtain the following expression for the sector boundary:

$$(x_{010})^2 = 2\left(\frac{k_D}{k_R}\right)x_{100} \qquad (12\text{-}14)$$

This is the equation of a parabola ($y^2 = cx$) and would thus explain the shape of the boundary between the (010) and (100) sectors. Any other combination of rate-controlling mechanisms would have produced a sector boundary of a different shape (Prob. 12-2).

If the rate-determining process for growth on a face of a sector-zoned crystal changes during the growth of the crystal, a discontinuity forms on the sector boundary. This can be illustrated by considering the shape of the sectors formed by growth zones in a plagioclase crystal grown in experiments by Lofgren (1974, Plate 3A; 1980, Fig. 9A). A plagioclase melt of An_{50} composition containing 10 wt % H_2O at a pressure of 0.5 GPa was cooled to successively lower temperatures in 50°C increments. While the temperature remained constant at each step, a normally zoned growth increment formed on each face of the crystal. The shape of the crystal at the end of each growth period is evident from the zoning (Fig. 12-13).

Consider first the sector boundary between (001) and (010). In thin section this boundary is a straight line over most of its length, indicating that the rate-determining growth step on both faces was the same or had the same rate dependency. During the final growth period, however, rates on the two faces changed differently, as evidenced by the fact that the outer corner of the crystal does not lie on the line passing through the corner at earlier stages of growth. This deflection of the sector boundary could result from the (001) face growing more slowly or the (010) face growing more rapidly. If it was the latter, however, we might expect to see an inflection of the boundary between (010) and (00$\bar{1}$) sectors at this same cooling step, and we do not. It is likely, therefore, that the growth rate on the (001) face slowed relative to that on the (010) face near 900°C, indicating a change in the rate-determining step for growth on the (001) face.

A change in the rate-determining step for growth on the (00$\bar{1}$) face can similarly be argued to have occurred when the temperature dropped below 1000°C. On this face, however,

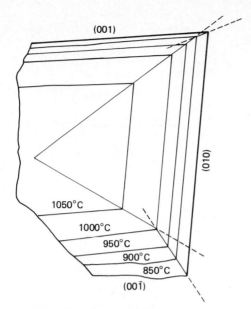

FIGURE 12-13 Sector zoning in plagioclase crystal grown in experiments by Lofgren (1974, Plate 3A; or 1980, Fig. 9A). Changes in the rate-determining mechanisms of growth on the various faces are indicated by the discontinuities in the lines drawn through the corners of the crystal at various stages of growth. See text for discussion.

served in the zoning (Tracy 1982) and grain size distribution (Kretz 1973) of this mineral. Most metamorphic garnets are strongly zoned, a feature not readily evident in thin section because garnet is isotropic and little color change accompanies the zoning. Compositional variation, however, is easily detected with electron microprobe analyses. Most garnets in pelitic rocks (rocks derived from shale), for example, have a manganese-rich core and iron-rich rim.

Of importance to interpreting nucleation and growth mechanisms from zoning is the fact that once a garnet has formed its composition cannot easily be changed, even by subsequent contact metamorphism. Diffusion rates in garnets are so low that steep compositional gradients show no signs of being lowered during most metamorphic events. As a first approximation, therefore, the composition at any point in a garnet crystal can be assumed to have remained unchanged since that part of the crystal grew.

Garnet can form by a number of reactions, the simplest of which is

$$\text{(chlorite)} \qquad\qquad \text{(quartz)}$$
$$(Fe, Mn)_9Al_3(Al_3Si_5)O_{20}(OH)_{16} + 4SiO_2$$
$$\qquad\text{(garnet)} \qquad\qquad \text{(water vapor)}$$
$$= (Fe, Mn)_9Al_6Si_9O_{36} + \quad 8(H_2O) \qquad (12\text{-}15)$$

In most rocks, chlorite and garnet would also contain magnesium, but for simplicity we will restrict the discussion to the two components Fe and Mn. Chlorite and garnet both belong to solid solution series; the reaction, therefore, takes place over a range of temperature (Fig. 12-14). The cryoscopic equation (Eq. 10-13) developed in Sec. 10-9 to express the equilibrium between a solid solution and a liquid solution is equally applicable to the equilibrium between two solid solutions. If the chlorite and garnet are assumed to form ideal solutions, we

the growth rate must have increased relative to that on (010) rather than decreased as it did on (001). Of course, because plagioclase is noncentrosymmetric, the growth-controlling mechanisms on (001) and (00$\overline{1}$) are not required by symmetry to be the same.

Considerable information on the nucleation and growth mechanisms of garnet crystals in metamorphic rocks is pre-

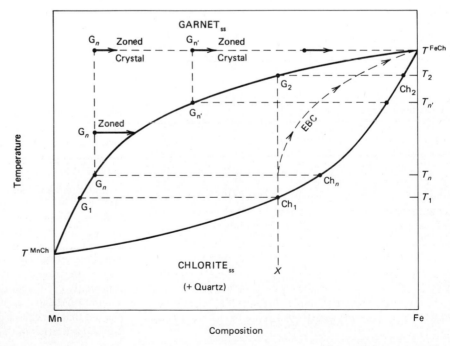

FIGURE 12-14 Temperature–composition plot of the reaction of chlorite solid solution with quartz to form garnet solid solution. See text for discussion.

can write

$$\ln \frac{X^{\mathrm{G}}\mathrm{Fe}}{X^{\mathrm{Ch}}\mathrm{Fe}} = \frac{\Delta H^{\mathrm{FeCh}}}{R}\left(\frac{1}{T^{\mathrm{FeCh}}} - \frac{1}{T}\right) \qquad (12\text{-}16)$$

and

$$\ln \frac{X^{\mathrm{G}}\mathrm{Mn}}{X^{\mathrm{Ch}}\mathrm{Mn}} = \frac{\Delta H^{\mathrm{MnCh}}}{R}\left(\frac{1}{T^{\mathrm{MnCh}}} - \frac{1}{T}\right) \qquad (12\text{-}17)$$

where ΔH^{FeCh} and ΔH^{MnCh} are the enthalpies of reaction to form garnet from pure Fe chlorite and pure Mn chlorite, respectively, and T^{FeCh} and T^{MnCh} are the temperatures of these reactions for the pure Fe and Mn end members, respectively. The resulting phase equilibrium diagram (Fig. 12-14) looks identical to the melting loops for olivine and plagioclase except that only solid solutions are involved in this case.

Under equilibrium conditions, heating of a rock containing quartz and chlorite with an Fe/Fe + Mn ratio of X in Figure 12-14 results in the formation of relatively Mn-rich garnet (G_1 in Fig. 12-14) at temperature T_1. As the temperature rises, the chlorite and garnet both become progressively enriched in Fe, reaching compositions Ch_2 and G_2, respectively, at temperature T_2 where, according to the lever rule, the last chlorite is consumed. But garnet crystals are zoned, and therefore equilibrium is clearly not achieved. As stated above, garnets, once formed, do not change their composition significantly. Consequently, the reaction from chlorite to garnet will approach complete fractionation. A more likely heating path for rock of composition X in Figure 12-14 would therefore be as follows. The rock must first be superheated to temperature T_n before garnet crystals nucleate. These crystals will again be Mn-rich (G_n) but not quite as rich as those first formed by equilibrium crystallization (G_1). With rising temperature the garnet becomes more Fe-rich, but the initial Mn-rich cores still remain because diffusion is unable to homogenize the crystals. A zoned crystal is therefore produced. Only the outer growing surface of the crystal has the equilibrium composition for a given temperature. This causes the effective bulk composition of the rock (EBC in Fig. 12-14) to become progressively enriched in Fe, and the garnet becomes zoned to rims of the pure Fe end member; chlorite in contact with the garnet also progressively changes its composition toward the Fe end member.

Because diffusion rates in garnet are so low and compositions remain essentially unchanged from the time of growth, the type of zoning that will result can be predicted quite accurately. The change from chlorite to garnet involves elements other than Fe and Mn (Mg and Ca, for example); indeed, the amount of Mn in the rock may be quite small. Let us assume that at a given temperature a garnet of some bulk composition is in contact with chlorite (Fig. 12-15a), and that the composition of the surface layer of garnet is in equilibrium with the chlorite. We will further assume that the surface layer of garnet has a concentration of Mn of $c_{\mathrm{Mn}}^{\mathrm{G}}$, which is related to the concentration of Mn in the chlorite, $c_{\mathrm{Mn}}^{\mathrm{Ch}}$, by the *Nernst distribution coefficient* (Sec. 13-11), $K^{\mathrm{G/Ch}}$, which is defined as

$$K^{\mathrm{G/Ch}} = \frac{c_{\mathrm{Mn}}^{\mathrm{G}}}{c_{\mathrm{Mn}}^{\mathrm{Ch}}} \qquad (12\text{-}18)$$

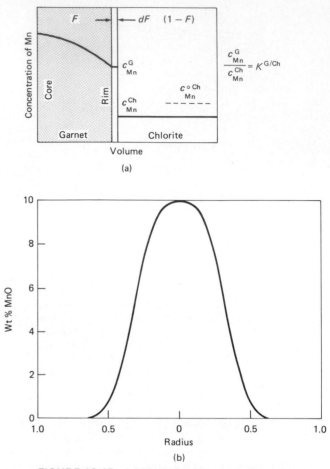

(a)

(b)

FIGURE 12-15   (a) Model for the growth of a zoned garnet crystal from chlorite plus quartz by Rayleigh fractionation of Mn. F is the fraction of garnet formed; $(1 - F)$ is the fraction of chlorite remaining; dF is a small growth increment, in which the concentration of Mn is related to the concentration of Mn in the chlorite by the Nernst distribution coefficient (see text for discussion). (b) Typical concentration profile of Mn across a garnet crystal that has formed by Rayleigh fractionation.

Because of the low diffusion rates, the Mn in each growth increment of garnet will be effectively removed from the system. To determine the zoning pattern that will develop in the garnet, we must be able to calculate how $c_{\mathrm{Mn}}^{\mathrm{G}}$ will change during growth of the crystal.

If we let the fraction of garnet formed (Eq. 12-15) at a given time be F, the amount of remaining chlorite will be $(1 - F)$ (Fig. 12-15). The total amount of Mn in the chlorite at this time will be $c_{\mathrm{Mn}}^{\mathrm{Ch}}(1 - F)$, which from Eq. 12-18, can be written as $(c_{\mathrm{Mn}}^{\mathrm{G}}/K^{\mathrm{G/Ch}})(1 - F)$. Note that $c_{\mathrm{Mn}}^{\mathrm{G}}$ refers to the concentration of Mn in the surface layer of garnet, for it is here only that equilibrium exists—the core of the crystal may have a very different composition. If the garnet grows by a small increment, dF, the amount of Mn that enters that increment must be equal to the amount that the chlorite loses: that is, $-d[(c_{\mathrm{Mn}}^{\mathrm{G}}/K^{\mathrm{G/Ch}})(1 - F)]$. The concentration in the growth increment, $c_{\mathrm{Mn}}^{\mathrm{G}}$, must therefore be

$$c_{Mn}^{G} = \frac{-d[(c_{Mn}^{G}/K^{G/Ch})(1-F)]}{dF}$$

On rearranging,

$$-K_{Mn}^{G/Ch}c_cG = \frac{dc_{Mn}^{G}}{dF} - c_{Mn}^{G} - F\left(\frac{dc_{Mn}^{G}}{dF}\right)$$

Multiplying through by dF and rearranging, we obtain

$$\frac{\dfrac{dc_{Mn}^{G}}{c_{Mn}^{G}}}{1-K} = \frac{dF}{1-F}$$

Integration of this expression gives

$$\frac{\ln c_{Mn}^{G}}{1-K} = -\ln(1-F) + \text{constant} \qquad (12\text{-}19)$$

When $F = 0$, $c_{Mn}^{G} = c_{Mn}^{\circ Ch}K^{G/Ch}$, where $c_{Mn}^{\circ Ch}$ is the initial concentration of Mn in chlorite before garnet started to form. The constant of integration is therefore

$$\text{constant} = \frac{\ln(c_{Mn}^{\circ Ch}K^{G/Ch})}{1-K}$$

Equation 12-19 then becomes

$$\ln c_{Mn}^{G} = \frac{-\ln(1-F) + [\ln(c_{Mn}^{\circ Ch}K^{G/Ch})]}{1-K}(1-K)$$

which simplifies to

$$c_{Mn}^{G} = c_{Mn}^{\alpha Ch}K^{G/Ch}(1-F)^{(K-1)} \qquad (12\text{-}20)$$

Equation 12-20 expresses how the Mn concentration in the surface layer of a garnet crystal changes as a function of the fraction of garnet formed, the Mn concentration in the initial chlorite, and the Nernst distribution coefficient for Mn between garnet and chlorite. After converting the fraction of garnet formed, which is a volume term, to crystal radius, we obtain a concentration gradient of the form shown in Figure 12-15b (Prob. 12-3). Gradients such as this are common in garnet crystals [e.g., Hollister (1966)].

This particular growth model, which is known as a *Rayleigh fractionation* process, has certain limitations. For example, it requires that diffusion of Mn to the garnet through the chlorite not be the rate-determining step. If Mn could not diffuse to the garnet rapidly enough, the chlorite near the garnet would become depleted in Mn, and the Mn content of the garnet would drop more rapidly than would otherwise happen. Eventually, when the Mn did diffuse to the garnet, it would produce a Mn-rich rim, a feature that is common to many garnets. Equation 12-20 also treats the Nernst distribution coefficient as a constant, but if the temperature changes during the reaction, as is likely during progressive metamorphism, the

distribution coefficient may change. More complicated growth models have been developed (Tracy 1982), but the Rayleigh fractionation model provides a simple explanation of the zoning process.

From Figure 12-14 it is clear that if during the growth of zoned garnet crystals—say, at temperature T_n—a second batch of crystals nucleates, their core composition would be poorer in Mn ($G_{n'}$) than the cores of the first-nucleated crystals (G_n). The core composition of the later-formed crystal could be used to determine at what stage in the growth of the first crystal the later one nucleated. This is illustrated in Figure 12-16a by the line joining the core composition of the late-formed crystal to the point of identical composition in the zoning profile of the early-formed crystal. Also, because the second generation of crystals would have had less time to grow than the first, they would be recognizable in the rock as the smaller crystals.

If we were to compare the zoning profiles in the first and second generations of garnet crystals, we would expect the zoning in the second crystals to match that in the outer part of

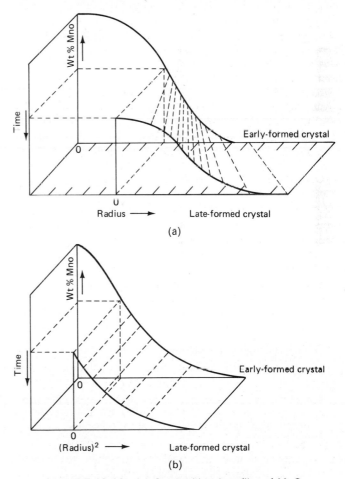

(a)

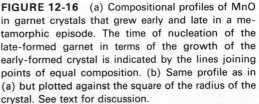

(b)

FIGURE 12-16 (a) Compositional profiles of MnO in garnet crystals that grew early and late in a metamorphic episode. The time of nucleation of the late-formed garnet in terms of the growth of the early-formed crystal is indicated by the lines joining points of equal composition. (b) Same profile as in (a) but plotted against the square of the radius of the crystal. See text for discussion.

the first crystals. Kretz (1973), who measured the compositional profiles across garnets of different size from a single hand specimen, found that despite similar compositions, the gradients were different. He found that early-formed crystals grew proportionately less toward the end of their growth than did later-formed crystals in this same growth period, where the growth periods were defined by identical compositional changes (Fig. 12-16a).

This correlation between diminishing growth rates and the time of nucleation can be explained only if diffusion were the rate-controlling crystal growth process (Eq. 12-6). If a phase boundary reaction had controlled growth, the rate would have been independent of the size of the crystals (Eq. 12-8), and if surface nucleation had controlled growth, the larger crystals would have grown faster (Eq. 12-9). With diffusion-controlled growth, the rate of change in the radius of a crystal is, according to Eq. 12-6, $dr/dt = K(D/t)^{1/2}$, where K is a constant and D is the diffusion coefficient. Upon integrating, we obtain

$$r = \text{constant} \times \sqrt{t}$$

which if squared and differentiated with respect to time gives

$$\frac{dr^2}{dt} = \text{constant} \qquad (12\text{-}21)$$

The squared term in Eq. 12-21 indicates that for diffusion-controlled growth the rate of increase of a crystal's surface area is a constant. If the zoning profiles in early- and late-formed garnet crystals in Figure 12-16a are plotted against the square of the radius, the profiles are seen to be the same. By this means, Kretz (1973) demonstrated that diffusion through the surrounding rock had been the rate-determining process for garnet growth in the sample he studied. He also showed, through the range of crystal sizes and core compositions, that nucleation of garnet had occurred not just at one time but throughout much of the growth period of garnet in that rock.

12-5 EQUILIBRIUM SHAPE OF CRYSTALS

Although the nucleation and initial growth of crystals are controlled by kinetic factors, with sufficient time and appropriate environment, crystals may develop forms that tend to minimize their surface free energy. Such shapes are independent of kinetics and are described as equilibrium forms. This process, first proposed by Gibbs (1875), can be expressed as

$$a_1\gamma_1 + a_2\gamma_2 + \cdots + a_n\gamma_n \text{ tends to a minimum} \quad (12\text{-}22)$$

where a_1 is the area of face number 1 and γ_1 is the surface free energy per unit area of that face. The surface free energy is defined as the change in free energy accompanying the creation of new surface area; that is,

$$\gamma_i = \left(\frac{\partial G}{\partial a_i}\right)_{T,P,n} \qquad (12\text{-}23)$$

Each face of a crystal has a surface free energy which is determined by the structure of the mineral on that face and the material with which that face comes in contact. Strongly anisotropic minerals, such as mica, have very different surface energies on different faces. By adjusting the areas of these faces, a crystal can minimize its total surface free energy.

Pierre Curie (1885) showed that for a crystal that had minimized its surface free energy, the perpendicular distance from a crystal face to the crystal's center (d_i) divided by the surface free energy of that face is a constant for that crystal; that is,

$$\frac{d_1}{\gamma_1} = \frac{d_2}{\gamma_2} = \cdots = \frac{d_n}{\gamma_n} = \text{constant} \qquad (12\text{-}24)$$

This expression later became known as *Wulff's theorem*, named after one of its chief proponents. It shows that the higher the surface free energy of a crystal face, the farther that face is from the center of the crystal, which in turn, because of geometrical considerations, means that this face is smaller than the other faces that have lower surface energies (Fig. 12-17a).

Consider, for example, the mineral staurolite, which in micaceous schists typically forms long prisms that have cross sections similar to the one shown in Figure 12-17a. The cross section shown in Figure 12-17b is also possible but is certainly not common. The difference between these two crystals, which have identical volumes (same areas in cross section), is in the relative sizes of the (110) and (010) faces. If the (110) face has a lower surface free energy than the (010) face, the crystal in Figure 12-17a would clearly have a lower total surface free energy than the one in Figure 12-17b. According to Wulff's theorem, $d_{110}/\gamma_{110} = d_{010}/\gamma_{010}$ (Fig. 12-17), or $\gamma_{010}/\gamma_{110} = d_{010}/d_{110}$; that is, the surface free energy on the (010) face relative to that on the (110) face is greater by a factor of d_{010}/d_{110}, which in this case is 1.7. The independent values of the surface free energies, however, cannot be determined from this relation.

In comparing the total surface free energies of the crystals in Figure 12-17, one might ask: Could the surface free energy be lowered still more by completely eliminating the (010) face? But note that as the (010) face is made smaller, the (110) face has to be extended, and the total surface area of the crystal actually increases. The more equant a crystal is, the smaller is its surface area for a given volume. The crystal in Figure 12-17b, for example, has a smaller surface area than the one in Figure 12-17a. To achieve the equilibrium shape, therefore, a crystal must strike a balance between eliminating high-energy faces and keeping the total surface area of the crystal small.

Little is known about the actual values of surface free energy on common minerals, but there is little doubt that they play an important role in determining the shape of many metamorphic minerals and the ways in which these minerals are intergrown. In a rock, equilibrium requires that the surface free energy of the rock as a whole tends toward a minimum. Wulff's theorem is, therefore, not directly applicable to most minerals because it considers a single crystal in isolation. In a rock, however, some mineral grains are bounded by faces that would not be expected if that grain were considered in isolation. It is unlikely, for geometrical reasons, that all minerals in a rock can be bounded by crystal faces. In striving for an equi-

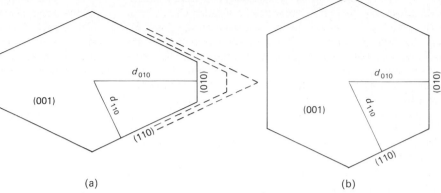

FIGURE 12-17 Cross sections of staurolite crystals showing common (a) and uncommon (b) relative sizes of crystal faces. The smaller size of the (010) face in common crystals indicates that this face has a higher surface free energy than does the (110) face. According to Wulff's theorem, the ratio of the surface free energies on (010) and (110) is given by the ratio of the lengths of the lines d_{010} and d_{110}.

(a)   (b)

librium texture, some minerals develop prominent crystal faces while others tend to fill in around the euhedral crystals. The effect of this is to decrease the surface area of the euhedral ones and increase that of the anhedral ones. Clearly, euhedral minerals must have greater surface free energies and thus need to be bounded by the faces that minimize their contribution to the total surface free energy of the rock. At the same time, however, the development of euhedral faces on one mineral means that another mineral with low surface energy is probably increasing its surface area and increasing the surface area of the grains in the rock as a whole. Equilibrium, then, must strike a balance between the development of crystal faces on some minerals and increasing the total grain surface area.

Based on petrographic experience, metamorphic minerals can be arranged in what is known as the *crystalloblastic series* (Table 12-1). A mineral can develop euhedral faces against any mineral below it in the series. The series must therefore reflect decreasing surface free energies from top to bottom.

Because surface free energies depend on the way in which crystal structures come together across grain boundaries, the total surface free energy of a rock can be lowered if grains are able to modify their relative orientations. Mica crystals, for example, when oriented with their basal planes parallel, as occurs in schists, create a lower surface free energy than when the crystals are randomly oriented. This factor must play some role in the development of schistosity, but the direction of preferred orientation in a schist is, of course, determined by the stress field extant during crystallization. Some minerals actually nucleate on others in special crystallographic orientations that minimize the surface free energy between them. Such growth is described as *epitaxial*. A special case of epitaxy involves the growth of exsolution lamellae in such minerals as pyroxene, amphibole, and feldspar (Robinson et al. 1971).

Many grain boundaries in rocks are not rational crystallographic planes. For example, the essentially monomineralic metamorphic rocks quartzite and marble typically consist of anhedral polygonal grains. Such rocks commonly have a striking texture known as *granoblastic-polygonal*, where grain boundaries meet at angles of approximately 120° to form triple junctions (Fig. 12-18). This texture, which is also common in annealed metals (Fig. 12-4), can be explained in terms of surface free energies.

Commonly, surface free energies are discussed in terms of *surface tension*, which is the force necessary to stretch a surface. Such usage seems natural when discussing the close packing of soap bubbles, but its application to rigid crystals, while valid, is not inherently obvious. Nonetheless, this means of treating surface free energies provides a simple means of analyzing the angles at which grains come together.

TABLE 12-1

Crystalloblastic series

Magnetite, Rutile, Sphene, Pyrite
Sillimanite, Kyanite, Garnet, Staurolite, Tourmaline
Andalusite, Epidote, Zoisite, Forsterite, Lawsonite
Amphibole, Pyroxene, Wollastonite
Mica, Chlorite, Talc, Prehnite, Stilpnomelane
Calcite, Dolomite, Vesuvianite
Cordierite, Feldspar, Scapolite
Quartz

FIGURE 12-18 Quartzite exhibiting granoblastic-polygonal texture. Most grain boundary intersections meet at angles of approximately 120°, and most grains are about the same size, indicating a close approach to an equilibrium texture. The width of field is 2 mm.

Consider three grains of quartz, each with different crystallographic orientations, meeting at a triple junction. The surface free energies on the three boundaries are γ_1, γ_2, and γ_3, and the angles between grains opposite these boundaries are θ_1, θ_2, and θ_3, as shown in Figure 12-19a. If we treat these surface energies as tensions pulling at the triple junction, their magnitudes and directions can be represented by vectors. Then, according to the *law of sines*,

$$\frac{\gamma_1}{\sin \theta_1} = \frac{\gamma_2}{\sin \theta_2} = \frac{\gamma_3}{\sin \theta_3} \qquad (12\text{-}25)$$

If the three surface tensions are equal, so must be the three sines, in which case $\theta_1 = \theta_2 = \theta_3 = 360°/3 = 120°$. Of course, surface tensions cannot literally pull the grain boundaries into this equilibrium shape. However, if the quartz is able to diffuse along grain boundaries, especially where aided by a fluid phase, solution and redeposition can result in the triple junction, adjusting its shape to the equilibrium form.

The angles between grain boundaries at triple junctions will be 120° only when the surface free energies on these boundaries are all the same. Based on the frequency of occurrence of such junctions in quartzite, marble, and some amphibolites, surface free energies on quartz, calcite, and amphibole cannot be strongly dependent on crystallographic orientation. Some minerals, such as mica, however, have very different surface energies on different faces, and these minerals may not develop 120° triple junctions, nor do minerals that are surrounded by grains with a very different surface free energy, such as magnetite in quartz.

Consider, for example, the situation illustrated in Figure 12-19b, where the mineral subtending the angle θ_3 at the triple junction is different from the mineral that forms the other two grains. The angles θ_1 and θ_2 are equal, because they involve the same mineral boundaries and hence the same surface energies. Consequently, $\theta_1 = \frac{1}{2}(360° - \theta_3)$. According to the sine law,

$$\frac{\gamma_3}{\sin \theta_3} = \frac{\gamma_1}{\sin \theta_1} = \frac{\gamma_1}{\sin(180 - \frac{1}{2}\theta_3)} = \frac{\gamma_1}{\sin \frac{1}{2}\theta_3}$$

from which it follows that

$$\frac{\gamma_3}{\gamma_1} = \frac{\sin \theta_3}{\sin \frac{1}{2}\theta_3} = 2\cos \frac{1}{2}\theta_3 \qquad (12\text{-}26)$$

This equation, then, relates the *dihedral angle*, θ_3, to the ratio of the surface energies on the two types of grain boundary. Kretz (1966) found that the dihedral angle formed by pyroxene grains contacting pairs of scapolite grains in a high-grade metamorphic rock is 128°. According to Eq. 12-26, the surface free energy on the scapolite–pyroxene boundaries is therefore 1.14 times that on the scapolite–scapolite boundaries.

We have seen from Eq. 12-25 that if surface free energies on all grain boundaries in a rock are equal, the angles between grain boundaries at triple junctions will tend toward 120° at equilibrium. If all grains have hexagonal shapes in cross section, this angular requirement can easily be satisfied. But with polygonal grains having more or less than six sides in cross section, 120° triple junctions can exist only if grain boundaries are curved (Fig. 12-20). Grains with fewer than six sides will be bounded by faces that are concave inward, and those with more than six sides will be bounded by faces that are concave outward.

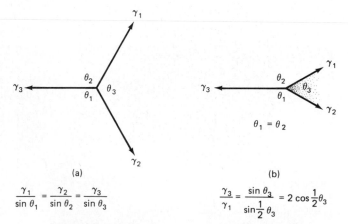

$$\frac{\gamma_1}{\sin \theta_1} = \frac{\gamma_2}{\sin \theta_2} = \frac{\gamma_3}{\sin \theta_3}$$

$$\frac{\gamma_3}{\gamma_1} = \frac{\sin \theta_3}{\sin \frac{1}{2}\theta_3} = 2\cos \frac{1}{2}\theta_3$$

FIGURE 12-19 (a) The angles between grain boundaries meeting at a triple junction are related through the law of sines to the surface free energies between the grains. These energies can be expressed as tensions acting on the triple point. At equilibrium these forces must balance. If the surface free energies on all three boundaries are the same, the angles between the boundaries will be equal, that is, 120°. (b) At a triple junction involving two different minerals, the ratio of the surface free energies between the like grains (γ_3) and the unlike grains (γ_1 and γ_2) is given by twice the cosine of half the dihedral angle subtended by the odd mineral.

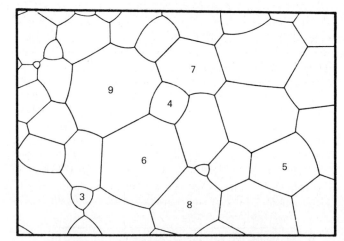

FIGURE 12-20 All grain boundaries in this illustration intersect at angles of 120°. Six-sided polygons tend to be surrounded by planar faces. Polygons with more than six sides are surrounded by concave outward faces; those with fewer than six sides have convex outward faces. Small grains with convex outward faces are, with time, consumed by larger grains with concave outward faces.

An important consequence of curved grain boundaries is that pressure differences develop across them which provide a driving force to eliminate small grains and make large ones grow. Consider for a moment the pressure difference between the inside (P_i) and outside (P_o) of a small gas bubble in a lava. This difference determines the force that must be applied if the bubble is to be inflated. The work done in changing the volume at this pressure difference, $(P_i - P_o)\,dV$, must equal the surface free energy times the increase in area of the bubble; that is,

$$(P_i - P_o)\,dV = \gamma\,dA \qquad (12\text{-}27)$$

We can differentiate the expressions for the volume ($V = \frac{4}{3}\pi r^3$) and area ($A = 4\pi r^2$) of a sphere with respect to radius to obtain $dV = 4\pi r^2\,dr$ and $dA = 8\pi r\,dr$. Substituting these into Eq. 12-27 gives

$$P_i - P_o = \frac{2\gamma}{r} \qquad (12\text{-}28)$$

This relation indicates that the pressure difference between the inside and outside of a bubble is inversely proportional to the radius of curvature of the bubble. Thus, when a bubble is very small, the pressure in the bubble is very much greater than that outside. This explains why bubbles have difficulty nucleating (Prob. 12-5).

Pressure differences across curved boundaries between crystalline phases are also given by Eq. 12-28. Because the radius of curvature of small grains is less than that of larger ones, the pressure in small grains is greater than that in large ones. From Eq. 7-32 we know that increased pressure raises the free energy of a phase $[(\partial G/\partial P)_T = \bar{V}]$. Thus small grains are metastable with respect to larger grains and will, as equilibrium is approached, be eliminated in favor of the larger ones. This process accounts for the general coarsening of metamorphic rocks during recrystallization.

In some rocks a directed pressure (stress) can modify the shapes of grains by causing solution or diffusion of material from points of high pressure to ones of low pressure. This is known as *Riecke's principle*. Its effects are most clearly seen in deformed clastic sedimentary rocks. Quartz grains, for example, on being forced together dissolve at points of contact; the dissolved silica is transported through pore fluids and precipitated in crystallographic continuity on the sides of the grains where the pressure is less (Fig. 18-1). Transfer of material from the sides of grains where the compressive stress is at a maximum to the sides where it is a minimum causes grains to become flattened, and the rock develops a prominent foliation normal to the direction of maximum compressive stress (see Sec. 18-2).

12-6 SURFACE FREE ENERGY AND WETTING OF CRYSTALS BY MAGMA

The way in which water beads on a freshly waxed and polished car but spreads out evenly on an unwaxed surface is a familiar example of the "wetting" ability of a liquid. It is wetting that causes water to rise in a glass capillary and mercury to sink.

The way in which a liquid wets a surface is determined by the tendency of the system to minimize its surface free energy.

The Si-rich liquid mesostasis of many tholeiitic basalts contains droplets of immiscible Fe-rich liquid (Philpotts 1982) that commonly nucleate and grow as hemispheres on the surface of plagioclase crystals (Fig. 12-21a). The fact that these droplets do not wet the surface of the plagioclase indicates that such spreading would increase the surface free energy of the rock as a whole. We can conclude, therefore, that the surface free energy between plagioclase and Si-rich liquid must be less than that between Fe-rich liquid and plagioclase. But can we tell from the shape of the droplets how different these surface energies are?

Figure 12-21b is a schematic representation of the point of contact of the boundary between the immiscible liquids and the plagioclase crystal. The surface free energies on the three different boundaries are γ_{Fe-Si} on the immiscible liquids boundary, γ_{Fe-Pl} on the Fe-rich liquid–plagioclase boundary, and γ_{Si-Pl} on the Si-rich liquid–plagioclase boundary. Each of these can be represented by a vector acting on the triple junction, which is assumed to be able to move to achieve equilibrium. Unlike the case considered in Figure 12-19, however, the face of the plagioclase crystal can be considered a rigid plane which the deformable liquid boundary is able to move along. If at

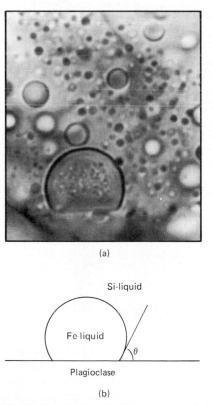

(a)

Si-liquid

Fe-liquid

θ

Plagioclase

(b)

FIGURE 12-21   (a) Droplet (20 μm diameter) of Fe-rich immiscible liquid on the surface of a plagioclase crystal in contact with Si-rich liquid in a tholeiitic basalt. (b) Schematic representation of the droplet in (a) showing the angle θ formed between the immiscible liquid boundary and the surface of the plagioclase crystal.

equilibrium the immiscible liquid boundary forms an angle of θ degrees with the crystal face, the surface tension on the two-liquid boundary produces a component of force of $\gamma_{Fe-Si} \cos \theta$ pulling in the same direction as the surface tension between plagioclase and Si-rich liquid. At equilibrium, the sum of these two forces must balance the surface tension between the Fe-rich liquid and the plagioclase; that is,

$$\gamma_{Fe-Si} \cos \theta + \gamma_{Si-Pl} = \gamma_{Fe-Pl}$$

or

$$\cos \theta = \frac{\gamma_{Fe-Pl} - \gamma_{Si-Pl}}{\gamma_{Fe-Si}} \qquad (12\text{-}29)$$

If $\gamma_{Fe-Pl} = \gamma_{Si-Pl}$, then $\theta = 90°$. If $\gamma_{Fe-Pl} < \gamma_{Si-Pl}$, then $\cos \theta < 0$ and $\theta > 90°$. In this case the Fe-rich liquid would tend to wet the plagioclase. If $\gamma_{Fe-Pl} > \gamma_{Si-Pl}$, then $\cos \theta > 0$ and $\theta < 90°$. In this case, the Si-rich liquid tends to wet the plagioclase. Because the angle θ on the Fe-rich droplet in Figure 12-21a is less than $90°$, we can conclude that $\gamma_{Fe-Pl} > \gamma_{Si-Pl}$ (Prob. 12-6).

The result that the surface free energy on the Si-rich liquid–plagioclase boundary is less than that on the Fe-rich liquid–plagioclase boundary is not surprising. The Si-rich liquid is highly polymerized and has a structure that is similar to that of the plagioclase. The amount of mismatch between the structures is therefore small and the surface free energy would

also be small. The Fe-rich liquid is less polymerized, and therefore its structure does not match as closely the framework structure of the plagioclase and a higher surface free energy results.

Some lamprophyres contain ocelli (small globules of felsic rock), which are probably formed as immiscible droplets of felsic liquid in mafic melt (Philpotts 1972). Where these droplets contact phenocrysts of amphibole or pyroxene, contact angles indicate that the mafic melt has a greater tendency to wet the surface of the ferromagnesian minerals than does the felsic melt (Fig. 12-22a). The surface free energy on the mafic liquid-crystal boundary is therefore less than on the felsic liquid-crystal boundary, which also would be expected from the structures of the phases involved.

A small pocket of mafic liquid trapped between the two hornblende phenocrysts in Figure 12-22a gives some indication of the relative surface free energies on different faces of the hornblende crystals. This pocket of liquid, which is shown schematically in Figure 12-22b, wets the (010) face of one crystal and the 011 face of the other. From the angle the droplet's surface makes with both crystal faces we can conclude that the mafic melt preferentially wet the hornblende crystal on both faces more than did the felsic melt. If it had not, the droplet would have a shape such as that shown in Figure 12-22c. If the hornblende showed only a slight preference to be wet by the mafic melt, a convex outward meniscus would still be possible

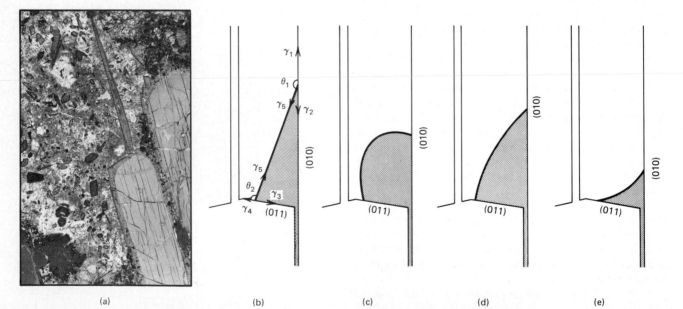

(a)   (b)   (c)   (d)   (e)

FIGURE 12-22   (a) Large, zoned hornblende phenocrysts protuding across the boundary of a felsic ocelli in a lamprophyre dike, Montreal, Quebec. The fine-grained mafic groundmass of the lamprophyric and the felsic material are interpreted to have formed as immiscible liquids. Note that the boundary of the ocelli, as marked by the presence of magnetite crystals in the mafic fraction, can be traced through the marginal zone of the hornblende phenocrysts. The width of field is 4 mm. (b) Schematic representation of the patch of "mafic liquid" trapped between the two large hornblende phenocrysts in (a). (c–e) illustrate other conceivable configurations that this patch of liquid could have taken had the wetting properties of the liquids with the hornblende been different. See text for discussion.

(Fig. 12-22d). The fact that the meniscus is almost straight indicates that on both crystal faces the hornblende has a strong preference to be wet by the mafic melt rather than by the felsic one. Even so, the meniscus can still have several different configurations. If the meniscus took the form shown in Figure 12-22e, we would conclude that the (011) face of the hornblende was wet more easily by the mafic melt than was the (010) face, and that the repulsion of the felsic liquid by the (010) face was less than it was by the (011) face. The actual shape of the meniscus (Fig. 12-22a and b) indicates that the reverse relations must be true. Using the nomenclature for the surface free energies on the different boundaries shown in Figure 12-22b, we can conclude that $\gamma_2 = \gamma_1 + \gamma_5 \cos \theta_1$, and $\gamma_3 = \gamma_4 + \gamma_5 \cos \theta_2$, from which it follows that

$$\frac{\gamma_1 - \gamma_2}{\gamma_4 - \gamma_3} = \frac{\cos \theta_1}{\cos \theta_2} \tag{12-30}$$

Because $\theta_1 > \theta_2$, the ratio of the cosines is greater than 1. Therefore,

$$(\gamma_1 - \gamma_2) > (\gamma_4 - \gamma_3) \tag{12-31}$$

The individual values of surface free energy on the different boundaries cannot be determined from Eq. 12-31. But because the meniscus is not convex outward, the surface free energies between mafic liquid and hornblende (γ_2 and γ_3) must be significantly less than those between felsic liquid and hornblende (γ_1 and γ_4). We can conclude from Eq. 12-31 that γ_1 is probably greater than γ_4. This would indicate that the structure of a highly polymerized felsic melt is able to attach more easily to the ends of double chains of silica tetrahedra exposed on the (011) face of the hornblende crystal than it is to the lengths of such chains exposed on the (010) prism faces.

The differences in wetting ability of mafic and felsic melts on different faces of hornblende (and pyroxene) crystals may explain the striking difference in morphology of these crystals in mafic and felsic rocks. In gabbros, hornblende and pyroxene commonly form stubby crystals, whereas in felsic rocks (granites, syenites) they tend to be much more acicular. Indeed, this difference is seen in the lamprophyre illustrated in Figure 12-22a, where the hornblende crystals in the mafic part of the rock have an *aspect ratio* (length/breadth) of approximately 3:1, whereas those in the ocelli have an aspect ratio of more than 10:1. One extremely long needle of hornblende extends in optical continuity from one of the stubby phenocrysts of hornblende into the ocelli.

If the mechanism controlling growth on the (010) and (011) faces on these crystals is a phase-boundary reaction, the greater ease with which the highly polymerized felsic melt attaches to the (011) face relative to its attachment on the (010) face would cause more rapid growth parallel to the c axis. In a mafic melt where the ratio of bridging to nonbridging oxygens requires that the liquid be largely polymerized into silica chains, structural units of melt may be able to attach themselves almost as easily on faces paralleling the c axis as they can on ones transverse to the c axis. Differences in growth rate on these faces would therefore not be so great, and the aspect ratio would be smaller.

Finally, wetting plays an important role in determining how easily melts formed by partial fusion in the upper mantle or lower crust are able to segregate and form bodies of magma that are large enough to rise (Sec. 22-4). Fusion of rock occurs first where grains of different composition come together (Fig. 10-5), for here is where the eutectic or peritectic mixtures of minerals occur. The initial melt phase can be distributed at grain boundaries in three different ways (Fig. 12-23). From Eq. 12-26, the dihedral angle produced by the liquid (θ) is expressed in terms of the surface free energy between the liquid and solids (γ_{LS}) and between the solids themselves (γ_{SS}) by

$$\cos \tfrac{1}{2}\theta = \frac{\gamma_{SS}}{2\gamma_{LS}} \tag{12-32}$$

If γ_{LS} is greater than γ_{SS}, the dihedral angle will be greater than 120°, and the liquid forms isolated pockets at four-grain corners (Fig. 12-23a). If γ_{LS} is less than γ_{SS} but greater than $\gamma_{SS}/\sqrt{3}$, the dihedral angle is between 120 and 60°, and the liquid partially penetrates along grain edges (Fig. 12-23b). If γ_{LS} is less than $\gamma_{SS}/\sqrt{3}$, the dihedral angle is less than 60°, and the melt is able to penetrate along the entire length of grain edges (Fig. 12-23c).

If partial melting gives rise to a liquid that does not wet the surface of grains ($\gamma_{LS} > \gamma_{SS}$), isolated pockets of melt formed at grain intersections (Fig. 12-23a) will not connect with each other until a large fraction of the rock has melted. Only then will the melt be able to segregate and form bodies of magma large enough to rise. On the other hand, if the melt easily wets the solids ($\gamma_{SS} > \gamma_{LS}\sqrt{3}$), any small amount of melting will

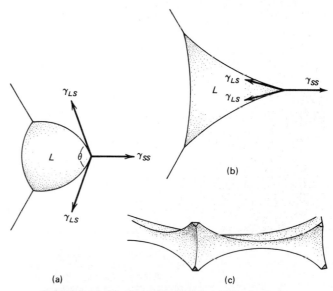

FIGURE 12-23 Partial fusion of rock first generates pockets of liquid at grain junctions where minerals of different composition come together. The shape of the pockets depends on the wetting of the minerals by the magma, which is indicated by the dihedral angle, θ. If this angle is greater than 120° (a), liquid does not penetrate along grain boundaries; if the angle is between 120 and 60°, partial penetration occurs (b); and if the angle is less than 60°, the liquid penetrates along the entire length of grain intersections (c).

form a continuous liquid phase along grain boundaries. In this case, magma may be extracted from the partially fused rock at very small degrees of partial melting. In this way the wetting ability of a magma plays an important role in determining the composition of bodies of magma (Sec. 22-4). The wetting ability of a magma also plays a role in determining how much residual magma can be expelled from the pore spaces between crystals that accumulate on the floor of a magma chamber (Sec. 13-5).

PROBLEMS

12-1. Plot a graph of the free energy of growth of a vapor bubble in a basaltic magma as a function of radius if the ΔG_{vol} for the formation of the bubble is -2.4×10^3 kJ m^{-3}, and the surface free energy of the bubble is 0.3 J m^{-2}. By differentiating Eq. 12-2, show that the critical radius is given by $r_c = -2\Delta G_{intf}/\Delta G_{vol}$, and determine the value of r_c. Does your answer agree with your graph?

12-2. If the growth on the (010) face of a crystal is controlled by a phase boundary reaction, and growth on the (100) face is controlled by surface nucleation, derive a mathematical expression for the sector boundary formed by the growth of these two faces. In integrating the growth equations, assume some finite size for the nucleus on which these faces grow. Draw a section through the crystal showing the sectors, and compare the result with the parabolic hourglass structure in Figure 12-12c. Would you be able to distinguish these two patterns in thin section? Would "martini glass" rather than "hourglass" be a better description of this zoning? [See, for example, Bryan (1972).]

12-3. Assuming a Rayleigh fractionation model for garnet growing from chlorite (+quartz), calculate the type of zoning that would result if the initial concentrations of MnO and FeO were 0.5 and 1.0 wt %, respectively, and the distribution coefficient between garnet and chlorite is 20 for MnO and 0.5 for FeO. Also, cal-culate the profile for an element whose concentration is initially 1.0 wt % and has a distribution coefficient (G/Ch) of 5.0.

12-4. Kretz (1966) found that the dihedral angle formed by grains of scapolite in contact with pairs of pyroxene grains in a metamorphic rock is 109°. What is the ratio of the surface free energies on the pyroxene–scapolite and pyroxene–pyroxene boundaries?

12-5. (a) What is the excess pressure in the bubble of Prob. 12-1 when it has the critical radius (0.25 μm)? The surface free energy is 0.3 J m^{-2}.

(b) If the molar volume of the vapor is 4×10^{-3} cm^3, what increase in the free energy of the vapor is caused by the excess pressure?

(c) If the entropy of the vapor phase is 280 J mol^{-1} K^{-1}, what degree of supercooling would be required to eliminate the increase in free energy due to the excess pressure in the bubble?

12-6. If the angle the surface of the Fe-rich droplet of immiscible liquid makes with the surface of the plagioclase crystal shown in Figure 12-21 is 64° and the surface free energy on the immiscible liquid boundary is 0.02 J m^{-2}, what is the difference between the surface free energies for the Fe-rich liquid–plagioclase boundary and the Si-rich liquid–plagioclase boundary?

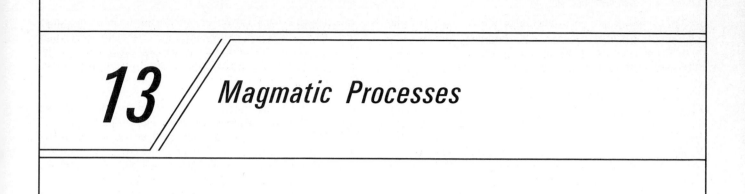

13 / Magmatic Processes

13-1 INTRODUCTION

Magmas that reach the Earth's surface to form lavas are highly varied, ranging in composition from ultramafic komatiites, through basalts and andesites, to rhyolites and feldspathoidal felsic rocks. Although the compositions of these lavas may not represent all magmas formed in the Earth (some may be too dense to rise to the surface, and others may require high pressures to keep volatile fluxes in solution), they do indicate the enormous diversity of magmas. Explaining the origin of this diversity has been the dominant goal of petrology.

Early in the history of the science most of the different magmas were thought to have independent origins; some were interpreted as the products of magma mixing, and still others the products of magma splitting (immiscible fractions). In this century, however, the trend has been to interpret the wide diversity of igneous rocks as being derivative from only a few primary magmas. The process by which these magmas are modified is known as *magmatic differentiation*. For example, N. L. Bowen (1928), who championed this new interpretation, argued that basalt was the primary magma from which other magmas were derived. So persuasive were his arguments that during the first half of this century, the interpretation that compositional variations in magmas might reflect primary variations in the source region was almost completely neglected. Although petrologists now recognize the importance of the source region, magmatic differentiation is still considered the major cause for variations in the composition of suites of igneous rocks.

Lavas erupted from a single volcano at different times can reasonably be interpreted to share a common origin. The same is true of different igneous rocks within a single pluton or even in separate, but closely related bodies, especially when the rock types share some striking geochemical signature. Igneous rocks that are related to a common source are said to be *comagmatic* or *consanguineous*. The latter term implies a clear genetic lineage; that is, one rock type is derived in some way from another. For example, on the basis of field evidence, geochemistry, or experimental work, a particular magma might be interpreted as being derivative from another magma, which we can refer to as a *parental magma*. That parent may, in turn, have been derived from another magma, and so on. As the lineage is traced backward toward the primary magma, magmas are said to become more *primitive*. Conversely, magmas late in the lineage are said to be *evolved* or *differentiated*. The *primary magma* itself is the one derived directly from the source region without being affected by any process of differentiation. Although easy to define, primary magmas are, in fact, very difficult to identify, because of the uncertainty in the composition of the source region and hence in the composition of magma with which it would be in equilibrium.

Given that a group of igneous rocks, and the magmas from which they were derived, are comagmatic, we may ask how their differentiation occurred. Undoubtedly, this is the most fundamental question in petrology today, yet it is far from being completely answered because, for the most part, magmatic processes occur at depths where they are not accessible to direct observation, or they take place at rates that are far slower than can be achieved in laboratory experiments. Nonetheless, igneous rocks preserve many textural and chemical characteristics that shed light on the differentiation processes involved in their formation. Few of these characteristics, however, are without multiple interpretations. Herein, then, lies the reason for much of the interest and excitement in studying petrology.

Many different processes have been invoked to account for magmatic differentiation. Some of these involve changes in the composition of magmas while entirely in the liquid state. The components of a magma, when placed in a potential field, may have to modify their concentrations to achieve an equilibrium distribution. The effect of the Earth's gravitational field in this respect is very small, but steep temperature gradients can cause a significant redistribution of elements by what is known as the Soret effect. Steep temperature gradients, however, can exist only at the extreme outer margins of magma bodies, and then only for short periods of time. The Soret effect is therefore likely to be only of local and transient importance. Some magmas are capable of splitting, into immiscible fractions, producing liquids of contrasting compositions. For the most part, however, such unmixing is restricted to magmas of evolved compositions and is not of importance to the fractionation of more primitive magmas. Thus, although differentiation does take place in the liquid state, it cannot be the main cause for the differentiation of igneous rocks.

Ever since the evidence was laid out so clearly by Bowen (1928) in his influential book *The Evolution of the Igneous Rocks*, most petrologists have accepted crystal-liquid fractionation as the major cause of igneous differentiation. As was evident from the discussion of phase diagrams in Chapter 10, the composition of a multicomponent liquid is changed dramatically by the crystallization of a mineral, especially when the composition of the mineral differs significantly from that of the liquid. What makes this process particularly attractive is that most lavas contain phenocrysts (magmas are not superheated), and the calculated composition of liquids that would be formed if the phenocrysts are removed commonly matches the composition of associated lavas. So convincing is this evidence, especially when documented for many different elements, that crystal fractionation is undoubtedly the dominant process of magmatic differentiation. How the crystals actually separate from the magma is, however, a hotly debated question.

As proposed by Bowen (1915b), crystals were thought to separate from their host magma by sinking or floating. This appears self-evident for minerals such as the ferromagnesian ones, which are considerably denser than the magmas in which they form. Other minerals, such as plagioclase in basaltic magma, however, may show no density contrast with the liquid. Furthermore, we now know that many crystallizing magmas behave as Bingham liquids, and thus even the dense ferromagnesian minerals may not be able to sink if they are unable to overcome the yield strength of the magma. For this reason, additional processes have been invoked, most of which involve the movement of magma. For example, *flowage differentiation* calls upon shear stresses in magma to help move crystals; *convecting* magma, on the other hand, can transport crystals in suspension to distant depositional sites; and *filter pressing* can separate liquid from crystals in much the same way that coffee is separated from coffee grinds. Each of these processes is likely to leave telltale evidence of its action.

Finally, the separation of a vapor phase from a magma can bring about marked differentiation, as evidenced by the high concentrations of certain rare elements in pegmatites (Sec. 11-6). Separation of a vapor phase, however, tends to occur in rather evolved magmas or during the final stages of crystallization of more primitive ones. It therefore cannot be responsible for the major trends of magmatic differentiation.

TABLE 13-1

Analyses of lavas from Kilauea, Hawaii*

	Analysis							
	1	2	3	4	5	6	7	8
SiO_2	48.05	48.43	47.92	48.21	49.16	49.20	49.71	50.10
TiO_2	2.04	2.00	2.16	2.24	2.29	2.57	2.68	2.71
Al_2O_3	10.33	10.70	10.75	11.37	13.33	12.77	13.65	13.78
Fe_2O_3	1.34	1.15	1.08	1.50	1.31	1.50	1.19	1.89
FeO	10.19	10.08	10.65	10.18	9.71	10.05	9.72	9.46
MnO	0.17	0.17	0.18	0.18	0.16	0.17	0.17	0.17
MgO	17.39	16.29	15.43	13.94	10.41	10.00	8.24	7.34
CaO	8.14	8.67	9.33	9.74	10.93	10.75	11.59	11.46
Na_2O	1.66	1.71	1.79	1.89	2.15	2.12	2.26	2.25
K_2O	0.36	0.35	0.44	0.44	0.51	0.51	0.54	0.57
P_2O_5	0.19	0.18	0.23	0.22	0.16	0.25	0.25	0.27
Total	99.86	99.73	99.96	99.91	100.12	99.89	100.00	100.00
M'^a	75.3	74.2	72.1	70.9	65.6	63.9	60.2	58.0
F	34.0	35.0	38.0	38.0	43.0	44.0	47.0	48.0
M	59.0	57.0	55.0	53.0	46.0	44.0	40.0	37.0
A	7.0	7.0	8.0	9.0	12.0	12.0	13.0	14.0
D.I.[b]	16.18	16.54	17.75	18.59	21.20	20.95	22.31	23.52
Q	0.00	0.00	0.00	0.00	0.00	0.00	0.00	1.11
Or	2.13	2.07	2.60	2.60	3.01	3.01	3.19	3.37
Ab	14.05	14.47	15.15	15.99	18.19	17.94	19.12	19.04
An	19.67	20.49	20.00	21.24	25.22	23.82	25.51	25.82
Cpx	15.60	17.14	19.95	20.65	22.66	22.61	24.74	23.87
Opx	21.12	21.12	14.87	15.99	14.45	18.66	16.92	18.28
Ol	21.04	18.57	21.20	16.51	9.98	6.21	3.13	0.00
Mt	1.94	1.67	1.57	2.17	1.90	2.17	1.73	2.74
Il	3.87	3.80	4.10	4.25	4.35	4.88	5.09	5.15
Ap	0.45	0.43	0.54	0.52	0.38	0.59	0.59	0.64

13-2 COMPOSITIONAL VARIATION IN SUITES OF VOLCANIC ROCKS

Lavas are important in the study of petrology because they provide samples of unequivocal magmatic liquids (or liquid plus phenocrysts). Many plutonic igneous rocks, although formed from magmas, are not, themselves, representative of the composition of liquids. They may, for example, be formed by the accumulation of early crystallizing minerals, as happens in the case of dunite, which is formed by the accumulation of olivine. Lavas from a single volcano may show progressive changes in composition both during a single eruptive episode or during successive eruptions. This compositional variation, then, is the most direct evidence we have of magmatic differentiation.

Almost any well-studied volcano can be used to illustrate the type of compositional variation found in volcanic rock suites. We will use, as an example, the Kilauea volcano, located on the southeastern side of the island of Hawaii. This basaltic shield volcano, which is very active (0.1 km^3 of magma per year), has been monitored continuously by the U.S. Geological Survey for many years. Not only are the compositions of its recent and historic eruptions well known, but geophysical monitoring of the volcano has provided much information about the magmatic plumbing and the movement of magma beneath the volcano.

As described in Sec. 4-3, seismic evidence indicates that the ultimate source of Kilauean lavas is at a depth of at least 60 km; from there magma rises into a shallow chamber directly beneath the summit caldera (Fig. 4-5). As this chamber fills, the summit of the volcano expands and does so until either a summit eruption occurs or the magma moves laterally into the East or South-West Rift Zones. Movement of magma into the rift zones is not always accompanied by an eruption, but when it is, the first magma to be extruded is not of the newest addition to the rift-zone but is derived from pockets of magma emplaced during earlier periods of activity.

Table 13-1 lists analyses of the full range of lavas erupted from Kilauea [mainly from Wright and Fiske (1971)]. These are all basaltic rocks ranging in composition from picrite (no. 1) to quartz tholeiite (no. 15). Rocks containing more than 7.0 wt % MgO contain phenocrysts of olivine (Fo_{87-85}), a typical analysis of which is given in no. 16. Rocks containing less than

TABLE 13-1 (Continued)

	Analysis							
	9	10	11	12	13	14	15	16^c
SiO_2	50.37	50.56	50.74	50.85	50.92	51.24	53.42	40.01
TiO_2	3.09	3.16	3.35	3.36	3.61	3.74	3.36	0.04
Al_2O_3	14.02	13.92	13.57	14.02	13.80	13.60	13.75	1.13
Fe_2O_3	1.88	1.78	1.36	1.90	1.85	1.87	1.96	0.30
FeO	10.07	10.18	10.63	10.44	10.71	11.19	10.45	12.33
MnO	0.17	0.18	0.18	0.18	0.19	0.18	0.18	0.17
MgO	6.75	6.33	6.16	5.68	5.46	5.12	3.92	44.77
CaO	10.39	10.24	9.94	9.71	9.45	9.03	7.75	1.23
Na_2O	2.35	2.61	2.69	2.77	2.80	2.81	3.34	0.01
K_2O	0.62	0.64	0.67	0.74	0.75	0.83	1.10	0.00
P_2O_5	0.32	0.33	0.37	0.38	0.40	0.41	0.59	0.00
Total	100.03	99.93	99.66	100.03	99.94	100.02	99.82	99.99
M'^a	54.4	52.6	50.8	49.2	47.6	44.9	40.1	86.6
F	51.0	52.0	53.0	53.0	54.0	56.0	56.0	22.0
M	34.0	32.0	31.0	29.0	28.0	26.0	21.0	78.0
A	15.0	16.0	17.0	18.0	18.0	18.0	24.0	0.0
D.I.^{b}	26.01	27.93	28.97	30.64	31.54	32.92	41.44	
Q	2.46	2.10	2.25	2.83	3.42	4.24	6.68	
Or	3.66	3.74	3.96	4.37	4.43	4.90	6.50	
Ab	19.89	22.09	22.76	23.44	23.69	23.78	28.26	
An	25.88	24.38	22.97	23.64	22.87	22.04	19.28	
Cpx	19.30	19.92	19.69	18.17	17.66	16.63	12.77	
Opx	19.51	18.32	18.84	17.48	17.41	17.67	15.74	
Ol	0.00	0.00	0.00	0.00	0.00	0.00	0.00	
Mt	2.73	2.58	1.97	2.75	2.68	2.71	2.84	
Il	5.87	6.00	6.36	6.38	6.86	7.10	6.38	
Ap	0.76	0.78	0.88	0.90	0.95	0.97	1.40	

*Data mainly from Wright and Fiske (1971).
^{a}M' is the magnesium number (100 × Mg/(Mg + Fe″)).
^{b}D.I. is the Thornton and Tuttle (1960) differentiation index.
^{c}Analysis 16 is of typical olivine phenocryst in Kilauean basalts.

7.0% MgO contain phenocrysts of plagioclase and augite. Below 5.5% MgO, orthopyroxene phenocrysts are present, which take the place of olivine phenocrysts. At still lower MgO contents (<4.0%), ilmenite becomes a phenocrystic phase.

From a cursory inspection of Table 13-1, which is arranged in order of decreasing MgO content (17.4–3.9%), certain relations are readily apparent. For example, with decreasing MgO content, the TiO_2 and alkali contents increase. The table, however, contains far too many data for any but the most obvious relations to be found by inspection. To simplify the task, petrologists, utilize plots known as *variation diagrams*, which graphically portray compositional trends between certain groups of elements or oxides in igneous rock suites.

One of the most widely used diagrams is the triangular *FMA* plot, which shows the simultaneous variation in the weight percentages of FeO (may also include ferric iron recalculated as FeO, that is, $0.9 \times Fe_2O_3$), MgO, and total alkalis ($Na_2O + K_2O$). The position of points in this diagram are determined by summing the FeO, MgO, and $Na_2O + K_2O$, and recalculating to 100%. For example, $F = 100 \times FeO/(FeO + MgO + Na_2O + K_2O)$.

The *FMA* diagram for the Kilauean rocks (Fig. 13-1) reveals a remarkable systematic variation, in which the iron and alkali contents increase as the magnesium content decreases. In addition, all but the least magnesian of the lavas (no. 15 in Table 13-1) lie on a straight line that radiates from the composition of the typical olivine phenocryst in these rocks. Based on our experience with ternary-phase diagrams and the lever rule (Sec. 10-14), this variation could be explained simply by subtraction of olivine from a magnesium-rich magma. The fact that the least magnesian lava falls off what is known as the *olivine control line* implies that for this rock, some additional factor must have played a role in its differentiation. Indeed, this lava is the only one in this group to contain ilmenite pheno-

crysts. Ilmenite plots near the *F* corner in Figure 13-1 and if subtracted from a basaltic magma, would displace derivative magmas to the iron-poor side of the olivine control line.

Although the evidence from Figure 13-1 for olivine fractionation in the Kilauean rocks may be persuasive, it is rash to draw such a conclusion based only on the parameters *F*, *M*, and *A*. If correct, other components in the rocks should vary in a predictable manner. For this reason it is useful to compare all the components in the rock series. A variation diagram in which this is done (Fig. 13-2) plots the weight percentage of each oxide in the rock against the atomic ratio of $100 \times Mg/(Mg + Fe'')$, which is determined by dividing the oxide weights in the analyses by their molecular weights. This ratio, which is known as the *magnesium number* and designated by M', is particularly useful as a *differentiation index* because, as we know from numerous phase diagrams in Chapter 10, ferromagnesian minerals that crystallize at high temperatures are more magnesian than those that crystallize at lower temperatures. Consequently, we can expect rocks with high magnesium numbers to have formed at higher temperatures and be more primitive than those with low magnesium numbers.

A number of important oxides in the Kilauean rocks are plotted against M' in Figure 13-2 along with the composition of typical phenocrystic olivine. For compositions with high values of M', the effect of olivine fractionation is readily apparent, with the concentration of all oxides (except MgO) increasing with decreasing M'. For M' values of less than 67, however, the rate of increase in CaO with falling M' decreases. This change occurs where augite first appears in the rock series as phenocrysts. Separation of augite phenocrysts from the magmas would diminish the increase in CaO caused by the separation of olivine alone. At slightly lower M' values, plagioclase also becomes a phenocrystic phase, and its separation can account for the Al_2O_3 content remaining constant and the CaO

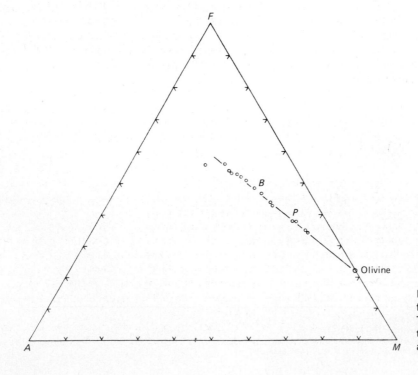

FIGURE 13-1 *FMA* variation diagram for Kilauean lavas (analyses given in Table 13-1). *B* and *P* mark compositions that may correspond to parental basalt and picrite magmas, respectively.

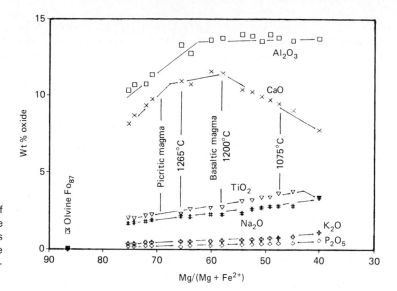

FIGURE 13-2 Variation diagram of weight percentages of oxides versus the magnesium number of Kilauean lavas (Table 13-1). Temperatures are those measured during eruptions or are laboratory-measured liquidus temperatures.

content falling in the less magnesian rocks. Oxides such as P_2O_5, K_2O, Na_2O, and TiO_2 do not enter the early-crystallizing minerals in the Kilauean basalts except plagioclase, which does incorporate some Na_2O, but even here sodium is concentrated in the residual melt. Because these oxides do not enter early-crystallizing minerals, they are described as *incompatible*. With decreasing M' the concentrations of incompatible elements steadily increase in the derivative liquids. The only exception is in the basalt with the lowest M', where ilmenite becomes a phenocrystic phase, and thus TiO_2 is no longer an incompatible component at this stage of differentiation (a crystalline phase is present into which TiO_2 can enter) and thus the TiO_2 content begins to fall.

Based on direct measurements of eruption temperatures and the laboratory determination of liquidus temperatures, we can label the M' variation diagram with typical magmatic temperatures. As would be expected, temperatures decrease with falling M'. Maximum eruption temperatures, which occur at the summit of Kilauea, are near 1200°C; flank eruptions are typically cooler than this. Picritic rocks, which contain high concentrations of olivine phenocrysts, may represent primitive liquids, or they may simply be basaltic liquids in which olivine phenocrysts have accumulated. Evidence indicates that primary magmas beneath Kilauea may be picritic in composition, in which case their liquidus temperature would have to be higher than the 1200°C eruption temperatures at the summit of Kilauea.

The hypothesized compositions of a primary picritic magma and of a basaltic parent magma are shown in Figures 13-2 and 13-1 (*P* and *B*). Compositions with M' greater than 70 are formed by magmas becoming enriched in olivine and are not representative of liquid compositions. Compositions with M' between 70 and 58 may also reflect accumulations of olivine in basaltic magma. Lavas with M' greater than 58 erupt primarily from the summit of Kilauea and have compositions that are determined largely by variations in the amount of olivine added or subtracted; that is, they lie on the olivine control line. Lavas with M' less than 58 erupt primarily along the rift zones, where magma has had time to cool to temperatures

where plagioclase and pyroxene have crystallized and played a role in determining the composition of lavas.

The M' index is particularly useful for studying compositional variations in basalts, because the ferromagnesian components constitute such a large fraction of these rocks. In granitic rocks, however, where there may be only a few percent ferromagnesian minerals, this index is likely to be less sensitive to differentiation, and thus a different index is required. Two of the most commonly used ones are the *Differentiation Index* (D.I.) of Thornton and Tuttle (1960), which is simply the sum of the normative minerals which are concentrated in residual magmas, that is, quartz, orthoclase, albite, nepheline, leucite, and kaliophilite (kalsilite), and the Harker diagram, which simply uses the weight percent of SiO_2. The resulting diagrams for the Kilauean rocks (Figs. 13-3 and 13-4) are very similar, and the explanation for the compositional trends they exhibit is the same as for the M' diagram.

Variation diagrams can be used in a quantitative way to determine possible relations between rock types. To do this, however, care must be used to plot conservative quantities, such as the weight percentages of the oxides. Consider, for example, Figure 13-5, in which analyses 2 and 8 of Kilauean lavas and analysis 16 of the phenocrystic olivine from Table 13-1 are plotted against their weight percent of MgO. With the exception of a small deviation in the amount of CaO, equivalent oxides in these three analyses lie on straight lines. This indicates that the intermediate analysis, no. 2, could be formed by combining the two extreme analyses, or, conversely, analysis 8 could be formed from no. 2 by subtracting no. 16. Furthermore, using the lever rule, the relative lengths of the lines between the three analyses indicate the proportion of phases involved. For example, analysis 2 could be formed by combining 24 wt % of olivine (line 2–8/line 8–16) with 76% of lava 8. The small deviation in the CaO plot is probably attributable to too high a value of CaO in the olivine analysis. Na_2O and P_2O_5, which have been omitted from Figure 13-5 for clarity, also lie on linear trends in this plot.

The lines in Figure 13-5 were visually fitted to the data. In this case the task was simple. But consider how much more

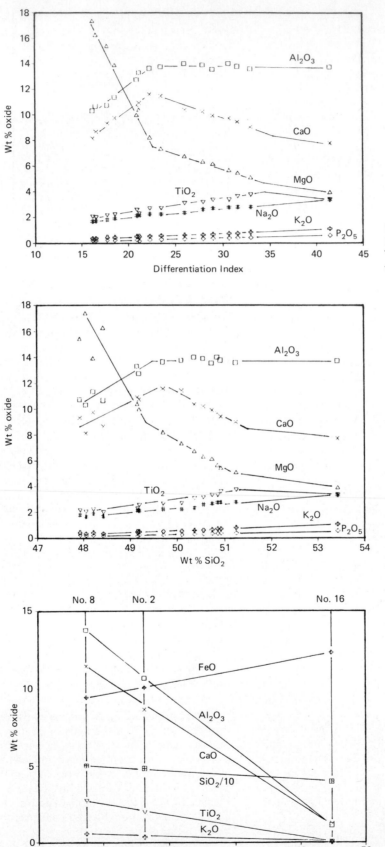

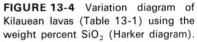

FIGURE 13-3 Variation diagram of Kilauean lavas (Table 13-1) using the Thornton and Tuttle (1960) Differentiation Index.

FIGURE 13-4 Variation diagram of Kilauean lavas (Table 13-1) using the weight percent SiO_2 (Harker diagram).

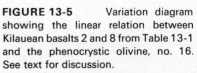

FIGURE 13-5 Variation diagram showing the linear relation between Kilauean basalts 2 and 8 from Table 13-1 and the phenocrystic olivine, no. 16. See text for discussion.

difficult it would be if we were trying to account for the derivation of one lava from another by subtracting some unknown ratio of analyzed olivine and pyroxene crystals or possibly a combination of olivine, pyroxene, and plagioclase crystals. Even if the supposed mechanisms were correct, unavoidable errors in the analyses would mean that no exact solution would exist. Only a best fit could be obtained. Calculations of this type are tedious and are therefore done on computers using least squares techniques to fit the data (Wright and Doherty 1970; Gray 1973).

Calculations such as these do not prove that mixing or separation occurred; they simply show that it is mathematically possible. Where olivine addition or separation is suspected, however, we have a means of testing the hypothesis. Roeder and Emslie (1970) have found that in a wide range of magmas and simpler silicate systems the composition of olivine can be related to the mole fractions of Mg and Fe″ in the melt with which the olivine is in equilibrium by a simple distribution coefficient, K_D, which remains relatively constant over the range of magmatic temperatures. The distribution coefficient, which has values between 0.30 and 0.35, is defined as

$$K_D - \frac{(X_{Mg}/X_{Fe''})^{liquid}}{(X_{Mg}/X_{Fe''})^{olivine}} \qquad (13\text{-}1)$$

Thus, if the mole fractions of Mg and Fe″ in the magma are known, we can calculate the composition of the olivine that would be in equilibrium with it, or, conversely, given the composition of an olivine, we can calculate the Mg/Fe ratio in the liquid from which it formed.

To use the olivine distribution coefficient to test the two possible mechanisms suggested by Figure 13-5, we have first to decide with which liquid the olivine crystals would have been in equilibrium. For example, if analysis 2 was formed by the addition of olivine crystals to analysis 8, the olivine crystals must have been in equilibrium with liquid 8. Using the Mg/Fe ratio of analysis 8, we calculate a value of K_D of 0.21, which is well outside the possible range of K_D values. We can conclude, therefore, that formation of lava 2 by the addition of olivine crystals with the composition of no. 16 to magma of composition 8 is not thermodynamically possible. Conversely, if we assume that lava 8 was formed from a picritic magma of composition 2 by the subtraction of olivine crystals, the olivine crystals would have to have been in equilibrium with a liquid of composition 2. This gives a K_D value of 0.45, which is equally as impossible. Presumably the true mechanism, then, lies somewhere between these two extreme hypotheses; that is, the original magma was not as magnesian as no. 2, which must have had some olivine accumulate in it but not as much as 24%. If we take the value of K_D to be 0.33, the magma with which olivine of analysis 16 would be in equilibrium would have a magnesium number of 68. This is the way in which the parental picrite magma in Figure 13-1 was determined [see Irvine (1979)].

Although rocks from only Kilauea have been discussed in this section, suites of rocks from other volcanoes reveal the same general features; that is, the compositional variation in lavas from a given volcano can be explained in terms of separation of phenocrystic phases from a parental magma. The con-

clusion that crystal fractionation must be responsible for the differentiation of these rocks seems inescapable. It is to the processes taking place within the magma chambers and conduits beneath volcanoes that we must now turn our attention in order to learn how this fractionation occurs.

13-3 CRYSTAL SETTLING IN MAGMA

Although many geologists had considered crystal settling as a possible mechanism of igneous differentiation, it was not until Bowen's 1915 paper on the "crystallization differentiation in silicate systems" that incontrovertible evidence of the process was documented. Bowen showed from experiments in the system forsterite–diopside–silica that crystals of olivine and diopside could sink (and crystals of tridymite float) a centimeter or two in periods of less than an hour. Although the experiments were carried out at high temperatures just below the liquidus where viscosities are much lower than they would be at typical magmatic temperatures, the magnitude of the effect was so great that Bowen concluded that crystal sinking (and to a lesser extent floating) must be of general importance to igneous differentiation.

It is not surprising that once crystal settling had been confirmed in the laboratory, field evidence for the process became more apparent. One of the most famous bodies in which crystal settling was believed to have occurred was the Palisades Sill, which forms the prominent escarpment on the west side of the Hudson River facing New York City. This sill, which is over 300 m thick, has a layer containing abundant olivine crystals approximately 15 m above its base. Olivine that was not trapped in the upper chilled margin was believed to have sunk and accumulated to form the olivine layer. Recent studies, however, have shown that the Palisades Sill has had a complicated history involving at least three injections of fresh batches of magma (Shirley 1987), and although gravitative settling may have played a role in its differentiation, a simple model of olivine settling is no longer believed to be valid (see Sec. 13-6). Nonetheless, the Palisades Sill is a large enough body that it would have taken several hundred years to solidify, and if olivine sinking occurs at a perceptible rate, some accumulation of olivine would be expected.

The rate at which a crystal sinks or floats depends on the difference between the buoyant force caused by the density contrast between the crystal and the magma ($\Delta\rho = \rho_s - \rho_m$) and the viscous drag of the liquid on the crystal. If we consider, for simplicity, a spherical crystal of radius r, the buoyant force on the crystal is

$$\text{buoyant force} = \tfrac{4}{3}\pi r^3 g\,\Delta\rho \qquad (13\text{-}2)$$

where g is the acceleration of gravity. The viscous drag resulting from the laminar flow of the liquid (viscosity η) across the surface of the crystal at a constant velocity of dz/dt, where z is measured downward, is given by *Stokes formula* (Turcotte and Schubert 1982a):

$$\text{drag force} = 6\pi\eta r \frac{dz}{dt} \qquad (13\text{-}3)$$

Equating these two forces, we obtain an expression for the *steady state* or *terminal* velocity (v_t) of the crystal:

$$v_t = \frac{dz}{dt} = \frac{2g\,\Delta\rho\,r^2}{9\eta} \tag{13-4}$$

which is commonly referred to as *Stokes law*. This law is valid only when the crystal moves slowly enough that no turbulent vortices develop. This condition is met when the Reynolds number is less than 1. The Reynolds number is defined as

$$\text{Re} = \frac{\rho_m 2 r v_t}{\eta} \tag{13-5}$$

For most geological cases this number is much smaller than 1. For nonspherical crystals, the factor $\frac{2}{9}$ in Eq. 13-4 can be changed to account for shape.

We can use Stokes law to determine how far crystals of olivine might have sunk in the Palisades Sill. Following Shirley (1987), we assume that the viscosity of the magma was 400 Pa s and its density was 2620 kg m^{-3}. If the olivine crystals, which have a density of 3500 kg m^{-3}, had a radius of 1 mm (note that grain size is normally given as grain diameter), their settling velocity would have been 4.8×10^{-6} m s^{-1}, or 1.7 cm h^{-1}, or 41.4 cm day^{-1}, or 151 m yr^{-1}. This is a significant settling velocity, and in the hundreds of years the intrusion took to solidify, olivine crystals should have been able to sink to the floor of the magma chamber. Note that the Reynolds number for these conditions is 6.2×10^{-8}, so the use of Stokes law is justified.

To calculate the distance traveled by a sinking crystal, it is necessary to integrate Eq. 13-4, which if we assume ρ and η to be constant, can be written as

$$\int dz = \frac{2g\Delta\rho}{9\eta} \int_0^t r^2\, dt \tag{13-6}$$

The radius of a crystal is unlikely to remain constant during sinking; instead, the crystal will either grow or dissolve. Crystals nucleate near the roof of the intrusion at some degree of undercooling, which allows them to sink into hotter magma without necessarily dissolving. Furthermore, the rise in pressure with increasing depth in a magma chamber causes the liquidus temperature to rise. The general tendency, then, is for crystals that form near the roof of the magma chamber to continue growing as they sink.

To evaluate the right-hand side of Eq. 13-6, we need to express the radius of a crystal as a function of time, which can be done by making use of the growth rate equations in Chapter 12 (Eqs. 12-7 to 12-10). For example, if the growth of olivine crystals were controlled by a phase boundary reaction, integration of Eq. 12-8 would give

$$\int_{r_i}^{r_t} dr = k_R \int_0^t dt \quad \text{from which} \quad r_t = r_{\text{initial}} + k_R t$$

The initial and final radii of the crystal might be determined, for example, from the size of crystals trapped in the upper chilled margin and in the layer of accumulation. Determining

the value of k_R is more difficult and requires that we know how long the crystal took to grow to its final size. A rough estimate might be obtained from the length of time required for the magma to solidify up to the layer of olivine accumulation (see Prob. 13-5). Equation 13-6 could then be rewritten as

$$\int dz = \frac{2g\Delta\rho}{9\eta} \int_0^t (r_{\text{initial}} + k_R t)^2\, dt \tag{13-7}$$

Even without precise knowledge of the effect of crystal growth rates on settling velocities, the simple application of Stokes law indicates that ferromagnesian minerals should sink in basaltic magmas with significant velocities that are on the order of centimeters per day. The question, then, is: Why do more sills not show evidence of crystal settling? Many sills of the same age, composition, and thickness as the Palisades Sill, but in other parts of the Mesozoic basins of eastern North America, show little, if any, evidence of crystal settling. Clearly, other factors must be involved.

McBirney and Noyes (1979) have drawn attention to the fact that magmas, on cooling into the crystallization range, become non-Newtonian liquids with a significant yield stress that increases with falling temperature, especially when plagioclase and pyroxene begin crystallizing. They found, for example, that a Columbia River basalt had a yield strength of 60 Pa at 1195°C. A dense crystal in such a magma would not be able to sink until it first grew large enough to overcome the yield strength of the magma. We can calculate this critical size for a spherical crystal of radius r by equating the gravitational force on the crystal (Eq. 13-2) with the shear force imposed by the yield strength (σ_y) acting on the surface of the crystal

$$\tfrac{4}{3}\pi r^3\, \Delta\rho\, g = \sigma_y 4\pi r^2$$

or for the case where the gravitational force overcomes the yield strength

$$\frac{r\,\Delta\rho\, g}{3} > \sigma_y \tag{13-8}$$

If the magma in the Palisades Sill had a yield strength of 60 Pa, the radius of an olivine crystal would had to have exceeded 2 cm (grain diameter 4 cm) for the crystal to sink. This grain size is an order of magnitude greater than occurs in the Palisade Sill. Clearly, no settling would have occurred in this intrusion if, or when, the yield strength of the magma reached significant values. Indeed, crystal settling in any magma chamber may be restricted to a short period while the magma is still at high temperatures and has Newtonian behavior.

We have, so far, discussed crystal settling in terms of stationary magmas. If, however, a magma is moving, either because of emplacement or convection, the flow may counteract, or add to, any movement due to gravitative settling of crystals. In extreme cases, convection could become turbulent, and then the effects of crystal settling would be completely neutralized by the movement of magma. As will be seen in the next section, convection in magma chambers is common, and it may therefore be another reason why simple crystal settling is not more prevalent in igneous bodies.

13-4 MAGMA CONVECTION

Magma in an isolated chamber may convect because of density differences resulting from thermal or compositional variations within the magma (Martin et al. 1987). This is known as *free convection*. If the magma chamber is still connected to a feeder conduit, influx of fresh magma may cause another type of convection known as *forced convection*. Forced convection is undoubtedly important in chambers that are repeatedly replenished with magma, such as those beneath the mid-ocean ridges, and the vorticity associated with this type of convection can be an effective means of mixing magmas (Campbell and Turner 1986). The effects of forced convection, however, are transient compared with those of free convection, which may last through a considerable part of the cooling history of an intrusion. The remainder of this discussion deals only with free convection; forced convection is discussed in Sec. 13-10.

Consider magma near a vertical wall of an intrusive body. Following intrusion, a temperature gradient, such as that shown in Figure 13-6, soon develops due to the conduction of heat into the surrounding rocks. Because magma contracts on cooling (Eq. 2-3), lower temperatures toward the contacts cause magma there to be denser than that a short distance into the intrusion. The system is therefore gravitationally unstable, and the dense magma near the wall sinks, while the hotter magma in the center rises, with the convection continuing as long as magma remains fluid enough to move.

The approximate rate at which magma sinks near a cool, vertical wall is given by

$$\frac{dz}{dt} = \left(\frac{g\Delta\rho\,kL}{\eta}\right)^{1/2} \tag{13-9}$$

where g is the acceleration of gravity, $\Delta\rho$ the density contrast that exists over a distance L resulting from the difference in temperature over that distance, k the thermal diffusivity, and η the viscosity. For example, if a temperature difference of

400°C exists across a distance of 2.0 m near the vertical contact of a dike of basaltic magma for which the coefficient of expansion is 6×10^{-5} °C^{-1}, the density contrast would be 64.8 kg m^{-3} (Eq. 2-3). If the thermal diffusivity, viscosity, and density of the magma are 10^{-6} m^2 s^{-1}, 100 Pa s, and 2700 kg m^{-3}, respectively, the convective velocity would be 0.4 cm s^{-1}. This is many orders of magnitude greater than the velocities at which ferromagnesian minerals would sink based on Stokes law.

The velocity calculated with Eq. 13-9 would, in reality, be too great, because the equation does not take into account the increase in viscosity that would accompany the decrease in temperature toward the contact. In addition, lower temperatures would be accompanied by higher degrees of crystallization, which might impart a yield strength to the magma (Eq. 13-24); this, then, would have to be overcome before convection could occur. These factors have been taken into account by Spera and others (1982) in calculating the velocity profile in the thermal boundary layer of a silicic magma chamber whose temperature is maintained constant by the intrusion of mafic magma into its base. As would be expected, the low temperatures near the contact produce a highly viscous crystal mush that is unable to sink, despite its negative buoyancy. Toward the inner side of the thermal boundary layer, however, magma becomes less viscous and yet has negative buoyancy. The downward convecting velocity of the magma therefore increases rapidly away from the contact, reaching a maximum of about 12 km yr^{-1} (137 cm h^{-1}) at 10 m from the contact on the inside of the thermal boundary layer. This velocity is still orders of magnitude greater than the rate at which crystals will sink in a silicic magma.

Magma cooling along a vertical wall is always gravitationally unstable. If, however, the contact is horizontal, as in a sill, the situation is more complex, and convection may, or may not, occur depending on a number of factors (Turner 1973). Consider a horizontal sheet of liquid between upper and lower boundaries that are maintained at constant temperatures T_1 and T_2 (Fig. 13-7). If $T_1 > T_2$, magma near the base of the sheet has a lower density than magma near the roof and so experiences a buoyant force. The magnitude of this force depends on the coefficient of thermal expansion of the magma, α, the temperature difference ($T_1 - T_2$), and the thickness of the sheet (d). This buoyant force causes magma near the base to rise and displace the denser magma near the roof. Resisting this movement, however, is the viscous drag of the magma. Also, the diffusion of heat through the magma diminishes the temperature difference that causes the density inversion. Thus the higher the thermal diffusivity, the less likely is a temperature (and hence density) difference to be maintained.

Just as the Reynolds number was defined as the ratio of the inertial and viscous forces, another dimensionless number, the *Rayleigh number* (Ra), can be defined as the ratio of the factors driving thermal convection to those opposing it.

$$\text{Ra} = \frac{\rho g d^3 \alpha (T_1 - T_2)}{\eta k} \tag{13-10}$$

When the Rayleigh number exceeds the critical value (Ra$_c$) of 1708 for the situation illustrated in Figure 13-7, the sheet of liquid becomes unstable and breaks up into a series of convective cells, known as Bénard cells, which have wavelengths of

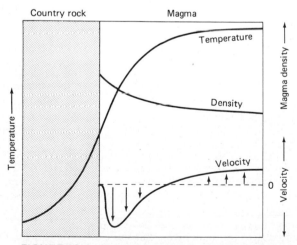

FIGURE 13-6 Temperature and density distribution near the vertical contact of an igneous intrusion. The higher density of the cooler magma near the contact relative to that farther into the intrusion produces gravitational instability, with magma convecting down the wall and up in the center of the intrusion.

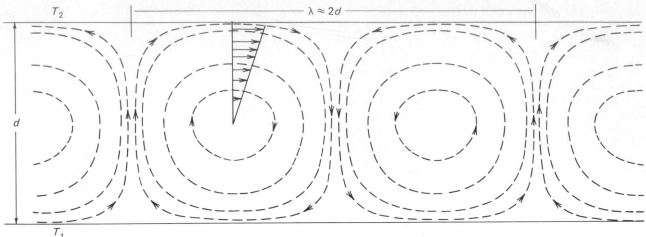

FIGURE 13-7 Bénard convection cells produced in a sheet of liquid that is gravitationally unstable because its base is kept at a higher temperature than its roof. In the third dimension, cells form rolls at low Rayleigh numbers, polygonal cells at intermediate values, and chaotic turbulent cells at high Rayleigh numbers. See text for discussion.

approximately twice the thickness of the sheet. If magma in the 330-m-thick Palisades Sill, for example, had a density of 2620 kg m^{-3}, a coefficient of expansion of 2.5×10^{-5} K^{-1}, a viscosity of 400 Pa s, and a thermal diffusivity of 4.4×10^{-7} m^2 s^{-1} (Shirley 1987), the Rayleigh number would be $10^{11} \times (T_1 - T_2)$. Consequently, with only a fraction of a degree difference in temperature between the bottom and top of the Palisades Sill, the Rayleigh number would exceed the critical value and convection would occur. This is likely to be the case in all but the thinnest sheets.

When the Rayleigh number just exceeds the critical value, the Bénard cells form long rolls in plan view. When the Rayleigh number exceeds 10^4 to 10^5, these rolls break up into polygonal convection cells, and at values of Ra in excess of 10^6, the convection becomes turbulent with chaotic unsteady motion (Sparks et al. 1984). Thus, in the Palisades Sill the convection would have most definitely been turbulent, at least during the early stages of convection. As the magma cooled and became more viscous and the sheet of remaining liquid became thinner, the Rayleigh number would have decreased and more regular, steady patterns of convection may have developed.

Once convection cells become established and achieve a steady velocity, the dissipation of potential energy caused by the density distribution in the magma exactly balances the viscous dissipation of mechanical energy. The potential energy ultimately results from the temperature distribution, and the convection cell provides a far more efficient and rapid means of transferring heat down the temperature gradient than does conduction. If no convection occurred in the sheet of liquid between the two surfaces illustrated in Figure 13-7, heat could be transferred only by conduction (assuming that radiation is unimportant), which according to Eq. 5-3 would be

$$J_z = -K \frac{dT}{dz} = \frac{K(T_1 - T_2)}{d} \qquad (13\text{-}11)$$

where J_z is the heat flux and K is the thermal conductivity of the magma. When the magma convects, heat is physically transported in the form of hot magma from the bottom to the top of the chamber. Thermal diffusion has then to transport the heat only a short distance through the roof. The heat flux through the roof of a convecting chamber is therefore greater than through the roof of a chamber with stationary magma; that is, $J_z > K(T_1 - T_2)/d$. The ratio of the actual heat flux (J_z) to that which would have been released had the magma been stationary gives a measure of the vigor of the convection and is known as the *Nusselt* number (Nu); that is,

$$\text{Nu} = \frac{J_z d}{K(T_1 - T_2)} \qquad (13\text{-}12)$$

When the Rayleigh number is greater than 10^5, which is usually the case for most magma bodies, the Nusselt number is given by $0.10\text{Ra}^{1/3}$ (Turner 1973).

Because convection transports magma from the hot lower parts to the cooler upper parts of a chamber, the temperature gradient in a convecting body of magma is far less than the gradient in a nonconvecting one. The thermal gradient in a convecting cell approaches the adiabatic gradient (see Prob. 7-7), which can be expressed as

$$\frac{dT}{dP} = \frac{T\bar{V}\alpha}{C_p} \qquad (13\text{-}13)$$

where $\bar{V}$ is the molar volume ($\bar{V} = 1/\rho$), and C_p is the heat capacity of the magma. But we know from Eq 1-1 that $dP = \rho g \, dz$, so Eq. 13-13 can be rewritten as

$$\frac{dT}{dz} = \frac{gT\alpha}{C_p} \qquad (13\text{-}14)$$

Again, using data for the magma of the Palisades Sill as an example, and assuming a temperature of 1373 K, and a heat capacity of $1.2 \text{ kJ kg}^{-1} \text{ C}^{-1}$, the temperature gradient in the convecting magma would approach 0.3 K km^{-1}. This is a remarkably low gradient compared with the typical geothermal gradient of 30 K km^{-1}.

The melting points of most minerals rise with increasing pressure. In a basaltic magma, the melting points of the early-crystallizing minerals increase, on average, about 3°C for every kilometer of depth (slightly less in granitic magma). This rise in melting points is approximately one order of magnitude greater than the rise in temperature with depth in a convecting basaltic magma (Fig. 13-8). Consequently, crystallization takes place near the base of a convecting cell where the pressure is greatest. This is illustrated schematically in Figure 13-8 for a superheated convecting magma that cools and eventually intersects the liquidus. Because the convecting magma is essentially the same temperature throughout, the magma first intersects the liquidus at the lowest point in the convecting cell. Some crystallization would occur in a narrow boundary layer near the roof where there is a steep temperature gradient. This material, being cool and laden with crystals, would be dense and thus gravitationally unstable. Periodically, plumes of this material would sink from the roof to the floor of the chamber. The result is that most accumulation of crystals in a convecting magma chamber takes place on the floor, either by direct crystallization or from high-density crystal-laden sluries that sink from the upper, cooler parts of the chamber.

The rate of cooling of a convecting magma body is far greater than that of a nonconvecting one. Because of the low-temperature gradient within a convecting cell, temperatures near the roof of a chamber are almost as high as those deeper in the body (Eq. 13-14). The resulting steep temperature gradient across the upper contact causes the heat flux from the magma body to be high. The rate of cooling of the body as a whole depends almost entirely on this heat flux through the roof. The length of time for a convecting magma chamber to crystallize completely is, according to Brandeis and Jaupart

(1987),

$$\text{convecting crystallization time} = \frac{15d^2}{\text{Ra}^{1/3}k} \quad (13\text{-}15)$$

whereas the time for a nonconvecting one is

$$\text{nonconvecting crystallization time} = \frac{d^2}{4k} \quad (13\text{-}16)$$

Thus for a convecting body, the length of time to crystallize is proportional to the thickness of the body (substitute Eq. 13-10 into Eq. 13-15), whereas for a nonconvecting body it is proportional to the square of the thickness.

The Rayleigh number for many magma bodies is well in excess of the critical value, and thus convection in magma chambers is common. The importance of convection depends on its rate relative to the rates of other processes. We have already seen that the rate of convection of even viscous granitic magma near a vertical wall is orders of magnitude greater than rates of crystal settling. The rate at which magma convects in a sheetlike body (Turcotte and Schubert 1982a) is given by

$$\text{convective velocity} = \frac{0.271k(\text{Ra})^{2/3}}{d} \quad (13\text{-}17)$$

If the Rayleigh number for the Palisades magma was 10^{11}, the convective velocity could have been as high as 3 km day^{-1}. Thus a packet of magma in a convective cell that occupied the entire thickness of the sill might have made two complete revolutions around the cell per day. Clearly, these rates are orders of magnitude greater than those calculated for crystal settling, and may actually be rapid enough to keep magma relatively homogenized during convection. Crystal concentrations can, however, develop in convecting cells. If crystals are denser than magma, they tend to concentrate in regions of convective upwelling, whereas if they are less dense, they concentrate in regions of convective downwelling (Marsh and Maxey 1985).

As stated at the beginning of this section, the density differences that drive convection are caused not only by variations in temperature, but by compositional differences. For example, a basaltic magma that first crystallizes plagioclase on cooling near the wall of a chamber produces a residual liquid that is more dense than the initial magma and would therefore sink. Conversely, if olivine or pyroxene were to crystallize first, the residual liquid would be less dense, so it would rise. Or consider the downward thermal convection of basaltic magma that causes partial melting of country rocks to form granitic melt, which, being less dense than the basaltic one, would rise. Similarly, if a boundary layer of magma assimilates water from country rocks, the less dense hydrated magma would rise. Crystallization of Fe at the inner core–outer core boundary produces liquid residue that is less dense than the overlying liquid. The resulting convection may generate the Earth's magnetic field. Any process, then, that changes the composition of a magma may induce convection. The volume of magma affected by this process at any one time is small because of low diffusion rates, but over the life span of an intrusion its cumulative effect

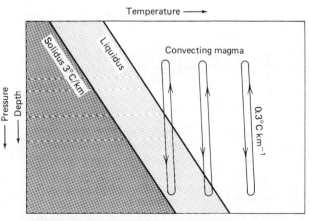

FIGURE 13-8 On cooling, a convecting magma, which has a shallow adiabatic temperature gradient, first intersects the steeper liquidus at its deepest part; convecting magmas consequently solidify from their base upward.

may be large. Magma in thin boundary layers along vertical walls of a chamber could, for example, float or sink and accumulate at the top or bottom of an intrusion to form large bodies of differentiated magma (Fig. 13-9a and b).

The density difference that drives thermal and compositional convection results from the diffusion of heat and chemical constituents, respectively. When both factors play roles, the convection is described as *double-diffusive convection* to indicate

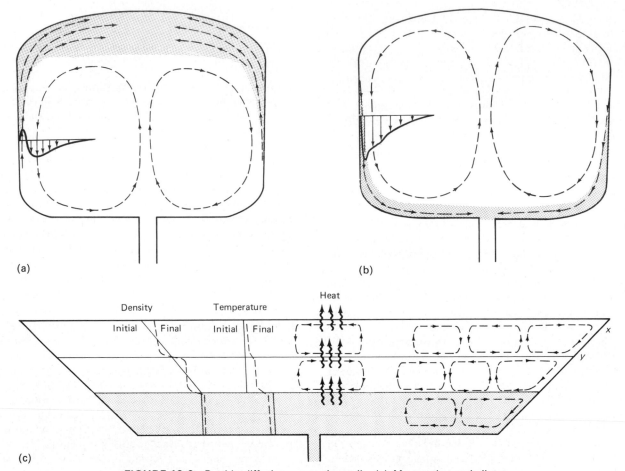

(a)

(b)

(c)

FIGURE 13-9   Double-diffusive convection cells. (a) Magma that assimilates water from wall rocks produces a low-density boundary layer that rises and accumulates in the upper part of the chamber, even though a wider thermal boundary layer causes magma to the inside of the compositional boundary layer to convect downward. (b) Magma cooling and convecting down the walls and crystallizing plagioclase onto the walls produces a dense residual liquid that can descend rapidly and pond on the floor of the chamber. (c) A laterally extensive sheet of magma which initially has almost constant temperature throughout and steadily increasing density with depth has a second more dense, hot magma emplaced at its base. Heat from this second intrusion rises into the base of the overlying magma, which then buoyantly rises, but as it does so, it eventually reaches magma of the same density (because of initial density gradient), and then it is forced to move laterally. As it does so, heat is transferred by diffusion into the overlying, less dense magma. Little, if any diffusion of chemical constituents takes place across the boundary because of the very low mass diffusion rates. Magma in this upper cell now convects, but on rising, it too encounters lower-density magma, which prevents it from rising farther. In this way, many separate sheets of convecting cells can form. Near the outer contact of the chamber, magma cooling at *x* becomes more dense and sinks, but on reaching point *y*, its density matches that of the main body of magma at this depth (because of initial density gradient), and the downward-convecting magma moves laterally away from the wall. Although stratified double-diffusive convection cells of this type can be produced with salt solutions in carefully adjusted laboratory experiments, their existence in magma chambers is yet to be proven.

that two diffusion coefficients are involved (Turner 1973). Heat diffuses about 10,000 times more rapidly than do chemical species ($k = 10^{-6} \ \text{m}^2 \ \text{s}^{-1}$, whereas $D = 10^{-10} \ \text{m}^2 \ \text{s}^{-1}$ as seen in Fig. 18-11). Consequently, thermal convection develops first in an intrusion, and thermal boundary layers are always much thicker than compositional ones. The effect of composition on magma density, however, is far greater than the thermal effect, because of the small coefficient of thermal expansion of magmas. Compositional changes in magma brought about by crystallization can therefore produce liquids of very different densities which can float or sink rapidly (Prob. 13-12).

Complex stratified sequences of Bénard cells (Fig. 13-9c) can form through double-diffusive convection (Turner 1973). Such layered convection cells have been found in seawater that has special temperature and salinity gradients. Although sequences of double diffusive convection cells have been invoked to explain some of the layering in large gabbroic intrusions (Irvine et al. 1983; Sparks et al. 1984; Wilson and Larsen 1985), the special conditions necessary for this type of convection seem unlikely to occur in typical magmas.

13-5 IGNEOUS CUMULATES

The compositions of certain igneous rocks differ significantly from those of known magmas that reach the surface as lavas. These rocks are enriched in early crystallizing minerals, and their textures (Fig. 13-10) provide clear evidence that crystals of these minerals were concentrated by accumulation. Rocks of this type are, therefore, referred to as *cumulates* (Wager et al. 1960). They were originally thought to form by crystals sinking to the floor of a magma chamber. This also accounted for the prominent horizontal or gently dipping layering that most cumulates exhibit (Fig. 13-12), which was interpreted to result from magmatic sedimentation. Although such a process may occur, it cannot be the only means by which cumulates form, because some contain plagioclase crystals that would certainly have floated in the magmas from which they crystallized. Regardless of the process by which crystals are concentrated in cumulates, fractional crystallization is clearly involved in the formation of these rocks.

Cumulates have a characteristic texture formed by two contrasting types of mineral grains (Irvine 1982). First, there are the more abundant *cumulus* grains, which tend to be euhedral and rest against each other to form a cumulus framework. Between these grains are minerals that form during a *postcumulus* period from the *intercumulus* liquid. A randomly stacked pile of crystals would have a porosity of 25 to 50%, and cumulates that contain this volume of postcumulus material are described as *orthocumulates* (Fig. 13-11). Some cumulates, however, contain far less postcumulus material, and these are termed *mesocumulates* if the percentage is between 7 and 25% and *adcumulates* if it is less than 7%. In mesocumulates, cumulus grains become enlarged by overgrowths, and many grains contact one another along mutual interference boundaries rather than at the point contacts that are more typical of cumulate grains in orthocumulates. In adcumulates, all cumulate grains

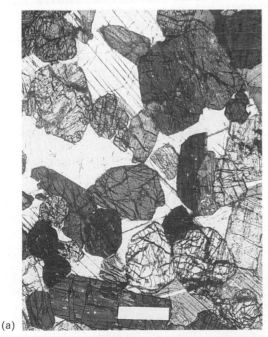
(a)

(b)

FIGURE 13-10 Photomicrographs of two rocks composed of pyroxene and plagioclase from the Great Dyke of Zimbabwe. In (a), pyroxene crystals accumulated from the magma, whereas plagioclase crystallized from the intercumulus liquid to form one single large poikilitic grain that cements the pyroxene grains in place. In (b), plagioclase was also a cumulus phase; many plagioclase laths aligned themselves parallel to the floor of the intrusion during accumulation. Crossed polars; scale bar is 1 mm.

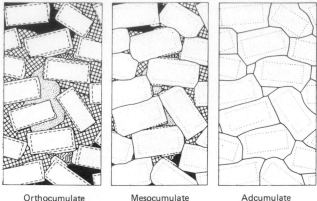

Orthocumulate   Mesocumulate   Adcumulate

FIGURE 13-11   Three different types of plagioclase cumulate. In each, plagioclase is the only cumulus phase. In the orthocumulte, intercumulus liquid crystallizes to an intergrowth of pyroxene, olivine, and magnetite; a small amount of plagioclase forms thin postcumulus rims on cumulus plagioclase grains. In the mesocumulate, overgrowths on cumulus grains are thicker, and the percentage of other postcumulus minerals is greatly decreased. In the adcumulate, the amount of postcumulus plagioclase added to cumulus grains is so great that no significant amount of other intercumulus minerals is present. [After Wager et al. (1960); published with permission of Oxford University Press.]

contact one another along boundaries that are formed either by mutual interference of overgrowths, or by recrystallization or pressure solution of the original grains.

An additional, less common type of cumulate known as a *crescumulate* has the cumulus grains grow normal to the layering in which they occur. The resulting texture is similar to the comb structure in pegmatites where minerals nucleate on the wall of the pegmatite and grow inward. Crescumulates can involve felsic as well as mafic minerals. In the steeply dipping marginal border group of the Skaergaard intrusion, sheets paralleling the contact of the intrusion contain bladed 2.5-cm-long plagioclase crystals oriented perpendicular to the contact. These crystals are believed to have grown during periods of quiescence of the magma; convection of magma down or up the wall would have deflected the crystals. Olivine crystals in some of the peridotite layers of the Rhum intrusion grew perpendicular to the layering, forming vertically elongated crystals that branch upward. This was originally described as a *harrisitic structure*. Apparently, during periods of slow accumulation of cumulus grains, olivine crystals resting on the floor of the intrusion that had their a-axis oriented approximately normal to the floor grew more rapidly than did those with other orientations (Wager and Brown 1967).

The common cumulus minerals are olivine, pyroxene, plagioclase, chromite, and magnetite, and the postcumulus minerals typically include pyroxene, plagioclase, magnetite, ilmenite, hornblende, and granophyric intergrowths of quartz and alkali feldspar. The compositional differences between the cumulus and intercumulus grains are consistent with an origin through fractional crystallization; that is, cumulus ferromagnesian minerals are more magnesian than postcumulus ones, and

cumulus plagioclase is more calcic than intercumulus plagioclase. Compositional zoning on the rims of cumulus grains is common, especially where there has been adcumulus overgrowths. This is particularly true of plagioclase and to a lesser extent of pyroxene. Olivine homogenizes rapidly enough that zoning is much less common. Cumulates are described in terms of their cumulate phases, which are given in order of decreasing abundance. Thus, describing a gabbro as a plagioclase-olivine cumulate tells the reader that this rock contains cumulate grains of plagioclase and lesser olivine, and that the other minerals, such as pyroxene and magnetite, occur as postcumulus grains.

The most common occurrence of cumulates is in layered gabbroic intrusions, but they also occur in some syenitic and granitic intrusions. An excellent review of these occurrences is given by Wager and Brown (1967). Much of what is known about cumulates comes from the study of tholeiitic layered gabbroic bodies, such as the Skaergaard intrusion of East Greenland (Wager and Deer 1939), the Stillwater complex of Montana (Hess 1960), the Muskox intrusion of the Northwest Territories, Canada (Irvine and Smith 1967; Irvine 1980), the Bushveld intrusion of South Africa and the Great Dyke of Zimbabwe (Rhodesia) (Jackson 1967), the Kiglapait intrusion of Labrador (Morse 1969), the Duke Island ultramafic complex of southwestern Alaska (Irvine 1974), and the Rhum complex of the Inner Hebrides, Scotland (Emeleus 1987). Cumulates are also found in alkaline gabbroic and syenitic rocks, as for example in those from the Gardar province of southwestern Greenland (Sørensen 1974b) and the Monteregian province of Quebec (Philpotts 1974).

The layering exhibited by most cumulates ranges in thickness from millimeters to meters and can result from variations in modal abundance of cumulus minerals, grain size, texture, or composition of minerals (Figs. 13-12 and 4-40). Because variations in mineral composition are normally not evident until analyses have been done, layering due to this is referred to as *cryptic layering*. Variations in modal abundance and grain size are commonly gradational, with coarse, dense minerals occurring at the base and finer, less dense minerals toward the top of layers. By analogy with similar variations in sedimentary beds, this *graded layering* was one of the first pieces of evidence used to argue that cumulates were formed by gravity settling. The variations responsible for the layering commonly repeat themselves over and over again to produce *rhythmic layering*. If the sequence of appearance of cumulus grains in passing up though rhythmic layers is identifiable as the fractional crystallization sequence of the minerals involved, the layering is said to be *cyclic*. Cyclic units are typically formed by injections of new batches of undifferentiated magma.

In large gabbroic intrusions, layers generally dip at shallow angles toward the center of the intrusion. Toward the margins, dips may increase and even become so steep that the layers become unstable and slump down onto the floor of the intrusion, where they disrupt the flat-lying layers. In other cases, channels may cut down through the cumulate layers to produce troughs that later are filled with other cumulates that commonly exhibit strong graded compositional layering. Such features are clear evidence of magmatic convection. The disruption or erosion of cumulate layers typically affects cumulates to depths of several meters, providing important evidence about the thickness of crystal mush on the floors of these intrusions.

(a)

(b)

(c)

FIGURE 13-12 Layering in rocks of the Kiglapait intrusion, Labrador. (a) Gabbro exhibiting millimeter- to centimeter-scale layering produced by concentrations of ferromagnesian minerals. (b) Graded layering; each layer grades from an olivine-rich base, through a pyroxene-rich central part, to a plagioclase-rich top before abruptly returning to the olivine-rich base of the next layer. (c) Troughs cut into layered gabbro are filled mainly with olivine.

Although the gently dipping cumulate layers in many intrusions have been interpreted as the products of gravity settling from either static or convecting magma, identical layering occurs on vertical walls of intrusions where gravity and density could not be a factor (Figs. 13-13 and 4-43). Such evidence led McBirney and Noyes (1979) to propose that layering in the Skaergaard intrusion formed by the nucleation and growth of minerals along the walls, roof, and floor of the intrusion in a static boundary layer in which the minerals were trapped in a Bingham liquid. Brandeis and others (1984) have shown that the rhythmic layering in cumulates can be developed in such a boundary layer by the interplay of crystal nucleation and the

dissipation of the heat of crystallization. Once nucleation occurs, the latent heat of crystallization raises the temperature and prevents further nucleation; crystal growth therefore takes place only on the already formed nuclei. The rate of this growth is determined by the rate at which heat is dissipated from the boundary layer. As the crystals approach their final size, which is determined by phase equilibria and the temperature, they grow more slowly, and the temperature is able to begin to fall. Eventually, a second batch of nuclei form and the process repeats. Differences in nucleation and growth rates produce graded modal abundances across each layer. Calculations indicate that layers produced by this process would be several

(a)

(b)

FIGURE 13-13 Vertical layering paralleling the walls of the Monteregian alkaline intrusion, at Mount Johnson, Quebec (Philpotts 1968). In (a), the layered gabbro is cut by a vertically plunging trough that is filled by a younger layered gabbro (center of intrusion to the right). In (b), layering is graded with respect to abundance of mafic and felsic minerals (center of intrusion to left). Because no tilting has occurred since formation, the compositional differences producing the vertical layering cannot be the product of gravity settling.

centimeters thick. This could explain the origin of the common rhythmic layering of cumulates, referred to as "inch-scale" layering (Fig. 13-12a).

Although the concept of cumulus and intercumulus minerals is simple, the distinction between such grains is often difficult in practice. If the intercumulus liquid crystallizes to distinct mineral grains, its identification is easy. Commonly, the intercumulus minerals nucleate at widely separate locations and grow around the cumulus minerals to form very large crystals that may contain many cumulus grains. These large crystals that grow late are known as *oikocrysts*, to distinguish them from phenocrysts, which grow early. A rock with oikocrysts has a *poikilitic* texture (Figs. 13-10 and 13-14). Some intercumulus liquid may crystallize as overgrowths on cumulus grains, and then the identification of the postcumulus material must depend on textural interpretations or the identification of compositional zoning in the regions of overgrowth. In still other cases, the intercumulus liquid may react with the cumulus grains, as happened in the peridotite from the Stillwater complex, where olivine was rimmed by orthopyroxene (Fig. 13-14).

When cumulates contain less than 25% intercumulus material, some process other than random packing of cumulus grains must be involved in their formation. For example, the intercumulus liquid may have been expelled from the crystal mush, or this liquid's composition may have changed so that cumulus minerals could continue crystallizing from it, or the initial cumulate pile may simply have formed with a lower percentage of intercumulus liquid. Petrologists do not yet agree on which of these mechanisms is the most important, but most believe that some change in composition of the intercumulus liquid takes place so as to allow the cumulus grains to continue growing—hence the term adcumulate. Of course, several mechanisms may operate simultaneously, any one of which may play the dominant role in a particular case.

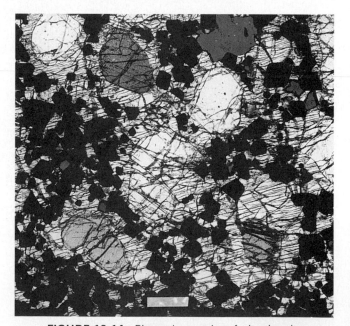

FIGURE 13-14 Photomicrograph of harzburgite from the Stillwater complex, Montana. The original outlines of cumulus olivine grains are marked by the position of chromite grains, which accumulated with olivine. Interstitial liquid reacted with olivine to produce postcumulus orthopyroxene, which poikilitically encloses all cumulus grains within this field of view in a single oikocryst (note common cleavage direction). Crossed polars; scale bar is 1 mm.

Originally, adcumulates were believed to form when compositional imbalances between the intercumulus liquid and the main body of magma were set up by the continued crystallization of cumulus minerals from the intercumulus liquid. For example, olivine crystallizing onto cumulus olivine grains would enrich the residual liquid in the other components of the melt. These higher concentrations would cause these components to diffuse up into the overlying magma. At the same time, olivine components in the main body of magma would diffuse down into the cumulate pile to replace the olivine components that had already crystallized onto the cumulate grains. With this two-way diffusion process the cumulate could eventually be converted into a monomineralic olivine rock—dunite. Because diffusion rates are so low, the distance between the intercumulus liquid and the interface with the main body of magma would have to be small for this process to operate. This, in turn, would require a slow rate of accumulation of cumulus grains. Morse (1986) estimates that rates should not exceed 0.5 cm yr^{-1} if adcumulus growth is to produce monomineralic rocks.

The change in composition of the intercumulus liquid that causes diffusion may also set up convection. Indeed, the change in density of a liquid resulting from crystallization (Fig. 3-1 and Prob. 13-12) is likely to provide a far greater driving force for exchange of intercumulus liquid than is diffusion. For convection to bring about an exchange with overlying magma, the residual liquid must be less dense than the overlying magma. This is true when crystallization of ferromagnesian minerals brings about enrichment of the residual liquid in felsic components, but when plagioclase is the cumulus mineral residual liquids are more dense and thus would remain in situ among the cumulus plagioclase grains on the floor of the intrusion. Convective removal of intercumulus liquid from plagioclase cumulates can therefore not be called upon to produce monomineralic layers of anorthosite in layered intrusions. This mechanism, however, may be an effective means of replenishing intercumulus liquid in mafic cumulates.

Irvine (1980) showed from whole-rock analyses and the composition of cumulus minerals that the interstitial liquid in cyclical cumulate units formed from repeated intrusions of batches of unfractionated magma in the Muskox intrusion was displaced upward by as much as 20 m. The first cumulus mineral to form from each batch of magma was olivine, followed by clinopyroxene, and then plagioclase. The emplacement of a new batch of magma is recorded in the magmatic stratigraphy by the abrupt appearance of olivine (and minor chromite) as the only cumulus phase. Olivine forming from each new pulse of unfractionated magma should have a higher magnesium content than that which formed in the fractionated magma at the top of the underlying unit. Irvine found, however, that the increase in magnesium invariably occurred above the stratigraphic level at which olivine appears as the only cumulus phase. He concluded, therefore, that the more iron-rich residual intercumulus liquid in the underlying rock was displaced upward into the new cumulate layer of olivine, where it reacted with the olivine to lower its magnesium content. He proposed that this *infiltration metasomatism* results from the upward migration of intercumulus liquid expelled from the underlying cumulates by compaction. Although the composition of the intercumulus liquid may be changed in this way, it does not necessarily produce adcumulates because this liquid still contains other components of the magma.

Compaction of cumulus grains can reduce porosity and thus produce a rock with an exceptionally high concentration of cumulus minerals, which would then be described as an adcumulate, even though no postcumulus material need have been added to the cumulus grains. The compaction process, which has been dealt with in detail by McKenzie (1984), results from the imbalance between the load pressure on the network of cumulus grains and the pressure on the intercumulus liquid. Mafic minerals and mixtures of mafic minerals and plagioclase form cumulates that have higher densities than the magma from which they crystallize. Compaction can therefore take place in these cumulates. The density of plagioclase cumulates, on the other hand, differ little from the density of the magma from which they form and may even be lower. Compaction could, therefore, not be a factor in the formation of plagioclase cumulates.

Where compaction is effective, porosity can be reduced to less than 1% by deformation and recrystallization of the cumulus grains (Sparks et al. 1985). This is because the interstitial liquid wets the cumulus grains and produces a small dihedral angle (see Sec. 12-6); the interstitial liquid therefore remains connected even when reduced to very small amounts. Compaction does, however, require a cumulate pile that is tens of meters thick to generate the necessary pressure differentials. Morse (1986) argues that such thicknesses are probably unusual. He also points out that the decrease in porosity that would be expected toward the base of a cumulate pile, because of the pressure differences, has not been documented. Shirley (1987), however, has made a convincing case for compaction being the main cause for differentiation of the last pulse of magma in the Palisades Sill.

Concentrations of early crystallizing minerals can occur when phenocryst-laden magma moves. One way in which this takes place is by *filter pressing*. Any constriction to flow may cause phenocrysts to be left behind. Changes in the width of dikes feeding magma into the east rift zone of Kilauea might, for example, be expected to filter out phenocrysts of olivine. Another process that comes into play when magma is in laminar flow and near the wall of a conduit is known as *flowage differentiation* (Komar 1972). The differential rate of flow of magma around phenocrysts at different distances from the wall of an intrusion produces a force that pushes the crystals apart (Fig. 13-15). Because the rate of change of velocity with distance increases toward the walls of a conduit (differentiate Eq. 3-7 or 3-13), the *grain dispersive pressure* reaches a maximum at the walls and decreases to a minimum at the center of the conduit. As a result, the concentration of phenocrysts is least near the walls and greatest near the center of conduits. Because large phenocrysts cause a greater distortion of the flow lines than do small ones, the effect is greater for large crystals (Ross 1986). This mechanism of producing rocks with concentrations of early crystallizing phases is most effective in narrow conduits or in boundary layers of convecting cells where velocity gradients are high.

All of these different processes of concentrating early crystallizing minerals may play roles in producing cumulates. But the apparent indifference of the cumulate-forming process

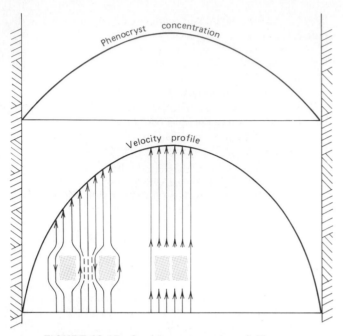

FIGURE 13-15 Crystals near margins of dikes are forced apart by differential laminar flow rates of magma around the crystals. Toward the center of dikes, shear strain decreases and so does the force dispersing the grains. As a result, phenocrysts tend to concentrate toward the centers of dikes that are emplaced by laminar flow.

to density contrasts between cumulus grains and magma and to the dip of the surface on which deposition occurs suggests that gravity does not play an important role. Instead, nucleation and growth of crystals on the walls of chambers seems to be the dominant process of formation, and diffusion is responsible for most of the small-scale compositional variation. Gravity does play a role in redistributing cumulates that slump from the walls or roof of a chamber onto the floor. With these controlling factors, it is not surprising that most cumulates form in large, slowly cooled plutonic bodies.

13-6 LIQUID IMMISCIBILITY

The unmixing of magmas into liquids of contrasting composition appealed to early petrologists as a simple mechanism by which magmas might differentiate. However, the field evidence advanced in support of this hypothesis, such as agate geodes in basalt, spherulites in rhyolite, and orbicular granites, which were all supposed to have formed as immiscible liquid globules, was easily shown by Bowen (1928) to be flawed. Furthermore, Bowen argued that the only liquid immiscibility field known at that time to exist in silicate systems (Fig. 10-9) was at too high a temperature to be of importance in the Earth's crust. This, combined with the success that fractional crystallization met with as a means of explaining magmatic differentiation, resulted in liquid immiscibility being almost completely ignored by petrologists during the first half of the twentieth century. The discovery of the low-temperature immiscibility field in the

system fayalite–leucite–silica (Fig. 10-24) by Roedder (1951) rekindled interest in this phenomenon, and by the time samples were returned from the Moon, petrologists were again receptive to finding evidence of immiscibility, which was present in a large fraction of the lunar samples (Roedder 1979).

The residual liquid formed by fractional crystallization of tholeiitic magma is commonly quenched to a glassy mesostasis between the plagioclase and pyroxene crystals of tholeiitic basaltic lavas. In the majority of these rocks this glass is now recognized to consist of two phases, one an iron-rich silicate glass of approximately pyroxene composition, and the other a silica-rich glass of granitic composition (Philpotts 1982). These glasses, which are quenched immiscible liquids, form globules of one in the other, with diameters reaching 50 μm (Fig. 13-16). The abundance of this two-phase glass varies from a few percent in primitive mid-ocean ridge basalts to approximately 30% in iron-rich basalts.

The compositions of pairs of immiscible liquids in these tholeiitic rocks (Table 13-2) match closely the compositions of conjugate liquids in the system fayalite–leucite–silica (compare Figs. 10-24 and 13-17). The compositional gap is slightly wider in basalts than in the simple system, due to the presence of additional components in the basalts, such as TiO_2, P_2O_5, and Fe_2O_3, that are known to expand the immiscibility field.

Figure 13-17 shows the composition of immiscible liquids in a wide range of tholeiitic lavas; it also shows the bulk compositions of the immiscible liquid pairs and the whole-rock compositions. The dashed line from the whole-rock compositions to the bulk compositions of the immiscible liquid pairs are fractional crystallization trends due to the crystallization mostly of plagioclase and pyroxene. Crystallization of these minerals causes the residual liquid to enter the two-liquid field at a point that depends mainly on the oxidation state of the magma and the stage at which magnetite starts crystallizing (Philpotts and Doyle 1983). If magnetite crystallizes late, the residual liquid remains iron-rich and the two-liquid field is reached rapidly, with approximately equal proportions of iron-

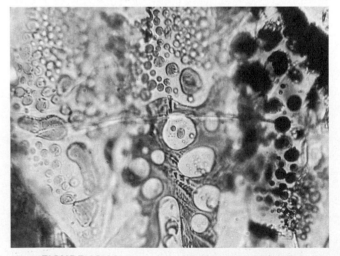

FIGURE 13-16 Iron-rich (dark) and silica-rich (clear) glasses formed as immiscible residual liquids in a tholeiitic basalt from Kilauea, Hawaii. Plane light; width of field is 100 μm.

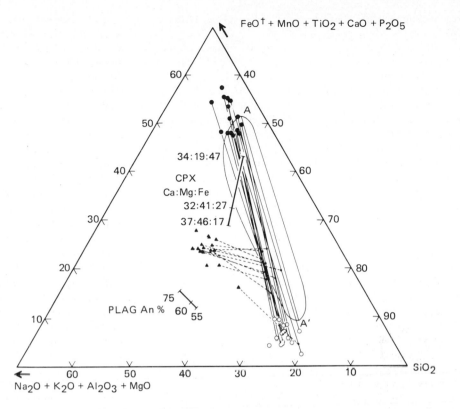

FIGURE 13-17 Plot of the composition of immiscible iron-rich (filled circles) and silica-rich (open circles) immiscible liquids preserved as glassy globules in a range of tholeiitic volcanic rocks (triangles) (Philpotts, 1982). The elliptical area is the two-liquid field from the system fayalite–leucite–silica (Roedder, 1951). The compositional range of plagioclase and of augite in these rocks is shown, with a mark on each solid solution line indicating the composition at which the magmas are believed to have unmixed. [After Philpotts (1982); published with permission of Springer-Verlag.]

rich and silica-rich immiscible liquids being formed. When magnetite crystallizes early, the residual liquid becomes enriched in silica and the two-liquid field is encountered at a later stage of crystallization, with a smaller fraction of iron-rich liquid being formed. If the oxidation state of the magma rises above the values associated with the nickel–nickel oxide buffer (Fig. 11-22), the fractionated liquid misses the immiscibility field completely.

Although liquid immiscibility produces melts of granitic composition (and of Fe-rich pyroxenite) in the mesostasis of tholeiitic basalts, is there evidence that immiscibility has played a role in the generation of significant amounts of differentiated tholeiitic rocks? It should be emphasized that the immiscibility field in tholeiitic basalts is encountered only in the fractionated liquid residue after at least 70% of the original magma has crystallized. Immiscibility, therefore, cannot be called upon simply to cause magmas to split into contrasting liquids upon cooling, as was done by early petrologists. Once a magma has solidified beyond 70%, immiscible liquids have little likelihood of being able to separate from one another to form sizable bodies of magmas, despite the significant density contrast between the two liquids. Perhaps if a crystal mush were to be compacted, some separation of the immiscible liquids might occur. Granophyric segregations toward the tops of bodies, such as the Palisades Sill, might form in this way.

If immiscibility is to produce sizable bodies of contrasting magmas, it appears necessary for a magma first to crystallize, fractionally, so that when unmixing does take place, the conjugate liquids are free to separate. These conditions exist during the final stages of crystallization of large layered gabbroic intrusions. Indeed, McBirney (1975) has shown through experiments that late-stage liquids in the Skaergaard intrusion entered the two-liquid field. The light silica-rich liquid rose to the roof of the residual lens of magma to form granophyre, while the dense

iron-rich liquid sank along with dense minerals to form an underlying zone of ferrodiorite. In this way liquid immiscibility can play a role in the differentiation of tholeiitic rocks, but only as an additional factor to the more important process of fractional crystallization.

Granitic magma can form from tholeiitic magma simply by fractional crystallization or by a combination of fractional crystallization and immiscibility. In Figure 13-17 these alternate paths involve either fractionating a magma directly from an initial tholeiitic composition (solid triangles) to the granitic residual liquid (open circles), or following one of the dashed lines into the two-liquid field and then producing the granite liquid through immiscibility. These granitic liquids have geochemical signatures that are diagnostic of the path followed.

Elements with large ionic radius and charge (incompatible elements) do not enter the early crystallizing minerals and, therefore, become concentrated in residual liquids (Sec. 13-11). Granitic magma formed through fractional crystallization should therefore be enriched in elements such as phosphorus and rare earths. Experiments, however, reveal that if a silica-rich liquid is in equilibrium with an immiscible iron-rich liquid, the incompatible elements partition strongly into the less polymerized iron-rich melt (Watson 1976; Ryerson and Hess 1978). Granitic magma formed through immiscibility would therefore have low concentrations of these elements. Despite the simplicity of this distinction, its application is often difficult. A granite cannot simply be said to be rich or poor in incompatible elements; this concentration must be given relative to either the parental magma or, in the case of suspected immiscibility, with the rock type that may have formed from the conjugate liquid. In many occurrences neither of these are known or their compositions are uncertain. Nonetheless, examples of comagmatic felsic and mafic rocks have been formed where this geochemical criterion of immiscibility has been satisfied (Vogel and Wilband 1978).

Although the evidence for liquid immiscibility is ubiquitous in the mesostasis of tholeiitic basalts, it is rare in alkaline basalts. This is due, in part, to the tendency of the mesostasis in alkaline basalts to crystallize rather than to quench to the glass, which would be necessary to preserve immiscible globules. Also, magnetite commonly crystallizes earlier in alkaline rocks than it does in tholeiitic ones, and thus residual liquids are not as enriched in iron and are less likely to encounter the immiscibility field. On the other hand, most alkaline rocks contain higher concentrations of the elements that expand the immiscibility field (TiO_2, P_2O_5, Fe_2O_3) than do tholeiitic ones. A small number of immiscible alkaline glasses have been found (Philpotts 1982). Their appearance is identical to those in tholeiitic rocks and their compositions are similar, except that both conjugate liquids contain less silica and more alkalis than do those in tholeiitic rocks (Table 13-2).

These few examples of alkaline immiscible liquids that are quenched to glasses are important because they provide a compositional link between the common immiscible glasses in tholeiites and the features in some lamprophyric dike and sill rocks known as ocelli (Philpotts 1976). Ocelli are globular felsic bodies that occur most commonly in lamprophyres of camptonitic composition (Chapter 6). They have diameters of several millimeters, but they are capable of coalescing to form larger globules (Figs. 4-50 and 13-18). They have been interpreted either as gas cavities that were filled with late crystallizing liquid, or as globules of immiscible felsic liquid in mafic alkaline magma. The partitioning of incompatible elements between ocelli and host lamprophyre favors the immiscibility origin for these structures. Calculated viscosities and densities of the ocellar phase based on the shape and distribution of ocelli within sills also favors the immiscibility hypothesis.

Ocelli are capable of rising and coalescing to form sheets near the top of sills. When these sheets attain thicknesses of about 1 cm, they become gravitationally unstable and may rise into the dense overlying crystal mush as diapirs (Fig. 3-10).

Some ocelli become large enough that they form small magmatic bodies in which gravitative crystal settling takes place as they cool and crystallize (Fig. 13-18). On this small scale, liquid immiscibility appears to play the dominant role in fractionating the lamprophyric magma. In larger intrusions, immiscibility may still be effective, but with slower cooling other processes may obscure its evidence.

Lamprophyric dike rocks, which are compositionally equivalent to alkali basaltic lavas, are common throughout most alkaline igneous provinces, and they probably reflect the compositions of magmas that formed the major intrusions in these regions. A striking feature of the compositions of plutonic rocks in these intrusions is that they tend to be either mafic or felsic, with very few intermediate rocks. Such a suite is illustrated in Figure 13-19 with rocks from the Cretaceous Monteregian province of Quebec. Also included are analyses of fine-grained dike rocks from this province, with those containing ocelli being shown with open circles. The fact that the ocellar dike rocks have compositions that correspond to the gap in the plutonic rock series is strong evidence that ocelli do indeed represent immiscible globules and that in the more slowly cooled plutons the immiscible liquids have time to separate from each other. This conclusion has been supported by the incompatible element partitioning between these rocks (Eby 1980).

One distinct group of alkaline rocks that may have been generated through liquid immiscibility are carbonatites. Such an origin has been widely accepted, in part, because of associated silicate rocks. Moreover, experiments have demonstrated a wide field of immiscibility (Fig. 13-20) separating the entire range of carbonatite compositions from silicate melts (Koster van Groos and Wyllie 1966; Kjarsgaard and Hamilton 1988).

It is surprising, in light of these experiments, that only a few rocks have been reported to contain immiscible droplets of carbonate liquid. Some lamprophyres contain carbonate ocelli,

TABLE 13-2

Average compositions of immiscible liquids preserved as glassy globules in tholeiitic and alkaline lavas*

| | Tholeiitic | | Alkaline | |
	Fe-rich Glass	Si-rich Glass	Fe-rich Glass	Si-rich Glass
SiO_2	41.5	73.3	37.1	65.4
TiO_2	5.8	0.8	8.2	1.0
Al_2O_3	3.7	12.1	5.1	13.9
FeO^t	31.0	3.2	28.9	4.0
MnO	0.5	0.0	0.5	0.0
MgO	0.9	0.0	1.9	0.5
CaO	9.4	1.8	8.7	2.3
Na_2O	0.8	3.1	1.1	4.0
K_2O	0.7	3.3	1.6	4.9
P_2O_5	3.5	0.07	3.4	0.5
Total	97.8	97.67	97.0	96.5

* After Philpotts (1982).

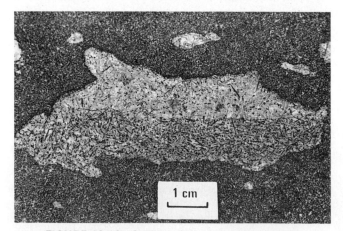

FIGURE 13-18 Settling of hornblende and plagioclase toward the base of an ocellus in a lamprophyric sill produced a lower zone of nepheline monzonite and an upper zone of nepheline syenite. The ocellus, in turn, was probably formed by coalescence of immiscible globules of felsic melt that rose and accumulated toward the top of the sill. Ste. Dorothée, Quebec.

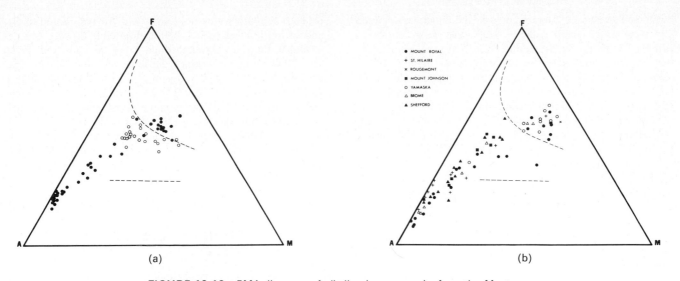

FIGURE 13-19 *FMA* diagrams of alkaline igneous rocks from the Montere-gian Province, Quebec. (a) Fine-grained dike rocks; ocellar ones are shown as open circles. Dashed line indicates probable extent of two-liquid field. (b) Plu-tonic rocks of the main intrusive bodies. The compositional gap in plutonic rocks corresponds to the composition of the ocellar dikes. Liquid immiscibility in the plutonic complexes may therefore have helped produce the striking division of these rocks into mafic and felsic. [From Philpotts (1976); published with per-mission of American Journal of Science.]

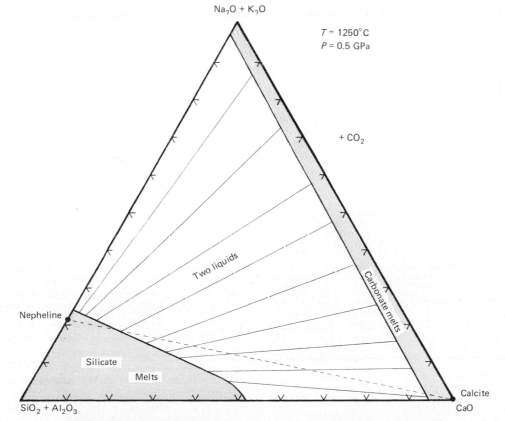

FIGURE 13-20 Isothermal, isobaric ternary diagram showing extent of the two-liquid field between carbonatite and silicate liquids. [After Kjarsgaard and Hamilton (1988).]

but these can also be interpreted as calcite amydules. Dawson and Hawthorne (1973) interpret small diapiric structures on carbonatite sheets in kimberlite, which appear identical to those of felsic ocelli sheets in lamprophyre sills (Fig. 3-10), as evidence of liquid immiscibility. Rankin and Le Bas (1974) found immiscible carbonate droplets in silicate melt inclusions in apatite crystals from an ijolite pegmatite in the Usaki complex of West Kenya. They were able to homogenize the carbonate and silicate liquids by heating the apatite crystals; the immiscible droplets, then, reappeared on cooling.

The concentration of carbonate in most silicate melts is small. Crystallization of non-carbonate-bearing silicates, however, will enrich a magma in carbonate, and if the immiscibility field is encountered before solidification has proceeded too far, an immiscible carbonate liquid will be able to segregate. The low viscosity of carbonate melts will allow them to migrate rapidly and perhaps separate completely from their silicate parent.

Sulfide liquids also can form as an immiscible fraction in silicate magmas. As indicated in Sec. 11-4, the solubility of sulfides in silicate melts is extremely small. Thus only a slight degree of fractional crystallization of silicate minerals from a sulfide-bearing silicate melt is usually all that is necessary to produce immiscible sulfide droplets, which if free to coalesce, will sink and segregate on the floor of a magma chamber.

An important factor to keep in mind when considering how rapidly immiscible liquids of any kind might be able to segregate is that immiscible droplets are able to coalesce. If initially all immiscible droplets in a magma were to nucleate and grow to a similar size, and if they were buoyant they would, according to Stokes law (Eq. 13-4), rise at the same rate. However, should any two touch and coalesce, the larger droplet, so formed, would rise more rapidly, which, in turn, would cause it to catch up with smaller ones. Further coalescence would

produce still larger globules that would rise still more rapidly, and so on. Furthermore, because immiscible droplets are deformable, they can travel through a crystal mush and bring about differentiation in much the same way that oil can separate and float on top of water in a porous sandstone. Figure 13-21 illustrates a felsic ocelli that was in the process of buoyantly rising around a sinking augite phenocryst in a 2-m-thick sill when solidification occurred. Other ocelli that managed to pass the sinking crystals accumulated near the top of the sill. The ability of immiscible droplets to coalesce and deform may allow differentiation to take place under conditions that would not permit crystal fractionation to occur.

13-7 DIFFUSION PROCESSES: SORET EFFECT

We normally think of diffusion taking place in magmas in response to compositional gradients set up, for example, by fractional crystallization (Sec. 13-5) or mixing of magmas (Sec. 13-10). But diffusion can also be induced by thermal gradients, in which case it is known as thermal diffusion or the *Soret effect*, so named in honor of the chemist who first demonstrated the phenomenon. When a solution is placed in a thermal gradient, its components redistribute themselves by diffusing up or down the gradient in order to establish an equal distribution of internal energy. For example, in gases, where the effect can be pronounced, the translational kinetic energy of the molecules ($\frac{1}{2}mv^2$) can be equalized by having heavier molecules move to the cooler end of the gradient (lower thermal velocity) and the lighter ones to the hotter end (higher velocity). In liquids and solids the situation is more complex because of the structure of the medium and the interactions between species. Qualitatively, however, the lighter components still migrate toward the hotter end, but the actual degree to which this occurs must be determined experimentally.

Walker and DeLong (1982) have investigated the Soret effect in a sample of mid-ocean ridge basalt placed in a thermal gradient of $330°C\,cm^{-1}$ at 1.0 GPa. They found that Si, Al, Na, and K were enriched toward the hot end of the charge (1480°C), and Fe, Mg, Ca, Mn, and Cr toward the cold end (1215°C). The differentiation was great enough to produce liquids having compositions of quartz-normative andesite at the hot end and nepheline-normative picrite at the cold end.

The diffusing species in these experiments are not likely to be individual cations or oxides; otherwise, the results would indicate that the mass of a species played no role in the differentiation. For example, K (atomic weight 39) and Na (23) are concentrated to a similar degree at the hot end, while Fe (56) and Mg (24) are concentrated to a similar degree at the cold end. The alkalis are probably coupled with alumina to form network-building units ($KAlO_4^{4-}$), which diffuse to the hot end along with polymerized SiO_4^{4-} groups. Walker and DeLong found that when there is insufficient alkalis to couple with alumina to form network groups, alumina diffuses to the cold end instead. The Fe, Mg, and Ca are probably coupled with silica to form chains and rings that have greater mass than the network units and so migrate to the cold end. The resulting differentiates have compositions similar to those formed through

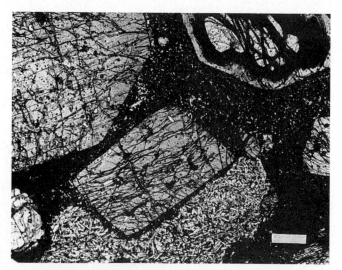

FIGURE 13-21 Photomicrograph of ocellus wrapped around the lower side of an augite phenocryst in a lamprophyric sill, Mount Bruno, Quebec. The ocellus is interpreted to have been an immiscible globule of felsic liquid, which was able to deform around the sinking crystal; other ocelli were able to float and concentrate near the top of the sill. Plane light; scale bar is 1 mm.

liquid immiscibility (Sec. 13-6). Thus a geochemical signature may serve only to distinguish fractional crystallization from liquid differentiation, not to characterize rocks as being uniquely the products of liquid immiscibility or the Soret effect.

The difference between Soret fractionation and crystal fractionation should be evident in the compositional variations of coexisting plagioclase and ferromagnesian minerals. With Soret diffusion, iron and magnesium both concentrate toward the cold end of the gradient, but the effect is greatest for the heavier iron. As a result, the Mg/Mg + Fe ratio is greatest at the hot end. But this is the end with the lowest Ca/Ca + Na ratio. Thus, when Soret-differentiated liquids crystallize, the resulting compositional variation in plagioclase and ferromagnesian minerals is very different from the trend formed by fractional crystallization, where Mg/Mg + Fe and Ca/Ca + Na vary sympathetically (Fig. 13-22). If fractional crystallization and Soret diffusion both affect a liquid, intermediate trends would result.

Differentiation of a magma by the Soret effect requires that a temperature gradient be maintained long enough for components to diffuse the distance over which the gradient exists. Values of thermal diffusivity in magmas are on the order of 10^{-6} m^2 s^{-1} (Sec. 5-2), whereas diffusion coefficients in basaltic magma are near 10^{-11} m^2 s^{-1}, and in granitic magma they are still less. Therefore, because heat diffuses much faster than chemical species, a thermal gradient would be dissipated long before the Soret effect could cause significant differentiation, unless some process, such as convection, were to replenish the heat and maintain the thermal gradient. If this occurs, the length of time necessary to establish a Soret gradient is given approximately by

$$\text{time}_{Soret} = \frac{d^2}{4D} \qquad (13\text{-}18)$$

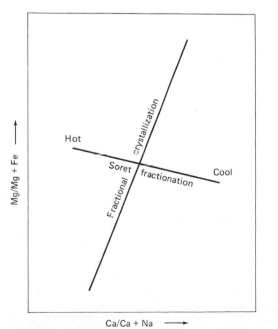

FIGURE 13-22 Compositional trends of melts fractionated by crystallization and by Soret diffusion expressed in terms of Ca/Ca + Na and Mg/Mg + Fe. [After Walker and DeLong (1982).]

where d is the distance over which species must diffuse and D is the diffusion coefficient for the mobile species. This equation is similar to Eq. 13-16 for the length of time for a nonconvecting body of magma to cool, the difference being that the diffusion coefficient is substituted for the thermal diffusivity. In Walker and DeLong's experiments, the gradient was established over a distance of 8 mm, and the diffusion coefficient of, for example, K in basaltic magma at 1300°C is 10^{-10} m^2 s^{-1}. Potassium in their experiments would therefore have required 44 h to establish an equilibrium gradient. The fact that they ran their experiments for 137 h assured that there was sufficient time for the gradient to develop.

The steep thermal gradient necessary to drive Soret diffusion can be maintained at the margin of an intrusion as long as the magma convects. However, convection dramatically shortens cooling times (Eq. 13-15), so that the Soret-effect will be most pronounced in those bodies where heat is periodically replenished by the intrusion of batches of new magma from the mantle, as is thought to happen in many large granitic bodies and in magma chambers beneath mid-ocean ridges. The fraction of magma in a chamber that occupies the thermal boundary layer at any one time is very small, but convection continually renews this magma. Spera and others (1982) have calculated that the entire volume of magma in a large intrusion (6000 km^3) can be cycled through a thin (20-m) boundary layer as many as a thousand times during the life span (10^6 years) of such a chamber. Thus, although Soret diffusion operates over only short distances, convection may allow it to affect large volumes of magma.

As long as the Soret effect produces magma near the wall of a chamber that has a density which causes it either to rise to the top or sink to the bottom of the chamber (Fig. 13-9a and b, respectively), convection will not rehomogenize this magma with the main body of magma. Large volumes of differentiated magma will then be able to accumulate slowly from the thin boundary layer. Hildreth (1981), for example, believes that Soret diffusion near the walls of chambers of convecting bodies of granitic magma is responsible for developing water-rich melt that convects upward and accumulates near the roof. In addition to water, many other components are concentrated in this buoyant boundary layer. Their accumulation results in a zoned body of magma, which, if erupted, can produce ash flows that preserve, in their stratigraphy, the record of the emptying of the chamber from the roof downward. The compositional variation in these rocks cannot be accounted for simply by fractional crystallization, which is evidence that some other process, such as Soret diffusion, must be responsible for their differentiation.

13-8 PNEUMATOLITIC ACTION

When a volatile-rich magma rises toward the Earth's surface, the decreasing pressure may allow a separate fluid phase to form, which, being lighter, rises more rapidly than the magma. The streaming of the volatile-rich phase through the magma depletes the lower part of the body in those elements that preferentially enter the fluid and transports them to the upper part of the chamber, where they may accumulate beneath an impervious roof or escape to form pegmatites and hydrothermal solutions. A fluid phase can also separate from a magma as a result of the crystallization of anhydrous minerals, which enriches the melt in water. This type of separation, known as *resurgent boiling*, may not be as effective in differentiating magma, especially if the fluid phase separates after a large fraction of the magma has crystallized (Sec. 11-6 and Prob. 11-9).

The fluid phase is rich in the volatile magmatic constituents, such as H_2O, CO_2, S, F, Cl, B, and P, but also contains less volatile elements that do not readily enter common rock-forming minerals but are not present in high enough concentrations to form their own minerals. These include Li, Be, Sc, Ga, Ge, As, Rb, Y, Zr, Nb, Mo, Ag, In, Sn, Sb, Cs, Ba, rare earth elements (REE), Hf, Ta, W, Re, Au, Pb, Bi, Th, and U. Some of these elements, on rising to the cooler upper parts of a magma chamber or on entering fractures in the roof rocks, precipitate from the fluid to form the common minerals of pegmatites. Thus, along with quartz, feldspar, and mica, minerals such as lepidolite (Li), spodumene (Li), beryl (Be), zircon (Zr), apatite (REE), scheelite (W), and uraninite (U, Th) form. Many of the other elements remain in the fluid and precipitate at much lower temperatures from hydrothermal solutions. Of these, the group that is characteristically associated with sulfides—the chalcophile elements (Cu, Zn, As, Mo, Ag, Au, and Bi)—is of great economic importance.

In addition to transporting certain elements, an immiscible fluid phase can play a role in differentiation by affecting other processes, such as convection or crystal settling. The presence of gas bubbles in a boundary layer of hydrous siliceous magma would increase the buoyant force and augment the segregation of a low-density melt that was being formed by Soret diffusion or fractional crystallization. On the other hand, gas bubbles rising through the Kilauea Iki lava lake significantly decreased the distance olivine crystals were able to sink and decreased the vigor of thermal convection.

Magmatic differentiation by volatile transfer may play a major role in determining the composition of carbonatites. Inspection of Figure 13-19 reveals that immiscible carbonate melts that coexist with most silicate liquids should contain significant amounts of alkalis. Indeed, immiscible phonolitic liquid (nepheline in diagram) can coexist only with a sodium carbonate melt. Although natrocarbonatites do occur, the majority of carbonatites are predominantly calcite-bearing rocks. Most calcite carbonatite intrusions, however, are surrounded by zones of intense alteration in which rocks are replaced by aegerine and orthoclase. This alkali metasomatism is known as *fenitization*, named after the locality where it was first identified—the Fen district of southeastern Norway. Fenitization is clear evidence that carbonatite melts give off considerable quantities of alkalis during crystallization. It seems likely, therefore, that volatile extraction is responsible for the low alkali contents of most carbonatites.

13-9 MAGMATIC ASSIMILATION

The composition of a magma rising from its source can be modified by interactions with material encountered on route to the final crystallization site. Contamination can result, for

example, from the incorporation and assimilation (i.e., melting) of xenoliths that come from the walls of the magma conduit or founder from the roof of a magma chamber. On the other hand, magma can encounter and mix with other bodies of magma, which may come from the same source at an earlier time or from completely different sources. In this section we deal with the assimilation of solid rocks by magma and leave to Sec. 13-10 the problem of magma mixing.

Evidence of contamination of magma through the assimilation of country rocks can be found at the margins of many intrusions. It is particularly striking when minerals are produced that could not possibly form if there had been simply fractional crystallization. Quartz, for example, can be found near the margins of nepheline-bearing intrusions that were contaminated by quartz-bearing sedimentary rocks. Conversely, nepheline can be found at the margins of some tholeiitic bodies where they intrude limestones. Despite this clear evidence of contamination, assimilation of crustal rocks cannot produce large quantities of contaminated magma because of the large heat of fusion necessary to melt country rocks and the limited amount of heat available in the magma.

One of the lowest-melting fractions obtainable from sedimentary rocks is generally of granitic composition. It is not surprising, therefore, to find that where magmas have assimilated sedimentary rocks, contaminated magmas are intermediate in composition between the initial magma and a granitic end component. This indicates that melting of xenoliths is not wholesale but is only of that fraction which requires the least amount of heat to fuse. Partially digested xenoliths of more refractory sedimentary rocks are consequently common in such contaminated magma. Most of the Cretaceous Monteregian intrusions in Quebec, for example, are surrounded by zones of contaminated magma (Philpotts 1974). The early mafic phase in these bodies is nepheline normative, but where the country rocks are siltstones and shales, a zone of *rheomorphic breccia* is present which grades from a granitic rock carrying abundant refractory xenoliths of quartz-rich and alumina-rich sedimentary rock in contact with the country rocks (Fig. 4-44) to a progressively more silica-undersaturated rock with fewer distinct xenoliths toward the interior of the intrusions.

Where intruded sedimentary rocks contain abundant shale, the granitic fraction causing contamination is peraluminous. Such a fraction may alter the abundance of the early crystallizing minerals. Augite, for example, may react with the silica and alumina derived from the sedimentary rocks to produce anorthite and orthopyroxene according to the reaction

(potential augite)   (contaminant in magma)

$$Ca(MgFe)Si_2O_6 + \quad Al_2O_3 + SiO_2 \quad \longrightarrow$$

$$\text{(anorthite)} \qquad \text{(opx)}$$

$$CaAl_2Si_2O_8 + (MgFe)SiO_3$$

Thus a magma that might initially have crystallized to a gabbro would, on being contaminated in this way, form a norite containing anorthite-rich plagioclase. The Haddo House complex of Aberdeenshire, Scotland, provides a classic example of this type of contamination, where a border zone of norite contains abundant xenoliths of aluminous metamorphic rocks in all stages of digestion (Read 1923). Longhi has argued that the early crystallization of orthopyroxene in most large layered gabbroic intrusions, such as the Stillwater and Bushveld complexes, is evidence that these mantle-derived magmas have assimilated continental crustal material. He points out that similar magmas in oceanic regions rarely crystallize orthopyroxene, and when they do, it is always preceded by Ca-rich pyroxene (Campbell 1985).

The effect of carbonate sedimentary rocks on silicate magmas is dramatic because of the reactive nature of calcite and dolomite with silica-rich melts. This can be illustrated with the reaction

(dolomite)   (albite in magma)

$$CaMg(CO_3)_2 + \quad NaAlSi_3O_8 \quad \longrightarrow$$

(diopside)   (nepheline)   (gas)

$$CaMgSi_2O_6 + NaAlSiO_4 + 2CO_2$$

In all such reactions CO_2 is liberated, and CaO and MgO from the carbonate react with silica in the magma to produce calcium magnesium silicates. This, in turn, reduces the amount of silica available to form silica-saturated minerals, such as the albite in the reaction above. These reactions are therefore commonly referred to as *desilication* reactions. Daly (1933) proposed that all alkaline igneous rocks were formed by the desilication of subalkaline magmas through limestone assimilation. Although this can occur on a small scale at the contact of intrusions (Tilley 1952), alkaline rocks are now known to have a primary source in the mantle.

The amount of magmatic assimilation that can take place depends on the thermal energy of the magma. Most magmas are not superheated; that is, their temperatures lie somewhere between the liquidus and solidus, and thus they would have a porphyritic texture if quenched rapidly enough. With such magmas, the heat needed to raise the temperature of the country rock to its melting point and then to supply the latent heat of fusion can come only from the crystallization of primary minerals. Thus the amount of assimilation that takes place is normally offset by an equivalent amount of crystallization. Large quantities of magma containing a significant fraction of assimilated material are therefore not likely to form.

Some magmas may be superheated and therefore capable of assimilating more material. Because the liquidus of dry magmas increases with increasing pressure, the decompression associated with the rapid rise of magma causes resorption of crystals and may generate magma that is well above its liquidus. Also, because the Earth was hotter early in its history, mantle-derived magmas were hotter in the Archean than they are today and were therefore capable of assimilating larger amounts of crustal rocks (Sparks 1986). The Archean ultramafic lavas known as komatiites, for example, were capable of assimilating three times as much crustal material as modern basalts.

Assimilation is greatest when the intruded rocks are near their melting point, and they have a large proportion of a low-melting fraction. The heat needed to melt this low-melting fraction of country rock is $\Delta H_m \times M_c$, where ΔH_m is the latent heat of fusion and M_c is the mass of country rock. This amount of heat withdrawn from the magma can be expressed as $C_p M_m \Delta T$, where C_p is the heat capacity of the magma, M_m the mass of

magma affected, and ΔT the change in temperature of the magma. Equating these two and rearranging gives

$$\frac{M_m}{M_c} = \frac{\Delta H_m}{C_p \Delta T} \qquad (13\text{-}19)$$

With typical values for ΔH_m and C_p of 400 kJ kg^{-1} and 0.8 kJ kg^{-1} C^{-1}, respectively, the ratio of M_m/M_c is $500/\Delta T$. Thus if 10% country rock were to be assimilated, the temperature of the magma would be lowered by 50°C. If the initial magma were not superheated by 50°C, the lowering of temperature would cause crystallization to occur, which in turn would liberate heat and maintain the magma at approximately the same temperature, provided that the heats of fusion and crystallization were similar (Prob. 13-13).

The way in which a magma rises through the crust may also play an important role in determining the amount of assimilation that occurs. If magma emplacement is through dikes or diapirs that force aside the country rocks, assimilation will normally be minimal. If the flow of magma in a dike is turbulent, contact temperatures are higher than when the flow is laminar, and thus assimilation is more likely. Turbulence occurs when the critical Reynolds number exceeds 2300 (Eq. 3-15). This is possible only for magmas with low viscosities, such as those of basaltic and komatiitic composition, and, then, only in dikes that are more than a few meters wide (see Fig. 3-8). Many mafic dikes, however, satisfy these conditions, and zones of granophyre between quartz and feldspar grains in the wall rocks attest to the partial melting caused by the intrusion of such dikes. The more primitive a magma is, the hotter it is, and the lower is its viscosity. Consequently, primitive magmas are more likely to assimilate crustal rocks than more evolved, cooler magmas. Huppert and Sparks (1985) estimate that komatiite magmas can be contaminated with up to 30% crustal material.

If magma rises through the crust by stoping or melting, assimilation of its roof rocks may be extensive. Near the surface of the Earth, where the roof of a magma chamber may be highly fractured, stoping of xenoliths is an efficient means of magma emplacement. If the xenoliths are denser than the magma, which is likely for all but the most mafic magmas, they will sink and possibly be assimilated, depending on their size. Deeper in the Earth, where rocks behave plastically, xenoliths are less likely to form. There the roof rocks may be melted in situ, as hot convecting magma continually removes the partially melted country rocks. These two styles of emplacement are illustrated in Figure 13-23.

Whether the magma melts the roof directly or melts sinking xenoliths, the necessary heat must come from the crystallization of the magma if there is no superheat. Convection within the magma keeps the temperature at the top and bottom of the chamber almost the same (Eq. 13-14), so crystallization takes place on the floor of the chamber, where the pressure is greatest. Simultaneous melting of roof rocks and crystallization on the floor means that the volume of magma could remain constant if there were no loss of heat to the surroundings. The result would be a volume of magma that rises through the crust, continuously changing its composition as it goes. At first the magma might be entirely mantle-derived, but as it ascends it would steadily be contaminated by crustal material. Because

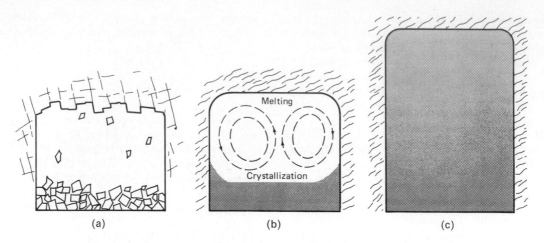

FIGURE 13-23 Upward migration of magma body through crust. (a) At shallow depths, fractured crust provides xenoliths which may sink through the magma, displacing it upward. Partial melting of xenoliths is accompanied by crystallization of magma. (b) At greater depths, where the crust is ductile, bodies of convecting magma can ascend by partially melting roof rocks while crystallizing refractory minerals on the floor. (c) Crust through which a convecting body of magma has passed becomes stratified by this "zonerefining" process, with the most refractory minerals concentrated toward the base and the least refractory ones toward the top.

the passage of this zone of melt through the crust is similar to the metallurgical process of zone refining, it is commonly referred to as *zone melting*. The distance a magma can rise by assimilating and crystallizing material is of course limited by the amount of heat lost to the surroundings. Oxburgh and McRae (1984) estimate that a mantle-derived magma, on being emplaced in the crust, may have enough energy to rise by zone melting a distance equivalent to the height of the magma chamber.

For magmas that are not superheated, assimilation causes crystallization of minerals with which the magma is already saturated. These minerals would crystallize in the normal course of events, even if there was no assimilation; assimilation simply causes them to crystallize earlier. Furthermore, the composition of the low-melting contaminating fraction may be similar to that of the final fractionated liquid from the original magma. Thus assimilation may simply bring on crystallization earlier and increase the amount of late stage differentiate. Because the precise quantities of the various rock types in an igneous intrusion are rarely known, the effects of assimilation may not be readily detectable from major and trace element chemistries. It is here that isotopic studies are particularly useful.

As shown in Sec. 21-3, the isotopic composition of the Earth's mantle and crust have steadily evolved along different paths throughout much of geologic time. As a result, magmas derived from the mantle have isotopic compositions that are quite distinct from those derived from the crust. Thus the amount of crustal contamination in a mantle-derived magma can be gauged from the degree to which the mantle values are disturbed. Of particular use in this regard are the isotopic ratios $^{87}Sr/^{86}Sr$ and $^{143}Nd/^{144}Nd$. In both pairs the isotope in the numerator is the product of radioactive decay, ^{87}Sr coming from ^{87}Rb and ^{143}Nd from ^{147}Sm. Because Rb is strongly fractionated into the Earth's crust, the $^{87}Sr/^{86}Sr$ ratio in crustal

rocks is higher than that in mantle-derived rocks. Conversely, Sm is not as strongly fractioned into the crust as Nd, and therefore the $^{143}Nd/^{144}Nd$ ratio in crustal rocks is lower than in mantle ones.

The Sr and Nd isotopic study by DePaolo (1985) of the Kiglapait layered intrusion in Labrador provides a good example of how isotopes can be used to determine the amount of assimilation. The initial magma was derived from the mantle and consequently has $^{87}Sr/^{86}Sr$ and $^{143}Nd/^{144}Nd$ ratios typical of that source. During crystallization of the intrusion, however, these ratios changed. Using the composition of the plagioclase as a measure of the extent of fractional crystallization, DePaolo showed that the Kiglapait chamber was first filled by several surges of uncontaminated mantle-derived magma. During this period isotopic ratios remained essentially constant. Once the chamber was no longer being replenished, however, the isotopic ratios steadily increased toward a crustal value (Fig. 13-24). A nearby Archean gneiss was the probable source of this contamination, and using its isotopic composition as the contaminating end member, the rate of assimilation was calculated to be only 1 to 4% of the rate of crystallization. The latent heat of crystallization would have been more than adequate to provide the heat necessary to cause this amount of assimilation. Physical processes, such as the rate of heat and chemical transfer through a boundary layer, the rate of magma convection, or the degree of fracturing of roof rocks, are therefore more likely to have controlled the rate of assimilation.

Isotopic ratios can give misleading impressions of the amount of assimilation that has taken place if used carelessly. Continental flood basalts the world over vary only slightly in composition, but their initial $^{87}Sr/^{86}Sr$ ratios vary widely. If this variability is due to different degrees of crustal assimilation, bulk compositions might be expected to vary much more than they do. The amount of contamination in these rocks must

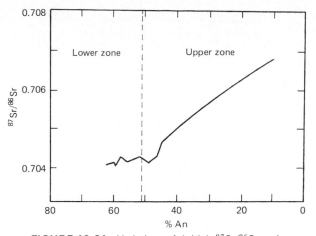

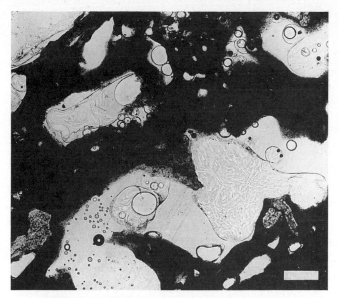

FIGURE 13-24 Variation of initial $^{87}Sr/^{86}Sr$ ratio with anorthite content of plagioclase in rocks of the Kiglapait layered intrusion, Labrador. [After DePaolo (1985).] During crystallization of the Lower Zone, several pulses of new magma maintained the initial $^{87}Sr/^{86}Sr$ ratio at a relatively constant mantle value. During crystallization of the Upper Zone, the magma was no longer replenished and was progressively contaminated by crustal material with a high $^{87}Sr/^{86}Sr$ ratio.

FIGURE 13-25 Intimately commingled basaltic and rhyolitic glasses from the Newberry volcano, central Oregon. Plane light; scale bar is 1 mm.

therefore be small, but it must involve a fraction that contains a disproportionately high concentration of strontium isotopes. Of course, the $^{87}Sr/^{86}Sr$ ratio of the assimilated rocks also plays an important role.

13-10 MIXING OF MAGMAS

Evidence for commingling of magmas is now recognized to be common. The magmas may be successive pulses from a common source, in which case their compositions may be similar. On the other hand, they may have completely different sources and be related in time only through a common thermal event; such magmas may have very different compositions. Most large magma bodies are not formed by a single injection of magma but are the result of numerous pulses, each of which may have slightly different compositions. In addition, magma in a chamber may cool, fractionally crystallize, or assimilate country rocks before the next pulse arrives. Successive batches may therefore encounter resident magmas that have had their compositions and physical properties modified by processes taking place in the chamber. Mantle-derived magmas bring heat into the crust, where partial melting can generate bodies of silicic magma. It is not unusual, therefore, to find mafic and silicic magmas commingling, as is clearly evidenced by the intimate mixtures of siliceous pumice and basaltic glass in some ash flows (Fig. 13-25), and the presence of pillowlike bodies of basaltic rock in some granites (Fig. 4-46).

The commingling of magmas may result in a range of products. Magmas may, for instance, not mix at all but simply remain as discrete bodies or xenoliths of one in the other. At the other extreme, they may mix completely to form a homogeneous hybrid magma. Whether magmas mix or not depends on how different their physical properties are. Density differences, for example, determine whether a second magma rises through or ponds at the bottom of an earlier magma. Viscosity contrasts determine how readily turbulent plumes of magma can bring about mixing. And the temperatures and volumes of the magmas and their solidification temperatures set further limits on the degrees of mixing. Studies by Huppert and Sparks (1980, 1988), Campbell and Turner (1986), and Sparks and Marshall (1986) are among a number of recent papers that quantify the effects these various factors have on magma mixing.

Two possible situations can result when one magma is intruded into another magma of different density. If the new magma enters at the base of a chamber in which the resident magma is more dense (possibly from cooling, for example), the new magma buoyantly rises in a plume to the top of the chamber, where it spreads laterally beneath the roof (Fig. 13-26c and d). If the new magma is denser than the resident magma, it may rise as a fountain into the chamber if intruded rapidly enough, but it then sinks and accumulates on the floor of the chamber (Fig. 13-26a and b).

If the rate of intrusion of new magma into a resident magma is great enough, turbulence within the plume or fountain causes entrainment of the resident magma and some intimately commingled or mixed magma is generated. The bulk density of this hybrid magma is intermediate between the other two magmas, and it therefore accumulates between them (Fig. 13-26b and d).

Campbell and Turner (1986) have shown that the amount of mixing in a turbulent fountain depends strongly on the velocity at which magma is intruded and the viscosity of the resident magma. They define a Reynolds number based on the viscosity of the resident magma as

$$Re_r \equiv \frac{2r\bar{v}\rho_r}{\eta_r} \qquad (13\text{-}20)$$

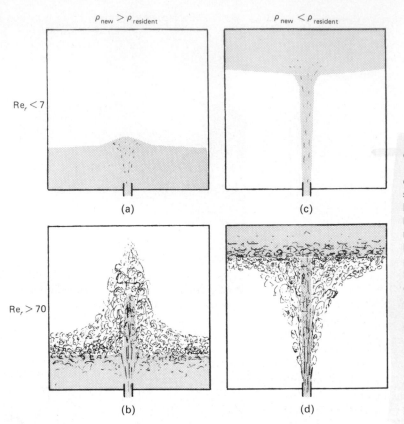

FIGURE 13-26 Four possible modes of replenishment of a magma chamber. (a) If the new magma has a greater density than the resident one, and intrusion velocities are low (Reynolds number < 7), the new magma ponds at the bottom of the chamber and there is no mixing. (b) If the new magma is denser but is intruded rapidly (Re > 70), it still ponds at the bottom of the chamber, but entrainment of the resident magma during fountaining produces a layer of commingled magmas between the two. (c) If the new magma is less dense than the resident one and intrusion takes place slowly, the new magma accumulates near the roof. (d) If intrusion of a less dense magma is rapid, entrainment of the resident magma in the fountain again produces a layer of commingled magmas between the two.

where r is the radius of the input pipe, $\bar{v}$ the mean input velocity, ρ_r the density, and η_r the viscosity of the resident magma. If the viscosity of the resident magma is more than eight times that of the new magma, then for values of $\text{Re}_r > 70$, vigorous mixing occurs along the margins of the fountain, but when $\text{Re}_r < 7$, almost no entrainment occurs. If basaltic magma, for example, is injected into granitic magma, the high viscosity of the granitic magma probably will keep the Reynolds number well below 70, so very little mixing of these magmas occurs at this stage.

If no mixing accompanies the emplacement of a hot dense magma beneath a cooler, less dense one, a stably stratified magma chamber is established, such as that illustrated in Figure 13-9c. Convection within the lower layer results in heat being transferred into the upper layer, but the density difference and low chemical diffusion rates allow little mixing of the two magmas to occur. This condition continues until the loss of heat from the lower layer results in crystallization; this, in turn, causes the composition and density of the residual liquid to change. If the lower layer is a picritic magma, crystallization of olivine causes the density of the residual liquid to decrease, and once it becomes less dense than the overlying magma, plumes of the fractionated lower liquid rise into the upper magma, and only then does mixing of the two layers take place.

When magmas with different temperatures commingle, heat diffuses so much more rapidly than material that thermal equilibrium is established long before significant homogenization of the magmas occurs. The fate of commingled magmas therefore depends on the physical state of the magmas after thermal equilibrium. If both are mainly liquid, homogenization

may occur, but if one is largely crystalline, no mixing will take place.

The equilibrium temperature of commingled magmas can be represented qualitatively in the mixing diagram of Figure 13-27. Consider two magmas, M_1 and M_2, with temperatures T_1 and T_2, respectively. If neither is superheated at first, their temperatures must fall between the liquidus and solidus for the respective compositions. If the heat capacities of all phases are the same, and we ignore, for the moment, heat associated with crystallization or melting, then the final temperature, T_f, of a mixture of these magmas falls on the line T_1–T_2 at a point determined by the proportions of the two magmas. As the amount of magma 1 increases, the lower is the final temperature of the mixture. However, because the compositions of the two magmas remain essentially unchanged during this period of thermal exchange, the solidus temperatures for both magmas remain the same. Thus, as the proportion of magma 1 increases, the final temperature of the mixture approaches the solidus temperature of liquid 2, and this melt shows higher and higher degrees of solidification. Indeed, once the magma becomes more than 60% crystalline, it behaves essentially as a solid, and no mixing with magma 1 can take place.

To determine quantitatively the final temperature of commingled magmas, we must take into account the heat capacities of the phases involved and the latent heat of crystallization of minerals that may form or be resorbed during mingling (Sparks and Marshall 1986). The heat liberated by either of the magmas on commingling with the other is given by

$$\Delta H = [-C_p(T_f - T) + (L \times \Delta F)]W \qquad (13\text{-}21)$$

FIGURE 13-27 When no crystallization or melting accompanies commingling of magmas, the final temperature of mixtures depends on the temperatures, proportions, and heat capacities of the magmas; final temperatures lie on the line between T_1 and T_2. If magmas contain crystals prior to commingling, resorption of crystals during heating of low-temperature magma results in absorption of heat, and crystallization during cooling of high-temperature magma causes liberation of heat. These additional heat effects are significant (Eq. 13-23) and typically cause the final temperature of mixtures to plot above the line T_1–T_2.

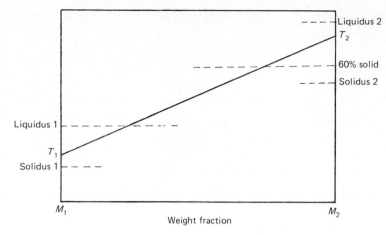

where C_p represents the heat capacities of all the liquids and solids, which are assumed to be equal, T_f the final temperature of the commingled mixture, T the initial temperature of the magma, L the latent heat of crystallization of minerals in the magma, ΔF the change in the fraction of crystals present in the magma as a result of the change in its temperature, and W the weight of the magma. If T_f is greater than T, then ΔH is negative; that is, heat will be taken in by the magma. If T_f is less than T, then ΔH is positive and heat must be liberated by the magma. When two magmas commingle, the heat given up by one must equal the heat taken in by the other; that is, the sum of the ΔH's must be zero.

$$[-C_p(T_f - T_1) + (L_1 \, \Delta F_1)]W_1 +$$
$$[-C_p(T_f - T_2) + (L_2 \, \Delta F_2)]W_2 = 0 \quad (13\text{-}22)$$

In this equation the subscripts 1 and 2 refer to magmas 1 and 2. If we divide both sides of the equation by $W_1 + W_2$, we can replace the actual weights of the two magmas by the weight fractions of the magmas in the mixture; that is, $X_1 = W_1/(W_1 + W_2)$ and $X_2 = W_2/(W_1 + W_2)$. Then, on rearranging Eq. 13-22, we can solve for the final temperature:

$$T_f = (X_1 T_1 + X_2 T_2) + \left(\frac{X_1 L_1 \, \Delta F_1}{C_p} + \frac{X_2 L_2 \, \Delta F_2}{C_p} \right) \quad (13\text{-}23)$$

This equation consists of two groups of terms; the first gives the contribution that the temperatures of the two magmas make to the final temperature, and the second group gives the temperature changes due to crystallization or resorption. Note that ΔF can be positive (crystallization) or negative (resorption), and that $X_2 = 1 - X_1$.

When two magmas commingle, the viscosity of the lower-temperature one decreases as its temperature rises toward the final equilibrium temperature, and the viscosity of the higher temperature magma increases as its temperature falls. The final viscosities of the two magmas depend on the equilibrium temperature, which is determined by the proportions of the two magmas (Eq. 13-23). If no crystallization takes place, the viscosities of the magmas can be expressed as a function of temperature using Eq. 2-9. If the cooling of the magma is accompanied by crystallization, the viscosity increases more than

it would due to cooling alone. If the fraction of solids present in the magma is F, the effective bulk viscosity of the magma, η_B, is related to the viscosity of the liquid, η_L, by

$$\eta_B = \eta_L (1 - 1.65F)^{-2.5} \quad (13\text{-}24)$$

Of course, as the low-temperature magma becomes hotter, crystals in it are resorbed, and the viscosity of the melt decreases more than it would due to heating alone. Thus, depending on the amount of crystallization or resorption that takes place, the final viscosities of the magmas vary as a function of the mixing proportions in a manner such as that shown in Figure 13-28.

If the two magmas have different compositions—silicic and mafic, for example—the viscosity of the lower-temperature

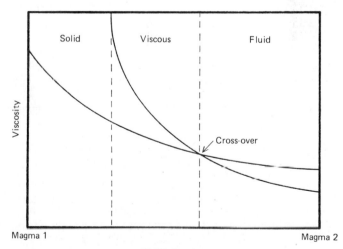

FIGURE 13-28 Variation in viscosity of commingled magmas as a function of equilibrium temperature, which depends on the proportions of the two magmas. Magma 1 is a low-temperature siliceous magma, and thus it initially has a high viscosity; on commingling and heating its viscosity decreases. The viscosity of magma 2, which is more mafic and hotter, increases as it cools and crystallizes; once it is 60% solidified it behaves essentially as a solid. [After Sparks and Marshall (1986).]

FIGURE 13-29 Mafic magma quenched to a pillow-like body in granite, Mount Desert Island, Maine. The mafic magma was more viscous than the granitic one at the equilibrium temperature, as indicated by the cuspate boundary on the pillow; sharp cusps point toward the more viscous liquid.

silicic magma is initially greater than that of the hotter mafic magma. On commingling and reaching the same temperature, however, the viscosity of the silicic magma may be greater or less than that of the mafic one, depending on the proportions of the two magmas (Fig. 13-28). Indeed, for one specific mixture, the viscosities are the same. This point is referred to by Sparks and Marshall (1986) as the cross-over of the viscosity curves. To the silicic side of this point, the viscosity of the mafic melt is greater than that of the silicic one. In the field, evidence for this viscosity cross-over is seen in the cuspate boundary developed on mafic pillows in granitic rock (Fig. 13-29); the sharp points on the cusps point toward the melt with the higher viscosity. As the proportion of the lower-temperature silicic melt increases, the viscosity of the mafic melt increases, especially as the mafic melt becomes more crystalline. From Eq. 13-24 we see that as the percentage of crystals in the mafic melt approaches 60%, the bulk viscosity approaches infinity; that is, the mafic phase behaves as a solid. Mixing cannot take place easily once the mafic melt reaches this stage. Small amounts of mafic magma in a silicic magma therefore behave as solid xenoliths.

On the basis of the relative viscosities following thermal equilibration, commingled magmas can be divided into three types (Fig. 13-28). The first involves mixtures containing a large percentage of the higher-temperature mafic melt, which after mingling is still more fluid than the silicic one. In the second, the proportion of mafic melt is less, and after thermal equilibration its viscosity is higher than that of the silicic one. In the third type the mafic magma becomes essentially solid during commingling.

Sparks and Marshall (1986) have calculated how these three types of commingling vary as a function of the composition of the higher-temperature mafic magma (Fig. 13-30a). The high solidus temperature of mafic magma with a high magnesium number causes the viscosity to rise rapidly when this magma mingles with a low-temperature siliceous magma. Consequently, the cross-over in the viscosity curves and the point

at which the percentage of crystals in the mafic melt reaches 60 occur at relatively low contents of admixed siliceous melt. As the magnesium number of the mafic magma decreases, so does its solidus temperature, and the cross-over and 60% solidification points occur at progressively higher contents of siliceous magma. Thus a mafic magma with a low magnesium number is more likely to remain fluid after commingling with siliceous magma than is one with a high number.

The composition of the siliceous magma can also vary, especially in water content. This is of particular importance because water both lowers the solidus temperature and decreases the viscosity. By reference to Figure 13-27 it can be seen that when the temperature of the siliceous magma (M_1) is lowered by the addition of water, the point at which the mafic magma becomes essentially solid is shifted to smaller admixtures of siliceous melt. Furthermore, the lowering of viscosity caused by the presence of water results in the cross-over in the viscosity curves occurring at more mafic mixtures. The effect of water in the siliceous magma is, therefore, to increase the range of mixed magma compositions in which the mafic fraction becomes essentially solid during commingling. This range would be decreased if the mafic magma also contained water.

The combined effects of variable magnesium number in the mafic magma and water content in the silicic magma on the commingling of magmas are illustrated in Figure 13-30b. It is believed that little homogenization of the magmas can take place if the mafic phase becomes essentially solid during commingling. If, however, it remains in the viscous or fluid fields, homogenization is possible if there is sufficient time. With this proviso then, we see that under all conditions hybridization is favored by having a large proportion of mafic magma in the mix, especially if it has a low magnesium number. On the other hand, the likelihood of homogenization decreases with increasing water content in the silicic magma.

Production of a homogeneous hybrid magma first involves the intimate commingling of two magmas, but eventually diffusion of components from one to the other must take place. The driving force for diffusion comes from the difference in the chemical potentials a component has in the two melts (Sec. 9-3). The chemical potential of a component, of course, depends on its concentration (Eq. 9-28), and components diffuse from regions of high concentration to ones of low concentration to equalize chemical potentials. The rate at which diffusion can homogenize magmas can be derived from *Fick's law:*

$$J_z = -D_i \frac{\partial c}{\partial z} \qquad (13\text{-}25)$$

where J_z is the flux of a component i diffusing in the z direction in response to a concentration gradient $\partial c/\partial z$ through a medium in which the diffusion coefficient for that component is D_i. This equation is identical in form to Fourier's law for heat transfer (Eq. 5-3), and just as we were able to determine rates of cooling from it (Eq. 5-11), we can determine rates of change of concentrations from Eq. 13-25; that is,

$$\frac{\partial c}{\partial t} = D \frac{\partial^2 c}{\partial z^2} \qquad (13\text{-}26)$$

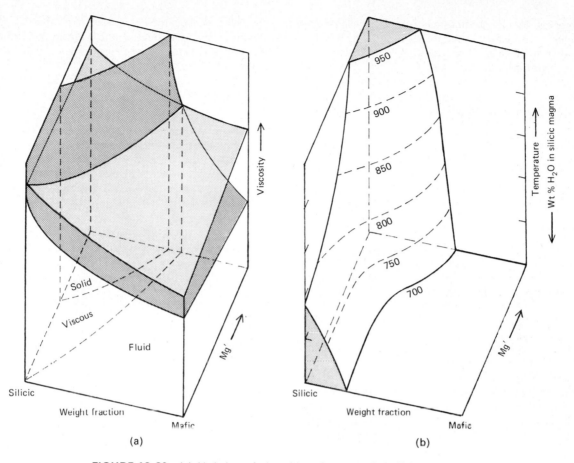

FIGURE 13-30 (a) Variation of viscosities of commingled silicic and mafic magmas as a function of the proportions of the two magmas and the magnesium number of the mafic magma. Viscosity of the mafic magma indicated by the darker-shaded surface and that of the silicic one by the lighter-shaded surface. Once the mafic magma is essentially solid, no mixing of the magmas is likely to occur. (b) The area marked solid in (a) is the shaded area near the top of this block diagram, which shows the effect of the addition of water to the mixing ability of silicic and mafic magmas. The higher the water content of the silicic magma, the lower is its temperature and the greater quenching effect it has on the commingled mafic melt. As the magnesium number of the mafic melt decreases, so does the temperature of the mafic melt and so the less likely is it to be quenched by the silicic melt. Commingled magmas falling within the shaded volume will be cool enough to cause the mafic melt to become essentially solid and no mixing of the magmas is likely to occur. [After Sparks and Marshall (1986).]

Values for D in silicate melts at magmatic temperatures are very small, being on the order of 10^{-10} to 10^{-14} m^2 s^{-1} (Hofmann 1980). The rate at which these diffusion coefficients cause compositional gradients to change with time can be calculated using the solutions given in Chapter 5 for thermal changes; simply read $\partial c/\partial t$ for dT/dt, and replace the thermal diffusivity with the diffusion coefficient. To obtain an approximate idea of these rates, we can calculate the time necessary to homogenize a sphere of liquid. We will define the homogenization time as that necessary to raise the concentration of the diffusing component at the center of the sphere to more than 98% of its concentration in the material surrounding the sphere, which is assumed to remain constant. Under these conditions $(Dt)^{1/2} = 0.75r$, where r is the radius of the sphere. If we con-

sider, for example, the diffusion of sodium into a sphere of basalt where the value of D_{Na} is 10^{-10} m^2 s^{-1} at 1200°C, 1.8 years would be required to homogenize the sodium if the radius were 10 cm, and 180 years if the radius were 1 m. At 1000°C, $D_{Na} = 10^{-11}$ m^2 s^{-1}, and then the 10-cm-radius sphere would take 18 years to homogenize, whereas the 1-m sphere would take 1800 years (Prob. 13-16).

The actual calculation of the homogenization of a sphere is far more complicated than this, for as sodium diffuses into the basalt, the change in composition would probably change the diffusion coefficient. Also, the inward diffusion of sodium would have to be coupled with the outward diffusion of other components in order to conserve mass. Nonetheless, the calculation does provide an order-of-magnitude estimate of the times and distances involved with diffusion. They clearly show why magmas must be intimately commingled to begin with if they are to produce homogeneous hybrids; diffusion then does not have as far to transport components.

An additional factor making homogenization difficult is that diffusion coefficients decrease exponentially with falling temperature according to the relation

$$D = D_0 e^{-E/RT} \qquad (13\text{-}27)$$

where D_0 is a constant, E an activation energy, R the gas constant, and T the absolute temperature (review the explanation for the Arrhenius relation in Eq. 2-9). If magma mixing brings about a significant lowering of temperature of one of the magmas, the distance components can diffuse in that melt will be dramatically reduced (Prob. 13-17).

According to Eq. 13-26, diffusion will be most rapid at first when the concentration gradients are steep; it will then decrease as the gradients diminish, especially if the temperature falls during this time and the diffusion coefficients decrease. It is unlikely, then, that diffusion alone can produce completely homogeneous hybrid magmas in the times available during typical magmatic episodes. Convective movements are probably necessary to help bring about complete homogenization.

A further problem with homogenizing magmas by diffusion is that even when the magmas have significantly different compositions, the chemical potentials or activities of components in the magmas may not be very different because of nonideality in the melts. This can be demonstrated by comparing the activity of SiO_2, for example, in basaltic and granitic magmas (Carmichael et al. 1970). If a tholeiitic basaltic magma contains crystals of olivine that are in the process of reacting to form orthopyroxene, we can write the following reaction for the magnesium components:

(olivine)   (magma)   (orthopyroxene)
$$Mg_2SiO_4 + \quad SiO_2 \quad = \quad 2MgSiO_3 \qquad (13\text{-}28)$$

The silica in this reaction is not pure quartz but is molten silica in solution in the magma. The forsterite is in solid solution in the olivine and the enstatite in the orthopyroxene. The equilibrium constant (K) for this reaction can be written in terms of the activities of these three components in their respective solutions; that is,

$$K = \frac{(a_{En}^{Opx})^2}{(a_{Fo}^{Ol})(a_{SiO_2}^{L})} \qquad (13\text{-}29)$$

where $a_{SiO_2}^{L}$ is the activity of silica in the magma.

In a basalt, the olivine and pyroxene are relatively magnesian, and both solid solutions are nearly ideal, so mole fractions (X_i) can be used in place of activities without introducing serious errors. Furthermore, from Eq. 9-56, $\ln K = -\Delta G/RT$, where ΔG is the free-energy change of the reaction. Equation 13-29 can therefore be rewritten to express the activity of silica in the magma as

$$\ln a_{SiO_2}^{L} = \frac{\Delta G}{RT} + 2 \ln X_{En}^{Opx} - \ln X_{Fo}^{Ol} \qquad (13\text{-}30)$$

The value of ΔG can be obtained from Table 7-1. If the mole fractions of forsterite and enstatite in olivine and orthopryoxene respectively are known, the activity of silica in the basaltic melt can be calculated.

In the granitic melt, the activity of silica might be determined by the presence of phenocrysts of quartz, which would buffer the silica activity through the following reaction:

(quartz)   (magma)
$$SiO_2 \quad = \quad SiO_2$$

and the activity of silica in the magma is given by

$$\ln a_{SiO_2}^{L} = -\frac{\Delta G}{RT} \qquad (13\text{-}31)$$

Using Eqs. 13-30 and 13-31, Carmichael et al. (1970) showed that the activity of silica increases from about $10^{-0.15}$ in basaltic magma at 1150°C to only $10^{-0.10}$ in granitic magma at 850°C (Prob. 13-18). This small increase in activity is surprising in view of the corresponding large increase in weight percent of silica, which goes from 48 to 75. This is clear evidence of the nonideal behavior of these melts. Indeed, as the mafic melt becomes more iron-rich, it approaches the immiscibility field (Fig. 13-17). Once it reaches the two-liquid field, melts with very different compositions coexist, but because the activities of components in both melts are equal, there is no driving force to cause homogenization. Therefore, although Figure 13-30b indicates that mafic melts with low magnesium numbers have physical properties that allow them to commingle more easily with silicic magmas than ones with high magnesium numbers, these same melts are closer to the two-liquid field, and the potential for homogenization is diminished.

Under appropriate conditions, magma mixing can produce completely homogeneous melts, in which the only evidence of mixing might come from isotopic data or assemblages of phenocrysts inherited from the parent magmas that are not in equilibrium with the hybrid magma. One of the best studied examples of such hybrid magmas is the Paricutín Volcano, Mexico (Wilcox 1954; McBirney et al. 1987). This volcano came into existence on February 20, 1943, and for nine years erupted over a cubic kilometer of basaltic andesite and andesite before the eruptive episode ended. The first 75% of the total erupted material varied little in composition, but it did contain xenoliths of siliceous basement rocks in all stages of fusion. The melts derived from these country rocks apparently migrated upward and collected near the roof of the chamber to form a density zoned and stratified magma chamber. The later-stage eruptions tapped the upper parts of this chamber to give a series of hybrid magmas that range in composition from the original basaltic andesite ($SiO_2 = 55$) to andesite ($SiO_2 = 60$). Part of the compositional variation is due to fractional crystallization of olivine and plagioclase, but McBirney et al. (1987) conclude that the most contaminated magma assimilated 20% siliceous material. In terms of Figure 13-30b, this admixture of siliceous melt in a mafic magma (basaltic andesite) with a relatively low magnesium number (60) is consistent with the mafic melt remaining liquid during commingling. It is not surprising, therefore, that this contaminated magma formed a homogeneous hybrid.

One of the most important consequences of magma mixing is that certain phase relations develop that cannot otherwise form by fractional crystallization of a single magma. Inspection of any multicomponent phase diagram (Chapter 10) reveals that melts become saturated with an increasing number of phases on undergoing fractional crystallization. A melt that starts crystallizing in the primary field of a single mineral soon becomes saturated with a second mineral; these two minerals then crystallize together until the liquid becomes saturated in a third, and so on, until the eutectic is reached where the number of phases is a maximum. A similar increase in the number of phases accompanies the fractionation of natural magmas. It is surprising, therefore, that some magmas which are multiply saturated can abruptly start crystallizing only one mineral. Such behavior is difficult to explain without resorting to magma mixing.

Fractional crystallization of basaltic magmas can, in part, be represented in the simple ternary system forsterite–anorthite–silica (Fig. 13-31). Consider a primary magma of composition X which, because it lies in the primary field of olivine, fractionates along the line $X1$ away from olivine until it reaches the cotectic with anorthite, whereupon it descends to the peritectic (P). With continued fractional crystallization the liquid descends the anorthite–orthopyroxene cotectic to the ternary eutectic (E).

If, during the fractional crystallization of magma X, a second pulse of magma mixes and homogenizes with it, the new

hybrid magma will have an intermediate composition that in general will not plot on a multiply saturated boundary but will, instead, fall in the primary field of just one mineral. Consider, for example, a resident magma that had fractionated to composition 2 on the anorthite–orthopyroxene boundary when a new pulse of primary magma with composition X mixes with it. The resulting magma would plot on the straight line joining compositions 2 and X, as long as no crystal resorption disturbs the composition. If only a small amount of magma X were added, the hybrid magma would, on cooling, fall in the orthopyroxene field. Accumulation of this mineral would then form a pyroxenite on top of the norite that would initially have been forming from magma 2. If the second pulse of magma were slightly evolved, for example with a composition of melt 1, and it mixed with a magma of composition 2, the hybrid magma would fall in the primary field of anorthite, and an anorthosite could form by accumulation.

Magma mixing may produce cumulates of economic value. Irvine (1977) has shown that layers of chromitite in mafic and ultramafic intrusions may owe their origin to precipitation from evolved magmas that mix with more primitive ones. Because the olivine–chromite cotectic is concave toward chromite, mixing of different magmas on this cotectic produces hybrid melts that fall in the primary field of chromite. The strong curvature on the sulfide–silicate liquid immiscibility field can also be responsible for the accumulation of magmatic sulfide deposits. If an evolved liquid near the ternary eutectic (e) in the system $FeO–FeS–SiO_2$ (Fig. 11-6) is mixed with less evolved liquid, for example of composition a in Figure 11-6, the hybrid magma would enter the two-liquid field, and a dense immiscible sulfide would separate and possibly segregate.

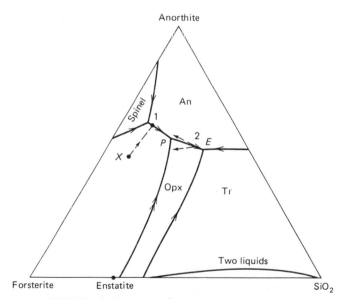

FIGURE 13-31 System forsterite–anorthite–silica at atmospheric pressure (Andersen 1915), showing the compositional trends produced by mixing a derivative magma 2 with either a less evolved magma, 1, or a primary magma, X. In the first case the hybrid magma moves into the primary field of anorthite, and a plagioclase cumulate can form. In the second case, the hybrid magma first moves into the field of orthopyroxene, and then into the field of olivine, allowing cumulates of these minerals to form.

13-11 TRACE ELEMENT FRACTIONATION BY MAGMAS

Magmatic processes and multicomponent phase relations are still too poorly understood for petrologists to explain precisely the major element variations in suites of differentiated igneous rocks. Trace element variations, on the other hand, are often simpler to account for. Unlike major elements, trace elements are not essential to the stability of the phases involved, and thus they play a relatively passive role. Nonetheless, their concentrations are affected by the magmatic processes.

When more than one phase is present, a trace element under equilibrium conditions partitions itself between the phases so that its chemical potential in each phase is the same; that is, for example, $\mu_i^\alpha = \mu_i^\beta$, where α and β are two of the coexisting phases (mineral, liquid, vapor). Making use of Eq. 9-34, we can write

$$RT \ln \frac{a_i^\beta}{a_i^\alpha} = -(\mu_i^{*\beta} - \mu_i^{*\alpha}) = -\Delta G^* \qquad (13\text{-}32)$$

where ΔG^* is the free-energy change for the transfer of element i from phase α to phase β at the particular pressure and temperature. Because trace elements are present in very small amounts, they can be expected to follow Henry's law (Eq. 9-36),

liquid is, according to Eqs. 13-37 and 13-19,

$$c_i^l = \frac{c_i^{ol}}{F} \qquad (13-42)$$

Thus the ratio of any two incompatible elements, a and b, is

$$\frac{c_a^l}{c_b^l} = \frac{c_a^{ol}}{F} \frac{F}{c_b^{ol}} = \frac{c_a^{ol}}{c_b^{ol}} = \text{constant} \qquad (13-43)$$

Clearly, the ratio of these two elements is independent of the fraction of liquid remaining and is simply equal to the initial ratio. Thus regardless of the degree of fractionation or the minerals crystallizing, the ratio of incompatible elements remains the same and serves as a signature of that lineage. It can be changed only by the assimilation of country rocks or magmas that have a different ratio.

The concentration of trace elements that have partition coefficients that differ significantly from unity can change by orders of magnitude during fractional crystallization. These elements can therefore be sensitive indicators of the degree of fractionation. Also, because different minerals have different distribution coefficients, the appearance of a new primary phase can produce a marked change in the trace element variation. Vanadium, for example, is an incompatible element in primary basalts and therefore becomes concentrated in fractionated liquids. However, it partitions strongly into magnetite, so that once this mineral starts crystallizing, the concentration of vanadium in the melt drops sharply. Similarly, strontium is an incompatible element in basaltic magma until plagioclase starts crystallizing.

The rare earth elements (REE, lanthanides La-57 to Lu-71) are particularly useful trace elements in the study of igneous petrogenesis. Each of these elements occurs in the trivalent state under normal magmatic conditions, except for europium, which occurs both as Eu^{2+} and Eu^{3+}, depending on the oxygen fugacity. They all occur in sixfold coordination, but as their atomic weight increases the ionic radius decreases from 1.03 Å in lanthanum to 0.86 Å in lutetium. Because of their identical charge and similar ionic radii, chemical processes do not discriminate strongly between the various REE, and when there is discrimination it varies systematically through the series; europium is the only exception, because of its different valence states. The combination of high charge and relatively large ionic radius make the REE incompatible in most early-crystallizing minerals. Their K_D values do, however, vary slightly—the larger the ionic radius, the more incompatible is the element. In addition, the abundance of each even-numbered REE is greater than that of adjacent odd-numbered ones. Plots of absolute abundance versus atomic number therefore have a distinct sawtooth appearance. This pattern is commonly smoothed by dividing the abundance of each REE by its abundance in a chondritic meteorite, which also shows similar even–odd fluctuations in abundance. The resulting chondrite-normalized plot (Fig. 13-33) provides a convenient means of viewing the overall enrichment in REE relative to a chondritic composition (presumably, a primordial composition from which the solar system formed) and the degree of fractionation of the light versus heavy REE.

The REE distribution coefficents between basaltic magma and the early-crystallizing minerals are sufficiently varied that REE abundance patterns in rocks can commonly be used to determine the minerals responsible for fractionation. In Figure 13-33, the distribution coefficients in Table 13-3 have been used to construct equilibrium distribution patterns of REE between the common primary minerals and a basaltic magma that is assumed initially to have had a chondritic abundance of REE; that is, $c^l/c_{\text{chondrite}} = 1$. Two sets of curves are given for each mineral. The lines marked with squares indicate the chondrite-normalized abundance of each REE in the basaltic magma following 75% crystallization ($F = 0.25$ in Eq. 13-39). The lines marked with X's indicate the abundance of each REE in the coexisting solid (Eq. 13-40).

Olivine, which is by far the most common primary mineral in basalts, almost completely excludes the REE, with the heavier (smaller ionic radii) ones being only slightly more compatible than the light ones. Early crystallization of olivine therefore invariably results in strong enrichment of REE in the melt, but the slope in the plot of chondrite-normalized abundance versus atomic number remains almost horizontal.

The REE are slightly less incompatible with respect to orthopyroxene, especially those toward the heavy end. Fractionation of orthopyroxene therefore results in the development of a prominent negative slope throughout the chondrite-normalized plot. Only the lightest three REE are incompatible with respect to augite; the remainder have distribution coefficients very near unity. As a result, fractionation of augite causes no change in the abundance of the REE except for La, Ce, and Nd, which become concentrated in the liquid. Separation of augite from a basaltic magma causes a distinct steepening of the REE pattern at the light end of the chondrite-normalized plot. Amphibole behaves in a similar way to augite.

All the REE exhibit approximately the same degree of incompatibility with respect to plagioclase, except for europium, which is significantly less incompatible than the rest. Eu^{3+} is probably as incompatible as the other trivalent REE, but Eu^{2+} is far less incompatible, substituting for Ca^{2+} in plagioclase. The more reducing the conditions, the greater will be the proportion of Eu^{2+} and the greater will be the difference in the europium distribution coefficient from those of the other REE. Because plagioclase always has a positive europium anomaly, that is, europium is relatively more abundant than the other REE in plagioclase, magmas from which plagioclase separates develop a negative europium anomaly. Such a negative anomaly exists in all lunar mare basalts, indicating that plagioclase fractionation was involved in their genesis.

Finally, garnet preferentially takes in REE with masses greater than europium and rejects those that are less. Consequently, basalt that has undergone garnet fractionation develops a strong negative slope to its REE pattern in the chondrite-normalized plot.

Basalt fractionation typically involves several minerals, making it necessary to calculate bulk distribution coefficients for the REE. Depending on the weight proportions of these minerals, various patterns develop in the chondrite-normalized plot (Probs. 13-21 and 13-22). Some of these patterns are sufficiently characteristic that they are diagnostic of the minerals causing the fractionation. In Figure 13-34, typical REE patterns

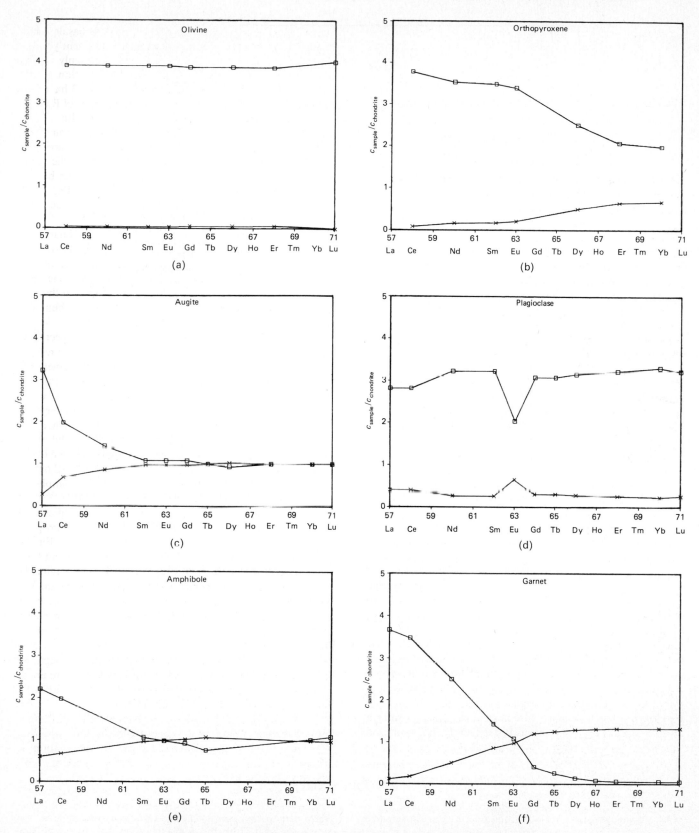

FIGURE 13-33 Chondrite-normalized abundances of the rare earth elements partitioned between common mineral (× 's) and basaltic melt (squares) following 75% equilibrium crystallization. The melt initially had a chondritic abundance of rare earth elements; that is, $c_{\text{sample}}/c_{\text{chondrite}} = 1$. See text for discussion.

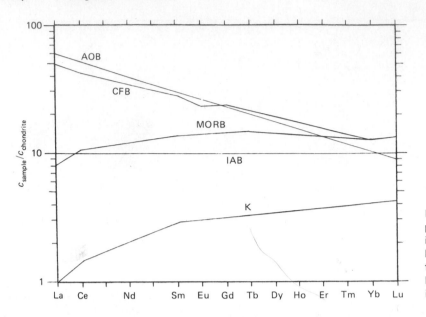

FIGURE 13-34 Chondrite-normalized plot of rare earth abundances in typical island arc basalts (IAB), alkali olivine basalts (AOB), tholeiitic continental flood basalts (CFB), mid-ocean ridge basalts (MORB), and Archean komatiites (K). See text for discussion.

are given for five major types of basalt (BVSP 1981). Each of these types shows varying degrees of REE enrichment relative to a chondritic composition, which is assumed to be similar to the primordial composition of the Earth. Such enrichment is consistent with the general incompatibility of REE. The pattern for each basalt, however, is distinct, indicating differences in the minerals involved in their fractionation or generation.

The simplest pattern is that of the island arc basalt, which shows a remarkably constant enrichment relative to chondrites of about 10 times over the entire range of REE. Reference to Figure 13-33 indicates that only olivine fractionation could give this flat REE pattern. Alkali olivine basalts also show a simple linear pattern but with strong enrichment in light REE. This enrichment could result from high degrees of fractional crystallization, but the high magnesium number and remarkably consistent REE pattern of these basalts argues against such an origin. Instead, the high enrichment is interpreted to result from very small degrees of partial melting in the source region, as discussed in Chapter 22. Also, because the light REE are strongly enriched relative to the heavier ones, garnet is likely to be a stable phase in that source region. Tholeiitic continental flood basalts also show enrichment in the light REE relative to the heavy ones, but the effect is not as pronounced as in the alkali basalts. Moreover, many continental flood basalts exhibit a distinct negative europium anomaly, indicating that plagioclase has been removed. It is significant that none of the other basalt types has a negative europium anomaly, even though most contain plagioclase phenocrysts. Clearly, no significant

amount of plagioclase can have been fractionated from these other basalts.

Two of the most peculiar, yet characteristic REE patterns are those of mid-ocean ridge basalts and Archean komatiites. Both are less enriched in light REE than heavy ones. Inspection of Figure 13-33 reveals that none of the common early-crystallizing minerals causes such enrichment. Fractional crystallization of magma derived from a source of chondritic composition cannot, therefore, be responsible for this pattern. This problem, however, has a simple solution. Consider a source containing augite that initially has a chondritic REE abundance. Partial melting of this rock gives rise to magma enriched in light REE (Fig. 13-33 augite) and leaves residual augite depleted in light REE. Subsequent melts derived from this depleted source will, themselves, be depleted in light REE relative to heavy ones. Mid-ocean ridge basalts and komatiites are therefore interpreted to come from a mantle source that has been depleted in incompatible elements by earlier periods of partial melting.

In closing this chapter it is important to emphasize that chemical variations in suites of igneous rocks are not due only to fractional crystallization. Another important factor is the partial melting and magma extraction process that takes place in the source region. This topic is dealt with in Chapter 22, where we will see that the Nernst distribution coefficients and Rayleigh and equilibrium fractionation models (Eqs. 13-37 and 13-39) are as applicable to partial melting as they are to fractional crystallization.

PROBLEMS

13-1. Using the data in Table 13-1, construct a triangular variation diagram for the weight percentages of CaO, Na_2O, and K_2O in the Kilauean lavas. Discuss the trend obtained in light of the variations found in the diagram involving the magnesium number (Fig. 13-1).

13-2. If the compositional trend in the Kilauean basalts from analysis 8 to 15 in Table 13-1 is due to assimilation of more siliceous rocks, use a Harker variation diagram to determine the composition and amount of the most siliceous material that could be added to analysis 8 to produce 15. Calculate the CIPW norm of the assimilated material, and comment on the likelihood of this assimilation mechanism being responsible for the compositional variation of lavas between nos. 8 and 15. In extending the trends to higher silica contents in the Harker diagram, certain oxides decrease, and where the first one reaches zero must be the most siliceous material that can be assimilated, because negative values of components are not possible.

13-3. The Kilauean lava whose analysis is given as no. 10 in Table 13-1 contains only 1.5% olivine phenocrysts; its analysis can therefore be treated essentially as that of a liquid. Using a K_D of 0.33, what forsterite content would you expect these phenocrysts to have?

13-4. Calculate the sinking velocities of olivine crystals with a density of 3500 kg m^{-3} and diameter of 2 mm and chromite crystals with a density of 5010 kg m^{-3} and diameter of 1 mm in a magma with a density of 2600 kg m^{-3} and Newtonian viscosity of 300 Pa s. Express your answer in cm day^{-1}. Check that Stokes law is applicable by calculating the Reynolds number in both cases. Comment on the relative importance of a mineral's density and grain size to rate of sinking. (Be careful with units and determining the radius of crystals.)

13-5. Stubby augite phenocrysts, with a density of 3410 kg m^{-3}, occur in the upper chilled margin of a 3.41-m-thick mafic alkaline sill and then are completely absent down to a distance 0.85 m above the base, where they form a cumulate layer. Augite phenocrysts in the upper chilled margin have diameters of 1 mm, whereas those in the cumulate layer have diameters of 2 mm. Calculations show that the magma took 7.22×10^5 s to solidify up to the 0.85-m level. This, then, was the length of time the crystals had to sink. The density of the magma was 2540 kg m^{-3}. Calculate the Newtonian viscosity of the magma assuming **(a)** that the crystal growth rate was controlled by a phase boundary reaction, and **(b)** that the crystal growth rate was controlled by a surface nucleation process. (*Hint:* First evaluate the constants in the growth rate equations 12-8 and 12-9.)

13-6. **(a)** Derive an expression for the distance through which a spherical crystal of radius r will sink in a magma if the crystal growth rate is diffusion-controlled and the viscosity and density of the magma remain constant. Use Eq. 12-6 for the crystal growth rate. Plot your equation in a graph of depth versus time.
(b) How do you think the velocity of the crystal will affect the growth rate of the crystal? How might the real sinking velocity of the crystal differ from that calculated in part (a)?

13-7. If the magma in Prob. 13-4 has a yield strength of 50 Pa, what minimum sizes of olivine and chromite grains would be required to cause crystal settling?

13-8. If a magma in which crystal sinking has occurred has a yield strength, the minimum size of cumulate crystal on the floor of the chamber is determined by the yield strength of the magma and the density contrast between the crystal and the magma (Eq. 13-7). Because we can measure the size and density of mineral grains in a sample of cumulate, we are left with two unknowns, the density and yield strength of the magma. If two or more minerals have accumulated together (Fig. 13-14), we can determine the density and yield strength of the magma, assuming, of course, that the yield strength played a role in determining the grain size. Later resorption of, or addition to, grains would disturb the relations. Olivine and chromite grains such as those illustrated in Figure 13-14 have diameters of 2.54 and 0.88 mm, respectively, and their densities are 3500 and 5010 kg m^{-3}, respectively. With the assumptions above, calculate the density and yield strength of the magma. Do the results

indicate that the proposed mechanism determining grain size has any validity?

13-9. A flood basalt is erupted rapidly and forms an extensive 50-m-thick lava lake. A 5-m-thick crust develops rapidly due to radiation cooling and water percolating into surface cracks. A 1-m-thick chill zone forms at the base of the lake, but the low thermal diffusivity of underlying rocks does not allow it to continue thickening rapidly. Following some cooling, the magma in the lake has an average viscosity of 10^3 Pa s, a density of 2700 kg m^{-3}, a thermal diffusivity of 5×10^{-7} m^2 s^{-1}, and a coefficient of thermal expansion of 5×10^{-5} K^{-1}. The surface of the lava lake is kept at the ambient temperature of 25°C, which in turn keeps the base of the crust (top of magma) at 1065°C. The top of the lower chill zone remains almost constant at 1150°C; heat lost into the underlying rocks is replaced by heat liberated by crystallization.
(a) Assuming that these conditions hold relatively constant for some period of time, will the magma convect?
(b) If convection does occur, what would the maximum convective velocity be?
(c) Approximately how long would the lava lake take to crystallize if there is convection, and how long, if there is not?
(d) Assuming that the convective velocity calculated in part (b) is valid for the first quarter of the crystallization period, how many revolutions might a neutrally buoyant plagioclase phenocryst make around a convecting cell in this time?

13-10. **(a)** If the heat loss from the convecting magma in the Palisades Sill was 4 W m^{-2} and the latent heat of crystallization was 400 kJ kg^{-1}, what was the rate of crystallization in kg m^{-2} day^{-1} (assume that crystallization takes place isothermally)?
(b) If crystallization takes place on the floor of the sill, and the cumulate minerals, which have a density of 3300 kg m^{-3}, form a layer with 40% porosity, what would the daily rate of rise of this layer be?
(c) At the rate determined in part (b), how long would it take the floor of the sill to rise to the olivine layer 15 m above the base?
(d) If the Rayleigh number for the convecting Palisades magma was 10^{11}, what approximately would the Nusselt number have been?
(e) Using the Nusselt number, calculate what the rate of rise of the cumulate layer would have been had the magma not convected. How long would it take in the nonconvecting magma to accumulate the same thickness as formed in one day in the convecting magma?

13-11. Draw a schematic velocity profile through the margin of a convecting basaltic magma, where partial melting of the country rocks has developed a boundary layer of buoyant granitic magma.

13-12. Using the method of Bottinga et al. (1982) (Sec. 2-3), calculate the densities of the following Kilauean lavas (Table 13-1) at the given temperatures, assuming that analyses are representative of liquids: no. 4, 1270°C; no. 5, 1265°C; no. 8, 1200°C; no. 9, 1160°C; no. 11, 1100°C; no. 13, 1075°C; no. 14, 1040°C; and no. 15, 1020°C. Plot the density of the lavas as a function of the magnesium number and comment on the shape of the graph.

13-13. Two spherical 1-m-diameter xenoliths, one of quartzite and the other of granite and both at 500°C, are incorporated into a basaltic magma. Both xenoliths absorb heat from the surrounding magma, but the granitic one, unlike the refractory quartz one, undergoes melting and thus absorbs additional heat. By the time the xenoliths reach 1000°C the granitic one is totally melted.

(a) If both xenoliths have a density of 2650 kg m^{-3}, their heat capacities (for solid and liquid) are 0.8 kJ kg^{-1} °C^{-1}, and the latent heat of fusion of the granite is 400 kJ kg^{-1}, calculate the total amount of heat withdrawn from the surrounding magma by each xenolith in raising its temperature to to 1000°C.

(b) If the heat withdrawn from the surrounding magma by the xenoliths results in crystallization of the basalt, what would the relative thicknesses of crystal mush be around the two xenoliths?

13-14. A basaltic magma at 1200°C and containing 10 wt % phenocrysts commingles with a 900°C rhyolitic magma containing 10 wt % phenocrysts of cristobalite. The weight proportions of basaltic to rhyolitic magma are 80 to 20. After commingling, all cristobalite phenocrysts are resorbed, whereas the percentage of phenocrysts in the basaltic magma increases to 30 wt %, which is made up of 60 wt % pyroxene and 40 wt % plagioclase. If the heat capacities of all phases are taken to be 0.8 kJ kg^{-1} °C^{-1} and the latent heats of fusion of cristobalite, pyroxene, and plagioclase are 135.8, 587, and 490 kJ kg^{-1}, respectively, what is the final temperature of the commingled magmas?

13-15. A basaltic and a rhyolitic magma commingle. Their initial temperatures are 1200 and 900°C, respectively, and neither contain any crystals at first. Their viscosities obey an Arrhenius relation (Eq. 2-9), where η_0 is 5×10^{-9} Pa s for the basalt and 10^{-13} Pa s for the rhyolite; their activation energies are 300 and 500 kJ mol^{-1}, respectively. On commingling, the basaltic magma cools and begins to crystallize, whereas the rhyolitic one becomes superheated. The mass fraction of crystals formed in the basaltic melt, F, is a linear function of temperature: $F = 4 - T°C/300$. The latent heat of crystallization is 500 kJ kg^{-1}, and the heat capacity of the magma is 0.8 kJ kg^{-1} K^{-1}. First, determine the temperature of the basaltic magma as a function of degree of crystallization, and then determine what admixture of rhyolitic magma would give that temperature. Second, plot the $\log_{10}$ of the viscosities of both magmas as a function of the weight fraction of rhyolite in the commingled mixture, making use of Eq. 13-24 to account for the crystallization of the basalt. Finally, at what fraction of rhyolite does the crossover in the viscosity curves occur?

13-16. (a) A granitic magma containing 5 ppm Rb and 5 ppm Cs is intruded by a second granitic magma containing 105 ppm Rb and 105 ppm Cs. Diffusion coefficients for Rb and Cs at 900°C, the temperature of both magmas, are $10^{-12.3}$ and $10^{-15.3}$ m^2 s^{-1}, respectively. Calculate how long it would take for the concentration of Rb in the Rb-poor melt to rise to 25 ppm at a distance of 1 cm from a planar interface between the two melts. (Modify Eq. 5-15 to solve this problem.)

(b) What is the concentration of Cs in this melt at this time?

(c) How near to the contact do you have to go to find a Cs concentration of 25 ppm?

(d) In light of the difference in the diffusion coefficients for Rb and Cs, which element would be best to use if you were looking for evidence that an apparently homogeneous magma had been formed by mixing of magmas with different trace element contents?

13-17. If, at 1200°C, one year is required for the sodium concentration in the core of a spherical globule of molten basalt immersed in granitic magma to rise to 98% of that in the surrounding granite, how long would it take to achieve the same result at 1100°C if the preexponential factor (D_0) in the Arrhenius relation describing the temperature dependence of the diffusion coefficient is 9.6×10^{-5} m^{-2} s^{-1} and the activation energy (E) is 163 kJ mol^{-1}?

13-18. Basaltic magma containing crystals of olivine (Fo$_{80}$) and orthopyroxene (En$_{77}$) commingles with rhyolitic magma containing quartz crystals. The magmas attain thermal equilibrium at 1100°C. The free energies of formation of pure forsterite, enstatite, quartz, and silica liquid are $-2,372,478$, $-1,684,380$, $-993,236$, and $-710,402$ J mol^{-1}, respectively, under these conditions. Assuming ideality in the olivine and orthopyroxene solid solutions, calculate the activity of silica in both melts, and comment on the potential for silica to homogenize by diffusion.

13-19. If a magma at the ternary eutectic in Figure 10-21 mixes with a more primitive one of composition l_1, what mineral would crystallize first from the hybrid magma, and what would happen to crystals of the other minerals that had been forming at the eutectic?

13-20. (a) Plot a graph similar to the one in Figure 13-32 showing the variation in concentration caused by Rayleigh fractionation of elements having distribution coefficients of 0.45 and 2.0, and the variation in concentration caused by equilibrium fractionation of elements having distribution coefficients of 0.35 and 3.0.

(b) Phosphorus in Kilauean magmas is completely incompatible, so that the ratio $c_{P_2O_5}^l/c_{P_2O_5}^{cl}$ in any magma is a measure of the degree of crystallization involved in its derivation from the parental magma. Assuming that the concentrations of P$_2$O$_5$, TiO$_2$, and MgO in the parental Kilauean magma are 0.18%, 2.3%, and 12.0%, respectively, plot in the graph prepared for part (a) the relative concentrations (c^l/c^{cl}) of TiO$_2$ and MgO as a function of the fraction of the parental magma crystallized for analyses 7 to 15 in Table 13-1.

(c) Is Ti as incompatible as P in these rocks? Explain why analysis 15 plots at such a different position from the other analyses.

(d) Do the trends for TiO$_2$ and MgO fit a Rayleigh or equilibrium fractionation model best?

(e) The early-crystallizing ferromagnesian minerals have a bulk MgO mineral/liquid distribution coefficient of approximately 5; plagioclase, on the other hand, completely excludes Mg ($K = 0$). Based on the fit of the data in part (b), what proportion of plagioclase to ferromagnesian minerals would provide the best bulk distribution coefficient?

13-21. Calculate the chondrite-normalized abundances of REE in a basaltic magma that undergoes 60% equilibrium crystallization to form solids consisting of 80% olivine and 20% augite. Use the distribution coefficients given in Table 13-3 and assume that the basalt initially had a chondritic abundance of REE. Plot results versus the atomic number of the REE.

13-22. Repeat Prob. 13-21, but for 80% equilibrium crystallization to form solids consisting of 50% olivine, 20% augite, and 30% plagioclase.

14 / Igneous Rock Associations

14-1 INTRODUCTION

Early in the development of petrology it was recognized that certain rock types are commonly associated, whereas others never occur together. Moreover, the common associations were seen to correlate with certain geologic settings. Today, with the insight provided by plate tectonic theory, most igneous rocks can be assigned to particular plate tectonic environments, each of which has its own distinctive thermal regime, magma source region, and crustal stress pattern. But not all rock associations can be explained through plate tectonics. Some magmatism in the Archean and even the Proterozoic was different from that of Phanerozoic time, and distinctive rock associations were formed that were never again repeated in later times.

Seismic evidence indicates that the lithosphere and upper mantle are essentially solid, although a small amount of liquid may exist in the low-velocity layer. The formation of large magma chambers and volcanic edifices is therefore a rare occurrence that requires special conditions. Yet the majority of crustal rocks are of igneous origin; thus these conditions must, on occasion, be met. The steady-state geotherm beneath a continent or ancient ocean floor (Sec. 1-6) does not come near the dry beginning of melting curve for mantle peridotite, at least not at the depths at which we believe magmas are generated. Therefore, either the geotherm must be raised or the beginning of melting curve lowered if magmas are to form. Both of these perturbations appear to occur most often as a result of convective movements in the mantle, which at the same time cause motion of the lithospheric plates. For this reason, magmatism and plate motion are directly related.

The greatest production of igneous rocks is at mid-ocean ridges where steepened geothermal gradients resulting from upward-convecting mantle cause melting at shallow depths. The degree of partial melting under these conditions is moderately high, so that incompatible elements that enter the first-formed liquid are diluted by the additional melt, and magmas of tholeiitic composition are formed. Despite the enormous production of igneous rocks in this environment, most are subducted and do not enter the geologic record preserved on continents. The relatively few instances where ocean floor has been obducted onto continents to form the ophiolite suite provide rare opportunities to study these ocean floor rocks.

The second most productive region of magmas is at convergent plate boundaries. Here, subduction of cold lithosphere lowers geothermal gradients, but melting is caused by the fluxing effect of water liberated from subducted oceanic crust. The magmas are of calc-alkali type and have compositions that are determined by the depth at which the fluxing water is released, which is controlled by the rate and angle of subduction.

Where convecting mantle rises beneath a continent prior to its rifting apart to form ocean floor, geothermal gradients are not as steep as those beneath mid-ocean ridges, and melting occurs at greater depth and involves a smaller degree of partial melting. This produces alkaline magmas with higher concentrations of incompatible elements. These magmas and their differentiation products therefore characterize rift valleys on continents. Similar magmas can form in mantle plumes that rise at locations not necessarily related to plate boundaries. Movement of the lithosphere over these "hot spots" produces trails of igneous rocks. If the plume is particularly hot, higher degrees of partial melting can produce transitional or tholeiitic rocks.

14-2 IGNEOUS ROCKS OF OCEANIC REGIONS

Beneath a thin veneer of sediment, the ocean floors, which comprise approximately 65% of the Earth's surface, are composed almost entirely of basaltic lavas. This is the most extensive occurrence of igneous rocks on Earth. While ocean floor is continually being created along oceanic ridges, equivalent volumes of ocean floor are being destroyed at subduction zones (Fig. 14-1). This creation and destruction of ocean floor is an extremely active process, as indicated by the age of ocean floors, which are Mesozoic or younger. More than 10 km^3 is erupted per year from the 65,000-km-long mid-ocean ridge system. This is an order of magnitude more than is erupted subaerial. The enormous production of igneous rocks within oceans accounts for approximately 60% of all heat lost from the Earth.

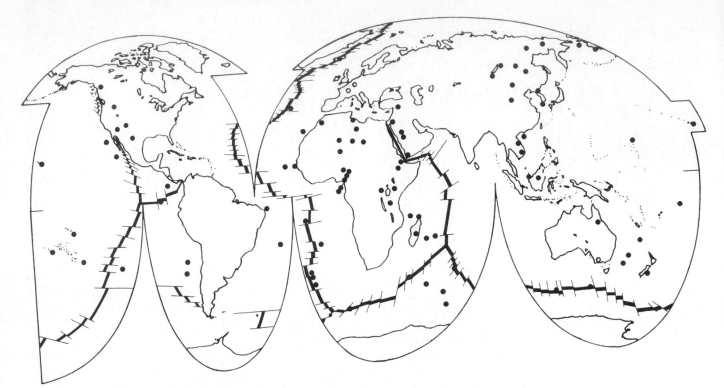

FIGURE 14-1   Mid-ocean ridge system where tholeiitic basalts (MORB) are being generated at a rate of more than 10 km³ yr⁻¹. Other centers of non-plate-margin volcanism (hot spots) are shown with dots. [Data mainly from BVSP (1981).]

Although most oceanic rocks form at mid-ocean ridges, some form within plates above long-lived "hot spots"—regions of thermal upwelling—whose positions appear to have remained relatively constant for long periods of time, despite the movement of overlying lithospheric plates (Wilson 1973). They consequently must have deep-seated origins well below the lithosphere. Some ocean floor rocks are generated on aseismic ridges. These are not clearly related to hot spots or spreading centers.

Mid-ocean Ridge Basalts (MORB)

The crest of most of the mid-ocean ridge system is marked by a 35-km-wide rift valley, along the center of which is a prominent fracture zone (Hekinian 1982). Lava erupting from this fracture either floods the rift valley or builds small volcanic edifices along the fracture. Some lava is erupted from faults that bound the rift valley. Along most of their length, oceanic ridges are segmented by transform faults. Within individual segments, the surface character of lava flows commonly changes from relatively smooth near the center, through lobate, to pillowed near the end of segments. This is thought to correspond to decreases in temperature and magma flux away from the center of segments.

Almost all rocks created at mid-ocean ridges are basalts, which have a small but significant range of composition. They are classified under the general name of *mid-ocean ridge basalt* (MORB) and are olivine tholeiites. They are either aphyric, or

contain olivine phenocrysts ($\pm$ Mg-Cr spinel) or plagioclase phenocrysts, or both; rarely do they contain augite phenocrysts. Olivine ranges in composition from Fo_{91} to Fo_{65} and plagioclase from An_{88} to An_{40}. The olivine–phyric lavas tend to erupt from the axial fracture zone, whereas the plagioclase–phyric ones occur more toward the edge of the rift valleys, suggesting the existence of zoned magma chambers beneath the rifts (Fig. 14-4). The olivine-bearing variety has a higher Mg number than the plagioclase one, indicating that differentiation and not simply crystal accumulation is involved in their genesis.

Because MORB have a restricted compositional range, their source must be relatively constant, and their degree of fractionation must be similar. Most have a Mg number of approximately 60, indicating that they have differentiated only slightly from a primitive magma, which would have a Mg number of 70. Also, the absence of a europium anomaly in the chondrite-normalized REE pattern of most MORB indicates that no significant amount of plagioclase could have been fractionated. When fresh, MORB have low contents of alkalis and other incompatible elements, and the light REE are strongly depleted. By contrast, concentrations of compatible elements such as Cr and Ni are high. As indicated in Sec. 13-11, the source from which MORB are derived must have been depleted in incompatible elements and enriched in compatible ones by repeated episodes of partial melting. Furthermore, the lack of a negative slope to the chondrite-normalized REE pattern (Fig. 13-34) indicates that their source does not contain garnet, and therefore it cannot be deeper than 65 km (Chapter 22). Toward topographic highs on ocean ridges, especially those capped by an oceanic island, MORB are less depleted in incompatible elements, have a flat REE pattern, and resemble more the tholeiites on ocean islands. In these regions magma may be tapping a deeper, less depleted source. The largest variation in the composition of MORB is due to alteration by circulating ocean water, which increases the alkali content, in particular Na, and converts plagioclase to albite and ferromagnesian minerals to chlorite. The altered rock, which is known as *spilite*, is discussed under ophiolites below.

Another minor rock type that occurs among these basalts is a plagiogranite or rhyodacite. This is believed to be formed as a final differentiation product of MORB magma. In coarsely crystallized pillows of MORB the residual mesostasis is seen to consist of immiscible globules of iron-rich glass in silica-rich glass (Philpotts 1982). The silica-rich glass matches closely the composition of the plagiogranite. Dixon and Rutherford (1979) have shown through experiments that these immiscible silica-rich liquids invariably develop after MORB is crystallized to about 95%.

Variations in the composition of MORB can be explained broadly in terms of fractionation of olivine and in some cases plagioclase, but in detail it breaks down (BVSP 1981). Because the compositions of olivine and plagioclase are so different from those of MORB, fractionation of these minerals brings about significant changes in the concentration of major elements in the fractionated liquids. The compositional range for major elements in MORB, however, is quite small compared with the range that would be predicted for the degree of fractionation indicated by changes in the concentration of incompatible elements. But even the degree of fractionation based on incompatible elements often cannot be made to match that based on the decrease in the concentration of highly compatible elements. Explaining the differentiation in terms of a single parental magma from which olivine, spinel, and plagioclase (the normal phenocrystic phases) have been separated is consequently very difficult, if not impossible, to do. The problem is alleviated somewhat by having several parental magmas, perhaps in separate magma chambers, that mix at different stages of fractionation (Rhodes et al. 1979). Problems still remain, however, until augite is added as a fractionating phase. Because this mineral's composition is much closer to that of MORB, its fractionation does not change the major element chemistry as much.

There is little doubt that augite must play a role in fractionating MORB, despite its rarity as a phenocrystic phase. Experiments indicate that with increasing pressure, augite becomes a primary phase on the liquidus of MORB. Rhodes et al. (1979) have pointed out that magma mixing could also be responsible for removing augite as a primary phase. It was pointed out in the discussion of Figure 13-31 that mixing of an evolved, multiply saturated magma of composition 2 with a primary magma of composition x caused resorption of plagioclase. This same mixing would cause resorption of augite. The phase diagram for forsterite–augite–SiO_2 is almost identical to that of Figure 13-31; simply replace the anorthite field with the augite field. Primary magma ($Mg' = 70$) is rarely erupted along ridges, but some may be intruded into the underlying magma chambers, where it would mix with evolved magma. The composition of the evolved magma may be determined in part by crystallization of augite, but when the primary magma mixes with it, augite disappears from the liquidus of the hybrid magma, which could then be erupted as MORB. This continual replenishing with primary magma could also explain why MORB does not become highly evolved. In the section on ophiolites we will see that beneath ocean ridges, there are indeed magma chambers in which clinopyroxene forms cumulate rocks.

Intraplate Oceanic Islands

Ocean floors are low areas on the surface of the Earth because they are underlain by oceanic crust, which, on average, is denser than continental crust. An oceanic island, especially one with considerable topographic elevation, is therefore a remarkable feature that normally requires active volcanism to survive. Some measure of the enormous amount of construction required to build one of these islands can be obtained by comparing the heights of Mauna Loa on the island of Hawaii, which is the largest volcanic edifice on Earth, and Mount Everest, which is the highest tectonically constructed feature on Earth. Mauna Loa towers 10 km above the ocean floor, whereas Mount Everest rises only 8.9 km above sea level.

Some oceanic islands are formed by volcanoes developed at convergent plate boundaries (Sec. 14-3), but the remainder are formed over what are believed to be rising mantle plumes, or "hot spots" (Wilson 1973), whose positions have remained relatively fixed over extended periods of time. The adiabatic rise of the mantle within a plume produces a partial melt, which buoyantly rises through the lithosphere to form a volcano on the ocean floor. If the plume is sufficiently active, the volcano grows and eventually becomes an island; otherwise, it remains a seamount. The thermal input from the mantle plume lowers densities and allows ocean floor to stand higher than it otherwise would. Because the lithosphere is continually moving relative to plumes, volcanoes (and islands) tend to develop in chains or ridges whose orientation defines the direction of plate motion. Once a volcano is moved off a hot spot by plate motion, it becomes extinct. With cooling and sinking of the lithosphere, the island is eventually planed off at wave base to form a truncated seamount, or atoll if the growth of fringing coral reefs is able to keep up with the rate of subsidence. The Hawaiian Islands–Emperor Seamount chain provides the best example of such a linear array of hot spot–generated volcanoes. Hot spots are not restricted to oceanic regions, but the majority occur there, and of these many are located at divergent plate boundaries (Fig. 14-1).

The volcanic rocks of ocean islands differ from MORB in having a wider range of composition, which in part is due to differentiation within near-surface magma chambers, but also to the composition of primary magmas, which contain higher concentrations of incompatible elements (Sec. 13-11). This difference is thought to be due to the magmas of oceanic islands having deeper sources than MORB; that is, the mantle rising in the plumes is not as depleted as the shallow mantle beneath mid-ocean ridges. A deep source is also required to account for the fixed position of hot spots beneath the moving lithospheric plates.

The rocks of oceanic islands are commonly both tholeiitic and alkaline (mainly Na), with tholeiitic rocks typically being early and alkaline ones late in the development of an island. Small oceanic islands tend to have only alkaline rocks, some of which are ultra-alkaline and silica poor; in this respect they resemble the ultra-alkaline rocks associated with rifting and mantle plumes on continents (Sec. 14-7). Despite the greater differentiation of oceanic island rocks compared to MORB, basaltic compositions still predominate. Table 13-1, for example, gives the typical range of tholeiitic rocks erupted from Kilauea. On Iceland, tholeiitic magmas differentiate all the way to rhyolite, which constitutes as much as 8% of the rocks. The most common differentiate of the alkaline magmas is trachyte, and in most differentiated suites intermediate rocks between basalt and trachyte are scarce.

The earliest crystallizing mineral in the tholeiitic lavas is a Cr–Al spinel. This is followed by olivine, which occurs only as phenocrysts and has a reaction relation with liquid to produce a Ca-poor pyroxene, either orthopyroxene or pigeonite. Olivine is consequently restricted to relatively magnesian compositions (Fo_{88-80}). Plagioclase and augite crystallize next and both can form small phenocrysts, but these are not as common as in the alkaline rocks. Because augite coexists with a Ca-poor pyroxene, it has the subcalcic composition ($Ca_{40-35}Mg_{50-42}Fe_{10-223}$) required of an augite on the pyroxene solvus (Fig. 10-32).

The tholeiitic rocks of ocean islands have higher concentrations of incompatible elements than do MORB. Their chondrite-normalized REE pattern does not show the depletion in light REE. Indeed, the REE pattern of most ocean island tholeiites is similar to that of alkali olivine basalts of ocean islands (Fig. 13-34) except that the overall concentration is slightly less in tholeiites. The steep negative slope to the REE pattern indicates that ocean island tholeiites are derived from mantle that is far less depleted than that from which MORB are derived, and that it probably contains garnet. This would indicate a depth of origin greater than 65 km (Chapter 22).

Alkali olivine basalt is normally the most abundant and most primitive member of the alkaline rocks. It differs from tholeiitic olivine basalt in having a lower activity of silica, which causes it to be nepheline normative. Some actually contain modal nepheline, in which case the rock is called a basanite. Spinel and olivine form the earliest phenocrysts, followed by plagioclase and augite. Olivine covers a much wider range of composition than it does in the tholeiites (Fo_{87-34}). This is because olivine in alkaline rocks does not react to form a Ca-poor pyroxene, and thus it can continue crystallizing to late stages. Ca-poor pyroxenes never form in alkaline rocks. These rocks only ever contain a single pyroxene, augite, which commonly contains a significant amount of Ti ($CaTiAl_2O_6$), which gives it a distinct pink or violet color in thin section. Augite may also contain significant amounts of Ca-Tschermak's molecule [$CaAl(AlSi)O_6$]. Both of these substitutions result from the low silica activities of these melts, as does the lack of a Ca-poor pyroxene (Ca-poor Pyx = Ol + SiO_2).

Stemming from alkali basalt is a series of differentiated rocks, in which, with decreasing color index, the plagioclase becomes progressively more sodic (Fig. 6-11a). The members of this series and their plagioclase compositions are: alkali basalt (labradorite), hawaiite (andesine), mugearite (oligoclase), benmoreite (anorthoclase), and trachyte (anorthoclase). The

series is critically undersaturated in silica at its mafic end and may remain so or become oversaturated toward its evolved end. Trachyte is normally the most differentiated member of the series, but in more extreme cases of differentiation phonolite (nepheline-bearing), pantellerite (Na-rhyolite), or comendite (K-rhyolite) may form. The development of both critically undersaturated and oversaturated melts from a common parental magma cannot be satisfactorily explained by fractional crystallization (Sec. 10-6 and Fig. 10-5). If both melts are indeed developed from a common parent, some process of differentiation affecting only the liquid must be involved, such as immiscibility, Soret diffusion, or vapor transport.

In most alkaline series intermediate rocks are scarce, and on many islands basalt and trachyte are the only members of the series present. The compositional range between basalt and trachyte is named the *Daly gap*, after Reginald Daly, who first drew attention to it. Numerous explanations have been advanced to explain the gap. Differences in the physical properties of the various alkaline magmas (density, viscosity) might increase the chances of basalt and trachyte reaching the surface. On the other hand, nonideal behavior in the liquids could result in sharp changes in the slope of the liquidus; this, in turn, would cause the rate at which liquid changed its composition with falling temperature to change rapidly. If the system is sufficiently nonideal, liquid immiscibility could be responsible for the gap. It has also been suggested that trachyte is not even derived from the basalt but is a separate product of partial melting in the mantle. This problem has yet to be resolved.

The trace elements of the alkaline rocks are very different from those of MORB. Alkaline rocks are much richer in incompatible elements and show no signs of having been derived from the strongly depleted mantle from which MORB has come. Their steep chondrite-normalized REE pattern indicates that they are derived from a source in which garnet is stable (> 65 km). Alkaline rocks contain higher concentrations of incompatible elements than do the tholeiites of oceanic islands. This difference cannot be accounted for by differentiation of the alkaline magmas, because the contents of compatible elements, such as Cr and Ni, are comparable to those in the tholeiites. Fractional crystallization of olivine or pyroxene would rapidly decrease the concentration of these elements. The high concentration of incompatible elements is therefore interpreted to result from small degrees of partial melting of a relatively undepleted mantle (Probs. 14-1 and 14-2). The smaller degree of partial melting would also explain why alkaline rocks are less abundant than tholeiitic ones and why they tend to form only small oceanic islands.

On some islands, extremely undersaturated alkaline rocks occur, in which the silica activity is so low that nepheline and melilite form instead of plagioclase. These are the nephelinites. They are typically younger than associated tholeiites or alkali basalts and commonly erupt explosively to form small cinder cones. They have very much higher concentrations of incompatible elements and a slightly steeper chondrite-normalized REE pattern than alkali basalts. They are consequently probably formed by still smaller degrees of partial melting. Nephelinites commonly contain mantle nodules whose mineralogy indicates they were derived from depths of at least 100 km.

These rocks are more common in alkaline complexes associated with rifts on continents and are discussed further there.

Aseismic Ridges

A small percentage of ocean floor rocks are formed along aseismic ridges, such as the Iceland–Faeroe Ridge (Hekinian 1982). It is not clear whether these ridges formed over hot spots or developed along fracture zones—leaky transforms. The rocks from these ridges resemble those of oceanic islands rather than MORB. They have a wide range of composition and have undergone significant differentiation. Like the rocks of oceanic islands, their trace element content indicates derivation of magmas from a mantle that has not been strongly depleted.

Ophiolite Suites

Despite the large number of studies in recent years of ocean floor rocks, the sampling of such a large fraction of the Earth's surface (65%) is still extremely small. Furthermore, the sampling, by necessity, is strongly biased toward the material within the first few meters of the ocean floor; it certainly is not representative of the entire oceanic crust. Deep drilling and exposures of rock in fracture zones where there is significant relief has shown that the pillowed basalts on the ocean floor are underlain by intrusive rocks, which like the basalts themselves, have undergone varying degrees of hydrothermal alteration.

Seismic velocities reveal that oceanic crust has a relatively simple three-tiered structure. Layer 1, which has the lowest velocity and averages 0.3 km thick, corresponds to the sedi-mentary veneer covering the lavas. Layer 2, with a compressional wave velocity of about 5 km s^{-1} and average thickness of 1.4 km, has been shown by drilling to be pillowed basalt. Layer 3, with a compressional wave velocity of 6.7 km s^{-1}, is about 5 km thick, and overlies the Mohorovičić (M) discontinuity, below which the seismic velocity abruptly increases to 8.1 km s^{-1}. Layer 3 is probably composed of intrusive rocks. The mantle immediately below the M discontinuity exhibits seismic anisotropy, which is probably due to preferred crystallographic alignment.

Although most oceanic crust is destroyed by subduction at convergent plate boundaries, some is preserved by being thrust onto continents (obduction) to form what is known as the *ophiolite suite* (Coleman 1977). This suite has three distinct members, serpentinized ultramafic rocks, hydrothermally altered pillowed basalts (spilites) and diabases, and pelagic sediment, which is characterized by radiolarian ribbon cherts. The pelagic sediments and pillow lavas have long been recognized as indicating a deep-water origin. Initially, however, ophiolites were thought to be autochthonous eugeosynclinal rocks. The association of the three types of rock is so common in most Phanerozoic mountain belts (Fig. 14-2) that it became known as the "Steinmann trinity," named after the petrologist who first drew attention to the association. Ophiolites, however, exhibit puzzling field relations, such as the lack of contact metamorphism around the ultramafic rocks, which were clearly formed at high temperature. These ultramafic bodies were called *alpine peridotites* to distinguish them from peridotites that do have metamorphic aureoles. Not until the advent of plate tectonics was the ophiolite suite recognized to be allochthonous and probably composed of rocks that were formed at mid-ocean ridges.

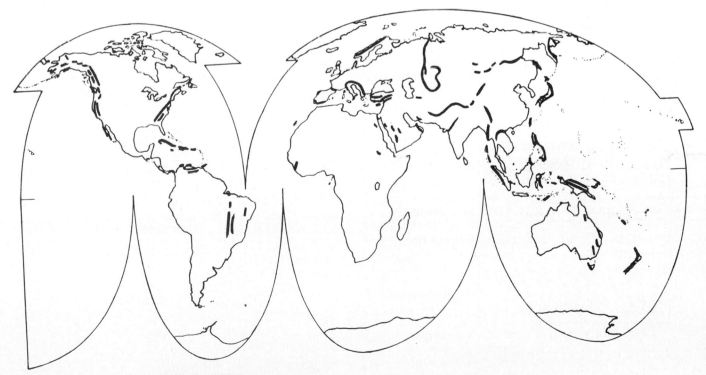

FIGURE 14-2 Distribution of ophiolite suites. [Data from Coleman (1977).]

Because ophiolites are tectonically emplaced, many of the contacts between its members are faulted, and much of the suite may be missing. Nonetheless, a number of well-preserved examples, such as those of Oman, Newfoundland, Papua, and the Troodos Massif in Cyprus, show that the ophiolite suite has a characteristic stratigraphy (Fig. 14-3), which consists of a lower relatively homogeneous ultramafic unit, overlain by a sequence of layered mafic and ultramafic rocks, which in turn may be overlain by a sheeted dike complex, which feeds overlying pillow lavas; these rocks are in turn overlain by the sedimentary rocks.

The lowest member is invariably a strongly deformed, serpentinized harzburgite (ol + opx). Primary layering, formed by variations in the abundance of olivine or spinel, is folded and cut by a deformational foliation. Toward the top, the foliation becomes parallel to the layering in the overlying rocks, which is thought to have originally been horizontal. The harzburgite also becomes richer in olivine toward the top, typically grading into a dunite. Irregular podlike bodies of chromite up to 1 km in length are common in the dunite (Thayer 1964). These rocks have been strongly deformed and recrystallized at high temperature. The olivine, which is unzoned and has a composition of Fo_{90-92}, has deformation kink bands and a strong preferred crystallographic orientation. The orthopyroxene is similarly unzoned and has a composition of En_{90}. Orthopyroxene never bears a reaction relation to olivine, indicating that these rocks may have equilibrated at pressures in excess of 0.5 GPa (Fig. 10-8). Spinel is variable in composition, generally becoming more chromium-rich toward the top. Clinopyroxene, which is a chrome diopside, never exceeds 5% and commonly is totally absent.

A narrow transitional zone ranging from tens to several hundred meters thick separates the lower ultramafic rocks from the overlying layered rocks (Nicolas and Prinzhofer 1983). The deformation, which is characteristic of the underlying harzburgite, dies out in this transition zone. The amount of clinopyroxene and plagioclase increases upward through the zone. Thus dunite (ol) at the base passes up through wehrlite (ol + cpx), to clinopyroxenite and troctolite (plag + ol). Pods and lenses of chromitite are common. Near the base of the transition zone olivine grains exhibit kink bands, but the chrome diopside and plagioclase that grow poikilitically or interstitially show no signs of deformation. With the decrease in amount of olivine, then, the degree of deformation decreases.

The transition zone terminates abruptly against the overlying layered gabbroic rocks, in which all signs of penetrative deformation are lacking. These cumulates are generally mafic at the base and grade upward to more feldspathic rocks. Olivine becomes progressively more iron-rich, reaching Fo_{70} in the uppermost gabbros. Orthopyroxene is present but far less abundant than in the underlying deformed ultramafic rocks. Diopsidic augite is the principal pyroxene. Toward the top of the layered gabbros, irregular intrusive bodies of plagiogranites occur. These are thought to be the final differentiation product of the gabbroic magma, which gave rise to the layered rocks and would be equivalent to the granophyres formed in differentiated tholeiitic sills on continents (Sec. 14-4).

In many but not all ophiolite suites, the layered gabbroic rocks are overlain by a sheeted dike complex. This unit consists of 100% dikes, that is, dikes intruding dikes, with no intervening screens of other rocks. Most dikes are less than 5 m wide. Many appear to have only one chilled margin. This is a consequence of repeated intrusions in the center of a single opening fissure. By studying statistically the asymmetry of chilled margins, the position relative to the main axis of spreading can be determined. The dikes preserve an ophitic texture, but most of their minerals have been hydrothermally altered. They were initially aphyric subalkaline diabases composed of plagioclase and augite; olivine and orthopyroxene are not reported. Although probably related to the underlying gabbros, they are not derived directly from them. The dikes cut the underlying gabbros but pinch out downward. The overlying lavas appear to have been fed by the dikes.

The pillow lavas are generally basaltic in composition but have been pervasively altered to mineral assemblages in the zeolite and greenschist metamorphic facies (Sec. 15-4). Thus the plagioclase has been converted to albite, and the mafic minerals have been transformed mainly to chlorite. These basaltic pillow lavas with albitic plagioclase were originally thought to be a special type of magma and were given the name *spilite*. Today, however, their peculiar composition is recognized to be the product of hydrothermal alteration (Amstutz 1974). Like the dikes, they initially were composed of plagioclase and augite; olivine is not common. The REE, which are less prone to change during hydrothermal alteration and weathering than most other elements, indicate that spilites are similar to MORB.

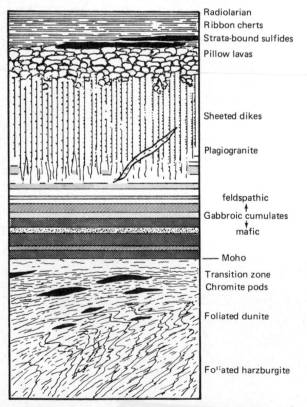

FIGURE 14-3 Schematic section through an ophiolite suite. Chilled margins on sheeted dikes are indicated by dots; spreading axis is to left. See text for discussion.

Radiolarian
Ribbon cherts
Strata-bound sulfides
Pillow lavas

Sheeted dikes

Plagiogranite

feldspathic
Gabbroic cumulates
mafic

Moho
Transition zone
Chromite pods

Foliated dunite

Foliated harzburgite

Toward the top of the pillowed succession, which typically is about 1 km thick, nonpillowed keratophyre and quartz keratophyre (Chapter 6) lavas occur. Like the basalts, these too are pervasively altered to low-grade metamorphic rocks. These lavas undoubtedly correlate with the plagiogranites.

Despite the pervasive low-grade metamorphism of the lavas and dikes, these rocks are not penetratively deformed. For this reason, the metamorphism is interpreted to result from hydrothermal alteration by circulating seawater near mid-ocean ridges. Exploration of mid-ocean ridges with submersibles has provided firsthand evidence of the effectiveness of the hydrothermal systems along spreading axes. Much of the cooling of the lavas and dikes near ridges is by circulating seawater. The entire volume of the oceans is estimated to cycle through these hydrothermal systems once every 8 million years. The circulating hot waters are certainly capable of converting the original basalt to spilite. Metals leached from basalts are known to precipitate from the hydrothermal solutions when they issue forth on the ocean floor as "black smokers." Meter-high chimneys of copper, iron, and zinc sulfides are built up around the hydrothermal vents. Strata-bound sulfide deposits in the pillow lavas and overlying sedimentary rocks of the ophiolite suite were probably formed by similar hydrothermal systems along ancient spreading axes. These sulfide deposits are of economic importance. Indeed, those associated with the ophiolites of the Troodos massif in Cyprus played a pivotal role in the development of the Chalcolithic Period and later Bronze Age.

The ophiolite suite is important because it provides a section through the oceanic crust. Whether these sections correspond to normal oceanic crust developed at mid-ocean ridges or is crust developed perhaps in small basins is not totally clear. MORB, for example, typically contains olivine phenocrysts followed by plagioclase phenocrysts; augite is a rare phenocrystic phase. Spilites and sheeted dike complexes, in contrast, commonly have augite phenocrysts, but olivine is rare. Nonetheless, the ophiolite suite is undoubtedly a slice of oceanic crust.

The boundary between the undeformed cumulate rocks and the underlying highly deformed ultramafic rocks is interpreted to have been the Mohorovičić discontinuity at the time these rocks were formed. Seismic velocities of the harzburgite match those of the mantle immediately below the M discontinuity, and the preferred orientation of olivine crystals in the harzburgite would certainly account for the seismic anisotropism of this mantle. The harzburgite is strongly depleted in incompatible elements and is interpreted to be the mantle from which MORB may have been derived; that is, it is a *restite*. The transition zone between harzburgite and overlying gabbroic rocks is interpreted to be mantle from which extraction of melt was incomplete; the clinopyroxene and plagioclase, which do not exhibit the deformation shown by the coexisting olivine in this zone, formed from this melt (Nicolas and Prinzhofer 1983).

Magma extracted from the harzburgite is believed to have collected at the base of the crust, where it formed the layered gabbroic rocks. The magma chamber may never have been very wide nor very deep and was probably repeatedly replenished. Where ophiolite suites contain sheeted dike complexes, the magma chamber may have had a continuous existence. Because the dikes have no intervening screens of other rocks, they represent 100% extension. Structures in the underlying layered rocks, however, show no evidence of significant extension. The magma chamber may therefore have been stretched laterally as new magma was emplaced. A zoned magma chamber, such as that shown in Figure 13-9, could produce the layered rocks, with the more mafic cumulates forming near the base and the more feldspathic ones near the top. As the magma chamber was replenished, the sidewalls of the chamber would move outward and magma on these walls would solidify to give the layered rocks (Fig. 14-4). At the same time, roof rocks would be pulled apart, providing the extension necessary for the emplacement of the sheeted dikes. A small, zoned magma chamber expanding and cooling laterally and being filled axially could eventually produce layered rocks of great lateral extent. En

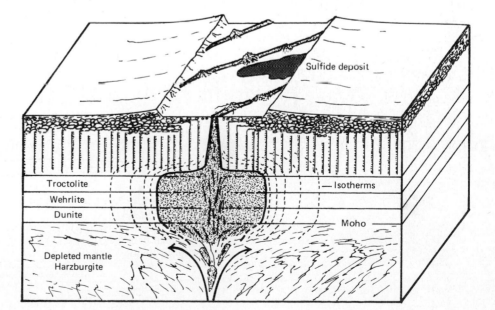

FIGURE 14-4 Possible shape of magma chamber beneath a mid-ocean ridge. Width of magma chamber is indicated by the width of the en echelon dike system. Vertical walls of chamber solidify as they are transported away from the ridge, producing layered gabbroic rocks that grade upward from mafic to feldspathic. The magma chamber, which must be zoned, is periodically replenished, at which time magma may be displaced into the axial dike. Splitting of this dike by repeated injections produces sheeted dikes with chilled margins (dots) that are symmetrically disposed on either side of the spreading axis. [Data mainly from BVSP (1981) and Nelson (1981).]

echelon fissure systems over active spreading axes have been interpreted to indicate that the underlying magma chambers are no more than 20 km and perhaps as little as 1 or 2 km wide (Nelson 1981; BVSP 1981).

14-3   IGNEOUS ROCKS ASSOCIATED WITH CONVERGENT PLATE BOUNDARIES

Convergent plate boundaries rank next to mid-ocean ridges in their rate of production of igneous rocks. Unlike MORB, most of which are doomed to a fate of subduction and disappearance, igneous rocks created at convergent boundaries survive by being accreted to continents. Indeed, the igneous process at convergent plate boundaries can be thought of as those primarily responsible for the building of continents.

Igneous activity does not occur right at convergent boundaries, but is located in the overriding plate some distance behind the boundary. The distance from this boundary to the igneous rocks depends on the angle of subduction of the downgoing plate. The top of this cold, brittle slab is the source of many earthquakes, which defines the so-called *Benioff seismic zone.* The descent of the slab into the mantle can be traced by mapping the depth of earthquake foci. Igneous activity begins once the slab descends to a depth of about 100 km. On the surface, the onset of igneous activity is marked by arcuate chains of regularly spaced volcanoes, which in oceanic

regions gives rise to *island arcs*, and in orogenic regions to mountain chains topped by large strato volcanoes (Fig. 14-5).

What actually initiates igneous activity is still uncertain. As will be seen in Sec. 21-4, isotopic studies indicate that the primary magmas must be derived largely from the mantle wedge overlying the Benioff zone, rather than from the subducted slab itself. But, then, why is the 100-km depth to the Benioff zone so critical to generating magmas? The most likely explanation is that melting is triggered by the release of water from the downgoing slab. Ocean floor rocks are highly altered and hydrated. As they are subducted, metamorphic reactions release this water, which then rises into the overlying mantle to act as a flux in generating melts (Sec. 22-3).

A wide range of igneous rocks is associated with convergent plate boundaries. Andesite is normally the most abundant volcanic rock, but many others, ranging from basalt (gabbro) to rhyolite (granite), are present and can predominate (Thorpe 1982). The proportions of various rock types depend on a number of factors, some of the most important of which are the types of plates involved in the convergence, the angle of subduction, and the rates of convergence. Oceanic plates tend to produce higher proportions of basaltic rocks, whereas continental ones produce more siliceous rocks. Both the angle and rate of subduction can affect the depth at which the water-releasing metamorphic reactions take place, and this, in turn, can affect the types of magmas formed.

It is commonly possible to recognize three distinct series of rocks, each of which covers most of the compositional range

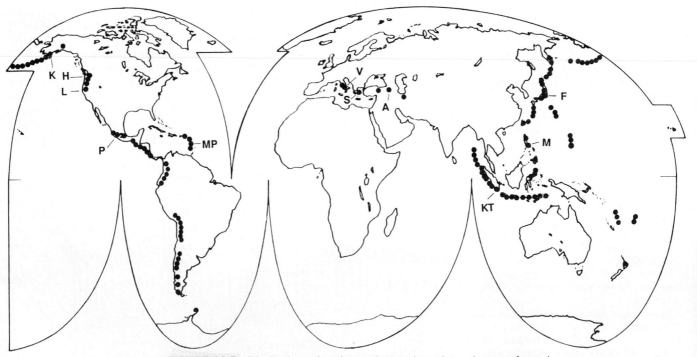

FIGURE 14-5 Distribution of active and recently active volcanoes formed above Benioff zones (individual dots may represent more than one volcano). Locations of the following well-known volcanoes are indicated: Katmai (K), St. Helens (H), Lassen (L), Paricutín (P), Pelée (MP), Vesuvius (V), Santorin (S), Ararat (A), Krakatoa (KT), Mayon (M), Fujiyama (F).

between basalt and rhyolite but differ in their potassium contents. The series with the lowest K contents is the *island-arc tholeiite*, which as the name implies, is found in oceanic regions but also occurs at continental margins, where rates of subduction exceed 7 cm per year. The *calc-alkali series*, which is characteristic of convergent boundaries involving continental crust (orogenic belts), has intermediate K contents. Finally, a *high-K series* occurs on continental crust. An extremely potassic series, the *shoshonites*, occurs in some belts.

The suites of rocks formed above Benioff zones commonly have distinct chronologies, as seen in the volcanic stratigraphy of dissected volcanoes. In general, the sequence goes from early mafic rocks to late siliceous ones and includes boninite, island-arc tholeiite, calc-alkali basalt, shoshonite, basaltic andesite, andesite, dacite, and rhyolite. Above any one subduction zone only part of the sequence may be present, and the proportions and compositions of individual rocks may differ depending on the type of convergent boundary (ocean–ocean or continent–continent, for example).

A continuous chemical range of rock types occurs between basalt and rhyolite (BADR in Fig. 14-6), and describing individual members of this series can create the false impression that distinct rock types exist. Terms such as "basaltic andesite" and "rhyodacite" (Chapter 6) have been introduced to bridge gaps between some of the main rock units. Although these terms increase the number of available choices, they belie the fact that the rocks form a continuous series. The descriptions below are therefore of typical samples from the midrange of the various rock types.

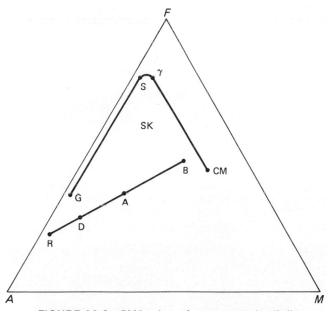

FIGURE 14-6 *FMA* plot of average calc-alkali rocks: basalt (B), andesite (A), dacite (D), rhyolite (R). In contrast, the differentiation trend of the tholeiitic Skaergaard intrusion (SK), which consists of chilled margin (CM), Upper Border Group (γ), Sandwich horizon (S), and transgressive granophyre (G), shows strong early enrichment in iron.

Normally, the earliest igneous rocks associated with convergent plate boundaries are basaltic in composition (BVSP 1981). Where the convergent plates involve oceanic crust, the basalts are island-arc tholeiites, whereas when continental crust is involved (orogenic belts), they are either calc-alkali basalts or high-K basalt, shoshonite. These basalts can be distinguished by their K_2O contents—<0.5% in the island-arc tholeiite, about 1% in the calc-alkali basalt, and >2% in the high-K basalt. Island-arc tholeiites may carry phenocrysts of plagioclase and pyroxene, but they are not strongly porphyritic rocks. They are distinct from tholeiitic continental flood basalts and MORB, as indicated by their chondrite-normalized REE patterns (Fig. 13-34), which point to derivation from a relatively undepleted mantle. In addition to occurring in island arcs, they are found on continental margins, where the rate of convergence exceeds 7 cm per year. Calc-alkali basalts have high alumina contents (about 18%), and consequently are also referred to as *high-alumina basalt* (Table 6-7). They are subalkaline and typically, quartz normative. Plagioclase is the dominant phenocryst phase, which can be as calcic as An_{90}. Orthopyroxene and augite can also occur as phenocrysts. These basalts are strongly porphyritic, with phenocrysts forming as much as 25% of the rock. The high-K basalt, shoshonite, is the K equivalent of an alkali olivine basalt, and some actually contain leucite in their groundmass. They have phenocrysts of olivine, labradorite, and augite. Because basaltic rocks tend to form early, they are likely to be concealed by later eruptions, and in the case of island arcs, these early rocks may all be submarine. Basaltic rocks may be more abundant than presently recognized.

In some island arcs, such as those of Bonin, Mariana, and Papua (Fig. 14-5), there are highly magnesian rocks which range from basalt (MgO = 12%), through basaltic andesite, to andesite in composition. High magnesium contents normally indicate the presence of abundant olivine, but most of these rocks are quartz normative and typically contain two pyroxenes. They are further characterized by very low contents of Ti, Zr, Y, and REE; in this respect they resemble MORB, but their silica contents are all in excess of 54%. This peculiar group of rocks is known as the *boninites*. They may be present in other island arcs but concealed by later rocks.

Andesite is the major volcanic rock formed above Benioff zones (Ewart 1976). Much andesitic volcanism is explosive, so that many andesites form tuffs and agglomerates. Most andesites are markedly porphyritic or vitrophyric, containing large, complexly zoned plagioclase phenocrysts (Fig. 14-7), which can be as calcic as anorthite (Prob. 14-3). Augite also commonly forms phenocrysts as does orthopyroxene. Olivine phenocrysts may occur, but not in rocks containing orthopyroxene phenocrysts. Orthopyroxene also occurs in the groundmass of orogenic andesites, where it may be rimmed by pigeonite. Andesites associated with island-arc tholeiites typically contain pigeonite in the groundmass rather than orthopyroxene. This is a consequence of the stronger iron enrichment in the tholeiitic rocks. Resorbed phenocrysts of hornblende can also occur. Titanomagnetite is also a common phenocrystic phase in orogenic andesites but not in island-arc andesites. High-K andesites may also contain phenocrysts of sanidine.

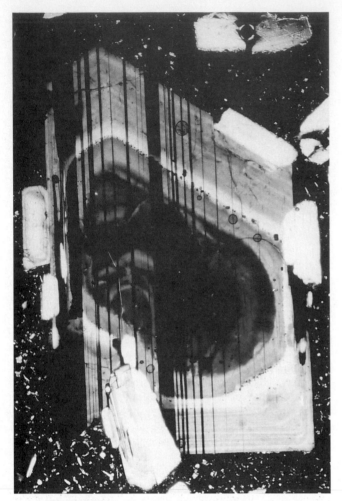

FIGURE 14-7 Complexly zoned plagioclase phenocryst in andesite from Mount St. Helens. Such phenocrysts may record magma mixing events, or changes
in the depth of the magma.

The groundmass of most andesites consists of a pale brown
glass or cryptocrystalline mesostasis. Many pyroclastic andesites are crystal tuffs, with phenocrysts surrounded by glass
shards.

The bulk composition of andesites can vary considerably
due to the presence of inclusions and xenoliths. Some of these
have glomeroporphyritic textures and may represent phases
that accumulated in the magma at depth. Others are true xenoliths. Many phenocrysts in andesites are not in equilibrium with
the melt in which they occur. This is most clearly seen in the
case of plagioclase (Fig. 14-7) but is also true of other phenocrystic phases. Magma mixing is the most satisfactory explanation for these complexities, but failure of phenocrysts to
reequilibrate with magma as it rises through the lithosphere
must also be a contributing factor.

With increasing silica content andesites grade into dacites and then rhyolites. These rocks tend to be glassier than
andesites and many are obsidians. With increasing silica, the
magmas also erupt more explosively, and the percentage of pyroclastic rocks increases until ash flows and ash falls constitute
most of the rocks at the rhyolitic end of the series. Dacites and
rhyolites are generally porphyritic or vitrophyric, and those
that are not pyroclastic are commonly flow-banded. Plagioclase phenocrysts, as calcic as bytownite, occur in most dacites
except those of the high-K series, which are no more calcic than
andesine. Phenocrysts of augite, hypersthene, hornblende, biotite, Fe–Ti oxides, and quartz may also be present. Sanidine
is a common phenocryst in high-K dacites. In the rhyolites,
plagioclase is still a common phenocrystic phase, but its
composition is normally more sodic than andesine. It is accompanied by phenocrysts of sanidine, and quartz, which commonly have embayed outlines.

Like andesites, dacites and rhyolites tend to be inhomogeneous, containing xenoliths or schlieren of more mafic volcanic rocks (Fig. 14-8). In many cases the mafic rocks were
liquid when incorporated in the siliceous magma (Fig. 13-25).
This is clear evidence, then, of magma mixing.

A strong correlation exists between the potassium content of volcanic rocks and the depth to the Benioff zone (Fig.
14-9), as demonstrated by Hatherton and Dickinson (1969)
and by a number of workers since. To separate spatial varia-

FIGURE 14-8 Fine-grained vesicular basalt inclusions in obsidian, Glass Mountain, Medicine Lake
Highlands, California. Eruption of obsidian may have
been caused by the influx of basalt at depth.

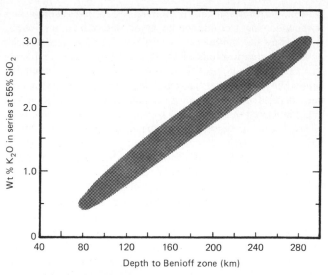

FIGURE 14-9 Plot of K_2O content of rocks containing 55% SiO_2 versus the depth to the Benioff zone. [After Hatherton and Dickinson (1969) and others.]

tions in K_2O from those due to differentiation, Hatherton and Dickinson restricted their comparisons to rocks containing 55 wt % SiO_2. At this concentration, which corresponds to a low-silica andesite (basaltic andesite), the 0.5% K_2O typical of the island-arc tholeiite series corresponds to a depth to the Benioff zone of about 80 km, whereas the approximate 1.0% K_2O of the calc-alkali series corresponds to a depth of 120 km, and the >2.0% K_2O in the high-K series corresponds to a depth in excess of 200 km. Other incompatible elements show this same correlation. The cause for this variation is still uncertain. It might be due to a systematic change in the composition of partial melts with depth, a change in the percentage of melting with depth, or a difference in the thickness through which melts rise and assimilate low melting fractions containing incompatible elements.

In addition to compositional variations across arcs, there can also be lateral variations. Rhyolites (and granite), for example, occur at the eastern end of the Aleutian arc where the overriding plate is of continental material, but they are almost totally lacking from the western end, where only oceanic plates are involved. Assimilation of crustal material would seem a likely explanation for this variation. In the Lesser Antilles, however, the variation occurs in the mafic members of the series, which change from tholeiitic in the north to alkaline in the south. Because these differences affect the most primitive members of the series, the differences must be generated in the mantle source regions.

The correlation of intrusive rocks with convergent boundaries is not as easily made as with volcanic rocks, because it is only in older belts that erosion has exposed the plutonic rocks. Nonetheless, these rocks constitute an important fraction of the material created at convergent boundaries. Plutonic equivalents of all the volcanic rocks can be found, ranging from gabbro, through diorite and granodiorite, to granite. Representatives of the low-, intermediate-, and high-K series can be recognized, but members of the low-K series, occurring mainly

in island arcs, are less likely to be exposed by erosion than those occurring in orogenic belts.

Although andesite is volumetrically the most abundant volcanic rock, its plutonic equivalent, diorite, is normally a minor constituent of the plutonic suite. Instead, granodiorite, or even granite, is the most abundant plutonic rock. These rocks form batholiths extending the entire lengths of arcs, but as pointed out in Sec. 4-12, these huge bodies are composites of many smaller intrusions. Compositional zoning, schlieren, xenoliths, or even pillows of one rock type in another are common and produce considerable inhomogeneity in all rock types. One feature that is prominent because of its absence is large magma chambers in which differentiation of basaltic magma has given rise to siliceous rocks. Moreover, there is no evidence of the bodies of ultramafic cumulates that would have to be formed if the siliceous magmas are products of differentiation of basaltic magma. Nor is there geophysical evidence to indicate that these dense rocks are hidden at depth in the crust.

Inclusions of mafic rocks in most granites and granodiorites signal an input of basaltic magma into the crust at depth. If basaltic magma were to rise just to the base of the crust, it could differentiate there, leaving dense cumulates in the underlying mantle. Once sufficient low-density siliceous magma had formed, it would rise as a diapir into the crust. Such a magma would bear a mantle signature. On the other hand, a basaltic magma emplaced at the base of the crust could raise temperatures to the point that crustal rocks would partially fuse, and, again, once sufficient low-density siliceous melt had formed, a diapir would rise. In this case, however, the siliceous magma would bear the imprint of the crustal rocks from which it formed.

White and Chappell (1983) have shown that granitic rocks can be divided, on the basis of chemistry, into two groups that may reflect these two extremes in origin. They name these the *I-* and *S-type granites*. I-type granites are relatively rich in Na and Ca and so have hornblende as the major mafic mineral. Initial $^{87}Sr/^{86}Sr$ ratios (Sec. 21-4) are less than 0.708, indicating a large input of mantle-derived magma. Porphyry copper deposits are associated with this type of granite. S-type granites are Na-poor, Al-rich rocks that contain normative corundum, which appears modally as muscovite. They contain biotite instead of hornblende. Xenoliths and schlieren of metamorphosed sedimentary rocks are common. Their initial $^{87}Sr/^{86}Sr$ ratios are greater than 0.710, which indicates a large component of continental crust. These granites are believed to have formed through partial fusion of sedimentary rocks. Tin deposits are associated with S-type granites.

I-type granites occur nearer convergent boundaries than S-type granites, suggesting that thicker crust may be necessary to cause partial melting of crustal rocks. S-type granites are water-rich, and as they rise toward the surface and the pressure decreases, they solidify (Sec. 11-6). This type of magma is therefore not as widely represented as the I-type among volcanic rocks. Other types of granites have been identified on the basis of geochemical traits, but these simply serve to emphasize how varied the conditions of formation of these rocks are.

Igneous rocks associated with convergent plate boundaries clearly form under a wide range of conditions. No single model of their genesis is therefore possible to make. The one common thread they all share is that the igneous activity is triggered by the subduction of an oceanic plate. If the overriding plate is also oceanic, magmas can be derived only from oceanic crust or underlying mantle. As will be shown in Sec. 22-3, these magmas must be derived largely from the mantle wedge overlying the Benioff zone. The subducted slab contributes mainly water to the magmas, which have a tholeiitic character. Because no continental crust is present, siliceous magmas must be differentiates of mafic magmas. Their presence in island arcs is therefore proof that differentiation is capable of producing such an extreme rock type. The quantities of siliceous rock, however, are small. In the South Sandwich Islands, for example, dacite and rhyolite constitute only 4% of the rocks (Baker 1968). This may mean that primary magmas are capable of producing only this small amount, but it may also indicate that without a thick continental crust through which to rise magmas have insufficient time to differentiate.

When the overriding plate is continental crust, the diversity of rock types increases as does the proportion of siliceous rocks. Based on their occurrence in island arcs, siliceous magmas can clearly be derived by differentiation from basaltic magma. Isotopic data, however, indicate that a large proportion of the siliceous rocks in orogenic zones are formed by partial melting of crustal rocks. Intermediate members of the series could therefore be the products of either differentiation or magma mixing.

The series formed by basalt, andesite, dacite, and rhyolite in the calc-alkali series of orogenic belts (BADR in Fig. 14-6) defines a relatively narrow trend in FMA diagrams, as would be expected of a differentiated series. This trend, however, differs from that commonly observed in known differentiated magma bodies in that intermediate members do not exhibit iron enrichment. If the series is the product of differentiation, the lack of iron enrichment can be accounted for by early crystallization of magnetite, which is a result of the higher Fe^{3+}/Fe^{2+} ratio in these rocks compared to that in tholeiites. But the evidence for magma mixing is strong and includes inhomogenieties in volcanic and plutonic rocks, complex zoning of phenocrysts, and the presence of phenocrysts that are out of equilibrium with their surrounding melt. The straight-line variation of BADR in the *FMA* diagram could therefore simply reflect mixing of basaltic and siliceous end members.

Numerous magmatic processes and a variety of different sources must be involved in the formation of igneous rocks associated with convergent plate boundaries. The normal tendency then of scientists to try and create the simplest model possible can lead to trouble in studying these rocks. Each occurrence should be evaluated in terms of its own specific environment.

14-4 CONTINENTAL FLOOD BASALTS AND ASSOCIATED ROCKS

Continental flood basalts are one of the most consistent rock types in the geologic record. Comparison, for example, of the Proterozoic Zig-Zag Dal basalts of eastern North Greenland (Kalsbeek and Jepsen 1984) and the Tertiary basalts of County Antrim, Northern Ireland (Patterson 1951), reveal similarities in the following: tectonic setting, associated sedimentary rocks; thicknesses and lateral extent of flows; internal structures of flows; and compositions of flows, including major and trace elements, and isotopes. These similarities are important, because they indicate that the source from which these rocks have been derived has not changed significantly over this period; also, similar quantities of heat (volumes of magma) have been released, and mechanisms of emplacement and differentiation have not changed.

A description of the morphology of flood basalts and their internal structures is given in Sec. 4-2 and will not be repeated here. One of their most remarkable features is the enormous cumulative volume of the flows in any one province, which can be in the hundreds of thousands of cubic kilometers. Individual flows can also have enormous volumes, some measuring hundreds of cubic kilometers. Eruptions of flood basalt take place rapidly from fissures or strings of vents along fissures. The feature, then, that sets continental flood basalt volcanism apart from other types is the very large volumes of magma that must be able to collect and erupt in short periods of time.

Flood basalts are formed in association with continental rifting (Fig. 14-10) and thus occur along passive continental margins that were previously successful branches of rift systems. They may also extend into continents along paleorifts that follow failed arms. Flood basalts are not restricted to rift systems, as are alkaline igneous rocks (Sec. 14-6), but can extend into surrounding regions. The lavas are interlayered with sedimentary rocks which are typically of the terrestrial red bed variety. Sedimentation precedes the volcanism, indicating that subsidence and crustal thinning are precursors to igneous activity. Subsidence continues during and after volcanism, with the result that the lowest stratigraphic members can eventually be well below sea level.

One of the earliest known occurrences of flood basalts is the 1.1-Ga-old Keweenawan basalts of Lake Superior (Fig. 14-10). These basalts are believed to have formed along a failed rift that is now marked by an enormous geophysical anomaly known as the *midcontinent gravity high* (BVSP 1981). A similar paleorift, but of early Mesozoic age, occurs within the Siberian Platform. Flood basalts were erupted in many regions during the Mesozoic because of the breakup of Pangea. First, with the opening of the central Atlantic, basalts erupted in many basins along the entire length of what is now the eastern seaboard of North America and the coast of Morocco, and then with the opening of the South Atlantic, the Parana basalts formed in Brazil and the Karoo basalts in South Africa. At the same time, basalts erupted in Antarctica and Tasmania. During the Cretaceous, the Deccan traps formed in northwestern India as it broke away from Gondwanaland. Still later (about 50 Ma), during the opening of the North Atlantic, flood basalts of the Thulean Province formed in northwestern Scotland, Antrim (Northern Ireland), and the southeast coast of Greenland. The most recent (Miocene 17-6 Ma), extensive continental flood basalts were erupted in the Columbia and Snake River plains of the northwestern United States (Fig. 4-1).

The most common type of flood basalt is a quartz tholeiite, which contains about 52% SiO_2 and has a magnesium

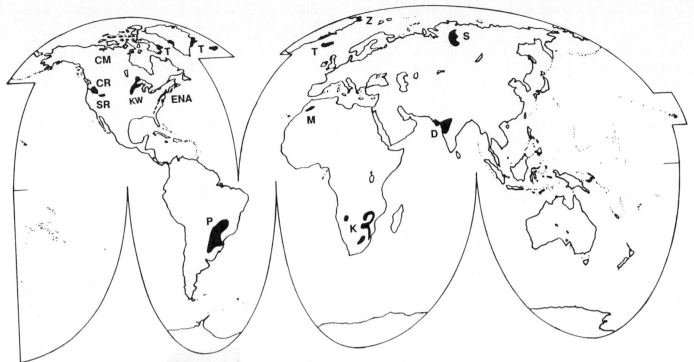

FIGURE 14-10 Distribution of continental flood basalts. Coppermine River (CM), Keweenawan (KW), Columbia River (CR), Snake River (SR), eastern North America (ENA), Parana (P), Karoo (K), Morocco (M), Thulean (T), Zig-Zag Dal (Z), Deccan (D), Siberia (S).

number of 55. Olivine tholeiites also occur, and in some provinces transitional and even alkali basalts are present. These rocks are either aphyric or only slightly porphyritic, carrying phenocrysts of olivine, plagioclase, and augite. The augite is commonly rounded and resorbed, as are rare phenocrysts of orthopyroxene. Regardless of the composition of the basalt, individual flows are remarkably homogeneous, and they rarely contain xenoliths of country rocks or cognate inclusions of cumulate material. Although vesicles occur near the base and top of flows, flood basalts do not appear to have formed from highly gas-charged magmas. Indeed, pyroclastic rocks are extremely rare. Most flows are massive, unless they are pillowed from having flowed into lakes that formed in the sedimentary basins. A few provinces have small volumes of dacitic or rhyolitic flows. These siliceous rocks occur late in the history of a province and are more common where central volcanic complexes develop.

Irregular sheetlike intrusions of diabase intrude the sedimentary rocks beneath the lavas. They can normally be chemically matched with specific volcanic units. These sheets are particularly abundant near the unconformity at the base of the associated sedimentary sequence. The Palisades Sill, for example, is emplaced at or near the base of the Triassic sedimentary rocks in the Newark Basin of New Jersey. These sheets rarely intrude the older rocks beneath the sediments, especially if the older rocks are deformed and metamorphosed. The sheets are normally composed of massive, medium-grained diabase. They rarely exhibit prominent cumulate layering but, nonetheless, do undergo considerable differentiation, which is marked by strong iron enrichment. Toward the top of most thick diabase sheets,

lenses of coarse-grained granophyre are formed from the residual liquids (Shirley 1987).

Dikes and dike swarms are also associated with flood basalts. Some dikes were undoubtedly feeders to the flows, but exposures of actual connections between dikes and flows are rare. Most dikes are less than 5 m wide, but their cumulative thickness in a swarm can be considerable. The orientation of dike swarms can therefore be used to determine directions of crustal extension during ancient episodes of rifting. Dikes, like the flows, show little variation in composition within individual dikes, but separate dikes can have compositions as varied as the flows. In contrast to the flows, the dikes are generally more porphyritic and may also contain xenoliths of country rock. Regional variations in dike composition have been found. For example, dikes associated with the Mesozoic basins of eastern North America are all quartz tholeiites north of Pennsylvania, but to the south, olivine tholeiites also occur (Weigand and Ragland 1970).

In most flood basalt provinces igneous activity is episodic. Each episode is marked by lavas which, although perhaps covering a wide range of composition, are nonetheless characterized by a geochemical signature that distinguishes them from lavas of other episodes. Periods of sedimentation or erosion may separate the episodes, but in some cases episodes may overlap slightly in time. Although the compositions of successive episodes may not form a clear evolutionary progression, lavas within an episode may follow an apparent differentiation trend. In many flood basalt provinces, three distinct episodes can be recognized, which commonly result in the lavas being grouped into lower, middle, and upper, as for example in the Zig-Zag Dal, Keweenawan, Newark, Deccan, and Antrim basalts.

The trace element and isotopic data for tholeiitic flood basalts indicate small but variable amounts of crustal contamination of magmas that are initially derived from a mantle that is only slightly depleted in incompatible elements relative to the bulk Earth (Sec. 21-4). This initial magma is probably similar to the tholeiitic magma of ocean islands (Thompson et al. 1983). The chondrite-normalized REE pattern of flood basalts (Fig. 13-34) has the same steep negative slope that characterizes the ocean island rocks, and indicates initial derivation from mantle containing garnet. The isotopic ratios $^{87}Sr/^{86}Sr$ and $^{143}Nd/^{144}Nd$ are variable, but their average is near that for the bulk Earth (Sec. 21-4). This initially led DePaolo and Wasserburg (1979) to invoke a relatively primitive undifferentiated source for these rocks, but subsequent work has shown that the source is somewhat depleted, and that mixing with crustal material gives compositions near that for the bulk Earth (BVSP 1981, p. 87). Isotopic ratios within a given province may show a progression toward more crustal values as flows become younger, but even within an individual unit there can be considerable variation. A positive correlation between the $^{87}Sr/^{86}Sr$ ratio and the SiO_2 content is also found, as might be expected if the higher ratios are associated with crustal contamination.

Experiments indicate that at low pressure tholeiitic flood basalts are multiply saturated, or nearly so, with olivine and plagioclase (BVSP 1981). This suggests that fractionation of olivine or plagioclase, or both, has taken place; that is, a magma is not likely to have a composition on the plagioclase–olivine cotectic unless it has been fractionated there by crystallizing the phase in excess of the cotectic. Geochemical mass balance calculations (Sec. 13-2 and Prob. 13-2) on these basalts commonly require that augite be included as a primary crystallizing phase if solutions with small residuals are to be obtained. Although augite is commonly not an early crystallizing phase in 1-atm experiments, it is at higher pressures (BVSP 1981). The amount of plagioclase fractionated from these basalts cannot be great; otherwise, they would have larger negative europium anomalies (Fig. 13-34).

If multiple saturation is due to low-pressure fractionation, it is natural to wonder what the composition of the primary magma was, where the fractionation took place, and what processes were involved. In answering these questions, it is important to keep in mind that the source of these lavas must be capable of repeatedly supplying enormous volumes of homogeneous magma through laterally extensive fissures.

There is no reason to suppose that the primary magma in all flood basalt provinces must be identical; inhomogeneities in the mantle, for example, could provide different source materials. But the compositional uniformity of flood basalts, in particular the quartz tholeiite, has led petrologists to seek a single primary magma, and they have narrowed the search to two likely candidates: quartz tholeiite and picritic basalt (Cox 1980). The most abundant, and commonly the most primitive, lava in flood basalt provinces is the quartz tholeiite. It is therefore a good candidate for a *parental* magma. However, its magnesium number (<60) is too low to be in equilibrium with olivine of the composition thought to exist in the upper mantle (Fo_{90}). It must therefore be a differentiate of a more primitive

magma, or the mantle beneath flood basalt provinces contains more iron-rich olivine than is currently believed. Some provinces contain picritic basalts, and these do have magnesium numbers (about 70) that are compatible with mantle olivine compositions; they could therefore be *primary* magmas. The objection to them being *parental* to the more evolved flood basalts is that they are absent from many provinces and even where present, are normally not abundant. Advocates for a primary picritic magma must therefore invoke plumbing systems which minimize the chances of this magma reaching the surface.

Regardless of which magma is primary, the compositional variation among flood basalts appears to be controlled by low-pressure fractionation involving olivine, augite, plagioclase, and, in some, orthopyroxene. These phases are fractionated in the associated sheetlike intrusions. But with the iron enrichment that normally characterizes this differentiation, residual liquids would have difficulty extruding rapidly in large volumes because their densities would be greater than that of the overlying sedimentary rocks. More important, dikes having the compositions of the differentiated lavas can be found cutting through these sills and associated sedimentary rocks. The source of these magmas must therefore be still deeper. Evidence of large magma chambers lying immediately beneath the sedimentary rocks is rare. The Keweenawan province may be an exception; there, the midcontinent gravity high points to large volumes of mafic rocks at depth. In other provinces where this evidence is lacking, magma reservoirs must be hypothesized to exist in the lower crust.

Cox (1980) proposes that these magma chambers form at the crust–mantle boundary in much the same way as those associated with ophiolite complexes (Fig. 14-4), the only difference being that they are overlain by continental crust rather than oceanic crust. Picritic magma, on reaching the base of the crust spreads laterally, floating the less dense crust above it. In regions of rapid crustal extension and fracturing, some picritic magma may reach the surface, but most is filtered out by the low-density crust. Some contamination of the magma with crustal material may take place at this stage. With cooling, the differentiating picritic magma becomes less dense and eventually is able to intrude the overlying rocks; at this point flood basalt volcanism is initiated. Reference to Figure 3-1 or your answer to Prob. 13-12 will reveal that the lowest density achieved by a differentiating tholeiitic magma occurs when the magnesium number is about 60. This, then, could explain the ubiquity of the quartz tholeiite lavas.

Further differentiation of the magma in this lower crustal chamber leads to iron enrichment and greater densities (Fig. 3-1). As the density of the magma approaches that of the overlying crust, the driving force for extrusion diminishes. Nonetheless, voluminous, high-iron flood basalts are common, even though their magmatic density would have been greater than that of the rocks they intrude. It may therefore be necessary with this model to provide some additional driving force for intrusion.

A point of considerable significance is the aphyric nature of most flood basalts. The mantle source and lower crustal magma chamber cannot be overheated—extra heat would simply result in more melting in the mantle or slower crystallization and differentiation in the chamber. The lack of phenocrysts must therefore be due to resorption of crystals during the rise and decompression of the magma. Providing a melt is relatively dry, which is indicated by the scarcity of hydrous minerals and pyroclastic rocks with flood basalts, rapid decompression is accompanied by melting. Flood basalts are certainly emplaced rapidly [see Chapter 3, and Shaw and Swanson (1970)], so no significant loss of heat can occur on route to the surface. Thus, as the magma rises (Fig. 14-11), its temperature approaches, and perhaps exceeds, the liquidus. While crystals are melting, the latent heat of fusion causes the temperature of the magma to fall, but once all crystals have been resorbed, the temperature remains essentially constant until cooling starts in the lava flow. The temperature of the magma would therefore reach its highest point above the solidus on the surface of the Earth. Up to this point, resorption of crystals and assimilation of xenoliths would take place.

In light of the relations illustrated in Figure 14-11, it is not surprising that flows are commonly less porphyritic than their feeder dikes. This is well illustrated by a 250-km-long diabase dike that fed the first lava in the Mesozoic Hartford basin of Connecticut (Fig. 4-26). Normal faulting associated with the subsidence of the basin has exposed the dike at levels ranging from an initial depth of 10 km to the point where the dike actually connects with the surface flow. The chilled margin of the dike in the deepest section contains zoned phenocrysts of euhedral plagioclase and orthopyroxene and rounded augite. These phenocrysts constitute 27% of the rock. Where the dike connects with the flow, chilled margins contain only 15% phenocrysts, including euhedral plagioclase, rounded orthopyroxene and augite crystals, and small euhedral crystals of olivine. The olivine apparently is a low-pressure phase in this dike, taking the place of orthopyroxene, which was clearly stable in the dike at the deeper section. In the lava flow itself, phenocrysts constitute only 5% of the rock, and these are euhedral

plagioclase and olivine with minor rounded augite crystals and extremely rare, highly embayed crystals of orthopyroxene.

If a flood basalt magma does contain crystals that are resorbed on rising toward the surface, a number of interesting effects ensue. First, as the percentage of crystals decreases, the effective bulk viscosity of the magma decreases (Eq. 13-24). This then allows the magma to rise more rapidly, which, in turn, increases the rate of resorption. Second, as crystals melt, the bulk volume of the magma increases, due to the approximate 10% volume expansion on melting. The increased volume would result in greater rates of intrusion, which, in vertical dikes, would cause more melting and thus more expansion. Both of these processes would have feedback effects that could contribute to the high rates of extrusion of flood basalts.

A further consequence of resorption is that the magma continually changes its composition to be in equilibrium with the minerals that are dissolving; that is, the magma is multiply saturated. Thus, instead of invoking a magma chamber in which fractional crystallization brings about multiple saturation, a rising, melting crystal mush could produce the same result. Moreover, should there be any separation of crystals from liquid on the way to the surface, the resulting compositional variations would look similar to those formed by fractional crystallization in a magma chamber. Finally, turbulence in a superheated dike or lava flow would provide ideal conditions for the fusion and assimilation of crustal xenoliths.

14-5 LARGE LAYERED IGNEOUS COMPLEXES

Grouped together under this category are rocks that may have very different origins, but they are all products of the differentiation of tholeiitic magma. At a number of times in Earth history, particularly during the Precambrian, conditions as varied as continental rifting and meteorite impact led to the development of enormous volumes of tholeiitic magma, which, on slowly cooling, differentiated into remarkable suites of cumulates, some of which are of great economic importance. Although size is not an essential criterion in this category, it is mostly in large intrusions that magmatic processes were able to produce the strongly differentiated rocks. Detailed descriptions of most of these occurrences are given by Wager and Brown (1967).

The form and structures of these bodies have been described in Sec. 4-10 and will not be repeated here. Suffice it to say that most of these intrusions are lopoliths. Some are circular to elliptical or lobate in plan, such as the Bushveld complex of South Africa (Fig. 14-13) or the Sudbury lopolith in Ontario (Fig. 14-31), whereas others are elongate, such as the Muskox intrusion in the Northwest Territories of Canada (Fig. 14-15) or the Great Dyke of Zimbabwe. Some that are circular to elliptical, such as the Sudbury body, were formed by meteorite impact, their shape being determined largely by the explosion crater, which was later filled with magma (Sec. 14-9). Perhaps most large circular to elliptical lopoliths have this origin. The elongate ones are associated with crustal extension, dike swarms, and flood basalts. The Muskox intrusion, for example,

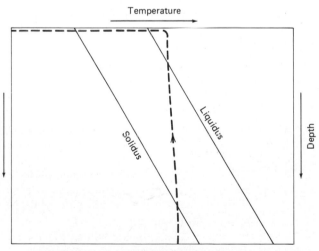

FIGURE 14-11 Possible path of anhydrous continental flood basalt magma as it rises to the surface. Decreasing pressure causes melting, and the magma may become totally liquid before reaching the surface. Because of rapid extrusion rates, magma reaches its highest point above the solidus on reaching the surface.

is overlain by the Coppermine River basalts and the Duluth gabbro at the west end of Lake Superior by the Keweenawan basalts.

Almost all large layered complexes are of Precambrian age. For example, the Stillwater complex in Montana is 3.2 Ga, the Great Dyke of Zimbabwe is 2.5 Ga, the Bushveld complex is 2.0 Ga, the Sudbury lopolith is 1.7 Ga, the Kiglapait intrusion, Labrador, is 1.3 Ga, and the Muskox and Duluth bodies are both 1.1 to 1.2 Ga. Layered complexes did form in Phanerozoic time, such as the Tertiary Skaergaard intrusion on the southeastern coast of Greenland (Fig. 14-12), but they are much smaller. The large volumes of the Precambrian bodies may reflect higher heat production early in the Earth's history. In addition, at least for those formed through meteorite impact, the flux of large meteorites would have been much greater during the Precambrian.

Whether beneath a rifting continent or a meteorite explosion crater, large-scale melting of the upper mantle produces tholeiitic magmas which rise to form lopoliths. Debates on the precise composition of these magmas center around the amount of normative olivine they contained and the degree to which they were contaminated by crustal rocks. Chilled margins containing pristine quenched magma do not occur on large intrusions. The finer-grained rocks that are found at contacts may have assimilated country rocks or been altered by circulating hydrothermal solutions. Commonly, nearby small dikes and sills that are thought to be related to the main intrusions are used to estimate initial magma compositions. These are all quite similar, containing about 50% SiO_2, 7 to 9% MgO, 11% CaO, and less than 0.3% K_2O. Their most variable constituent, Al_2O_3, ranges from 15 to 19%. More olivine-rich picritic magmas have been proposed for intrusions containing large volumes of olivine cumulates, but these can be explained with less olivine-rich magmas if the chambers remain open and are repeatedly replenished and flushed with primary magma.

Crystallization in these lopoliths takes place from the floor upward, to form gently dipping layers of cumulates. Toward sidewalls dips can become steeper and even vertical. Evidence of slumping and cross-bedding in the layered rocks attests to their partially molten state for some time following accumulation. Other discordances in the layering can be explained as a result of magma convection. Some form of convection probably takes place in all of these bodies and may play an important role in promoting differentiation (Sec. 13-4).

Because olivine and pyroxene are the first minerals to crystallize from these magmas, dunites, peridotites, and pyroxenites form the lowest layers. Once plagioclase starts crystallizing, noritic and gabbroic rocks form, and then, as the plagioclase becomes more sodic, these rocks grade into iron-rich ferrodiorites. Toward the top of intrusions, granophyre is common. It may be derived by differentiation from the mafic magma or from assimilation and partial melting of country rocks.

The actual order of crystallization of minerals depends on the precise composition of the magma. Three of the most common sequences are (I) olivine–orthopyroxene–plagioclase–clinopyroxene, (II) olivine–clinopyroxene–plagioclase–orthopyroxene, and (III) olivine–orthopyroxene–clinopyroxene–plagioclase. The quaternary phase diagram for the system forsterite–diopside–anorthite–silica (Fig. 10-36a) shows how slight changes in the composition of a primary magma plotting near the peritectic can bring about these different sequences (Prob. 14-5). Because of the peritectic, olivine does not continue crystallizing from the differentiating magma but is replaced by calcium-poor pyroxene. If a magma has a composition that causes minerals to appear as in sequence I, the order of formation of cumulate rocks would be dunite, harzburgite, bronzitite, norite, and gabbro (refer to Fig. 6-3 for rock classification). With sequence II, the rocks would be dunite, wehrlite, olivine gabbro, and two-pyroxene gabbro. With sequence III, they would be dunite, harzburgite, lherzolite, websterite, and two-pyroxene gabbro. The only difference in these three sequences is in the stage at which orthopyroxene crystallizes, which as pointed out in Sec. 13-9, may reflect different degrees of crustal contamination, orthopyroxene appearing earlier in the more contaminated magmas.

Crystallizing with olivine as a primary phase is chromite. This mineral is normally an accessory phase because magmas contain only trace amounts of chromium. But in most large lopoliths chromite-rich layers have repeatedly formed by a process of accumulation that probably involves magma mixing (Sec. 13-10). Chromitite layers vary in thickness from centimeters to meters, the thicker ones being of economic value. They occur in olivine and orthopyroxene cumulates, but once clinopyroxene starts crystallizing, the small amount of Cr in the magma is taken into the pyroxene structure, and chromite no longer forms. Layers of almost pure magnetite rock also form in these intrusions but normally not until intermediate stages of fractionation.

Accompanying the succession of rock types are variations in the composition of minerals, the ferromagnesian ones becoming more iron-rich, and the plagioclase, once it starts crystallizing, becoming more sodic. Detailed inspection of these trends commonly reveals reversals, which are most likely the result of intrusion of fresh batches of magma. These new surges are also typically associated with the formation of an olivine or chromite cumulate. Most large intrusions have evidence of numerous periods of replenishment and mixing during their early stages of crystallization. This causes them to continue accumulating ultramafic rocks rather than fractionating to more evolved differentiates.

Once replenishment of the magma ceases, fractionation of the remaining liquid progresses to form gabbro, ferrodiorite, and eventually granophyre. At the ferrodiorite stage, magnetite and apatite are typical cumulus phases, and they can produce layers of extremely iron-rich rocks. The iron-rich magmas at this stage of differentiation may enter a two-liquid field (Fig. 13-17), and an immiscible siliceous liquid separates to form granophyre. If the two-liquid field is not encountered, the fractionating magma will eventually attain a granophyric composition in any case. This late-stage siliceous liquid remains molten to lower temperatures than the associated mafic fraction and commonly forms irregular transgressive sheets.

The upper part of these intrusions can be simple or complex, depending largely on the size of the intrusion. In smaller bodies, rocks accumulating on the floor of the intrusion finally meet a narrow zone of rocks that solidifies down from the roof, the residual liquid being trapped between the two advancing crystallization fronts. Granophyre typically occurs in this zone. In large intrusion, so much heat is liberated through the roof that overlying rocks are partially melted. The siliceous melts derived in this way then mix with the granophyres formed by fractionation of the mafic magma and complex relations develop.

We will summarize three different layered intrusions. The Tertiary Skaergaard intrusion, while small compared with most Precambrian bodies, is the best-studied intrusion in the world, and its history is relatively simple—intrusion of a single batch of magma that differentiated in a closed chamber. At the other end of the spectrum is the enormous Bushveld complex, the world's largest igneous body—it formed through the mixing of several pulses of magma. Finally, the Muskox intrusion formed by repeated surges of magma which, each time, flushed out most of the old fractionated magma.

Skaergaard Intrusion, East Greenland

The Skaergaard intrusion is one of a number of bodies on the east coast of Greenland that formed during the initial opening of the North Atlantic (Wager and Brown 1967). Many diabase dikes and flood basalts found in the vicinity of the intrusion formed during this same period of crustal extension. The body itself developed near the unconformity separating the Tertiary basalts from the underlying Precambrian gneisses (Fig. 14-12). The magma made room for itself by lifting a conical pluglike body of gneiss and basalt, as evidenced by a roof pendant of gneiss at the stratigraphic level of the basalts. The resulting

form of the intrusion is that of a cone, which is elliptical in plan with maximum and minimum axes of 10 and 7.5 km.

The rocks of the intrusion form three distinct groups (Wager and Deer 1939): a Marginal Border Group, a Layered Series, and an Upper Border Group. The *Marginal Border Group* is a 200- to 300-m-thick zone of generally finer-grained olivine gabbro that crystallized from the wall inward. Layering, which parallels the contacts, is formed by growth of different minerals on the wall of the intrusion or by flowage of convecting magma. In some layers plagioclase crystals up to 2.5 cm long grew out from the wall into stationary magma to form a crescumulate texture, which Wager and Deer refer to as the perpendicular feldspar rock. Xenoliths of country rock occur in this zone. The compositions of the minerals change systematically through this zone in a manner that mimics the change found in the layered series of rocks. For this reason it is thought that the rocks of the Marginal Border Group accumulated on the walls of the intrusion at the same time that the Layered Series accumulated on the floor.

The main part of the intrusion is formed by the *Layered Series*, which consists of over 2700 m of layered gabbroic rocks that accumulated on the floor of the intrusion. This series is divided into three zones on the basis of whether or not the rocks contain olivine. In the *Lower Zone* magnesian olivine (Fo_{67-53}) is a cumulus phase (Fig. 14-13). Toward the top of this zone, olivine develops a reaction relation to form pigeonite ($Ca_9Mg_{56}Fe_{35}$), which on cooling changed to inverted pigeonite. The top of the Lower Zone is marked by the disappearance of olivine, and the overlying *Middle Zone* consists only of layered two-pyroxene gabbro. Magnetite also is a cumulus

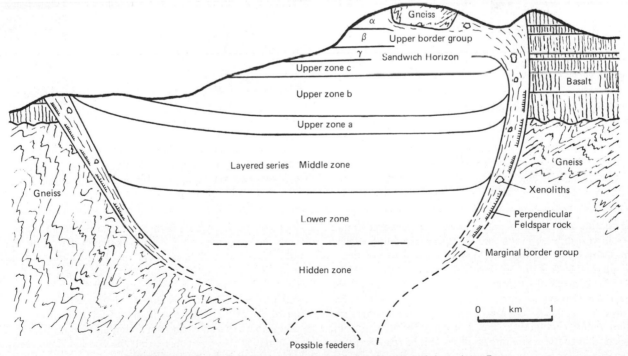

FIGURE 14-12 Simplified cross section of the Skaergaard intrusion, East Greenland. No vertical exaggeration. [Data mainly from Wager and Brown (1967).]

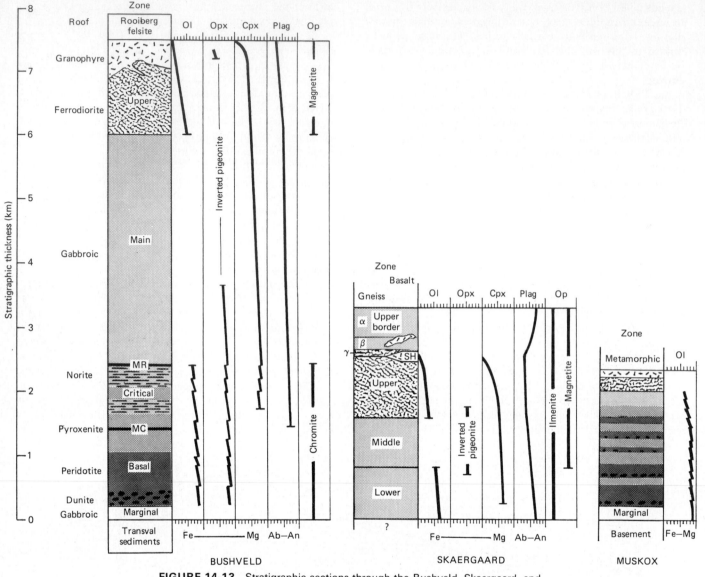

FIGURE 14-13 Stratigraphic sections through the Bushveld, Skaergaard, and Muskox intrusions. Layers in the Basal and Critical Zones of the Bushveld and in the Muskox intrusions symbolize many thin layers. The average compositional trends of minerals are shown accurately, but the minor sharp compositional fluctuations, which reflect the entry of new batches of magma, are shown only symbolically. Main Chromitite layer (MC); Merensky Reef (MR); Sandwich Horizon (SH). [After Wager and Brown (1967) and Irvine (1979).]

phase in this zone. The magma became progressively more iron-rich with differentiation and when the pigeonite reached a composition of $Ca_9Mg_{45}Fe_{46}$ olivine once again became a stable phase, but this time with a composition of Fo_{40} (Prob. 14-6). The reappearance of olivine marks the beginning of the *Upper Zone*. Olivine continues to change its composition, becoming pure fayalite at the top of the Upper Zone, where the residual liquid encountered the downward solidifying roof. Plagioclase in the Upper Zone is all less calcic than An_{45}, so these rocks are ferrodiorites. Apatite is also a cumulus phase in this zone. A fourth zone, referred to as the *Hidden Layered Series*, underlies the Lower Zone rocks. It would presumably contain earlier crystallizing minerals, examples of which are found in the outer parts of the Marginal Border Group (olivine with a composition of Fo_{81}, for example). Based on the best estimate of the original composition of the magma and the integrated composition of the exposed rocks, Wager and Deer calculated that the hidden rocks constituted as much as 70% of the intrusion. The parameters on which such a calculation are based are prone to serious error and the calculation does not take into account the effects of assimilation. Recent drilling and geophysical surveys indicate that the Hidden Zone constitutes only a small fraction of the intrusion.

The *Upper Border Group*, which solidified downward from the roof, can also be divided into three zones, which mineralogically match the three zones in the Layered Series. In the

Upper Border Group, however, the division must be based on plagioclase composition, because contamination by xenoliths of granite gneiss prevented olivine from crystallizing in many of the rocks. The zone nearest the roof, α, contains plagioclase more calcic than An_{53}; the next zone, β, contains plagioclase between An_{53} and An_{44}, and the lowest zone, γ, contains plagioclase less calcic than An_{44}.

Toward the end of crystallization a lens of residual liquid became trapped between the upward accumulating Layered Series and the downward solidifying Upper Border Group. This lens, which is referred to as the *Sandwich Horizon*, contains the most evolved liquids in the Skaergaard. Olivine is pure fayalite, and the clinopyroxene became so iron-rich that it reached the high-temperature stability field of iron wollastonite, which on cooling inverted to a polygonal aggregate of hedenbergite ($Ca_{43}Mg_0Fe_{57}$). This zone contains considerable quantities of granophyre, but some of the siliceous liquid rose to form transgressive sheets in the overlying rocks.

The Skaergaard intrusion was formed by a single pulse of magma that underwent extreme fractionation in a closed chamber. The differentiation resulted from fractional crystallization of a convecting magma that deposited minerals mainly on the floor of the intrusion. Wager and Deer (1939) believed that minerals separated by sinking from the convecting magma currents as they crossed the floor of the intrusion. More recently, McBirney and Noyes (1979) have interpreted the layering to result largely from the nucleation and crystallization of different minerals directly onto the solidification front.

Bushveld Complex, South Africa

The Bushveld complex covers an area of approximately 66,000 km^2 and contains an 8-km-thick sequence of layered igneous rocks (Wager and Brown 1967). In detail it consists of four lobes, which contain similar but not identical series of rocks. The lowest parts of the intrusion, marked by the Main Chromitite layer and the Merensky Reef (Fig. 14-14), are absent from the northern and southern lobes but are present in the eastern and western lobes. At times the lobes were separate,

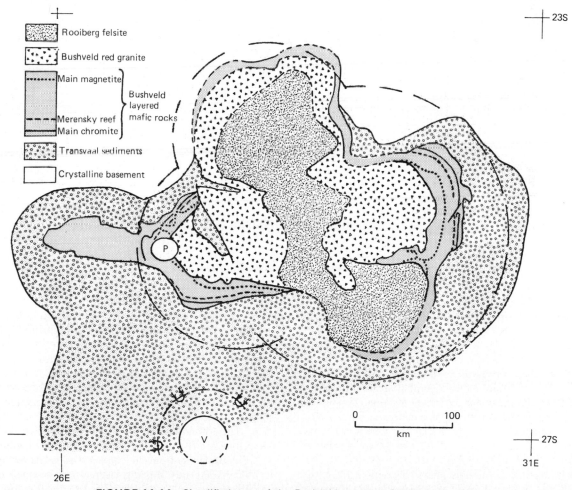

FIGURE 14-14 Simplified map of the Bushveld complex, South Africa. The Vredefort Ring (V) and three of the lobes of the Bushveld (dashed lines) may have been caused by meteorite impact. The Pilansberg (P) plug is a younger alkaline intrusion (see Fig. 4-33). [Based on maps in Wager and Brown (1967) and Rhodes (1975).]

but at others they were connected, as evidenced by prominent stratigraphic marker layers that can be traced throughout. Magma levels must have fluctuated considerably, possibly as a result of the fresh influxes of magma that gradually built the intrusion to its final size. Repeated surges of magma produce more complex sequences of rocks, and in this respect the Bushveld is different from the Skaergaard.

The rocks immediately beneath the Bushveld are sediments of the Transvaal System. Like the lopolith, they are bowed down into a large basinlike structure, which could have resulted from the loading of the dense igneous rocks or the withdrawal of magma from beneath the crust to form the lopolith. The upper contact of the body is complex and in most areas is obscured by the intrusion of the roughly contemporaneous Bushveld red granite, which overlies most of the lopolith. In places, however, the mafic rocks contact overlying Rooiberg felsite, which is also approximately the same age as the Bushveld. As shown in Sec. 14-9, this felsite has been interpreted as shock-melted rock formed by a meteorite impact that simultaneously triggered the mafic magmatism (Rhodes 1975). Considerable assimilation of the felsite produced hybrid rheomorphic granophyres, which are difficult to distinguish from granophyres derived from the mafic magma.

The lowest part of the intrusion consists of approximately 100 m of noritic diabase (Marginal Group in Fig. 14-13), which contains approximately 5% normative olivine. This presumably formed by relatively rapid cooling of the original undifferentiated magma and may therefore be representative of the parental magma. Its high content of orthopyroxene may, however, indicate that it was contaminated with sedimentary rocks.

Immediately above the Marginal Group is a 1200-m-thick *Basal Series* composed of interlayered dunite, harzburgite, and bronzitite. The minerals in these rocks change little in composition throughout the series, olivine ranging from Fo_{88} at the base to Fo_{86} at the top, and orthopyroxene from En_{87} to En_{83} over this same interval. However, the mineral compositions do fluctuate, indicating additions of fresh batches of magma. These additions prevented the magma from fractionating to more evolved compositions.

Toward the top of the Basal Series numerous thin layers of chromitite appear. These are precursors to the *Main Chromitite* layer (Steelpoort), which is one of the prominent stratigraphic markers of the intrusion. Between it and another prominent stratigraphic marker, the Merensky Reef, is a remarkable sequence of 1000 m of finely layered cumulates forming the *Critical Series*. Approximately 60 m above the base of this series plagioclase appears as a cumulus phase, and at 300 m augite also appears. Many monomineralic adcumulate layers and reversals in the fractionation trends of minerals point to an unstable magma that was being repeatedly replenished. The layers are predominantly bronzitite near the base of the series but become mainly norites toward the top. Many chromitite layers occur throughout the series.

Olivine remains a cumulus phase throughout the Critical Series in the western lobe of the Bushveld, but it ceased crystallizing at the top of the Basal Series in the eastern lobe. This difference indicates that the magmas in these two parts of the intrusion had slightly different compositions, assuming

the pressures were the same. But the Main Chromitite layer and the Merensky Reef occur in both the western and eastern lobes. Their formation therefore cannot be extremely sensitive to magma composition. Instead, magma mixing is most likely responsible for the precipitation of these layers, especially since they are associated with reversals in mineral composition trends.

The top of the Critical Series is marked by the *Merensky Reef*, the world's most important source of platinum. This layer, which varies in thickness from 1 to 5 m, is composed of cumulus bronzite poikilitically enclosed by plagioclase with a thin chromite layer at its base. The platinum occurs with disseminated sulfides. Irvine et al. (1983) believe that the ore was formed as a result of magma mixing. Fractionation of a sulfur-poor magma had brought about enrichment in platinum when a new sulfur-rich magma was intruded. During mixing an immiscible sulfide liquid formed into which the platinum strongly partitioned. The dense sulfide liquid then sank to the floor of the intrusion.

Above the Merensky Reef is the *Main Zone*, 3600 m of poorly layered gabbroic rocks. In addition to lacking the striking fine scale layering of the Critical Series, the minerals show a steady upward change to more fractionated compositions. Apparently, during this stage, no new pulses of magma entered the chamber, so fractionation was able to take its course.

The appearance of magnetite as a cumulus phase marks the beginning of the *Upper Zone*, which consists of 1500 m of ferrodiorites. Numerous magnetite cumulate layers occur near the base of this zone, one of which, the Main Magnetite layer, forms a prominent stratigraphic marker that can be traced throughout the eastern and western lobes. No downward-solidifying upper border group exists as it does in the Skaergaard. Instead, the ferrodiorite grades upward into a granophyric zone that contains assimilated roof rocks.

Muskox Intrusion, Northwest Territories, Canada

The Muskox intrusion is an elongate body (Fig. 14-15) emplaced at the unconformity between a basement of granitic and metamorphic rocks and overlying sediments and basalts of the Coppermine River, which were erupted during the same general magmatic episode as the Muskox (Irvine 1979). A dike, which presumably served as a feeder, extends for 60 km to the south of the intrusion. But the layered ultramafic and mafic rocks that form the main part of the intrusion are not cut by this dike. The magma that formed all but the lowest layered rocks must therefore have have been intruded laterally from a source farther to the north.

The 150-m-wide dike and 200-m-wide marginal zone of the intrusion are composed of bronzite gabbro, which grades to peridotite away from the contact. Inside the marginal zone is 1800 m of gently dipping layered series, which consist of 25 cyclic units, each of which formed by the fractional crystallization of a new batch of magma. Each cyclic unit consists of a number of cumulate layers, the mineralogy of which was determined by the order of crystallization. In the lower part

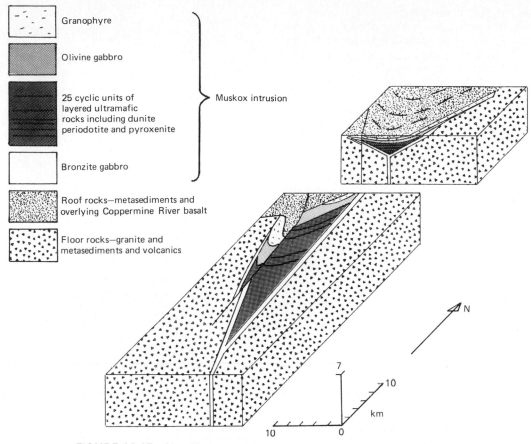

FIGURE 14-15 Simplified perspective view of the Muskox intrusion, Northwest Territories, Canada [From Irvine (1980); reprinted with permission of Princeton University Press.]

of the intrusion, the sequence olivine–clinopyroxene–plagioclase–orthopyroxene was common, producing dunite, olivine clinopyroxenite, olivine gabbro, and two-pyroxene gabbro. Most cyclic units, however, do not contain the complete suite of rocks, because influxes of fresh magma interrupted the fractionation process. With time, the sequence of crystallization changed to olivine–clinopyroxene–orthopyroxene–plagioclase and then to olivine–orthopyroxene–clinopyroxene–plagioclase toward the top of the layered series. Consequently, cumulus orthopyroxene appears at progressively earlier stages toward the top of the intrusion, indicating that later magmas may have been contaminated more with crustal rocks. Chromitite layers occur in the dunite, but they are absent where clinopyroxene is a cumulus phase.

Through a complete cyclic unit, the magnesium number of the cumulates decreases by about 10%, but injections of new magma and displacement of old kept restoring the magnesium number to essentially its original value throughout most of the layered series. Thus olivine, for example, changes from Fo_{85} to only Fo_{80} in the first 1200 m. However, during the accumulation of the final 600 m, when the magma was sufficiently fractionated to form gabbroic rocks, the composition of the olivine progressed to Fo_{60}. The layered series become progressively more siliceous toward the top, with a prominent sheet of granophyre containing an abundance of xenoliths of the overlying metasedimentary rocks capping the intrusion.

Irvine (1979) estimates that the Muskox contains more than 50% olivine. Because magmas do not contain this high a percentage, accumulations from repeated surges of less olivine-rich magma must be invoked. From each surge, ultramafic and mafic cumulates were deposited on the floor of the intrusion, and the fractionated residual liquid was displaced by the next surge of magma, erupting possibly as the Coppermine River basalts. Despite the great thickness of the layered series, the actual volume of magma in the intrusion at any one time may have been quite small. The repeated pulses of magma would, however, have brought in the heat needed for the large-scale melting and assimilation of roof rocks.

14-6   CONTINENTAL ALKALINE ROCKS

Apart from Benioff zone-related magmatism and continental flood basalts, igneous activity on continents is extremely limited and restricted to rift valleys and local hot spots (Sørensen 1974a; Bailey 1974). The volumes of magma emplaced in these environs is small compared with those formed in the other two environments. Continental flood basalts are also related to rifting, but they are more common along successful branches of rift systems, whereas alkaline magmatism predominates along the failed arms normally preserved within continents. A wide range of rock types, some of which have unusual mineralogy, commonly occur together in alkaline provinces. It is not surprising, therefore, that petrologists have devoted considerable effort to working out the petrogenesis of these rocks. As a result, alkaline rocks have received a degree of attention and a share of the rock nomenclature that far outweighs their volumetric significance. Despite these efforts, we are still a long way from fully understanding these rocks.

Alkaline igneous rocks of Archean age have not yet been found. There are, however, numerous Proterozoic examples, such as those of the Gardar province of southwest Greenland (1.2 Ga) and the Kapuskasing–James Bay belt of Ontario (1.1 Ga). The prevalence of alkaline rocks with an age of 0.57 Ga around the perimeters of many shield areas points to a worldwide period of rifting at this time. The fact that the beginning of the Paleozoic era, with its new life forms, is normally taken to be 0.57 Ga may not be entirely a coincidence. During the Paleozoic, alkaline rocks were formed along numerous rift systems, such as the Carboniferous Midland Valley of Scotland and the Permian Oslo graben. During the Mesozoic, alkaline rocks were formed along failed arms of the rift system that caused the breakup of Pangea, such as those of New England, the coastal regions of Brazil, and the Benue trough in Cameroun. In the Cenozoic, alkaline rocks were formed along the Rhine graben, the Rio Grande rift in New Mexico, and the East African rift system.

Although eroded paleorift systems provide excellent exposures of both volcanic and plutonic rocks, only in modern rift systems such as that of East Africa can the relation between tectonism and magmatism be precisely determined. In East Africa, well-developed erosion surfaces of known age provide convenient datum plains with which to monitor tectonic movements during rifting. These erosion surfaces indicate that the East African rift system now occupies the crest of an uplifted ridge, along which are a number of culminations or domes. These domes have diameters on the order of 1000 km and central uplifts of a couple of kilometers (Gass 1970; Baker and Wohlenberg 1971; Bailey 1974). Major branches or triple junctions in the rift system are located at the centers of these domes. The Afro-Arabian dome, for example, is centered on the Afar triangle at the northern end of the rift system, and the Kenya dome is centered on the triple junction to the east of Lake Victoria (Fig. 14-16).

These domes may be surface expressions of upwelling mantle plumes, as suggested by the lower-density and lower seismic velocity of the underlying mantle. Melting of the mantle appears to be localized beneath these domes, because, although eruptions can occur anywhere along the rift system, igneous activity is centered on the domes, with each dome having its own individual history. In the case of the Kenya dome, for example, igneous activity began in the early Miocene at about the same time as doming began. Large central volcanoes, composed essentially of nephelinite with minor basanite, phonolite, and trachyte, developed near the triple junction at the center of the dome. Next, fissure eruptions of phonolite issued from fractures formed during the initial rifting. The grabens at this stage were not prominent topographic features, and flows were able to extend well beyond the rift valley. Although composed of phonolite, they behaved very much like flood basalts, with individual flows extending laterally as much as 250 km. In the

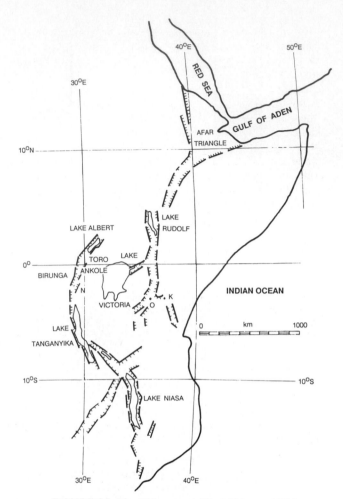

FIGURE 14-16   Rift system of East Africa and Ethiopia. Oldoinyo Lengai (O); Kilimanjaro (K); Nyiragongo (N).

mid- to late Pliocene and early Pleistocene, numerous central volcanoes developed on the flanks of the dome, such as Mount Kilimanjaro, which was built largely of alkali basalt, trachyte, and phonolite. Present-day activity is localized mainly within the rift and is highly variable in character. Pantellerite, for example, erupted recently near the center of the dome, and natrocarbonatite has erupted at Oldoinyo Lengai at the southern end of the dome (Fig. 14-16). The association of igneous activity with uplift and rifting is clear in this case, and the domes play an important role in determining the location and composition of magmas. In general, rocks become more alkaline away from the center of domes, probably reflecting smaller degrees of melting in the mantle beneath these parts. Also, magmatism generally becomes less alkaline with time.

In addition to compositional differences within a dome, there may be larger-scale regional differences. For example, rocks of the Kenya dome are sodic, whereas those of the next dome to the west, which is associated with the western branch of the rift system, are highly potassic. Even here there are regional differences. For example, many of the rocks in the Birunga district are leucite basanites and leucite nephelinites, whereas in the Toro and Ankole districts to the north (Fig.

14-16) potassium is so abundant that in addition to leucite, kalsilite is present and feldspar is absent; this produces rare rock types, such as ugandite and mafurite (Table 6-4). Farther to the north, rocks associated with the Afro-Arabian dome are far less alkaline, and indeed some are even tholeiitic, in particular those associated with the actual generation of ocean floor in the Gulf of Aden and the Red Sea (Gass, 1970). Voluminous alkali basalts, covering an area of about 800,000 km^2, erupted first from the center of this dome prior to the separation of the Arabian and African plates. A number of central volcanoes then developed along the rift valleys, erupting transitional basaltic rocks and peralkaline silicic differentiates. Finally, with continental breakup, tholeiitic lavas began erupting in the Afar triangle and on the new ocean floor.

These regional differences may depend on the vigor of individual mantle plumes. The more active, hot ones may undergo greater degrees of partial melting, which produces more tholeiitic magmas, whereas the less active, cooler ones may undergo smaller degrees of partial melting, which produces more alkaline magmas. Some differences may reflect lateral variations in the composition of the mantle, but melting at different depths in a vertically zoned mantle could also produce these differences.

In other areas, regional variations appear to be induced by differences in crustal rocks. The Monteregian province of southern Quebec is located at the intersection of three grabens, which formed early in the Paleozoic and are now occupied by the valleys of the Ottawa and St. Lawrence rivers and Lake Champlain (Figs. 4-42a and 14-19). The St. Lawrence and Champlain grabens succeeded in rifting apart to become the ocean that later closed to form the Appalachian Mountains. The Monteregian intrusions, which were emplaced in the Cretaceous, are localized by faults extending eastward from the Ottawa graben across the floor of the old triple junction into the Appalachian fold belt. Seismic studies reveal a steady increase in the thickness of the crust in this direction. The composition of the Monteregian rocks also changes systematically in that direction, with the degree of silica saturation increasing toward the east. At the western end of the province, feldspar is completely lacking from the rocks, which are carbonatite, kimberlite, and alnöite. Feldspar first appears near Montreal, but nepheline remains a major constituent. Not until the Appalachians are reached does quartz become a significant component of any of the rocks. Along with increasing silica activity, the size of the intrusions steadily increases (Fig. 4-42a). In this case it is tempting to invoke crustal contamination as the cause for regional compositional differences.

In many alkaline provinces, intrusions that are close to one another and clearly related by tectonic setting have very different rock types. Let us take the Monteregian province again as an example and examine the rock types encountered in going eastward through the province (Philpotts 1974). At the western end, the first main intrusive complex is a carbonatite (no. 5 in Fig. 4-42a). Next, is a series of small plugs of alnöite and kimberlite (no. 7). Mount Royal (no. 10), in the center of Montreal, is composed of essexite and nepheline monzonite. St. Helen's Island (no. 11) in the St. Lawrence River is a diatreme breccia composed of Paleozoic and Precambrian rock fragments with no igneous matrix. Mount Bruno (no. 12) is composed entirely of feldspathic peridotite, as is Rougemont (no. 16). Olivine in Mount Bruno ranges in composition from Fo$_{84}$ to Fo$_{76}$ and the plagioclase from An$_{60}$ to An$_{10}$. The olivine in Rougemont is more iron-rich, ranging from Fo$_{78}$ to Fo$_{74}$, but the plagioclase is extremely calcic and ranges from An$_{98}$ to An$_{80}$. Between these two peridotite intrusions is Mount St. Hilaire (no. 13), an alkali gabbro and zirconium-rich sodalite nepheline syenite complex, which has one of the world's largest assortment of rare minerals. Farther to the east, Mount Yamaska (no. 17) is composed almost entirely of gabbroic rocks. Finally, Brome (no. 18) and Shefford (no. 19) are composed of gabbroic rocks with nepheline and quartz syenites.

Differences in these intrusions cannot be attributed to variable depths of erosion through a common type of magma body. Fundamental compositional differences distinguish each intrusion. Also, because these bodies are vertical pluglike stocks with steeply dipping contacts between rock types, rapid vertical changes in composition cannot be expected. Each intrusion seems to have its own separate source, for there is no geophysical evidence for a large magma reservoir underlying the entire province. Of course, the intrusions were not all emplaced at precisely the same time, which might account for some of the compositional variability. Different depths of origin of the magma in juxtaposed intrusions may also be responsible for differences.

In other alkaline provinces, there may be much less variability in the rock types. In the Permian Oslo graben, for example, 80% of the lavas consist of "rhomb-porphyry," an intermediate alkaline rock that is just saturated (Oftedahl 1960). The plutonic rocks consist predominantly of larvikite (plutonic equivalent of the rhomb-porphyry) and syenite, with smaller amounts of peralkaline granite. Only 15% of the lavas are of alkali basalt, and an extremely small percentage of the intrusive rocks is composed of its intrusive equivalent, essexite. Similarly, in the Kenya dome in the East African rift, phonolite forms a very large percentage of the lavas. Geophysical evidence in both of these regions points to the existence of large elongate mafic magma chambers beneath the rifts. These chambers may have provided sites in which differentiation could produce the large quantities of intermediate and felsic magmas (Ramberg 1976; Baker et al. 1978).

In most alkaline provinces there is a noticeable scarcity of intermediate rock types in both volcanic and plutonic series; that is, they exhibit a Daly gap. Volcanic rocks, for example, may consist predominantly of alkali basalt and trachyte, and plutonic ones of essexite and syenite. Figure 13-19a, which illustrates the compositional range of Monteregian plutonic rocks, shows such a gap between rocks of gabbroic and nepheline monzonite composition. A similar gap exists between nephelinite and phonolite in the rocks of the Kenya dome. Baker et al. (1978) believe that this gap is a result of density filtering by the crust. Some dense mafic magmas may initially have risen directly to the surface, but those that did not must have had to reside in lower crustal chambers until fractionation lowered their density to the point that they could buoyantly rise through the crust. In the Monteregian province, however, where rocks as different as peridotite and syenite are juxtaposed, density does not seem to have been a critical factor. There, liquid immiscibility may have played a role in generating the gap (Sec. 13-6).

Some continental alkaline rocks do not appear to be associated with rift valleys, but instead form elongate belts that may trace the paths of continents over mantle plumes. For example, extending through New England, and cutting obliquely across the trend of the Appalachian mountain belt, are the alkaline intrusions of the White Mountains (Fig. 14-19). This 300-km-long chain, when extrapolated to sea, appears to connect with the New England seamount chain, which can be traced all the way to the mid-Atlantic ridge. On land, however, the ages of the intrusions, which range from 185 to 100 Ma, do not follow the simple linear progression that would be expected if a plate moved over a plume (Foland and Faul 1977). A similar belt of Jurassic age extends for 1300 km in a northerly direction through Nigeria and Niger (Bowden and Turner 1974).

These rocks typically form ring dike complexes, many of which have foundered blocks of rhyolitic lavas in their center (Fig. 4-29). Elongate chains of intersecting ring dike complexes testify to the migration of magmatic activity with time (Fig. 4-32). The rocks are predominantly oversaturated, with granites being the most abundant. Quartz syenites are also common and may predate granites during cycles of activity. Many of the quartz-rich rocks are peralkaline. The granites of northern Nigeria contain economic deposits of tin, which occurs as the mineral cassiterite. Unlike most tin deposits, these contain no tourmaline. Mafic rocks, both younger and older than the granites, form a minor but essential part of this type of alkaline rock association. Many granites contain xenoliths of basaltic rocks, indicating that mafic magmas existed at depth, despite their scarcity near the surface. While extremely uncommon, critically undersaturated syenites and mafic rocks can occur. Some of the intrusions in Niger contain early intrusive sheets of anorthosite, one of the rare Phanerozoic occurrences of this rock type (Sec. 14-8).

The geochemistry of continental alkaline rocks is not very different from that of the alkaline rocks of oceanic islands. They appear to have been formed by small amounts of partial melting of relatively undepleted mantle that contains garnet. In general, alkaline magmas are rich in volatiles, in particular water. Amphibole is therefore a common mineral in the plutonic rocks, even in ones of mafic composition. Alkaline magmas have high concentrations of incompatible elements, which can become further concentrated by fractionation. Many of the felsic members of the alkaline series on continents show signs of some crustal contamination. This is particularly noticeable for magmas that resided in large crustal chambers for any length of time.

14-7   ULTRA ALKALINE AND SILICA-POOR ALKALINE ROCKS

Associated with many but not all alkaline provinces are rocks with exceptionally high alkali contents and low silica contents. They include several varieties of lamprophyre, kimberlite, and carbonatite, and their volcanic equivalents. They constitute a very small percentage of the rocks in alkaline provinces. Nonetheless, in the case of kimberlite and carbonatite, they are of great economic importance as sources of diamond and niobium, respectively. Kimberlites and some lamprophyres are also of great petrologic interest because they contain an abundance of xenoliths ripped from the conduit walls during the rapid ascent of these magmas. Some of these xenoliths provide samples of the mantle down to depths of at least 200 km.

Alkaline Lamprophyres

As a group, lamprophyres are porphyritic melanocratic hypabyssal rocks that contain euhedral phenocrysts of either amphibole or biotite but never feldspar (Table 6.6). Included under such a classification are lamprophyres belonging to two different associations. First, there are the calk-alkaline lamprophyres, including minette, kersantite, vogesite, and spessartite. These are commonly associated with granitic rocks. Second, there are the alkaline lamprophyres, including sannaite, camptonite, monchiquite, polzenite, and alnöite. It is this second group we discuss here.

The most common alkaline lamprophyre is *camptonite*. It contains phenocrysts of kaersutitic (high TiO_2) or barkevikitic (high Fe) amphibole, titanaugite, and possibly olivine, and biotite in a fine-grained groundmass of amphibole, titanaugite, plagioclase, and iron-titanium oxides, with minor amounts of feldspathoid and apatite. Calcite is also a common constituent, occurring both as a primary and secondary mineral; it may fill vesicles or form ocelli. Chemically, camptonites are equivalent to alkali basalts, which explains why they are so common in alkaline provinces. *Sannaites* are similar to camptonites, but plagioclase is subordinate to potassium feldspar in the groundmass.

Ocelli are a characteristic feature of the groundmass of camptonites (Figs. 12-22 and 13-21). These millimeter- to centimeter-sized globules of felsic rock are interpreted to have formed as droplets of immiscible liquid (Sec. 13-6). Toward the top of sills, ocelli may coalesce to form sheets of felsic rock (Fig. 3-10). Thus on a small scale we see the development of contrasting rock compositions that are similar to those of the alkali basalts and trachytes that produce the Daly gap in many alkaline provinces on continents and ocean islands.

With decreasing silica activity, camptonites pass into *monchiquites*. The phenocrysts are similar to those in camptonites, but calcic plagioclase is lacking from the groundmass of monchiquites. The calcium and aluminum that would form plagioclase in more siliceous rocks enter titanaugite as the Ca-Tschermak's molecule ($CaAl_2Si_2O_8 - SiO_2 = CaAl_2SiO_6$). These lamprophyres also contain ocelli of carbonate and nepheline monzonite (Fig. 13-18). *Fourchite* is similar to monchiquite but lacks olivine. The volcanic equivalent of monchiquite is nephelinite.

At still lower silica activities, melilite becomes an important mineral, taking the place of pyroxene ($2CaMgSi_2O_6 - \frac{3}{2}SiO_2 = Ca_2MgSi_2O_7 + \frac{1}{2}Mg_2SiO_4$). This gives rise to *alnöite*, which is composed essentially of phlogopite, melilite, olivine, titanaugite, monticellite, iron-titanium oxides, and calcite, with accessory nepheline, apatite, and perovskite. The presence of perovskite instead of sphene is a clear indication of the low silica activity of these rocks (Prob. 14-8). Alnöites are remarkable in that they are both rich in magnesium and potassium, a characteristic shared with kimberlites. In other groups of rocks, the higher the magnesium content, the lower is the potassium content. Alnöites must be derived from a source containing phlogopitic mica. With this in mind, it is interesting to consider what would happen to an alnöitic magma once it reached high-enough levels in the crust for the mica to become unstable. Phlogopite, for example, breaks down as follows:

$$\underset{\text{(phlogopite)}}{2KMg_3AlSi_3O_{10}(OH)_2} = \underset{\text{(kalsilite)}}{KAlSiO_4} + \underset{\text{(leucite)}}{KAlSi_2O_6}$$

$$+ \underset{\text{(olivine)}}{3Mg_2SiO_4} + \underset{\text{(vapor)}}{H_2O} \quad (14\text{-}1)$$

The reaction products are essential minerals in the rare volcanic rock types *ugandite* (olivine–melilite–leucite) and *mafurite* (olivine–pyroxene–kalsilite) found in the Toro and Ankole districts on the East African rift. Alnöites could therefore be the hypabyssal equivalents of these lavas (Probs. 14-9 and 14-10).

Alnöites appear to have been emplaced rapidly and in some cases even explosively, for they commonly contain fist-sized nodules of mantle and lower crustal rocks; they may even be associated with diatreme breccias. Their content of phlogopite and calcite certainly indicates the presence of H_2O and CO_2. In these respects alnöites are similar to kimberlites, with which they are commonly associated.

Kimberlite

Kimberlite is a rare and unusual alkaline rock that has the distinction of being the primary source of diamonds. It occurs in narrow dikes and diatremes, where it forms the matrix surrounding fragments of the rocks passed through on the way to the surface (Sec. 4-8). Kimberlites are restricted to cratons, with the most widespread occurrences being in the shields of southern Africa and Siberia, but they are also found in North America, Brazil, Australia, and India. In many of these areas, kimberlites have been emplaced repeatedly over a time span ranging from the Proterozoic to Cenozoic; this indicates the existence of a very stable source for these magmas. Although their age may correspond to alkaline igneous activity in nearby rift systems, kimberlites themselves are not confined to rifts. Their locations do, however, appear to be determined by prominent fractures. Kimberlites and associated rocks are treated in detail in the book by Mitchell (1986). Shorter review articles are given by Meyer (1979) and Cox (1978).

The igneous part of a kimberlite is normally severely altered by hydration and carbonation, making it difficult to ascertain the rock's original composition. Nonetheless, they are clearly volatile-rich, potassic ultramafic rocks composed of phenocrysts of olivine (commonly serpentinized) and possibly phlogopite, clinopyroxene, magnesian ilmenite, magnesian garnet, and orthopyroxene, in a groundmass of serpentine, phlogopite, monticellite, magnetite, perovskite, chlorite, calcite, and apatite. The groundmass has essentially the composition of alnöite; kimberlite then could simply be an alnöitic liquid with phenocrysts of olivine and phlogopite. Many of the phenocrysts in kimberlites are xenocrysts formed by the disaggregation of

mantle nodules. Three distinct generations of olivine are commonly present: rounded megacryst (about Fo_{85}), phenocrysts ($> Fo_{90}$), and groundmass (Fo_{90-85}).

Kimberlites contain up to 50% xenoliths and xenocrysts derived from the mantle and crust. The mantle-derived ones fall into two classes: the eclogites, which are normally less abundant, and the more abundant ultramafic nodules, which have variable amounts of olivine but generally can be characterized as garnet lherzolite. Eclogite nodules are composed essentially of omphacitic clinopyroxene (NaAl) and pyrope garnet. Chemically, they have the composition of basalt, and experiments by Yoder and Tilley (1962) indicated that they have a small melting range. These fragments, therefore, are not pieces of residual mantle, which would be expected to have a wide melting range. Instead, they appear to be the high-pressure crystallization product of basaltic magma or metamorphosed basaltic rocks that have been subducted.

The ultramafic nodules include, in addition to garnet lherzolite, garnet harzburgite, garnet pyroxenite, spinel peridotite, and dunite. These come from a wide range of depths, as indicated by the presence of spinel in some and garnet in others. The assemblage plagioclase + olivine, which occurs in basalts at low pressure, is replaced at higher pressures by spinel + pyroxene according to the reaction

$$\underset{\text{(olivine)}}{2Mg_2SiO_4} + \underset{\text{(plagioclase)}}{CaAl_2Si_2O_8} = \underset{\text{(opx)}}{2MgSiO_3}$$

$$+ \underset{\text{(cpx)}}{CaMgSi_2O_6} + \underset{\text{(spinel)}}{MgAl_2O_4} \quad (14\text{-}2)$$

At still higher pressure, spinel is replaced by garnet according to the reaction

$$\underset{\text{(opx)}}{2MgSiO_3} + \underset{\text{(cpx)}}{CaMgSi_2O_6} + \underset{\text{(spinel)}}{MgAl_2O_4}$$

$$= \underset{\text{(garnet)}}{CaMg_2Al_2Si_3O_{12}} + \underset{\text{(olivine)}}{Mg_2SiO_4} \quad (14\text{-}3)$$

These reactions give rise to the stability fields shown in Figure 14-17.

Because kimberlites are emplaced rapidly, minerals in the mantle nodules do not have time to reequilibrate on rising to the surface—diamond, for example, is not converted to graphite. By determining the pressures and temperatures at which the minerals in these nodules equilibrated, the geothermal gradient in the mantle through which the kimberlite passed can be determined. Boyd and Nixon (1975) have done this for nodules in South African kimberlites. Garnet lherzolite nodules contain both Ca-rich and Ca-poor pyroxenes, making it possible to estimate the temperature of equilibration from the pyroxene solvus (Sec. 10-20 and Fig. 10-32). Once the temperature is known, the pressure is estimated from the Al_2O_3 content of orthopyroxene coexisting with garnet. The results of their study (Fig. 14-17) indicate that the nodules formed at pressures and temperatures close to the geothermal gradient that would be expected beneath a stable craton (Sec. 1-6) and at pressures just above and below the diamond–

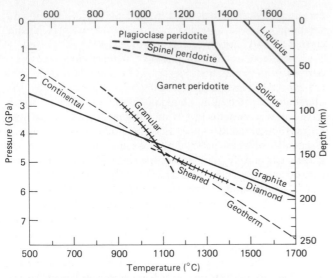

FIGURE 14-17 Equilibrium temperatures and pressure of sheared and granular lherzolite nodules in kimberlites from South Africa [After Boyd and Nixon (1975).] Stability fields of plagioclase-, spinel-, and garnet-peridotites and their anhydrous liquidus and solidus are from Kushiro and Yoder (1966), and the diamond–graphite equilibrium boundary is from Bundy et al. (1961). The continental geotherm is from Figure 1-7 and Prob. 1-4.

graphite equilibrium boundary. This is consistent with the fact that only some of the kimberlites contain diamond.

Boyd and Nixon's study also revealed that the garnet lherzolite nodules fall into two distinct classes, those having a coarse-grained granular texture and those with finer-grained porphyroclastic and mosaic textures, which indicate shear. The granular nodules all come from depths between 100 and 150 km, whereas the sheared ones come from depths of 150 to 200 km. They interpret the boundary between these two types as being the top of the asthenosphere. The two types have slightly different compositions, the granular ones being more depleted in incompatible elements (except K) than the sheared ones. This suggests that the shallower mantle has had melt extracted from it.

Another significant difference is that the granular nodules define a distinctly shallower geotherm than the sheared ones. Because a sharp inflection in the geotherm could not be a steady-state phenomenon, Boyd and Nixon interpret the steeper gradient in the mantle from which the sheared nodules came to be the result of a thermal perturbation in the rocks below 150 km. They suggest that the heating could result from the shearing itself. It is also possible that the perturbation was related to the magmatic event responsible for the production of the kimberlites, which must have come from a depth of at least 200 km.

With contents of SiO_2 of about 30% and of MgO of about 23%, kimberlites are clearly ultramafic rocks. They differ from other ultramafic rocks, however, in having high concentrations of TiO_2, K_2O, P_2O_5, and other incompatible elements. Their high Cr and Ni contents indicate that they have not undergone any significant fractionation of olivine. They must therefore be derived by small degrees of partial melting

of fertile mantle, that is, mantle from which little if any melt has previously been extracted. Zone melting may also have played a role in developing the high concentrations of incompatible elements.

An interesting, but rather rare group of volcanic rocks, the *lamproites*, may be related to kimberlites. In northwestern Australia, for example, they occur with kimberlites and also contain diamonds and mantle xenoliths. Lamproites are potassium- and magnesium-rich rocks that typically contain leucite and phlogopite. Olivine-rich varieties, which have compositions approaching kimberlite, are the ones that contain diamond.

Carbonatites

The most silica-undersaturated rocks found in alkaline provinces are the carbonatites. The term *Karbonatite* was introduced by Brögger in 1921 to denote carbonate rocks from the Fen district of southern Norway, which he believed were of igneous origin. The idea of magmatic carbonates met with immediate opposition from Bowen, who thought that the carbonates were of a replacement origin. Despite experimental evidence that such melts could exist at low temperatures and pressures (Wyllie and Tuttle 1960), a magmatic origin for carbonatites was not universally accepted until carbonate lavas were witnessed erupting from the Oldoinyo Lengai volcano in Tanzania (Dawson 1962). Reviews of the history of ideas relating to carbonatites and a coverage of related field and experimental studies can be found in the books by Heinrich (1966) and Tuttle and Gittins (1966).

Carbonatites are found in most alkaline provinces throughout the world, but like kimberlites, they appear to be restricted to continental regions, with a few possible exceptions (Cape Verde Islands, for example). They are closely associated with rift valleys and can have ages from Proterozoic to modern, as for example: Paloboro in the Transvaal (Proterozoic); Mountain Pass, San Bernardino, California (1.1 Ga); Alnö Island in the Baltic, Sweden (0.56 Ga); Fen, southern Norway (0.56 Ga); Minas Gerais, Brazil (Jurassic); Oka, Quebec (Cretaceous); Magnet Cove, Arkansas (Cretaceous); Bearpaw Mountains, Montana (Eocene); Kaiserstuhl, Baden, Germany (Miocene); and Oldoinyo Lengai (modern—1960).

Most carbonatites occur in small ring complexes (Fig. 4-42b) associated with silica-poor rocks composed essentially of nepheline and clinopyroxene but also containing variable amounts of melilite, melanite garnet, perovskite, apatite, and opaques. *Ijolite*, which is composed of nepheline and aegerine diopside, grades to *melteigite* (pyroxene rich) and *urtite* (nepheline rich). Rocks composed largely of melilite and titanaugite are called *melilitites*, and they grade from *okaite* (melilite rich) to *jacupirangite* (augite rich). Nepheline syenite may also be present. The volcanic equivalent of ijolite is nephelinite, and at the world's only active carbonatite volcano, Oldoinyo Lengai, it is interlayered with carbonatitic tuffs and capped by the recent natrocarbonatite lavas. These occurrences leave no doubt as to the close association of carbonatitic and nephelinitic magmas.

In many carbonatite complexes, the concentrically arranged rock types become progressively poorer in silica toward the core, which is commonly occupied by carbonatite, but this pattern can be disturbed by multiple intrusions. A typical succession of rock types from rim to core would consist of nepheline syenite, ijolite, and carbonatite, with all rocks being cut by lamprophyric dikes of alnöite or fourchite. The carbonatite itself may also be concentrically zoned from an older, outer zone of calcite carbonatite (sövite), followed by a zone of dolomite carbonatite (beforsite) and a younger core of ankerite or siderite carbonatite. Based on experimental studies, such a sequence is consistent with the fractional crystallization of a carbonate melt.

The typical carbonatite is composed of approximately 75% carbonate (calcite, dolomite, ankerite, siderite), with lesser amounts of clinopyroxene, phlogopite, alkali amphibole, apatite, magnetite, olivine, monticellite, perovskite, and pyrochlore [$(Ca, Na)_2(Nb, Ta)_2O_6(Oh, F)$]. Carbonatites are highly variable in composition, even down to the scale of a hand specimen, where modal variations almost always define a flow layering. Carbonate grains may be elongate parallel to the layering, but many have been recrystallized into smaller polygonal grains. Original grain sizes may still be evident, however, from the distribution of other minerals, such as apatite, which outlines the original grains. Phlogopite crystals may be zoned in shades of pale green with orange-red rims.

The rocks surrounding virtually all carbonatite complexes have undergone intense sodium metasomatism; that is, their compositions have been changed by the addition of sodium. Siliceous country rocks may be converted to aegerine-bearing syenites that are termed *fenites*. Zones of *fenitization* are typically hundreds of meters wide. The prevalence of such metasomatized rocks around carbonatite complexes indicates that large volumes of alkali-bearing solutions are given off during cooling.

Ancient intrusive carbonatites are composed predominantly of calcite, whereas modern volcanic ones contain abundant sodium carbonate. Rainwater rapidly dissolves sodium carbonate, so it is not surprising that natrocarbonatite lavas are unlikely to be preserved in the geologic record. However, the lack of evidence for sodium carbonate in plutonic complexes has puzzled petrologists. One explanation is that carbonatite magmas do indeed have high concentrations of sodium carbonate, but during solidification and cooling, hydrothermal solutions remove the sodium from the carbonatite (leaving it composed essentially of calcite) and transporting the sodium into the country rocks to form fenites. This idea was strongly favored when it appeared from experimental studies that only sodium carbonate melts could separate as immiscible liquids from silicate magmas. In view of the recent work by Kjarsgaard and Hamilton (1988) showing that the immiscibility field extends to low alkali compositions (Fig. 13-20), it is not necessary for all carbonatites to be initially sodium-rich and then be leached of alkalis. They can form directly as calcite carbonatites.

Some carbonatites are of economic value as ores of niobium and rare earths. Initially, carbonatites were mined only for iron and limestone, which is used for cement and as a flux in smelting iron. Their potential as the major source of Nb was not recognized until the 1950s. Although most carbonatites contain accessory amounts of pyrochlore, some contain zones with high concentrations (about 0.4% Nb_2O_5). These zones are independent of the shape of the carbonatite in which they occur, and they may even be discordant. The Nb enrichment probably takes place during a late magmatic or deuteric stage. REE and Th can also be present in economic concentrations, occurring mainly in perovskite, but also to a lesser extent in pyrochlore. The more highly differentiated ankeritic and manganiferous carbonatites typically contain the highest concentrations of REE.

Carbonatites, as a group, have exceptionally high concentrations of Ti, Nb, Zr, REE, P, F, Ba, Sr, and Th. These elements are abundant in alkaline magmas in any case, but it appears that during the process of generating carbonatites they are further concentrated, perhaps through strong liquid–liquid partitioning into an immiscible carbonate liquid (Sec. 13-6). Once fractional crystallization of this liquid takes place, Sr and Nb are depleted by entering early-crystallizing carbonate and pyrochlore, respectively, and the residual liquid becomes enriched in the other elements.

Strontium isotopic data clearly demonstrate that carbonatites are derived from the same mantle source as the associated alkaline rocks, both sharing low initial $^{87}Sr/^{86}Sr$ ratios (0.701 to 0.704). They are clearly not the product of limestone assimilation, for sedimentary limestones have initial values of about 0.709. Carbonatites are probably derived from a variety of alkaline magmas. Dawson and Hawthorne (1973) have shown that at least on a small scale, carbonatite can form as an immiscible liquid from kimberlitic magma, as occurred in the Benfontein Sill, South Africa, where carbonatite formed small diapirs that rose into the kimberlite. Many alnöites contain abundant calcite, and fourchites, monchiquites, and camptonites contain carbonate ocelli, which may be quenched immiscible droplets (Prob. 14-11). The carbonate globules trapped in apatite crystals in the Usaki complex of West Kenya (Rankin and Le Bas 1974) show that immiscible carbonate liquids can separate from ijolitic magmas. It is likely, therefore, that carbonatite liquids can separate from silicate melts over a wide range of compositions. The experimental results of Kjarsgaard and Hamilton (1988) indicate that silicate melts encountering the immiscibility field at an early stage of fractional crystallization (low Na) would separate calcium-rich carbonate melts (Fig. 13-20), whereas more evolved silicate melts would separate more sodium-rich carbonate melts.

14-8 SPECIAL PRECAMBRIAN ASSOCIATIONS

One of the tenets of geology, first formulated by Hutton and widely publicized by Playfair and Lyell, was the doctrine of *uniformitarianism—the present is the key to the past*. This principle has helped geologists unravel much of the geologic record. Conditions during the earliest part of the Earth's history, however, were distinctly different from those during later time. As a result, we find that certain rocks were formed only during the Precambrian and at no other time. Without modern analogs, these unique Precambrian rocks present a special petrologic challenge.

Radioactive elements were more abundant in the Archean than in later time, and they would have generated more heat, thus creating a steeper geothermal gradient. This, in turn, would have resulted in a thinner lithosphere (Tarling 1980). The geotherm would have intersected the beginning of melting curve for mantle materials over a wider range of pressures, allowing melting to occur at both shallower and greater depths than in more recent times. Mantle convection would have been more vigorous, and lithosphere was probably created and destroyed more rapidly. On being subducted, thin oceanic lithosphere may not have retained its water long enough to flux the mantle above the deep Benioff zone, thus decreasing the quantity of calc-alkali rocks formed, especially the more potassic varieties. Bombardment by meteorites would have been more intense, with the larger ones creating huge impact structures. Unlike the Moon, the Earth, with its active tectonism, has tended to destroy the evidence of these ancient impacts, but some are still preserved, especially in stable Precambrian shields.

Archean Crust

Shield areas on each of the main continents have a nucleus of Archean rocks (>2.3 Ga), some of which, despite their age, are surprisingly well preserved. Indeed, some sedimentary and volcanic rocks have hardly been metamorphosed and provide a reasonably good picture of the surface conditions at the time. The igneous rocks provide some insight into the thermal status of the upper mantle and lithosphere. These rocks indicate that the Archean world differed significantly from that of later times.

First, the Earth's atmosphere had a very different composition. Gases liberated by the earliest volcanoes developed an atmosphere composed largely of H_2O, CO_2, CO, SO_2, H_2S, and N_2 (Sec. 11-1). This atmosphere lacked oxygen until the end of the Archean. Ultraviolet light from the sun caused some dissociation of water vapor to hydrogen and oxygen, but this small amount of oxygen would have combined with elements such as iron ($2Fe + O_2 = 2FeO$) to form oxides and silicates. Not until the appearance of green plants and photosynthesis was oxygen produced in sufficient quantity to oxygenate the atmosphere. Stromatolitic limestones formed from algae that were capable of carrying out this process are found in rocks as old as 3.0 Ga, and they are believed to have produced sufficient oxygen in the atmosphere by about 2.7 Ga for primitive oxygen-respiring bacteria to develop. These algae become particularly abundant in the geologic record by 2.3 Ga, which is normally taken as the beginning of the Proterozoic era.

The lack of oxygen in the atmosphere throughout most of the Archean allowed iron to be transported in surface waters in the soluble ferrous state. Today, ferrous iron is rapidly oxidized by the atmosphere to the insoluble ferric state, which gives the brown and red colors to weathered rock surfaces and soils. The greater solubility of iron in Archean times was undoubtedly a factor in making sedimentary iron formations a common rock type. These are interlayered with cherts, to form *banded iron formations*, rocks containing millimeter- to centimeter-thick layers of hematite, jasper, or chert. Iron formations become progressively rarer during the Proterozoic and are virtually nonexistent in Phanerozoic time. Other sedimentary rocks found in Archean terrains are stromatolitic limestones and clastic sedimentary rocks, which are important because

they indicate that some granitic crustal material was present and available for weathering. Unlike Phanerozoic sedimentary sequences, Archean ones are never thick, indicating the lack of long-lived subsiding basins.

Archean sedimentary rocks occur in small basins with a variety of volcanic and intrusive rocks, which have typically been metamorphosed to the greenschist facies (Sec. 15-4). They are consequently referred to as *greenstone belts*. These belts form short arcuate outcrop patterns that have a general parallelism within a given province, but they lack the long continuous patterns of Proterozoic and Phanerozoic mountain belts. The thinner sequences of sedimentary and volcanic rocks, the discontinuous nature of the fold belts, and shorter wavelengths of folds have been interpreted as indicating that the lithosphere was thinner at the time and thus buckled more easily. Greenstone belts have also been interpreted as deformed basins that were initially formed by meteorite impact (Green 1972). On the Moon, large impact structures have been flooded with basalt to form maria. On Earth, such structures would have been metamorphosed and deformed.

The igneous rocks of greenstone belts, which typically constitute more than 50% of a belt, are generally basaltic, but other compositions ranging from ultramafic through basaltic to andesitic and rhyolitic are also present. Some lavas of ultramafic compositions, known as komatiites, are found only in Archean terrains (see next section). In general, the basaltic and ultramafic rocks form the lower part of the succession, with intermediate and siliceous ones forming the upper part.

Between the greenstone belts are vast stretches of granitic rocks, which form plutons and gneissic diapirs. In contrast to the greenstone belts, these granitic rocks are intensely metamorphosed at upper amphibolite or granulite facies grades. They are characterized by a low K/Na ratio; that is, they are trondjhemites or tonalites. They are typically gray and constitute the so-called *gray gneisses*, which are so characteristic of Archean terrains. At Isua, western Greenland, it is on a dome of such a gray gneiss (3.7 Ga) that the oldest rocks so far found on Earth are draped (3.8 Ga).

The K/Na ratio in igneous rocks has increased through time (Engel et al., 1974). The preponderance of trondjhemites in the Archean is responsible for the low K/Na ratio during that time, but toward the end of the Archean the K/Na ratio increases rapidly as a result of the widespread emplacement of more normal granites. The Archean gray gneisses seem to be part of a calc-alkali suite. Why more potassic varieties did not develop is a puzzle. Perhaps insufficient continental crust, which typically concentrates the strongly incompatible potassium, had not yet formed. On the other hand, in view of the distribution of Na and K rocks above modern Benioff zones, perhaps melting occurred at too shallow a depth.

The major and trace element chemistry of the gray gneisses of the Archean basement of Finland indicate that they must have been derived by the partial melting of tholeiitic basalt (Martin 1987). They have low initial $^{87}Sr/^{86}Sr$ ratios of 0.702 to 0.701, which indicates derivation from the mantle and not from a reworked crust. The REE data, however, do not allow generation of the gray gneisses directly from the mantle. Instead, a period of REE enrichment during the formation of a tholeiitic basalt is necessary. This basalt must next have been metamorphosed to a garnet-bearing amphibolite and then partially melted. A residue of hornblende, garnet, plagioclase, clinopyroxene, and ilmenite must have existed in order to generate the major and trace elements of the gray gneisses.

The presence of hornblende and garnet in the source from which the gray gneisses came is of great importance and indicates a fundamental difference between these rocks and later calc-alkali ones. The geothermal gradient must have been much steeper in the Archean for subducted basalt to be metamorphosed to garnet amphibolite and then melt before dehydration could occur. Today, hydrated ocean floor basalts lose their water to the overlying mantle wedge long before the basalts can actually melt.

Komatiites

The ultramafic lavas known as komatiites are unique to the Archean greenstone belts. They were first recognized in Zimbabwe and then described in detail from the Komati River, in the Barberton Mountainland of South Africa (Viljoen and Viljoen 1969). They are now also recognized from the Archean terrains of Canada, Western Australia, and Finland. Apart from high MgO contents, their most striking feature is a texture found in the upper part of many flows, consisting of long criss-crossing sheafs of olivine crystals that radiate or fan out downward into the flow. This is referred to as the *spinifex* texture because of its resemblance to inverted tufts of an Australian grass by that name. Komatiites have proved to be of economic importance, because some contain nickel sulfide deposits.

Komatiites are defined as ultramafic volcanic rocks that contain at least 18% MgO (Arndt and Nisbet 1982); many contain as much as 33% MgO. They occur with tholeiitic basalts but never constitute more than 10% of the volcanic rocks. One of the best preserved successions of komatiite lavas is in Munro Township, Ontario, where they form part of the Abitibi greenstone belt (Pyke et al. 1973). Here 60 ultramafic flow units are exposed over a stratigraphic thickness of 125 m, with the thickest being 15 m, and the average, 3 m. Most flows have a characteristic stratigraphy, which is illustrated in Figure 14-18. The top of each flow is marked by prominent fracturing. Immediately beneath this is the spinifex zone, with olivine blades increasing in length downward and reaching a maximum length of 1 m in the thickest flow. The olivine blades, which have a marked skeletal form (Fig. 12-7a), constitute approximately 50% of the rocks, with the intersticies being occupied by skeletal clinopyroxene and altered glass. The spinifex zone terminates abruptly against a fine- to medium-grained peridotite that forms most of the lower part of the flow. Immediately beneath the spinifex zone in some flows is a zone containing elongate olivine dendrites with a strong foliation parallel to the flow. Even in the main lower peridotite part of the flow a weak foliation is commonly present. Toward the base of the peridotite zone the flow develops a knobby weathering.

The spinifex texture, with its skeletal olivine crystals which nucleate and grow down from the upper quenched surface of the flow, is an ideal example of crystal growth controlled by the rate at which heat is dissipated from the growth surface (Sec. 12-3). These olivine quench crystals constitute important

FIGURE 14-18 Steeply dipping 1-m-thick komatiite flow, Munro Township, Ontario. The top of the flow, which is at the extreme left, is marked by a prominent zone of fracture. Beneath this is dark-weathering peridotite with spinifex texture (long blades of olivine), which terminates abruptly against a fine- to medium-grained pale-weathering peridotite. Toward the base of the flow is a dark, knobby-weathering peridotite. The hammer is lying on the lower contact of the flow, which is most clearly marked by the beginning of the fractured flow top of the underlying flow. (Photograph by J. M. Duke.)

textural evidence that komatiites are indeed ultramafic lavas and are not basaltic ones in which olivine crystals were concentrated. Olivine in the spinifex zone definitely grew as quench crystals in the melt. The fact that the blades of olivine are not broken or bent and can extend down through as much as two-thirds of the thickness of a flow indicates that little, if any, differential movement occurred in this part of the flow during the growth of the olivine blades. This has been taken as evidence that the olivine grew in situ. However, the velocity profile calculated in Prob. 2-9 indicates that little differential movement occurs in the upper part of a flow in which there is a temperature gradient; most of the movement is concentrated in the central hot zone. The spinifex zone may have been rafted along on the hotter interior, which may have formed the strongly foliated peridotite immediately beneath the spinifex zone. Some lavas of komatiitic affinity have been found in post-Archean rocks, but these generally have lower MgO contents. They may, however, have formed under conditions that approached those under which komatiites formed. Perhaps on rare occasions the Phanerozoic geotherm locally behaved like that in the Archean.

Experiments indicate that komatiitic magma has a high liquidus temperature, in excess of 1650°C at atmospheric pressure (Green 1975). It is not surprising then that such hot magma, when erupted on the Earth's surface, would experience rapid cooling both by radiation and conduction, and that quench crystals would likely develop, at least in the upper part of flows. Passage of such hot magma through a continental crust is likely to have caused partial melting and assimilation. Numerous geochemical studies have shown komatiites to have highly variable concentrations of incompatible elements. This could indicate derivation from inhomogeneous mantle, but it is more likely that komatiites all come from depleted mantle but have suffered variable degrees of contamination while passing through the crust [see, for example, Barley (1986)]. Huppert and Sparks (1985) calculate that the rate of meltback of conduit walls by komatiitic magma could have been as high as tens of meters per day. These hot, low-viscosity magmas, on erupting, would have flowed turbulently across the Earth's surface (Prob. 14-12). Turbulent flow maintains a lava at a relatively constant temperature throughout, which causes considerable heating of the underlying rock. Huppert et al. (1984) have shown that this heat would cause sediment beneath a komatiite flow to melt, and the lava would eventually flow in a thermal erosion channel of its own making. At Kambalda in Western Australia such

thermal erosion has led to the development of nickel sulfide ores at the base of the komatiite flows. The assimilated sediment contained sulfur, which on entering the komatiite lava, combined with iron and nickel to form a dense immiscible sulfide liquid that sank to the bottom of the channel.

The existence of such high-temperature lavas is strong evidence that the Archean geothermal gradient was steeper than that in later times. Today, partial melting of peridotitic mantle generates basaltic to picritic magmas, not komatiitic ones. To increase the MgO content to the levels in komatiites, the degree of partial melting would have to be increased to more than 50%. This, however, presents a problem, because melt instantly segregates from residual solids once the degree

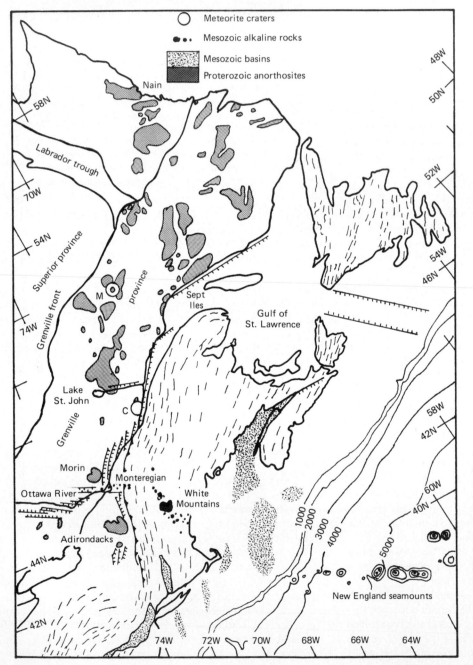

FIGURE 14-19   Massif-type anorthosites of the southeastern part of the Canadian Shield. Those near Nain have ages of 1.45 Ga, whereas many of those south of the Grenville Front have ages as young as 1.1 Ga, with the exception of the Sept Iles anorthosite, which is only 0.5 Ga. The New England Seamounts may mark the trace of a hot spot that earlier formed the Mesozoic igneous rocks of the White Mountains. The Mesozoic basins formed during crustal extension associated with the opening of the Atlantic. Also shown are two large meteorite impact structures, Manicouagan (M) and Charlevoix (C).

of partial melting exceeds 30%; the 50% melt required to generate komatiitic liquid in this way could therefore not be achieved. Takahashi and Scarfe (1985), however, have found from experiments that the normative olivine content of the anhydrous melt that first appears above the solidus of mantle peridotite systematically increases with increasing pressure. At low pressure this melt is basaltic, but at 4 GPa it contains 20% MgO, and at 7 GPa it contains >30% MgO. They conclude therefore that basaltic magmas are generated at depths of less than 100 km, whereas komatiitic ones are formed between 150 and 200 km.

Despite the presence of komatiites in Archean greenstone belts, basalts are still by far the most abundant volcanic rock. The conditions in the upper mantle must therefore have been capable of producing both basaltic and komatiitic magmas. With a steeper geothermal gradient, the depth range over which the geotherm would intersect the peridotite solidus would have been greater than in later times, thus allowing basalts to form at shallow depth and komatiites at greater depth. In post-Archean times the shallower geotherm would have intersected the solidus over a shorter depth interval and apparently at depths too shallow to produce high-magnesium melts.

Anorthosites

Anorthosites are rocks composed essentially of plagioclase feldspar. They occur as cumulate layers in large, layered gabbroic intrusions, where they typically have compositions in the bytownite range. They also occur in large Proterozoic massifs, where there are no mafic or ultramafic rocks; these typically have an intermediate plagioclase in the labradorite or andesine range. These bodies, which rival the large layered lopoliths and granite batholiths for size, represent a remarkable and unique period of magmatism that occurred between 1.45 and 1.1 Ga. The author knows of only two Phanerozoic massif-type anorthosites. One Cambrian body at Sept Iles, Quebec, looks identical to nearby Proterozoic ones (Fig. 14-19). It occurs on the margin of the Shield and is associated with alkaline magmatism and rifting. The other body is of Silurian age and occurs in the alkaline ring complexes of Niger. Anorthosites also occur in Archean terranes, where they typically form lenses in high-grade gneisses. Some of these may be part of metamorphosed layered gabbroic intrusions, whereas others may have formed by the metamorphism of sedimentary rocks. They differ from the Proterozoic anorthosites in having extremely calcic plagioclase. Anorthosites are believed to make up much of the lunar highlands. They were possibly formed early in the Moon's history by plagioclase crystals floating and accumulating on the top of basaltic magma. It is the Proterozoic massif-type anorthosites that are dealt with in this section.

Proterozoic anorthosites lie in a belt that extends across Pangea from Scandinavia, through southern Greenland, Labrador, Quebec, and New York State (Fig. 14-19), and beneath the sedimentary cover across the craton to Wyoming and California. A second, less well defined belt may extend from Virginia, through Africa, Madagascar, and India, to Australia. A collection of works describing these bodies has been edited by Isachsen (1968), and partisan reviews have been written by Morse (1982), Duchesne (1984), and Emslie (1985).

Most massif-type anorthosites in the Grenville province form batholithic bodies that appear to have been emplaced diapirically (Philpotts 1981). They consequently do not reveal floors, and thus we do not know what underlies them. Gravity surveys do not indicate the presence of ultramafic rocks beneath anorthosites, but this is not to say that such rocks might not occur at much greater depth and were simply too dense to rise with the anorthosite. Most anorthosites are emplaced into upper amphibolite or granulite facies metamorphic rocks. The marginal zones of these bodies may be intensely deformed and recrystallized as a result of the diapiric emplacement, but toward their cores, primary textures are preserved that show massif anorthosites to have formed initially as igneous rocks. In Labrador, north of the Grenville front, anorthosites appear to have formed in situ rather than being emplaced diapirically.

Within most anorthosite massifs the rock types have a distinct primary stratigraphy (Fig. 14-20). If the body has been emplaced diapirically, these rocks may be disposed in annular zones with the oldest in the core. The lowest exposed rock is invariably anorthosite, containing only small amounts of orthopyroxene, augite, or olivine. Variations in the proportions of these minerals can produce layering on the scale of tens of centimeters to meters in thickness. This layering originally appears to have been horizontal. More commonly, the anorthosite appears massive. One of the most characteristic features of the Proterozoic anorthosites is their coarse primary grain size (Fig. 14-21). Plagioclase crystals several centimeters long are common, and they may even reach a meter. This coarse grain size is not due to pegmatitic fluids. Indeed, all indications are that anorthosites formed under anhydrous conditions. Plagioclase crystals are commonly of two different sizes. The largest

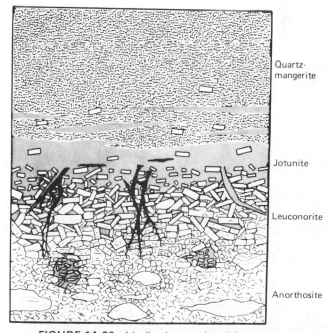

Quartz-mangerite

Jotunite

Leuconorite

Anorthosite

FIGURE 14-20 Idealized stratigraphic section through a typical massif-type anorthosite. The black dikes and sheets are iron-titanium oxide–apatite rock. [After Philpotts (1981); published with permission of Canadian Mineralogist.]

(a)   (b)

FIGURE 14-21   (a) Single meter-sized crystal of plagioclase from the Nain anorthosite, Labrador. Reflection of light is from (010) cleavage. (b) Single megacryst of aluminous bronzite from Lake St. John anorthosite, Quebec. The crystal has kink bands as a result of deformation.

crystals, which typically have lengths of several centimeters, occur separately or in clumps that are surrounded by centimeter-length laths or granules that are typically intergrown with ferromagnesian minerals. The granular plagioclase, which is prevalent in the Grenville anorthosites, is interpreted as being *protoclastic*, that is, broken and recrystallized prior to complete solidification of the magma. This texture indicates that these anorthosites were emplaced as a crystal mush. The primary plagioclase is invariably darkened by small oriented inclusions of iron oxide. Recrystallization associated with later tectonic activity, which presumably took place at lower temperature, expels these inclusions and turns the plagioclase white. Anorthosites also commonly contain megacrysts of aluminous bronzite (Fig. 14-21b), which contains between 7 and 9% Al_2O_3 and indicates crystallization at high pressures. Xenoliths of country rocks are rare, but some anorthosites enclose meter-sized blocks of other anorthosite that differs slightly in mineralogy or texture.

The content of ferromagnesian minerals (opx normally most abundant) in anorthosites increases upward as the rock grades into *leuconorite*, which in many bodies is the main rock type. Undeformed samples of this rock have a prominent alignment of plagioclase laths parallel to any compositional layering that is present. Toward the top of the leuconorite the abundance of pyroxene, apatite, and iron-titanium oxides increases rapidly, and the grain size decreases to form a ferrodioritic rock, *jotunite*. Dikes of ferrodiorite may cut the underlying leuconorite. Occurring in this zone and in the underlying leuconorite may be sills and dikes of magnetite–ilmenite (rutile)–apatite rock (*nelsonite*), which in some areas has been mined for ilmenite. The jotunite zone, which is commonly only tens of meters thick, marks the top of the anorthositic part of an intrusion; below this plagioclase is the dominant mineral; above it alkali feldspar dominates. A variety of rock types may lie above the jotunite, but in general they have the composition of a hypersthene bearing quartz monzonite (*quartz mangerite*); quartz syenite and hypersthene granite (*charnockite*) may also be present. The contact between the

jotunite and more siliceous rocks is abrupt, but sheets of jotunite continue to occur for some distance above the main contact.

Plagioclase in the anorthositic rocks typically has compositions between An_{60} and An_{30}, but within a given body it rarely varies by more than a few percent. For example, in the world's largest anorthosite body, which covers an area of 25,000 km^2 north of Lake St. John, Quebec (Fig. 14-19), all plagioclase falls in the labradorite range. To the southwest, plagioclase in the Morin anorthosite, located directly north of Montreal, is all in the andesine range. Plagioclase does become slightly more sodic in the jotunites, reaching An_{45} in those associated with labradorite anorthosites and An_{30} in those associated with andesine anorthosites. The composition of the plagioclase remains essentially the same in the quartz mangerites. Dark megacrysts of andesine, measuring up to several centimeters in diameter, commonly occur in quartz mangerites near anorthosite bodies.

By contrast, pyroxenes, which are minor components in most anorthositic rocks vary widely in composition. Orthopyroxenes show a complete range of composition from about En_{75} in the most primitive anorthosites to En_{30} (inverted pigeonite) in the quartz mangerite. This enrichment in iron takes place progressively from leuconorite, through jotunite, to quartz mangerite. The iron enrichment is sufficient in many quartz mangerites for the iron-rich pigeonite (now inverted pigeonite) to be replaced by the assemblage fayalite plus quartz. Coexisting augite in this rock series changes steadily from about $Ca_{47}Mg_{43}Fe_{10}$ in anorthosites to ferroaugite in quartz mangerites. The fact that the most magnesian orthopyroxene in anorthosites is only En_{75} indicates that these rocks form from a magma with a magnesium number that is lower than that of basaltic magmas in layered gabbroic intrusions which typically are capable of precipitating orthopyroxene as magnesian as En_{85}.

One of the major problems to unraveling the origin of massif-type anorthosites is the relation of the quartz mangerites to the other rocks. Are they fractionation products of the magma that produced the anorthosites, or are they crustal rocks melted by the heat given off by this magma? The answer to this question is critical to determining the composition of the parental magma of anorthosites. Because most anorthosite bodies lack chilled margins, the only way of estimating the composition of the parental magma is from a weighted average of all comagmatic rocks. This, of course, is difficult to do in the case of a body emplaced diapirically, because some of the original rocks may not be represented at the present level of erosion. Nonetheless, if this possibility is ignored, the problem remains of whether the quartz mangerites belong to the series. If they are omitted from the weighted average, the parental magma would have the composition of leuconorite; if they are included, the magma would be granodioritic.

The continuity to the variation in the composition of the major minerals in passing from anorthosite, through leuconorite and jotunite, to quartz mangerite suggests these rocks are comagmatic. But it is possible that a mantle-derived magma could rise into the crust by assimilating material at its roof and crystallizing plagioclase and pyroxene at its base. The rise of such a zone of melt into the crust could produce a progressive change in the composition of the minerals accumulating on the floor, even though most of the siliceous fraction of the rock series would be derived from crustal rocks.

There is a simple test of whether quartz mangerites are fractionation products of an anorthositic magma or are partial melts of crustal rocks. Plagioclase is the primary cumulus phase in anorthosites and leuconorites, so fractionation products should show the effects of the early separation of this mineral. Because europium preferentially enters plagioclase, separation of this feldspar should produce negative europium anomalies in fractionated liquids (Sec. 13-11 and Fig. 13-33). Although some quartz mangerites have distinct negative europium anomalies, the majority do not. We can conclude, therefore, that although quartz mangerite can be a differentiation product of the magma from which anorthosites formed, the large volumes of quartz mangerite associated with most anorthosites are the products of partial melting of crustal rocks. Initial $^{87}Sr/^{86}Sr$ ratios support this conclusion.

Two of the most puzzling features of anorthosites is the unzoned nature of its plagioclase crystals and the relatively constant composition of plagioclase within a given body. Plagioclase crystals are always enriched in anorthite relative to the liquid with which they are in equilibrium. As crystallization proceeds the liquid and the crystals become more albitic (Fig. 10-12), and if equilibrium is maintained, the crystals should become homogeneous throughout. Diffusion rates in plagioclase, however, are so slow that once a plagioclase crystallizes, its composition is essentially fixed; that is, crystals have little chance of homogenizing, and as a result, plagioclase is commonly zoned (Fig. 14-7). The fact that even the largest plagioclase crystals in anorthosite are remarkably free of zoning calls for explanation.

Homogeneous plagioclase crystals can grow if the proportion of liquid is much greater than the proportion of crystals forming from it. This, however, cannot be the case for anorthosite massifs, because not only are individual crystals unzoned, but the plagioclase within a given complex shows little compositional variation. The process by which anorthosites form must therefore involve some mechanism of buffering the plagioclase at a relatively constant composition, but this composition can be different for different anorthosite complexes. Such a mechanism might, for example, involve special phase relations, or perhaps a physical process of magma replenishment or mixing.

As an example of one mechanism that could produce plagioclase of essentially fixed composition, we will examine the possible role of liquid immiscibility (Philpotts 1981). Rocks of the anorthosite–quartz mangerite series, when plotted in variation diagrams such as those based on silica content, differentiation index, or magnesium number (see Sec. 13-2), show a continuous spectrum of compositions. When these same data are plotted in terms of the three groups of components $SiO_2–(FeO + MnO + TiO_2 + CaO + P_2O_5)–(MgO + Al_2O_3 + Na_2O + K_2O)$, the series is seen to have a distinct fork to it (Fig. 14-22). Rocks from anorthosite through leuconorite to jotunite are marked by strong enrichment in iron and other high field-strength cations. The rocks from leuconorite through mangerite and quartz mangerite to charnockite, on the other hand, show a decrease in iron and an increase in silica. Thus there seems to be two contrasting trends of fractionation, one toward iron-rich residues (Fenner trend) and the other toward silica-rich residues (Bowen trend). When compared to the liquid immiscibility field in the system fayalite–leucite–silica (Roedder 1951) and its expanded form resulting from the addition of 3 wt % TiO_2 and 1% P_2O_5 (Freestone 1978), these two trends skirt the two liquid field, as if they were composed of conjugate liquids. Experiments (Philpotts 1981) reveal that quartz mangerite and jotunite produce immiscible liquids on melting, just as do the granophyre and ferrodiorite of the Skaergaard (McBirney 1975).

Figure 14-23 shows the results of an experiment where a homogeneous, high-temperature fused mixture of jotunite and quartz mangerite was held at 995°C at atmospheric pressure. Plagioclase (An_{50}), augite, and pigeonite have crystallized, and the remaining melt has unmixed into silica-rich and iron-rich liquids. In this one experiment we have the necessary phases to form the entire anorthosite series: that is, anorthosite from the plagioclase, jotunite from the iron-rich liquid, and quartz mangerite from the siliceous liquid. As plagioclase and pyroxene crystallize in this phase assemblage, the liquids do not change their compositions significantly; instead the two liquids simply change their proportions. As a consequence, the plagioclase crystallizes with a constant composition, buffered by the two liquids. Note that in this model, the quartz mangerites could be derived by fractionation from the parental liquid of the anorthosites or by melting of crustal rocks. The buffering of the plagioclase requires only that two liquids be present, regardless of their ultimate source. Mechanisms other than liquid immiscibility could also produce relatively constant composition plagioclase. Any successful model for the generation of massiftype anorthosites must be capable of doing this.

Anorthosites remain one of petrology's greatest puzzles. With Phanerozoic igneous rocks, a particular geochemistry can be associated with a general tectonic setting—alkaline rocks with rifting, depleted tholeiites with ocean floor spreading, and calc-alkali rocks with subduction. Anorthosites do not fit into any of these categories. Estimates of the composition of the parental magma vary widely, from basalt to anorthosite and granodiorite. The amount of crustal contamination is also hotly

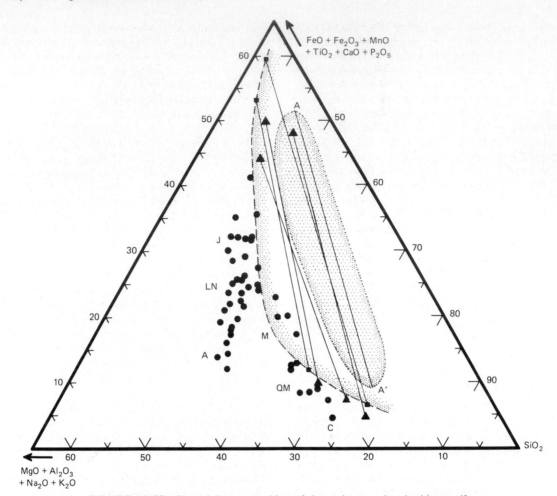

FIGURE 14-22   Plot of the composition of the rocks associated with massif-type anorthosites; A, anorthosite; LN, leuconorite; J, jotunite; M, mangerite; QM, quartz mangerite; C, charnockite. The oval shaded area is the two-liquid field from the system fayalite–leucite–silica (Roedder, 1951), and to its left is the expanded two-liquid field resulting from the addition of 3 wt % TiO_2 and 1 wt % P_2O_5 (Freestone, 1978). Tie lines connect conjugate immiscible liquids in experiments (squares) or andesites (triangles). [After Philpotts (1981); published with permission of Canadian Mineralogist.]

debated. But overshadowing all else is the problem that they formed almost exclusively in the short interval between 1.45 and 1.1 Ga, and then, only in restricted belts. What unique conditions could have led to their formation?

Isotopic studies reveal anorthosites to have highly variable initial $^{87}Sr/^{86}Sr$ ratios, but there appears to be a distinct input of magma from the mantle. However, anorthosites do not resemble any other mantle-derived rocks, in that they are so feldspathic and have such low magnesium numbers. A mantle-derived magma would therefore have to fractionate before it could produce anorthosites. Experiments indicate that the eutectic between diopside and anorthite (Fig. 10-4) shifts to feldspar-rich compositions at high pressures (Yoder 1965; Presnall et al. 1978). Fractionation of a mantle-derived basaltic magma in the lower crust could lead to the developement of this feldspar-rich residual liquid. It is here that the aluminous megacrysts of bronzite might form. The feldspar-rich eutectic liquid would have a low density and would rise into the crust. As the

pressure decreased, the liquidus field of plagioclase would expand, and plagioclase would become a prominent cumulus phase. As the magma rose, more and more plagioclase would form, converting the magma into a plagioclase mush. Such a medium would be capable of transporting the megacrysts of bronzite, which otherwise could not have been raised by a liquid magma (Prob. 14-13).

All anorthosites have a strong positive europium anomaly, which is to be expected in view of the partition coefficients of the REE between plagioclase and magma (Sec. 13-11). These positive anomalies, however, must have their negative counterparts somewhere in the geologic record if the REE are to preserve their overall normall abundance distribution. Furthermore, the fact that anorthosites have positive europium anomalies indicates that anorthosites must be cumulates and cannot be liquid residues from a fractionation process. For example, if anorthosites formed from an anorthositic liquid that was generated from a basaltic magma by the separation

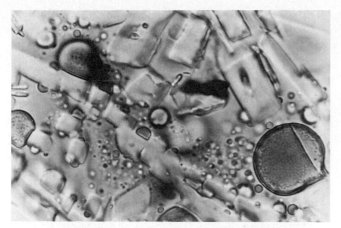

FIGURE 14-23 Immiscible iron-rich and silica-rich liquids coexisting with plagioclase of An_{50} composition formed by cooling a homogeneous fused mixture of jotunite and quartz mangerite to 995°C at low pressure under an oxygen fugacity buffered by nickel—nickel oxide. Diameter of large iron-rich globule is 15 μm. [After Philpotts (1981); published with permission of Canadian Mineralogist.]

of mafic minerals at depth there would be no reason for it to have a europium anomaly. However, an anorthositic liquid developed from the melting of a previous plagioclase cumulate would have a positive europium anomaly. But this simply pushes the problem of the origin of the anorthosite back one step. The question then is: Where is the residue from the anorthosite cumulate that has the negative europium anomaly? So far, only a small number of quartz mangerites and jotunites have such anomalies. Perhaps other quartz mangerites had negative anomalies but large-scale assimilation of crustal rocks may have masked them. It is also possible that the residual liquid with the negative anomaly was more dense than anorthosite and sank. A similar relation would exist if anorthosites formed by accumulation of floating plagioclase at the top of a magma chamber. The stratigraphy of anorthositic rocks (Fig. 14-20), however, is not consistent with this model. Another complicating factor is that apatite crystallizes with a negative europium anomaly, which causes residual liquids to become enriched in europium relative to the other REE. This would tend to cancel the effect of plagioclase crystallization. Apatite becomes extremely abundant in jotunites, and thus rocks more evolved than this will have their REE patterns modified by this additional factor.

Anorthosites in the southern part of the Canadian shield form a prominent belt extending from the coast of Labrador to the Adirondacks in New York State (Fig. 14-20), but within this belt anorthosites are of several ages. The distribution, sizes, shapes, and petrology of the anorthosites throughout this region are so similar that the entire belt appears to be one igneous province. Yet the anorthosites in Labrador have ages of 1.45 Ga, whereas most in the Grenville province are only 1.1 Ga or even 0.5 Ga if we include the body at Sept Iles, Quebec. Perhaps anorthosites were formed at depth throughout this region at 1.45 Ga and then reactivated and intruded to higher levels at later dates in the Grenville province. In fact, even in the Grenville, anorthosites may be of several ages.

Anderson and Morin (1968), for example, argue on the basis of the presence of blocks of labradorite anorthosite in andesine anorthosite that andesine anorthosites are formed by the partial melting of labradorite anorthosites. Duchesne (1984) has pointed out that structural relations in the Grenville anorthosites are not consistent with them simply being older anorthosites that were tectonically reactivated during the Grenville orogeny; they must involve melting at this time.

The volumes of magma involved in the formation of anorthosite complexes are enormous; for example, anorthosites constitute 20% of the rocks in the belt extending from Labrador to the Adirondacks. They must therefore reflect a major thermal event (or episode). Are there any other significant occurrences in the geologic record that might give some indication of what this event was? In the vicinity of Lake Superior, Keweenawan basalts and the Mid-Continent gravity high were developing between 1.2 and 1.1 Ga, as were the Coppermine River basalts and the Muskox intrusion in the Northwest Territories. Many Keweenawan basalts and diabases actually contain large xenoliths of anorthosite. To the northeast of Lake Superior, alkaline igneous rocks were formed between Kapuskasing and James Bay at 1.1 Ga, and just to the east of Labrador, alkaline rocks of the Gardar province in southwest Greenland were formed at 1.2 Ga. Crustal extension therefore seems to have become active shortly after the emplacement of the first Proterozoic anorthosites and remained active during the emplacement of the Grenville ones. In fact, Emslie (1985) interprets the anorthosite suite as a rift-related group of rocks. Widespread rifting at this time might reflect some major overturning of the mantle which has never been repeated on the same scale. But this is pure speculation, and the solution to the anorthosite problem awaits new insight. This solution, when it comes, is likely to be of considerable significance to our understanding of the history of the Earth.

14-9 METEORITE-IMPACT-GENERATED ROCKS

Inspection of the surface of the Moon and many other planets reveals that cratering by meteorite impact has been a common occurrence in the solar system. The flux of meteorites reaching the surface of the Moon, and presumably its nearest planetary neighbor, the Earth, has, however, decreased dramatically with time, as evidenced by the difference in the frequency of craters on the older lunar highlands and the younger maria. Most of the ancient surface of the Earth has been destroyed by tectonic processes, and none older than 3.8 Ga has yet been found. Nonetheless, it is in the cratonic areas, which have remained tectonically stable since the Proterozoic, that we have the greatest chance of finding evidence of meteorite impacts.

Meteorite craters are of two types. Small meteorites, on passing through the Earth's atmosphere, are slowed by friction, which causes them to glow brightly, and most are completely evaporated before reaching the surface. Those that do penetrate cause little damage on landing, and the size of the crater, which may be up to several meters in diameter, depends on the material on which they fall. By contrast, large meteorites are not perceptibly slowed by the atmosphere and they reach the Earth's

FIGURE 14-24   Meteor Crater, Arizona. (Photograph courtesy of U.S. Geological Survey.)

surface with essentially the velocities they had in space, which are typically on the order of 20 km s^{-1}. The energy dissipated on impact is so great that it causes an explosion, which excavates the crater (Fig. 14-24). These hypervelocity impact structures are the ones likely to be recognized in the geologic record, for they produce distinctive fracture patterns, rock types, and even minerals.

When a meteorite hits the Earth it has a kinetic energy that depends on its mass and velocity, that is, $\frac{1}{2}mv^2$. Following impact, this energy is dissipated both as shock waves that fracture and displace the impacted crust to form the crater, and as heat, which vaporizes the meteorite and melts the underlying rock on the crater floor. Approximately 20% of the energy goes into heating and perhaps even more with very large impacts. Considerable insight into the formation of large explosion craters has been gained from underground nuclear testing [see, for example, Short (1966)]. If we define the distance from the explosion out to the point where the rock is completely crushed as R, the rock is vaporized to a radius of $0.05R$, melted to a radius of $0.07R$, and fractured to a radius of $2R$. The diameter of a crater is related to the kinetic energy of the meteorite by the relation

$$\text{diameter} = 9.47 \times 10^{-3}(\tfrac{1}{2}mv^2)^{0.294} \qquad (14\text{-}4)$$

where all values are in SI units (m, kg, s). The depth R to which

complete crushing of the rock occurs is approximately one-third the crater diameter.

Craters, whether formed by nuclear explosion or hypervelocity impact, have a common structure (Fig. 14-25). Because they are formed by explosion, they are almost perfectly circular and are not elongated parallel to the direction of flight of the meteorite. If the impacted rocks have marked anisotropic physical properties (prominent joints or schistosity, for example), fracturing during the explosion may produce a noncircular form. Meteor Crater, Arizona, for example, tends to be rectangular (Fig. 14-24). Explosion craters are quite shallow, the depth to the floor being only one-tenth their diameter. Material ejected during the explosion forms a blanket surrounding the crater. The size of the ejecta decreases rapidly away from the rim. If the crater penetrates stratified rocks, their fragments are deposited in the ejecta blanket in reverse stratigraphic order. Flat-lying sedimentary rocks may be tipped up on end or even completely overturned by the force of the explosion. The floor of the crater is covered with breccia and melted rock (Fig. 14-27). In large impact structures this zone of melt may be hundreds of meters thick. Beneath the breccia is a zone of intense fracturing. Certain rocks (quartzite, limestone, for example) may develop a type of fracture known as *shatter cones*, which are believed to be diagnostic of impact structures; they have not been found associated with volcanic explosions. Shatter cones are conical fractures, with their apex pointing toward the ex-

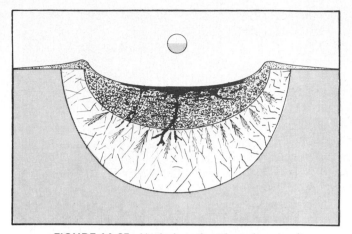

FIGURE 14-25 Vertical section through meteorite explosion crater. Melt (black) formed by fusion of impacted rock mixes in all proportions with the underlying breccia. Crater depth is typically one-tenth of the crater diameter, and the depth to which rock is totally crushed is one-third the diameter. Intense fracturing extends twice as deep as the crushed zone. Shatter cones fan out away from the explosion. The size of the meteorite shown assumes it to be a stony variety with a density of 3500 kg m^{-3} traveling with a velocity of 20 km s^{-1}.

plosion. They commonly form interpenetrating sets as seen in Figure 14-26.

The steep walls of very large craters are gravitationally unstable and collapse along normal faults into the crater. Where the toes of these slump structures converge toward the center of the crater, they develop a central uplift. In extremely large craters, central uplift may continue long after the initial slumping as a result of the lithosphere isostatically trying to compensate for the lack of mass in the crater and zone of broken rock.

The melt formed by impact rapidly quenches to a glass, which in some ancient impact structures has been mistakenly identified as a volcanic rock. Impact glasses, however, differ in significant ways from volcanic ones (Dense, 1971). First, they are extremely inhomogeneous and contain abundant fragments of the surrounding rocks, many of which exhibit shock features (see below). With increasing number of fragments, these glasses grade into breccia (Fig. 14-27). Because impact glasses are superheated they never contain phenocrysts, whereas most volcanic glasses do. Impact glasses also contain streaks of glass that correspond in composition to almost pure minerals, such as *lechatelierite*, formed from fused quartz or high-pressure silica polymorphs. These glasses testify to the high temperatures associated with hypervelocity impacts. Some glasses contain spherules of iron, which may be of meteoritic origin. Melt ejected from craters cools rapidly by radiation as it passes through the atmosphere, raining down as small glassy spheres. With very large explosions, melt may be ejected into space, and when it returns through the atmosphere, frictional heating and melting of the surface sculptures the glass particles into aerodynamic forms. These are known as *tektites*.

The high pressures generated by the explosions, which momentarily may exceed 10^3 GPa, produce new minerals and a variety of deformation structures. The high pressures do not last long enough to develop stable high-pressure mineral assemblages, and those transformations that do occur are limited to ones requiring no diffusion, that is, simple polymorphic transformations. Nor is there time for large crystals to be formed, and most identification of high-pressure forms is made by x-ray diffraction.

Both of the high-pressure polymorphs of quartz, *coesite* and *stishovite*, have been found at impact structures. Coesite is stable only above 2 GPa at low temperature, and though it

FIGURE 14-26 Shatter cones in Trenton limestone from the Charlevoix impact structure, Quebec. Shatter cones fan out in the direction away from the blast.

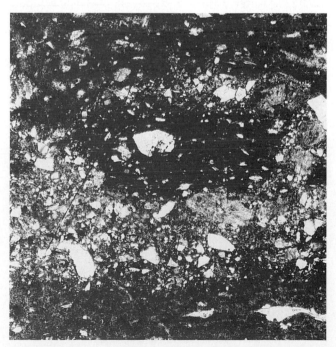

FIGURE 14-27 Impact melt from the Manicouagan impact structure, Quebec.

may form in high-pressure metamorphic rocks, it would normally invert to quartz on returning slowly to the surface. It has been found, however, in some Alpine high-pressure metamorphic rocks, so its presence cannot be taken as positive proof of meteorite impact. The polymorph stishovite, however, requires pressures in excess of 7 GPa (corresponding to a depth of 220 km), and it has been found only at meteorite craters. Stishovite may melt to lechatelierite after a shock wave passes. Some meteorite fragments, such as those of Cañon Diablo, which formed Meteor Crater, Arizona, contain diamond, which requires about the same pressure as stishovite to form.

The high strain rates associated with the passage of a shock wave can produce disorder and dislocations in minerals. Plagioclase, for example, is converted at pressures above 5 GPa to a glassy pseudomorph known as *maskelynite* (Fig. 14-28). When examined in plane light under the microscope, maskelynite resembles normal plagioclase, retaining its crystal shape, cleavage, and twin lamellae. But under crossed polars it is isotropic. Also, it does not diffract x-rays. Maskelynite therefore resembles a glass, but its refractive index, while being less than those of plagioclase of the same composition, is greater than that of glass formed by fusing the plagioclase. Also, heating of maskelynite at 900°C for 2 h converts it back to a single crystal of plagioclase (Bunch et al. 1967). Clearly, maskelynite lacks long-range order (hence no x-ray pattern), but unlike fused glass it retains the short-range order that allows it easily to revert back to plagioclase. Commonly, the transformation to maskelynite is selective, affecting only parts of grains or perhaps only one set of twin lamellae. This shows that crystallographic orientation with respect to the path of the shock wave is critical to the formation of maskelynite.

One of the most common microscopic features produced by meteorite impact is sets of planar features in quartz (Fig. 14-29). They occur as fine, closely spaced, crystallographically controlled multiple sets of lamellae. They may be visible because of decorations (minute inclusions), differences in refrac-

FIGURE 14-28   Maskelynite (isotropic plagioclase glass) developed in anorthosite from the central uplift of the Manicouagan impact structure, Quebec. Note that only one of the two twinned directions in the lower grain has been converted to maskelynite, the remainder of the grains are still plagioclase.

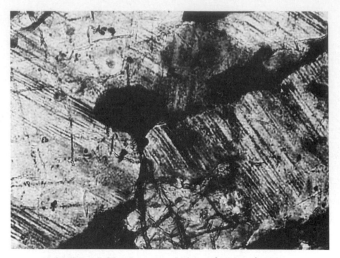

FIGURE 14-29   Decorated planar features in quartz from the Charlevoix impact structure, Quebec.

tive index or crystallographic orientation, or actual cleavages. In their study of meteorite craters in the Canadian Shield, Robertson et al. (1968) were able to classify these planar features on the basis of crystallographic orientation into five different types, which could be correlated with the intensity of deformation (Table 14-1). The least deformed quartz has planar features parallel {0001}. With greater deformation planar features develop at 23° to the basal plane ({$10\bar{1}3$}) and then at 80° ({$22\bar{4}1$}). By this stage the basal features are no longer present. Next, lamellae develop at 32° to the basal plane ({$10\bar{1}2$}). Finally, at the highest degree of deformation parts of the quartz grains become isotropic. The planar features become more closely spaced with increasing intensity of deformation, and the refractive indices of quartz decrease. These five classes of planar feature can be used to map zones of deformational intensity around old craters.

The diameters of impact structures range from hundreds of meters to possibly as much as 100 km. Figure 14-30 shows the range of sizes of impact structures found in the Canadian Shield. They range from 2.3 km for the Holleford structure in southern Ontario to 67 km for the Manicouagan structure in Quebec (see also Fig. 14-19). Structures with diameters of less than 30 km form simple craters, their precise form depending on the depth of erosion or amount of younger sedimentary fill. Structures with diameters greater than 30 km are complicated by a central uplift. In the case of the Clearwater Lakes, which were probably formed by the impact of a pair of meteorites, we see that the western structure, with a diameter of 32 km, has a central uplift which forms a ring of islands composed of melt rock and breccia. The eastern structure, with a diameter of 27 km, lacks a central uplift. The large Manicouagan structure has a central uplift consisting of anorthosite and other basement rocks from beneath the crater. Melt rock that would have originally covered the crater floor now occurs only around the central uplift.

The zone of rupture associated with impact structures having diameters of more than 50 km would extend into the lower crust, and the zone of fracture would penetrate to the upper mantle. The resulting decompression on the underlying mantle could cause large-scale melting if the geotherm were

TABLE 14-1

Planar features in quartz formed by meteorite impact*

Type[a]	Crystallographic Orientation	Number of Lamellae per Millimeter
A	$\{0001\}$	100–250
B	$\{10\bar{1}3\}$; $\{01\bar{1}3\}$; $\pm\{0001\}$	200–500
C	$\{22\bar{4}1\}$; $\{10\bar{1}3\}$; $\{01\bar{1}3\}$; $\{0001\}$ rare	400–1000
D	$\{10\bar{1}2\}$; $\{01\bar{1}2\}$; $\{22\bar{4}1\}$; $\{10\bar{1}3\}$; $\{01\bar{1}3\}$	>1000
E	Same as in D, but with isotropic patches in which there are no planar features.	>1000

* After Robertson et al. (1968).
[a] Type A is least deformed and type E is most deformed.

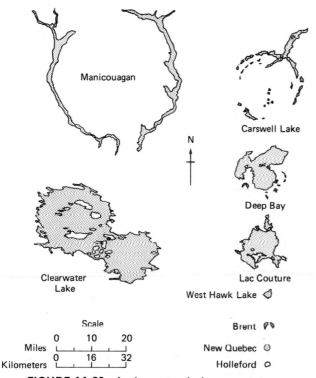

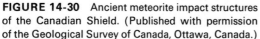

FIGURE 14-30 Ancient meteorite impact structures of the Canadian Shield. (Published with permission of the Geological Survey of Canada, Ottawa, Canada.)

near solidus temperatures prior to impact. Large-scale melting would favor the formation of tholeiitic magma, which would then rise into the crater and pond beneath the less dense breccia and melt rock formed by the initial impact.

Such a mechanism was proposed by Dietz (1964) for the origin of the 1.7-Ga-old Sudbury lopolith, Ontario (Fig. 14-31). Since its formation, this body has been deformed by the Grenville orogeny (1.1 to 1.0 Ga), and thus some interpretation is involved in reconstructing its original form. The lopolith is now elliptical in plan, but prior to Grenville deformation it may have been circular. The presence of shatter cones, most of which radiate and plunge outward from the body, is strong evidence of meteorite impact (Guy-Bray et al. 1966). Along the southern side of the lopolith, numerous small, irregular veins and stockworks contain a remarkable rock known as the Sudbury breccia. Its fragments, which are mostly of local rocks, but may include some exotic ones, range in size from large

boulders to microscopic particles. This was probably an impact breccia intruded into the underlying rocks. The magma that rose into the crater solidified to form norite. This rock contains a large percentage of interstitial granophyre, which increases in abundance toward the top, where it forms a continuous sheet (micropegmatite). The granophyre, in turn, intrudes the Onaping tuff, which forms the roof of the intrusion. This rock was originally interpreted as a welded tuff of volcanic origin. However, it contains many fragments of the country rocks, and some quartz fragments contain the diagnostic planar features (French 1972). It seems most likely that this tuff was generated as an impact melt. The crater was later filled with sedimentary rocks, which now form the Onwatin slate and the Chelmsford Arkose. Most of the world's nickel comes from the ore deposits associated with the Sudbury lopolith. Although Dietz proposed that this nickel may have come from the original meteorite, the high Cu/Ni ratio of the Sudbury ores (about 1:1) makes this highly unlikely because this ratio in iron–nickel meteorites is about 1:500. The ores, which sank as immiscible sulfide liquids to the base of the body, are therefore probably of terrestrial origin.

On a still larger scale, meteorite impact has been proposed as the triggering mechanism for the magmatism that gave rise to the Bushveld complex, South Africa (Rhodes 1975). This body and the nearby Vredefort Ring (Fig. 14-14) may have been formed by four simultaneous hypervelocity impacts. Although multiple impacts may seem highly improbable, the Clearwater Lakes (Fig. 14-29) show that they have occurred.

The Vredefort structure has numerous features supporting an impact origin (Dietz 1961). Sediments of the Transvaal System, which underlie the Bushveld intrusion and normally have shallow dips, are turned completely up on end and are even overturned around the Vredefort structure. This produces a ring syncline 60 km in diameter (Fig. 14-14). The core of the structure, which has a diameter of 40 km, is occupied by the Vredefort granite. The structure therefore appears to be a classic diapir. However, shattered rock is abundant, and both the granite and the overlying sediments are cut by black glassy veins of *pseudotachylite*, which is formed by the in situ fusion of the host rock (Philpotts 1964). Granophyre, which forms still larger veins, is also thought to be formed by in situ fusion. All of the quartzite units surrounding the structure have shatter cones. Their orientation varies depending on the dip of the sediments, but if bedding is restored to its original horizontal position, the shatter cones all point toward the center of the structure. This evidence indicates that following a hypervelocity

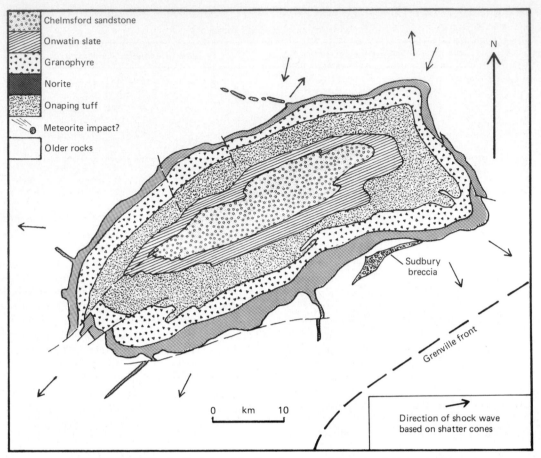

FIGURE 14-31   Sudbury lopolith, Ontario, a possible meteorite impact structure. [Map from various sources; shatter cone orientations from Guy-Bray et al. (1966).]

impact, which caused melting of the sedimentary and granitic rocks (pseudotachylite and granophyre), upward doming destroyed the crater and overturned the surrounding rocks.

The Bushveld complex is very much like the Sudbury lopolith, with noritic rocks capped by a granophyre which is intrusive into overlying volcanic rock, the Rooiberg Felsite. Rhodes interprets the felsite as being formed from an impact melt. Its age is indistinguishable from that of the Vredefort granophyre and the Bushveld complex itself (1.95 to 2.0 Ga).

No shatter cones have been found around the Bushveld, but Rhodes claimed that the lobes of the complex are formed by magma flowing into ring synclines around each of the three impact structures. A central uplift exposes highly deformed and brecciated rocks from beneath the complex. The evidence for an impact origin for the Bushveld, while tantalizing, is far from positive at this stage and will require further documentation. It seems quite likely, however, that early in the Earth's history impact structures of this dimension would have formed.

PROBLEMS

14-1. Two basalts with different concentrations of the trace elements Ni and P might be related by fractional crystallization or different degrees of partial melting. The distribution coefficient (solid/liquid) for the compatible Ni is 10, and that for the incompatible P is zero. Calculate the factors by which the Ni and P in the less evolved composition (lower P) would have to be multiplied to give those in the more evolved composition if the two basalts are related by **(a)** one being a primary magma and the other its differentiation product following 5% crystallization, and **(b)** one being a 1% partial melt and the other a 6% partial melt. Assume equilibrium fractionation and melting

(review Eq. 13-39). **(c)** In light of your answers to parts **(a)** and **(b)**, how might you distinguish chemical changes in rock series produced by fractional crystallization from those produced by partial melting? **(d)** What effect would Rayleigh rather than equilibrium fractionation have?

14-2. The concentrations of Ni and Rb in typical primitive ocean island basalt are 120 and 5 ppm, respectively, and in primitive alkali basalt they are 100 and 30 ppm, respectively. Assuming that these lavas are formed by partial fusion of an identical source and that the distribution coefficient (solid/liquid) for Ni is 10 and for Rb is 0, what degrees of equilibrium partial fusion would produce the given compositions, and what are the concentrations of these elements in the source rock? [*Hint*: If the source rock was totally melted, the concentration of an element in the melt (c_i^{ol}) would be the same as in the source rock.]

14-3. Carefully study the zoning in the plagioclase phenocrysts from Mount St. Helens shown in Figure 14-7, and list, in chronological order, all growth and resorption episodes visible. Give a possible explanation for the uneven distribution of inclusions on the one surface marked by inclusions (review Sec. 12-4 and Fig. 12-13). Finally, what geological events may be recorded in the zoning of this crystal?

14-4. In tabular form, contrast the compositions and mineralogy of typical MORB, ocean island alkali basalt, continental flood basalt, and calc-alkali basalt.

14-5. Using Figure 10-36a, show how three magmas of only slightly different composition could produce the following sequences of crystallization:

    I   olivine–orthopyroxene–plagioclase–clinopyroxene

    II   olivine–clinopyroxene–plagioclase–orthopyroxene

    III olivine–orthopyroxene–clinopyroxene–plagioclase

14-6. In many tholeiitic layered intrusions, early-crystallizing olivine, which is forsteritic, is replaced in intermediate rocks by calcium-poor pyroxene (opx or pigeonite) and then reappears in late-stage rocks as a fayalitic variety. Assuming complete fractional crystallization, select an initial magma composition in the phase diagram for FeO–MgO–SiO₂ that would produce such a sequence.

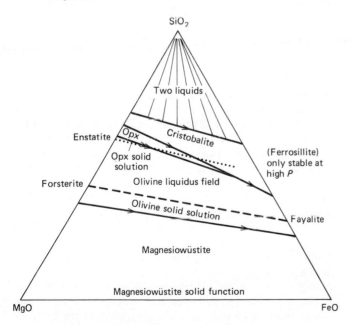

14-7. The most magnesian olivine in the Muskox intrusion is $Fo_{89.5}$. Using a K_D for the partitioning of magnesium and iron between liquid and solid of 0.31 (Eq. 13-1), calculate the magnesium number of the magma with which this olivine must have been in equilibrium.

14-8. Alnöite commonly contains perovskite rather than sphene, which occurs in the more siliceous lamprophyres. Perovskite

reacts with magma to form sphene if the silica activity is high enough according to the reaction

$$CaTiO_3 \text{ (perovskite)} + SiO_2 \text{ (magma)} = CaTiSiO_5 \text{ (sphene)}$$

The presence of perovskite therefore sets an upper limit on the activity of silica in alnöites. Calculate the maximum silica activity that an alnöitic magma can have at 1100°C. The free energy of formation in kJ mol⁻¹ of silica liquid at this temperature is -663.996, of perovskite is -1277.885, and of sphene is -1958.551.

14-9. Many of the volcanic rocks of the Toro and Ankole districts of the East African rift contain both kalsilite and leucite, which therefore defines the silica activity in these lavas through the reaction

$$KAlSiO_4 \text{ (kalsilite)} + SiO_2 \text{ (magma)} = KAlSi_2O_6 \text{ (leucite)}$$

According to Robie et al. (1978), the free energies of formation of these materials from the elements at 1100°C in J mol⁻¹ are $-1,560,611$ for kalsilite, $-663,906$ for silica in magma, and $-2,262,666$ for leucite. Calculate the activity of silica in a lava containing these minerals at 1100°C. Assume leucite and kalsilite to be pure phases.

14-10. In view of the possibility that ugandite and mafurite might be related to alnöitic magma through the breakdown of phlogopite at low pressure, is the silica activity calculated in Prob. 14-9 compatible with the presence of perovskite in alnöitic magma (Prob. 14-8)?

14-11. (a) Calculate the silica activity in a carbonatitic magma containing crystals of pure diopside and monticellite at 1100°C. Use thermodynamic data from Table 7-1.

    (b) If a carbonatitic magma formed as an immiscible liquid from an alnöitic magma, how would the silica activities in these two liquids compare?

    (c) In light of the silica activity calculated in part (a) and the possible silica activities in alnöites (Prob. 14-8), could carbonatite form as an immiscible liquid from an alnoitic magma?

14-12. The viscosity of a komatiitic lava near its liquidus temperature is 0.5 Pa s. If the density of the magma is 2800 kg m⁻³, how thick would an extensive sheet flowing down a 2° slope have to be to become turbulent? (*Hint*: Review Prob. 3-14 and the critical Reynolds number in Sec. 3-6.)

14-13. Could megacrysts of bronzite, such as the one in Figure 14-21b, be transported from depth into the crust by magma? Assume that the megacryst has a diameter of 20 cm and a density of 3400 kg m⁻³. We do not know the density of the magma, but it cannot be far from 2500 kg m⁻³. The viscosity of the magma is unknown and will depend strongly on composition. Let us assume two different values, one typical of a basalt, 50 Pa s (a), and the other of an intermediate composition, 1000 Pa s (b). Calculate the sinking velocity of the crystal using Stokes law (Eq. 13-4), and check that the conditions are appropriate for the equation (Eq. 13-5). In light of your calculations, what conclusion can you draw from the presence of bronzite megacrysts in anorthosites?

14-14. The Manicouagan impact structure has a diameter of 67 km. Assuming that it was formed by a common stony meteorite with a density of 3500 kg m⁻³ and was traveling with a velocity of 20 km s⁻¹, (a) what was its kinetic energy, and (b) what was its diameter?

15 / Metamorphic Reactions and Metamorphic Facies

15-1 INTRODUCTION

In Chapter 14 we saw that igneous rocks vary widely in composition, with most being in the basalt to rhyolite range. The surface processes of weathering, transportation, and sedimentation serve to further differentiate these compositions into such extreme rock types as pure quartz sandstone (orthoquartzite), alumina-rich shale (pelite), and carbonate rocks (limestone, dolostone). This wide assortment of igneous and sedimentary rocks provides the starting materials (*protoliths*) from which metamorphic rocks are formed.

Metamorphism is the sum of all changes that take place in a rock as a result of changes in the rock's environment, that is, changes in temperature, pressure (directed as well as lithostatic), and composition of fluids. The changes in the rock may be textural, mineralogical, or chemical. In its broadest sense, metamorphism includes the entire spectrum of changes that take place between the zone of weathering and the zone in which melting gives rise to magmas. Traditionally, however, the low-temperature changes associated with weathering and the lithification and diagenesis of sediments have been omitted from the study of metamorphism. Typical metamorphic reactions take place at temperatures above 150 to 200°C. At the highest temperatures metamorphism gives way to magmatic processes where partial melting produces *migmatites*, mixed igneous–metamorphic rocks. Average continental crust starts melting under water-saturated conditions around 1000°C at low pressures, but this temperature decreases with increasing pressure, dropping to 650°C at 0.5 GPa (Fig. 11-10). The water-saturated beginning of melting of granite marks the upper temperature limit of metamorphism in many regions, because the latent heat of fusion provides an enormous heat sink. If the activity of water is less than 1, the melting curve is raised (Fig. 11-11), and thus higher metamorphic temperatures may occur.

When sedimentary rocks are buried, either through subsidence and filling of sedimentary basins or by overthrusting of other crustal rocks, their temperature rises. This is due in part to the geothermal gradient and in part to the blanketing effect of crustal rocks, which normally have relatively high concentrations of heat-producing radioactive elements (review Prob. 1-5). The metamorphism resulting from this type of burial and heating is associated with convergent plate boundaries and develops on a regional scale; it is therefore referred to as *regional metamorphism*. It is normally accompanied by tectonism, with the result that the metamorphic rocks develop prominent foliated fabrics—slate, schist, gneiss. Heat can also be introduced into the crust by bodies of magma, which, on cooling, liberate their heat content, and, on crystallizing, liberate the latent heat of crystallization. Metamorphism resulting from this heat is localized around igneous intrusions and is referred to as *contact metamorphism*. Most contact metamorphic rocks are not strongly deformed during metamorphism, and they therefore lack the foliated fabric of regionally metamorphosed rocks.

When studying metamorphic rocks, petrologists attempt to determine the conditions that brought about the metamorphism and then, from these, to deduce the major geologic events that have affected a region. The mineralogy, textures, and structures of metamorphic rocks can all record important information about the conditions prevalent during the formation of these rocks, and one of the main tasks of metamorphic petrology is to learn how to read this record. We have already seen, for example in Sec. 1-6, that many metamorphosed pelites contain one or more of the Al_2SiO_5 polymorphs, andalusite, kyanite, or sillimanite, yet the calculated steady-state continental geotherm (Fig. 1-7) lies entirely within the kyanite stability field. Metamorphic rocks containing andalusite or sillimanite must therefore have formed at temperatures well above the steady-state geotherm. In addition, rocks containing all three polymorphs or just kyanite and sillimanite must have formed at pressures of 0.376 GPa or higher, that is, at depths greater than 13 km. These rocks are exposed on the surface of the Earth in regions where the present crust is 35 km thick. At the time of metamorphism, then, the crust must have been at least 50 km thick. We can conclude from these simple observations that such metamorphic rocks must form in a thickened continental crust with a steepened geothermal gradient. These requirements are met only near convergent plate boundaries above Benioff zones.

In this and subsequent chapters we will see how petrologists go about reading the metamorphic record. We start by examining a simple model field area that has undergone only one episode of metamorphism. This will allow us to see how field data are collected, how they are analyzed, and what metamorphic conditions they record. The principles introduced are developed further in subsequent chapters.

15-2 MODEL METAMORPHIC TERRANE

Figure 15-1 is a map of an idealized metamorphic terrane underlain by folded and faulted metamorphosed sedimentary rocks. Two different types of geological boundary are shown in this map. First, there are the primary boundaries between different rock bodies, in this case sedimentary bedding planes separating mappable formations and fault contacts. In other areas these boundaries might also include igneous contacts. Second, there are boundaries between different metamorphic zones that have experienced different intensities of metamorphism, as evidenced by their mineralogy.

Most primary boundaries are easily recognized, for they separate rocks of distinctly different composition, for example, limestone/sandstone or shale/granite. Primary features, such as sedimentary bedding and igneous textures, may also help distinguish rock types, but with increasing metamorphic grade such features become obscured by the growth of new minerals or the local redistribution of elements, and they may be completely obliterated at the highest grades of metamorphism. Sedimentary layering is one of the most difficult features to identify with certainty in highly metamorphosed rocks, because millimeter- to centimeter-scale layering, which resembles sedimentary bedding, can develop during metamorphism at pronounced angles to bedding. The thickness of secondary layers tends to be rather constant, whereas that of sedimentary layers is commonly variable. In addition, most metamorphic layering is produced by abrupt changes in mineral abundance; for example,

one set of layers may consist essentially of pure quartz and the alternating set of muscovite. Changes in composition within sedimentary beds, by contrast, tend to be more gradational. In some cases, however, it is necessary to look for compositional layering on the scale of an outcrop, or larger, before a decision can be made as to the primary or secondary origin of layering.

Boundaries between most metamorphic zones are not as evident in the field as the primary boundaries, and commonly they can be identified only after close inspection of the rocks with a hand lens or even in a thin section. Metamorphic zones are identified on the basis of the mineral assemblage in the rocks. As will be evident from the discussion below, these assemblages are determined both by the initial bulk composition of the protolith and by the temperature, pressure, and fluid composition during metamorphism. The boundary between metamorphic zones is named an *isograd*, the implication being that all points along a boundary were exposed to the same intensity or grade of metamorphism. The term was introduced by Barrow (1893) in the mapping of the metamorphic rocks of the southeastern Scottish Highlands. Because mineral assemblages on either side of most isograds can be related through simple reactions, an isograd is inferred to represent the trace of the intersection of the equilibrium reaction surface (in terms of temperature, pressure, and fluid composition) with the present erosion surface. Isograds are named with an index mineral or mineral assemblage that appears on the higher-grade side of the boundary.

Two different types of protolith form the metamorphic rocks shown in the map of Figure 15-1. Extending from west to east across the map area is a folded and faulted bed of quartzite (stippled in Fig. 15-1) formed from a feldspathic sandstone (arkose). It is overlain and underlain by muscovite schists (lined in Fig. 15-1), which were initially shales (pelites). The compositions of the protoliths are sufficiently different that the rock types are easily distinguished at all metamorphic grades, the metasandstone being composed predominantly of quartz and the metapelite of muscovite.

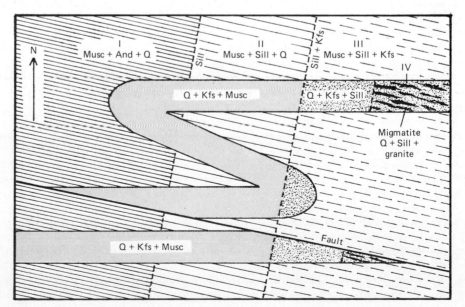

FIGURE 15-1 Idealized geologic map of a model metamorphic terrane containing two protoliths, sandstone (stippled) and shale (lined). These rocks were folded and faulted prior to metamorphism. The shale was converted to a muscovite schist and the sandstone to a quartzite during metamorphism. The intensity of metamorphism increases from west to east, with four metamorphic zones being developed (Roman numerals), each with its own characteristic mineral assemblages, as shown on the map. The boundaries between zones, which are called isograds, mark the intersection of reaction surfaces in P–T–a_{H_2O} space with the erosion surface. The isograds are labeled with the new mineral or mineral assemblage appearing on the high-temperature side of the isograd.

In tracing these rock units through the map area, we find systematic changes in their mineralogy, which allow us to define metamorphic zones. As will be evident from the discussion below, these mineralogical changes indicate increasing intensities of metamorphism from west to east.

Let us first examine the changes in the pelitic schist. In the western corner of the area this rock consists of muscovite, quartz, and the Al_2SiO_5 polymorph andalusite. In progressing eastward, andalusite disappears from the metapelite, and the polymorph sillimanite takes its place. The metamorphic boundary separating these two zones can therefore be named the sillimanite isograd. Farther to the east, quartz vanishes from the metapelite, and potassium feldspar appears. The boundary between these zones marks another isograd. It would be tempting to name it the K-feldspar isograd, but we will postpone naming it until the mineralogical changes in the other rock unit are examined.

The metasandstone in the western part of the area is composed of quartz, minor potassium feldspar, and muscovite. On following this unit to the east we reach the point where the sillimanite isograd is identified in the pelitic rock. But because the metasandstone has no andalusite to be converted to sillimanite, this isograd cannot be mapped in this unit. We see, therefore, that isograds are discontinuous boundaries that occur only in rocks of appropriate composition. For an isograd to be useful, then, it should occur in a rock that has a wide distribution. Pelitic rocks are particularly useful in this respect, and they contain minerals that take part in many reactions.

Where the metasandstone reaches the second isograd in the pelitic rock a mineralogical change does take place, with sillimanite appearing and muscovite disappearing. Recall that in the metapelite, potassium feldspar appears and quartz disappears at this isograd. As will be evident from our analysis of the mineral assemblages in these zones, both sillimanite and potassium feldspar are created at this isograd, but in the metasandstone, which already contains potassium feldspar below the isograd, we recognize only sillimanite as a new mineral, and in the metapelite, which already contains sillimanite below the isograd, only potassium feldspar is recognized as a new mineral. The isograd is named the sillimanite–K-feldspar isograd, because both minerals are produced by the metamorphic reaction that occurred in both rock types along this line.

In the extreme eastern part of the area, the metasandstone develops stringers of coarse-grained granitic material parallel to the foliation of the rock, thus making it a migmatite. This rock indicates that the temperature, pressure, and fluid composition during metamorphism in this part of the region reached the beginning of melting of granite. On the basis of mineralogy, then, we are able to divide the rocks of this area into four different zones, which are indicated on the map with Roman numerals. Zone I, in the western part, is recognized by the presence of andalusite with quartz and muscovite in the pelitic schist. Zone II is also distinguished by the mineral assemblage in the pelitic unit, with sillimanite taking the place of andalusite. The third zone, however, is marked by mineralogical changes in both the metapelite and metasandstone, quartz vanishing and K-feldspar appearing in the metapelite, and sillimanite appearing in the metasandstone. The fourth zone, marked by the presence of migmatite, is recognized only in the metasandstone.

These divisions are purely descriptive and require no interpretation of the data. The resulting map showing the distribution of the primary rock units and metamorphic zones is the raw material from which we must interpret the metamorphic history of the region.

15-3 INTERPRETATION AND REPRESENTATION OF MINERAL ASSEMBLAGES

The different mineral assemblages in the four metamorphic zones of Figure 15-1 indicate that during metamorphism the protoliths were exposed to different intensities of metamorphism, with those in zone IV reaching the melting point of granite. The mineral assemblages can be related through reactions. Some of these are easily deduced, such as the polymorphic transformation of andalusite to sillimanite between zones I and II. The sillimanite isograd can therefore be interpreted as the intersection of the andalusite → sillimanite reaction curve with the present erosion surface. Other reactions are more difficult to deduce, such as the one taking place at the second isograd. Here sillimanite and potassium feldspar are produced at the expense of muscovite and quartz. These minerals can be related by the simple reaction

$$
\begin{array}{ccc}
\text{(muscovite)} & \text{(quartz)} & \text{(K-feldspar)} \\
KAl_2AlSi_3O_{10}(OH)_2 + & SiO_2 & = KAlSi_3O_8 \\
& & \text{(sillimanite)} \quad \text{(vapor)} \\
& & + Al_2SiO_5 + H_2O \quad (15\text{-}1)
\end{array}
$$

Identifying such reactions becomes increasingly difficult, with larger numbers of minerals and more widely varied rock types. It is necessary therefore to have techniques by which we can systematically analyze mineral assemblages and deduce what metamorphic reactions have taken place.

One of the simplest ways of systematizing the data on mineral assemblages in metamorphic zones is to use the type of diagram introduced by Miyashiro (1973). This is a simple bar graph with lines used to indicate the presence of a mineral. Figure 15-2 lists all of the minerals found in the model metamorphic terrane of Figure 15-1. Each of the four zones is represented by a column. A horizontal line in any of these columns indicates the presence of the mineral beside which the line is drawn. Because the mineral assemblages in the metapelite and metasandstone are different in each zone, they are distinguished in the graph with solid and dashed lines, respectively. In such a diagram, the changes in mineral assemblages in going from one zone to another are readily apparent. For example, in the metapelite, quartz is present in zones I and II but not zones III and IV, and conversely, potassium feldspar is not present in zones I and II but is in zones III and IV. At a glance, then, we see that the reaction that took place between zones II and III must have consumed quartz and produced potassium feldspar. Unfortunately, this diagram does not indicate all of the minerals involved in the reaction, only those that are either totally consumed or formed for the first time. The Miyashiro

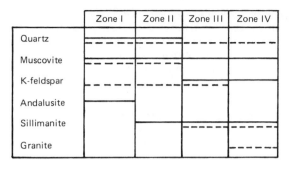

	Zone I	Zone II	Zone III	Zone IV
Quartz				
Muscovite				
K-feldspar				
Andalusite				
Sillimanite				
Granite				

——————— Metapelite

— — — — — Metasandstone

FIGURE 15-2 Minerals present in the metapelite and metasandstone in the four different metamorphic zones of the area shown in Figure 15-1.

diagram, therefore, while being useful in organizing field observations does not allow us quickly to deduce the reactions that have taken place.

Differences in the mineral assemblages in the metasandstone and metapelite in any one zone must be related to differences in bulk composition. In zone I, for example, the metasandstone is not sufficiently aluminous to contain the andalusite found in the aluminous metapelite. In analyzing changes in mineral assemblages between zones, we must therefore be able to take into account the differences in bulk composition of the rocks. This is done by making use of compositional diagrams similar to those used for phase diagrams in Chapter 10. When dealing with metamorphic rocks, these diagrams are normally plotted in terms of mole fractions rather than weight fractions. Because mole fractions are not conservative measures, the lever rule (Sec. 10-3) cannot be used to read off quantities from these diagrams. The use of mole fractions, however, does simplify the plotting of metamorphic minerals in these diagrams.

To construct such a diagram, we must first select appropriate components with which to describe all of the minerals in the map area. The minerals and their formulas are:

Quartz	SiO_2
Andalusite, sillimanite	Al_2SiO_5
Potassium feldspar	$KAlSi_3O_8$
Muscovite	$KAl_2AlSi_3O_{10}(OH)_2$

It will be recalled from Sec. 8-5 that components are defined as the minimum number of compositional building units required to describe all the phases present. Thus rather than selecting the five elements Si, Al, K, H, and O as components, we can reduce this number by noting that certain elements invariably go together. For example, the mineral formulas can be written in terms of the oxides SiO_2, $AlO_{3/2}$, $KO_{1/2}$, and $HO_{1/2}$ rather than the elements, because the first four elements are invariably combined with oxygen; that is, none of these elements occurs in its native state. This reduces the number of components from five to four. Note that the oxides are not written in their conventional way with whole numbers, but are

expressed in terms of single cations; this greatly simplifies plotting minerals. One further simplification can be made with respect to the potassium component. Neither feldspar nor muscovite contain much $KO_{1/2}$, and if the oxide is used as one of the components, the mineral assemblages of interest will be compressed into the potassium-poor part of the diagram. If, instead, $KAlSi_3O_8$ is used as the potassium component, we can obviously make feldspar, and muscovite can be formed by combining $KAlSi_3O_8$ with $2AlO_{3/2}$ and $2HO_{1/2}$. The four components we will use to describe the minerals are, therefore, SiO_2, $AlO_{3/2}$, $KAlSi_3O_8$, and $HO_{1/2}$.

Four components can be plotted using a tetrahedron (Fig. 15-3a), with each apex representing one of the components (see Sec. 10-21). The minerals in the map area can now easily be plotted in terms of mole fractions. Andalusite and sillimanite, for example, plot one-third of the way along the line from $AlO_{3/2}$ toward SiO_2, and muscovite plots on the $AlO_{3/2}$–$KAlSi_3O_8$–$HO_{1/2}$ face of the tetrahedron. Quartz and feldspar both plot at apexes.

If the minerals in a metamorphic rock represent an assemblage that was at equilibrium at the time of metamorphism, they can be joined by *tie lines* in the tetrahedral plot. In zone I, for example, quartz coexists with andalusite and muscovite in the metapelite, and with potassium feldspar and muscovite in the metasandstone. Lines can therefore be drawn from quartz to each of these minerals (Fig. 15-3a). Of course, if quartz coexists with andalusite and muscovite, andalusite must coexist with muscovite, and these minerals can also be joined with a tie line. The result is the *tie triangle*, quartz–andalusite–muscovite, which indicates that this is a stable assemblage in the pelitic unit of zone I. The other tie triangle that we know exists from the mineral assemblage in the metasandstone in zone I is quartz–muscovite–K-feldspar.

Inspection of Figure 15-3a indicates that in addition to the three-phase tie triangles, there are four-phase tie tetrahedra. For example, from the diagram it would appear possible to have the coexistence of quartz–andalusite–muscovite–K-feldspar. This assemblage is not recorded from the area, but this is because no rocks have appropriate compositions. Two other possible four-phase assemblages are quartz–andalusite–muscovite–water and quartz–K-feldspar–muscovite–water. If water were present in the rocks at the time of metamorphism, these two assemblages are the ones found in the metapelite and metasandstone, respectively. Water is not retained as a separate phase in metamorphic rocks, except as small amounts trapped in fluid inclusions. We will, however, assume, for the moment, that water was present.

Each of the four-phase assemblages in Figure 15-3a forms a volume in the tetrahedron that does not overlap with the volumes occupied by the other phase assemblages. Thus, at the pressures and temperatures under which the mineral assemblages of zone I formed, any rock–water mixture having a bulk composition that plots in the volume quartz–andalusite–muscovite–water would have formed these phases in the appropriate proportions. This was the case for the metapelite. Similarly, a rock–water mixture having a bulk composition in the volume quartz–muscovite–K-feldspar–water would have formed these phases in the appropriate proportions. This was the case in the metasandstone. Clearly, andalusite could not

which many metamorphic mineral facies occur is so great that they most likely do represent near-equilibrium assemblages. The occurrence of zoned metamorphic minerals, however, is evidence, at least on a small scale, that the approach to equilibrium is not complete.

Eskola (1920) found that contact metamorphic rocks around a granite at Orijarvi, Finland, also have simple mineral assemblages, but that these are different from those around the Oslo plutons. He concluded therefore that equilibrium was probably attained in both areas, but that the conditions of metamorphism must have been different. This led him to propose the metamorphic facies concept, which he generalized to cover all conditions in contact and regional metamorphism (see Sec. 15-7).

The metamorphic facies concept has been introduced here through the example of a model metamorphic terrane with simple rock types so that the principles can easily be understood. The real world, however, is more complex, and the metamorphic facies concept, while still valid, has a number of pitfalls for the unwary. Much of metamorphic petrology is devoted to working around these difficulties. Two of the problems will be mentioned here briefly as examples, but their solutions will be left to later chapters.

It is normally assumed that sharp gradients in temperature, pressure, and fluid phase composition are not likely to have existed on the scale of an outcrop, and therefore all rocks in an outcrop would have experienced essentially the same temperature, pressure, and fluid phase composition. Rocks with different mineral assemblages must, according to the metamorphic facies concept, have different bulk compositions. This we check by plotting the mineral assemblages in a facies diagram to make certain that no tie lines cross. If tie lines do cross, the assumption about the lack of gradients across the outcrop may be invalid. Although temperature and pressure are most unlikely to vary much over short distances, the fluid-phase composition might, in which case it cannot be treated as a homogeneous environmentally controlled variable. In Chapter 16 we see that in some rocks the fluid-phase composition can vary markedly over short distances as a result of buffering reactions in adjacent layers of different mineralogy.

When the metamorphic facies concept is applied to rocks with more complex compositions than those of our model terrane, crossing tie lines may not be an indication of different metamorphic conditions but simply a product of the graphical analysis. Because the mineral facies diagrams of Figure 15-3c need only three components to describe all the mineral assemblages in the model terrane, they are rigorously correct. But when a rock requires more components, the simplifying assumptions made to portray the mineral assemblages graphically may invalidate the facies diagram for distinguishing the effects of bulk composition from those of metamorphic conditions on the mineral assemblage.

15-5 SIMPLE PETROGENETIC GRID

We will now return to the interpretation of the metamorphic record preserved in the rocks of the model terrane. The mineral facies diagrams allow us to systematize the relations between mineral assemblages and bulk composition and to work out the reactions relating one facies to another. The isograds are interpreted as the intersection of the $P-T-a_{H_2O}$ reaction surfaces with the erosion surface. The positions of most of these reaction surfaces are known from either experimental studies or calculations based on thermodynamic data. When these are plotted in a $P-T$ diagram for a given activity of water, they form a network of lines known as a *petrogenetic grid* (Fig. 15-4). The grid is divided into areas that are bounded by two or more reactions, and within any one area only one arrangement of mineral tie lines is possible; that is, each area contains a separate mineral facies. By locating the mineral facies of the model terrane on the $P-T$ diagram, the possible limits of pressure and temperature in the region during metamorphism can be determined for any given activity of water.

The four metamorphic zones are separated by three reactions: the polymorphic transformation of andalusite to sillimanite, the reaction of quartz with muscovite to form sillimanite and potassium feldspar, and the melting of quartz and potassium feldspar to form a granitic melt. The first reaction involves only a polymorphic transformation and therefore is not dependent on the composition of the fluid phase. But the other two reactions involve water and are consequently dependent on the activity of water in the fluid phase.

The Al_2SiO_5 polymorphs, andalusite, kyanite, and sillimanite, are extremely useful minerals to have in metamorphic rocks when trying to determine the $P-T$ conditions of metamorphism. First, they are common constituents of metapelites. Second, their composition is almost exactly that of the simple formula; thus substitutions do not significantly affect their stability ranges, except for Mn, which can expand the field of andalusite into those of sillimanite and kyanite (Strens 1968). Finally, because the change from one to another is a simple polymorphic transformation, the fluid phase has no effect on their relative stability relations. This is an extremely important point because often the activity of water during metamorphism is not precisely known. The $P-T$ diagram for the Al_2SiO_5 polymorphs was calculated in Prob. 8-3. At 501°C and 0.376 GPa, all three polymorphs coexist at an invariant or triple point (Holdaway 1971). Three univariant lines radiate from the invariant point dividing $P-T$ space into three divariant fields, with kyanite occupying the high-pressure one, andalusite the low-pressure one, and sillimanite the high-temperature one.

The reaction of quartz with muscovite to produce sillimanite and potassium feldspar is typical of many metamorphic reactions in that it liberates a volatile phase, in this case water. At low pressures, the ΔV of the reaction is very large, and thus the slope of the reaction curve in $P-T$ space ($dP/dT = \Delta S/\Delta V$, Eq. 8-3) is shallow and positive when the water activity is 1. Because of the compressibility of the vapor phase, however, the ΔV of the reaction decreases sharply with increasing pressure, with the result that the slope of the reaction curve increases, becoming steep but still positive above 0.2 GPa. This reaction curve, which was calculated in Prob. 8-8, is shown in Figure 15-4a for a water activity of 1. If the activity of water is less than 1, for example due to the presence of CO_2 in the fluid phase, the reaction is driven to the right in order to try and build up the water pressure, and thus the stability of quartz plus muscovite is pushed to lower temperatures. As the water

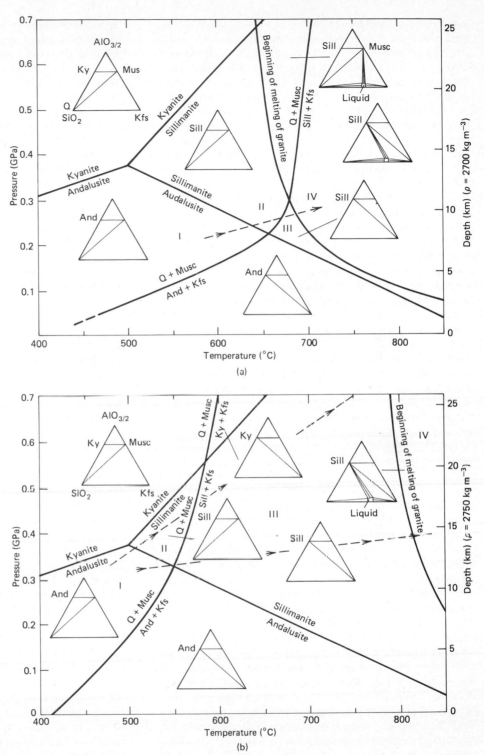

FIGURE 15-4 Petrogenetic grids for rocks in the model terrane of Figure 15-1 shown for an activity of water of 1.0 in (a) and of 0.3 in (b). The possible ranges of conditions in the four zones are shown by heavy dashed lines. The stability fields of the Al_2SiO_5 polymorphs are from Holdaway (1971); the beginning of melting curves for granite are from Clemens and Wall (1981); and the stability of muscovite + quartz is from Kerrick (1972).

activity decreases from 1, the slope of the dehydration reaction steepens and then becomes negative as it pivots about its intersection with the temperature axis (Fig. 11-14). In Figure 15-4b the reaction is shown for a water activity of 0.3.

The beginning of melting curve of quartz and potassium feldspar is also sensitive to the activity of water because of the solubility of water in silicate melts. The melting curve in Figure 15-4 is actually for a granitic composition rather than for the quartz–K-feldspar eutectic, which is only poorly known. The quartz–K-feldspar eutectic would occur at slightly higher temperatures. The melting curve in Figure 15-4a, which is for a water activity of 1, is almost a mirror image of the dehydration reaction, having a shallow negative slope at low pressures but steep negative ones at high pressures. With decreasing water activity, the melting curve steepens as it pivots about its intersection with the temperature axis, eventually having a positive slope at low water activities (Fig. 11-14). Figure 15-4b shows the melting curve for a water activity of 0.3. The liquid field is shown as a single phase plotting at the quartz–feldspar eutectic in the facies diagrams because during metamorphism it was a homogeneous liquid, but on cooling it would have crystallized to a near-eutectic mixture of quartz and feldspar. Note that once melting occurs in a rock with the bulk composition of the metasandstone, all potassium feldspar enters the liquid, so that the rock still contains only three phases, quartz, sillimanite, and liquid.

With these reactions plotted on the P–T diagram, we now draw a facies diagram in each one of the areas on the petrogenetic grid. It is then a simple task to identify which ones correspond to those found in the model terrane, and to determine the possible range of metamorphic P–T conditions. The P–T stability range of the mineral assemblages of zone I are limited by the reactions of andalusite to kyanite, andalusite to sillimanite, and quartz + muscovite to andalusite + K-feldspar. This sets an upper limit on the pressure in this zone of 0.376 GPa regardless of the activity of water. The upper temperature limit is given by the intersection of the dehydration reaction with the andalusite-to-sillimanite reaction. The position of this intersection depends on the activity of water. If the fluid phase is pure water ($a_{H_2O} = 1$), the maximum temperature for this zone would be 650°C (Fig. 15-4a). This point of intersection also sets a lower limit on the pressure in this zone at the sillimanite isograd of 0.22 GPa. For activities of water less than 1, the intersection of these two reactions shifts to lower temperatures and higher pressures. For an activity of 0.3, for example, it occurs at 550°C and 0.32 GPa (Fig. 15-4b).

The stability field of zone II is limited by the reactions of sillimanite to andalusite, sillimanite to kyanite, quartz + muscovite to sillimanite + K-feldspar, and, at high water activities, by the beginning of melting of granite. This is a large area on the P–T diagram, but at least under high-water activities the possible range of conditions is limited by the position of zone III on the diagram. The isograd separating zones II and III is limited to the dehydration reaction between the sillimanite-to-andalusite reaction and the beginning of melting curve. This limits the conditions on this isograd to being between 655°C at 0.22 GPa and 680°C at 0.3 GPa if the activity of water is 1 (Fig. 15-4a). If the activity is less than 1, the shift in the dehydration reaction to lower temperatures causes the intersection with the beginning of melting curve (which shifts to higher temperatures) to be at very much higher pressures. At low enough water activities, the dehydration reaction intersects the sillimanite-to-kyanite reaction before intersecting the beginning of melting curve. At a water activity of 0.3, the upper stability limit of zone II is 0.55 GPa and 585°C (Fig. 15-4b).

The upper stability limit of zone III, which is the beginning of melting of granite, is strongly dependent on the composition of the fluid phase. At high activities of water, the upper pressure limit of zone III is set by the intersection of the beginning of melting curve with the dehydration reaction (0.3 GPa and 680°C at $a_{H_2O} = 1$). But as the activity of water decreases, this point of intersection rapidly moves to much higher pressures as the dehydration curve shifts to lower temperatures and the beginning of melting curve to higher temperatures. At low

activities of water, the upper pressure limit of this zone is determined by the intersection of the beginning of melting curve with the sillimanite-to-kyanite reaction.

We are now in a position to estimate the metamorphic conditions recorded in the rocks of the model terrane. Pressure and temperature gradients must be smooth functions. Even though faulting or igneous intrusion might temporarily produce discontinuities, the flow of rock and diffusion of heat would soon eliminate such irregularities. So in addition to selecting regions in the petrogenetic grid that match the mineral facies from the field, temperatures and pressures must be chosen that allow for smooth variations from one zone to the next.

The deduced metamorphic conditions in the model terrane are most tightly constrained if the activity of water was 1. Zone I would then have formed at temperatures near 600°C and 0.2 GPa, and zone IV at about 700°C and slightly less than 0.3 GPa (dashed line in Fig. 15-4a). If the activity of water was less than 1, the temperature range represented by the four zones greatly increases and so also may the pressure range. For example, if the activity of water was 0.3, zone I would have formed below 500°C and at a pressure of approximately 0.3 GPa, whereas zone IV would have formed near 800°C and at pressures that could have been anywhere between 0.3 GPa and possibly 1 GPa. Clearly, without more knowledge of the activity of water, the conditions of metamorphism cannot be better defined with the present data. In Chapter 16 we will see that there are a number of common metamorphic reactions from which the activity of water can be determined.

Finally, having determined the possible range of metamorphic conditions in the model terrane, the question remains as to their significance. The simplest interpretation is that they represent a fossil geotherm. But if they do, they give unrealistically steep temperature gradients (dT/dz, shallow slope in Fig. 15-4). There can, of course, be no doubt that high temperatures were reached during metamorphism, but the interpretation that conditions in successive metamorphic zones represent geothermal gradients makes the implicit assumption that the mineral assemblages that record the information formed simultaneously. In Chapter 20 we will see that successive metamorphic zones are probably formed at slightly different times, and this produces an apparent geotherm that is oversteepened.

15-6 *ACF* AND *AKF* DIAGRAMS

The model metamorphic terrane considered in previous sections was designed to be compositionally simple, so that the principles of analysis of mineral assemblages could be illustrated without encumbrance from complexities resulting from the graphical representation of systems containing more components. Most rocks, however, contain several more components than those used in Figure 15-3c. The major oxides needed to describe the essential composition of most igneous and sedimentary rocks includes SiO_2, Al_2O_3, Fe_2O_3, FeO, MgO, CaO, Na_2O, K_2O, H_2O, and CO_2. Because graphical representations are limited to three dimensions, all of these oxides cannot be treated as separate components. However, by restricting the types of rocks that are plotted and by making a few approximations, the number of components can be reduced to three, which can therefore be represented in triangular diagrams. Two such diagrams are the common *ACF* and *AKF* plots, both of which were introduced by Eskola (1920) in his study of the rocks of the Orijarvi district of Finland.

Diagrams for plotting mineral assemblages can be greatly simplified if the compositional range of the rocks is restricted. For example, basaltic rocks contain very little potassium, and therefore this element need not be included. Such rocks can therefore be described essentially in terms of SiO_2, Al_2O_3, CaO, ferromagnesian components, Na_2O, and the volatile constituents H_2O and CO_2. Moreover, in many ferromagnesian minerals FeO and MnO freely substitute for MgO. The number of components can therefore be reduced by treating these three oxides as a single component, which is designated *F*, for ferromagnesian. A further simplification is to plot only that alumina which forms minerals other than alkali feldspar; in this way, alkali feldspar (mainly albite in basalt) can be ignored. To do this, the alumina is reduced by an amount equal to the number of moles of $Na_2O + K_2O$ in the rock. This simplification assumes that muscovite is not present, which is reasonable for basaltic rocks. Also, because small amounts of ferric iron substitute for alumina in minerals, Fe_2O_3 is added to the alumina. If we further assume that H_2O and CO_2 are environmentally controlled variables, they need not be considered as components. Basaltic rocks can therefore be expressed approximately in terms of the following four components, which are expressed in mole percentages:

$$SiO_2$$
$$A = Al_2O_3 + Fe_2O_3 - (Na_2O + K_2O)$$
$$C = CaO$$
$$F = FeO + MgO + MnO$$

These four components can be plotted in a tetrahedron (Fig. 15-5a). But most metamorphosed basalts contain small amounts of quartz. The mineral assemblages thus form tetrahedral volumes extending from the silica apex of the tetrahedron. If we restrict the plots to rocks containing quartz, the diagram can be further simplified by projecting all minerals and rock compositions from quartz onto the *ACF* face of the tetrahedron. Mineral assemblages appearing on the resulting triangular plot (Fig. 15-5b), which is the common *ACF* diagram, must all include quartz. A systematic approach to reducing the variables to a number that can be plotted is given in Sec. 17-3.

Clearly, many assumptions and simplifications are made in reducing a rock to the three *ACF* components. Such plots, therefore, do not have the thermodynamic rigor of the triangular plot of Figure 15-3. Nonetheless, the *ACF* diagram is well suited for portraying generalized mineral assemblages in metamorphosed igneous rocks, especially those of basaltic composition, and metamorphosed limestones, dolostones, and impure carbonate rocks. It is not suitable for rocks formed from aluminous sediments. When metamorphosed, metapelites contain a number of ferromagnesian minerals, such as staurolite, chloritoid, garnet, cordierite, and chlorite that tend to be either iron-rich or magnesium-rich; that is, iron and magnesium do not substitute freely for each other and therefore cannot be treated as a single component. When plotted in the *ACF* diagram, these minerals commonly form four-phase, rather than three-phase assemblages and may also have crossing tie lines. Such rocks require a completely different plot (Sec. 17-2).

To plot the composition of a rock in the *ACF* diagram, the chemical analysis, which is normally presented in terms of weight percent, is recast in mole proportions by dividing the weight percentage of each oxide by its molecular weight. The mole proportions of the *A*, *C*, and *F* components are then recalculated to 100%. If accurate modal data are available,

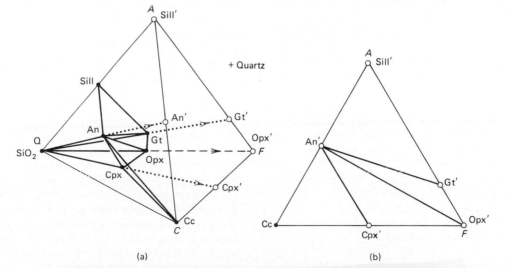

(a) (b)

FIGURE 15-5 *ACF* diagram for rocks containing quartz. (a) Tetrahedral plot of SiO_2–*A*–*C*–*F*, where $A = Al_2O_3 + Fe_2O_3 - (Na_2O + K_2O)$, $C = CaO$, and $F = FeO + MgO + MnO$. (b) For rocks containing quartz, compositions can be projected from quartz onto the *ACF* face. Mineral assemblages can then be plotted on the triangular *ACF* triangle. Phases shown are quartz (Q), sillimanite (Sill), anorthite (An), clinopyroxene (Cpx), orthopyroxene (Opx), garnet (Gt), and calcite (Cc). Projected compositions of phases are shown with a prime.

corrections can be made to the mole proportion of CaO for any accessory apatite or sphene present in the rock, and to FeO for accessory ilmenite and magnetite (see norm calculation, Sec. 6-3, for calculation). Normally, the amounts of these minerals are small, and they do not seriously affect the *ACF* plot.

The positions of the minerals in the *ACF* plot can be deduced from their formulas. For convenience, their compositions in terms of *A*, *C*, and *F* are given below:

	A	*C*	*F*
Mineral			
Anthophyllite, Cummingtonite, Orthopyroxene Talc, Serpentine	0	0	100
Actinolite, Tremolite	0	28.5	71.5
Hornblende	<25	~28.5	71.5
Diopside, Dolomite	0	50	50
Calcite, Wollastonite	0	100	0
Grossularite, Andradite	25	75	0
Vesuvianite	14	72	14
Epidotes	43	57	0
Anorthite	50	50	0
Al_2SiO_5, Pyrophyllite	100	0	0
Staurolite	67	0	33
Chloritoid (Fe), Cordierite (Mg)	50	0	50
Spesartine, Almandine, Pyrope	25	0	75
Chlorite	10–35	0	90–65

Most rocks form, in addition to quartz, a three-phase assemblage in the *ACF* diagram (accessory minerals are neglected). Some minerals, such as hornblende, however, have such a wide range of composition that some bulk compositions may fall in two-phase or even single-phase fields in *ACF* diagrams. The typical mineral assemblages formed under the range of pressures and temperatures encountered during metamorphism are shown in Figure 15-7.

Another less commonly used plot is the *AKF* diagram. The principles involved in its construction are similar to those for the *ACF* diagram, except that it is used for rocks containing excess alumina and silica, which have the minerals muscovite, quartz, and plagioclase. Quartz and plagioclase are not plotted; instead, the alumina content is reduced by the amount that would enter feldspar. The three components, expressed in mole fractions, are

$$A = Al_2O_3 - (CaO + Na_2O + K_2O)$$

$$K = K_2O$$

$$F = FeO + MgO + MnO$$

Serious problems arise from grouping MgO and FeO as one component, because many of the ferromagnesian minerals in the rocks for which this diagram is designed (metapelites) do not freely substitute MgO for FeO. For this reason a diagram in which MgO and FeO are treated separately is preferred (see Sec. 17-2).

15-7 FACIES CLASSIFICATION OF METAMORPHIC ROCKS

One of the earliest attempts to classify metamorphic rocks on a regional scale was by Barrow (1893) working in the southeastern Scottish Highlands. He mapped zones of progressively metamorphosed rock that he named after readily identifiable minerals; these were chlorite, biotite, garnet, staurolite, kyanite, and sillimanite, listed in order of increasing intensity of metamorphism. He named the outer part of a zone, where the index mineral first appeared, as an isograd. Mapping of isograds is still used as a convenient field method for studying metamorphic rocks and for giving a general idea of the intensity of metamorphism. But the precise relation between an isograd and the intensity of the factors causing metamorphism is complex and can be determined only through the detailed analysis of the entire mineral facies.

Eskola (1920), working on the contact metamorphic rocks of the Orijarvi region of Finland, was the first to appreciate fully the broad relations between mineral assemblages, rock composition, and the pressures and temperatures of metamorphism. He noted that within regions that were small enough to have experienced the same conditions of metamorphism throughout, mineral assemblages were determined only by rock composition, but that rocks from different areas, covering similar ranges of composition, could contain very different mineral assemblages, which he attributed to different conditions of metamorphism. Eskola incorporated these observations into the metamorphic facies concept. He defined a *metamorphic mineral facies* as comprising all the rocks that have originated under temperature and pressure conditions so similar that a definite chemical composition results in the same set of minerals (Eskola 1920).

Eskola originally proposed five different metamorphic facies, but today about 11 are recognized, covering the entire spectrum of possible metamorphic conditions (Fig. 15-6). These are named after a characteristic mineral or common rock type in each facies. Experimental studies provide the data necessary to determine the approximate pressures and temperatures over which each of the facies is stable, but the boundaries between the facies are strongly dependent on the activity of water, which in Figure 15-6 has been taken to be equal to 1.

The 11 metamorphic facies fall into three general series on the *P–T* diagram. One series extends from moderate to high temperatures at low pressures and includes the albite–epidote hornfels, hornblende hornfels, pyroxene hornfels, and sanidinite facies. These facies are normally restricted to contact metamorphic aureoles, or even xenoliths in mafic igneous rocks in the case of the sanidinite facies. (A hornfels is a hard, brittle contact-metamorphic rock.) Another series, which includes the zeolite, greenschist, amphibolite, and granulite facies, is marked by increasing temperature and pressure. This series is typical of many regionally metamorphosed orogenic belts. A third series, which includes the prehnite–pumpellyite, blueschist, and eclogite facies, shows very little increase of temperature despite a marked rise in pressure. This series is the product of regional metamorphism near convergent plate boundaries. Some rocks are metamorphosed under *P–T* conditions that fall between these three series, but they are less common.

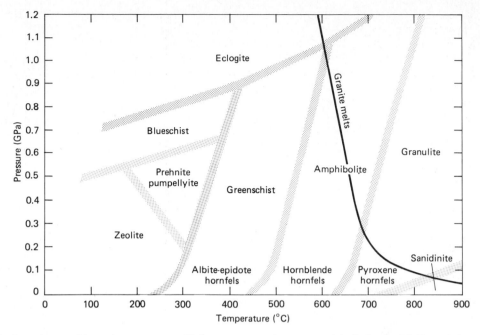

FIGURE 15-6 Approximate pressures and temperatures under which various metamorphic mineral facies form. Rocks are assumed to be in equilibrium with water at the same pressure as the load pressure. Facies are shifted to lower temperatures if partial pressures of water are less than the load pressure. Boundaries between facies are broad zones in which a number of important reactions take place (Fig. 15-7). The melting curve for peraluminous (S-type) granite under water-saturated conditions is from Clemens and Wall (1981).

Figure 15-7 shows *ACF* plots of mineral assemblages in each of the metamorphic facies. To illustrate how these vary, we will consider the mineralogical makeup of a typical basalt in each facies. Remember that in addition to the minerals in the *ACF* diagram, quartz is also present (Prob. 15-3). In the pyroxene hornfels facies, a tholeiitic basalt would have essentially the same mineralogy it had when it crystallized from the magma: that is, plagioclase, clinopyroxene, and orthopyroxene (+quartz); we will assume that the basalt has a composition near the center of this mineral tie triangle. At lower temperatures and under water-saturated conditions, hornblende is stable in the hornblende hornfels facies. The bulk composition of the basalt would plot in the two-phase region plagioclase + hornblende. At still lower temperatures, in the albite–epidote hornfels facies, calcic plagioclase is not stable; instead, it is replaced by albite + epidote, and the coexisting amphibole is actinolite rather than hornblende. The albite–epidote hornfels facies is the lowest-temperature facies of contact metamorphism, but around many intrusions nothing lower than the hornblende hornfels facies is formed.

Much lower-temperature metamorphic facies are formed during regional metamorphism, the lowest being the zeolite facies. The basaltic rock in this facies would consist of chlorite (possibly with smectite in a mixed layer structure), prehnite, and laumontite, or chlorite, prehnite, and calcite, or even chlorite, calcite, and dolomite, depending on the exact composition of the basalt. In addition, they would of course contain quartz and probably analcite. With increase of metamorphic grade to the greenschist facies, basaltic rocks are composed of chlorite, epidote, and actinolite (plus albite and quartz). These rocks form dark green-colored schists, from which the facies derives its name. In the amphibolite facies plagioclase and the aluminous amphibole, hornblende, are stable and are the major phases in basaltic rocks along with minor amounts of either clinopyroxene or almandine garnet. In the granulite facies, hornblende is not stable, its place being taken by clinopyroxene and orthopyroxene.

If the sequence of metamorphic facies had followed the low-temperature high-pressure series, the zeolite facies would have been followed by the prehnite–pumpellyite facies, in which the basalt might have consisted of prehnite, pumpellyite, and chlorite (plus quartz and albite). In the blueschist facies, the basalt would consist of the sodic amphibole, glaucophane, lawsonite, and aragonite. Glaucophane imparts a prominent blue color to these rocks, which gives the facies its name. At still higher pressures, in the eclogite facies, kyanite coexists with clinopyroxene. A basalt would therefore contain these two minerals plus an almandine or grossular garnet.

It will be noted that in some facies, a basalt produces a mineral assemblage that is not unique to that facies. For example, the assemblage plagioclase + orthopyroxene + clinopyroxene is found in the granulite, sanidinite, and pyroxene hornfels facies. A rock containing this assemblage could therefore not be assigned to any one of these facies unless associated rocks of different composition had unique mineral assemblages.

The boundaries between the various facies in Figure 15-7 are determined by important metamorphic reactions. These boundaries do not, however, produce a petrogenetic grid. Inspection of the petrogenetic grid created for the model metamorphic terrane in Figure 15-4 reveals that facies diagrams in adjoining areas of the grid differ by only one reaction. This is certainly not the case for adjoining areas in Figure 15-7. Compare, for example, the facies diagrams for the greenschist and amphibolite facies. Many tie lines must be broken or switched in going from one to the other; that is, several reactions must take place to convert the assemblages of the one facies into those of the other (Prob. 15-5). These reactions are unlikely to occur at precisely the same pressures and temperatures. We can therefore conclude that within the upper part of the greenschist facies or lower part of the amphibolite facies a number of reactions must occur to transform the assemblages of the greenschist facies shown in Figure 15-7 to those of the amphibolite facies.

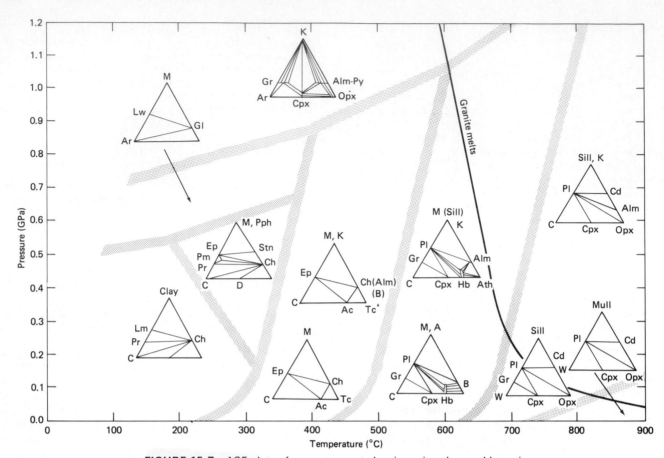

FIGURE 15-7 *ACF* plots of common quartz-bearing mineral assemblages in the metamorphic facies. Boundaries and conditions are the same as in Figure 15-6. Minerals plotted include andalusite (A), kyanite (K), sillimanite (Sill), muscovite (M), mullite (Mull), grossularite (Gr), almandine (Alm), pyrope (Py), orthopyroxene (Opx), clinopyroxene (Cpx), calcic plagioclase (Pl), epidote (Ep), lawsonite (Lw), laumontite (Lm), pumpellyite (Pm), prehnite (Pr), calcite (C), aragonite (Ar), dolomite (D), wollastonite (Wo), actinolite (Ac), hornblende (Hb), glaucophane (Gl), anthophyllite (Ath), talc (Tc), biotite (B), chlorite (Ch), cordierite (Cd), pyrophyllite (Pph), and stilpnomelane (Stn). The beginning of melting curve for water-saturated peraluminous granite is from Clemens and Wall (1981).

The discovery of rocks that formed in the transitional region of *P–T* space between the facies led to the division of the main facies into subfacies. As knowledge of metamorphic rocks grew, the number of reactions that had to be accounted for in this way also grew. It became apparent that subfacies would soon become too numerous to be useful. Clearly, the only solution was to create a petrogenetic grid that did contain all of the possible reactions. With modern experimental and theoretical data such diagrams can now be constructed. The facies classification, however, is still used, as originally defined, but only for broad regional classifications of metamorphic rocks.

PROBLEMS

15-1. If a shale containing 5.0 wt % water is metamorphosed to an entirely anhydrous set of minerals (granulite facies), calculate the volume of fluid formed, relative to the volume of metamorphic rock, at 800°C and 0.7 GPa if the density of water under these conditions is 770 kg m^{-3} and that of the rock is 2900 kg m^{-3}.

15-2. Metamorphic rocks at two localities have the following mineral assemblages:

Locality A: quartz + enstatite + phlogopite
                  quartz   + phlogopite + orthoclase

Locality B: quartz + enstatite   + orthoclase
                enstatite + phlogopite + orthoclase

The metamorphic facies concept can be used to determine whether metamorphic conditions at these two localities were the same or different. Begin by constructing mineral facies diagrams for the two localities. Although five oxides (SiO_2, MgO, $KO_{1/2}$, $AlO_{3/2}$, $HO_{1/2}$) are needed to describe the compositions of all the phases, careful grouping of the oxides and treatment of water as an environmental variable allows the number of components to be reduced to three.

(a) List three possible sets of three components that can be used to describe the minerals from the two localities on an anhydrous basis (water is treated as an environmental variable rather than as a component).

(b) Construct a triangular mineral facies diagram for each locality, using one set of the components listed in part (a). Could these assemblages belong to the same facies? Explain your answer.

(c) Write a balanced tie-line switching reaction that relates the mineral facies of the two localities.

(d) Which locality would have experienced the highest temperature during metamorphism, and why?

15-3. In *ACF* diagrams only assemblages containing quartz are plotted. This allows SiO_2 to be omitted from the plot. Because quartz is present in all assemblages, it is available as a reactant or product to balance tie-line switching reactions relating *ACF* diagrams for adjacent mineral facies. To illustrate this, write balanced reactions relating the following low- and medium-pressure subfacies of the granulite facies. Start by drawing an *ACF* diagram for each subfacies.

Low-P: anorthite + diopside + enstatite   + quartz
          anorthite   + enstatite + pyrope + quartz

Medium-P: anorthite + diopside   + pyrope + quartz
                     diopside + enstatite + pyrope + quartz

15-4. From the *ACF* diagrams in Figure 15-7, list the major mineral assemblages formed in an impure limestone containing small amounts of MgO and Al_2O_3.

15-5. How many reactions are required to convert the mineral facies diagram for the greenschist facies to that of the amphibolite facies in Figure 15-7?

16 // Mineral Reactions Involving H₂O and CO₂

16-1 INTRODUCTION

The majority of metamorphic reactions involve dehydration or decarbonation. The large increase in entropy that accompanies the liberation of a volatile phase from a mineral ensures that rising metamorphic temperatures will favor reactions that produce a separate vapor phase. The properties of this phase are critical in determining which metamorphic reactions take place and under what conditions they occur. We have already seen in the simple example of the model metamorphic terrane treated in Chapter 15 that very different conclusions about the conditions of metamorphism can be reached depending on the assumptions made about the composition of the fluid phase (Fig. 15-4). The purpose of this chapter is to outline some of the important principles governing metamorphic reactions that involve a volatile phase.

The fluid phase in most metamorphic rocks is composed essentially of H_2O and CO_2. Both of these are initially derived almost entirely from the atmosphere (meteoric). Water is incorporated by minerals, such as the clays, during the weathering of rocks or the diagenesis of sediments. Carbon dioxide may also be similarly incorporated with the formation of calcite. But the largest amount of CO_2 enters metamorphic rock as calcite of biological origin formed from the shells of organisms. Both H_2O and CO_2 are trapped directly as pore fluid, but during compaction and diagenesis of sediment most of this is expelled. The fluid phase in metamorphic rocks is therefore derived largely from the breakdown of minerals rather than from the initially trapped pore fluid. The composition of the fluid can therefore vary considerably depending on the composition of the host rock.

Under typical metamorphic conditions the fluid is above its critical point; that is, it is above the temperature and pressure at which liquid and gas can be distinguished. The critical point for water, for example, is at 374.1°C and 22.12 MPa, and that of carbon dioxide is at 31°C and 7.38 MPa. Thus, at the bottom of a 2.26-km-deep ocean, erupting lava would heat water, but it could not cause it to boil. In pore fluids, where the pressure is likely to approach that of the lithostatic pressure, water is above its critical point at depths greater than 1 km.

These supercritical fluids are highly compressible compared with incompressible minerals. Pressure therefore causes dramatic changes in the ΔV of reactions involving fluids, which in turn affects the temperatures at which the reactions take place. To determine the volume of a fluid as a function of temperature and pressure it is necessary to have what is known as an equation of state. The ideal gas law ($PV = nRT$) is one such equation, but under the conditions prevailing during metamorphism deviations from this law are large, and therefore other equations must be used.

The composition of the fluid is critical in determining metamorphic mineral equilibria. In Chapter 15 we saw how a change in the acitivity of water in the fluid dramatically affects the P–T conditions under which a dehydration reaction takes place. Similarly, the activity of CO_2 in the fluid affects the equilibria of decarbonation reactions. Many metamorphic reactions involve both H_2O and CO_2 as reactants and as products. For these reactions, variations in the composition of the fluid can completely change the sequence of appearance of minerals with increasing temperature. It is therefore necessary to know how the activities of these components vary with composition in these fluids and how they affect mineral equilibria.

16-2 P–V–T BEHAVIOR OF FLUIDS

Equation 8-10 in Sec. 8-3 describes how changes in the fugacity of a fluid affect the free-energy change of a reaction involving condensed phases and a gas. That equation was used in Prob. 8-8 to calculate the dehydration reaction of quartz plus muscovite to produce sillimanite plus K-feldspar. For simplicity, water was assumed to behave ideally for this calculation. Because of the importance of this reaction, and others like it, we now need to make the calculation, taking into account that water does not behave ideally under metamorphic conditions.

In deriving Eq. 8-10 it is necessary to evaluate the integral of $V\,dP$ at a fixed temperature, which requires a knowledge of the functional relation between V and P at a given temperature. An equation relating mass, volume, temperature, and pressure is known as an *equation of state*. The simplest such equation is the *ideal gas law,*

$$\bar{V} = \frac{RT}{P} \qquad (16\text{-}1)$$

which states that the product of the molar volume ($\bar{V}$) and pressure (P) is a constant for a given absolute temperature T. The constant of proportionality R, the *gas constant*, can be expressed in many different units (see Table 16-1). The great value of the ideal gas law lies in its independence of the composition of the gas.

Although many gases approach ideal behavior at low pressures, deviations from ideality become large at high pressures. At low pressures, nonideal gases can have larger or smaller molar volumes than would be predicted from the ideal gas law, but at high pressures the molar volumes are larger than predicted. Smaller volumes indicate that attractive forces between molecules must overcome the repulsive ones due to normal thermal vibrations. Larger volumes, on the other hand, indicate the existence of repulsive forces.

An obvious flaw in the ideal gas law is the prediction that the volume of a gas decreases to zero at the absolute zero of temperature. But gases must have a finite volume at absolute zero, and most of course, liquefy well above this, and then their volume changes very little with falling temperature. The ideal gas law can be modified by assigning a value of b to the finite volume that would exist at absolute zero. The equation then becomes

$$\bar{V} = b + \frac{RT}{P} \qquad (16\text{-}2)$$

Although this equation can account for volumes greater than those predicted by the ideal gas law, it cannot account for those that are smaller because b is positive. In the latter case, the pressure that causes the gas to expand must be diminished by an attractive force between the molecules, which is known as van der Waals force. The pressure must therefore be reduced by this amount. To do this we first rearrange Eq. 16-2:

$$P = \frac{RT}{\bar{V} - b} \qquad (16\text{-}3)$$

From this we subtract an amount $a/\bar{V}^2$, where a is a constant and the $\bar{V}^2$ term is introduced to account for the fact that attractive forces balance out within the gas and produce an imbalance only on the outer walls of the volume. The resulting equation is

$$P = \frac{RT}{\bar{V} - b} - \frac{a}{\bar{V}^2} \qquad (16\text{-}4)$$

which is known as *van der Waals equation.* Values of the constants a and b for H_2O and CO_2 are given in Table 16-1.

Although van der Waals equation provides an adequate fit of calculated P–V–T relations to experimental data for many gases at low to moderate pressures and temperatures, under the more extreme conditions encountered during metamorphism, there are significant deviations. Consequently, van der

TABLE 16-1

Constants used in equations of state for gases

Gas constant R = 8.3144 J mol^{-1} K^{-1}	(SI units)
8.3144 Pa m^3 mol^{-1} K^{-1}	(SI units)
83.144 bar cm^3 mol^{-1} K^{-1}	
8.3144 × 10^7 ergs mol^{-1} K^{-1}	
1.9872 cal mol^{-1} K^{-1}	
0.082054 liter atm mol^{-1} K^{-1}	

Coefficients in van der Waals equation (Eq. 16-4)[a]

	a(Pa m^6 mol^{-2})	b(m^3 mol^{-1})
H_2O	0.5537	3.049 × 10^{-5}
CO_2	0.3640	4.267 × 10^{-5}

Coefficients in modified Redlich–Kwong equation of Kerrick and Jacobs (1981)[a]

For H_2O
b = 2.9 × 10^{-5} m^3 mol^{-1}
c = [290.78 − (0.30276 × T) + (1.4774 × 10^{-4} × T^2)] × 10^{-1}
d = [−8374 + (19.437 × T) − (8.148 × 10^{-3} × T^2)] × 10^{-7}
e = [76600 − (133.9 × T) + (0.1071 × T^2)] × 10^{-13}

For CO_2
b = 5.8 × 10^{-5} m^3 mol^{-1}
c = [28.31 + (0.10721 × T) − (8.81 × 10^{-6} × T^2)] × 10^{-1}
d = [9380 − (8.53 × T) + (1.189 × 10^{-3} × T^2)] × 10^{-7}
e = [−368654 + (715.9 × T) + (0.1534 × T^2)] × 10^{-13}

[a] All values in equations are for SI units.

Waals equation has been modified to improve its fit to the data. One such modified version is the Redlich–Kwong equation,

$$P = \frac{RT}{\bar{V} - b} - \frac{a}{\sqrt{T}\bar{V}(\bar{V} + b)} \qquad (16\text{-}5)$$

which differs from van der Waals equation only in the formulation of the term for the attractive force, which is made a function of temperature. This equation has been further modified to give what is the best fit for the high-pressure data for H_2O and CO_2; this is the modified Redlich–Kwong equation of Kerrick and Jacobs (1981):

$$P = \frac{RT(1 + y + y^2 - y^3)}{\bar{V}(1 - y)} - \frac{a}{\sqrt{T}\bar{V}(\bar{V} + b)} \qquad (16\text{-}6)$$

where $y = b/4\bar{V}$, and a, which is both a function of pressure and temperature, is given by

$$a = c + \frac{d}{\bar{V}} + \frac{e}{\bar{V}^2} \qquad (16\text{-}7)$$

where the value of $\bar{V}$ is in cubic meters and the values of c, d, and e are given in Table 16-1. Note that Eq. 16-6 reduces to the ideal gas law at high temperatures where the volumes are large, because the second term on the right-hand side approaches zero and y also approaches zero. Equation 16-6 provides good agreement between calculated and measured molar volumes up to 0.8 GPa and between 300 and 925°C for water and 400 to 700°C for CO_2, with the poorest fit being between 400 and 500°C for water. The equation can be used to extrapolate to more extreme conditions with reasonable accuracy (approximately several percent).

Once equations of state are known, the integral of $\bar{V}\,dP$ in Eq. 8-10 can be evaluated and the free-energy change of a reaction involving a gas under any desired conditions can be determined. If the equation of state is the ideal gas law, RT/P is substituted for $\bar{V}$ and the integral gives $RT \ln P$ (Eq. 8-6). For real gases, however, it is necessary to integrate more complicated expressions, such as the modified Redlich–Kwong equation. This will not be done here, but for those interested, the integration is given in the paper by Kerrick and Jacobs (1981).

It will be recalled from Eq. 8-7 that a function known as the fugacity, f, was introduced so that the simple form of the integral of $\bar{V}\,dP$ for ideal gases could be retained for nonideal ones; that is, the integral is $RT \ln f$. By equating the actual integral of $\bar{V}\,dP$ obtained using the equation of state with $RT \ln f$, a fugacity coefficient, γ ($f = \gamma P$, Eq. 8-8), is obtained. Kerrick and Jacobs (1981) give the following expression for the fugacity coefficients of H_2O and CO_2 using the values for the constants given in Table 16-1:

$$\ln \gamma = \frac{8y - 9y^2 + 3y^3}{(1-y)^3} - \ln Z - \frac{c}{RT^{3/2}(\bar{V} + b)}$$
$$- \frac{d}{RT^{3/2}\bar{V}(\bar{V} + b)} - \frac{e}{RT^{3/2}V^2(V + b)}$$
$$+ \left(\frac{c}{RT^{3/2}b} \ln \frac{\bar{V}}{\bar{V} + b}\right) - \frac{d}{RT^{3/2}b\bar{V}}$$
$$+ \left(\frac{d}{RT^{3/2}b^2} \ln \frac{\bar{V} + b}{\bar{V}}\right) - \frac{e}{RT^{3/2}2b\bar{V}^2}$$
$$+ \frac{e}{RT^{3/2}b^2\bar{V}} - \left(\frac{e}{RT^{3/2}b^3} - \ln \frac{\bar{V} + b}{\bar{V}}\right) \quad (16\text{-}8)$$

where Z, the *compressibility factor*, is given by $P\bar{V}/RT$. Despite the length of this equation, it provides a simple means of calculating the fugacity coefficient for any desired molar volume and temperature (Prob. 16-2).

Fugacities can be calculated from the fugacity coefficients for any given pressure. Figure 16-1 shows the calculated fugacities of H_2O and CO_2 at 600°C up to a pressure of 1 GPa (Prob. 16-2). Water and CO_2 behave quite differently. Below a pressure of 0.88 GPa, at this temperature, the fugacity of water is less than the pressure of water, that is, $\gamma_{H_2O} < 1$. Above this pressure, however, the reverse is true, that is, $\gamma_{H_2O} > 1$. The pressure at which $\gamma_{H_2O} = 1$ decreases with rising temperature. At 1000°C, for example, it occurs at 0.4 GPa. This means that in the lower crust at the temperatures generated during metamorphism, water behaves very nearly ideally.

The fugacity of CO_2, on the other hand, is invariably greater than the pressure of CO_2, and markedly so at high pressures. For example, at 1 GPa and 600°C, the fugacity of CO_2 is 44 GPa. Such large deviations from ideality significantly

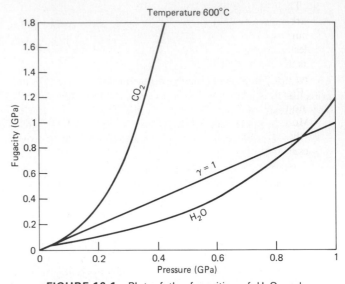

FIGURE 16-1 Plot of the fugacities of H_2O and CO_2 as a function of pressure at 600°C based on the modified Redlich–Kwong equation of Kerrick and Jacobs (1981). Ideal behavior is shown by the line marked with a fugacity coefficient of $\gamma = 1$.

affect mineral equilibria (Prob. 16-3). A decarbonation reaction proceeding under a pressure of 1 GPa would experience a fugacity (effective pressure) of CO_2 of 44 GPa. This means that the equilibrium temperature is much higher than would be expected from ideal behavior.

Equations 16-6 and 16-8 allow us to calculate the fugacities of pure H_2O and CO_2 over a wide range of conditions. But because metamorphic fluids may contain both H_2O and CO_2, it is necessary to know the fugacities of these gases in mixtures. Many gases at high pressures form ideal solutions, even though separately they do not obey the ideal gas law. For such gases, the fugacity of a component is given by

$$f_i = X_i F_i \quad (16\text{-}9)$$

where X_i is the mole fraction of component i and F_i is the fugacity that component i would have if pure and at the same temperature and total pressure as the mixture. This is a simple relation that allows us to obtain approximate values for the fugacities in mixed gases. Water and CO_2, however, exhibit nonideal mixing, so for more precise determinations more complex mixing laws must be used [see Kerrick and Jacobs (1981)].

16-3 METAMORPHOSED SILICEOUS CARBONATE ROCKS

Metamorphosed impure limestones and dolomites have long attracted the attention of petrologists because of the large number of mineral assemblages that can form in such a relatively simple system. These rocks can be represented essentially in terms of three components, CaO, MgO, and SiO_2, and a CO_2–H_2O fluid phase. Alumina may also be a component if the carbonate rocks initially contain shale.

The simplest mineralogy is found in metalimestones (= marble) that contain silica. One of the most common modes of occurrence of silica in such rocks is as chert nodules. The nodules are particularly useful because they form recognizable objects that can be traced through various metamorphic grades, making it possible to study the mechanisms of successive reactions. Figure 16-2 shows the progression of reaction rims around chert nodules in the contact metamorphic aureole of the Christmas Mountains gabbro, Big Bend region, Texas (Joesten 1976). Near the intrusion the limestone changes from gray to white as a result of recrystallization (Fig. 16-2a). The first metamorphic

mineral to form is wollastonite, which appears as a rim separating chert from calcite (Fig. 16-2c). Quartz and calcite in the unmetamorphosed limestone (Fig. 16-2b), on being heated, react to form wollastonite and liberate CO_2. Nearer the intrusion, wollastonite and calcite are no longer stable together and react to form tilleyite, with a further liberation of CO_2 (Fig. 16-2d). Still nearer the contact, tilleyite and wollastonite react to form spurrite, and more CO_2 is liberated (Fig. 16-2e). At still higher temperatures, the minerals rankinite and larnite might also have formed had the bulk composition of the rock been more siliceous (Fig. 16-2f and g).

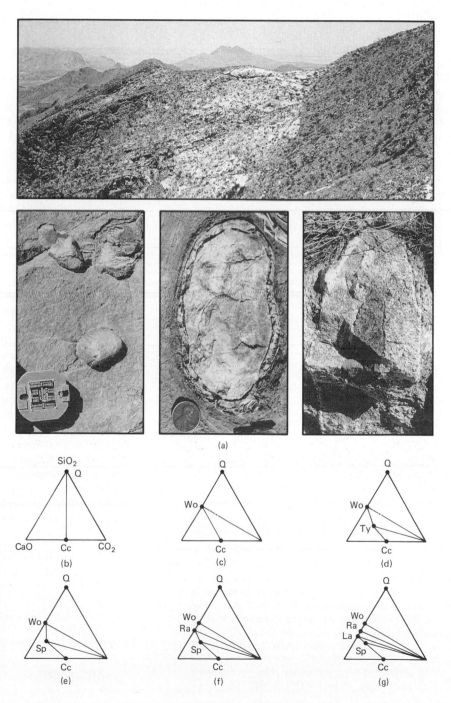

FIGURE 16-2 (a) Contact metamorphic aureole of Christmas Mountains gabbro, Big Bend Region, Texas (Joesten and Fisher 1988). Heat from the gabbro (on right) recrystallized and whitened limestone out to 50 m from contact. Chert nodules in limestone reacted with enveloping calcite to produce concentric layers of calc-silicate minerals. (b) Unmetamorphosed chert nodules in limestone 132 m from contact. The mineral facies diagram below shows the stable assemblage of calcite (Cc) + quartz (Q) + fluid (CO_2) in this unmetamorphosed rock. (c) Chert nodule rimmed by wollastonite (Wo), which formed by reaction between quartz and calcite at 100 m from contact. (d) From 23 to 15 m from contact a reaction zone of tilleyite (Ty) formed between calcite and wollastonite. (e) Within 13 m of the contact, tilleyite became unstable and spurrite (Sp) formed instead. (f,g) Had the temperature risen higher and the rock been of appropriate composition, two more reactions would have occurred, producing rankinite (Ra) and larnite (La). (Photographs by R. L. Joesten.)

The reactions taking place between chert and limestone all involve the generation of a CO_2 vapor. They are as follows:

(calcite) (quartz)   (wollastonite) (vapor)

$$CaCO_3 + SiO_2 = CaSiO_3 + CO_2 \quad (16\text{-}10)$$

(calcite)   (wollastonite)

$$3CaCO_3 + 2CaSiO_3$$
$$\qquad\qquad\qquad \text{(tilleyite)} \qquad \text{(vapor)}$$
$$= Ca_3Si_2O_7 \cdot 2CaCO_3 + CO_2 \quad (16\text{-}11)$$

(tilleyite)   (wollastonite)

$$Ca_3Si_2O_7 \cdot 2CaCO_3 + CaSiO_3$$
$$\qquad\qquad\qquad \text{(spurrite)} \qquad \text{(vapor)}$$
$$= Ca_4Si_2O_8 \cdot CaCO_3 + CO_2 \quad (16\text{-}12)$$

The effect of metamorphism is therefore to transfer CO_2 progressively from the minerals to the vapor phase. Apart from small amounts of fluid trapped in minerals as fluid inclusions or on grain boundaries, the large volumes of CO_2 produced by the reactions above are no longer present in the rocks. Therefore, another effect of metamorphism is to cause decarbonation.

If the carbonate rocks also include dolomite, the assortment of metamorphic minerals that can form is still larger. The composition of these minerals, projected into the ternary diagram $CaO\text{-}MgO\text{-}SiO_2$, is shown in Figure 16-3. It should be noted that although all of these minerals may occur in meta-carbonate rocks, the bulk composition of sedimentary carbonates is normally restricted to the left half of the diagram; that is, these rocks initially consist of a mixture of calcite, dolomite, and quartz. Metamorphosed ultramafic igneous rocks, however, have compositions in the right half of the diagram. With so many minerals, the number of three-phase assemblages that can be drawn is large, and working out which are possible and in what sequence they should appear with increasing metamorphic grade is no small task. In 1940, Bowen proposed that these mineral assemblages are formed by 13 decarbonation reactions, which lead to the appearance of 10 minerals in a definite order; from low to high temperature they are tremolite, forsterite, diopside, periclase, wollastonite, monticellite, akermanite, spurrite, merwinite, and larnite. He suggested the following mnemonic to help remember the sequence:

Tremble, for dire peril walks,
Monstrous acrimony's spurning mercy's laws.
(Bowen 1940)

Subsequently, additional reactions have been suggested. Tilley (1948), for example, found that talc appeared before tremolite. Bowen, on being asked by Tilley what this would do to the mnemonic, suggested stuttering at the beginning—ta. . . tremble, for dire. . . .

Not only have additional reactions been proposed, but Bowen's actual sequence of reactions has been shown to be only one of several. Indeed, diopside almost everywhere appears before forsterite. The cause for the different sequences is obvious when the reactions are studied in terms of a mixed $CO_2\text{-}H_2O$ fluid phase, but discussion of this aspect is postponed to the next section.

Despite the limited applicability of Bowen's decarbonation series, his 1940 paper was important, because it contained a proposal for the construction of a *petrogenetic grid*. He visualized that a $P\text{-}T$ grid would be formed by the intersection of the univariant decarbonation reactions with reactions involving only solid phases, which would likely have different slopes. Subsequent work has shown this to be the case, and today a detailed grid has been worked out in terms, not only of pressure and temperature, but also fluid-phase composition [see, for example, Skippen (1974)].

For the moment we will consider only one of Bowen's decarbonation reactions, leaving the others to the next section. The reaction of quartz and calcite to form wollastonite and CO_2 (Eq. 16-10) provides a good example of one of the many reactions in which only CO_2 is liberated. The free-energy change of the reaction at any given pressure and temperature can be expressed by Eq. 8-9, that is,

$$\Delta G_{r,T}^P = \Delta G_{r,T}^0 + \Delta V_{r,T,c}(P - 10^5) + RT \ln \frac{f}{10^5}$$

where P and f are measured in pascal, making it necessary to subtract the pressure of the reference state (atmospheric pressure = 10^5 Pa). Note that $\Delta V_{r,T,c}$ is for the condensed phases only, which in this reaction is -1.9692×10^{-5} m^3 mol^{-1}. For fugacity, we can substitute γP. The equation then becomes

$$\Delta G_{r,T}^P = \Delta G_{r,T}^0 + \Delta V_{r,T,c}(P - 10^5) + RT \ln \frac{\gamma P}{10^5} \quad (16\text{-}13)$$

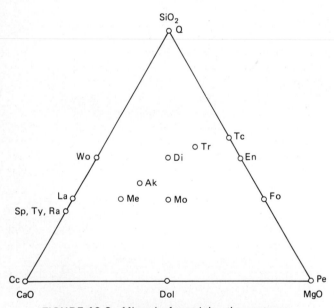

FIGURE 16-3 Minerals formed by the metamorphism of siliceous limestone and dolomite plotted in terms of mole proportions of $CaO\text{-}MgO\text{-}SiO_2$. Talc (Tc) and tremolite (Tr) contain H_2O, and calcite (Cc), dolomite (Dol), spurrite (Sp), and tilleyite (Ty) contain CO_2. The other minerals are enstatite (En), forsterite (Fo), periclase (Pe), diopside (Di), monticellite (Mo), akermanite (Ak), merwinite (Me), wollastonite (Wo), larnite (La), rankinite (Ra), quartz (Q).

At 700°C and atmospheric pressure, for example, $\Delta G_{r,T}^P = \Delta G_{r,T}^0 = -57818$ J mol^{-1} based on the data in Table 7-1. Because $\Delta G_{r,T}^P$ is negative, the reaction proceeds under these conditions and generates CO_2. To reestablish equilibrium at this temperature, $\Delta G_{r,T}^P$ must be reduced to zero by increasing the pressure [$(\partial\,\Delta G/\partial P)_T = \Delta V$, which is positive]. When $\Delta G_{r,T}^P = 0$ then,

$$-\Delta G_{r,T}^\circ = \Delta V_{r,T,c}(P - 10^5) + KT\ln\frac{\gamma P}{10^5} \quad (16\text{-}14)$$

The value of γ depends on P, which is yet to be determined. We solve the equation by estimating a value for γ and then calculating P. If the chosen value of γ does not match with the determined pressure, a new value is selected. After several iterations, matching values of γ and P are found. For the conditions considered here (700°C), a solution is obtained for a pressure of 0.117 GPa, a fugacity of 0.161 GPa, and a fugacity coefficient of 1.37. This point lies very near the experimentally determined equilibrium curve for this reaction (KJ in Fig. 16-4). Note that had CO_2 been assumed to behave ideally, the calculated equilibrium point would have plotted well above the experimentally determined curve at 0.183 GPa (I in Fig. 16-4).

This decarbonation reaction resembles a dehydration reaction, in that it has a positive slope that is shallow at low pressures and steepens with increasing pressure as the evolved

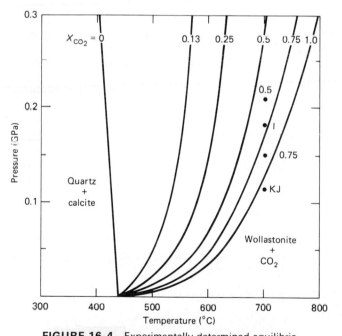

FIGURE 16-4 Experimentally determined equilibria for the reaction quartz + calcite = wollastonite + CO_2 for fluid compositions ranging from pure CO_2 (X_{CO_2} = 1) to pure water (X_{CO_2} = 0). Data from Greenwood (1967b) and Harker and Tuttle (1956). The four points refer to calculated equilibria for the reaction at 700°C (see text for discussion). Point KJ is based on the fugacity data of Kerrick and Jacobs (1981); point I assumes ideal behavior of CO_2; points 0.75 and 0.5 are calculated using the Kerrick–Jacobs data for mole fractions of CO_2 in the fluid of 0.75 and 0.5.

vapor phase becomes less compressible. Because the fugacity coefficient of CO_2 is invariably greater than 1 and becomes progressively larger with increasing pressure (Fig. 16-1), the effective pressure" (fugacity) of the CO_2 experienced by the reaction is greater than the actual pressure. As a result, the increased temperatures required to drive the reaction at higher pressures are greater than those for dehydration reactions, for which the fugacity coefficient of H_2O is less than 1. The slopes of decarbonation reactions on P–T diagrams are consequently shallower than those of dehydration reactions.

In calculating this decarbonation reaction, the vapor phase has been assumed to be pure CO_2. If it is not, then, according to Le Châtelier's principle, the reaction proceeds at lower temperatures. Equation 16-4 is still applicable, except that higher pressures are required at any given temperature to generate the equilibrium fugacities, because the vapor is not pure CO_2. If the fugacity of CO_2 in a mixed vapor phase obeys Eq. 16-9, the activity of CO_2 is given by its mole fraction, that is, $a_{CO_2} = X_{CO_3}$. Thus, according to Eq. 16-9, the fugacity of CO_2 in a mixed vapor is $X_{CO_2}F_{CO_2}$, where F_{CO_2} is the fugacity CO_2 would have if it were pure and under the same total pressure, P, as the system. Equation 16-14 can thus be written as

$$-\Delta G_{r,T}^0 = \Delta V_{r,T,c}(P - 10^5) + RT\ln\frac{\gamma PX_{CO_2}}{10^5} \quad (16\text{-}15)$$

where γ is the fugacity coefficient CO_2 would have if it were pure and at pressure P. This equation, then, can be used to calculate the equilibrium pressure of a decarbonation reaction for any given composition of the fluid phase, provided that the fugacity of CO_2 obeys Eq. 16-9.

Consider again, for example, the reaction of quartz and calcite to produce wollastonite. The fugacity of CO_2 at 700°C was determined to be 0.161 GPa when the vapor is pure CO_2. This same fugacity is required for equilibrium when the activity of CO_2 is less than 1; the difference is that the fugacity must be generated as a fraction of the fugacity of pure CO_2 at a higher pressure and, therefore, with a different γ. It is necessary again to solve Eq. 16-15 iteratively for the new conditions, which gives a pressure of 0.15 GPa when $a_{CO_2} = 0.75$ and 2.1 GPa when $a_{CO_2} = 0.5$ (Fig. 16-4). Differences between the calculated and experimental data arise because of the assumption that H_2O and CO_2 mix ideally.

The experimentally determined equilibria for the reaction of quartz and calcite to form wollastonite and CO_2 are shown in Figure 16-4 for several mole fractions of CO_2 in the vapor. The maximum stability of quartz + calcite occurs when X_{CO_2} = 1. With decreasing CO_2 concentrations, the reaction curve steepens and has a negative slope at very low mole fractions. This means that when the fluid is almost pure H_2O, which might occur, for example, in a thin marble layer embedded in a thick pelitic sequence of rocks, increasing pressure causes wollastonite to form at progressively lower temperatures. This P–T diagram, which is typical of any decarbonation reaction, illustrates an important point, that is, the presence of a high-temperature mineral, such as wollastonite, provides little indication of actual metamorphic temperatures unless the composition of the fluid phase is known. At 0.3 GPa, for example, the temperature of appearance of wollastonite could be as low as

400°C if $X_{CO_2} = 0$ or as high as 800°C if $X_{CO_2} = 1$. In the next section we will see that certain mineral assemblages provide a record of the mole fraction of CO_2 during metamorphism.

16-4 THERMODYNAMICS OF MINERAL REACTIONS WITH H_2O–CO_2 FLUIDS

Dehydration and decarbonation reactions in the presence of a mixed fluid phase involve three intensive variables, P, T, and fluid composition, which can be conveniently represented in a three-dimensional diagram (Fig. 16-5a). A decarbonation reaction is plotted in this diagram for illustrative purposes. In the presence of pure CO_2 this reaction appears on the right-hand face of the diagram. This face is shown separately in Figure 16-5b. The positions of the reaction in equilibrium with fluids containing mole fractions of CO_2 less than 1 are given by the dashed lines in the block diagram. These are lines of intersection of the reaction surface with planes parallel to the P–T face. The dashed lines can be projected onto the P–T face of the diagram to give the familiar curves for a decarbonation reaction at various mole fractions of CO_2 in the fluid phase

(Fig. 16-5b). Isobaric (dotted lines in Fig. 16-5a) and isothermal sections can also be taken through the diagram (Fig. 16-5c and d, respectively).

To determine the conditions under which a decarbonation (or dehydration) reaction will take place, the shape of the reaction surface in terms of P, T, X_{H_2O}, X_{CO_2} must be known. The surface, of course, represents the locus of points in P–T–fluid composition space where the free-energy change of the reaction is zero (by definition of equilibrium). By breaking the free-energy change of the reaction into its component parts and then expressing these as functions of P, T, and fluid composition, we generate the equation needed to describe the reaction surface. Although this derivation may appear complex because of the number of terms involved, it is, in fact, simple. The derivation is worth working through carefully because it employs many of the basic thermodynamic relations developed in Chapters 8 and 9. However, you may wish to go directly to the result, which is given in Eq. 16-16.

Consider a reaction of an assemblage of minerals, A, to produce a new assemblage B with the liberation of m moles of H_2O and n moles of CO_2.

$$A = B + mH_2O + nCO_2$$

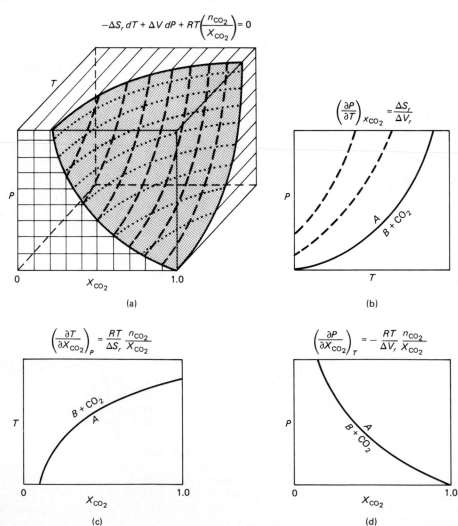

$$-\Delta S_r\, dT + \Delta V\, dP + RT\left(\frac{n_{CO_2}}{X_{CO_2}}\right) = 0$$

(a)

$$\left(\frac{\partial P}{\partial T}\right)_{X_{CO_2}} = \frac{\Delta S_r}{\Delta V_r}$$

(b)

$$\left(\frac{\partial T}{\partial X_{CO_2}}\right)_P = \frac{RT}{\Delta S_r}\frac{n_{CO_2}}{X_{CO_2}}$$

(c)

$$\left(\frac{\partial P}{\partial X_{CO_2}}\right)_T = -\frac{RT}{\Delta V_r}\frac{n_{CO_2}}{X_{CO_2}}$$

(d)

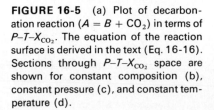

FIGURE 16-5 (a) Plot of decarbonation reaction ($A = B + CO_2$) in terms of P–T–X_{CO_2}. The equation of the reaction surface is derived in the text (Eq. 16-16). Sections through P–T–X_{CO_2} space are shown for constant composition (b), constant pressure (c), and constant temperature (d).

To avoid excessive subscripting, H_2O in the fluid will be symbolized by H and CO_2 by C. At equilibrium

$$\Delta G_r = G_B + mG_H + nG_C + G_{mix} - G_A = 0$$

Note that in addition to the free energies of the various phases, a free-energy term resulting from the mixing of the fluid components in included. If the reactants and products are in equilibrium on the reaction surface at a point represented by $(T°, P°, X_H°, X_C°)$, how will ΔG_r change with P, T, and X? In Sec. 1-6 the temperature at any depth in the Earth was determined by expanding T in a Taylor series as a function of depth, starting from the known temperature on the Earth's surface. Similarly, ΔG_r can be expanded in a Taylor series as a function of P, T, and X from the starting point $(T°, P°, X_H°, X_C°)$. Only the first two terms of the series need be used in the expansion, because eventually the expression is used to determine the conditions at a point that is separated from the first by only dP, dT, dX. In the following derivation the superscript° indicates a value at the initial point. Note that $X_C = 1 - X_H$ in the binary fluid phase.

The free energy of the solids is a function only of T and P. For the solids constituting assemblage B the Taylor series can be written as

$$G_B(T, P) = G_B° + \left(\frac{\partial G_B}{\partial T}\right)_{P,X} (T - T°) + \left(\frac{\partial G_B}{\partial P}\right)_{T,X} (P - P°)$$

but $(\partial G_B/\partial T)_P = -S_B$ and $(\partial G_B/\partial P)_T = V_B$, so

$$G_B(T, P) = G_B° - S_B(T - T°) + V_B(P - P°)$$

Similarly, for A,

$$G_A(T, P) = G_A° - S_A(T - T°) + V_A(P - P°)$$

The free energy of the fluid is a function of T, P, and fluid composition. Also, at constant T and P, the free energy of the fluid is given by $G_f = m\mu_H + n\mu_C$. By expressing the free energy of the fluid in terms of chemical potentials the free energy of mixing in the fluid is accounted for. We will assume ideal mixing in the fluid, so $\mu_i^f = \mu_i^* + RT \ln X_i^f$ (Eq. 9-28). The free energy of the fluid is therefore

$$G^f(T, P, X_H, X_C) = G^f + m\left(\frac{\partial \mu_H}{\partial T}\right)_{P,X_H,X_C}(T - T°)$$
$$+ n\left(\frac{\partial \mu_C}{\partial T}\right)_{P,X_H,X_C}(T - T°)$$
$$+ m\left(\frac{\partial \mu_H}{\partial P}\right)_{T,X_H,X_C}(P - P°)$$
$$+ n\left(\frac{\partial \mu_C}{\partial P}\right)_{T,X_H,X_C}(P - P°)$$
$$+ m\left(\frac{\partial \mu_H}{\partial X_H}\right)_{T,P,X_C}(X_H - X_H°)$$
$$+ n\left(\frac{\partial \mu_C}{\partial X_C}\right)_{T,P,X_H}(X_c - X_C°)$$

Making use of Eq. 9-28, we expand the partial derivatives to give

$$m\left(\frac{\partial \mu_H}{\partial T}\right)_{P,X_H,X_C} = m\left(\frac{\partial}{\partial T}(\mu_H^* + RT \ln X_H)\right)_{P,X_H,X_C}$$
$$= m\left(\frac{\partial \mu_H^*}{\partial T}\right)_{P,X_H,X_C}$$
$$+ m\left(\frac{\partial}{\partial T}(RT \ln X_H)\right)_{P,X_H,X_C}$$

Recall from Eq. 9-2 that $\mu_H \equiv (\partial G/\partial n_H) = \bar{G}_H = \bar{H}_H - T\bar{S}_H$, which on being differentiated with respect to T gives $-\bar{S}_H$. Making this substitution yields

$$m\left(\frac{\partial \mu_H}{\partial T}\right)_{P,X_HX_C} = -m\bar{S}_H + mR \ln X_H$$

The partial derivative with respect to P can be expanded as follows:

$$m\left(\frac{\partial \mu_H}{\partial P}\right)_{T,X_H,X_C} = m\left(\frac{\partial}{\partial P}(\mu_H^* + RT \ln X_H)\right)_{T,X_H,X_C}$$
$$= m\left(\frac{\partial \bar{G}_H}{\partial P}\right)_{T,X_H,X_C} = m\bar{V}_H$$

The partial derivative with respect to X, on being expanded, gives

$$m\left(\frac{\partial \mu_H}{\partial X_H}\right)_{T,P,X_C} = m\left(\frac{\partial}{\partial X_H}(RT \ln X_H)\right)_{T,P,X_C} = \frac{mRT}{X_H}$$

For the carbon dioxide component of the fluid, we similarly obtain

$$n\left(\frac{\partial \mu_C}{\partial T}\right)_{P,X_H,X_C} = -n\bar{S}_C + nR \ln X_C$$

and

$$n\left(\frac{\partial \mu_C}{\partial P}\right)_{T,X_H,X_C} = n\bar{V}_C$$

and

$$n\left(\frac{\partial \mu_C}{\partial X_C}\right)_{T,P,X_H} = \frac{nRT}{X_C}$$

We let $(T - T°) = \delta T$, $(P - P°) = \delta P$, $(X_H - X_H°) = \delta X_H$, $(X_C - X_C°) = \delta X_C$, and $\delta X_H = -\delta X_C$. We now combine all the terms to obtain the free energy of the fluid.

$$G^f = G^{f°} - m\bar{S}_H\,\delta T + (mR \ln X_H)\,\delta T - n\bar{S}_C\,\delta T$$
$$+ (nR \ln X_C)\,\delta T + m\bar{V}_H\,\delta P + n\bar{V}_C\,\delta P$$
$$+ \frac{mRT}{X_H}\,\delta X_H + \frac{nRT}{X_C}\,\delta X_C$$
$$= G^{f°} - (m\bar{S}_H + n\bar{S}_C)\,\delta T + R(m \ln X_H + n \ln X_C)\,\delta T$$
$$+ (m\bar{V}_H + n\bar{V}_C)\,\delta P + RT\left(\frac{n}{X_C} - \frac{m}{X_H}\right)\delta X_C$$

Let $(m\bar{S}_H + n\bar{S}_C) = S_f$ and $(m\bar{V}_H + n\bar{V}_C) = V_f$. Note from Eq. 9-20 that $R(m \ln X_H + n \ln X_C) = -S^f_{mix}(m + n)$ because $m = X_H(m + n)$ and $n = X_c(m + n)$. Substituting these into the expression for G^f gives

$$G^f = G^{f\circ} - [S_f + S^f_{mix}(m + n)]\delta T + V^f \delta P$$
$$+ RT\left(\frac{n}{X_c} - \frac{m}{X_H}\right)\delta X_C$$

We now combine the expressions for the solids and the fluids to obtain ΔG_r. Note that $\Delta S_r = S_B + [S_f + S^f_{mix}(m + n)] - S_A$, and $\Delta V_r = V_B + V_f - V_A$,

$$\Delta G_r = G_B - G_A + G^f$$
$$\Delta G_r = G_B^\circ - S_B\,\delta T + V_B\,\delta P - G_A^0 + S_A\,\delta T - V_A\,\delta P$$
$$+ G^{f\circ} - [S_f + S^f_{mix}(m + n)]\delta T + V^f\,\delta P$$
$$+ RT\left(\frac{n}{X_C} - \frac{m}{X_H}\right)\delta X_C$$

$$\Delta G_r = \Delta G_r^\circ - \Delta S_r\,\delta T + \Delta V_r\,\delta P + RT\left(\frac{n}{X_C} - \frac{m}{X_H}\right)\delta X_C$$

But for two points on the P, T, X_C surface that differ by only δP, δT, and δX_C, $\Delta G_r \Rightarrow \Delta G_r^\circ$. Therefore,

$$-\Delta S_r\,\delta T + \Delta V_r\,\delta P + RT\left(\frac{n}{X_{CO_2}} - \frac{m}{X_{H_2O}}\right)\delta X_{CO_2} = 0 \quad (16\text{-}16)$$

Equation 16-16 describes the divariant surface for any reaction evolving m moles of H_2O and n of CO_2 in terms of T, P, and X_{CO_2}, where mixing in the fluid is ideal. The slope of the reaction can be determined for any particular section through the P-T-X_{fluid} diagram of Figure 16-5. The three important sections are for constant composition, constant pressure, and constant temperature. At *constant composition*, $\delta X_{CO_2} = 0$, and Eq. 16-16 reduces to

$$\left(\frac{\delta P}{\delta T}\right)_X = \frac{\Delta S_r}{\Delta V_r} \quad (16\text{-}17)$$

which is the Clapeyron equation (Eq. 8-3). At *constant pressure*, $\delta P = 0$, and Eq. 16-16 reduces to

$$\left(\frac{\delta T}{\delta X_{CO_2}}\right)_P = \frac{RT}{\Delta S_r}\left(\frac{n}{X_{CO_2}} - \frac{m}{X_{H_2O}}\right) \quad (16\text{-}18)$$

Finally, at *constant temperature*, $\delta T = 0$, and Eq. 16-16 reduces to

$$\left(\frac{\delta P}{\delta X_{CO_2}}\right)_T = -\frac{RT}{\Delta V_r}\left(\frac{n}{X_{CO_2}} - \frac{m}{X_{H_2O}}\right) \quad (16\text{-}19)$$

The univariant reaction in any one of these sections can be obtained by integrating the appropriate equation (Prob. 16-7).

Five different types of reaction are possible involving H_2O and CO_2 (Greenwood 1967a). These are shown in an iso-

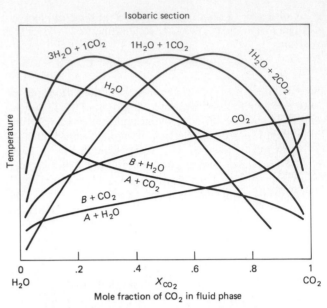

FIGURE 16-6 Isobaric T-X_{CO_2} plot showing the general shape of univariant lines for reactions involving H_2O, CO_2, or both. The number of moles of H_2O and CO_2 involved in a reaction is shown on the side of the univariant line on which it is evolved. See text for discussion.

baric, T-X_{CO_2} section in Figure 16-6. First is the simple decarbonation reaction ($m = 0$, $n > 0$), which has a positive slope and reaches its maximum temperature at $X_{CO_2} = 1$. A dehydration reaction ($m > 0$, $n = 0$), on the other hand, has a negative slope and reaches a temperature maximum at $X_{H_2O} = 1$. Some reactions evolve both H_2O and CO_2 ($m > 0$, $n > 0$), in which case the reaction passes through a temperature maximum at a fluid composition equivalent to the composition of the mixture of gases evolved by the reaction. In other reactions, H_2O may be evolved while CO_2 is consumed ($m > 0$, $n < 0$). Such reactions occur at progressively higher temperatures as the mole fraction of water approaches one. Conversely, reactions that evolve CO_2 while consuming H_2O ($m < 0$, $n > 0$) occur at progressively higher temperatures as the mole fraction of CO_2 approaches one.

All five types of reaction have been studied experimentally (see Greenwood 1967b, 1976; Skippen 1974). Some examples are as follows:

(dolomite)   (quartz)   (diopside)   (vapor)
$$CaMg(CO_3)_2 + SiO_2 = CaMgSi_2O_6 + 2CO_2 \quad (16\text{-}20)$$

(tremolite)   (calcite)   (quartz)
$$Ca_2Mg_5Si_8O_{22}(OH)_2 + 3CaCO_3 + 2SiO_2$$
                (diopside)   (vapor)
$$= 5CaMgSi_2O_6 + H_2O + 3CO_2 \quad (16\text{-}21)$$

(dolomite)   (quartz)   (vapor)
$$5CaMg(CO_3)_2 + 8SiO_2 + H_2O$$
          (tremolite)   (calcite)   (vapor)
$$= Ca_2Mg_5Si_8O_{22}(OH)_2 + 3CaCO_3 + 7CO_2 \quad (16\text{-}22)$$

Each of these reactions is shown in a separate plot at the top of Figure 16-7. The reader should check that the slope of these reactions in each T–X_{CO_2} diagram is appropriate for the type of reaction.

If reactions 16-20 and 16-21 are plotted in the same diagram (Fig. 16-7), they intersect at a temperature slightly in excess of 500°C at a fluid composition of $X_{CO_2} = 0.96$ for the pressure of this diagram (0.2 GPa). Along the isobaric univariant curve for reaction 16-20, three minerals coexist with the fluid: dolomite, quartz, and diopside. Along the curve for reaction 16-21, tremolite, calcite, quartz, and diopside coexist with the fluid. At the point of intersection, five minerals—tremolite, calcite, quartz, diopside, and dolomite—coexist. These phases

can be described in terms of the three components CaO–MgO–SiO_2 (Fig. 16-3) plus a fluid phase, which is not included among the components but is, instead, treated as an environmentally controlled intensive variable. The modified phase rule ($\phi + f = c + 3$) indicates that with five minerals, there is only one degree of freedom, which is used to define the pressure of the isobaric T–X_{CO_2} section. This assemblage of minerals therefore creates an isobaric invariant point.

Schreinemakers rules (Sec. 8-5) can now be used to complete the array of univariant lines about this invariant point in terms of the two intensive variables, T and X_{CO_2}. Reaction 16-21 is the dolomite-absent reaction, which is designated [Dol]. Reaction 16-20 is the tremolite- and calcite-absent reaction

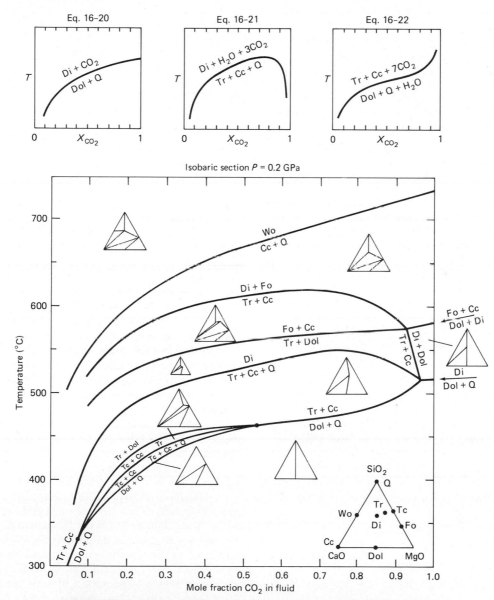

FIGURE 16-7 Isobaric T–X_{CO_2} diagram for several reactions that occur in metamorphosed siliceous dolomitic limestone at a pressure of 0.2 GPa. Shown separately at the top of the diagram are three reactions, 16-20, 16-21, and 16-22. When combined, these reactions create the three lowest-temperature invariant points in Figure 16-8. [After Skippen (1974) and Greenwood (1976).]

([Tr, Ct]). The other reactions that must pass through this point can be determined by noting where the minerals involved in the reactions plot in the CaO–MgO–SiO_2 diagram (Fig. 16-3). These reactions can then be tabulated according to the mnemonic scheme used in the Schreinemakers method for deriving the sequence of reactions around the invariant point.

(Tr)	(Ct)	(Q)	|Dol|	(Di)
(Dol)	(Q)	|Tr, Ct|	(Di)	
(Ct)	(Tr)	|Di|	(Q)(Dol)	
(Tr)	(Ct)	|Q|	(Di)(Dol)	

Note that only four reactions exist at this invariant point because of degeneracy caused by the collinearity of quartz, diopside, and dolomite in the CaO–MgO–SiO_2 plot.

The gas must appear on the high-temperature side of reactions 16-20 and 16-21. By labeling these two reactions in Figure 16-7, we see that only that part of reaction 16-20 ([Tr, Ct]) between the invariant point and the CO_2 side of the diagram is stable; on the water-rich side of the invariant point, the reaction is made metastable by reaction 16-21 ([Dol]). According to the mnemonic scheme, the diopside-absent reaction lies on one side of the tremolite–calcite–absent reaction, and the dolomite- and quartz-absent reactions on the other. The only arrangement of univariant lines around the invariant point that satisfies the mnemonic scheme is shown in Figure 16-7. The slopes of the lines can be determined from the balanced reactions and Eq. 16-18. The diopside-absent reaction is Eq. 16-22, which has a sigmoid shape in the T–X_{CO_2} diagram because H_2O is consumed and CO_2 is evolved. The quartz-absent reaction can be balanced as follows:

(tremolite)   (calcite)
$$Ca_2Mg_5Si_8O_{22}(OH)_2 + 3CaCO_3$$
(diopside)   (dolomite)   (vapor)
$$= 4CaMgSi_2O_6 + CaMg(CO_3)_2 + H_2O + CO_2 \quad (16\text{-}23)$$

Because this reaction produces 1 mol each of H_2O and CO_2, it rises from the invariant point to a temperature maximum at X_{CO_2} of 0.5. Once all univariant lines are correctly placed, facies diagrams are drawn in each isobaric divariant field. This serves to check that Schreinemakers rules are not violated and allows interpretation of metamorphic assemblages.

Before discussing the mineral assemblages found in the various divariant fields in Figure 16-7 we will examine the invariant point created by introducing a talc-producing reaction.

(dolomite)   (quartz)   (vapor)
$$3CaMg(CO_3)_2 + 4SiO_2 + H_2O$$
(talc)   (calcite)   (vapor)
$$= Mg_3Si_4O_{10}(OH)_2 + 3CaCO_3 + 3CO_2 \quad (16\text{-}24)$$

Intersection of this isobaric univariant reaction with reaction 16-22 produces two new invariant points involving the minerals talc, calcite, quartz, tremolite, and diopside. One invariant point occurs at 463°C and $X_{CO_2} = 0.53$ and the other at 340°C and

$X_{CO_2} = 0.08$. The reason the univariant lines intersect at two points is that the curvature of reaction 16-22 is greater than that of reaction 16-24 (Prob. 16-7). Three other reactions that must pass through these invariant points are

(talc)   (calcite)   (quartz)
$$5Mg_3Si_4O_{10}(OH)_2 + 6CaCO_3 + 4SiO_2$$
(tremolite)   (vapor)
$$= 3Ca_2Mg_5Si_8O_{22}(OH)_2 + 2H_2O + 6CO_2 \quad (16\text{-}25)$$

(talc)   (calcite)
$$5Mg_3Si_4O_{10}(OH)_2 + 6CaCO_3$$
(tremolite)   (dolomite)
$$= Ca_2Mg_5Si_8O_{22}(OH)_2 + CaMg(CO_3)_2$$
(vapor)
$$+ H_2O + CO_2 \quad (16\text{-}26)$$

(talc)   (dolomite)   (quartz)
$$5Mg_3Si_4O_{10}(OH)_2 + 2CaMg(CO_3)_2 + SiO_2$$
(tremolite)   (vapor)
$$= Ca_2Mg_5Si_8O_{22}(OH)_2 + 4CO_2 \quad (16\text{-}27)$$

These reactions, with the exception of 16-27, are shown in Figure 16-7. For simplicity, reaction 16-27 is omitted because it involves only compositions outside the range of siliceous carbonate rocks. It can, however, occur in metamorphosed ultramafic igneous rocks (Prob. 16-8).

This T–X_{CO_2} diagram explains why different sequences of minerals can form as a result of progressive metamorphism under different fluid compositions. The lowest-temperature field in Figure 16-7 contains the mineral assemblage typical of unmetamorphosed siliceous carbonate rocks, that is, calcite + dolomite + quartz. With rising temperature, the first metamorphic mineral to form depends on the fluid composition. At 0.2 GPa, if $X_{CO_2} > 0.53$, tremolite forms, whereas if $X_{CO_2} < 0.53$, talc forms, followed by tremolite. Bowen's decarbonation series, which starts with the appearance of tremolite, must therefore have been based on occurrences where the fluid was CO_2-rich, whereas in the area where Tilley found talc appearing before tremolite, the fluid must have been H_2O-rich. Note that tremolite again forms first if the fluid is extremely water-rich. In addition to accounting for differences in the first-appearing minerals, the diagram provides an explanation for the common occurrence of the sequence tremolite, diopside, forsterite, wollastonite, with progressive metamorphism.

Twenty-five stable reactions involving a fluid relate the minerals quartz, calcite, dolomite talc, tremolite, diopside, forsterite, and enstatite (Skippen 1974). In addition, there are numerous reactions that do not involve a fluid and hence are unaffected by the composition of the fluid. Such reactions plot as horizontal lines in isobaric T–X_{CO_2} diagrams. If pelitic rocks are also present, a large number of dehydration reactions can occur. These have negative slopes in the T–X_{CO_2} plot. The resulting petrogenetic grid provides many mineral assemblages that are indicative of fluid compositions.

When the composition of the fluid is taken to be an environmentally controlled intensive variable, we are implicitly invoking communication of the rock with a large reservoir of fluid with fixed or buffered composition. Under such conditions we read the effects of increasing metamorphic temperature in the T–X_{CO_2} diagram by following a vertical line at the fixed composition of the fluid. For example, in Figure 16-8, which is for a total pressure of 0.2 GPa, we can follow the path taken by a rock that is heated in equilibrium with an externally buffered fluid phase having a composition of $X_{CO_2} = 0.75$ (dotted line in Fig. 16-8). Tremolite first appears at 475°C, and because the fluid composition must remain constant the temperature also remains fixed, while the tremolite-producing reaction goes to completion. Once one of the reactants is exhausted the temperature of the rock is free to rise until it reaches the diopside-producing reaction at 550°C. Again the temperature remains constant while this reaction goes to completion. This steplike increase in temperature continues as the rock reaches successive reactions involving the phases remaining from previous reactions. Note that if the fluid has a different composition, the reactions take place at different temperatures, and therefore the steps on the heating curve are also at different temperatures. Mapping the first appearance of a mineral formed by such a reaction would therefore not provide a reliable indication of metamorphic temperatures unless independent evidence indicated the composition of the fluid was everywhere the same.

This type of metamorphism requires that large volumes of fluid be able to migrate freely. If the fluid flux is restricted in any way, reactions evolving H_2O and CO_2 in proportions that differ from that in the fluid can develop local compositional inhomogeneities. In the extreme case, the fluid composition can be controlled entirely by local reactions. In nature, conditions probably fall between these two extremes.

The temperatures at which metamorphic reactions occur, and even the types of reactions that take place, can be quite different if the composition of the fluid is controlled by the rock rather than by an external buffer. Consider again a rock under a total pressure of 0.2 GPa being heated in equilibrium with a fluid that initially has $X_{CO_2} = 0.75$, but this time the fluid is internally buffered. Tremolite again first forms at 475°C, but now, with the reaction consuming H_2O and liberating CO_2 (Eq. 16-22), the composition of the fluid becomes richer in CO_2 (dashed line in Fig. 16-8). As it does so, the temperature rises in order to keep the rock on the isobaric univariant reaction line. Reactants are gradually consumed as the rock follows the reaction curve, and once one of them is exhausted, the rock is free to leave the reaction curve, with the temperature rising along a vertical line marking the new fluid composition. This path is followed until the next reaction involving the remaining phases is reached, whereupon the rock follows that univariant line with the composition of the fluid changing as it does so. Depending on the slope of the next reaction curve, the fluid can become enriched in either CO_2 or H_2O as the reaction proceeds

When reactions taking place in the rock internally buffer the composition of the fluid, the temperature of the rock does not remain constant but rises gradually as the composition of the fluid changes. Once the reaction reaches an invariant point, however, the composition of the fluid becomes fixed and the temperature does remain constant. In Figure 16-8 two internally buffered paths are shown. The one indicated with the dashed line runs out of reactants before reaching the invariant point. Its time–temperature path consequently has a gradual slope with no plateaus; even inflections caused by reactions occur at

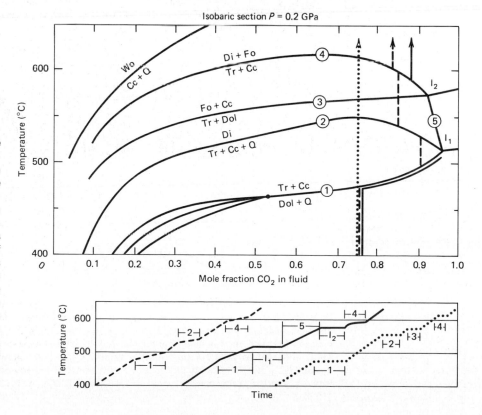

FIGURE 16-8 Isobaric T–X_{CO_2} diagram (same as Fig. 16-7) showing three possible paths followed during progressive metamorphism of a rock initially in equilibrium with a fluid having a mole fraction of CO_2 of 0.75. If the rock is externally buffered with a reservoir having $X_{CO_2} = 0.75$, the path followed is indicated by the dotted line. If the rock is internally buffered, the composition of the fluid is modified by the reactions taking place; the solid line shows such a path for a fluid that is driven all the way to the invariant point, whereas the dashed line is for a rock that exhausts one of the reactants before reaching the invariant point (in this case dolomite). A schematic representation of the rise of temperature along these three paths during metamorphism is given below. See text for discussion.

different temperatures, depending on local abundances of minerals. The other path, shown by a solid line, reaches the invariant point, which then produces a distinct plateau on the time–temperature plot that is independent of the bulk composition of the rock. As will be shown below, this latter path is the more common for internally buffered assemblages. Only these isobarically invariant assemblages can be used for mapping lines of equal metamorphic temperature (isograds).

The degree to which a reaction can buffer and change the composition of the fluid depends on the proportions in which H_2O and CO_2 are generated by the reaction relative to their mole fractions in the fluid. If we let the number of moles of H_2O and CO_2 in the fluid be N_{H_2O} and N_{CO_2}, respectively, then, as shown by Greenwood (1975), the mole fraction of CO_2 in the fluid is

$$X_{CO_2} = \frac{N_{CO_2}}{(N_{H_2O}^0 + N_{CO_2}^0) + A(N_{CO_2} - N_{CO_2}^0)} \quad (n \neq 0) \quad (16\text{-}28)$$

where $A = (m + n)/n$, m *and* n being the numbers of moles of H_2O and CO_2 produced by the reaction, respectively, and $N_{H_2O}^0$ and $N_{CO_2}^0$ are the initial numbers of moles of H_2O and CO_2, respectively, in the fluid. This equation relates the mole fraction of CO_2 in the fluid to the proportion of H_2O and CO_2 produced by the reaction (the A term), the initial composition of the fluid ($N_{H_2O}^0 + N_{CO_2}^0$), and the extent to which the reaction has progressed ($N_{CO_2} - N_{CO_2}^0$).

For a reaction such as $MgCO_3 + H_2O = Mg(OH)_2 + CO_2$, the value of A is 0, so Eq. 16-28 becomes $X_{CO_2} = N_{CO_2}$; that is, there is a one-to-one relation between the number of moles put into the fluid by the reaction and the mole fraction of CO_2 in the fluid. Figure 16-9a shows how the X_{CO_2} in the fluid changes during reaction. For example, if the fluid was initially pure H_2O, the fluid would follow the path labeled

($N_{CO_2}^0 = 0$); if the fluid initially had $X_{CO_2} = 0.2$, it would follow the line labeled $N_{CO_2}^0 = 0.2$. The buffering capacity of a reaction can be given quantitative expression by differentiating Eq. 16-28 with respect to N_{CO_2} (Prob. 16-9). For the magnesite-to-brucite reaction, dX_{CO_2}/dN_{CO_2} is a constant and equal to 1; that is, the buffering capcity of the reaction remains constant regardless of how long it takes place.

For a reaction such as calcite + quartz = wollastonite + CO_2, the value of A is 1, so Eq. 16-28 becomes $X_{CO_2} = N_{CO_2}/(1 + \Delta N_{CO_2})$, where $\Delta N_{CO_2} = N_{CO_2} - N_{CO_2}^0$, which is a measure of the progress of the reaction. In this case, dX_{CO_2}/dN_{CO_2} is a function of the progress of the reaction, with the value decreasing with time. Thus, in Figure 16-9b we see that for an initially pure H_2O fluid, $dX_{CO_2}/dN_{CO_2} = 1$, but as the reaction progresses, the derivative decreases (slope steepens in plot of N_{CO_2} versus X_{CO_2}) and the reaction causes progressively less change in the composition of the fluid. Notice also that the rate of change of composition of the fluid is less for those fluids that initially have a higher mole fraction of CO_2.

For a reaction such as Eq. 16-2, which produces $1H_2O$ and $3CO_2$, the change in composition of the fluid is similar to that for the wollastonite reaction, except that the fluid composition to which the reaction converges is $X_{CO_2} = 0.75$ rather than pure CO_2 (Fig. 16-9c). Here again, the change in composition of the fluid is greatest for early stages of reaction and for fluids with compositions that deviate most from the composition being evolved by the reaction.

Let us determine the amount of reaction that would have to take place along the reaction curve for $5Dol + 8Q + H_2O = Tr + 3Cc + 7CO_2$ (Eq. 16-22) in Figure 16-8 to change the mole fraction of CO_2 in the fluid from its initial value of 0.75 to a value of 0.9, where one of the reactants (we will assume dolomite) is exhausted and the fluid leaves the univariant line. Because the reaction consumes 1 mol of H_2O ($m = -1$) and

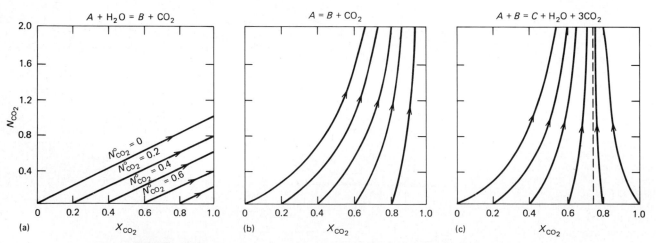

FIGURE 16-9   Change in the mole fraction of CO_2 in a fluid caused by metamorphic reactions that involve H_2O and CO_2, where N_{CO_2} is the number of moles of CO_2 entering the fluid from the reaction. In (a), the reaction consumes $1H_2O$ and evolves $1CO_2$. A fluid initially having $X_{CO_2} = 0.2$, follows the line marked $N_{CO_2}^0$ as the reaction proceeds. For the reaction in (b), only CO_2 is evolved, and the reaction in (c) evolves $1H_2O$ and $3CO_2$. [After Greenwood (1975).]

evolves 7 mol of CO_2 ($n = 7$), $A = \frac{6}{7}$. If we start with 1 mol of fluid that has $X_{CO_2} = 0.75$, it initially contains 0.75 mol of CO_2 ($N^0_{CO_2}$). From Eq. 16-28 we find that the number of moles in the fluid in its final state is 1.406, which is an increase of 0.66 mol of CO_2; at the same time 0.094 mol of H_2O and 0.47 mol of dolomite are consumed. Because dolomite is exhausted by the reaction, the volume of rock affected must have initially contained this amount of dolomite, which is equivalent to 0.087 kg. This means that for 1 mol of initial fluid, 0.087 kg of dolomite is consumed in changing the composition of the fluid from a mole fraction of 0.75 to 0.90. From this we see that if rocks buffer the fluid, variations in modal abundances brought about by progressive metamorphism could be used to estimate the amount of fluid that passed through the rock during metamorphism. We will return to this point in Chapter 18.

Greenwood (1975) has shown that rocks have enormous buffering capacities; that is, only small amounts of reactants need be consumed to cause major changes in the composition of the fluid. Indeed, he concludes that in most internally buffered rocks the changes are great enough to move fluid compositions along univariant lines all the way to isobaric invariant points, as was illustrated in the second internally buffered reaction path in Figure 16-8. On reaching invariant points, fluid compositions and temperatures become fixed, changing again only after one of the minerals has been eliminated. The rock then rises along a univariant line to the next invariant point where again the fluid composition and temperature remain constant until one phase is exhausted. During metamorphism a rock, consequently, spends a considerable fraction of its time at invariant points, and it is here that the most conspicuous changes in mineral assemblages occur.

The univariant lines and invariant points in the $T-X_{CO_2}$ diagram can be thought of as playing the same role for metamorphic fluids as the liquidus lines and eutectic and peritectic points do for magmas. The range of liquid compositions in phase diagrams might suggest that igneous rocks should have a much wider range of composition than they do, but fractionation causes these liquids to converge on the invariant points, which then determine the composition of common igneous rocks. Similarly, isobaric univariant metamorphic reactions move fluid compositions to invariant points where the assemblage of minerals buffers the fluid at a fixed composition.

One might have expected from drawing lines of constant fluid composition on $T-X$ diagrams that many different sequences and temperatures of first appearance of minerals could be found in metamorphic terranes. Instead, a relatively simple picture emerges, with much less variability, indicating that internal self-buffering is more common during metamorphism than external buffering. Progressive metamorphism tends to follow a few paths through $T-X$ diagrams moving from one invariant point to the next.

Isograds based on the first appearance of minerals in rocks that were externally buffered have little significance if the composition of the fluid is unknown. Similarly, the first appearance of minerals formed by isobaric univariant reactions in rocks that were internally buffered cannot provide unique indications of temperature. But reactions taking place at isobaric invariant points, to which the composition of the fluid is driven rapidly by the univariant reactions in the internally buffered rock, do record unique temperatures and fluid compositions for a given pressure. Isograds based on the appearance of minerals formed by these reactions are therefore the most reliable.

PROBLEMS

16-1. Using the data in Table 16-1, calculate pressure versus molar volume graphs for water using the ideal gas law, van der Waals equation, and the modified Redlich–Kwong equation for a temperature of 750°C and for molar volumes between 6×10^{-5} and 11×10^{-5} m³.

16-2. Starting with the spreadsheet used for solving Prob. 16-1, enter Eq. 16-8 for the fugacity coefficient of water, using separate columns for each group of terms in the equation. Then calculate the pressure, fugacity, and density of H_2O for molar volumes between 1.89×10^{-5} and 63×10^{-5} m³ at 600°C (873 K). Plot graphs of fugacity versus pressure and density versus pressure.

16-3. In Prob. 8-8 the equilibrium pressures of water for the reaction of quartz + muscovite = sillimanite + K-feldspar + H_2O were calculated, assuming ideal behavior, for 600, 650, 700, 750, and 800 K to be 0.26, 1.12, 3.98, 12.44, and 38.75 MPa, respectively. However, H_2O is not an ideal gas, so the calculated pressures are in fact fugacities. Using the spreadsheet from Prob. 16-2, select molar volumes for H_2O that give the fugacities above for the appropriate temperatures, and by so doing determine the actual equilibrium pressures. Graph the equilibrium curve for the reaction both for real and ideal behavior, noting how the fugacity affects the results. From the change

in the fugacity coefficient of H_2O with pressure, how might you expect the real curve to be positioned with respect to the ideal curve at high pressures?

16-4. The object of this problem is to calculate the equilibrium pressure for the reaction quartz + calcite = wollastonite + CO_2 at 750°C. The procedure is outlined for 700°C in Sec. 16-3. First, starting with the same spreadsheet used in Prob. 16-2 for H_2O, change the appropriate coefficients to those for CO_2 given in Table 16-1. Then create a table of fugacity coefficients and pressures for CO_2 over a range of pressures at 750°C (1023 K). Next, calculate the free-energy change of the reaction and the volume change of the condensed phases for 750°C and atmospheric pressure from data in Table 7-1. By an iterative process, select a value of γ and calculate P using Eq. 16-13 until the values of γ and P match values in the tables you created for γ and P.
(a) What is the equilibrium P and the value of γ at 750°C?
(b) What would the equilibrium P be if CO_2 were assumed to behave ideally?

16-5. Using the results from Prob. 16-4, calculate the equilibrium pressure for the reaction quartz + calcite = wollastonite + CO_2 at 750°C for the condition where the activity of CO_2 is buffered by the environment at (a) 0.75 and (b) 0.5.

16-6. Starting with Eq. 16-18, show that in an isobaric $T-X_{CO_2}$ diagram, a reaction that evolves both H_2O and CO_2 has a temperature maximum when the ratio of the mole fractions of H_2O and CO_2 in the fluid are in the same proportions as the numbers of moles of H_2O and CO_2 produced by the reaction.

16-7. Starting with Eq. 16-18, determine a general equation for a reaction involving H_2O and CO_2 in terms of T and X_{CO_2} at constant P. Then plot graphs for the following reactions, assuming that ΔS_r remains constant over the temperature interval considered. You can use any value for ΔS_r, but you may wish to experiment with how this value affects the shape of the reaction curves.
  (a) $A = B + 5H_2O + 3CO_2$
  (b) $A = B + H_2O + 7CO_2$
  (c) $A + H_2O = B + 2CO_2$
  (d) $A + H_2O = B + 5CO_2$

16-8. The invariant point at 463°C and $X_{CO_2} = 0.53$ in Figure 16-7 is missing one univariant line which was omitted because it involves a reaction that could not take place in the compositional range of metamorphosed sedimentary carbonate rocks. It can, however, occur in metaperidotites. Determine the missing reaction, and using Schreinemakers rules, draw the array of univariant lines around this invariant point. Check your answer by constructing facies diagrams for each divariant field. Make certain that the missing reaction has the correct slope for the gases evolved.

16-9. Repeat Prob. 16-8 for the invariant point at 575°C and $X_{CO_2} = 0.92$.

16-10. Using Eq. 16-28, determine how many moles of tremolite would have to be consumed by the Tr + 3 Cc = 4Di + Dol + H_2O + CO_2 reaction (Eq. 16-23) in order to move the univariant assemblage from the invariant point at 512°C and $X_{CO_2} = 0.96$ to the invariant point at 575°C and $X_{CO_2} = 0.92$.

16-11. The minerals enstatite (En), talc (Tc), magnesite (Mag), and quartz (Q) are related by the following reactions:

[En]	$3Mag + 4Q + H_2O = Tc + 3CO_2$
[Tc]	$Mag + Q = En + CO_2$
[Q]	$Mag + 3Tc = 4En + H_2O + CO_2$
[Mag]	$Tc = 3En + Q + H_2O$

  (a) Show each reaction on a separate $T-X_{CO_2}$ diagram, labeling each curve with reactants and products on the correct side.
  (b) These reactions meet at an isobaric invariant point at $X_{CO_2} = 0.9$ on the isobaric $T-X_{CO_2}$ diagram. Plot all four reactions on a $T-X_{CO_2}$ diagram, and use Schreine-makers rules to obtain the correct arrangement of stable and metastable curves around the invariant point. Check your answer by placing mineral facies diagrams in each divariant field.

17 / Mineral Reactions Among Solid Solutions

17-1 INTRODUCTION

Solutions are encountered in many different forms in petrology. A magma, for example, is a liquid silicate solution; the fluid phase in a metamorphic reaction is a supercritical solution composed essentially of H_2O and CO_2; and many of the common metamorphic minerals are solid solutions involving substitution of Fe for Mg. In Chapter 16 we have seen how important the composition of the fluid phase is to determining metamorphic mineral equilibria. In this chapter we see that solid solutions play an equally important role. Unlike fluids, which from the petrographic viewpoint no longer exist in the rock, solid solutions constitute many of the essential minerals of a metamorphic rock, and in their compositions is preserved valuable information about the metamorphic history of a rock.

In constructing ACF and AKF diagrams, FeO and MgO are assumed to behave identically in order to minimize the number of components that have to be plotted. Although Fe does substitute for Mg in many metamorphic minerals, it is erroneous to believe that these elements have the same effect on mineral equilibria, even when the solid solutions behave ideally. We have already seen in Chapter 10 that olivine, for example, forms an almost ideal solid solution between fayalite (Fe_2SiO_4) and forsterite (Mg_2SiO_4); yet when olivine melts, the liquid is invariably richer in Fe than the coexisting solid (Fig. 10-11). Similarly, in metamorphic rocks, garnet, for example, is invariably more iron-rich than coexisting chlorite (Fig. 12-14). To account for such compositional differences, FeO and MgO must be treated as separate components.

Mineral assemblages can be graphically analyzed only if the number of components can be reduced to four, or preferably three. The necessity of treating FeO and MgO as separate components therefore introduces problems. However, by placing restrictions on the types of rocks plotted and the conditions to which diagrams apply, clear graphical expressions of mineral assemblages can be developed. The rules for constructing these diagrams are given in this chapter.

The variance of mineral assemblages is increased by solid solution. Thus a metamorphic reaction that would occur at a specific temperature under a given pressure if it involved minerals of fixed composition takes place over a temperature range when the minerals form solid solutions. By analogy with igneous systems, these two cases are equivalent to reactions taking place at a peritectic, such as that of olivine + liquid = enstatite (Fig. 10-7), or along continuous reaction curves such as the liquidus and solidus in the olivine system (Fig. 10-11). We will see that metamorphic reactions can also be classified as being either discontinuous or continuous.

Although the added variance caused by solid solutions appears to reduce our ability to determine metamorphic conditions from mineral assemblages, the reverse is actually the case. If equilibrium is to be achieved, when a rock contains several minerals belonging to solid solution series, elements involved in substitutional exchange must partition themselves between these minerals in accordance with the thermodynamic laws governing solutions (Chapter 9). Because partition coefficients are functions of temperature and, in some cases, pressure, the compositions of some coexisting minerals can be used as geothermometers or geobarometers.

In no group of rocks are the effects of solid solution on mineral equilibria more evident than in those of pelitic composition. This chapter therefore concentrates on these rocks, but the principles dealt with are applicable to rocks of any composition.

17-2 PETROGRAPHY OF PELITIC ROCKS

Rocks formed from mud are known as shales or pelites. They constitute approximately two-thirds of all sedimentary rocks. They are composed essentially of clay minerals and variable amounts of extremely fine-grained detrital material (largely quartz). The average shale contains 62% SiO_2, 17% Al_2O_3, 7% FeO[t], 3.7% K_2O, 2.4% MgO, 1.5% CaO, 1% Na_2O, and up to 5% H_2O; all other components, and there may be many, are normally present in minor or trace amounts. Given this composition, it is not surprising that most of these rocks are metamorphosed to quartz muscovite schists. Despite their small compositional range, metapelites contain a large number of minerals in addition to muscovite and quartz, many of which

345

form assemblages with limited stability ranges. These rocks are, therefore, particularly sensitive indicators of metamorphic conditions, and for this reason they have been studied more intensively than any other group of metamorphic rocks. Barrow (1893), for example, used the first appearance of minerals in metapelites to map isograds in the southeast Scottish Highlands.

Shales, by definition, are extremely fine-grained, with most particles being less than 5 μm in size. The total surface area in a given volume of rock is consequently extremely large, and this increases the ease with which metamorphic reactions take place. The deformation that accompanies regional metamorphism also promotes reaction by inducing strain, which increases the free-energy change of reactions (Sec. 12-2). Regional metamorphism consequently starts as low as 200 to 300°C, whereas contact metamorphism, which lacks the strain energy, does not normally begin until at least 400°C.

Regionally metamorphosed pelites are converted first into *slate* by the growth of muscovite and chlorite grains that have their (001) plane oriented preferentially normal to the principal compressive stress axis (Fig. 17-2a). This *foliation plane* is typically at some angle to the bedding, and as the degree of recrystallization increases and the rock passes from a slate through a *phyllite* to a *schist*, the bedding may become transposed and eventually totally obscured by the metamorphic foliation. Micas growing in contact metamorphosed pelite, on the other hand, have random orientations. The resulting rock, known as a *hornfels*, is characterized by extreme toughness, which is caused by the interlocking randomly oriented crystals.

Characteristic sequences of minerals appear in progressively metamorphosed pelites, just as they do in metamorphosed siliceous limestones and dolomites. Barrow (1893) found that the sequence chlorite, biotite, almandine garnet, staurolite, kyanite, and sillimanite marked successively higher grades of metamorphism in the southeastern Highlands of Scotland (Fig. 17-1). The conditions that produce such a succession are commonly referred to as Barrovian-type metamorphism. Although Barrow's sequence is common, it is not unique. Indeed, in the Buchan area, just to the northeast of the Barrovian type locality (Fig. 17-1), the sequence is chlorite, biotite, cordierite, andalusite, sillimanite. In the central part of the Abukuma Plateau of Japan, muscovite, chlorite, and biotite all occur together at the lowest grade of metamorphism, followed successively by andalusite, cordierite, and sillimanite. Miyashiro (1961), who refers to these different sequences as *metamorphic facies series*, has shown that they are produced by metamorphism under different pressures; of the three mentioned here, the Barrovian is formed at the highest pressures, and the Abukuma type at the lowest.

Typical examples of pelitic rocks, as seen under the microscope, are shown in Figure 17-2. These have been arranged approximately so that temperature increases from left to right and pressure (depth) increases downward. The typical Barrovian sequence is illustrated along the diagonal from top left to lower right. Contact metamorphism at shallow depth produces minerals seen at the top right. High-pressure, low-temperature metamorphic minerals are shown along the left side of the figure. Note that many of the minerals form porphyroblasts (large crystals), which are particularly useful when it comes to mapping isograds in the field.

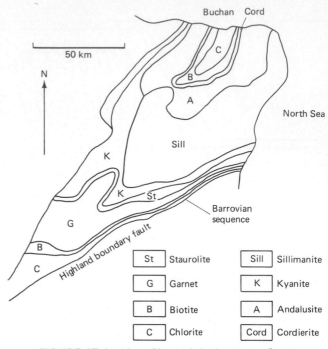

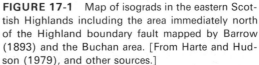

FIGURE 17-1   Map of isograds in the eastern Scottish Highlands including the area immediately north of the Highland boundary fault mapped by Barrow (1893) and the Buchan area. [From Harte and Hudson (1979), and other sources.]

17-3 GRAPHICAL REPRESENTATION OF MINERAL ASSEMBLAGES IN SYSTEMS OF FOUR OR MORE COMPONENTS

Most metapelites are quartz muscovite schists, which contain three additional, essential minerals from the list given in Table 17-1. Clearly, a large number of components is necessary to describe these phases. If the mineral assemblages are to be graphically analyzed, the number of components must be reduced to a manageable number while retaining those that show significant changes in mineral assemblages. The important question is: Which components need to be plotted, and which can be ignored? A mineral facies diagram is used to show how the bulk composition of a rock affects the mineral assemblages that form at any given pressure, temperature, and possibly, fluid composition. The components needed for the plot are therefore those whose variation in concentration causes changes in mineral assemblages; those that do not cause changes can be ignored. Four simple rules allow us to distinguish these two types of component. In applying these rules we are following a procedure introduced by J. B. Thompson (1957) for the graphical analysis of pelitic schists.

The variance of an assemblage of minerals is defined as the number of variables that can be independently changed without changing the assemblage (Eq. 10-11). Numerically, this is equal to the total number of variables minus the number of independent equations relating the variables. For each phase present, a Gibbs–Duhem equation (Eq. 9-7) can be written,

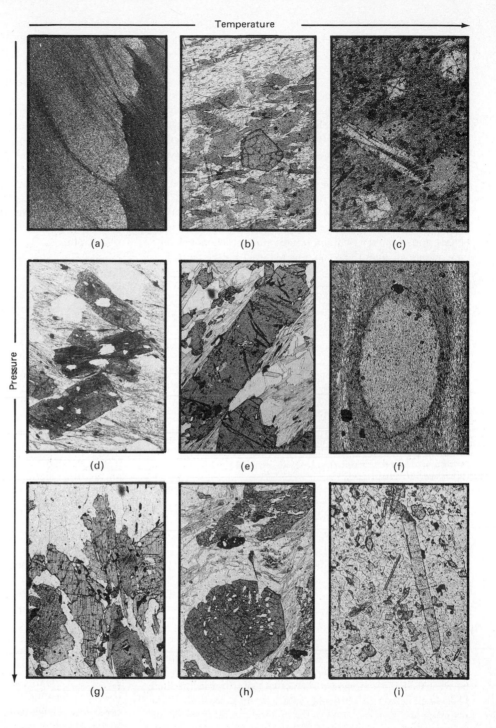

FIGURE 17-2 Photomicrographs of typical minerals in metapelites. All samples contain quartz and muscovite except for (i), which contains quartz and K-feldspar. The width of each field is 4 mm. All are under plane-polarized light. Minerals are arranged approximately with temperature of formation increasing from left to right and pressure increasing from top to bottom. (a) Graded silt bed (top of bed to left) in chlorite-bearing slate, Waterville formation, south-central Maine. (b) Biotite and garnet in Waterville formation. (c) Andalusite porphyroblasts in hornfels of the Skiddaw granite, English Lake District. Concentrations of inclusions along sector boundaries in andalusite produce the variety known as chiastolite. (d) Chloritoid porphyroblasts, Leeds, Quebec. (e) Staurolite in Gassetts Schist, Vermont. (f) Cordierite porphyroblast, Waterville Formation, south-central Maine. (g) Kyanite in Gassetts Schist, Vermont. (h) Garnet with core of quartz inclusions, Bronson Hill anticlinorium, south-central Massachusetts. (i) Sillimanite, Aberdeenshire, Scotland.

which describes the relations between the various intensive variables in any phase in internal, homogeneous equilibrium (i.e., no compositional gradients). Thus for a phase α, we can write

$$S^\alpha \, dT - V^\alpha \, dP + \sum n_i^\alpha \, d\mu_i = 0$$

The facies diagram in which such a phase is to be plotted will be for a given temperature and pressure. This leaves only the chemical potentials of the various components as variables. The

following rules show which of these chemical potentials can be independently varied, and thus which ones play a role in determining mineral equilibria.

RULE 1

Components that occur as pure phases, such as SiO_2 in quartz, TiO_2 in rutile, or Fe_2O_3 in hematite, need not be plotted.

The Gibbs–Duhem equation for any silicate, S, has the following form:

$$S^S dT - V^S dP + n_i^S d\mu_i + n_{SiO_2}^S d\mu_{SiO_2} = 0$$

If quartz (Q) is present, then $\mu_{SiO_2} = \mu_{SiO_2}^Q$, but $\mu_{SiO_2}^Q$ is simply the molar free energy of quartz, which is a function only of temperature and pressure. The chemical potential of silica is therefore not an independent compositional variable. Another way of expressing this is to say that the rock is saturated in quartz. Any addition or subtraction of SiO_2 from the rock simply varies the amount of quartz but does not change the chemical potential of SiO_2; variation in the amount of silica does not, therefore, affect the stability of the other minerals. A rock is saturated in all those components that occur as pure phases, and because variations in their amounts cannot change mineral assemblages these components need not be plotted.

RULE 2

Components whose chemical potentials are controlled external to the system need not be plotted.

This rule applies mostly to components that appear in the fluid phase, such as H_2O and CO_2, but it could include others if they are sufficiently mobile. The Gibbs–Duhem equation for a fluid containing H_2O is

$$S^f dT - V^f dP + n_i^f d\mu_i + n_{H_2O}^f d\mu_{H_2O} = 0$$

But $\mu_{H_2O}^f = \mu_{H_2O}^{f*} + RT \ln(f/10^5)$. Now $\mu_{H_2O}^{f*}$ is a function only of pressure and temperature which is fixed for any given facies diagram. If the fugacity of H_2O is buffered by an external reservoir, then, $\mu_{H_2O}^f$ is not an independent variable. It is common in such cases to express $\mu_{H_2O}^f$ in terms of activity:

$$\mu_{H_2O}^f = \mu_{H_2O}^{f*} + RT \ln(a_{H_2O}^f)$$

from which it follows that

$$a_{H_2O}^f = \frac{\exp(\mu_{H_2O}^f - \mu_{H_2O}^{f*})}{RT} \tag{17-1}$$

Thus a pure H_2O fluid has $a_{H_2O}^f = 1$, and if the fluid contains a mixture of components, $a_{H_2O}^f < 1$. Although H_2O need not be treated as a component if the activity of water in the fluid phase is buffered at a particular value, we must remember that the graphical representation of the mineral assemblages will be for only that one particular activity of H_2O.

RULE 3

Components occurring in only one phase need not be plotted if we also ignore the phase in which the component occurs.

Some components occur in only one phase, for example ZrO_2 in zircon, Na_2O in albite, CaO in plagioclase, and P_2O_5

in apatite. For each one of these phases a separate Gibbs–Duhem equation can be written, but each equation introduces just one new variable, that of the chemical potential of the component that occurs just in that phase. From the standpoint of the phase rule, the additional variable is balanced with the additional equation, and the variance of the system is unchanged by including these phases and components. Rule 3 perhaps offers the greatest opportunity to reduce the number of components erroneously. Before a component is eliminated by this rule, you must be certain that the component occurs in only one phase. Na_2O, for example, might be present, not only in albite, but in the hard-to-detect paragonite component of muscovite, and CaO might occur in both plagioclase and epidote. If such second phases are present, the elimination of the component may result in assemblages that apparently violate the phase rule.

RULE 4

Components that are not sufficiently abundant to stabilize a mineral can be ignored.

This rule applies to the trace components in rocks. As long as an element's concentration is so low that it does not affect the stability of minerals, it need not be treated as a separate component. Instead, its amount can be added to the major element for which the trace component substitutes. Thus Ni^{2+} and Co^{2+} can be added to Fe^{2+}, and Ba^{2+} can be added to Ca^{2+}. This grouping, however, must be done with care. Mn^{2+}, for example, is grouped with Fe^{2+}, but if garnet is present, Mn^{2+} preferentially enters this phase, thus stabilizing it. Zn^{2+}, on the other hand, preferentially enters staurolite, helping to stabilize that mineral. And Sr^{2+} is known to stabilize aragonite with respect to calcite. These elements are normally present in such small amounts that their stabilizing effect is negligible. But one must remain alert to the possibility that rocks with higher-than-normal concentrations of these elements may be present, and then apparent violations of the phase rule may occur. Mn-rich rocks, for example, are not uncommon in many shale sequences associated with volcanic rocks.

Let us now apply these rules to reducing the number of components in a metapelite. Most pelitic schists are composed essentially of quartz + muscovite and three other minerals from Table 17-1. At the highest grades of metamorphism, muscovite is not stable; instead, metapelites are composed of quartz + K-feldspar and three other minerals. In addition to the essential minerals, most metapelites contain a number of accessory minerals, such as hematite, magnetite, ilmenite, rutile, pyrrhotite, pyrite, graphite, zircon, tourmaline, and apatite. The formulas of the essential minerals are given in Table 17-1.

The minerals in a typical pelitic schist are composed of the following components: SiO_2, TiO_2, Al_2O_3, Fe_2O_3, FeO, MnO, ZnO, MgO, CaO, Na_2O, K_2O, ZrO_2, B_2O_3, H_2O, C, and S. This is a long list, but we will now proceed to reduce it to only four components.

Most pelitic rocks contain quartz, that is, they are saturated in silica. Thus, according to rule 1, SiO_2 need not be plotted as long as only rocks containing quartz are considered. Carbon can similarly be eliminated if it occurs as graphite,

TABLE 17-1

Common minerals in metapelites and their formulas

Quartz	SiO_2
Muscovite	$KAl_2(AlSi_3O_{10})(OH)_2$
Chlorite	$(Mg,Al,Fe)_6(Al,Si)_4O_{10}(OH)_8$
Biotite	$K(Mg,Fe)_3(AlSi_3O_{10})(OH)_2$
Almandine	$(Fe,Mg)_3Al_2Si_3O_{12}$
Spessartine	$Mn_3Al_2Si_3O_{12}$
Staurolite	$(Fe,Mg)_2Al_9Si_{3.75}O_{22}(OH)_2$
Andalusite, kyanite, sillimanite	Al_2SiO_5
Mullite	$3Al_2O_3.2SiO_2$
Chloritoid	$(Fe^{2+},Mg,Mn)_2(Al,Fe^{3+})Al_3O_2(SiO_4)_2(OH)_4$
Cordierite	$(Mg,Fe)_2Al_4Si_5O_{18}$
Orthopyroxene	$(Mg,Fe)_2Si_2O_6$
Orthoclase	$(K,Na)AlSi_3O_8$
Albite	$NaAlSi_3O_8$
Paragonite	$NaAl_2(AlSi_3O_{10})(OH)_2$

TiO_2 if it occurs as rutile, and Fe_2O_3 if it occurs as hematite. We eliminate H_2O as a component by rule 2, expressing it rather as an intensive environmental variable. We must not forget, however, that the mineral facies plot will be restricted to a certain activity of H_2O.

Under rule 3, the following components are eliminated because each occurs in only one phase: CaO (plagioclase), Na_2O (albite), ZrO_2 (zircon), B_2O_3 (tourmaline), and S (pyrrhotite). If the rocks do not contain rutile and hematite (rule 1 above), TiO_2 and Fe_2O_3 can be eliminated under this rule if the rocks contain ilmenite and magnetite.

The trace components MnO and ZnO are eliminated by adding their amounts to FeO, keeping in mind that any significant amount of these elements may cause problems in plotting rocks containing garnet and staurolite.

By these rules, the 16 components needed to describe all the minerals in a pelitic schist are reduced to only four significant ones, Al_2O_3–FeO–MgO–K_2O (*AFMK*), which can be plotted in a tetrahedron (Fig. 17-3a). Most of the minerals plot on the *AFM* face of the tetrahedron with the exception of muscovite, biotite, and K-feldspar. One further simplification can be made; that is, to project points plotting within the tetrahedron onto a face of the tetrahedron. This we do according to the following rule.

RULE 5

If a component occurs in two or more phases, and one phase is common to all assemblages, project the composition of all phases with this component from the composition of the common phase onto a face of the tetrahedron or some other convenient plane.

Let us see how this rule applies to the tetrahedron for pelitic rocks. Biotite and muscovite both contain the component K_2O. Furthermore, most pelitic rocks contain muscovite (ignore for the moment the high-temperature rocks in which

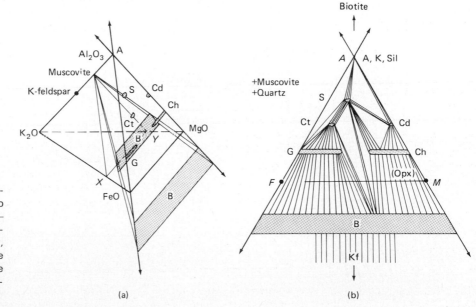

FIGURE 17-3 Projection of quartz- and muscovite-bearing metapelites onto the AFM face of the Al_2O_3–FeO–MgO–K_2O tetrahedron from muscovite. Minerals plotted are Al_2SiO_5 polymorphs (A), staurolite (S), chloritoid (Ct), cordierite (Cd), garnet (G), chlorite (Ch), biotite (B), muscovite (M), and potassium feldspar (Kf). [After Thompson (1957).]

(a)   (b)

K-feldspar forms in place of muscovite). Thus, if we restrict the plot to rocks containing muscovite (+quartz), compositions in the tetrahedron can be projected from the common phase, muscovite, onto the *AFM* face. This projection is shown graphically in Figure 17-3a. To project a point quantitatively onto the *AFM* face, the K_2O content of the point must be reduced to zero by subtracting the composition of the point from which the projection is made—muscovite in this case ($K_2O \cdot 3Al_2O_3$). The alumina content of the projected point is therefore obtained by reducing the initial value by the alumina removed with K_2O as muscovite, that is, $Al_2O_3 - 3K_2O$. To plot the projected point on the *AFM* face, the new alumina content plus the FeO and MgO are recalculated to a total of one. The alumina value then becomes

$$\text{alumina index} = \frac{Al_2O_3 - 3K_2O}{(Al_2O_3 - 3K_2O) + FeO + MgO} \quad (17\text{-}2)$$

Biotite, for example, has a composition of $K_2O \cdot 6(FeO,MgO) \cdot Al_2O_3$ in the tetrahedron. Its alumina index is therefore $(-2/(-2 + 6) = -0.5)$. Consequently, biotite does not fall within the *AFM* triangle but plots below its base, as would be expected from the graphical projection from the tetrahedron. Points plotting in the subtetrahedron bounded by muscovite, biotite, and the two points *x* and *y* project onto the *AFM* plane below biotite at distances that approach $-\infty$ as compositions approach the plane muscovite-*x*-*y*, which is parallel to the *AFM* face. Compositions plotting in the subtetrahedron muscovite-*x*-*y*-K_2O, rather than being projected *from* muscovite, can be projected onto the *FMA* face only *through* muscovite (negative projection), intersecting the face above *A*. K-feldspar projects

through muscovite to fall right on the *A* apex of the *AFM* triangle.

All of the essential minerals in metapelites can be plotted in the *AFM* projection (Fig. 17-3b), and their range of compositions in terms of FeO and MgO can be shown clearly. Staurolite allows little substitution of Mg for Fe and plots as a small area near the *AF* side of the diagram. Chloritoid allows a greater, but still limited substitution of Mg, with Mg-rich compositions forming cordierite instead. Garnet and chlorite have a complete range of compositions, but garnets are always Fe-rich relative to coexisting chorite. The Mg-rich garnet, pyrope, is stable only at high pressures. Stilpnomelane tends to be Fe-rich, whereas biotites have a complete range of Fe/Mg values; they also have variable alumina contents.

At the highest grades of metamorphism, quartz and muscovite react to form K-feldspar and an aluminum silicate. For such rocks, K-feldspar is the common phase, and so compositions in the tetrahedron are projected onto the *AFM* face from feldspar. Because most minerals lie in the *AFM* plane, the new projection does not change their positions. Biotite, however, plots at a slightly higher (less negative) value than in the muscovite projection (Prob. 17-1).

17-4   VARIANCE IN PELITIC MINERAL ASSEMBLAGES

Figure 17-4 shows a typical *AFM* mineral facies diagram for a pelitic schist. Note that because of rules 1 to 5, the diagram is for a given temperature, pressure, and activity of H_2O, and

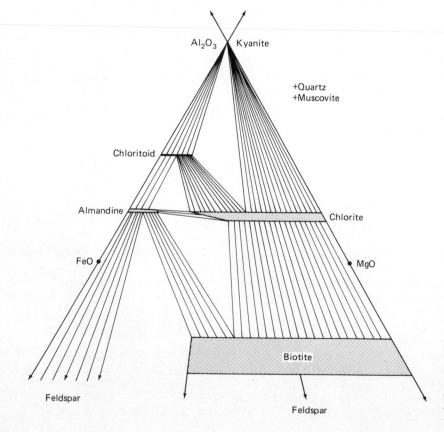

FIGURE 17-4 *AFM* plot showing possible three-phase and two-phase assemblages in metapelites. Note that all phases coexist with quartz and muscovite and a fluid phase with a fixed activity of water.

all mineral assemblages must coexist with quartz and muscovite. The diagram is different from those discussed in Chapters 15 and 16 because many of the minerals exhibit solid solution. We consider next the effect of solid solution on the variance, or degrees of freedom, of a mineral assemblage.

If the mineral assemblages depicted in Figure 17-4 occur in rocks over an extensive geographic region, the facies diagram can be assumed to hold true over some range of T, P, and a_{H_2O}. Thus, from the viewpoint of the phase rule ($\phi + f = c + 3$, the numeral 3 referring to T, P, and a_{H_2O}), three intensive variables were able to vary without causing changes in the mineral assemblages. The number of components is only three because all others have been eliminated according to rules 1 to 4. Consequently, for the variance we have $f = 6 - \phi$.

The variance of a rock containing three minerals in an *AFM* diagram (plus quartz and muscovite) is 3. But three degrees of freedom are required to define the T, P, and a_{H_2O} of the diagram, leaving no degrees of freedom for varying mineral compositions. Thus in any three-phase assemblage, such as chloritoid + garnet + chlorite, the composition of each mineral is fixed for a given T, P, and a_{H_2O}. In this case, variations in the bulk composition of the rock result in variations in the proportions of minerals but not in their compositions. Note that minerals in some three-phase assemblages have fixed compositions regardless of the pressure and temperature (kyanite, for example). Others belong to solid solution series, and although their compositions are fixed for a given T, P, and a_{H_2O}, under different conditions their compositions are fixed at different values. This allows the composition of minerals in some three-phase assemblages to be used for geothermometry and geobarometry.

If only two minerals are present in an *AFM* plot, the variance is 4. Again, three degrees of freedom are needed to specify the T, P, and a_{H_2O}, leaving one degree of freedom for composition. If the two minerals are chloritoid and chlorite, for example, both belong to Fe–Mg solid solution series, and chlorite even exhibits some variability in its Al content. However, only one degree of freedom exists among these compositional variables. Chlorite coexisting with chloritoid must contain its maximum amount of Al; that is, it must plot on the Al side of the chlorite field. If the one degree of freedom is used to specify the Fe/Mg ratio in the chlorite, the composition of coexisting chloritoid is then determined, with the composition being indicated by the chloritoid composition at the end of the tie line coming from the specified chlorite composition. All two-phase regions in Figure 17-4 are therefore ruled with tie lines indicating the compositions of coexisting phases. Note that if one of the two phases has a fixed composition, such as kyanite, the composition of the other phase cannot be varied independently. In this case, the bulk composition of the rock determines the composition of the other phase; that is, tie lines radiate from the phase of fixed composition through the bulk composition of the rock to the other phase.

Finally, if a rock contains only one of the minerals in an *AFM* diagram the variance is 5, that is, three environmental degrees of freedom and two compositional ones. Thus any two compositional variables can be specified, and then the third is determined. When only one mineral is present, its composition is determined simply by the bulk composition of the rock, and it can indicate little about the metamorphic conditions, other

than the rock must have crystallized in the stability field of that mineral.

The mineral assemblages in an *AFM* diagram consequently divide the diagram into three-phase triangles, two-phase regions ruled with tie lines, and one-phase regions. The three-phase assemblages have the smallest variance and therefore preserve the most information about metamorphic conditions.

17-5 ISOGRADS IN PELITIC ROCKS

An isograd was defined in Sec. 15-2 as a line on the surface of the Earth marking the first appearance of a metamorphic mineral or mineral assemblage. We will now see that such first appearances can result from either discontinuous or continuous reactions, but only the discontinuous reactions are of practical use in mapping. These will be illustrated with reactions responsible for the appearance of staurolite in metampelites.

Discontinuous Reactions. A discontinuous reaction brings about a change in the topology of a mineral facies diagram. This can result from the appearance or disappearance of a mineral. Such reactions are said to be *terminal* to that particular mineral. The topology can also be changed by a tie-line switching reaction; these are *nonterminal* because the particular minerals, when separate, have stability ranges above and below that of the reaction. Both of these reaction types have already been encountered in Chapters 15 and 16. We will now study examples involving staurolite.

Consider first the reaction

$$almandine_{ss} + chlorite_{ss} + kyanite$$
$$= staurolite_{ss} + quartz + H_2O \quad (17\text{-}3)$$

where the subscript ss indicates the mineral belongs to a solid solution series. Balancing such a reaction exactly requires knowledge of the Fe/Mg ratio in the various phases, which will be discussed in the next section. A P–T plot of this reaction is given in Figure 17-5. Because H_2O is evolved, the activity of H_2O plays an important role in determining the position of the reaction curve, which in this case is shown for $a_{H_2O} = 1$. On the low-temperature side of the reaction, almandine + chlorite + kyanite form a three-phase triangle in the *AFM* diagram. Note that the composition of staurolite plots within this triangle, so that when the reaction temperature is reached, staurolite forms from these three minerals. This produces three new phase triangles: almandine + staurolite + kyanite, kyanite + staurolite + chlorite, and almandine + staurolite + chlorite. At a given pressure and activity of H_2O, this reaction takes place at only one temperature. The reaction marks the first true appearance of staurolite; that is, it is a terminal reaction on the low-temperature side of the staurolite stability field. Note that the reaction affects all rocks with bulk compositions lying in the triangle almandine + chlorite + kyanite. Staurolite therefore appears in a wide range of rock compositions as a result of this reaction. This is an important consideration when looking for reactions that can be mapped easily.

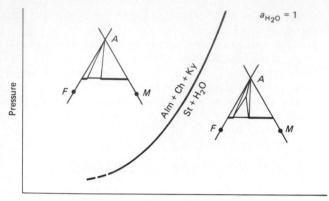

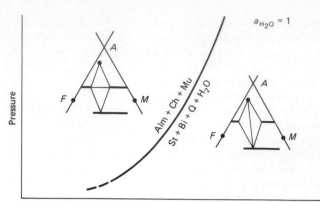

FIGURE 17-5 *P–T* plot of the discontinuous reaction almandine + chlorite + kyanite = staurolite + quartz + H_2O for $a_{H_2O} = 1$. This reaction is terminal to staurolite, because staurolite is not stable at temperatures below this reaction.

FIGURE 17-6 *P–T* plot of the tie-line-switching nonterminal reaction of almandine + chlorite + muscovite = staurolite + biotite + quartz + H_2O for $a_{H_2O} = 1$.

An example of a nonterminal discontinuous reaction is

$$\text{almandine}_{ss} + \text{chlorite}_{ss} + \text{muscovite}$$
$$= \text{staurolite}_{ss} + \text{biotite}_{ss} + \text{quartz} + H_2O \quad (17\text{-}4)$$

which is shown in Figure 17-6. Note that in this reaction muscovite and quartz are both needed to balance the reaction; but remember that these minerals are available in the rock even though they are not plotted in the *AFM* diagram. Again, because the reaction evolves H_2O, the a_{H_2O} must be specified. No new minerals are formed by the reaction; instead, the reaction

involves a tie-line switch (hence, nonterminal). At temperatures below the reaction, staurolite is stable only in rocks having bulk compositions above the almandine–chlorite tie line, but once the reaction occurs, rocks with bulk compositions below this line also contain staurolite. The reaction therefore greatly expands the range of rock compositions that can contain staurolite. This reaction would therefore also be a convenient one to use for mapping purposes.

Continuous Reaction. As in magmatic systems, a continuous metamorphic reaction occurs when some of the reactants and products belong to solid solution series. Consider

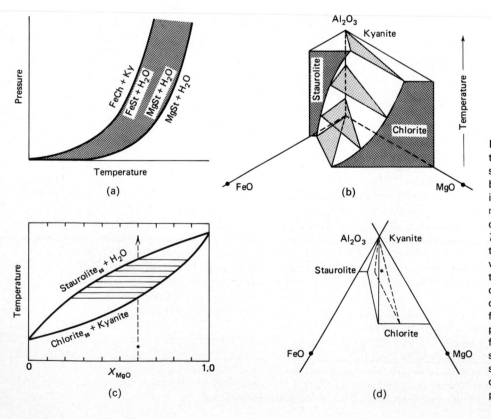

FIGURE 17-7 (a) *P–T* plot of the continuous reaction kyanite + chlorite = staurolite + quartz + H_2O. The reaction begins at low temperatures involving the iron-rich components and progresses to more magnesian compositions with increasing temperature. (b) Perspective *T–AFM* plot showing how the composition of staurolite and chlorite coexisting with kyanite change with increasing temperature. (c) *T–X_{MgO}* projection of compositions of staurolite and chlorite coexisting with kyanite, showing path followed by progressively metamorphosed rock having $X_{MgO} = 0.6$ (see text for discussion). (d) *AFM* plot showing swing of three-phase triangle kyanite–staurolite–chlorite to more magnesian compositions with increasing temperature.

the reaction

$$kyanite + chlorite_{ss} = staurolite_{ss} + quartz + H_2O \quad (17\text{-}5)$$

Both chlorite and staurolite form sold solutions, and without specifying the Mg/Fe ratio of one of the minerals, the reaction is not completely defined. As will be evident from the discussion in Sec. 17-6, the Mg/Fe ratios of both minerals continuously change as the reaction proceeds with increasing temperature (hence, a continuous reaction). In a *P–T* diagram the reaction is no longer a univariant line but is a broad zone (Fig. 17-7a). The reaction first starts with Fe-rich chlorite reacting with kyanite to form Fe-rich staurolite. As the temperature rises, the chlorite and staurolite become more magnesian. This causes the three-phase triangle kyanite + staurolite$_{ss}$ + chlorite$_{ss}$ to swing to more magnesian compositions by pivoting on its kyanite apex (Fig. 17-7b). These compositional changes are best seen in projections onto the *T–X*$_{Mg}$ plane (Fig. 17-7c) or the *AFM* plane (Fig. 17-7d).

The temperature at which staurolite is produced by a continuous reaction depends on the bulk composition of the rock. Let us assume that the Mg/(Mg + Fe) ratio in the rock

is 0.6 (asterisk in Fig. 17-7c and d). At low temperature, this composition lies in the two-phase region kyanite–chlorite. As the temperature rises, the three-phase triangle kyanite–chlorite–staurolite swings to more magnesian compositions, and eventually the kyanite–chlorite side of the triangle passes through the bulk composition of the rock, and staurolite first appears. Only an infinitesimal amount of staurolite forms at first, which would likely go undetected in the rock, at least to the unaided eye. As the temperature rises and the three-phase triangle continues swinging to more magnesian compositions, the amount of staurolite increases and that of chlorite decreases as both minerals adjust their compositions. Eventually, the three-phase triangle swings far enough that its kyanite–staurolite side passes the bulk composition of the rock; all chlorite is consumed at this stage and the reaction is complete. The rock is then free to continue heating until the next reaction is encountered.

Let us now examine the significance of the staurolite isograd in terms of these three reactions. We will assume that in a sequence of pelitic rocks three different units have compositions indicated in the AFM diagrams of Figure 17-8 by points *x*, *y*, and *z*. Unit *x* plots in the three-phase triangle kyanite–almandine–chlorite at low temperature. When the temperature

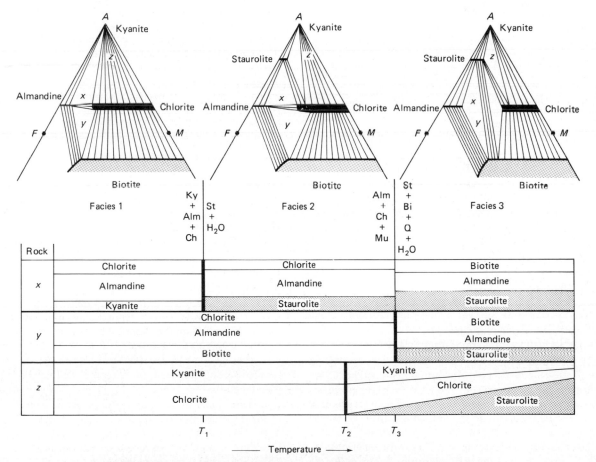

FIGURE 17-8 *AFM* diagrams showing two discontinuous reactions and one continuous reaction for the formation of staurolite. The bar graphs represent schematically the modes of three different rocks, having compositions that are indicated in the *AFM* diagrams by points *x*, *y*, and *z*. The temperature of first appearance of staurolite (i.e., the staurolite isograd) is different in the three rock types, as indicated by the heavy vertical lines. See text for discussion.

reaches T_1, staurolite forms by the discontinuous reaction of Eq. 17-3, and the new assemblage for this rock composition becomes staurolite–almandine–chlorite. Staurolite makes an immediate entrance at this temperature, with its abundance determined by the position of the bulk composition of the rock within the staurolite–almandine–chlorite triangle. As long as the rock is sufficiently aluminous, staurolite would be an easily recognized phase, and this reaction would make an ideal isograd, marking the trace of the T_1 isotherm at a given pressure and activity of water.

With rising temperature, rock x eventually reaches T_3, where the second discontinuous reaction takes place (Eq. 17-4). The switch in tie lines changes the assemblage of minerals to staurolite–almandine–biotite above this temperature. Biotite makes an immediate entrance at this temperature, and chlorite suddenly disappears. The amount of staurolite also slightly increases. This reaction could be used to map a biotite isograd.

Rock y, being less aluminous than rock x, consists of the assemblage almandine–chlorite–biotite at low temperatures. Staurolite does not appear in this rock at T_1 because it lacks the appropriate reactant minerals. Not until the tie-line switch at T_3 does staurolite appear, and then it does so abruptly, which would make this a mappable reaction. But note that this appearance of staurolite is different from that mapped by the reaction at T_1.

Rock z plots in the two-phase field of kyanite and chlorite at low temperature, and not until the three-phase triangle swings to more magnesian compositions does staurolite make its first appearance at T_2, but even then the amount is very small at first, increasing only with rising temperature. An isograd based on this continuous reaction would therefore be difficult to detect. The compositional range encompassed by the three-phase triangle is not large and therefore the range of rocks in which this reaction is likely to take place is small. Moreover, the temperature T_2 of first appearance of staurolite depends on the bulk composition of the rock. Thus, even if the first appearance of staurolite is detectable, its presence cannot be used as an indicator of temperature unless the bulk composition of the rock is known. This reaction is therefore an impractical one on which to base an isograd.

From the reactions above, it is clear that if an isograd is based simply on the first appearance of a mineral, it cannot be an accurate indicator of temperature, even if the pressure and a_{H_2O} are known, because the minerals can form by numerous reactions. The staurolite isograd defined in this way would occur at significantly different temperatures in rocks x, y, and z, because of their different compositions (heavy vertical line in lower part of Fig. 17-8). To avoid this compositional problem, the isograd must be based on a discontinuous reaction. These reactions generally affect a large range of rock composition. Also, an isograd should be based on the first appearance of a *mineral assemblage* rather than the first appearance of a *single mineral*. Thus in the examples above, two separate staurolite isograds could be mapped, one involving staurolite + almandine + chlorite, and the other staurolite + almandine + biotite. Clearly, if these two reactions had been lumped together and mapped as the first appearance of staurolite, the isograd would not necessarily indicate a line of equal grade of metamorphism.

17-6 EFFECT OF P AND T ON REACTIONS AMONG SOLID SOLUTION PHASES

Reactions involving solid solution phases can be of the continuous or discontinuous type. Because so many metamorphic minerals form solid solutions, in particular those in pelitic rocks, it is important to have an understanding of the effects of temperature and pressure on reactions involving these minerals. For a complete discussion of this topic, see the papers by A. B. Thompson (1976a and b).

The ratio $Mg/Mg + Fe (X_{Mg})$ in coexisting minerals in pelitic rocks decreases in the order cordierite > chlorite > biotite > chloritoid > staurolite > garnet. For example, in Figure 17-7c we saw that chlorite is more magnesian than coexisting staurolite. Knowing the relative distribution of Fe and Mg between these minerals allows us to deduce what reactions can take place among the minerals simply from their positions in the *AFM* plot. The direction in which these reactions proceed with increasing temperature can then be determined by noting on which side of the reaction H_2O is released (high-entropy side = high-temperature side).

In Sec. 10-9 the substitution of Mg for Fe in fayalite was seen to raise the melting point of olivine, and conversely, the substitution of Fe for Mg in forsterite lowered the melting point (Fig. 10-11). Although this is not a metamorphic reaction, the derivation of the cryoscopic equation, which quantified these effects, was perfectly general and is applicable to solid as well as liquid solutions. For example, in Sec. 10-10, polymorphic changes among the pyroxenes were seen to follow the same type of relation. We can generalize the observations by stating that the temperature of a reaction taking place between pure end members is raised by a component that is more soluble in the low-temperature phase or phases and is lowered by a component that is more soluble in the high-temperature phase or phases. Magnesium, for example, is more soluble in olivine$_{ss}$ than in olivine liquid, so the melting point of fayalite rises with substitution of Mg. Similarly, the temperatures of metamorphic reactions are raised if Mg substitutes more readily into the phases on the low-temperature side of the reaction and lowered if Mg is more soluble in the high-temperature phases. Knowing the relative distribution of Fe and Mg between the minerals of pelitic rocks allows us, then, to predict the $T–X_{Mg}$ relation of reactions involving these minerals. The effect of pressure on the reactions can be deduced if, in addition to knowing the Fe–Mg partitioning among phases, the ΔV's of the end-member reactions are known.

Let us consider the effects of temperature and pressure on a few specific metamorphic reactions. This will be done using pseudobinary $T–X_{Mg}$ and $P–X_{Mg}$ diagrams with compositions of minerals in the *AFM* plot being projected from the A apex of the diagram. We have already seen in Figure 17-7c the effect of Mg substitution for Fe on the continuous reaction of chlorite + kyanite to produce staurolite + quartz + H_2O (Eq. 17-5). Staurolite must be on the high-temperature side of this reaction, and because Mg is more soluble in chlorite than in staurolite, the reaction temperature rises as X_{Mg} increases (shown also in Fig. 17-9a). Also, as X_{Mg} increases, so does the pressure at which the reaction takes place (Fig. 17-9b).

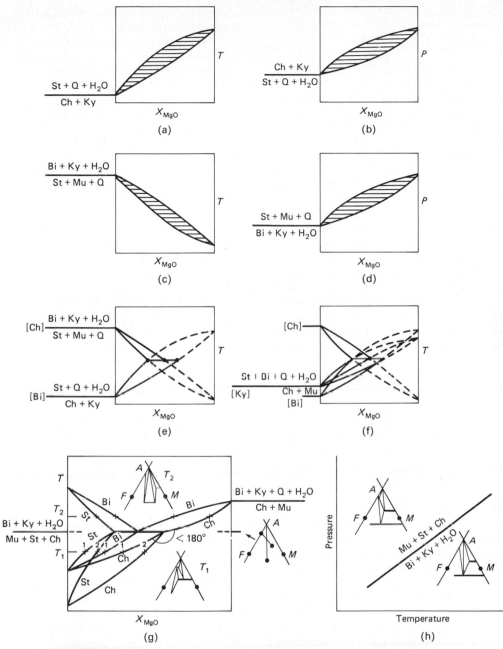

FIGURE 17-9 (a) Isobaric T–X_{MgO} diagram for the continuous reaction chlorite + kyanite = staurolite + quartz + H_2O. (b) Same reaction as in (a) plotted in P–X_{MgO} diagram. (c) Isobaric T–X_{MgO} diagram for the continuous reaction staurolite + muscovite + quartz = biotite + kyanite + H_2O. (d) Same reaction as in (c), plotted in P–X_{MgO} diagram. (e) Reactions in (a) and (c) combine to give an isobaric invariant assemblage where they intersect involving muscovite + staurolite + chlorite = biotite + kyanite + H_2O. Each reaction is made metastable on the magnesian side of the invariant point by the other reaction. (f) Extending from the invariant point is a reaction involving chlorite + muscovite = staurolite + biotite + quartz + H_2O. (g) A fourth and final reaction extending from the invariant point involves chlorite + muscovite = biotite + kyanite + quartz + H_2O. At the isobaric invariant point, the compositions of staurolite, chlorite, and biotite are fixed, whereas above and below this temperature the compositions vary continuously along the isobaric univariant continuous reaction curves. (h) P–T diagram for the reaction muscovite + staurolite + chlorite = biotite + kyanite + H_2O. See text for discussion.

Consider next the reaction

staurolite + muscovite + quartz

$$= \text{biotite} + \text{kyanite} + H_2O \quad (17\text{-}6)$$

In this reaction staurolite appears on the low-temperature side, but because Mg is more soluble in biotite than in coexisting staurolite, the reaction temperature decreases with increasing X_{Mg} (Fig. 17-9c). Increasing X_{Mg}, however, raises the pressure at which the reaction takes place (Fig. 17-9d).

If these two reactions (Eqs. 17-5 and 17-6) are plotted in the same T–X_{Mg} diagram (Fig. 17-9e), they intersect to create an isobaric invariant assemblage of the four phases chlorite + kyanite + staurolite + biotite. From the modified phase rule $(\phi + f = c + 3)$, we have $4 + f = 3 + 3$, and thus there are two degrees of freedom; these are required to define the pressure and a_{H_2O} of the T–X_{Mg} diagram.

Four independent reactions must extend from this invariant point. Each one involves three of the four minerals present at the invariant point. Two of these are already known; the chlorite-absent reaction is Eq. 17-6 and the biotite-absent reaction is Eq. 17-5. The kyanite-absent and staurolite-absent reactions can be deduced from the positions of the phases in the *AFM* diagram. They are:

[Ky]   chlorite$_{ss}$ + muscovite

$$= \text{staurolite}_{ss} + \text{biotite}_{ss} + \text{quartz} + H_2O$$

[St]   chlorite$_{ss}$ + muscovite

$$= \text{biotite}_{ss} + \text{kyanite} + \text{quartz} + H_2O$$

The temperature of the kyanite-absent reaction must increase with increasing X_{Mg} because Mg is more soluble in chlorite than in coexisting staurolite, and chlorite appears on the low-temperature side of the reaction. Also, at the invariant point, the reaction involves the coexistence of three solid solution phases—chlorite, staurolite, and biotite—the compositions of which are defined by the points of intersection of the other two reaction lines. Three lines must extend from these points to indicate the projected compositions of the minerals forming the three-phase triangle in the *AFM* diagram. Because the kyanite-absent reaction involves the coexistence of staurolite and quartz, this reaction must occur in the stability field of staurolite + quartz; that is, the reaction must occur above the biotite-absent reaction because it is only there that staurolite + quartz is stable (Fig. 17-9f).

The staurolite-absent reaction involves the coexistence of the solid phases chlorite and biotite, the compositions of which are determined by the temperature at which the chlorite- and biotite-absent reactions intersect. Because Mg partitions into chlorite more than into biotite and chlorite is on the low-temperature side of the reaction, increasing X_{Mg} causes the temperature of the reaction to rise. However, because the assemblage chlorite + muscovite occurs on the low-temperature side of both the kyanite- and staurolite-absent reactions, the slope of the staurolite-absent reaction must be less than that of the kyanite-absent reaction (Fig. 17-9g) so as not to violate the Schreinemakers rule stating that *the angle between univari-*

ant lines bounding a divariant field at an invariant point must be less than 180° (Sec. 8-4).

The point of intersection of these four isobaric, univariant, continuous reactions is an isobaric, invariant, discontinuous reaction involving

staurolite + chlorite + muscovite

$$= \text{biotite} + \text{kyanite} + \text{quartz} + H_2O \quad (17\text{-}7)$$

At the isobaric invariant point, the compositions of staurolite, chlorite, and biotite are fixed; above or below this temperature, however, the compositions of these minerals vary along the lines representing the projections of the various three-phase triangles. At temperature T_1 (Fig. 17-9g) the compositions in the two three-phase triangles (St + Bi + Ch and St + Ch + K) can be read off the T–X_{Mg} diagram and plotted on an *AFM* diagram (inset in Fig. 17-9g). As the temperature rises the compositions of staurolites 1 and 2 and chlorites 1 and 2 approach each other and eventually become the same at the invariant point where the tie-line-switching reaction of staurolite + chlorite = kyanite + biotite takes place. Above the invariant point, at temperature T_2 for example, we can again read off the T–X_{Mg} diagram the composition of the phases coexisting in the three-phase triangles and plot them in an *AFM* diagram (Fig. 17-9g).

Although the temperature and compositions of the phases are fixed at the isobaric invariant point, under other pressures or a_{H_2O} different values would be established. A diagram similar to Figure 17-9g can be constructed for the intersection of the four reactions in terms of P–X_{Mg} at constant T and a_{H_2O} [see Thompson (1976a)]. Finally, the reaction of Eq. 17-7 can be projected into P–T space to become the univariant reaction of Figure 17-9h.

17-7   PETROGENETIC GRID FOR PELITIC ROCKS

In Sec. 17-5 we saw that isograds based simply on the first appearance of a mineral with no regard to associated minerals is likely to produce a poor map of metamorphic grade. The metamorphic facies concept proposed by Eskola is superior, because it bases the metamorphic grade of a rock on assemblages of minerals that are stable together under a given set of conditions. But here we saw that numerous metamorphic reactions separate each of the main facies, and attempts to divide the facies into subfacies became unwieldy. The solution then was to construct a P–T grid which includes all of the reactions that could take place. Having discussed the types of reactions that can occur among minerals of solid solution series, we are in a position to see how a petrogenetic grid for pelitic rocks could be constructed.

A petrogenetic grid is based on discontinuous reactions because these are independent of the composition of the rock and are functions only of pressure, temperature, and a_{H_2O}. In the remainder of this section we assume that $a_{H_2O} = 1$, which is probably a reasonable assumption for thick successions of pelitic rocks where few, if any, carbonate rocks are present. If we restrict the petrogenetic grid to reactions involving the seven common minerals in the *AFM* diagram (Al$_2$SiO$_5$, staurolite, chloritoid, cordierite, garnet, chlorite, and biotite), it is possible to determine how many discontinuous reactions relate all of these phases. A discontinuous reaction, such as that of Eq. 17-7, involves four minerals in the *AFM* diagram (+ quartz + muscovite). To determine how many such reactions can be written between the seven minerals, we need to know how many different combinations of minerals, taken four at a time, can be formed regardless of the order in the combination. According to the *rule of combinations, the number of combinations of n different items taken r at a time is n!/r! (n − r)!.* Thus, for the seven minerals, the number of combinations of four are 7!/4! (7 − 4)! = 35. This, however, treats Al$_2$SiO$_5$ as a single mineral. The number of reactions is therefore much greater, for all those involving Al$_2$SiO$_5$ must be written as separate reactions for the different Al$_2$SiO$_5$ polymorphs. At high temperatures, muscovite reacts with quartz, and K-feldspar becomes the common phase from which *AFM* projections are made. Therefore, reactions involving muscovite at low temperatures must be written as different reactions involving K-feldspar at high temperature. Clearly, the number of reactions is large and the construction of a complete petrogenetic grid for pelitic rocks is beyond the scope of this book. A number of authors have constructed such grids [see, for example, Albee (1965), Hess (1969), Kepezhinskas and Khlestov (1977), and Harte and Hudson (1979)], and a composite of these efforts is given in Figure 17-11.

For purposes of illustration, consider just part of this grid. If we start with the univariant discontinuous reaction of Eq. 17-7 and add to it the mineral cordierite (Cd), we create an invariant assemblage for a given a_{H_2O}. Experiments and thermodynamic calculations indicate that this invariant point occurs at about 520°C and 0.3 GPa (Fig. 17-10a). It would therefore lie in the stability field of andalusite. We will not concern ourselves for the moment with distinguishing the particular Al$_2$SiO$_5$ polymorph involved in the reactions; instead, we will refer to them all by the letter A.

Radiating from the invariant point must be five univariant reactions, one for each group of four minerals that can be formed from the five minerals at the invariant point. These reactions can be deduced from the position of minerals in the *AFM* diagram. We label them according to the absent phase.

[Cd]	St + Ch = Bi + A + H$_2$O
[St]	Ch = Bi + A + Cd + H$_2$O
[Ch]	St + Cd = Bi + A
[Bi]	Ch + A = St + Cd + H$_2$O
[A]	Ch = St + Bi + Cd + H$_2$O

In each of these reactions quartz and muscovite are available for balancing.

We already know the position of the cordierite-absent reaction from Figure 17-9h. All but the chlorite-absent reaction liberate water, and therefore they would be expected to have positive slopes. The chlorite-absent reaction, however, has a negative slope. Using Schreinemakers rules, we then arrange the univariant lines around the invariant point (Fig. 17-10). The

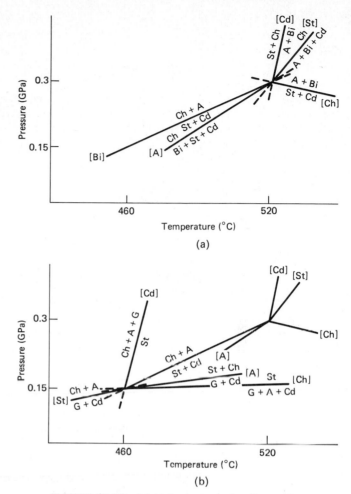

FIGURE 17-10 (a) Univariant reaction lines radiating from the invariant point involving the minerals staurolite, chlorite, biotite, kyanite, and cordierite (+quartz and muscovite). (b) A second invariant point is formed on the biotite-absent univariant reaction coming from the first invariant point by the addition of the mineral garnet. Univariant reactions radiating from this invariant point intersect those radiating from the first to create other invariant points. In this way a petrogenetic grid is developed.

reader should check that the positions of these reactions do indeed agree with Schreinemakers rules. The slopes of these reactions can be calculated from the Clapeyron equation and the ΔV's and ΔS's of the reactions.

On extending to lower temperatures, the biotite-absent reaction intersects the reaction of Eq. 7-3 (Fig. 17-5) to create a new invariant point involving garnet (G), chlorite, staurolite, cordierite, and Al$_2$SiO$_5$ (A). This invariant point, which is near 460°C and 0.15 GPa (Fig. 17-10b) must involve the following reactions:

[G]	Ch + A = St + Cd + H$_2$O
[A]	St + Ch = G + Cd + H$_2$O
[St]	Ch + A = G + Cd + H$_2$O
[Cd]	Ch + A + G = St + H$_2$O
[Ch]	St = G + A + Cd + H$_2$O

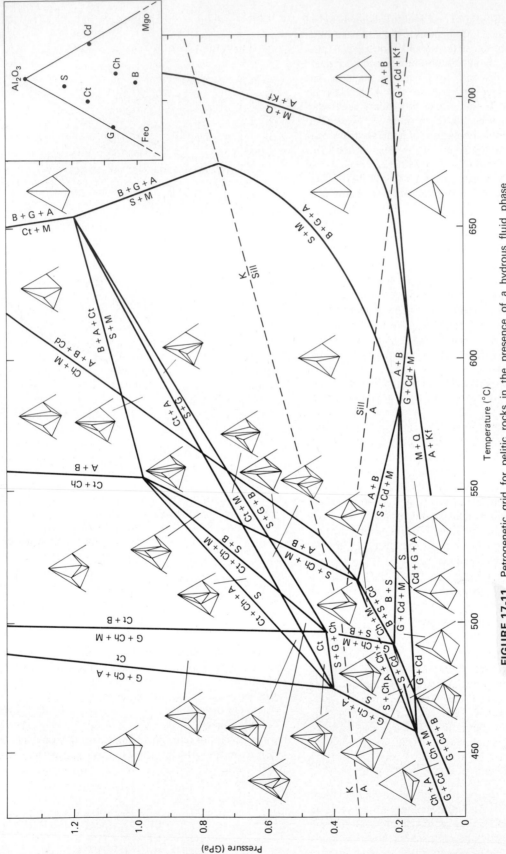

FIGURE 17-11 Petrogenetic grid for pelitic rocks in the presence of a hydrous fluid phase ($a_{H_2O} = 1$). The minerals shown in the *AFM* diagrams are kyanite, andalusite, and sillimanite (A), cordierite (Cd), chlorite (Ch), biotite (B), garnet (G), chloritoid (Ct), and staurolite (S). In addition, all rocks contain quartz (Q) and muscovite (M), except at high temperature, where they contain quartz and K-feldspar (Kf). [Modified from petrogenetic grids by Albee (1965), Hess (1969), and Kepezhinskas and Khlestov (1977). Al_2SiO_5 phase relations (A, K, Sill) are from Holdaway (1971).]

All of these reactions produce H_2O and must therefore have positive slopes on the $P-T$ diagram. Again, using Schreinemakers rules, the univariant reactions are distributed around the invariant point to produce the array shown in Figure 17-10b.

Some of the univariant lines radiating from this second invariant point will, on being extended, intersect other univariant lines radiating from the first invariant point. For example, the Al_2SiO_5-absent reaction from the second invariant point will intersect the Al_2SiO_5-absent reaction from the first. Similarly, the chlorite-absent reactions from both invariant points will intersect. A new invariant point is created at each point of intersection, and five univariant lines will radiate around each of these, unless degeneracy occurs. These lead to new intersections, and $P-T$ space is filled rapidly with a petrogenetic grid, such as that shown in Figure 17-11.

A petrogenetic grid can be a useful tool with which to analyze metamorphic mineral assemblages, because it should include all possible mineral assemblages if it has been constructed properly. It must, however, be used with care. The grid can be applicable to only one activity of H_2O, which in Figure 17-11 is $a_{H_2O} = 1$. If the activity of H_2O varies throughout a metamorphic region, as might happen if carbonate rocks were interlayered with pelites, the simple petrogenetic grid could not be used. Other components, in particular Mn, can also cause changes in the grid. Apart from these problems, the grid provides a useful means of explaining the sequence of successive mineral assemblages in progressively metamorphosed rocks, and it also provides a means of determining the approximate temperatures and pressures represented by particular mineral assemblages.

17-8 EXCHANGE EQUILIBRIA AND THE DISTRIBUTION COEFFICIENT

Although the petrogenetic grid provides a means of setting limits on the temperatures and pressures under which a given mineral assemblage can form, these limits are commonly rather wide. A rock containing the divariant assemblage staurolite + chloritoid + kyanite could be stable from 475 to 650°C and from 0.4 to 1.2 GPa (Fig. 17-11). Many of the minerals constituting these assemblages are solid solutions, and their compositions may vary considerably across a divariant field. If we can determine how Fe and Mg partition themselves between coexisting minerals as a function of temperature and pressure, it is conceivable that precise metamorphic temperatures and pressures could be determined from analyses of the minerals.

Continuous reactions can be thought of as resulting from two reactions proceeding simultaneously, one between the iron components and the other between the magnesium components. Reaction 17-5, for example, can be written as

$$41Al_2SiO_5 + 4Fe_5Al_2Si_3O_{10}(OH)_8$$
$$= 10Fe_2Al_9Si_4O_{23}(OH) + 13SiO_2 + 11H_2O$$

$$41Al_2SiO_5 + 4Mg_5Al_2Si_3O_{10}(OH)_8$$
$$= 10Mg_2Al_9Si_4O_{23}(OH) + 13SiO_2 + 11H_2O$$

If the second equation is subtracted from the first, we obtain

10 Mg-staurolite + 4 Fe-chlorite

$$= 10 \text{ Fe-staurolite} + 4 \text{ Mg-chlorite} \quad (17\text{-}8)$$

This is known as an Fe–Mg *exchange reaction*. The compositions of the coexisting staurolite and chlorite depend on how far the reaction progresses to the right or left. At equilibrium, the compositions of the two minerals are related through the equilibrium constant for the reaction.

It will be recalled from Eq. 9-5 that the free energy of any assemblage is given by $G = \sum_i n_i \mu_i$. At equilibrium, the free energies of the left- and right-hand sides of Eq. 17-8 must be equal; that is,

$$10\mu_{MgSt} + 4\mu_{FeCh} = 10\mu_{FeSt} + 4\mu_{MgCh}$$

On rearranging, this gives

$$10(\mu_{FeSt} - \mu_{MgSt}) - 4(\mu_{FeCh} - \mu_{MgCh}) = 0$$

But the chemical potential of each component can be written in terms of the activity of that component according to Eq. 9-34. For example, for the Fe-staurolite component we can write

$$\mu_{FeSt} = \mu_{FeSt}^* + RT \ln a_{FeSt}$$

where μ_{FeSt}^* is the chemical potential of pure Fe-staurolite at the given pressure and temperature. Substituting these expressions for each of the chemical potential terms and rearranging, we obtain

$$10(\mu_{FeSt}^* - \mu_{MgSt}^*) - 4(\mu_{FeCh}^* - \mu_{MgCh}^*)$$
$$= -10 \, RT \ln \frac{a_{FeSt}}{a_{MgSt}} + 4 \, RT \ln \frac{a_{FeCh}}{a_{MgCh}}$$

But $\mu_{FeSt}^* = \bar{G}_{FeSt}$. Thus the left-hand side of this equation is simply the free-energy change of the exchange reaction at the given pressure and temperature; that is,

$$\Delta G_{exch} = -RT \ln \frac{a_{FeSt}^{10} a_{MgCh}^4}{a_{MgSt}^{10} a_{FeCh}^4} \quad (17\text{-}9)$$

This free-energy change is known as the Fe–Mg *exchange potential*.

To relate the exchange potential to the composition of the minerals, we must relate the activities to mole fractions. From Eq. 10-14 it is known that if component i is mixed ideally on n equivalent sites in a mineral, the activity is given by $a_i = X_i^n$. Using this model, the activity of the Fe-staurolite component in staurolite becomes $a_{FeSt} = X_{FeSt}^2$, and for chlorite, $a_{FeCh} = X_{FeCh}^5$. Making these substitutions into Eq. 17-9 gives

$$\Delta G_{exch} = -RT \ln \frac{X_{FeSt}^{20} X_{MgCh}^{20}}{X_{MgSt}^{20} X_{FeCh}^{20}}$$

which on rearranging gives

$$-\Delta G_{exch} = 20RT \ln \frac{X_{FeSt}/X_{MgSt}}{X_{FeCh}/X_{MgCh}} \quad (17\text{-}10)$$

The ratio of the mole fractions of Fe and Mg in the coexisting phases is known as the *distribution coefficient*, which is designated K_D. In the example above, the distribution coefficient is for Fe and Mg between coexisting staurolite and chlorite and is therefore indicated as K_{DFe-Mg}^{St-Ch}. The distribution coefficient is a measurable property of a rock and simply requires careful analyses of coexisting minerals. These are normally obtained by electron microprobe analyses of juxtaposed grains.

Equation 17-10 relates the distribution coefficient to the free-energy change of the exchange reaction between pure end members, the data for which are available from thermodynamic tables. Because ΔG_{exch} can be expressed as a function of pressure and temperature, the distribution coefficient can be used to interpret the pressures and temperatures under which coexisting minerals achieved their compositions. To do this we make use of Eq. 8-9, which for an exchange reaction that does not liberate a gas becomes

$$\Delta G_{exch} = \Delta H^0_{exch} - T\,\Delta S^0_{exch} + (P - 10^5)\,\Delta V^0_{exch}$$

where P is measured in pascal. For the general case, Eq. 17-10 can be written as

$$\ln K_D = -\frac{\Delta H^0_{exch}}{mRT} + \frac{\Delta S^0_{exch}}{mR} - (P - 10^5)\frac{\Delta V^0_{exch}}{mRT} \quad (17\text{-}11)$$

where m is a coefficient based on the activity model and the coefficients in the reaction ($m = 20$ in the example given in Eq. 17-10). Whether a reaction can be used to indicate metamorphic temperatures (geothermometer) or pressures (geobarometer) depends on the magnitude of each term in this equation.

The volume change resulting from most exchange reactions is very small ($\Delta V_{exch} \approx 0$), so Eq. 17-11 reduces to the equation of a straight line ($y = mx + b$), where y is $\ln K_D$; m, the slope of the line, is $\Delta H^0_{exch}/mR$; x is $1/T(K)$; and b, the intercept on the y axis, is $\Delta S^0_{exch}/mR$. Such an equation can be used as a geothermometer if the thermodynamic data are available for the end members, or the exchange reaction can be calibrated through experimental studies or against other geothermometers.

One exchange reaction that has received considerable attention involves coexisting garnet and biotite [see, for example, Thompson (1976b), Ferry and Spear (1978), and Hodges and Spear (1982)]. The exchange reaction can be written as follows:

(phlogopite)   (almandine)

$KMg_3AlSi_3O_{10}(OH)_2 + Fe_3Al_2Si_3O_{12}$

(annite)   (pyrope)   (17-12)

$= KFe_3AlSi_3O_{10}(OH)_2 + Mg_3Al_2Si_3O_{12}$

The expression for the distribution coefficient for the reaction is

$$K_{DFe-Mg}^{Bi-G} = \frac{X_{FeBi}/X_{MgBi}}{X_{FeG}/X_{MgG}} \quad (17\text{-}13)$$

which can be related to temperature and pressure (Eq. 17-11) by

$$\ln K_D = -\left(\frac{\Delta H^0_{exch}}{3R}\right)\frac{1}{T} + \frac{\Delta S^0_{exch}}{3R}$$
$$-\left[(P - 10^5)\frac{\Delta V^0_{exch}}{3RT}\right] \quad (17\text{-}14)$$

where the numeral 3 is for the three sites in the garnet and in the biotite structures involved in the exchange. The ΔV^0_{exch} is small, so that the terms in square brackets can be dropped without introducing serious error; the remaining expression for $\ln K_D$ can then be plotted as a linear function of $1/T$ (Fig. 17-12). This exchange reaction has been calibrated experimentally by Ferry and Spear (1978), who give the following equation for $\ln K_D$:

$$\ln K_D = -\left(\frac{52108}{3R}\right)\frac{1}{T} + \frac{19.51}{3R}$$
$$-(P - 10^5) \times \frac{0.238 \times 10^{-5}}{3RT} \quad (17\text{-}15)$$

where all values are expressed in SI units.

The biotite–garnet exchange reaction is a useful geothermometer, especially in view of the common occurrence of this assemblage in low to high grades of metamorphism. As can be seen from Eq. 17-15, the ΔV^0_{exch} is small, so that even if the metamorphic pressure is unknown, a reasonably accurate temperature ($\pm 50°C$) can be obtained. Serious errors are introduced, however, if large amounts of Fe^{3+} are present, and in the upper amphibolite facies and granulite facies, biotites can

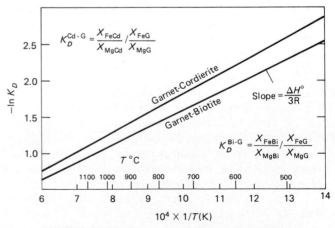

FIGURE 17-12   Plots of $-\ln K_D$ versus $1/T(K)$ for the Fe–Mg exchange reactions between coexisting garnet–biotite and garnet–cordierite. See text for discussion. [After Thompson (1976b).]

be rich in Ti, F, and Cl, which can seriously affect the calibration of the geothermometer (Essene 1982).

For a reaction to be a good geobarometer, the ΔV of reaction should be large. This eliminates simple ion-exchange reactions, which have very small ΔV's. Dehydration and decarbonation reactions have large ΔV's, but these reactions are also functions of a_{H_2O} and a_{CO_2}, respectively, and these are commonly not known exactly. Reactions involving solids and having large ΔV's must therefore be used.

One of the most widely used geobarometers in metapelites involves the common assemblage plagioclase + garnet + Al_2SiO_5 + quartz (Ghent 1976). It is based on the reaction

(anorthite)   (grossular)
$$3CaAl_2Si_2O_8 = Ca_3Al_2Si_3O_{12}$$

(kyanite)   (quartz)
$$+ 2Al_2SiO_5 + SiO_2 \quad (17\text{-}16)$$

The grossular forms a component in garnet, which is predominantly almandine in pelitic rocks, and the anorthite forms a component in plagioclase, which is normally albite rich. An equilibrium expression for the reaction can be written in terms of the activities of these components in the two minerals:

$$\ln \frac{a^G_{gross}}{(a^{Pl}_{an})^3} = -\left(\frac{62{,}640}{R}\right)\frac{1}{T} + \frac{160.76}{R}$$
$$- (P - 10^5) \times \frac{24.07 \times 10^{-5}}{RT} \quad (17\text{-}17)$$

Comparison of this equation with that for the biotite–garnet exchange (Eq. 17-15) shows that the ΔV term in Eq. 17-17 is two orders of magnitude greater. For this reason, the reaction makes a good geobarometer. Note that the ratio of the activities also depends strongly on temperature. Thus, to use this geobarometer, the temperature must be known independently. This could be obtained, for example, from the biotite–garnet geothermometer. The difficulty in using this geobarometer, however, arises from having to know the activities of grossular in garnet and anorthite in plagioclase. These components are normally present in small amounts in their respective phases, so serious errors can be introduced by assuming they behave ideally. Nonetheless, if we make this assumption, $a^G_{gross} = (X^G_{gross})^3$, and $a^{Pl}_{an} = X^{Pl}_{an}$, and then

$$\ln \frac{a^G_{gross}}{(a^{Pl}_{an})^3} = \ln \frac{(X^G_{gross})^3}{(X^{Pl}_{an})^3} = 3 \ln K_D \quad (17\text{-}18)$$

Various solution models for garnet and plagioclase have been used to refine this geobarometer [see Ghent et al. (1979), Ghent and Stout (1981), and Hodges and Crowley (1985)]. The amount of grossular component in the garnet of a typical pelitic rock is less than 5 mol %. Small errors in the analysis can therefore lead to large errors in the pressure determination. Also, errors in the temperature determination are translated into errors in pressure.

A more accurate geobarometer is based on the less common assemblage garnet + rutile + Al_2SiO_5 + ilmenite + quartz

(Bohlen et al. 1983), which involves the reaction

(almandine)   (rutile)
$$Fe_3Al_2Si_3O_{12} + 3TiO_2$$

(ilmenite)   (Al silicate)   (quartz)
$$= 3FeTiO_3 + Al_2SiO_5 + 2SiO_2 \quad (17\text{-}19)$$

This reaction is almost independent of temperature, proceeding to the left at pressures greater then 1 GPa (Fig. 17-13). In rocks, ilmenite forms a solid solution with hematite, and almandine forms a solid solution with other garnets, which allows the reaction to take place at lower pressures. In this case the equilibrium constant for the reaction can be expressed as

$$K = \frac{(a^{Ilm_{ss}}_{Ilm})^3 (a^{K,A,S}_{Al_2SiO_5})(a^Q_{SiO_2})^2}{(a^{G_{ss}}_{Alm})(a^{Ru}_{TiO_2})^3} \quad (17\text{-}20)$$

But rutile, quartz, and Al_2SiO_5 polymorphs form essentially pure phases, and thus they have unit activities. Ilmenite and hematite do not form an ideal solid solution series; indeed, they are separated by a large solvus at low temperatures. In pelitic rocks, however, ilmenite rarely contains more than 15 mol % hematite, so that ilmenite can be treated as an ideal solution without introducing serious error. Several solution models have been proposed for garnet. Bohlen et al. use a regular solution model (Sec. 9-6) with a temperature-dependent interchange energy. Using these models for the solid solution phases and the experimental data for the end-member reaction (Eq. 17-19), Bohlen et al. have calculated the position of lines of constant $\log_{10}K$ for this reaction on a P–T diagram (Fig. 17-13). Note that these lines have a very shallow slope, so that metamorphic temperatures need not be known accurately for this geobarometer to be used. Temperature uncertainties of $\pm 50°C$ result in

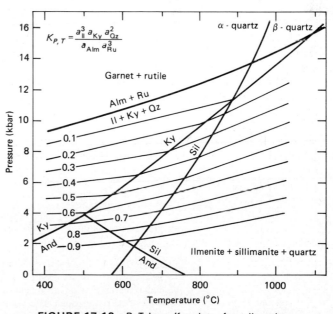

FIGURE 17-13 P–T–$\log_{10}K$ plot for ilmenite–Al_2SiO_5–quartz–almandine–rutile geobarometer. [From Bohlen et al. (1983); published with permission of American Mineralogist.]

maximum errors in inferred pressure of only 0.05 GPa. Note also that the lines of constant K have inflections where they cross the univariant lines for the transformations between the various polymorphs of Al_2SiO_5 and of quartz.

Geobarometry of pelitic rocks can be based purely on mineral assemblages. The simplest example involves the Al_2SiO_5 triple point. The location of this invariant point can be traced through a terrane by mapping the boundary between regions in which andalusite occurs and those in which kyanite transforms directly to sillimanite during progressive metamorphism. If the Al_2SiO_5 polymorphs are pure, this boundary corresponds to an isobar with a pressure of 0.375 GPa. Carmichael (1978) defines such a line as a bathograd; that is, a *bathograd* is a mapped line that separates occurrences of a higher-pressure assemblage from occurrences of a lower-pressure assemblage.

Carmichael uses the intersection of two dehydration reactions with the univariant lines marking the polymorphic transformations of Al_2SiO_5 to define four additional invariant assemblages that can be used to map bathograds. The dehydration reactions involve the disappearance of staurolite and muscovite from quartz-bearing pelites. The lower-temperature terminal reaction is

quartz + muscovite + staurolite

$$= biotite + garnet + Al_2SiO_5 + H_2O$$

Where this reaction crosses the andalusite–sillimanite transformation line (Fig. 17-14a), both polymorphs are stable,

creating an invariant assemblage. Mapping the occurrence of this assemblage, or what is more likely in practice, mapping the boundary between regions that have the higher-pressure assemblage from those with the lower-pressure assemblage delineates a bathograd whose pressure is approximately 0.33 GPa. Another invariant point exists at 0.48 GPa, where the same dehydration reaction crosses the kyanite–sillimanite transition line.

The higher-temperature terminal reaction is

quartz + muscovite = K-feldspar + Al_2SiO_5

At low pressures, Al_2SiO_5 forms the polymorph andalusite, whereas at higher pressures it forms sillimanite. The bathograd marking the boundary between these two zones corresponds to a pressure of 0.22 GPa. At higher pressures still, this reaction would intersect the kyanite–sillimanite transformation line, but the reaction is made metastable above a pressure of approximately 0.375 GPa by the beginning of melting of granite. Thus, above this pressure quartz + muscovite + Na-feldspar breaks down to form K-feldspar + Al_2SiO_5 + water-saturated granite melt. This reaction intersects the kyanite–sillimanite transition line at about 0.72 GPa (Fig. 17-14a).

Clearly, the position of all these points of intersection depend on the a_{H_2O}, which is assumed to be one in Figure 17-14a. The a_{H_2O} must be lowered considerably to cause a significant lowering of reaction temperatures. Because the activity of water is believed to be moderately high in all but the

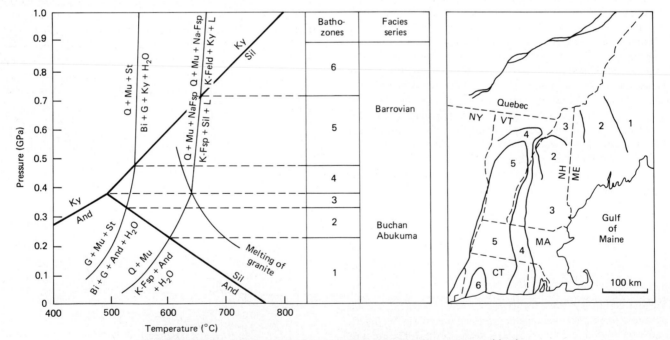

FIGURE 17-14 Intersection of two dehydration reaction curves with the Al_2SiO_5 polymorphic transformation lines and the Al_2SiO_5 triple point itself provide a convenient means of distinguishing six pressure (depth) zones which are referred to as bathozones. The boundaries between the bathozones, which are termed bathograds, are shown for New England. They show that the greatest uplift has occurred in western Connecticut and Massachusetts and in Vermont. [After Carmichael (1978); published with permission of the American Journal of Science.]

highest-grade metamorphic terranes, the diagram for $a_{H_2O} = 1$ should provide reasonably accurate pressures.

The four points of intersection of the dehydration reactions with the Al_2SiO_5 transformation lines and the Al_2SiO_5 triple point itself create five boundaries that can be used to map bathograds. The five boundaries divide $P–T$ space into six zones, which Carmichael refers to as *bathozones* (Fig. 17-14a). These zones can be correlated with Miyashiro's (1961) metamorphic facies series. For example, the Abukuma type corresponds to the boundary between bathozones 1 and 2, whereas the Barrovian type falls in the upper part of zone 5.

Figure 17-14b shows the distribution of the bathozones in the northern Appalachians of New England. Regions falling in the deepest bathozones must have undergone the greatest amount of uplift and erosion since the time of metamorphism. The distribution of the bathograds in New England resemble a northerly-plunging antiform extending through western Connecticut and Massachusetts and central Vermont, with the greatest uplift occurring in Connecticut. This region corresponds closely to a large gravity high and is approximately along the suture between North America and the Avalon Terrane.

A modification of the biotite-garnet geothermometer of Thompson (1976b) and Ferry and Spear (1978) has been developed by Spear and Selverstone (1983) so that it can be used as a geobarometer as well. It takes advantage of the fact that the zoning so commonly exhibited by garnet crystals preserves a record of a range of metamorphic temperatures and pressures. In the Ferry-Spear geothermometer, the value of K_D is used in Eq. 17-11 to solve for T. In the Spear-Selverstone method T is not solved for directly but only values of ΔT and ΔP are determined as functions of mineral compositions. If a specific temperature and pressure are known for one pair of compositions, the remainder can then be determined from the calculated values of ΔT and ΔP.

Spear and Selverstone's method involves expressing P and T as functions of n linearly independent compositional variables, where n is the variance of the system (i.e., P or $T = f(X_1, X_2 \ldots X_n)$). As temperatures and pressures change, so do the compositions of coexisting biotite and garnet. The total differentials of T and P can be expressed as

$$dT = \sum \left(\frac{\partial T}{\partial X_i}\right)_{X_{j \neq i}} dX_i \quad \text{and} \quad dP = \sum \left(\frac{\partial P}{\partial X_i}\right)_{X_{j \neq i}} dX_i$$

Values of $(\partial T/\partial X_i)_P$, $(\partial P/\partial X_i)_T$, and $(\partial P/\partial T)_X$ can be solved for using molar entropies and volumes of all phases in the assemblage and a model for mixing in the solid solutions.

Consider, for example, the common assemblage in pelitic rocks of garnet + biotite + Al_2SiO_5 + muscovite + quartz + H_2O (Fig. 17-15). This forms a divariant assemblage that is bounded at low temperatures by the discontinuous reaction

$$garnet + chlorite = biotite + Al_2SiO_5$$

and at high temperatures by the discontinuous reaction

$$muscovite + quartz = K\text{-feldspar} + Al_2SiO_5$$

as indicated in Figure 17-15. Within the divariant field, the compositions of both garnet and biotite vary with temperature and pressure. This would show in an *AFM* plot as shifts in the tie triangle garnet-biotite-Al_2SiO_5. Spear and Selverstone (1983) have calculated the compositions of coexisting garnet and biotite, assuming both to be ideal binary solutions; the composition of the garnet is expressed as mole fraction of almandine, X_{Alm}, and that of biotite as mole fraction of annite, X_{Ann}.

The isopleths (lines of constant composition) for coexisting garnet and biotite form a series of intersecting lines

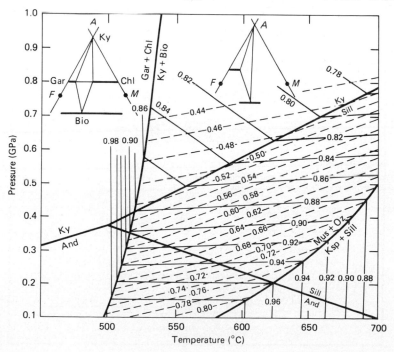

FIGURE 17-15 $P–T$ plot with isopleths giving the mole fractions of almandine in garnet (solid lines) and annite in coexisting biotite (dashed lines) in pelitic rocks. [Simplified from Spear and Selverstone (1983); published with permission of Springer-Verlag.]

that allow the pressure and temperature of equilibration to be read for any coexisting pair (Fig. 17-15). Note that with increasing pressure, both the garnet and biotite become progressively more magnesian. Because the isopleths for the two minerals intersect at high angles in the kyanite stability field, P–T conditions can be determined more accurately there than in the other fields, where the angle of intersection is small. Within the sillimanite and andalusite fields, the garnet isopleths have little slope to them, that is, they are functions almost entirely of pressure.

Graphs such as the one shown in Figure 17-15, and others that take into account the effects of Mn, Ca, Na, K on the garnet and biotite compositions (Spear and Selverstone 1983), are extremely useful in unraveling the pressure-temperature-time histories of rocks, and we shall return to them in Chapter 20. Although assumptions about the solution models for the minerals may cause the absolute temperatures and pressures to be inaccurate, the range of temperatures and pressures are only slightly affected. Use of the method does require that biotite grains can be found included in zoned garnet, and that diffusion has not modified their composition since inclusion. Careful electron microprobe traverses across grain boundaries can usually resolve the extent to which compositions have been modified. Experience shows that zoned garnet crystals do indeed preserve long histories of changing P–T conditions (see Sec. 20-5).

In closing, let us examine briefly two field areas where geothermomenters and geobarometers have been used extensively to give a regional picture of metamorphic conditions. Both areas are in New England. In south-central Maine, pelitic rocks of the Silurian Waterville formation underwent a Buchan-type metamorphism during the Devonian Acadian orogeny. Ferry (1980) has used eight different geothermometers/geobarometers to estimate the conditions during this metamorphic episode. To the west, in central Massachusetts, rocks ranging in age from late Precambrian to lower Devonian were also metamorphosed during the Acadian orogeny but under Barrovian-like conditions. Geothermometers and geobarometers have been used by Robinson et al. (1982) and Tracy et al. (1976) to unravel the complex history of this region.

Rocks of the Waterville formation in south-central Maine were deformed into a series of tight isoclinal folds. They appear to have undergone only one period of metamorphism. Many porphyroblasts crosscut the schistosity (Fig. 17-2), indicating that much of the metamorphism followed the deformation that caused the schistosity. Mapping by Osberg (1971) has shown that the metamorphic grade increases from chlorite zone in the north to sillimanite zone in the south. The following succession of isograds has been mapped: biotite, garnet, staurolite + andalusite, staurolite + cordierite, and sillimanite (Fig. 17-16). Intrusive bodies of synmetamorphic quartz monzonite occur in the highest-grade zones.

Ferry (1980) analyzed minerals in approximately 200 samples of pelitic schist from the area shown in Figure 17-16 in order to determine the pressures and temperatures during metamorphism. Above the garnet isograd, the ubiquity of the assemblage garnet + biotite makes this mineral pair the most convenient to use for geothermometry. The isotherms determined using these minerals are shown in Figure 17-16. Temperatures rise from about 430°C at the garnet isograd to 570°C

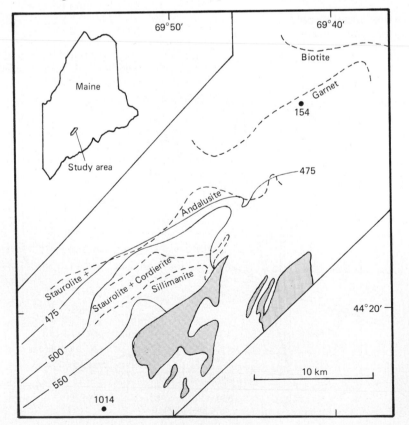

FIGURE 17-16 Map of south-central Maine showing isograds (dashed lines) and isotherms (solid lines, °C) determined using the garnet–biotite geothermometer. Stippled areas are synmetamorphic quartz monzonite bodies. Two numbered dots refer to samples discussed in Prob. 17-6. [Simplified from Ferry (1980); published with permission of American Mineralogist.]

in the southern part of the region. Pressures throughout the region are all very near 0.35 GPa. The range of $P–T$ conditions recorded in these rocks is shown in Figure 17-17 along with the Al_2SiO_5 phase diagram for reference. Ferry also calculates the composition of the fluid phase by determining the amount of paragonite in muscovite. The fugacity of H_2O can be related to the paragonite content of the muscovite through the reaction

(muscovite$_{SS}$)   (quartz)

$NaAl_3Si_3O_{10}(OH)_2 + SiO_2$

          (plagioclase$_{SS}$)   (and/or sill)   (vapor)

          $= NaAlSi_3O_8 + Al_2SiO_5 + H_2O$ (17-21)

Using thermodynamic data for the reaction at temperatures and pressures obtained from geothermometers and geobarometers, the fugacity of H_2O needed to produce the observed concentration of paragonite in muscovite is calculated. The values obtained indicate that the a_{H_2O} was between 0.5 and 1.

In central Massachusetts the geologic history is more complex than in south-central Maine. At about 400 Ma the rocks were folded into nappes which were overturned to the west. These were then backfolded (overturned to the east) and at about 375 Ma the rocks were again strongly deformed as gneiss domes buoyantly rose in the Bronson Hill anticlinorium (Fig. 17-18). Before and during the nappe stage, sheetlike intrusions of gabbro, tonalite, and granite were emplaced. Sillimanite pseudomorphs after andalusite indicate an early low-pressure metamorphism in the eastern part of the area. Peak metamorphic conditions were attained during the backfolding

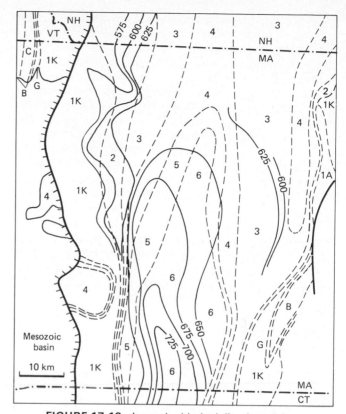

FIGURE 17-18 Isograds (dashed lines) and isotherms (solid lines, °C) in the Bronson Hill anticlinorium and Merrimack synclinorium of central Massachusetts [after Robinson et al. (1982)]. The metamorphic zones are chlorite (C), biotite (B), garnet (G), andalusite–staurolite (1A), kyanite–staurolite (1K), sillimanite–staurolite (2), sillimanite–muscovite (3), sillimanite–muscovite–K-feldspar (4), sillimanite–K-feldspar (5), and garnet–cordierite–sillimanite–K-feldspar (6).

stage with various retrograde reactions taking place during doming and unloading. The metamorphic isograds were deformed along with the primary rock units by the backfolding and doming.

The area is bounded to the west by the border fault of the Mesozoic Hartford and Deerfield basins. The lowest-grade rocks occur immediately east of this fault and directly north of the Deerfield basin. Their grade increases eastward toward the Bronson Hill anticlinorium through zones of chlorite, biotite, garnet, kyanite + staurolite, sillimanite + staurolite, sillimanite + muscovite, sillimanite + muscovite + K-feldspar, sillimanite + K-feldspar, and garnet + cordierite + sillimanite + K-feldspar. Farther to the east the grade decreases toward the Merrimack synclinorium through many of the same zones, but instead of a kyanite + staurolite zone, andalusite + staurolite is present.

The complex history of this region makes interpretation of geothermometers and geobarometers difficult. Fortunately, diffusion rates in garnets are sufficiently low that zoned crystals can preserve records of a range of metamorphic conditions. Zoned garnets fall into three main types. The first type,

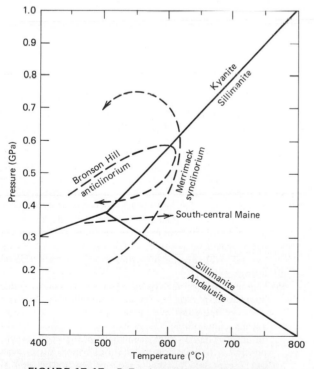

FIGURE 17-17 $P–T$ plot showing metamorphic conditions in south-central Maine and central Massachusetts. The Massachusetts rocks experienced higher pressures than those in Maine (see Fig. 17-14).

which occurs mostly in the lowest-grade zones, has spessartine–enriched cores with rims enriched in almandine–pyrope components. We have already seen in Sec. 12-4 and Figure 12-15 that such zoning is a natural consequence of Rayleigh fractionation of Mn into garnet as it grows during progressive metamorphism. The fact that zoning has not been eliminated by later metamorphism indicates how low diffusion rates in garnets are, especially in the lower metamorphic grades. The second type is almost homogeneous except for a narrow, continuous spessartine-enriched rim. These are common in the higher grades. If they originally had zoned cores, the temperature and duration of metamorphism was sufficient to homogenize the crystals. The spessartine–enriched rim is probably a product of retrograde reaction. Because the rim is not dependent on the proximity to minerals with which the garnet might have reequilibrated, the retrograde reaction probably involved a grain boundary fluid. The third type is similar to the second, except that the spessartine–enriched rims are discontinuous, occurring only where garnet contacts biotite or cordierite with which it has reacted, as evidenced by zoning in those minerals. This type is most common in the highest-grade rocks, where it probably indicates that the rocks lacked a grain boundary fluid phase that could have helped bring about retrograde reactions.

By paying careful attention to the type of zoning in garnet crystals, their compositions have been used in conjunction with that of other minerals to determine temperatures at various stages in the history of these rocks. Figure 17-18 shows isotherms representing the peak temperatures reached by the rocks. Temperatures rise from about 600°C in the kyanite + staurolite zone to 650 to 700°C in the highest-grade zones. Pressures increase over this range from about 0.5 to 0.7 GPa.

Rocks in different parts of the region reached their peak conditions via different paths. Rocks in the Merrimack synclinorium (eastern part) passed through the andalusite field (Fig. 17-17), whereas those in the Bronson Hill anticlinorium (central part) passed through the kyanite field (see also Sec. 20-5).

A striking feature of the isotherms in both the Maine and Massachusetts areas is that they do not deviate greatly from the pattern produced by the isograds. An isograd in these regions can therefore be thought of as a line of almost constant metamorphic temperature. This is found to be the case in many other metamorphic terranes. Thus, although rocks in different regions may be metamorphosed at different pressures, the different metamorphic zones reflect largely different temperatures. In the Waterville formation, for example, the pressure recorded in the rocks in passing from chlorite to sillimanite zones is a constant 0.35 GPa; yet the temperatures rise from 430°C to 570°C. Regional metamorphism therefore bears a striking resemblance to contact metamorphism (constant P, variable T), but on a larger scale.

Finally, a word of caution regarding interpretation of the record preserved by geothermometers/geobarometers and mineral assemblages throughout a metamorphic terrane. Although points recording identical temperatures can be joined by lines that we label "isotherms", nothing proves that all rocks on an isotherm attained that temperature simultaneously. Isotherms and isograds need not be isochronous. Nor need two rocks that fall on different isotherms, for example, one at 600°C and the other at 700°C, have attained those temperatures at the same time. In Chapter 20 we will see that time is an important variable, and that the records of temperature and pressure in rocks are likely to vary systematically in age across a region.

PROBLEMS

17-1. (a) Derive a formula for calculating the alumina index of compositions in the Al_2O_3–FeO–MgO–K_2O tetrahedron when projected onto the AFM plane from K-feldspar.

   (b) What is the alumina index of biotite in the AFM plane when projected from K-feldspar? Use the biotite formula given in Table 17-1.

17-2. A suite of metagabbros and metaperidotites contain the following mineral assemblages:

$$Fo + En + An + Gt$$
$$Fo + En + An \qquad + Di$$
$$Fo \qquad + An + Gt \qquad + Sp$$
$$Fo \qquad + An + \qquad + Di + Sp$$
$$Fo + En \qquad + Gt \qquad + Sp$$

where the abbreviations refer to

Forsterite	(Fo)	Mg_2SiO_4
Enstatite	(En)	$MgSiO_3$
Anorthite	(An)	$CaAl_2Si_2O_8$
Diopside	(Di)	$CaMgSi_2O_6$
Garnet	(Gt)	$CaMg_5Al_4Si_6O_{24}$
Spinel	(Sp)	$MgAl_2O_4$

The problem is to construct a triangular mineral facies diagram by projecting points in the CaO–MgO–$AlO_{3/2}$–SiO_2 tetrahedron onto the CaO–$AlO_{3/2}$–SiO_2 face from the phase common to all assemblages.

   (a) First, plot all minerals in the tetrahedron, and join coexisting phases with tie lines. Because olivine is common to all these assemblages, the composition of the other minerals can be projected onto the CaO–$AlO_{3/2}$–SiO_2 face from the composition of olivine, thus reducing the graphical representation of the assemblages to a triangle. Draw, as best you can, the approximate positions of the projected points of the minerals. Note that the projection of spinel falls outside the triangle.

   (b) Although the projected compositions of minerals can qualitatively be plotted graphically, it is necessary to have a means of doing this more accurately. We will now derive a general equation for the projection of any point within a tetrahedron onto a desired face from a given point. The composition of the point from which the projection is to be made can be expressed in terms of the four components A, B, C, and D as

projection point: $\qquad a_pA + b_pB + c_pC + d_pD$

where the lowercase letters are the mole fractions of each component. The composition of the point to be projected can similarly be defined as

$$\text{point } X: \quad a_X A + b_X B + c_X C + d_X D$$

If point X is to be projected onto the BCD face, all A component must be removed from composition X by subtracting an appropriate amount, n, of the projection composition; that is, $a_X A - na_p A = 0$, in which case $n = a_X/a_p$. Now derive the general formula for the composition of the projected point in terms of B, C, and D.

(c) Using the formula from part (b), plot all five mineral assemblages on the triangular projection, joining coexisting phases with tie lines. Care must be taken with the projection of spinel. The general formula is still valid; the problem is to interpret correctly the significance of a negative coefficient.

17-3. (a) Draw a Thompson AFM diagram showing the three-phase tie triangles and two-phase tie lines for the following assemblages ($+$muscovite $+$ quartz)

staurolite $+$ biotite $+$ almandine

staurolite $+$ biotite $+$ kyanite

biotite $+$ kyanite $+$ cordierite

(b) Write a qualitative exchange reaction (no need to balance) relating staurolite–biotite–kyanite ($+$muscovite $+$ quartz). Determine which way this reaction would proceed with increasing temperature by noting on which side H_2O is liberated.

(c) Draw a schematic isobaric T–X_{Mg} diagram showing variation in the composition of staurolite and biotite Fe–Mg solid solutions in the assemblage staurolite $+$ biotite $+$ kyanite as a function of rising temperature. Discuss the variation in composition and modal proportions of staurolite and biotite as a function of T using this diagram.

(d) Show how the compositions of staurolite and biotite coexisting with kyanite change with rising temperature by showing the shift of the staurolite–biotite–kyanite tri-angle with temperature.

17-4. The purpose of this problem is to see how several continuous reactions can form a discontinuous reaction that terminates the stability field of staurolite at high temperatures.

(a) Start by drawing two schematic T–X_{Mg} diagrams for the following continuous reactions:

$$\text{garnet} + \text{muscovite} = \text{biotite} + Al_2SiO_5 + H_2O$$

$$\text{staurolite} + \text{biotite} = \text{garnet} + \text{muscovite} + H_2O$$

Note that the second reaction requires three lines in the T–X_{Mg} diagram to show the compositions of the three co-existing solid solution phases.

(b) These two reactions intersect to create an isobaric invariant assemblage for a given a_{H_2O}. Draw the composition loops for the two reactions in the same T–X_{Mg} diagram, and then determine which parts of these loops must be metastable. Deduce what other two continuous reactions must intersect at this invariant point.

(c) Complete the T–X_{Mg} diagram by placing the other two reactions with appropriate relative slopes. Check that your construction does not violate Schreinemakers rules.

(d) Finally, select arbitrary temperatures above and below the invariant temperature, and show in AFM diagrams the compositions of the phases at these temperatures. What is the maximum temperature to which staurolite is stable in this isobaric diagram?

17-5. In Chapter 16 we saw that low-temperature metamorphism of siliceous dolomite can give rise to the following minerals: tremolite, diopside, forsterite, calcite, and quartz, which can be expressed in terms of the three components CaO–MgO–SiO_2 if the composition of the fluid is controlled by the environment. How many phases are needed to create an isobaric invariant assemblage in such a case? How many reactions can be written involving these minerals? Would you expect degenerate reactions to occur with these minerals?

17-6. The following table gives the mole fractions of Fe and Mg in coexisting garnet and biotite in pelitic rocks studied by Ferry (1980) in south-central Maine. Sample number 154 is from near the garnet isograd in one of the lowest-grade parts of the region, whereas number 1014 is from the highest-grade part of the region (Fig. 17-15). Geobarometers indicate that the pressure at the time of metamorphism was 0.35 GPa throughout the region. Using the Ferry and Spear (1978) calibration of the garnet–biotite geothermometer, determine the temperatures recorded by the minerals in these two rocks.

Sample Number	Location Isograd	Garnet		Biotite	
		X_{Fe}	X_{Mg}	X_{Fe}	X_{Mg}
154	Garnet	0.680	0.044	0.490	0.290
1014	Sill-Ksp	0.710	0.088	0.457	0.323

17-7. Recalculate the compositions of the garnet and biotite in sample 1014 from the previous problem in terms of mole fraction almandine and annite, assuming them to be simple binary Fe–Mg solutions. Using the isopleths in the P–T diagram of Spear and Selverstone (Fig. 17-15), determine the pressure and temperature of metamorphism of this rock. Is your answer consistent with the values obtained in the previous problem? Sample 1014 is from the sillimanite–K-feldspar isograd. In view of the position of the reaction defining this isograd in Figure 17-15, what can you conclude about the acitivity of H_2O during metamorphism in south-central Maine?

18 / Material Transport during Metamorphism

18-1 INTRODUCTION

In igneous petrology the question of the movement of magma is clearly of importance, because the site in which a magma solidifies is rarely the site in which the magma is generated, and between these two locations many important processes can occur that play an important role in determining the eventual composition of the igneous rock. Material transport, therefore, is obviously an important aspect of the magmatic process. By contrast, there is a tendency when first dealing with metamorphic rocks to think of them as rather inert bodies, which despite mineralogical changes, undergo little if any change in bulk composition apart from the loss of volatiles. Material transport does not, therefore, appear to be as important as in igneous processes. A moment's reflection, however, indicates that material transport must play significant roles in determining not only the bulk composition of metamorphic rocks but also the rates and mechanisms of metamorphic reactions, which in turn play important roles in determining the textures of metamorphic rocks.

Approximately 5 wt % H_2O must be driven from a shale to convert it to a high-grade metapelite. Conversely, several wt % H_2O must be added to an anhydrous basalt to convert it to a chlorite schist. Clearly, metamorphism in these cases must involve the transport of considerable quantities of H_2O. Similarly, CO_2 released by decarbonation reactions must also be transported away from the rock. At the temperatures and pressures normally extant during metamorphism, H_2O forms a supercritical fluid, which has a density similar to that of water at room temperature and pressure. This hot fluid, in equilibrium with a rock, is certainly not pure H_2O but, instead, contains considerable amounts of dissolved ions. The loss or addition of such a fluid phase during metamorphism can therefore change the bulk composition of a rock in ways other than simply modifying its H_2O content.

When a metamorphic rock undergoes significant changes in composition, the process of change is referred to as *metasomatism*. In extreme cases, metasomatism can completely change the composition of a rock, as, for example, in the conversion of limestone to magnetite ore bodies at the contact of igneous intrusions. Despite clear evidence for such major compositional changes, metasomatism was the subject of much controversy during the first half of this century. Unlike the analysis of metamorphic mineral assemblages, which is firmly based on thermodynamic principles, metasomatism, which was invoked to explain certain field relations (see Fig. 4-27b), provided very little that could be tested. Commonly, the rock that was supposed to have been replaced no longer existed, nor did the solutions that brought about the change. The petrologic community became polarized into those who believed that metasomatism was capable of effecting enormous compositional changes over large regions and those who thought its effects were minor. The debate culminated with the controversy over the origin of granite (Read 1957), with metasomatists arguing in favor of a metasomatic origin (*granitization*), and magmatists arguing for an igneous origin. The experimental work of Tuttle and Bowen (1958) decided the question in favor of the magmatists, and metasomatism fell out of favor. Indeed, a number of metamorphic petrology textbooks went so far as to ignore the term completely. More recently, however, metasomatism has experienced a comeback, but this time it is better routed in physical-chemical principles (Hofmann et al. 1974). Isotopic studies have also provided means of monitoring changes in rock composition. Metasomatic processes in the mantle are now thought to be important precursors to certain types of igneous activity, as discussed in Chapter 22.

Most metamorphic reactions were shown in Chapter 17 to result from rising temperature, regardless of the metamorphic pressure. In Chapter 5 it was shown that rocks are poor conductors of heat. Thus the rate of metamorphism will be increased by any process that increases the rate of heat transfer. As long as fluids travel more rapidly than heat diffuses, the flux of fluids can play an important role in heating or cooling bodies of rock.

Transport of material over short distances is necessary for metamorphic reactions to proceed. Inspection of the photomicrographs in Figure 17-2 reveals that considerable redistribution of material is required to transform the mineral assemblages of one metamorphic facies to those of another. The textures of metamorphic rocks on either side of an isograd commonly indicate that the steps involved in even the simplest reaction, such as a polymorphic transformation, are often complex and involve considerable movement of material. Several different transport mechanisms may be involved in a single reaction, such as transfer by a fluid phase, diffusion through solids, and diffusion along grain boundaries. Each mechanism has a certain rate constant, the slowest of which controls the overall rate of the metamorphic reaction.

In this chapter we examine the evidence for mass transfer during metamorphism and the mechanisms by which it can occur. Knowledge of the transport mechanisms is of importance to understanding the kinetics of metamorphism. If we are to interpret properly the record preserved in a metamorphic rock, it is necessary to know the rates of metamorphic reactions relative to other important geologic rates, such as those of plate motion and conductive heat transfer.

18-2 EVIDENCE FOR MASS TRANSFER DURING METAMORPHISM

Before examining the evidence for material transport during metamorphism, let us consider the processes involved in the lithification or diagenesis of sediment. Freshly deposited marine sediment can contain as much as 50% pore fluids. With burial and compaction, this is soon reduced to about 30%, at which stage the sediment would consist, in the simplest case, of close-packed spheres. The pore fluid in the compacted sediment remains a continuous phase connected through the intergranular channelways. The fluid is therefore under hydrostatic pressure. The sedimentary grains, however, in addition to experiencing the pressure from the surrounding fluid must support the weight of the overlying sedimentary grains. This creates stresses on the grains at their points of contact. As indicated in Sec. 12-5, stress increases the solubility of minerals, and consequently, the grains dissolve at the points of contact (Fig. 18-1). The pore solution, however, becomes saturated in the dissolving mineral, and further dissolution leads to redeposition on the parts of grains protruding into pore spaces. In this way the porosity of a sedimentary rock is reduced almost to zero at the base of a thick sedimentary sequence. For this reason, fluids involved in metamorphic reactions must come from minerals rather than from trapped pore fluids. In addition, once the porosity has been reduced almost to zero, fluid flow through the rock becomes difficult, and then the hydrostatic and lithostatic pressures become essentially the same.

The stress causing dissolution of grains can also be of tectonic origin. Such stresses are likely to be considerably greater than those due to load, and because they are ultimately the result of plate motions, they tend to be approximately horizontal. In most cases, the rock being deformed already has had its porosity reduced to a minimum, and therefore no large pores are present into which dissolved material can be transported. If tension fractures open, they provide sites for redeposition, but these are soon filled and become veins. Material dissolved by the stress soon saturates the extremely small amount of pore fluid remaining in the rock, thus preventing further dissolution. If stresses continue to build and the pore fluid cannot be changed, deformation of the rock must occur by some process other than dissolution; this could be by a process as slow as dislocation creep or as rapid as thrust faulting. However, if the intergranular fluid is flushed out of the rock with fluid that is undersaturated in the phase being dissolved, the stress on the rock is released by dissolution.

FIGURE 18-1 Grains of sand in this rock dissolved at points of contact as a result of pressure solution due to compaction alone. Original outlines of grains are marked by inclusions of clay particles that adhered to the grains. The dissolved silica reprecipitated as cement, which grows in optical continuity on the original grains (note that birefringence of white grain is the same as that of surrounding cement).

Fluids that could cause such flushing are likely to be common in orogenic belts as products of metamorphic dehydration reactions. Under some circumstances, these fluids may travel along grain boundaries, but as flow and dissolution continue, undulous interconnecting channels develop along which the flow is concentrated. Once established, the channels become the main sites for dissolution. In hand specimens they are marked by a buildup of insoluble residues. Closely spaced channels slowly migrate toward each other and become one as the intervening rock is dissolved away. This results in the development of *solution cleavage* (Geiser and Sansone 1981), which is a prominent feature of many carbonate rocks in orogenic belts. Figure 18-2 shows an example of such cleavage in a limestone composed largely of foraminifera. The tests or shells of this type of foraminifera have a flat elliptical cross section. Where the tests abut against the solution cleavage, which is marked by zones of dark insoluble residue, they have square truncated terminations as a result of dissolution. Solution cleavage may be responsible for limestones in orogenic belts decreasing their volume by as much as 40%.

Solution cleavage also develops in quartz-bearing rocks. Nowhere is its effect more noticeable than in the development of what is referred to as *crenulation schistosity*. The schistosity of many metapelites is cut by a later schistosity that deforms or crenulates the early schistosity into small regularly spaced folds (Fig. 18-3). Unlike the early schistosity, which involves a parallelism of almost every grain in the rock, the later crenulation schistosity develops on planes that are millimeters apart.

FIGURE 18-2 Foraminifera limestone with solution cleavage marked by seems of clay-rich insoluble residue. These forams normally have rounded terminations; their abrupt termination against the clay seems is clear evidence of dissolution. The width of field is 10 cm.

layers begin to get folded by the crenulation schistosity the layers of quartz decrease in thickness, until they completely vanish in the main foliation plane of the later schistosity. The crenulation folds are therefore produced by the shortening of the rock normal to the later schistosity as a result of pressure solution of quartz. The concentration of muscovite in the crenulation schistosity is formed simply as an insoluble residue.

Figure 18-4 illustrates how the geometry of crenulation schistosity develops as a result of dissolution of quartz from rock on either side of the newly developed schistosity (Prob. 18-1). As will be shown below, fluid capable of dissolving quartz is not likely to exist along all grain boundaries because of wetting problems, and as a result dissolution is most rapid where channels develop, and these become the new schistosity planes. The geometry of the crenulation schistosity does not indicate what happens to the dissolved quartz. It might, for example, be completely removed from the rock, but it could also be added to the already existing quartz layers, making them thicker. Such growth, however, commonly leads to the growth of "beards" on grains, but rarely can the volume loss due to dissolution be accounted for by local redeposition (Etheridge et al. 1983). Regardless of the ultimate fate of the dissolved quartz, the transport of material appears to have played an important role in the deformation of this rock, and possibly in determining its composition.

Further evidence of material transport during metamorphism is provided by the movement of material into the pull-aparts between boudins in high-grade metamorphic rocks (Fig. 18-5). The *boundinage structure* is formed when a competent layer embedded in less competent rock necks down as a result of being stretched parallel to its length. The flow of the less competent rock may actually segment the competent layer

The folding of the early schistosity by the crenulation schistosity was initially interpreted as resulting from a slip parallel to the later schistosity; for this reason it has also been referred to as *strain-slip cleavage*. Inspection of Figure 18-3, however, reveals that the crenulations are not the result of simple folding of the earlier schistosity. The early foliation is defined by layers that are alternately quartz- and muscovite-rich. Where these

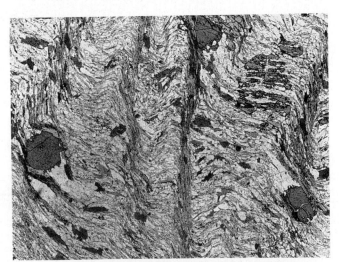

FIGURE 18-3 Deformation of an early schistosity (top left to lower right) by a later one gives rise to crenulation schistosity. The later schistosity is marked by a decrease in the abundance of quartz in the crenulation planes. Offset of the earlier schistosity by the crenulation schistosity is most likely the result of dissolution of quartz from within the crenulation zones rather than of shear. The width of field is 15 mm.

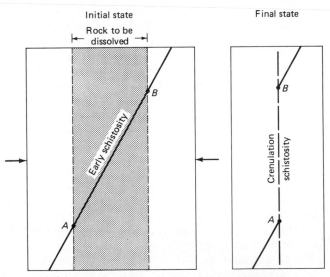

FIGURE 18-4 Geometry of crenulation schistosity developed by dissolution of rock composed of mica and quartz. Within the shaded area, all quartz is removed by solution, leaving mica as an insoluble residue on the newly formed crenulation schistosity. The early schistosity appears offset between points *A* and *B* on the crenulation schistosity, but the displacement of the rock has only been in a direction normal to the plane of the crenulation schistosity.

FIGURE 18 5 Coarse crystals of quartz and biotite in the necked-down regions between boudins formed by the migration of these minerals into zones of low pressure. Where quartz has grown, the adjoining rock is depleted in quartz, and thus it is darker; where biotite has formed, the adjoining rock is depleted in biotite and is lighter. See text for discussion.

into separate "boudins" (French for "sausage"). The pressure in the regions between the boudins is low relative to that elsewhere in the rock. If a fluid phase is present, the pressure differential can result in the dissolution of a mineral in regions of higher pressure with reprecipitation in regions of low pressure. If a fluid phase is not present, this same transfer of material can occur through diffusion but the process is very much slower, as will be shown in Sec. 18-3. Quartz is the most common mineral to find in the pull-aparts between boudins. In high-grade metamorphic rocks, feldspar, micas, amphiboles, and tourmaline may also migrate into these low-pressure zones. Quartz and biotite have grown between the boudins shown in Figure 18-5.

The change in composition of the competent layer as it approaches the pull-aparts between the boudins in Figure 18-5 is particularly interesting. The competent layer is composed essentially of quartz, feldspar, and biotite. Where the layer ap-

proaches the lower pull-apart it becomes darker because of a decreased quartz content. It seems reasonable to conclude that the quartz in the pull-apart must have come from this quartz-depleted part of the boudin. At the top end of this same boudin, the layer becomes lighter because of a decreased biotite content. At this pull-apart, biotite crystals formed instead of quartz. It appears that the depletion in biotite toward the upper end of this boudin was caused by the migration of biotite into the upper pull-apart. We are faced with the interesting situation of different minerals being involved in migration at opposite ends of the same boudin. Temperatures, pressures, and fluid compositions could not have differed significantly over such short distances; besides, similar mineralogical variations are found at other boudins along the same layer. The most likely explanation is that differential pressures developed solutions that were supersaturated in both quartz and biotite in the pull-aparts; were quartz nucleated first, the concentration gradient of quartz in the fluid steepened, and transfer of quartz occurred; where biotite nucleated, the concentration gradient in biotite was steepened, causing biotite to migrate. If such a conclusion is correct, it indicates that the factors causing migration may be rather subtle.

Other examples of the transport of material during metamorphism could be given, but these few suffice to show that material does actually move. It is now necessary to examine the various ways in which material can actually be transported.

18-3 MECHANISMS OF MASS TRANSFER

Material has two basic means of transport. An atom can behave passively, being moved only when its surroundings move, or it can move independently of its surroundings in response to a force developed by a potential gradient that affects that particular element. The first type of movement is referred to as *advection* or *infiltration* and the second as *diffusion*. Ions carried in solution by a fluid that migrates through a rock would be an example of advection, whereas diffusion of an ion down a concentration gradient in a solution would be an example of the second type. Both forms of transport play important roles in metamorphism and both may operate on an atom simultaneously. Their rates, however, are very different, and thus they are capable of acting over very different distances in the time available during metamorphism. We will start by examining advection.

Under most circumstances the fastest means of transporting material through a rock is as a component of a fluid that is migrating along grain boundaries or fractures. The transport rate is determined by the size of the channels, the viscosity of the fluid, and the pressure differential causing flow. The equations governing such flow are similar to those derived in Chapter 3 for the flow of magma through conduits. The differences are that the fluids have much lower viscosities, and the conduits are very much smaller.

The viscosities of supercritical H_2O and CO_2 fluids between 400 and 600°C at moderate pressures fall between 0.1 and 0.2 mPa s (Walther and Orville 1982). These viscosities are not particularly sensitive to changes in temperature and pressure. Viscosities may also vary with dissolved solids, but these variations are also likely to be small compared with other factors affecting flow.

As indicated in Sec. 18-2, lithification of sedimentary rocks eliminates most porosity, and any that does remain is essentially eliminated by recrystallization in the lowest-grade metamorphic rocks. All rocks, however, do retain a small but finite porosity simply because crystal structures of adjoining grains cannot fit together perfectly. The degree of mismatch along grain boundaries depends on the disparity in the structures and orientations of juxtaposed grains. If lattice planes in adjoining crystals match, the grain boundary is said to be *coherent*; if only some lattice planes match, it is semicoherent; and if none match, it is *incoherent*. Figure 18-6 illustrates possible grain boundaries in a monomineralic rock composed of crystals having a structure based on close-packed spheres. The *intergranular region* consists in part of open spaces, which could be filled with a fluid phase, and zones of disordered structure (Brady 1983). The width of these zones can range from about 0.1 to 100 nm (1 nanometer = 10^{-9} meter). A monomolecular layer of H_2O or CO_2 might be adsorbed onto the walls of the grain boundary channels, and this fluid would not be free to move advectively. Such layers would be approximately 0.5 nm wide. Consequently, advective flow would occur only in channels that are more than 1 nm wide. When the grain bound-

ary width is less than this, movement would be by surface diffusion.

Equation 3-14 can be used to estimate the average flow velocity of fluid rising in a channel of width W.

$$\bar{v} = -\frac{1}{12\eta}\left(\rho_f g + \frac{dP}{dz}\right)W^2$$

Let us assume that the fluid has a density of 1 Mg m^{-3}, a viscosity of 1 mPa s, and the channel has a width of 100 nm. To calculate the average velocity, we must know the pressure gradient. Walther and Orville (1982) assumed that a schist would have a vanishingly small tensile strength, and thus the pressure gradient would simply be that resulting from the buoyancy of the fluid phase. This provides a lower limit for the pressure gradient. If the rock does have a tensile strength and devolatilization reactions are taking place, pressure gradients might be as high as 10 MPa m^{-1} (Etheridge et al. 1984). We will assume, for the moment, that buoyancy is the only force acting on the fluid, in which case the terms in parentheses become $(\rho_f - \rho_r)$, where ρ_r is the density of the rock, which we can take to be 2.8 Mg m^{-3}. This produces a pressure gradient of only 0.02 MPa m^{-1}. The calculated velocity is then 0.5 m yr^{-1}, which is two orders of magnitude greater than the

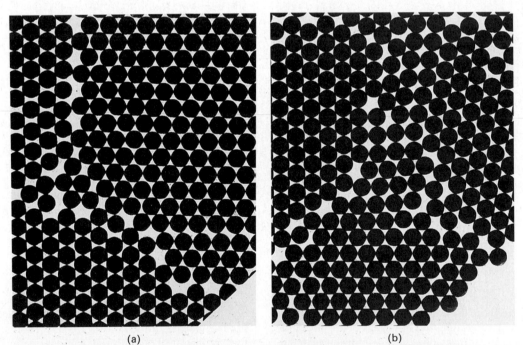

(a)   (b)

FIGURE 18-6 Grain boundaries are regions of misfit between the regular crystal structures of adjoining grains. In this two-dimensional model, crystal structures are represented by regions of close-packed spheres. The amount of open space in intergranular regions depends on the degree of misfit across boundaries. The horizontal boundary in (a) is a dislocation in the crystal occupying the right half of the field. The boundary between this crystal and the one to the left is *semicoherent* because only every second row of atoms can be traced across the boundary; such a boundary is more open than a *coherent* one in which each row can be traced across the boundary. Boundaries in (b) are all *incoherent* and are marked by irregular, open intergranular regions. The more incoherent the boundary, the more space there is for intergranular fluids.

rate at which regional metamorphic isograds are likely to advance through metapelites (see Sec. 18-8). If the channel width is reduced to 1 nm, the average flow velocity becomes 46 μm yr^{-1}, that is, slightly more than the thickness of a petrographic thin section. This velocity is two orders of magnitude less than the rate at which regional metamorphism is likely to advance. If flow occurs along schistosity planes or prominent fractures, flow rates would be very much greater (Prob. 18-3).

If the rate of advance of metamorphic isograds is faster than the rate at which fluid can flow out of the rock, devolatilization reactions would cause the pressure on the fluid to rise above that of the lithostatic pressure. How much excess pressure could be generated in this way would depend on the tensile strength of the rock. It will be recalled from Sec. 3-3 that partial fusion of rocks can also generate excess pressures due to the volume expansion on melting. It was pointed out that rocks were able to withstand no more than 0.03 GPa of excess pressure. This pressure, however would be sufficient to increase greatly the flow rate of a fluid. Etheridge et al. (1984) estimate that pressure gradients of as much as 10 MPa m^{-1} can be generated in this way. This would cause fluid to flow through a 100-nm-wide channel at 263 m yr^{-1} and through a 1-nm-wide channel at 2.6 cm yr^{-1}. If fluid were not able to escape from the rock as fast as it was produced, the resulting increase in pressure would force open grain boundaries or cause hydrofracturing of the rock. The increased width of the channels would then easily accommodate the increased flux of fluids and thus lower the pressure on the fluid phase. In this way a metamorphic rock would behave as its own pressure release valve, which would prevent fluid pressure ever exceeding lithostatic pressure by more than the tensile strength of the rock (about 0.03 GPa).

In Figure 18-6 it will be noted that channels between grains become more open where they meet other grain boundaries at triple junctions. Thus, although flow might occur between pairs of grains, most should occur in the channels between groups of three grains. Flow in these channels is better described by the Hagen–Poiseuille law for pipe flow (Eq. 3-9). But such an equation is not a convenient means of describing the flow of fluid through a rock because of the tortuous paths the channels must follow (Fig. 18-7). Instead, we use *Darcy's law*, which states that the flux of fluid in a given direction through a permeable medium, is proportional to the negative pressure gradient on the fluid in that direction; that is,

$$J_X = -\frac{\mathbf{K}}{\eta}\frac{dP}{dx} \qquad (18\text{-}1)$$

where the constant of proportionality, **K**, is the *permeability*, which has units of square meters, and η is the viscosity. Comparison of Darcy's law with the Hagen–Poiseuille law for flow through a pipe (Eq. 3-9),

$$J_X = -\frac{\pi r^4}{8\eta}\frac{dP}{dx}$$

reveals that the permeability is a measure of the average size of the through-going channels. The permeability of unfractured metamorphic and igneous rocks increases with decreasing grain size (higher percentage of channels) but is about 10^{-18} m^2.

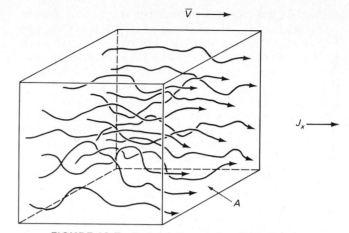

FIGURE 18-7 Paths followed by fluid flowing through rock are tortuous and complex. It is more convenient, therefore, to refer to the flux of fluid in a given direction (x); flux is the volume of fluid passing through a reference area A in a unit of time. Dividing flux by porosity, which is the fraction of the area through which fluid travels, gives the average velocity.

The average flow velocity through a rock can be calculated from Darcy's law. The flux of fluid, J_X, is the volume of fluid passing a reference area, A, in a unit time, that is, V/At (Fig. 18-7). This flux represents an average over the area, but in detail the fluid migrates through channels that constitute only a small fraction of the area A. This fraction is the *porosity*, which is designated by the letter ϕ. If the porosity were 100%, the fluid would cross the reference area at every point, and the average velocity, $\bar{v}$, would be given by the flux (m^3/m^2 s = m/s). But the channels constitute only the fraction ϕ of the reference area, so the average flow must be proportionately greater in these channels to account for the flux, that is,

$$\bar{v} = \frac{J_X}{\phi} = -\frac{\mathbf{K}}{\phi\eta}\frac{dP}{dx} \qquad (18\text{-}2)$$

The porosity of unfractured metamorphic rocks is less than 1%. If we assume a porosity of 0.1% and a permeability of 10^{-18} m^2, and we take the pressure gradient to be the minimum due only to buoyancy, the average flow velocity is 0.5 m yr^{-1}, which is of the same order of magnitude as the calculated flow rate along a 100-nm-wide channel. Again, if the pressure gradient were greater than that due to buoyancy, proportionately higher flow rates would result. In any case, the calculation demonstrates that fluids are able to move advectively at significant rates through narrow channels.

In the examples above, the fluid has been assumed to migrate freely between grains or along grain edge channels. It will be recalled from Sec. 12-6, however, that whether a fluid flows between grains or not depends on the wetting property of the fluid, which can be expressed in terms of the surface free energy between the fluid and the solid (γ_{FS}) relative to the surface free energy between the solids themselves (γ_{SS}). The dihedral angle formed by the fluid in contact with two grains (Fig. 12-23) depends on the relative values of the two surface free energies (Eq. 12-32). Only when the dihedral angle is less than 60° is the fluid able to penetrate along the entire length of channels between grain edges. When the dihedral angle is greater than 60°, the channels close off and the fluid is isolated in pockets at four grain junctions. This criterion applies when the fluid constitutes less than 1% of the volume of the rock. When there is 2% fluid, the critical angle becomes 65°. Dihedral angles are not a factor when dealing with low-temperature groundwater flow, because the solids are unable to change their shapes. Under metamorphic conditions, however, mineral grains in contact with a fluid would be expected to adjust their shapes so as to produce equilibrium dihedral angles by dissolving and reprecipitating the solids.

Experiments reveal that the size of the dihedral angle produced with supercritical H_2O–CO_2 fluids depends on the composition of the fluid and the minerals involved but in almost all cases is greater than 60°. The dihedral angle formed with quartz varies from 57 to 90°, depending on whether the fluid is H_2O- or CO_2-rich, respectively. The dihedral angle for olivine ranges from 65° for H_2O-rich fluid to 90° for CO_2-rich fluid (Watson and Brenan 1987). Preliminary results for clinopyroxene, orthopyroxene, and feldspar indicate dihedral angles that are considerably larger than 60°. The dihedral angle formed with calcite is larger still, being about 140° for H_2O fluids and about 150° for CO_2 fluids (Hay and Evans, 1988). With the exception of the case of pure H_2O fluid in contact with quartz, these angles are all greater than 60°, and thus fluids in most metamorphic rocks cannot exist as thin continuous films or channels between grains but, instead, must occur as isolated pockets at four-grain and three-grain junctions (Fig. 18-8). It seems likely, therefore, at least for H_2O-rich compositions, that fluids will penetrate channels along grain edges on quartz grains.

We can conclude that although advective flow can be a rapid means of moving material through channels, the per-

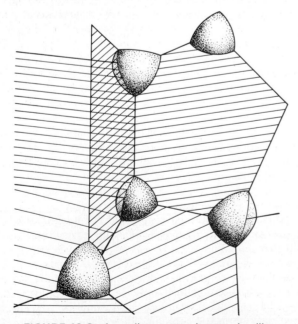

FIGURE 18-8   According to experiments, the dihedral angle between minerals and most fluids in metamorphic rocks is greater than 60°. Fluid therefore occurs as isolated beads at junctions between four grains and is not free to flow through the rock.

meability of most metamorphic rocks is essentially zero because of the failure of fluids to wet mineral grains. During devolatilization reactions, however, the evolution of volatiles causes a buildup in fluid pressure, which forces apart grains by *hydrofracturing* (Etheridge et al. 1984). Both the increased pressure gradient and the dilated channels result in rapid advective movement. This, however, releases the pressure that opens the channels to flow. Therefore, once reaction is complete and the fluids been expelled, grain boundary channels heal themselves, and any remaining fluid is trapped as isolated pockets. The permeability of the rock once again becomes essentially zero. Fluid flow in a metamorphic rock is therefore likely to be a pulsating phenomenon driven by devolatilization reactions.

The large dihedral angles measured in experiments for fluids in contact with metamorphic minerals presents a paradox when considered in light of the extremely low porosity of most metamorphic rocks. Although much of the fluid produced by devolatilization reactions may be able to escape advectively during periods of fluid overpressure, residual pockets of fluid would be expected to occur at four-grain junctions, especially when the fluids are CO_2-rich and/or the grains are carbonates. Considering the large volumes of fluid that are evolved during metamorphism, it is surprising that the rocks are not vesicular. Some minerals, such as quartz, do, however, contain small (< 5 μm) fluid inclusions; these are typically much smaller than those in many igneous rocks. There is little doubt that most of the metamorphic fluid phase must be effectively purged from the rocks by some process.

It is possible that minor constituents may be present in natural fluids that cause H_2O–CO_2 mixtures to wet metamorphic minerals (dihedral angle $< 30°$). Very small amounts of impurities can dramatically affect the surface properties of phases. A small drop of detergent, for example, is all that is necessary to make a whole basin of water wet greasy dishes. Watson and Brenan (1987) found that addition of NaCl to water decreases the dihedral angle with quartz from 57° to 40°. The general lack of fluid inclusions in metamorphic rocks is strong evidence that some wetting agent is present in natural fluids. If no such wetting agent exists, it is necessary to call on surface diffusion to purge the rocks, but as will be shown below, this is a slow process.

Etheridge et al. (1983) have suggested that the intergranular fluid might thermally convect during metamorphism. Large-scale circulation would have a significant effect on the thermal evolution of a metamorphic terrane. Convection would flatten the temperature gradient, and mineral assemblages reflecting a wide range of pressures at relatively constant temperatures would result. The transfer of near-surface waters to depth within convection cells would not only cause a lowering of temperatures but would bring about retrograde hydration reactions.

In Sec. 13-4 we saw that a dimensionless number, the Rayleigh number (Eq. 13-10), could be used to indicate whether a cooling sheet of magma would undergo convection. The Rayleigh number was defined as

$$Ra = \frac{\rho g \, d^3 \alpha (T_1 - T_2)}{\eta k}$$

where ρ is the density of the magma, g the acceleration of gravity, d the depth or thickness of the sheet, α the coefficient of thermal expansion, T_1 and T_2 the temperatures at the bottom and top of the sheet, respectively, η the viscosity of the magma, and k the thermal diffusivity of the magma. This equation can be modified so that it is applicable to convection in a porous medium by replacing the d^3 term with the product of the permeability and depth ($\mathbf{K} \times d$), which has units of cubic meters. The Rayleigh number for convection in a porous medium therefore becomes

$$\text{Ra} = \frac{\rho g \mathbf{K} \, d\alpha (T_1 - T_2)}{\eta k} \tag{18-3}$$

The critical Rayleigh number above which convection takes place in this situation is 40 (Prob. 18-4).

We can use Eq. 18-3 to determine what permeability a metamorphic terrane would require for the intergranular fluid to convect. If the fluid has a density of 800 kg m^{-3}, a viscosity of 1 mPa s, a coefficient of thermal expansion of 10^{-3} K^{-1}, and a thermal diffusivity of 6×10^{-7} m^2 s^{-1}, and the temperature rises 700 K between the surface of the Earth and a depth of 25 km, the permeability would have to exceed 1.7×10^{-16} m^2 for convection to occur. This permeability is at least two orders of magnitude greater than is normally found for metamorphic rocks (10^{-18} m^2). During periods in which fluid pressures are raised by devolatilization reactions permeabilities probably do exceed this critical value. Fluid overpressures, however, are likely to be short lived and would not aid convection significantly. Etheridge et al. (1983) suggest that material carried in solution by ascending fluids would precipitate and create a low-permeability cap over a metamorphic terrane. This they believe would maintain the pressure necessary to keep the permeability above the critical value for convection in the deeper parts of the metamorphic terrane. However, if devolatilization reactions produce sufficient pressure to dilate grain boundaries at depth, it seems likely that these same pressures could fracture the cap rock.

Equation 18-3 is strictly applicable only to fluids that are initially under hydrostatic conditions. This clearly would not be the case if devolatilization reactions were taking place. The equation also does not consider the compressibility of the fluid nor its change in composition with depth. In Sec. 13-4 we saw that convection in magmas is more dependent on composition than it is on temperature because of the small coefficient of thermal expansion of magma (10^{-5} K^{-1}). Metamorphic fluids have larger coefficients of expansion (10^{-3} K^{-1}), and their compositional variations are smaller than those of magmas. Nonetheless, density differences resulting from variations in the content of dissolved solids would exist. In Sec. 18-4 we will see that the solubilities of most minerals (carbonates excepted) increase with increasing pressure and temperature. The densities of metamorphic fluids would therefore be expected to increase with increasing depth in a metamorphic terrane. The compositional effect would thus tend to counteract thermal convection. England and Thompson (1984) believe that the absence of large-scale retrograde metamorphism, and the lack of evidence for low geothermal gradients that would be produced by circulating fluids is evidence that convection is not common in metamorphic terranes.

Clearly, more must be known about the properties of rocks and fluids under metamorphic conditions before this question can be resolved. On balance, however, it appears likely that advective movement of metamorphic fluids is unidirectional toward the surface of the Earth, and that the movement is driven by fluid pressures that periodically exceed the lithostatic pressure as a result of devolatilization reactions.

We now deal with the other mechanism for the transfer of matter, that of *diffusion.* Even though a material may be at rest, its atoms are in thermal motion. Although this motion is random, in the presence of a concentration gradient, it results in a net transfer of atoms down the gradient. This motion is known as diffusion.

In a gas, thermal motion causes atoms to travel in a zigzag path as a result of collisions with other atoms. The distance traveled between collisions, which is known as the *mean free path* (λ), is inversely proportional to the diameter (d) of the atoms and to their numbers (n); that is,

$$\lambda = \frac{1}{\pi \, d^2 n \sqrt{2}} \tag{18-4}$$

Because the value of n decreases with increasing temperature, the mean free path increases with increasing temperature. Increased pressure, on the other hand, increases the number of atoms in a given volume, and the mean free path decreases. At atmospheric pressure and room temperature, the mean free path of an atom in gas is about 1 μm.

The distance traveled by atoms in solids is much less than that in gases and is normally restricted to jumps from one site in the structure to another; the individual jumps are therefore on the order of 0.3 nm. Crystals are not perfect, regular structures. Most contain large numbers of defects, which produce regions where the atoms are not as closely packed or where vacancies may actually exist in the structure. Most diffusion involves these defects. For example, if a crystal structure contains vacancies, diffusion occurs when an atom or ion jumps from an adjoining site into the vacancy. At the same time, the vacancy moves to the site from which the atom jumped. This mechanism can therefore be considered as the diffusion of vacancies through the structure. Because grain boundaries have many vacancies and other defects, diffusion along grain boundaries is more rapid than diffusion through crystals (Brady 1983). Some diffusion may not involve defects. For example, diffusion might simply involve the exchange in position of two atoms. This would obviously be energetically more difficult than the vacancy transfer mechanism because the atoms would have to squeeze past each other.

Because thermal vibrations are random, the mean free path of an atom in a gas or the jump direction in a crystal can be in any direction. But when a concentration gradient exists, statistically more jumps take place down the gradient than up, resulting in a net transfer of atoms down the gradient. Consider a zoned crystal with a compositional gradient in the x direction for some diffusing species of atom of dc/dx, where c is measured in number of atoms per unit volume (Fig. 18-9). Visualize the crystal being divided into thin slices that are Δx thick, where Δx is the length of individual jumps of the diffusing atoms. On either side of the shaded slice (Fig. 18-9), we have concentrations of c_1 and c_2, where $c_1 > c_2$. If the jump frequency is f, then in a short time interval dt, the number of jumps at the high-concentration end will be $c_1 A \, \Delta x \, f \, dt$, where A is the

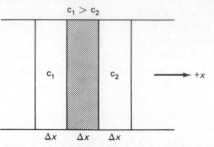

$c_1 > c_2$

c_1 c_2 $\longrightarrow +x$

Δx Δx Δx

FIGURE 18-9 A substance through which there is a compositional gradient can be thought of as consisting of small slabs, Δx thick, each with its own composition. Random jumps of atoms from one slab into the next are just as likely to occur to the left as they are to the right. If the concentration of a diffusing atom in a slab is C_1 and that in another slab is C_2, where $C_1 > C_2$, more jumps will occur from the high-concentration end into an intervening slab than will occur from the low-concentration end. This results in a net transfer of material down the concentration gradient as a result of diffusion.

area of a slice measured normal to x. Of these jumps, however, as many are likely to be to the right as to the left, and therefore the number of jumps down the gradient will be $\frac{1}{2}(c_1 A \Delta x f \, dt)$. Jumps may occur in directions other than along the x axis, but these need not concern us because only jumps in the x direction (+ or −) can contribute to diffusion in that direction. The number of jumps that take place up gradient from the low concentration end will be $\frac{1}{2}(c_2 A \Delta x f \, dt)$. The net flux of atoms passing the unit area in the short time interval in the $+x$ di-

rection is

$$J_x = \tfrac{1}{2}(c_1 \Delta x f) - \tfrac{1}{2}(c_2 \Delta x f) = \tfrac{1}{2}\Delta x f (c_1 - c_2) \quad (18\text{-}5)$$

Because Δx is very small, $c_1 - c_2 = -\Delta x (dc/dx)$. Thus

$$J_x = -\frac{1}{2}\Delta x^2 f\left(\frac{dc}{dx}\right) \quad (18\text{-}6)$$

The group of terms $(\tfrac{1}{2}\Delta x^2 f)$ is defined as the *diffusion coefficient*, D, which has units of $m^2\,s^{-1}$. Equation 18-6 can then be expressed as

$$J_x = -D\frac{dc}{dx} \quad (18\text{-}7)$$

This is known as *Fick's first law*. Because chemical potential is related to concentration, the diffusion flux can also be expressed in terms of the chemical potential gradient; that is,

$$J_x = -D\frac{d\mu}{dx} \quad (18\text{-}8)$$

in which case the diffusion coefficient has units of $mol^2\,(m\,s\,J)^{-1}$.

Fick's first law is identical in form to other transport laws (see Sec. 3-6). Indeed, the derivations and solutions for heat flow equations (Chapter 5) can be used for problems involving diffusion simply by replacing the temperature gradient with the concentration gradient and the thermal diffusivity with the diffusion coefficient. For example, to determine the rate of change

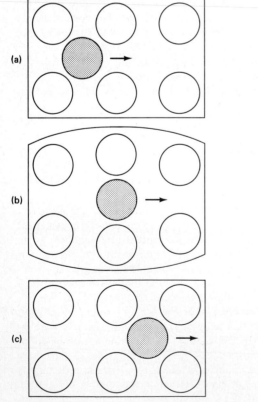

(a)

(b)

(c)

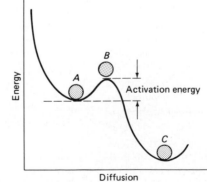

Energy

B

A

Activation energy

C

Diffusion

FIGURE 18-10 Diffusion of an atom through a structure is an activated process; that is, when an atom moves from one position to another, it passes through an intermediated stage where the energy of the system is momentarily increased. In this example the atom forces apart the structure of the crystal. The magnitude of the activation energy depends on the actual mechanism by which the atoms move. The number of atoms possessing sufficient energy to make the move is increased by raising temperatures.

of composition as a result of diffusion, we can follow a parallel derivation to that used for Fourier's equation (Eq. 5-11) for the change of temperature with time. The resulting equation for diffusion down a gradient in the x direction is

$$\frac{dc}{dt} = D\left(\frac{d^2c}{dx^2}\right) \tag{18-9}$$

which is known as *Fick's second law*. The solution to any diffusion problem must satisfy this equation.

In comparing the transport of heat and matter, it is interesting to note that values of thermal diffusivity in most geological materials is about 10^{-6} m^2 s^{-1}, whereas the diffusion coefficient in most minerals at metamorphic temperatures is less than 10^{-20} m^2 s^{-1} (Fig. 18-11). Even diffusion along grain boundaries, which may be several orders of magnitude greater than that through crystals, is still very much slower than the diffusion of heat.

The jump of an atom from one position to another in a crystal structure is resisted by the surrounding atoms; this is illustrated in Figure 18-10. An atom must therefore have extra energy in order to make the jump, and only atoms with this energy are able to diffuse. Diffusion is therefore referred to as an *activated process*; that is, the atoms must possess an activation energy before they can jump. Like other activated processes, diffusion obeys an Arrhenius-type relation;

$$D = D_0 \exp\left(-\frac{E_D}{RT}\right) \tag{18-10}$$

where D_0 is a constant, E_D an activation energy, R the gas constant, and T the absolute temperature. The preexponential term, D_0, can be thought of as the diffusion coefficient at infinitely high temperature, and the exponential term gives the fraction of the attempted jumps that have sufficient energy to complete the jump. A plot of the log of the diffusion coefficient versus $1/T$ gives a straight line from whose slope we can obtain the activation energy (Fig. 18-11). An extensive list of values of D_0 and E_D for silicate minerals and glasses is given by Freer (1981).

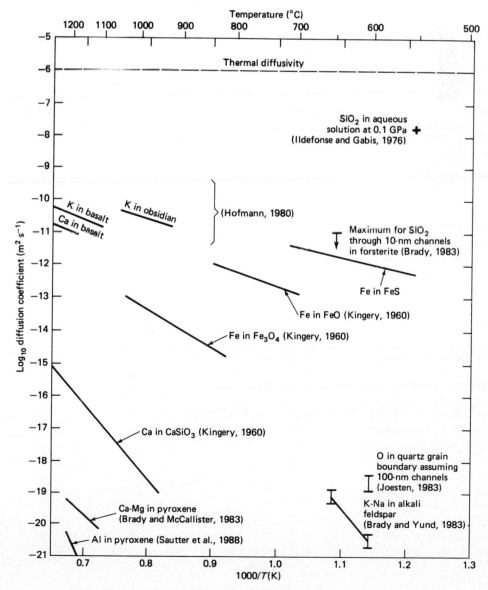

FIGURE 18-11 Plot of $\log_{10}$ of the diffusion coefficients for a variety of atoms in some minerals, solutions, and magmas versus $1000/T$(K). For comparison, the typical value of thermal diffusivity in rocks and magmas is included.

When an atom moves from one site to another there is a momentary increase in volume (Fig. 18-10). We might suspect therefore that increased pressure would decrease diffusion rates. If the momentary increase in volume is defined as an activation volume (ΔV_a), the effect of pressure on the diffusion coefficient at constant temperature can be expressed as

$$D = D_p \exp\left(-\frac{P \,\Delta V_a}{RT}\right) \qquad (18\text{-}11)$$

where D_p is a constant. The effect of pressure on diffusion is very small compared with that of temperature. During periods of deformation, however, directed stresses may induce defects in crystals, which in turn increase diffusion rates.

Fick's second law, as expressed in Eq. 18-9, is of limited use because it applies only when the diffusion coefficient is constant. But in most systems D is compositionally dependent, in which case Eq. 18-9 becomes

$$\frac{dc}{dt} = \frac{d}{dx}\left[D\left(\frac{dc}{dx}\right)\right] \qquad (18\text{-}12)$$

One instance where the diffusion coefficient is unaffected by composition is the case of *self-diffusion*, that is, the diffusion of a substance through itself. This is normally measured by introducing and tracking a radioactive isotope of the element in question. Self-diffusion is important to the migration of grain boundaries and the establishment of equilibrium textures in metamorphic rocks (Sec. 12-5).

In many crystals diffusion of one ion must be balanced by a counter diffusion of another ion in order to maintain electrical neutrality. The diffusion of these ions cannot, therefore, be treated as independent process. In such cases, an *interdiffusion coefficient* is defined, which is the effective diffusion for the coupled constituents. Brady and Yund (1983), for example, define a K–Na interdiffusion coefficient for alkali feldspars. The flux of K in an alkali feldspar is then given by

$$J_K = -D_{\text{K}-\text{Na}}\,\frac{dc_K}{dx}$$

where the interdiffusion coefficient is approximately 10^{-21} $\text{m}^2\,\text{s}^{-1}$ at 600°C (Fig. 18-11).

When diffusion occurs in multicomponent systems, interdiffusion coefficients can be written for each pair of constituents. These are referred to as cross coefficients and are represented by D_{ij}, which refers to the diffusion of component i through a gradient of j. If the gradient is expressed in terms of chemical potential, the coefficient is represented by L_{ij}. These terms are then incorporated into Onsager's extension of Fick's equation by summing the effects of all the components (Fisher 1973); this gives

$$J_i = \sum_{j=1}^{n} L_{ij}\left(-\frac{du_i}{dx}\right) \qquad (18\text{-}13)$$

In many metamorphic rocks diffusion is believed to have occurred predominantly along grain boundaries or in the intergranular region. This is based on the knowledge that intergranular regions are zones of misfit between adjoining grains, where abundant defects, vacancies, and impurities all serve to increase diffusion rates. Moreover, reaction rims on minerals in metamorphic rocks commonly completely surround grains, indicating that diffusion is probably more rapid along boundaries than it is into grains (lattice diffusion). Of course, in a rock, the effective diffusion coefficient reflects the sum of both lattice and intergranular diffusion. Although intergranular diffusion may be very much faster than lattice diffusion, the intergranular regions may constitute such a small fraction of the volume of a rock that the effective diffusion coefficient may be little higher than the lattice diffusion coefficient.

For purposes of evaluating a rock's effective diffusion coefficient, the rock can be thought of as consisting of two phases: mineral grains through which there is lattice diffusion (D^L), and intergranular regions where diffusion occurs along boundaries (D^B). The effective diffusion coefficient, D^{LB}, depends on the volume fractions of the two phases (V_L and V_B) and the tortuosity and orientation of the channels with respect to the diffusion direction. Brady (1983) has shown that with the exception of diffusion normal to schistosity, various grain geometries give effective diffusion coefficients that are only slightly less than the maximum possible value, which is obtained when diffusion parallels the schistosity and $D^{LB} = V_L D^L + V_B D^B$. The main factor determining the effective diffusion coefficient is therefore the volume fraction of the intergranular region, which is determined by the grain size of the rock.

Brady (1983) obtains a maximum limit for the effective diffusion coefficient by assuming the intergranular regions to be occupied by aqueous fluid in which the diffusion coefficient for SiO_2 is 2.4×10^{-8} $\text{m}^2\,\text{s}^{-1}$ at 550°C and 0.1 GPa (Ildefonse and Gabis 1976). If the intergranular regions have widths of 10 nm and the average grain size is 0.1 mm [$V_B = 10$ nm/ (10 nm + 0.1 mm)], and lattice diffusion is insignificant, then $D^{LB} = V_B D^B = 2.4 \times 10^{-12}$ $\text{m}^2\,\text{s}^{-1}$. This maximum value is, however, six to seven orders of magnitude greater than that determined for oxygen diffusion through alkali feldspar in hydrothermal experiments. The maximum value is probably too high because of the nature of the intergranular regions. If an aqueous phase is present on the grain boundaries, instead of being a continuous sheet it may resemble more a braided stream, with points of contact between adjoining grains being equivalent to bars in the stream. This would reduce still more the volume fraction of the intergranular region. Of course, the experiments on the wetting of metamorphic minerals by aqueous fluids (Watson and Brenan, 1987) suggest that fluids may not penetrate grain boundaries, in which case surface diffusion may be considerably slower. Joesten (1983), for example, calculated the intergranular diffusion coefficient of oxygen through polycrystalline quartz from the coarsening of chert nodules in the contact metamorphic aureole of the Christmas Mountains (Fig. 16-2); his data indicate that if the intergranular zones are 10 nm wide, the diffusion coefficient would be 2×10^{-19} $\text{m}^2\,\text{s}^{-1}$ at 600°C (Fig. 18-11). We can conclude that although surface diffusion may be orders of magnitude more rapid than lattice diffusion, it is still a slow process.

We will now examine a few simple diffusion problems in order to show that the solutions used in Chapter 5 for heat flow are equally applicable to material diffusion. First, let us consider two grains of alkali feldspar having compositions Or_1

and Or_2 ($Or_1 > Or_2$) that are in contact along a planar grain boundary. The temperature of the grains is raised to T K when the rock in which they occur becomes a xenolith in a diabase dike; this causes the feldspars to diffuse into one another. We will assume that the feldspar grains are sufficiently large that despite diffusion near their junction, feldspars with compositions of Or_1 and Or_2 remain at some distance from the boundary between the two grains. These boundary conditions are illustrated in Figure 18-12.

This problem is similar to that of magma cooling in the vicinity of a plane contact, the solution for which is given in Eq. 5-13. In that equation, the temperature scale was adjusted so that the country rock temperature was zero; that is, T/T_0 is actually $T - T_c/T_0 - T_c$. We can write an equivalent equation for diffusion as follows:

$$\frac{Or - Or_2}{Or_1 - Or_2} = \frac{1}{2} + \frac{1}{2}\,\text{erf}\left(\frac{x}{2\sqrt{D_{K-Na}t}}\right) \quad (18\text{-}14)$$

where Or is the orthoclase content of the alkali feldspar at a distance x from the boundary between the grains, with positive values of x being measured in the feldspar with the initial composition Or_1; D_{K-Na} is the interdiffusion coefficient, which is assumed to be independent of composition in the range dealt with; and t is time. This equation shows that the rate of advance of a diffusion front for a given composition is proportional to the square root of time (review the discussion of Eq. 5-13). Problem 18-6 provides the data necessary to quantify the rate of homogenization of the feldspar.

In the problem above it was specified that feldspars with the initial compositions Or_1 and Or_2 must continue to exist at some distance from the grain boundary for the solution of Eq. 18-14 to be valid. In practice, however, this condition cannot exist indefinitely. Consider, for example, an exsolution lamella with a composition of Or_1 in a host phase of composi-

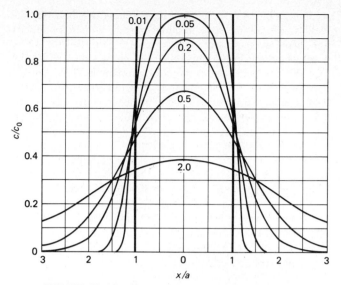

FIGURE 18-13 Changes in the composition of an exsolution lamella as a result of diffusion into a host crystal. The lamella has a half-width of a. Time is indicated on the gradients by the dimensionless term Dt/a^2. The gradients satisfy Eq. 18-15. See text for discussion.

tion Or_2 (Fig. 18-13). The distance x is measured from the center of the lamella, which has a half-width of a. We will assume that the lamella is far enough from other lamellae that the host is able to remain at a composition of Or_2 at some distance from the lamella. This problem is similar to that of the cooling of a dike (Eq. 5-20). We can write the equivalent equation for diffusion of matter as

$$\frac{Or - Or_2}{Or_1 - Or_2} = \frac{1}{2}\left[\text{erf}\left(\frac{a - x}{2\sqrt{Dt}}\right) + \text{erf}\left(\frac{a + x}{2\sqrt{Dt}}\right)\right] \quad (18\text{-}15)$$

The compositional gradients across the lamella are shown in Figure 18-13 for different times represented by the dimensionless term Dt/a^2.

Problems 18-6 and 18-7 should make it clear that diffusion cannot be responsible for large-scale transport of material during metamorphism. Advection is the only mechanism capable of transporting material significant distances. Diffusion, of course, must play a role at the level of individual grains involved in metamorphic reactions, but here distances of typically less than 1 mm are involved. We conclude, therefore, that infiltration metasomatism rather than diffusion metasomatism is the principal mechanism of transferring matter in metamorphic rocks.

18-4 DISSOLUTION OF MINERALS IN SUPERCRITICAL H_2O

Because the pore space in metamorphic rocks is so small, the fluid phase formed during dehydration and decarbonation reactions must be largely lost. The amounts of other material lost from the rock during any such reaction depends on the

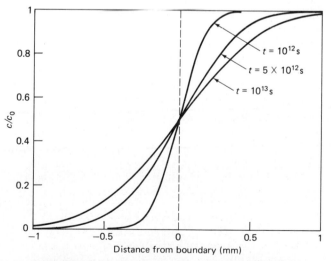

FIGURE 18-12 Compositional gradients developed across a planar boundary for three different times (in seconds) as a result of diffusion of a constituent with a diffusion coefficient of 10^{-20} m^2 s^{-1}. These gradients satisfy Eq. 18-14. See text for discussion.

solubilities of the various minerals in the fugitive fluids. In this section we examine the solubilities of some common minerals and the effects of pressure and temperature on their solubility. A detailed discussion of this topic is given by Fyfe et al. (1978).

Quartz is the most common vein mineral at most metamorphic grades. We can conclude, therefore, that quartz is the most soluble rock-forming mineral in most metamorphic fluids, a conclusion supported by experimental studies. Although quartz is relatively insoluble in surface waters, its solubility in supercritical H_2O increases significantly with both increasing temperature and pressure (Fig. 18-14). For example, at the temperature and pressure of the Al_2SiO_5 triple point (500°C and 0.375 GPa), H_2O can contain up to 0.6 wt % SiO_2; this increases to 2.7 wt % at 800°C at 0.375 GPa and to 5.5 wt % at 800°C and 0.8 GPa. At about 1 GPa and 1200°C, the system quartz–H_2O has a second critical endpoint where the solution contains about 10 wt % SiO_2. Because of the effect of temperature and pressure on the solubility of quartz, silica-saturated hydrous fluids rising toward the Earth's surface invariably become supersaturated, which eventually results in precipitation of quartz. This undoubtedly accounts for the abundance of quartz veins in low-grade metamorphic terranes.

Other common rock-forming minerals are not as soluble as quartz in H_2O under most metamorphic conditions. Although H_2O is able to dissolve 0.25 wt % SiO_2 at 500°C and 0.1 GPa, it can dissolve only 0.08 wt % albite or microcline, and 0.06 wt % enstatite. Like quartz, the solubilities of these minerals increase with temperature and pressure. At pressures in excess of 0.5 GPa and at temperatures above 600°C, the solubility of alkali feldspar may equal or exceed that of quartz. This could explain why quartz veins in upper amphibolite facies rocks may also contain feldspar. Such veins, however, could result from melting under these conditions (see Sec. 22-6 and Fig. 22-17). Vidale (1974), for example, found that veins in

pelitic rocks in Dutchess County, New York, and adjoining Connecticut exhibit a systematic variation in composition with metamorphic grade. Quartz veins are found in all metamorphic grades; quartz + albite veins, however, occur only in a narrow zone just below the biotite isograd; quartz + plagioclase (An_{20-50}) veins occur above the staurolite isograd; and quartz + plagioclase + orthoclase veins occur only above the sillimanite + K-feldspar isograd. Quartz + calcite veins also occur in rocks up to the staurolite isograd. These vein compositions probably reflect changes in mineral solubilities with temperature, but they also are controlled by the stability of minerals in the host rock. At low temperatures and high pressures albite is more soluble than quartz, with the result that albite is the most common vein mineral in blueschists.

The solution of most complex silicates takes place incongruently. For example, solutions in equilibrium with alkali feldspar are slightly enriched in alkalis relative to the feldspar composition. Dissolution of feldspars therefore leads to a residue enriched in aluminous minerals (muscovite, Al_2SiO_5, etc.).

Next to quartz, calcite is the most common vein-forming mineral. Its solubility, however, decreases with increasing temperature, unlike that of quartz. Calcite's solubility does increase with increasing pressure, but this effect is small compared with that of temperature. At the moderate to high pressures and temperatures experienced during metamorphism, calcite and dolomite are relatively insoluble in H_2O fluids. As these fluids migrate toward the surface, their ability to dissolve carbonates continuously increases with falling temperature. Thus fluids released by dehydration reactions at depth could well have the ability to generate solution cleavage in overlying limestones, as long as the temperature of the fluid continues to drop.

When a solution is in equilibrium with a mineral assemblage, its composition is fixed by the chemical potentials of the components in the solid phases (Helgeson 1967). Consider, for example, a low-temperature metamorphic rock consisting of the four minerals albite, orthoclase, muscovite, and quartz. These minerals can be related through three different exchange reactions with a hydrous fluid; these are as follows:

$$\text{albite} + K^+ = \text{orthoclase} + Na^+ \tag{18-16}$$

$$3\,\text{orthoclase} + 2H^+ = \text{muscovite} + 6\,\text{quartz} + 2K^+ \tag{18-17}$$

$$3\,\text{albite} + K^+ + 2H^+ = \text{muscovite} + 6\,\text{quartz} + 3Na^+ \tag{18-18}$$

where the ions are in the hydrous solution. Thermodynamic data indicate that the equilibrium constants for these three reactions are 10, 5×10^7, and 10^{11}, respectively, at 300°C. These equilibrium constants allow us to determine the activities of K^+, Na^+, and H^+ in equilibrium with the mineral assemblage. For Eq. 18-16 the equilibrium constant is written as

$$K = \frac{[a\text{Or}][a\text{Na}^+]}{[a\text{Ab}][a\text{K}^+]} = 10 \tag{18-19}$$

At these low temperatures the amount of solid solution in the two feldspars is minimal, so their activities can be taken to be unity. Equation 18-19 therefore reduces to

$$[a\text{Na}^+] = 10[a\text{K}^+] \tag{18-20}$$

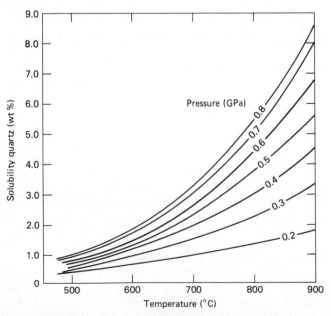

FIGURE 18-14 Solubility of quartz in supercritical H_2O as a function of pressure and temperature. [After Anderson and Burnham (1965).]

Because this exchange reaction does not involve H^+ ions, it does not define the hydrogen ion activity. Indeed, according to this reaction, coexisting albite and orthoclase could form at any hydrogen ion concentration. Reaction 18-17, however, does involve H^+ ions. The equilibrium constant for this reaction is written as

$$K = \frac{[a\text{Musc}][a\text{Qtz}]^6[a\text{K}^+]^2}{[a\text{Or}]^3[a\text{H}^+]^2} = 5 \times 10^7 \qquad (18\text{-}21)$$

Assuming unit activities for the minerals, we obtain

$$[a\text{K}^+] = 7071[a\text{H}^+] \qquad (18\text{-}22)$$

Finally, the equilibrium constant for Eq. 18-18 is

$$K = \frac{[a\text{Musc}][a\text{Qtz}]^6[a\text{Na}^+]^3}{[a\text{Ab}]^3[a\text{K}^+][a\text{H}^+]^2} = 10^{11} \qquad (18\text{-}23)$$

from which it follows that

$$\left(\frac{a\text{Na}^+}{a\text{H}^+}\right)^3 = 10^{11}\left(\frac{a\text{K}^+}{a\text{H}^+}\right) \qquad (18\text{-}24)$$

These equilibria are best represented on an activity diagram where the $\log_{10}(a\text{Na}^+/a\text{H}^+)$ is plotted against the $\log_{10}(a\text{K}^+/a\text{H}^+)$ in the fluid (Fig. 18-15). According to Eq. 18-22, the

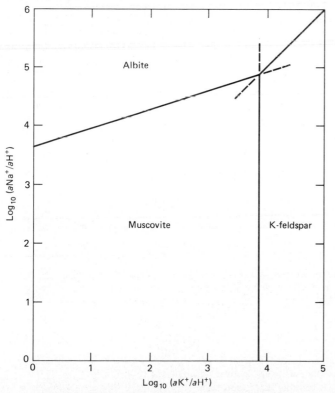

FIGURE 18-15 Plot of the stability fields of albite, K-feldspar, and muscovite in terms of activities of Na^+, K^+, and H^+ ions in coexisting fluid.

coexistence of muscovite and orthoclase occurs when the ratio of $[a\text{K}^+]/[a\text{H}^+] = 7071$; this ratio plots as a vertical line at $\log_{10}(a\text{K}^+/a\text{H}^+) = 3.8$. The line marking the coexistence of orthoclase and albite has a positive slope with $\log_{10}(a\text{Na}^+/\text{H}^+) = 1 + \log_{10}(a\text{K}^+/a\text{H}^+)$. The muscovite–orthoclase line becomes metastable on the albite side of the albite–orthoclase reaction, and the albite–orthoclase reaction becomes metastable on the muscovite side of the muscovite–orthoclase reaction. Finally, the third reaction (Eq. 18-18) passes through this same point of intersection, with a slope given by Eq. 18-24.

The intersection of these three reactions on the activity diagram is an isothermal, isobaric invariant point that defines the ratios of ions in the fluid in equilibrium with orthoclase, muscovite, and albite (+ quartz). In this fluid $\log_{10}(a\text{Na}^+/a\text{H}^+) = 4.9$ and $\log(a\text{K}^+/a\text{H}^+) = 3.8$ at 300°C. Fluids emanating from a rock containing these minerals would have to have these ions in these ratios at this temperature, that is, the composition of the fluid is buffered by the rock.

Although Eq. 18-17 represents the bulk equilibrium between the solution, feldspar, and muscovite, the (001) surfaces of muscovite establish their own equilibrium with the solution according to the reaction

$$\text{KAl}_3\text{Si}_3\text{O}_{10}(\text{OH})_2 + \text{H}^+ \rightleftharpoons$$
$$\text{HAl}_3\text{Si}_3\text{O}_{10}(\text{OH})_2 + \text{K}^+ \qquad (18\text{-}25)$$

Winch (1975) has suggested that during deformation, the exposure of new (001) surfaces on muscovite drives this reaction to the right, thus increasing the $a\text{K}^+/a\text{H}^+$ ratio in the solution. This, in turn, drives reaction 18-17 to the left with the formation of feldspar (Fig. 18-15). The growth of feldspar porphyroblasts in strongly deformed micaceous rocks may result from this surface reaction on muscovite.

18-5 MASS TRANSFER MECHANISMS AT A METAMORPHIC ISOGRAD

In previous chapters metamorphic rocks have been discussed in terms of the development of mineral assemblages having the minimum free energy for a given set of conditions. Thermodynamics, however, does not provide information on how such assemblages are achieved. For this we must have kinetic information on the various paths by which a reaction can take place. Often, reaction paths are extremely circuitous; this is simply because more direct paths are kinetically slower.

The amount of material transport that accompanies a reaction depends on the reaction mechanism. Balanced reactions can be written by equating mineral assemblages above and below an isograd, but this does not tell us how the reaction actually took place. To determine this, we must interpret the textural changes in the rocks across the isograd.

Consider, for example, the reaction that defines the sillimanite isograd in a Barrovian metamorphic series. Below the isograd, rocks of appropriate composition contain kyanite, whereas those above contain sillimanite. The reaction is, therefore, the simple transformation between Al_2SiO_5 polymorphs, and one might expect that sillimanite would simply make its appearance by replacing kyanite. This, however, rarely happens. In Barrow's type locality, for example, Chinner (1961) has shown that sillimanite appears first in biotite crystals, not in kyanite. He concludes that sillimanite nucleated more rapidly in biotite than in kyanite, and thus the reaction proceeded by transferring Al_2SiO_5 from the kyanite to nearby biotite grains. Exactly how the Al_2SiO_5 moved is uncertain, but it might have involved advection or diffusion through an aqueous pore fluid.

The first appearance of sillimanite in biotite is a common phenomenon and has been investigated by other workers. Carmichael (1969) has proposed that aluminum is relatively immobile during metamorphic reactions, and if this is the case, the sequence of textural changes across an isograd can be used to deduce the reaction mechanism. He does not believe that the reaction at the sillimanite isograd involves the transfer of aluminum. Instead, he proposes that three related reactions occur, which keep aluminum fixed while other components are transferred through the pore fluid.

The three reactions each take place in separate locations, which are determined by the positions of the minerals that contain the immobile aluminum. Thus at kyanite grains the following reaction takes place:

$$3 \text{ kyanite} + 3 \text{ quartz} + 2K^+ + 3H_2O \rightleftharpoons$$
$$2 \text{ muscovite} + 2H^+ \quad (18\text{-}26)$$

and at biotite grains the reaction is

$$\text{biotite} + Na^+ + 6H^+ \rightleftharpoons$$
$$\text{albite} + K^+ + 3(Mg, Fe)^{2+} + 4H_2O \quad (18\text{-}27)$$

and where muscovite and albite grains are close, the reaction is

$$2 \text{ muscovite} + \text{albite} + 3(Mg, Fe)^{2+} + H_2O \rightleftharpoons \text{biotite}$$
$$+ 3 \text{ sillimanite} + 3 \text{ quartz} + K^+ + Na^+ + 4H^+ \quad (18\text{-}28)$$

The sum of these three reactions is simply *kyanite = sillimanite*, the polymorphic transformation. By having the transformation proceed via the three reactions the textural changes in the rock can be explained (Fig. 18-16). Kyanite is typically rimmed by muscovite at the sillimanite isograd, and biotite is embayed by plagioclase. Finally, sillimanite and biotite are produced together as reaction products at the expense of muscovite and plagioclase. Note that in this interpretation sillimanite and biotite are both products of reaction, whereas in Chinner's interpretation sillimanite is the only product, which grows epitaxially on preexisting biotite.

Carmichael's interpretation hinges on the assumption that aluminum is essentially immobile during metamorphism. This is a reasonable assumption. Direct measurements of the diffusion coefficient indicate that at metamorphic temperatures aluminum is unlikely to move far (Fig. 18-11). The rocks themselves provide textural evidence of aluminum's immobility. Metapelites commonly preserve fine-scale compositional layering of aluminum despite the growth of aluminous porphyroblasts whose dimensions are considerably larger than the thickness of the layers. Variation in the abundance of aluminum in the rock results in fluctuations in the abundance of inclusions in the porphyroblasts (Fig. 18-17).

Of course, some migration of aluminum must occur in order to convert fine-grained shale into coarse-grained metapelite. The actual distance aluminum has moved in a rock can be judged from how small a volume can be selected that still has the average aluminum content of the rock. A garnet crystal, for example, will likely have a higher concentration of aluminum than will the rock as a whole. Aluminum must therefore have migrated to the garnet crystal as it grew, and by so doing depleted the surroundings in aluminum. Around the garnet we therefore typically find aluminum-poor minerals, such as quartz. By including the aluminum-poor grains surrounding aluminum-rich grains, a minimum volume can be determined in which the bulk aluminum composition is identical to that of the rock as a whole. This, then, is a measure of the dis-

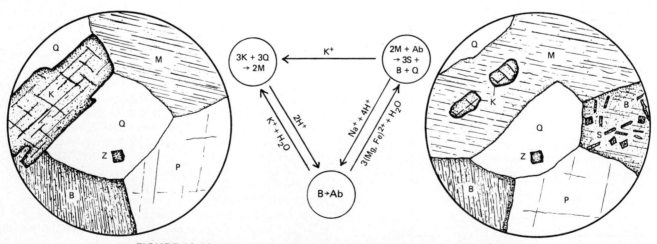

FIGURE 18-16 Textural changes across the sillimanite isograd. Kyanite (K) associated with quartz (Q), muscovite (M), biotite (B), plagioclase (P), and zircon (Z) below the isograd is transformed into sillimanite (S) at the isograd by a complex ion-exchange reaction. See text for discussion. [After Carmichael (1969).]

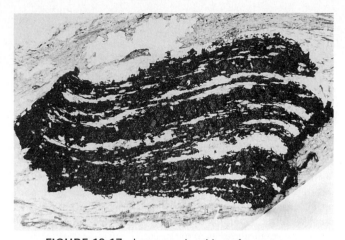

FIGURE 18-17 Large porphyroblast of garnet containing many layers of quartz, which are preserved from the original compositional layering of the rock. The alumina, rather than diffusing to form a single homogeneous garnet, remained essentially in situ, with garnet forming in those layers that were initially rich in aluminum. Such textures are evidence of the relative immobility of aluminum. The width of field is 15 mm.

which is shown graphically in the *AFM* plot of Figure 18-18, can be written as

$$9 \text{ staurolite} + \text{muscovite} + 5 \text{ quartz} \rightleftharpoons$$
$$17 \text{ sillimanite} + 2 \text{ garnet} + \text{biotite} + 9H_2O \quad (18\text{-}29)$$

Although sillimanite is produced by this reaction, it is never found in contact with the reactant staurolite; instead, it is intergrown with quartz and biotite. This indicates that the actual reaction mechanism is more complex than the simple net reaction. A number of textures in rocks at this isograd provide evidence of what that mechanism might be. Again, ion-exchange reactions are written assuming that aluminum is immobile relative to other components involved in the reaction.

At the isograd, staurolite and quartz are replaced by albite and biotite. Commonly, small isolated grains of staurolite in plagioclase or biotite have common crystallographic orientation, indicating that originally they were part of the same large staurolite porphyroblast. This texture suggests that the following reaction might have occurred:

$$9 \text{ staurolite} + 90 \text{ quartz} + 12K^+ + 24Na^+ + 27(Mg, Fe)^{2+}$$
$$+ 48H_2O \rightleftharpoons 24 \text{ albite} + 12 \text{ biotite} + 90 \text{ } H^+ \quad (18\text{-}30)$$

Also, wherever sillimanite forms an intergrowth with biotite and quartz, plagioclase is not present. This suggests the following reaction:

$$43 \text{ albite} + 7K^+ + 18(Mg, Fe)^{2+} + 7H_2O \rightleftharpoons 17 \text{ sillimanite}$$
$$+ 6 \text{ biotite} + \text{muscovite} + 91 \text{ quartz} + 43 \text{ } Na^+ \quad (18\text{-}31)$$

Furthermore, garnet crystals have salients where they contact plagioclase or biotite and reentrants where they contact quartz.

tance over which aluminum was able to migrate. Carmichael estimates that this distance is no more than 0.2 mm in lower-amphibolite-grade rocks, which is an order of magnitude less than the distance other components are able to migrate.

Let us apply this principle of writing reactions in which aluminum remains constant to a reaction that is more complex than the simple polymorphic transformation considered above. Carmichael has studied the textural changes across the isograd terminating staurolite in quartz-bearing pelites. This reaction,

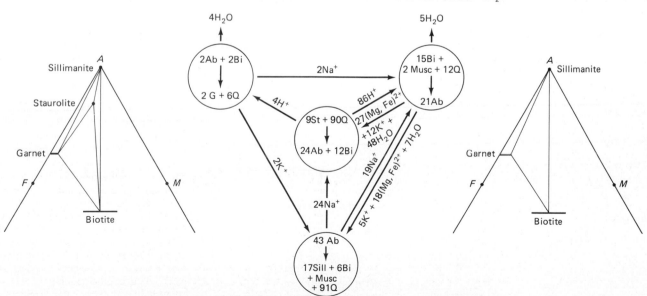

FIGURE 18-18 Possible ion-exchange reactions associated with the terminal reaction for staurolite. See text for discussion. [After Carmichael (1969).]

This suggests the reaction

albite + biotite + $2H^+$ ⇌

$\quad\quad\quad$ garnet + 3 quartz + $Na^+ + K^+ + 2H_2O$ (18-32)

The final reaction necessary to complete the net reaction is

15 biotite + 2 muscovite + 12 quartz + $21Na^+ + 86H^+$ ⇌

$\quad$ 21 albite + $17K^+ + 45(Mg, Fe)^{2+} + 60H_2O$ (18-33)

The sum of the four reactions (18-30 to 18-33) is the net reaction (18-29) describing the termination of the stability field of stuarolite. This complex set of ion-exchange reactions, which is illustrated in the flow diagram of Figure 18-18, is only one of many possible paths that could be followed, but this one does provide an explanation for the textural changes across this isograd.

$\quad\quad$ Two important conclusions can be drawn from this discussion. First, regardless of the simplicity of a metamorphic reaction, the actual mechanism by which a reaction proceeds may be complex and involve circuitous reaction paths that are quite different from the net reaction. The subreactions forming these paths are favored because they provide the fastest route by which the net reaction can proceed. Second, these kinetically favored reaction paths typically involve considerable exchange of material, presumably either by diffusion or advection through an intergranular fluid. Indeed, the cation exchange is so important to the overall progress of the reaction that were a fluid phase not available to transfer the cations or increase diffusion rates, the reaction might not take place or at least it would be forced to proceed by a more sluggish mechanism. As will be seen in Chapter 20, once the fluid phase has been expelled from a metamorphic rock, reaction rates may become negligible. This, in part, is responsible for the preservation of metamorphic mineral assemblages.

18-6 METASOMATIC ZONATION

In the reactions considered in Sec. 18-5, considerable exchange of material was necessary to account for the reaction mechanisms, but no change in the bulk composition of the rock took place, except perhaps for the loss of volatiles. Many reactions, however, involve a change in rock composition, and these are referred to as *metasomatic*. The change may involve large volumes of rock, such as that of a contact metasomatic ore body of magnetite, or it may involve small volumes of rock, such as millimeter-thick zones separating chert nodules or pelitic layers from enclosing limestone. The scale of metasomatic processes therefore varies considerably.

$\quad\quad$ A rock undergoes a change of composition in order to eliminate chemical potential gradients (Eq. 18-8). These gradients may result from compositional differences inherited from the protolith, or they may be induced by changes in the composition of the fluid phase. Chemical potential differences also produce reactions between minerals, with the result that many metasomatic rocks are characterized by series of sharply defined reaction layers, which are characterized by rocks with relatively small numbers of minerals. The metasomatic transfer of material and the presence of reaction layers are clear evidence that metasomatic rocks as a whole do not represent equilibrium products. Equilibrium thermodynamics, however, can be used to interpret the mineral assemblages in these rocks as long as sufficiently small volumes of rock are considered. The idea of *local equilibrium* is central to the interpretation of metasomatic mineral assemblages proposed by Korzhinskii (1970) and Thompson (1959).

$\quad\quad$ Let us consider the metasomatic layers that develop around a chert nodule in a contact metamorphosed limestone, such as that of the Christmas Mountains, Texas, shown in Figure 16-2 (Joesten 1976; Joesten and Fisher 1988). On being heated, coexisting quartz and calcite become unstable, and a reaction layer of wollastonite forms between them. For simplicity, we will assume that the fluid phase is pure CO_2 ($X_{CO_2} = 1$). With time, the incompatible phases, quartz and calcite, are separated by a progressively thickening layer of wollastonite (Fig. 18-19). The equilibrium assemblage quartz + wollastonite + CO_2 occurs on one boundary of this reaction layer, and the equilibrium assemblage calcite + wollastonite + CO_2 occurs on the other boundary. Thus, although the system as a whole is not at equilibrium, local equilibrium exists at the boundaries on either side of the reaction layer. Diffusion through the reaction layer allows more wollastonite to form, but it does not change the equilibrium assemblages on either side of the reaction layer.

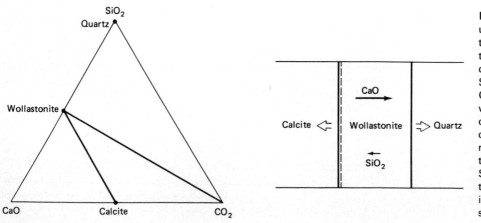

FIGURE 18-19 Heating of chert nodules in limestone results in the formation of a reaction rim of wollastonite on the nodules. Growth of the wollastonite depends on the diffusion of CaO and SiO_2 through the reaction rim. Because CaO diffuses more rapidly than SiO_2, wollastonite grows more rapidly on the quartz side of the rim than it does on the calcite side; the lengths of the two arrows indicate the relative magnitudes of the diffusion coefficients for CaO and SiO_2. The original boundary between the chert nodule and the limestone is indicated by the dashed line. [After Joesten and Fisher (1988).]

Diffusion through the reaction layer is caused by the chemical potential gradient set up by the compositional difference between the chert and the calcite (Joesten 1977). All minerals can be considered to be in local equilibrium with an intergranular phase, which is largely the fluid but would also include the disordered regions on grain boundaries; we will refer to this as the fluid phase. The chemical potential of components in the fluid are determined by the chemical potentials of components in the local minerals. These chemical potentials can be defined using simple reactions between the minerals and the fluid. They are as follows:

(mineral)   (fluid components)

$$CaCO_3 \rightleftharpoons CaO + CO_2 \tag{18-34}$$

$$CaSiO_3 \rightleftharpoons CaO + SiO_2 \tag{18-35}$$

$$SiO_2 \rightleftharpoons SiO_2 \tag{18-36}$$

If each mineral is in local equilibrium with the fluid, we can use Eq. 9-5 to write

(mineral)   (fluid)

$$\mu_{calcite} = \mu_{CaO}^f + \mu_{CO_2}^f \tag{18-37}$$

$$\mu_{wollastonite} = \mu_{CaO}^f + \mu_{SiO_2}^f \tag{18-38}$$

$$\mu_{quartz} = \mu_{SiO_2}^f \tag{18-39}$$

The stability of these minerals can, therefore, be expressed in terms of the chemical potentials of CaO, SiO_2, and CaO. Because the fluid phase is taken to be pure CO_2, the chemical potential of CO_2 is determined only by temperature and pressure. Thus, at a given temperature and pressure, only the chemical potentials of CaO and SiO_2 are variables. We therefore show the stability relations between the minerals in an isothermal, isobaric plot of μ_{CaO} versus μ_{SiO_2} (Fig. 18-20).

For a fluid to be in equilibrium with quartz, $\mu_{SiO_2}^f$ must have the value defined by Eq. 18-39. At a temperature of 700°C and a pressure of 35 MPa, which corresponds to the depth at which wollastonite layers developed around the chert nodules in the contact metamorphic aureole of the Christmas Mountains intrusion (Joesten 1976), $\mu_{SiO_2}^f$ would, according to the data in Table 7-1 and Eq. 7-45, be -977 kJ mol^{-1}. This chemical potential plots as a vertical line in Figure 18-20. For a fluid to be in equilibrium with calcite, μ_{CaO}^f must have a value defined by Eq. 18-37. From the data in Table 7-1, $\mu_{calcite}$ is -1345 kJ mol^{-1}. The $\mu_{CO_2}^f$ is calculated from the data in Tables 7-1 and 16-1, and Eq. 8-7 to be -574 kJ mol^{-1}. Consequently, the μ_{CaO}^f in the fluid in equilibrium with calcite under these conditions must be $\mu_{calcite} - \mu_{CO_2}^f$; that is, -771 kJ mol^{-1}. This plots as a horizontal line in Figure 18-20. Finally, fluids in equilibrium with wollastonite can have a range of chemical potentials of CaO and SiO_2 as long as they satisfy Eq. 18-38. For example, $\mu_{wollastonite}$ is -176 kJ mol^{-1}, and if we take the chemical potential of SiO_2 to be that which would be in equilibrium with quartz (-977 kJ mol^{-1}), μ_{CaO}^f is -781 kJ mol^{-1}. On the other hand, if the μ_{CaO}^f is that which is in equilibrium with calcite (-771 kJ mol^{-1}), $\mu_{SiO_2}^f$ is -988 kJ mol^{-1}. These

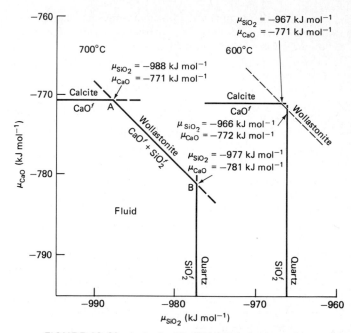

FIGURE 18-20 Isobaric ($P = 35$ MPa), isothermal ($T = 700$°C) plot of the chemical potentials of CaO and SiO_2 in fluids in equilibrium with calcite, wollastonite, and quartz. Chemical potentials of CaO and SiO_2 are uniquely defined where calcite and wollastonite are in contact and where wollastonite and quartz are in contact. Differences in chemical potential between these two boundaries drive the diffusion through the wollastonite layer. At 600°C the wollastonite saturation surface only just intersects the surfaces for calcite and quartz. At temperatures below this, wollastonite is unstable with respect to calcite and quartz, and then the wollastonite saturation surface does not produce a stable intersection with the saturation surfaces of calcite and quartz. [After Joesten and Fisher (1988).]

two points define a line in Figure 18-20 indicating fluids with chemical potentials of CaO and SiO_2 that would be in equilibrium with wollastonite. From the stoichiometry of Eq. 18-35, the slope of this line must be 45°. Indeed, the slope of any reaction line in an isothermal, isobaric μ_i versus μ_j plot can be determined from the stoichiometry of the reaction as follows:

$$\frac{d\mu_i}{d\mu_j} = \frac{-v_j}{v_i} \tag{18-40}$$

where v_i and v_j are the stoichiometric coefficients for the species i and j, respectively, in the balanced reaction, with the coefficients being positive if they are products and negative if they are reactants.

Note in Figure 18-20 that the line representing the equilibrium between fluid and calcite is metastable once it crosses the line marking the fluids coexisting with wollastonite, because here the chemical potential of SiO_2 in the fluid is sufficiently high to react with CaO to produce wollastonite. Furthermore, the wollastonite line is metastable once it crosses the calcite line, because at these high values of μ_{CaO} calcite forms in place of CaO. Similar arguments apply to the intersection of the wollastonite and quartz lines. The intersecting lines delineate a *saturation surface*, which must be a convex polyhedron. Within the surface, fluid is undersaturated with respect to solids, but at the surface, the chemical potentials are high enough to precipitate a mineral or minerals. At point *A*, where the calcite and wollastonite saturation surfaces intersect, fluid with a $\mu_{CaO}^f = -771$ kJ mol^{-1} and $\mu_{SiO_2}^f = -988$ kJ mol^{-1} is in equilibrium with both calcite and wollastonite; at point *B*, where $\mu_{CaO}^f = -781$ kJ mol^{-1} and $\mu_{SiO_2}^f = -977$ kJ mol^{-1}, fluid is in equilibrium with wollastonite and quartz.

At different temperatures (and pressures) the position of the saturation surface and the points of intersection of individual saturation lines change. In Figure 18-20 a saturation surface is also shown for 600°C. Note that at this temperature, the calcite saturation line is approximately the same as that at 700°C, but the quartz saturation line shifts to higher values (less negative). More important, however, the wollastonite saturation line also shifts to higher values and almost does not produce a stable intersection with the calcite and quartz saturation lines. At temperatures just below 600°C (at $P = 35$ MPa), the wollastonite saturation line plots above the intersection of the calcite and quartz saturation lines and is, therefore, entirely metastable. This simply means that these conditions are below the stability field of wollastonite, and instead, calcite + quartz is stable.

We are now in a position to evaluate the driving force causing diffusion through the wollastonite layer separating the calcite and quartz at 700°C. On one side of this layer, coexisting calcite and wollastonite buffer the chemical potentials of SiO$_2$ and CaO at -988.0 and -770.7 kJ mol^{-1}, respectively (*A* in Fig. 18-20), whereas on the other side, coexisting quartz and wollastonite buffer these potentials at -977.4 and -781.3 kJ mol^{-1}, respectively (*B* in Fig. 18-20). The chemical potentials of SiO$_2$ and CaO both decrease by 10.6 kJ mol^{-1} through the wollastonite layer. The gradients of these two components are therefore equal but in opposite directions.

If the Onsager diffusion coefficients for CaO and SiO$_2$ through the intergranular channels in the wollastonite were the same, the identical chemical potential gradients would cause equal amounts of CaO and SiO$_2$ to diffuse through the wollastonite layer. This would result in equal amounts of growth on both sides of this layer. Joesten and Fisher (1988), however,

have shown that almost all of the growth takes place on the wollastonite/quartz boundary; that is, CaO diffuses much more rapidly than SiO$_2$. Separate diffusion coefficients for CaO and SiO$_2$ cannot be determined from the field data. The thickness of reaction layers is certainly controlled by diffusion rates, but these rates would have varied as the rocks were heated and then cooled during the magmatic episode, and diffusion coefficients may also have varied as a function of composition. The layer thickness therefore represents the cumulative growth of the wollastonite formed while the rock was above 600°C. Joesten and Fisher, however, were able to extract an Onsager diffusion ratio of $L_{CaO}/L_{SiO_2} = 42$. With this ratio, the reactions on the two boundaries of the wollastonite layer can be written as follows (see Fig. 18-19) at the calcite/wollastonite boundary,

$$26 \text{ calcite} + 0.6\text{SiO}_2 \Longrightarrow$$
$$0.6 \text{ wollastonite} + 25.4\text{CaO} + 26\text{CO}_2 \quad (18\text{-}41)$$

and at the wollastonite/quartz boundary,

$$26 \text{ quartz} + 25.4 \text{ CaO} \Longrightarrow$$
$$25.4 \text{ wollastonite} + 0.6\text{SiO}_2 \quad (18\text{-}42)$$

Nearer the igneous contact, where temperatures exceeded 941°C, the mineral tilleyite formed between calcite and wollastonite around the chert nodules (Fig. 16-2). The position and orientation of the saturation surface for this mineral in the $\mu_{CaO}-\mu_{SiO_2}$ diagram can be evaluated from the stoichiometry of the reaction relating the mineral to its components:

$$\text{Ca}_5\text{Si}_2\text{O}_7(\text{CO}_3)_2 \Longrightarrow 5\text{CaO} + 2\text{SiO}_2 + 2\text{CO}_2 \quad (18\text{-}43)$$

from which it follows that $d\mu_{CaO}/d\mu_{SiO_2} = -\frac{2}{5}$. Its slope is shallower than that of the saturation surface of wollastonite (Fig. 18-21). Consequently, the tilleyite layer forms between the calcite and wollastonite; this is also evident from the position of the mineral with respect to calcite and wollastonite in the CaO–SiO$_2$–CO$_2$ mineral facies diagram (Fig. 16-2d). This shallower slope also means that the chemical potential difference of CaO across this layer is less than it is across the wollastonite layer. It is not surprising, therefore, to find that although the wollastonite layer grows mainly at the boundary with quartz,

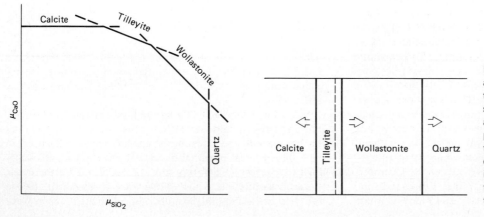

FIGURE 18-21   At temperatures above 941°C at $P = 35$ MPa, the saturation surface for the mineral tilleyite intersects the saturation surfaces for calcite and wollastonite in a μ_{CaO} versus μ_{SiO_2} plot. Consequently, tilleyite forms a reaction rim between the wollastonite and calcite on metamorphosed chert nodules in limestone. The dashed line represents the original calcite/chert boundary. [After Joesten and Fisher (1988).]

considerable growth of the tilleyite layer occurs at the boundary with calcite (Fig. 18-21).

Although two components (plus CO_2) are necessary to describe the minerals in the various layers around the chert nodules in the limestone, individual reaction layers are composed of just one mineral. For metasomatically layered rocks in general, the maximum number of phases in any local thermodynamic system (individual layer) is equal to the number of components necessary to define all phases present in all the layers less the number of independent compositional gradients. In the case of the chert nodules, two components, CaO and SiO_2, are necessary to describe all the phases (we have assumed that a fluid phase of pure CO_2 is present), and there is one independent compositional gradient, either CaO or SiO_2. Consequently, only one phase is present in each layer. At layer boundaries, reactions buffer gradients, so that there are no independently variable gradients, and consequently, two minerals can coexist.

18-7 ESTIMATING VOLUMES OF FLUID FLUX DURING METAMORPHISM

The discussion in previous sections of this chapter should make it clear that mass transfer is an important aspect of metamorphism. Of the various transport mechanisms, advection of the fluid phase is by far the most effective. Not only is advection capable of changing the composition of rocks, it may, if great enough, be able to transfer sufficient heat to become a primary agent causing metamorphism. This topic is dealt with in Sec. 18-8. But first it is necessary to be able to estimate the volumes of fluid flux during metamorphism. This can be done with certain rocks by using modal data and our knowledge of the controls of fluid composition over mineral equilibria (Rice and Ferry 1982).

In Sec. 16-4 we saw that dehydration and decarbonation reactions occur under a wide range of conditions. At one extreme, the composition of the fluid is controlled entirely by local reactions, whereas at the other extreme the fluid composition is externally buffered. In the first case, the fluid is generated by reactions taking place in the rock in question; no fluid is introduced, but some may be lost. In the second case, the externally buffered fluid is introduced, in all probability, by advection, because this is the only transport mechanism likely to be rapid enough. Natural reactions can fall anywhere between the two extremes. The important question is whether we can determine the extent of buffering due to infiltration. If this can be done, the actual flux of externally derived fluids can be determined. This, in turn, allows us to estimate the amount of thermal energy the fluids may have transported.

In metamorphic terranes where rocks have been internally buffered, mineral assemblages form divariant zones (univariant isobaric zones) where modal abundances vary continuously as a result of continuous reactions (dashed and solid paths in Fig. 16-8). These zones are separated by abrupt changes in modal abundance resulting from discontinuous reactions at isobaric invariant points. Also, because the fluid composition is buffered by local mineral assemblages, initial differences in rock composition may result in different fluid compositions.

These differences produce chemical potential gradients in the fluid components over relatively short distances, such as those between layers of different composition in an outcrop. On the other hand, the chances of finding isobaric univariant reaction assemblages in rocks that are externally buffered are very small. The vertical paths followed by such buffered assemblages through T–X_{CO_2} diagrams (dotted line in Fig. 16-8) indicate that isobaric divariant assemblages are far more likely to be found. Also, fluid infiltration eliminates chemical potential gradients in the volatile constituents that may initially have existed. Consequently, layers with different bulk compositions have mineral assemblages that are in equilibrium with a single fluid phase.

It will be recalled from Sec. 16-4 that the degree to which a reaction can buffer and change the composition of a fluid is given by Eq. 16-28, which for a binary H_2O–CO_2 fluid can be written as

$$X_{CO_2} = \frac{N_{CO_2}}{N^0_{H_2O} + N^0_{CO_2} + A(N_{CO_2} - N^0_{CO_2})}$$

where X_{CO_2} is the mole fraction of CO_2 in the fluid at a given time, N_{CO_2} and N_{H_2O} are the number of moles of CO_2 and H_2O in the fluid at that time, $N^0_{CO_2}$ and $N^0_{H_2O}$ are the initial numbers of moles of CO_2 and H_2O in the fluid, and $A = (m + n)/n$ where m and n are, respectively, the numbers of moles of H_2O and CO_2 produced by the reaction. A measure of the progress of the reaction is given by the last term in the equation, $(N_{CO_2} - N^0_{CO_2})$, which is simply the number of moles of CO_2 produced or consumed by the reaction. To account for differences in the amount of CO_2 resulting from the stoichiometry of the reaction, the $N_{CO_2} - N^0_{CO_2}$ term is divided by the stoichiometric coefficient for CO_2 in the reaction (n) to give what is referred to as a *reaction progress variable*, ξ, which is defined as

$$\xi \equiv \frac{N_{CO_2} - N^0_{CO_2}}{n} \tag{18-43}$$

We can then substitute ξn for $N_{CO_2} - N^0_{CO_2}$ and $\xi n + N^0_{CO_2}$ for N_{CO_2} in Eq. 16-28, and recalling that $A = (m + n)/n$, we obtain

$$X_{CO_2} = \frac{X^0_{CO_2}(N^0_{H_2O} + N^0_{CO_2}) + \xi n}{(N^0_{H_2O} + N_{CO_2}) + \xi(m + n)} \tag{18-44}$$

which can be rearranged to give ξ

$$\xi = \frac{(N^0_{CO_2} + N^0_{H_2O})(X^0_{CO_2} - X_{CO_2})}{X_{CO_2}(m + n) - n} \tag{18-45}$$

This equation then relates the change in the composition of the fluid to the reaction progress variable, from which we can determine the change in the modal abundances of minerals.

Reactions in many metamorphic rocks are likely to occur under conditions intermediate to those of complete internal or external buffering. Mineral assemblages in these reactions still buffer the fluid composition, but some fraction of the fluid is derived from an external source. In such cases the extent of infiltration may be estimated from changes in the modal abundances of minerals. If no fluid were introduced, the amount of reaction necessary to establish an equilibrium fluid composition for any given change in temperature along an isobaric univariant reaction can be calculated from the stoichiometry of the reaction (review Prob. 16-10). But this amount of reaction produces a predictable change in the modal abundances of the reactant and product minerals if we have an estimate of the porosity of the rock. Differences between the predicted and actual change in the modal abundances indicate the degree of infiltration. For example, if the externally derived fluid had, coincidentally, the desired equilibrium composition, no reaction would be required whatsoever, and the modal abundances of the minerals would not change. On the other hand, if the infiltrating fluid was poorer in the component to which the buffered fluid must progress, an additional amount of reaction would be required, and the change in the modal abundances of the minerals would be larger than expected. To determine precisely the quantity of infiltrating fluid, the reaction progress must be related to the composition of the fluid and the abundances of minerals involved in the reaction.

Consider, for example, a rock containing quartz and calcite that has a porosity of 1% and is in equilibrium with a CO_2–H_2O fluid in which $X_{CO_2} = 0.01$ (Rice and Ferry 1982). On being heated to 470°C at 0.35 GPa ($P_{total} = P_{CO_2} + P_{H_2O}$), the following reaction starts:

$$\text{quartz} + \text{calcite} = \text{wollastonite} + CO_2 \qquad (18\text{-}46)$$

As reaction progresses with rising temperature, the evolved gas enriches the fluid in CO_2, causing the assemblage to follow the isobaric univariant reaction curve in the T–X_{CO_2} diagram (Fig. 18-22). Regardless of whether or not externally derived fluid is added, the fluid must have the composition given by the isobaric univariant curve if quartz + calcite + wollastonite are to be in equilibrium at any given temperature. Thus, at 600°C, the fluid must have a composition of $X_{CO_2} = 0.086$.

Using Eq. 18-45, we now calculate how much quartz and calcite would have to react to change the fluid from $X_{CO_2}^0 = 0.01$ to $X_{CO_2} = 0.086$ in rising from 470°C to 600°C. If we treat a rock volume of 1000 cm³ (a large hand specimen), the pore fluid occupies 10 cm³, and with a composition of $X_{CO_2} = 0.01$, would contain 0.447 mol of H_2O ($N_{H_2O}^0$) and 0.005 mol of CO_2 ($N_{CO_2}^0$). Substituting these values into Eq. 18-45 gives

$$\xi = \frac{(0.005 + 0.447)(0.01 - 0.086)}{0.086 \times (1) - 1} = 0.0376$$

From Eq. 18-43 we therefore obtain

$$N_{CO_2} - N_{CO_2}^0 = \xi n = 0.0376$$

Thus only 0.0376 mol of CO_2 need be produced to effect the change in fluid composition, and because of the stoichiometry of the reaction, 0.0376 mol each of quartz and calcite would be consumed and 0.0376 mol of wollastonite would be formed. The molar volume of wollastonite is 39.93 cm³ (Table 7-1). Thus the volume of wollastonite produced in the reaction is

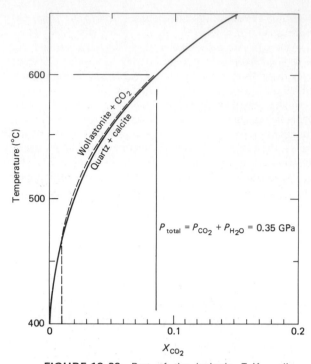

FIGURE 18-22 Part of the isobaric T–X_{CO_2} diagram for the reaction quartz + calcite = wollastonite + CO_2. The dashed line indicates the path followed by a fluid that initially has $X_{CO_2} = 0.01$ and is then buffered by the reaction as the temperature rises to 600°C. See text for discussion.

1.5 cm³, which is a 0.15% change in modal abundance. Clearly, very little wollastonite need be formed if the rock is internally buffered.

Rumble et al. (1982) have described the contact metamorphism of a siliceous limestone in west-central New Hampshire in which wollastonite is formed under a pressure of 0.35 GPa and a maximum temperature of 600°C. Near the igneous contact, the metamorphosed limestone contains 70% wollastonite, an amount far greater than would be expected if the rock were internally buffered, even if the porosity had been considerably higher than 1%. This indicates infiltration by an H_2O-rich fluid, the excess wollastonite being formed when the reaction was establishing the equilibrium fluid composition ($X_{CO_2} = 0.086$).

If again we deal with 1000 cm³ of this New Hampshire rock, 70 modal % of wollastonite corresponds to 700 cm³, or 17.53 mol of wollastonite. From the stoichiometry of the reaction (Eq. 18-46), 17.53 mol of CO_2 must also have been produced. Despite this large production of CO_2, the composition of the fluid at 600°C was only $X_{CO_2} = 0.086$. The CO_2 must therefore have been considerably diluted with a CO_2-poor fluid. If this fluid is assumed to have been pure H_2O, we can write

$$X_{CO_2} = 0.086 = \frac{N_{CO_2}}{N_{CO_2} + N_{H_2O}^{inf}} = \frac{17.53}{17.53 + N_{H_2O}^{inf}}$$

where $N_{H_2O}^{inf}$ is the number of moles of H_2O that infiltrated the rock during metamorphism. Thus 186.3 mol of H_2O must have passed through the rock during metamorphism. The molar volume of H_2O at 600°C and 0.35 GPa is 19.6 cm^3 (Eq. 16-6). Consequently, 3652 cm^3 of H_2O must have infiltrated the rock during the reaction. If we add to this the 871 cm^3 of CO_2 released by the reaction (17.53 × 49.7 cm^3), the volume of the fluid phase was 4523 cm^3. Thus the ratio of the fluid to rock volumes was 4.5:1. If the fluid had not been pure H_2O and had contained some CO_2, still larger volumes of fluid would have been required to infiltrate the rock.

Reactions that take place at isobaric invariant points provide a still better measure of the extent of infiltration. Again we consider a reaction treated by Rice and Ferry (1982). A rock containing anorthite, calcite, quartz, zoisite, tremolite, and diopside in equilibrium with a CO_2–H_2O fluid is isobarically

invariant (Fig. 18-23a). At 0.35 GPa the fluid has a composition of $X_{CO_2} = 0.094$ and the temperature is 481°C. The invariant point is produced by the intersection of four isobaric univariant reactions. The reaction at the invariant point can be described in terms of any two of these reactions, as long as they are linearly independent. For example,

tremolite + 3 calcite + 2 quartz
$$= 5 \text{ diopside} + 3CO_2 + H_2O \quad (18\text{-}47)$$
and

$$2 \text{ zoisite} + CO_2 = 3 \text{ anorthite} + \text{calcite} + H_2O \quad (18\text{-}48)$$

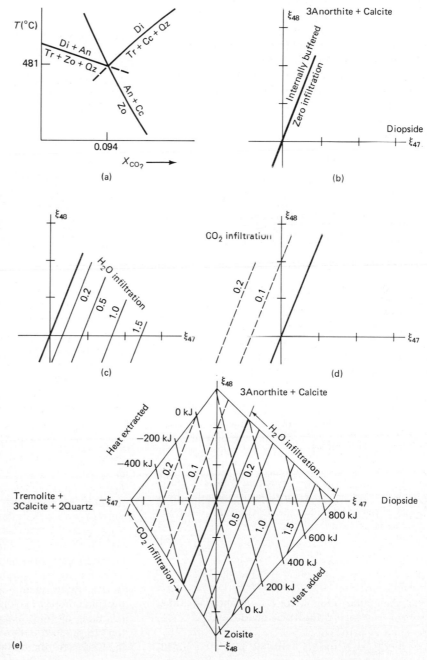

FIGURE 18-23 Reaction taking place at the isobaric invariant point in the T–X_{CO_2} diagram (a) can be expressed as the sum of two reactions: tremolite + calcite + quartz = diopside + 3CO$_2$ + H$_2$O and zoisite + CO$_2$ = anorthite + calcite + H$_2$O (Eqs. 18-47 and 18-48). Because fluid composition is fixed at the invariant point, reactions 18-47 and 18-48 must progress at rates that produce $\dot{C}O_2$ and H$_2$O in the proportions required by the invariant point. Relative rates of progress of these two reactions are given by the solid line in (b), which plots the reaction progress variable ξ (see text for definition) of the one reaction against that of the other. If pure H$_2$O infiltrates the rock, the reaction producing the larger amount of CO$_2$ (18-47) must progress more rapidly if the fluid composition is to remain at the invariant composition. Volumes of H$_2$O infiltrating 1000 cm^3 of rock are shown multiplied by 1000 in (c). If pure CO$_2$ fluid infiltrates the rock, the reaction producing the most water must progress more rapidly to maintain the invariant fluid composition. Volumes of infiltrating CO$_2$ (× 1000) are shown by dashed lines in (d). These figures are combined into a single diagram in (e). Negative values of reaction progress variables indicate the reactions above occurring in reverse. The amount of heat released or taken in by the reaction is indicated by dashed lines.

Reaction progress variables for these two reactions (ξ_{47} and ξ_{48}) are defined as (Eq. 18-43)

$$\xi_{47} = \frac{N_{CO_2} - N^0_{CO_2}}{3}$$

and

$$\xi_{48} = \frac{N_{CO_2} - N^0_{CO_2}}{1}$$

If the rock is internally buffered, the volatiles produced by the reactions must be in the proportions defined by the isobaric invariant point; that is $X_{CO_2} = 0.094$. The number of moles of CO_2 and H_2O produced by each reaction can be expressed in terms of the reaction progress variables. Thus reaction 18-47 produces $3\xi_{47}$ moles of CO_2 and ξ_{47} moles of H_2O, and reaction 18-48 produces $-\xi_{48}$ moles of CO_2 (negative sign indicates CO_2 is consumed) and ξ_{48} moles of H_2O. The mole fraction of CO_2 in the fluid at the invariant point can therefore be written as

$$X_{CO_2} = 0.094 = \frac{3\xi_{47} - \xi_{48}}{(3\xi_{47} - \xi_{48}) + (\xi_{47} + \xi_{48})}$$

$$= \frac{3\xi_{47} - \xi_{48}}{4\xi_{47}} \qquad (18\text{-}49)$$

from which it follows that $\xi_{47}/\xi_{48} = 0.381$. Thus for the mole fraction of CO_2 to remain fixed at 0.094, reactions 18-47 and 18-48 must progress in the proportions $0.381 : 1$. Graphically, this can be represented by a straight line in a plot of ξ_{47} versus ξ_{48} (Fig. 18-23b).

Because of the stoichiometry of reactions 18-47 and 18-48, the number of moles of diopside (N_{Di}) produced is $5\xi_{47}$, and the number of moles of anorthite (N_{An}) is $3\xi_{48}$. Thus $\xi_{47} = N_{Di}/5$, and $\xi_{48} = N_{An}/3$, which can be substituted into Eq. 18-49 to give the ratio of diopside to anorthite that must be produced for the fluid to retain its composition of $X_{CO_2} = 0.094$; that is,

$$\frac{N_{Di}}{N_{An}} = 0.635 \qquad (18\text{-}50)$$

If the rock is infiltrated by fluid with a composition different from that of the isobaric invariant point, the production of diopside and anorthite will differ from the ratio given by Eq. 18-50.

If a rock in which tremolite + calcite + quartz + zoisite is reacting to form diopside and anorthite is infiltrated by a fluid composed of pure H_2O, the mole fraction of CO_2 in the fluid at the isobaric invariant point is

$$X_{CO_2} = 0.094 = \frac{3\xi_{47} - \xi_{48}}{4\xi_{47} + N^{inf}_{H_2O}} \qquad (18\text{-}51)$$

where $N^{inf}_{H_2O}$ is the number of moles of H_2O infiltrating during the reaction. On rearranging, Eq. 18-51 gives

$$N^{inf}_{H_2O} = 27.915\xi_{47} - 10.638\xi_{48} \qquad (18\text{-}52)$$

Equation 18-52 can be expressed in terms of volume of H_2O by multiplying both sides of the equation by the molar volume of H_2O, which is 22.11 cm³ at 481°C and 0.35 GPa. This gives

$$V^{inf}_{H_2O} = 617.20\xi_{47} - 235.21\xi_{48} \quad cm^3 \qquad (18\text{-}53)$$

Equation 18-53 defines straight lines in a plot of ξ_{47} versus ξ_{48} (Fig. 18-23c). In the limiting case where $V^{inf}_{H_2O} = 0$, Eq. 18-53 becomes Eq. 18-49. The values of $V^{inf}_{H_2O}$ plotted in Figure 18-23c have been multiplied by 10^{-3}. Because we are dealing with a rock volume of 1000 cm³, the values on the lines give the ratio of the volume of infiltrating fluid to volume of rock. Thus a point on the line labeled 1.0 indicates that 1000 cm³ of H_2O would have infiltrated the rock during metamorphism. Conversely, if the modal abundance of diopside and anorthite indicate that $\xi_{47} = \xi_{48} = 1$, then 400 cm³ of pure H_2O could have infiltrated the rock during metamorphism. Had the infiltrating fluid not been pure H_2O, a larger volume of fluid would have been necessary to effect the same change.

Identical relations can be developed for infiltration by a fluid composed of pure CO_2. The mole fraction of CO_2 in the fluid at the isobaric invariant point is

$$X_{CO_2} = 0.094 = \frac{3\xi_{47} - \xi_{48} + N^{inf}_{CO_2}}{4\xi_{47} + N^{inf}_{CO_2}} \qquad (18\text{-}54)$$

On rearranging the equation and converting moles of CO_2 ($N^{inf}_{CO_2}$) to volume by multiplying by the molar volume of CO_2, which is 48.98 cm³ at 481°C and 0.35 GPa, we obtain

$$V^{inf}_{CO_2} = 51.85\xi_{48} - 136.07\xi_{47} \quad cm^3 \qquad (18\text{-}55)$$

Equation 18-55 defines straight lines in a plot of ξ_{47} versus ξ_{48} (Fig. 18-23d). These are parallel to the lines for infiltration by H_2O fluids but plot on the opposite side of the line marking the internally buffered reaction. In the limiting case where $V^{inf}_{CO_2} = 0$, Eq. 18-55 becomes Eq. 18-49. The values of $V^{inf}_{CO_2}$ plotted on the lines in Figure 18-23d have also been multiplied by 10^{-3}; these values express the ratio of the volume of the infiltrating fluid to rock volume (1000 cm³). Again, the volumes of infiltrating fluid are only minimum values. If the fluid were not pure CO_2, larger volumes would be required to effect the same change.

In describing the reaction progress variables for reactions 18-47 and 18-48, the reactions were assumed to progress to the right; that is, diopside and anorthite were reaction products. But the reactions could progress to the left and then diopside and anorthite would be reactants. Rather than define new reaction progress variables for the reverse reactions, we can simply use negative values of the progress variables for the forward reactions. Thus, in Figure 18-23, each ξ_{47} versus ξ_{48} plot extends to negative values of the two variables, dividing each plot into four quadrants. In the first quadrant, diopside and anorthite are both products of the isobaric invariant reaction. In the second quadrant, anorthite becomes a reactant (negative), and diopside is joined by zoisite as products. In the third quadrant, diopside also becomes a reactant along with anorthite to produce tremolite, calcite, quartz, and zoisite. Finally, in the fourth quadrant, anorthite again becomes a product, along with tremolite, calcite, and quartz, while diopside and zoisite are reactants. Each quadrant, therefore, expresses a different way in which the mode of the rock could change at the isobaric invariant reaction point, depending on the nature of the infiltrating fluid. Figure 18-23b to d have been combined into a single diagram in Figure 18-23e. The lines representing the minimum volumes of infiltrating H_2O or CO_2 fluids are applicable in each of the quadrants and indicate the composition of fluids that would give rise to the particular modal changes.

Because Figure 18-23 is constructed for 1000 cm³ of rock, there is a limit to how far the reactions can progress before going to completion. For example, reaction 18-47 is complete when the entire 1000 cm³ of rock is converted to diopside. Because the molar volume of diopside is 66.09 cm³ (Table 7-1), 1000 cm³ of diopside would correspond to 15.13 mol. According to the stoichiometry of reaction 18-47, this amount of diopside is produced once the reaction progress variable, ξ_{47}, reaches a value of 3.026. Conversion of the entire volume of rock into a mixture of anorthite and calcite in a 3:1 molar ratio limits ξ_{48} to a value <2.947. Conversion of the volume entirely into zoisite limits $\xi_{48} > -3.662$. Finally, conversion of the volume entirely into tremolite, calcite, and quartz in the molar ratio of 1:3:2 limits $\xi_{47} > -2.330$.

In addition to buffering the fluid composition, the isobaric invariant reaction also buffers the temperature. At 0.35 GPa the temperature of the invariant point is 481°C (Rice and Ferry 1982). If heat is added to the rock, the reaction must proceed in a direction that consumes heat in order to maintain the equilibrium temperature, and if heat is removed from the rock, the reaction must proceed in a direction that liberates heat. Both reactions 18-47 and 18-48 are endothermic; that is, ΔH_r is positive (review Sec. 7-3), and the reactions proceed to the right with addition of heat. The ΔH_r for reaction 18-47, as written, is 283.529 kJ; that is, 283.529 kJ must enter the rock in order to produce 5 mol of diopside, which is equivalent to the reaction progress variable increasing by one. The ΔH_r for reaction 18-48, as written, is 70.308 kJ for each unit increase in the reaction progress variable. The amount of heat taken in or given out by each reaction is therefore simply $\Delta H_r \xi$, and when both reactions occur together, as they must at the isobaric invariant point, the amount of heat taken in by the rock is given by

$$Q = 283.529\xi_{47} + 70.308\xi_{48} \quad \text{kJ} \quad (18\text{-}56)$$

Values of constant Q define straight lines in the ξ_{47} versus ξ_{48} plot (Fig. 18-23e). The line representing the progress of the reaction with no addition or loss of heat ($Q = 0$) must pass through the origin, for this point represents no reaction whatsoever. The slope of this line, and all others of constant Q, is simply $\xi_{47}/\xi_{48} = -70.308/283.529$.

Equation 18-56 expresses only the amount of heat involved in the progress of a metamorphic reaction. Most of these reactions, however, take place because the temperature of the rock has been raised by the influx of heat. The total heat budget for a metamorphic episode must therefore include a term for the heat needed to change the temperature of a rock. This, of course, is given by the product of the heat capacity, C_p, and

the change of temperature, ΔT (Eq. 7-12). The total amount of heat taken in by a metamorphic rock during a metamorphic episode which involves i separate reactions and a total change of temperature of ΔT is given by

$$Q_{\text{total}} = Q_{\text{in}} - Q_{\text{out}} = C_p \, \Delta T + \sum_i \Delta H_i \, \xi_i \quad (18\text{-}57)$$

Figure 18-23e provides a useful means of determining, from modal data, the conditions under which a reaction took place. Modal data are plotted in the figure by converting them to reaction progress variables. Thus, in 1000 cm³ of rock, the volume percentage (mode) of diopside can be converted to number of moles, which in turn can be converted to the reaction progress variable for reaction 18-47; that is,

$$\xi_{47} = \frac{\text{number of moles of diopside}}{5}$$

We can similarly express the modal abundance of anorthite as

$$\xi_{48} = \frac{\text{number of moles of anorthite}}{3}$$

Before the modal data are plotted in Figure 18-23e, some knowledge of the reaction history of the rock is required. If the figure is to indicate the amount and type of fluid infiltration and the amount of heat added to or subtracted from the rock, only those minerals produced at the invariant point must be plotted. A certain percentage of these minerals may have been present in the rock prior to the reaction. This information can be obtained by careful modal analysis of the rocks both below and above the isograd marking the isobaric invariant reaction. Studies such as this have been carried out by Ferry (1983) on rocks in south-central Maine (Fig. 17-15). Here changes in the modal abundances of impure carbonate rocks resulting from the progressive metamorphism from the biotite through the diopside zones indicate infiltration by two rock volumes of an H_2O fluid and addition of about 1.25 kJ cm⁻³ of rock. This amount of heat would have been sufficient to raise the temperature of the rock almost 300°C had there been no reaction. This shows how important the reaction is to the buffering of temperature.

18-8 EFFECT OF FLUIDS ON METAMORPHISM

In previous sections we have seen how fluids affect specific types of metamorphic reactions. We have not, however, considered their effects on the overall metamorphic process. In a regional metamorphic terrane or in a contact aureole, numerous reactions may occur at the same time in different places. Most of these liberate a fluid phase which is likely to migrate upward and affect other reactions occurring above. The fluid transports heat with it, and by so doing modifies the thermal gradient. The enthalpy of the reactions, which are mostly but not entirely endothermic, also affect the gradient. Predicting precisely a temperature–time path for any given rock in a metamorphic terrane is clearly a complex process, and one that is

fillings of quartz are common in low-grade metamorphic terranes [see, for example, Etheridge et al. (1984)] and attest to this pulsing flux of fluids.

In this scenario, a devolatilization reaction is a pulsating phenomenon, the frequency of which is determined by the tensile strength of the rock. The temperature at which the reaction progresses vigorously is also determined by the tensile strength of the rock. Because most rocks are relatively weak, devolatilization reactions are unlikely to exceed equilibrium temperatures by more than a few degrees (Walther and Wood 1984). In contrast, reactions that do not evolve a fluid and have small ΔG's, such as andalusite $\rightarrow$ sillimanite, may overstep equilibrium temperatures by as much as 40°C during rapid metamorphic events (contact metamorphism). Putnis and Holland (1986) have argued that the presence of sector trilling in the

common metamorphic mineral cordierite, which results from an hexagonal-to-orthorhombic transformation, is evidence that cordierite-forming reactions may exceed equilibrium temperatures by as much as 80°C (the higher-temperature hexagonal form is stable only at temperatures above the equilibrium reaction temperatures).

We may conclude that metamorphism is far from being the passive reaction of rocks to a raised geothermal gradient. Because large volumes of fluid are typically produced, metamorphism must involve a complex interaction between the dissipation of thermal, chemical, and mechanical energy. Working out the relation between these and finding features in rocks that provide a record of the metamorphic history will keep petrologist busy for many years to come.

PROBLEMS

18-1.   From the geometry of the crenulation schistosity in Figure 18-4, calculate the volume percent of the initial rock that must have been dissolved in order to create a crenulation schistosity that offsets the earlier schistosity by 1 mm if the schistosities intersect at 30° and the crenulation schistosity planes are spaced 5 mm apart.

18-2.   If the mica defining the crenulation schistosity planes is the insoluble residue left after quartz has been dissolved, its thickness should be a measure of the amount of quartz removed. If the rock on either side of the crenulation schistosity planes is composed of 30% muscovite and 70% quartz, and the rock in the crenulation schistosity planes is 100% muscovite, calculate what volume percent of the initial rock must have been removed by solution if the crenulation schistosity planes are 0.5 mm thick and are spaced 5 mm apart.

18-3.   What is the average flow velocity of fluid buoyantly rising through a 500-nm-wide planar fracture if the fluid's viscosity is 1 mPa s and its density is 1 Mg m^{-3}, and the density of the surrounding rock is 2.8 Mg m^{-3}? (Express your answer in m yr^{-1}.) Is flow laminar or turbulent?

18-4.   Because gold is such a minor constituent of the Earth's crust, its concentration in ore deposits is evidence of enormous enrichment. The solubility of gold in hydrothermal waters is low (about 1 to 100 ppb), so that large volumes of fluid are needed to transport the quantities of gold in ore bodies. One way in which this may occur is through deep convection of meteoric water in the upper crust (Nesbitt 1988). For convection to occur, however, rocks must be sufficiently permeable (Eq. 18-3). In the upper crust, where rocks are brittle, this permeability might be provided by fractures. If the upper temperature limit for brittle behavior is 400°C and the temperature gradient is 30°C km^{-1}, calculate the minimum permeability necessary for convection. The water can be assumed to have a density of 1 Mg m^{-3}, a viscosity of 1 mPa s, a coefficient of thermal expansion of 10^{-3} K^{-1}, and a thermal diffusivity of 10^{-6} m^2 s^{-1}.

18-5.   From the plot of $\log_{10}D$ versus $1/T$(K) (Fig. 18-11), calculate the activation energy for diffusion of Ca in wollastonite.

18-6.   Two large alkali feldspar crystals, one of Or$_{100}$ composition and the other of Or$_{60}$ composition, are in contact along a planar surface. If they are suddenly heated to 600°C by being incorporated as a xenolith in a body of magma, how many

years would it take the Or$_{100}$ crystal to change its composition to Or$_{90}$ 1 mm from the initial contact between the crystals? Feldspar of both Or$_{100}$ and Or$_{60}$ composition continue to exist at some distance from the initial boundary. The K–Na interdiffusion coefficient is 3×10^{-21} m^2 s^{-1}.

18-7.   If an alkali feldspar of Or$_{90}$ composition contains widely spaced 10-μm-thick exsolution lamellae of Or$_{10}$, how many years would it take at 650°C, where the K–Na interdiffusion coefficient is 10^{-19} m^2 s^{-1}, for the composition in the center of a lamella to reach a composition of Or$_{85}$? Assume that the lamellae are far enough apart that host crystal of composition Or$_{90}$ is always present.

18-8.   Figure 18-15 indicates the stability fields of muscovite, K-feldspar, and albite with respect to the activities of K$^+$, Na$^+$, and H$^+$ ions in coexisting fluid at 300°C. At high activities of H$^+$, the mineral kaolinite [Al$_2$Si$_2$O$_5$(OH)$_4$] becomes stable. Kaolinite is related to albite and muscovite by the following ion-exchange reactions:

(1) $2Ab + H_2O + 2H^+ = kaolinite + 4SiO_2 + 2Na^+$

$$K = 6.31 \times 10^8$$

(2) $2Musc + 3H_2O + 2H^+ = 3\,kaolinite + 2K^+$

$$K = 5 \times 10^4$$

Assuming unit activities for the minerals and H$_2$O, and using the equilibrium constants given, determine the position of the kaolinite stability field in Figure 18-15, and determine the $\log(a\text{Na}^+/a\text{H}^+)$ and $\log(a\text{K}^+/a\text{H}^+)$ in the fluid in equilibrium with albite, muscovite, and kaolinite at this temperature.

18-9.   If periclase (MgO) and quartz are heated together, two reaction layers form between them, one of forsterite in contact with periclase, and the other of enstatite in contact with quartz. The rate of thickening of these layers depends on the rate of diffusion of Si and Mg through the layers, which in turn depends on the chemical potential gradients across the layers. First, construct a saturation surface for these minerals in a μ_{MgO} versus μ_{SiO_2} diagram for 700°C and 35 MPa using the data in Table 7-1.

(a)   What is the chemical potential decrease for SiO$_2$ and MgO across the forsterite layer and across the enstatite layer?

(b) Assuming that diffusion coefficients in forsterite and enstatite are similar, would you expect the forsterite or enstatite layer to thicken faster?

18-10. Impure carbonate rocks below the isograd marked by the isobaric invariant assemblage anorthite + calcite + quartz + tremolite + diopside + zoisite contain no diopside or zoisite, but those above the isograd contain 42.96 vol % diopside and 35.33% zoisite. If the reaction occurred at a pressure of 0.35 GPa ($T = 481°C$), the molar volumes of H_2O and CO_2 would be 22.11 and 48.98 cm^3, respectively. The molar volumes of diopside and zoisite are 66.09 and 135.9 cm^3, respectively.

(a) What values of the reaction progress variables for reactions 18-47 and 18-48 would account for the modal change?

(b) What volume and composition of fluid is produced by these reactions in 1000 cm^3 of rock? Does the fluid composition produced by the reactions match that of the isobaric invariant point?

(c) What minimum volume and composition of fluid must have infiltrated the rock during reaction?

(d) What was the total fluid to rock volume ratio for the reaction?

(e) How much heat was taken in by the rock during the reaction?

(f) If the reaction had not taken place, how much would this heat have raised the temperature of the rock if the heat capacity per 1000 cm^3 of rock is 3765 J°C^{-1}?

(g) Finally, plot the reaction progress variables in Figure 18-23e, and check that your calculations in parts (c) to (e) are correct.

19 // Deformation and Textures of Metamorphic Rocks

19-1 INTRODUCTION

In Chapters 15 to 18 we have seen how the mineralogical composition of a metamorphic rock is used to determine the pressure, temperature, and fluid composition during metamorphism. But this is only part of the record these rocks preserve. Their textures are also a source of valuable information and in most cases are diagnostic of a particular type of metamorphism—contact versus regional, for example. Textures are important in working out the timing of metamorphic events, especially in regionally metamorphosed rocks. Indeed, the interpretation of metamorphic textures is an important part of structural geology and some books have been devoted entirely to this topic. Only a brief introduction can be given here, but for more detailed coverage reference can be made to texts by Spry (1969), Turner and Weiss (1963), Hobbs et al. (1976), and Suppe (1985).

Many of the textures in metamorphic rocks result from the recrystallization of existing minerals or the growth of new ones. The general principles governing these processes were dealt with in Chapter 12, but little attention was given to the effects of deformation, other than to indicate that strain energies provide an additional driving force for recrystallization. Regional metamorphism is normally accompanied, at some stage in its development, by strong deformation, which results in the formation of a number of characteristic textures. It is these textures with which this chapter is mainly concerned. Some of the textural terms have already been used in previous chapters, but it is worth systematically reviewing these in light of what they can reveal about the deformation history of a rock.

Textures in regionally metamorphosed rocks can develop prior to, during, or after the main period of deformation. Rocks may even undergo more than one period of metamorphism, in which case textures may be complicated by relict ones from earlier episodes. Radiometric age determinations from the cores and rims of zoned crystals have provided absolute ages for multiple metamorphic events in some areas. Normally, however, only relative ages are determined from interpretations of both the textures of the rock and the outcrop-scale structures, such as folds, faults, and deformed primary features. Once unraveled, this chronology, in conjunction with the interpreted pressure, temperature, and fluid composition during metamorphism, allows us to estimate pressure–temperature–time paths for a rock during a metamorphic episode. This topic is dealt with in Chapter 20.

Metamorphism normally involves the growth of grains. These may form as new minerals produced by metamorphic reactions, or they may result from the recrystallization of old ones; less commonly, the process simply involves grain breakage. We first examine the textures developed by the growth of new minerals in the presence of a stress field, leaving to later the textures developed by recrystallization and grain breakage.

19-2 METAMORPHIC FOLIATION

The most prominent textural feature exhibited by regionally metamorphosed rocks is a planar fabric, which is referred to generally as *foliation*. It is produced mainly by the parallel or subparallel arrangement of platy and elongate minerals. When fine-grained pelitic sedimentary rocks undergo regional metamorphism, muscovite and chlorite are among the first minerals to form. These grow with a preferred orientation which produces a prominent foliation referred to as *slaty cleavage* (Fig. 19-1). The micaceous minerals in slate are too fine-grained to be visible with the unaided eye. As they coarsen, the foliation planes develop a sheen and generally become slightly less regular. This rock is referred to as *phyllite*. Individual crystals are still too small to be visible to the unaided eye in phyllite. When they do become visible, the rock is called a *schist*. Schists can also form from mafic igneous rocks; these typically contain abundant chlorite and amphibole. In sedimentary rocks of silty and sandy composition and in intermediate-to-felsic igneous rocks fewer micaceous or needly minerals are formed during metamorphism. Those that are present tend to be concentrated into layers that parallel the foliation defined by the micas or amphiboles. Quartz and feldspars tend to concentrate in the

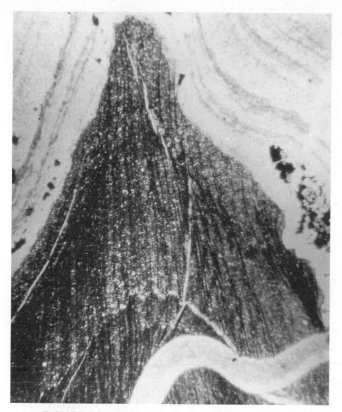

FIGURE 19-1 Slaty cleavage parallel to axial plane of fold in interlayered slate and limestone. Foliation planes are marked by concentrations of clay minerals and graphite. Note refraction of slaty cleavage near competent carbonate vein which has also been folded. The width of field is 1 cm.

even if nuclei form with random orientations, the unequal growth rates result in the development of a preferred orientation, with the majority of platy and elongate minerals lying in the plane normal to the maximum compressive stress (Fig. 19-2). In addition, pure shear tends to rotate platy and elongate minerals into this same plane. This is particularly true of clay minerals, which, during compression and dewatering, develop a strong preferred orientation. The growth of micaceous minerals from these clays during metamorphism is therefore likely to result in the nucleation of grains that already have a strong preferred orientation. Pressure solution can also cause quartz grains to become flattened in the plane of foliation.

Foliation may develop parallel to shear planes, but this does not appear to be common. This is not to say that shearing does not occur along foliation planes. Indeed, once formed, a foliation is difficult to eliminate from a rock, and the foliation provides planes of weakness on which later movement can occur. As will be seen below, there is considerable evidence for shear on foliation planes in metamorphic rocks.

In folded rocks, foliation parallels or nearly parallels the axial plane of folds (Fig. 19-1). In slaty and schistose rocks the foliation is almost everywhere parallel to the axial plane. Where more competent, quartz- or calcite-rich layers are interlayered with slate or schist, the foliation may fan across or converge toward fold axes (Fig. 19-3). This produces a refraction of the

alternate layers. Such rocks are referred to as *gneiss*. No hard-and-fast rule separates schists from gneisses. If micaceous or needly minerals predominate, the rock is a schist; if the granular minerals predominate, it is a gneiss. In general, schists tend to pulverize when hit with a geological hammer, whereas gneisses fracture.

When first developed in the lowest grades of metamorphism, foliation can be shown to coincide with the plane of maximum extension, as defined by deformed primary features, such as fossils and ooids in limestone, pebbles in quartzite, and reduction spots in slates. The reduction spots are particularly useful because they are initially spherical, and their physical properties, apart from color, are identical to those of the surrounding rock. Consequently, they form ideal passive strain markers. During deformation they are deformed into flattened ellipsoids, with their plane of maximum flattening coinciding with the slaty cleavage.

Several different mechanisms can cause minerals to become oriented during metamorphism. Minerals that form plates, blades, or needles have growth rates that clearly vary with crystallographic direction. Nuclei of these minerals that are oriented with their directions of maximum growth rate perpendicular to the maximum principal compressive stress can be expected to grow more rapidly than those with other orientations. Thus,

FIGURE 19-2 Schistosity defined mainly by alignment of muscovite and chlorite, but quartz grains are also flattened parallel to foliation. Some biotite grains (dark, stippled) are not aligned parallel to foliation and must have grown later when the stress conditions were different. The foliation plane is normal to the direction of maximum compression as shown by the strain ellipse. The width of field is 3 mm; crossed polars.

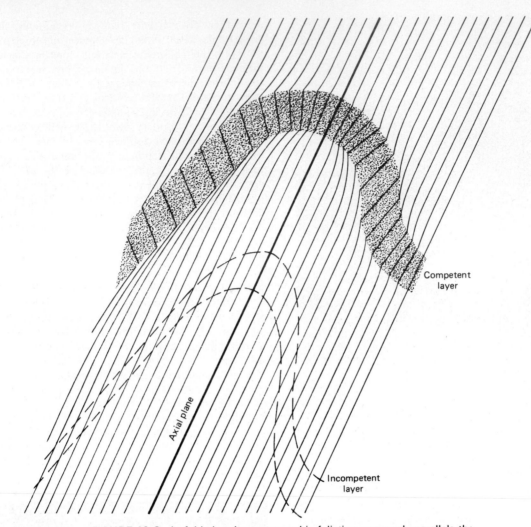

FIGURE 19-3   In folded rocks, metamorphic foliation commonly parallels the axial plane of folds. Competent layers of rock, however, tend to buckle during folding, which rotates axial plane foliation into fanlike patterns across fold axes.

foliation planes as they cross from one layer into another. But even here the foliation planes were probably parallel to the axial plane when first developed, but competent layers of rock buckle during folding and so rotate the foliation planes. Within folds, then, the foliation again generally parallels the plane of maximum flattening.

When flattening occurs in rocks that contain rigid grains, such as cubes of pyrite in slate or garnet crystals in schist (Fig. 19-4), zones of relatively low pressure develop on the sides of the crystals that are normal to the maximum compressive stress. These zones become the locus of precipitation of the mineral that is most soluble under pressure in the surrounding rock; this is normally quartz. This produces what are referred to as *pressure shadows*. Because most of these taper to a point away from the rigid grain, they resemble a goatee and thus are also referred to as *beards* (Fig. 19-4).

Within foliation planes, elongate minerals may be aligned so as to produce a *lineation*. Linear aggregates of equant grains, elliptical reduction spots in slate, and stretched pebbles in quartzite (Fig. 19-5) are other features capable of defining lineations. Some rocks may exhibit lineation but no foliation. In folded rocks, lineations lie in the axial-plane foliation, where they either parallel fold axes (*b* lineation) or are perpendicular to them (*a* lineation). In zones of strong shear, such as near thrust faults, lineations may parallel the tectonic transport direction (*a* lineation).

Foliation and lineation are described as *penetrative* structures because they are pervasive throughout the rock. Slaty cleavage, for example, is visible under the microscope down to the level of individual grains. A fault, by contrast, is *nonpenetrative* because it forms a discrete feature. Penetrative structures are useful because they provide, from a hand specimen or thin section, information on the orientation of larger nonpenetrative structures such as axial planes and plunges of folds that may not be exposed.

The foliation plane in many schists is folded into small crenulations which have a wavelength of a few millimeters to about a centimeter. These small folds define a *crenulation linea-*

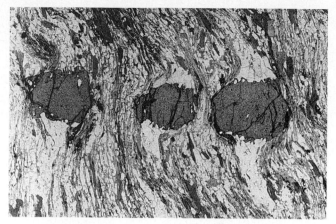

FIGURE 19-4 Rigid grains, such as these garnet crystals in schist, are commonly flanked by tapering wedges of quartz, which because of their shape are named beards. Quartz is thought to migrate to regions of low pressure, so beards are also referred to as pressure shadows. Note quartz grains in the beards are intergrown with the garnet in the rims of these crystals. Because the cores of the garnet crystals contain no quartz inclusions, the beards probably developed late in the growth of the garnet. Bolton schist, Connecticut. The width of field is 7 mm; plane light.

tion. These lines actually mark the intersection of a second, more widely spaced schistosity with the first schistosity. In cross section, the early schistosity is seen to be bent into sigmoid curves by the later *crenulation schistosity* (Fig. 18-3). Within the bends the amount of quartz is substantially less than in the surrounding rock, presumably as a result of removal by pressure solution (Sec. 18-2). This results in the crenulation schistosity being marked by mica-rich layers. This new compositional layering must not be mistaken for bedding. Crenulation schistosity can result from successive episodes of metamorphism, but in many cases it is produced by a single episode during which the orientation of the rock changes with respect to the stress field, possibly as a result of folding.

While platy and elongate minerals become aligned during deformation to form a foliated fabric, other minerals that exhibit no morphological anisotropy may become crystallographically aligned; quartz and calcite are two common examples. Such preferred orientation is not as readily apparent as the morphological alignment but is nonetheless an important part of the metamorphic reorganization of a rock. The quartz grains in the quartzite illustrated in Figure 19-5 have a strong preferred crystallographic orientation, which is made evident under the microscope when a first-order red interference filter is inserted under crossed polars and all of the quartz grains turn blue (or all turn yellow). The grains cannot, therefore, be randomly oriented, for there would then be equal numbers of blue and yellow grains.

Precise determination of preferred crystallographic orientation requires time-consuming measurements, which are normally made optically. The orientation of some easily identifiable crystallographic direction is determined using a universal

stage. In quartz and calcite, the optic axis (c axis) provides such a direction. A large number of grains are measured in thin section, and the results are plotted in an equal-area stereographic projection. The density of points in the plot is then contoured to give a statistical measure of the preferred orientation of the grains. In some fine-grained rocks, x-ray diffraction can be used to determine preferred orientations.

Many different patterns of preferred orientation have been found. Plots of c axes of quartz and calcite grains in stereographic projections form girdles or bands along which there may be one or more maxima. These patterns are classified according to the symmetry they exhibit—either orthorhombic, monoclinic, or triclinic. The pattern shown by the quartzite of Figure 19-5 is orthorhombic. The c axes define two girdles at 90° to each other and at 45° to the foliation plane. The two girdles intersect to produce a maximum that coincides with the lineation of the stretched quartz pebbles.

Why mineral grains that show no morphological anisotropy and, in the case of quartz, little anisotropy of any physical properties should undergo reorientation during deformation and recrystallization is still not entirely understood. There is probably no one single cause, but the mechanisms by which rocks undergo plastic deformation must play a role. Plastic deformation is that which takes place without brittle failure. It occurs by translation gliding, twin gliding, and recrystallization. Both gliding mechanisms are related to crystallographic directions, and for them to play roles in deformation they must be appropriately oriented with respect to the principal stress directions or be rotated into such orientations. Recrystallization in a stress field may also bring about preferred orientations, if certain orientations of nuclei effect a greater lowering of free energy than others.

Translation gliding takes place in crystals along specific planes known as *slip planes*. These are normally characterized by a close-packed structure. The *slip direction* also tends to be one of either close packing or rows of similar ions or ions of similar charge. Gliding can be thought of as the simple shear of one part of the crystal past another (Fig. 19-6). Slip does not occur along the entire length of a slip plane at one time; this would result in a fracture and the grain would loose cohesion (brittle failure). Instead, a defect is generated in the structure (see beneath the ⊥ on the slip plane in Fig. 19-6), which migrates along the slip plane and so transmits the displacement through the crystal. The morphology of the crystal is modified by this gliding, but the crystallographic orientation remains unchanged. The effectiveness of translation gliding depends on the ease with which dislocations are formed and move through the structure.

Twin gliding can also be thought of as resulting from simple shear, with each successive layer of atoms above a twin plane gliding past the one below. In contrast to translation gliding, twin gliding results in rotation of the structure (Fig. 19-6). These twins, which are referred to as *deformation twins*, may obey the same laws as, and be indistinguishable from, growth twins or transformation twins. In some cases, however, a distinction can be made. Many deformation twins do not extend all the way through a crystal but instead form wedge-shaped polysynthetic lamellae that extend in from grain boundaries or

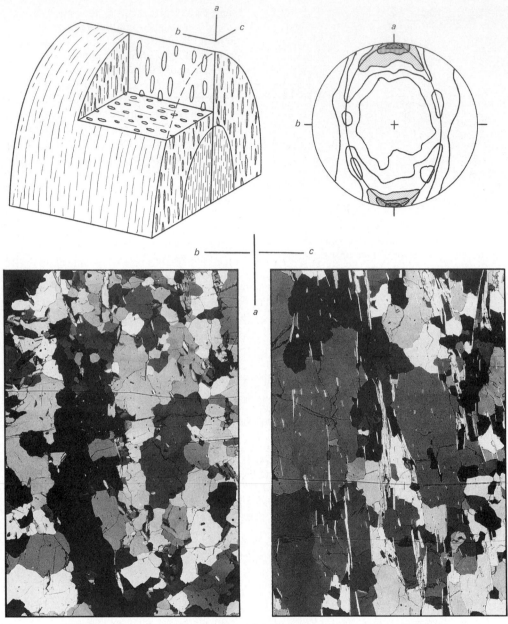

FIGURE 19-5 Folded quartzite containing quartz pebbles which are flattened in the foliation plane (*a–b*) and elongated parallel to *a*. Photomicrographs illustrate appearance of rock in sections parallel to the *a–b* and *a–c* planes (width of each field is 5 mm; crossed polars). Insertion of a first-order red interference filter causes grains to turn all blue or all yellow, indicating a strong preferred crystallographic orientation. This is shown in the stereographic plot of optic axes of quartz grains, most of which plot along two great circles at 45° to the foliation plane (*a–b*), which intersect to form a strong maximum parallel to *a*, the direction of pebble alignment. Clough quartzite, Connecticut.

are concentrated in parts of grains that are bent (Fig. 19-7a). Deformation twins also commonly form intersecting sets of lamellae (Fig. 19-7b). Deformation twins are common in calcite, dolomite, plagioclase, cordierite, and diopside.

In Chapter 12 we saw that deformation promotes recrystallization by introducing strain energy. We also saw that recrystallization attempts to minimize the free energy of a rock by reducing the surface area of grains and by eliminating boundaries that have high curvature; this results in grain coarsening (Sec. 12-5). The surface energy on grain boundaries also determines the dihedral angle that develops between grains during recrystallization. Clearly, a mineral grain in a metamorphic rock cannot be treated independently of its neighboring grains when considering size and shape because of the interfacial en-

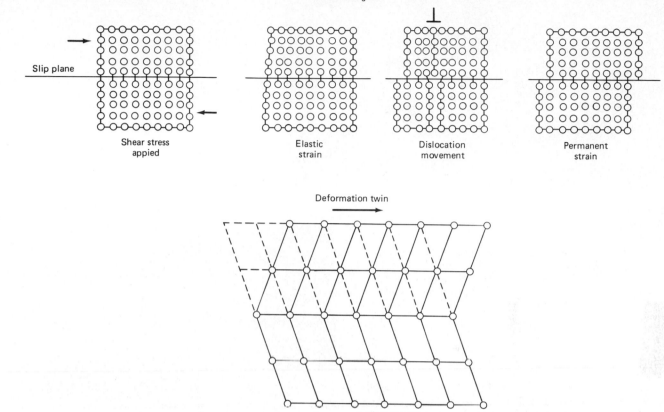

FIGURE 19-6 Crystalline material deformed by translation gliding and twin gliding. In translation gliding a dislocation (see below ⊥ on slip plane) migrates along a slip plane through the structure. In twin gliding each successive layer of atoms above the twin plane glides past the one below. Twin gliding changes the crystallographic orientation, whereas translation gliding does not.

ergies involved. Similarly, these energies may play a role in determining the orientation of juxtaposed grains. The surface energies of grain boundaries on anisotropic minerals vary with crystallographic orientation. Thus two grains of equal size and shape may not have the same surface energy if their crystal structures are oriented differently with respect to the structure of surrounding grains. Also, if the grains are under stress, anisotropy of the elastic properties will result in differences in the free energy of the grains. This could also result in grains of certain orientation being favored during recrystallization over those of other orientations.

In summary, then, the foliation visible in hand specimen is only the outward expression of the almost total internal reconstitution of the fabric of a regionally metamorphosed rock. Although the alignment of platy and elongate minerals is the most pronounced aspect of this fabric, crystallographic alignment of all minerals may be strong. The fabric, whether measured in the outcrop or under the microscope, can be used to interpret regional structures because the plane of foliation normally corresponds to the plane of maximum extension, which during early stages of deformation probably is normal to the maximum compressive stress.

19-3 PORPHYROBLASTS

Next to foliation, the most characteristic feature of many metamorphic rocks is the presence of certain minerals whose grain size is significantly larger than that of the rest of the rock. These large grains are known as *porphyroblasts*, and the rock is said to have a *porphyroblastic texture*. The suffix *-blastic* indicates that the texture is of metamorphic origin. When used as a prefix, *blasto-* indicates that the texture is inherited from an earlier rock. For example, *blastoporphyritic* would apply to metamorphosed porphyritic igneous rock. Many of the metamorphic index minerals in pelitic rocks form porphyroblasts; see, for example, garnet, staurolite, cordierite, Al_2SiO_5 polymorphs, and chloritoid in Figure 17-2. The fact that these minerals form porphyroblasts aids in their recognition in the field.

The grain size of rocks depends largely on the number of nuclei formed, but factors such as the relative times of nucleation and the abundances of the various minerals must also play roles (Bell et al. 1986). The tendency for large grains to grow at the expense of small ones might also be expected to contribute to the growth of porphyroblasts. However, this mechanism, while capable of eliminating small grains, does not

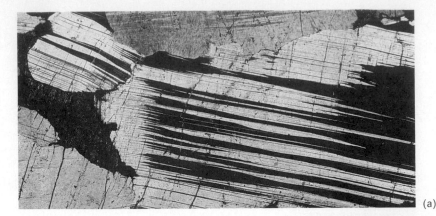

(a)

FIGURE 19-7 (a) Wedge-shaped and discontinuous albite twins in Precambrian anorthosite from Lake St. John, Quebec, were formed by deformation and are localized where grains are bent. (b) Many deformation twins form intersecting sets, as seen in this Precambrian Grenville marble from the Adirondacks, New York. The width of each field is 15 mm; crossed polars.

(b)

lead to the development of very large ones. This is because the excess pressures that drive this process are produced by the curvature on grain boundaries, and these are significant only on small grains (Eq. 12-28). As a result, porphyroblasts are not found in recrystallized monomineralic rocks, such as quartzite and marble. In multimineralic rocks, the number of nuclei formed of each mineral is unlikely to be the same, and thus variation in the grain size among the different minerals is to be expected.

The number of nuclei formed of a mineral depends on the distance over which diffusion is able to transport the nutrients and the nature and size of the critical nucleus (Sec. 12-2). How far material is transported depends very strongly on whether or not an intergranular fluid is present. Nucleation is likely to be heterogeneous, with some preexisting mineral providing the substrate on which the new one grows. In view of the discussion of reaction mechanisms in Sec. 18-5, prediction of the actual site of nucleation may be difficult. The greater the critical radius of the nuclei, the less chance there is for the nuclei to form, and the fewer there will be. The critical radius is proportional to the interfacial energy on the crystal (Eq. 12-2). Consequently, minerals that rank high in the *crystalloblastic series* (Table 12-1) are expected to have large critical radii and thus are likely to form porphyroblasts.

Some minerals high in the crystalloblastic series do not form porphyroblasts, such as magnetite and rutile, but these have simple structures and compositions, which probably al-

lows nucleation to occur more readily. By contrast, some minerals low in the crystalloblastic series, such as cordierite and feldspar, also form porphyroblasts. These porphyroblasts tend to be ovoid in shape rather than being bounded by crystal faces (Fig. 17-2f). Perhaps the complexity of the structures of these minerals is what prevents large numbers of nuclei forming.

When porphyroblasts grow, they make room for themselves by either replacing material (diffusional exchange of material from one place to another) or by physically pushing material aside. Of course, the growth of new minerals is accompanied by the disappearance of others, and this forms part of the diffusional exchange, as discussed in Sec. 18-5. Although some porphyroblasts appear to have pushed apart foliation planes during growth, these can equally well be interpreted as foliation planes collapsing around the porphyroblast during growth. Indeed, when trails of inclusions are preserved in porphyroblasts that mark the position of the original foliation planes, the later interpretation is normally found to be correct. Thus, during growth of the porphyroblast the spacing between foliation planes decreases, perhaps simply because of flattening of the rock or removal of quartz by solution, and as a result the foliation planes now wrap around the porphyroblast. In such cases the porphyroblast is normally bordered by pressure shadows (Fig. 19-4). Most porphyroblasts appear to grow principally by a volume-for-volume replacement. The large cordierite porphyroblast in Figure 17-2f, for example, grew without significantly deflecting the layering in the surrounding rock.

Many porphyroblasts incorporate inclusions of other minerals during their growth, giving rise to a *poikiloblastic* or *sieve* texture. In some porphyroblasts inclusions may be concentrated into crystallographic planes, as in the variety of andalusite known as chiastolite (Fig. 17-2c). In most, however, the inclusions are distributed throughout the porphyroblast, occupying the same positions that they did prior to the growth of the porphyroblast (Fig. 17-2). Their distribution commonly reflects compositional differences inherited from sedimentary bedding or earlier foliation planes (Fig. 18-17). The inclusions are commonly of one mineral. Garnet and staurolite, for example, typically contain quartz inclusions. This association is so common that the inclusions must be interpreted as a by-product of the reactions that produce the garnet and staurolite, or they are reactants that were present in excess.

The presence of inclusions in porphyroblasts indicates that minimizing surface free energies during metamorphism is not a high-priority process. Some porphyroblasts are composed almost entirely of inclusions, with the host mineral forming only a thin crystallographically continuous intergranular network (Fig. 19-8). Despite the ability of such crystals to enclose inclusions during most of their growth period, they are commonly bounded by inclusion-free margins that exhibit well-developed crystal faces. The fact that a porphyroblast can grow as a thin film along grain boundaries can be interpreted as indicating that the dihedral angle between the porphyroblast and the other minerals was zero at the time of growth; that is, the porphyroblast wetted the other minerals. Why, then, would the porphyroblast expel these minerals toward the end of the growth period? We have seen from experiments (Watson and Brenan 1987) that the composition of an intergranular fluid can significantly affect the dihedral angle between minerals. It is not unreasonable to suppose that toward the end of the growth period of a porphyroblast, the composition of a fluid phase may have changed, or the fluid phase may have disappeared altogether.

Variable growth rates may also be a factor in producing poikiloblastic crystals with inclusion-free rims. If growth was originally rapid and then slowed toward the end, diffusion may have had insufficient time at first to remove the inclusions. For growth to be rapid, however, the equilibrium conditions must be overstepped by some considerable degree, and then the frequency of nucleation would increase. Because poikiloblastic crystals can occur as widely scattered crystals, this explanation seems less probable.

Because the diffusion rates of most constituents through rocks at metamorphic temperatures, even when aided by an intergranular fluid phase, are so low, large porphyroblasts must take a long time to grow. This is particularly true of those containing aluminum, because of this component's relative immobility. If Carmichael's (1969) estimate of 0.3 mm for the limit of aluminum mobility in the rocks of the Whetstone Lake area of southeastern Ontario is typical, then, with a diffusion coefficient of 10^{-19} m^2 s^{-1}, aluminous porphyroblasts would require 14,000 years to grow ($x^2 = 2Dt$). The diffusion coefficient is probably several orders of magnitude less than this, and the growth time would be correspondingly greater. Recently, Christensen et al. (1988) have obtained a range of Rb/Sr absolute ages on garnet porphyroblasts from the Ottauquechee Formation of southeast Vermont, which was metamorphosed during the Devonian Acadian orogeny. These porphyroblasts, which are up to 3 cm in diameter and are poikiloblastic, have an age span of 9 Ma from their core to rim, indicating a very slow growth rate.

Many porphyroblasts, in particular those of garnet, are compositionally zoned, the origin of which is discussed in Sec. 12-4. Because diffusion rates in garnet are so low, compositions remain essentially unchanged once they are formed, at least at temperatures below 650°C (upper amphibolite facies). Thus the compositional range from core to rim of garnet porphyroblasts may preserve a lengthy record of the conditions of progressive metamorphism in all but granulite facies rocks (see Chapter 20).

In addition to growing over extended periods, porphyroblasts may form late in the metamorphic history of a region. Because porphyroblasts are large and commonly widely scattered (a few per thin section), their orientation with respect to foliation planes in the rock may go undetected. Commonly, porphyroblasts are aligned and define a foliation that may be at a considerable angle to earlier foliations. This is particularly noticeable with biotite porphyroblasts that may grow in a completely different orientation from earlier biotite grains. Observation of such facts is important in determining the relative ages of minerals in a rock.

FIGURE 19-8 Euhedral porphyroblasts of neptunite (Na$_2$FeTiSi$_4$O$_{12}$) in aegerine-bearing hornfels (protolith = shale) from the contact metamorphic aureole of the Cretaceous nepheline syenite intrusion of St. Hilaire, Quebec. Many grains of quartz and albite were included during the early growth of these crystals, making them poikiloblastic. During final stages of growth inclusion-free rims were able to form. See text for discussion. The width of field is 7 mm; plane light.

19-4 METAMORPHIC TEXTURES RESULTING FROM SHEAR AND ROTATION

Despite the general tendency for grain size to increase during metamorphism, a number of important rocks are formed by a reduction in grain size. These rocks typically have a strong foliation and may be compositionally layered. They occur in zones of intense deformation and may be associated with faults, in particular, thrust faults. There is little doubt that their fine grain size is related to the deformation. A clear and well-illustrated treatment of these rocks is given by Higgins (1971).

Many of the textural terms used to describe these rocks reflect early interpretations in which the fine grain size was thought to result from grain breakage. As a group, they are referred to as being *cataclastic*, a term defined by the American Geological Institute Glossary of Geology as "pertaining to a texture found in metamorphic rocks in which brittle minerals have been broken and flattened in a direction at a right angle to the pressure stress." Many of these rocks, however, are formed at depths that are too great to allow for brittle failure. It is now clear that much of the reduction in grain size results from plastic deformation and recrystallization during which the rock does not lose cohesion.

The transition from brittle to ductile behavior in rocks depends on confining pressure, temperature, strain rate, fluid phase, and mineralogical composition (Suppe 1985). In continental rocks, the transition is controlled largely by the presence of quartz, which is one of the first of the common minerals to become ductile with increasing depth (Tullis and Yund 1987). Because the brittle–ductile transition occurs at different temperatures with different minerals, a polymineralic rock may exhibit both forms of deformation. At the strain rates ($< 10^{-14}$ s^{-1}), temperatures, and pressures that most metamorphic rocks form, deformation is predominantly ductile. However, in the lowest-grade rocks, and even in high-grade rocks exposed to high strain rates, brittle deformation can occur.

The grain size of rocks undergoing strong deformation is reduced regardless of whether the deformation is brittle or ductile. With brittle behavior, grains are fractured. In shallow fault zones *breccias* develop, and if there is sufficient movement and comminution, rock can be reduced to an extremely fine powder known as *fault gouge*. With plastic deformation grain size is reduced by the generation and migration of defects and by recrystallization. In Chapter 12 we saw that the number of sites for nucleation and growth of new unstrained grains increases as the amount of strain in a rock increases. Consequently, rocks that have experienced high degrees of strain become fine-grained if they recrystallize. During progressive metamorphism rising temperatures promote recrystallization. But deformation itself causes heating and in extreme cases results in melting (see the discussion of pseudotachylite below). Recrystallization is therefore likely to occur in most zones of intense deformation.

In a polymineralic rock, some minerals become strained more easily than others, and these are likely to recrystallize more rapidly. Growth rates during recrystallization also differ from mineral to mineral, because recrystallization depends on the migration of grain boundaries, which, in turn, is determined by diffusion rates. In the brittle–ductile transition zone, ductile minerals may deform and recrystallize around the more brittle ones. Consequently, the various minerals in a rock have their grain sizes reduced to different extents by deformation. Regardless of the mode of deformation, the larger residual grains are referred to as *porphyroclasts*.

In many medium- to high-grade metamorphic quartzofeldspathic rocks, quartz grains deform rapidly into flattened

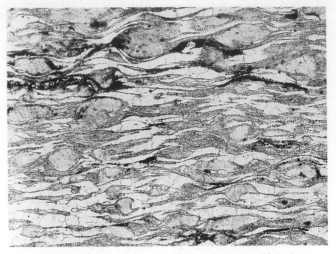

FIGURE 19-9 Strongly deformed augen (flaser) granite gneiss, Grenville Township, Quebec. Porphyroclasts of deformed perthitic feldspar are surrounded by extremely fine-grained feldspar and lenses of recrystallized quartz. The width of field is 3 mm; plane light.

sheets or lenses (Fig. 19-9), whereas feldspar grains may retain their original shape or show only small amounts of strain (Simpson 1985). The quartz in the sheets recrystallizes to unstrained grains. Such rocks are known as *flaser gneiss*. With increased deformation, feldspar also undergoes recrystallization. This starts with the rims of grains recrystallizing to very fine aggregates of unstrained grains that commonly have a strong preferred crystallographic orientation (Fig. 19-9). Commonly, the cores remain as large strained porphyroclasts within aggregates of fine-grained feldspar and quartz sheets. Because of the eyelike shape of relict feldspar grains, this rock is referred to as *augen gneiss* after the German word for an eye (*auge*). The term *phyllonite* is used for those rocks in which the grain size has been reduced to that which typifies a phyllite.

In zones of intense shearing, as for example near thrust faults, the original grains of a rock can be stretched into long thin streaks. With such large amounts of strain, recrystallization takes place easily, and because of the large number of nucleation sites, the resulting rock is extremely fine-grained and usually dark colored. Because the reduction in grain size was originally thought to have formed by the grinding or milling of material in shear zones, these rocks were named *mylonites* from the Greek word for a mill (*mylon*). In addition to being fine-grained, they are characterized by what is referred to as a *fluxion* or flow structure, a penetrative foliation which may or may not be accompanied by compositional layering and which is commonly visible only under the microscope.

Mylonites can be classified as protomylonite, mylonite, and ultramylonite, depending on the degree to which grain size is reduced (Higgins 1971). *Protomylonite* contains more than 50% megascopically visible porphyroclasts. This rock may resemble a metaconglomerate if the porphyroclasts are large enough. In *mylonite*, porphyroclasts constitute from 10 to 50% of the rock and are generally larger than 0.2 mm. *Ultramylonite* contains less than 10% porphyroclasts, which are smaller than 0.2 mm, and most are reduced to fine-grained streaks. The rock generally becomes darker as grain size decreases.

Figure 19-10 illustrates this progression in rocks from the Honey Hill thrust zone of southeastern Connecticut. The protolith is a coarse-grained, massive gabbro. Toward the fault zone its pyroxene is converted to amphibole, which recrystallizes more rapidly than the plagioclase. The resulting protomylonite contains porphyroclasts and lenses of plagioclase in fine-grained amphibole (Fig. 19-10a). Nearer the fault, the size of the porphyroclasts decreases, and the foliation of the mylonite is folded (Fig. 19-10b). In the fault zone itself, the rock is reduced to an ultramylonite, in which almost no porphyroclasts remain, but a strong compositional layering of amphibole and plagioclase attest to their former existence. Extremely flattened folds are evident in this layering (upper left of Fig. 19-10c). The rock is so fine grained (examine light-colored layers in Fig. 19-10c) that it resembles flint in hand specimen. The term *flinty-crush-rock* has, in fact, been used for these rocks in the Scottish Highlands.

Although some of the layering seen in ultramylonites is inherited from compositional differences in the protolith, completely homogeneous rocks can become layered during deformation. Apparently, differences in mechanical properties of minerals can result in cataclastic metamorphic differentiation. The layering and fine grain size of ultramylonites have led to their misidentification as volcanic rocks and siliceous sedimen-

tary rocks. In hand specimen, the distinction may be difficult to make.

The temperature of rocks can be raised by deformation, but the effect is usually small. If the maximum deviatoric stress that a metamorphic rock can withstand is 0.03 GPa, and the rock is strained at a rate of 10^{-14} s^{-1}, the rate of working is 3×10^{-7} Pa s^{-1}. But 1 Pa = 1 J m^{-3}. The rate of working is therefore 3×10^{-7} J m^{-3} s^{-1} or 0.3 μW m^{-3}. If the rock has a heat capacity of 0.9 kJ kg^{-1} K^{-1} and a density of 2.8 Mg m^{-3}, its heat content would be 2.5 MJ K^{-1} m^{-3}. If the deformation lasted for 1 Ma (3×10^{13} s) and the heat was completely retained by the rock, the temperature rise would be

$$\Delta T = \frac{3 \times 10^{-7} \, (\text{J m}^{-3} \, \text{s}^{-1}) \times 3 \times 10^{13} \, (\text{s})}{2.5 \times 10^{6} \, (\text{J K}^{-1} \, \text{m}^{-3})} = 3.6 \text{ K}$$

On a regional scale, therefore, deformation is not likely to cause significant heating. However, if deformation is concentrated into narrow zones, deviatoric stresses in excess of 0.1 GPa may cause appreciable heating (Graham and England 1976). Indeed, in extreme cases temperatures may be raised to the melting point of rock and the liquid formed injected into fractures, where it is quenched rapidly to a dark aphanitic or glassy rock known as *pseudotachylite* because of its resemblance to volcanic glass (tachylite).

Pseudotachylite forms narrow dikelets that are rarely more than 1 cm wide (Fig. 19-11). In the field, they resemble

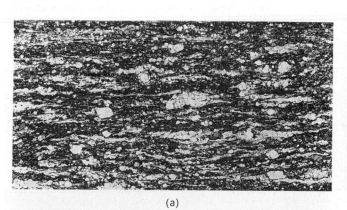

(a)

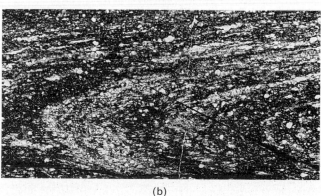

(b)

(c)

FIGURE 19-10 Stages in the development of mylonite from a coarse-grained gabbro in the Honey Hill thrust fault of southeastern Connecticut. (a) *Protomylonite:* On approaching the fault, pyroxene is converted to amphibole (dark), which recrystallizes rapidly to fine grains. Plagioclase resists deformation and survives as porphyroclasts surrounded by lenses of fine-grained recrystallized plagioclase, which is streaked out to form a fluxion structure. (b) *Mylonite:* In this section plagioclase porphyroclasts are greatly reduced in size, and fluxion layering is folded. (c) *Ultramylonite:* In the thrust fault itself the grain size is extremely fine, and porphyroclasts are visible only under the microscope. Layering is much more tightly folded than in (b), indicating the higher degree of strain. The width of each field is 16 mm; plane light.

FIGURE 19-11 Pseudotachylite cutting quartz monzonite, Paillé Lake, Quebec. Frictionally fused rock was injected into fractures formed by brittle failure and quenched to dark aphanitic material carrying abundant rock fragments (Philpotts 1964). The width of field is 1 cm; plane light.

rapidly quenched diabase. In thin section, however, their mylonitic origin is clear. They contain many small fragments of the surrounding rocks embedded in a dark aphanitic or glassy groundmass. They may or may not have a fluxion structure. In Figure 19-11, for example, the pseudotachylite extending through the center of the photograph between the two pieces of country rock has a foliation, whereas that forming the dikelets on the left and the main body on the right are devoid of flow structure; in the latter areas fragments appear to be freely suspended in the opaque groundmass.

The term *pseudotachylite* was introduced by Shand (1917) to describe dark aphanitic rock that forms networks of veins cutting the granite around the Vredefort ring structure, South Africa (Fig. 14-14). Because these veins did not appear to be associated with faults and most occur as isolated networks, he concluded that they must have been formed by shock and that fusion was caused by "incandescent gases." The Vredefort ring structure is now recognized to be most likely a meteorite impact structure (see Sec. 14-9), in which case pseudotachylite is, indeed, a shock-induced glass (Dietz 1961). Despite the meteorite impact origin of the original pseudotachylite, most other occurrences are associated with faults (Philpotts 1964).

Although deformation in fault zones containing pseudotachylite may be largely ductile, the dilational fractures filled by pseudotachylite indicate that occasionally brittle fracture does occur when strain rates exceed the levels that can be coped with by ductile flow. The fact that ultramylonite can be fused to form pseudotachylite through frictional heating indicates that strain rates can be extremely high. This is further supported by the fact that pseudotachylite has essentially the same composition as the rock in which it occurs. This indicates that the heating must have been sufficiently intense to cause total rather than partial melting. In addition, quartz fragments in pseudo-

tachylite may be surrounded by extremely silica-rich glass that has fusion temperatures well above those normally encountered in crustal magmatic systems. Moreover, these high temperatures must be generated in rocks that are at relatively low temperatures in order for the pseudotachylite to quench to a glass. Such short but intense thermal pulses can only be produced by the conversion of mechanical energy to heat. Other means of transferring heat in or out of these zones would be too slow to allow glass to form.

Pseudotachylite and ultramylonite can occur in zones where there is repeated movement. Inspection of Figure 19-12 reveals evidence of several periods of deformation, with some marked by mylonite and others by pseudotachylite. At times deformation on this fault was ductile, but at others it was clearly brittle when pseudotachylite was formed. It seems likely that glassy pseudotachylite is the petrologic record of fault movements associated with large earthquakes (McKenzie and Brune 1972).

Some rigid minerals, when embedded in a ductile groundmass, escape being strained during deformation. Garnet crystals, for example, in muscovite or chlorite schist can remain undeformed while the surrounding schist undergoes considerable shear. In such cases, the only evidence of deformation near the garnet might be the presence of quartz pressure shadows that are rotated by the shearing motion of the surrounding schist (Fig. 19-4). If the shearing is accompanied by progressive metamorphism, however, the garnet may grow during deformation, and if it includes minerals from the surrounding schist, the porphyroblast may preserve a record of the deformation.

In Chapter 18, aluminum was argued to be relatively immobile during metamorphism, as evidenced by the preservation of compositional differences in the rock within garnet porphyroblasts. For example, sedimentary bedding or compositional layering parallel to foliation planes can be preserved in

FIGURE 19-12 Mylonite and pseudotachylite, Paillé Lake, Quebec (Philpotts 1964). The pseudotachylite is a deep brown isotropic glass, which toward the margin of the dike and around feldspar fragments has become a deep red devitrified glass. The width of field is 1 cm; plane light.

FIGURE 19-13 Layers of quartz included in a garnet porphyroblast were rotated in a counterclockwise direction by shearing of the schist past the crystal, southeastern Vermont (Rosenfeld 1968). Because the layers are not bent, rotation must have occurred after garnet growth ceased. See text for discussion. The width of field is 25 mm; plane light.

a garnet porphyroblast as variable amounts of nonaluminous inclusions. Preservation of older fabrics within porphyroblasts gives rise to the *helicitic* texture.

Figure 19-13 illustrates a garnet porphyroblast through which quartz layers can be traced. In the surrounding rock these layers parallel the schistosity, which must therefore have existed prior to the growth of the garnet. The quartz layers in the porphyroblast now make an angle of 45° with the foliation in the surrounding schist. The rotation, which was in a

counterclockwise direction, must have occurred after the garnet had finished growing because the quartz layers are planar in the porphyroblast. If, for example, the garnet were to resume growing now, the newly incorporated quartz layers would be added at the ends of the planar ones at an angle of 45°. The planar layers must therefore have been included prior to any rotation.

From only this one thin section we can deduce a considerable amount of history. First, a metamorphic foliation developed prior to the appearance of garnet. This was probably an axial plane schistosity that developed normal to the maximum compressive stress. The metamorphic grade increased to the point that garnet formed, but there was no change in the attitude of the schistosity. The grade never rose above the garnet zone, and following termination of the growth of the garnet, a shearing motion rotated the porphyroblast counterclockwise. This late deformation may have been associated with tectonic processes that prevented the metamorphic grade from increasing above the garnet zone.

Figure 19-14 also illustrates garnet porphyroblasts with a helicitic texture, but here the quartz layers trace out a complex pattern of rotation. As in the previous example, the inclusions parallel an early foliation, which must therefore have existed prior to the growth of the garnet. The central part of each porphyroblast has relatively planar layers of inclusions. There must, consequently, have been an early period of growth in which there was little if any rotation. However, when the porphyroblasts reached approximately half their final size, the schistosity became shear planes that caused the garnets to rotate in a counterclockwise direction as they continued to grow. The layers of quartz in the surrounding schist continued to be included in the garnet, but once included they were rotated by the garnet. By the time the garnet finished growing, the original foliation planes in the core of the porphyroblasts had been rotated through 180°. Other garnets in this same rock unit

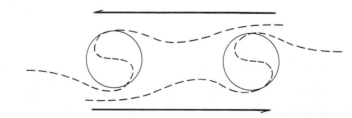

FIGURE 19-14 Layers of quartz in garnet porphyroblasts from southeastern Vermont (Rosenfeld 1968) have a helicitic texture, indicating a counterclockwise rotation of 180°. Because garnets were growing during rotation, layers become folded during inclusion. The width of field is 30 mm; plane light.

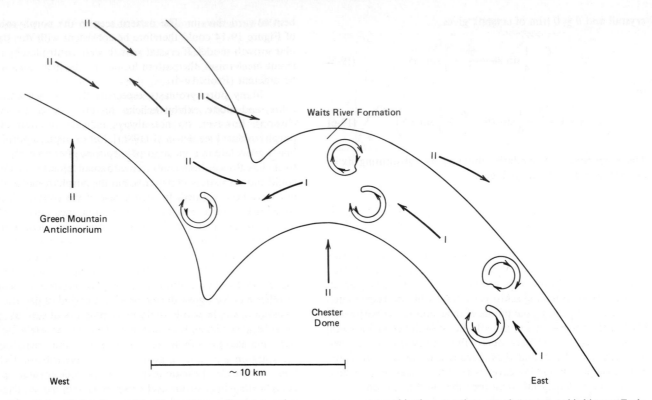

Waits River Formation

Green Mountain
Anticlinorium

Chester
Dome

⊢————————————⊣ ~ 10 km

West   East

Mountain anticlinorium and the Chester Dome. Garnets continued to grow during this period but were rotated about different axes. Rotation was caused by the rise of the dome and the eastward flow of material off the anticlinorium. List the evidence preserved in the rotated garnets that support this history. Early rotation directions are indicated by the inner circles and the later rotation direction by the outer circles.

20 // Pressure–Temperature–Time Paths in Regional Metamorphic Rocks

20-1 INTRODUCTION

Most metamorphic rocks have mineral assemblages that can be shown to have been in equilibrium at elevated pressures and temperatures. With certain assemblages, these conditions can be specified quite closely using thermodynamic and experimental data. The important question that we must now address is: To what point in the history of a metamorphic rock do these conditions relate? The protoliths of many metamorphic rocks have their origins on the surface of the Earth as sediment or lava. With burial, increasing pressure and temperature lead to the development of metamorphic minerals, which undergo successive changes as pressure and temperature continue to change. Then, through erosion and uplift, the rocks return to the Earth's surface. The rocks consequently experience changes in pressure and temperature, which are both functions of time. The question, then, is: What part of this pressure–temperature–time (P–T–t) path is recorded in the final metamorphic rock?

The fact that metamorphic rocks on the Earth's surface today have mineral assemblages that were stable at elevated pressures and temperatures indicates that they are now out of equilibrium (metastable) under the conditions on the Earth's surface. It is remarkable, then, that rocks that can so easily become metastable during unloading can, under other conditions, approach thermodynamic equilibrium so completely. Clearly, the rates of metamorphic reactions must change considerably with changing conditions, and kinetics must play an important role in determining what is preserved in a metamorphic rock.

The approach to equilibrium at one set of elevated temperatures and pressures is not always complete in metamorphic rocks, as evidenced by the presence of zoned minerals. The zoning in many porphyroblasts, in particular of aluminous minerals such as garnet, staurolite, and feldspar, records ranges of pressure and temperature during the growth of the mineral. This information, in conjunction with P–T conditions indicated by the mineral assemblage of the rock allows partial P–T–t paths to be constructed for some metamorphic rocks. As might be expected, metamorphic rocks can follow a variety of different P–T–t paths, depending on the specific tectonic conditions. The fact that many of these paths can be identified using analyses of zoned porphyroblasts means that metamorphic petrologists can make important contributions to the understanding of the development of mountain belts.

20-2 KINETICS OF METAMORPHIC REACTIONS

Metamorphic reactions can be divided into *prograde* and *retrograde* depending on whether they result from increasing or decreasing temperature, respectively. It is commonly believed that overstepping of equilibrium conditions by prograde reactions is small; that is, prograde reactions occur with ease. This accounts for the close approach of many metamorphic rocks to thermodynamic equilibrium. Retrograde reactions, on the other hand, tend to be sluggish and in most metamorphic rocks do not occur to any great extent, unless special conditions prevail.

The rate at which a metamorphic reaction progresses depends on both thermodynamic and kinetic factors. The thermodynamic potential must be large enough to overcome kinetic factors—this involves overstepping the equilibrium conditions by some finite amount. Kinetic factors are particularly sensitive to temperature, but the presence and composition of fluids can be equally important. Strain can also affect the rate of a reaction.

The driving force for a metamorphic reaction is provided by the lowering of free energy of the reaction. For a reaction to occur spontaneously, ΔG must be negative. With changing temperature and pressure, the change in ΔG is given by (Eqs. 7-33 and 7-34)

$$\left(\frac{\partial \Delta G}{\partial T}\right)_P = -\Delta S \quad \text{and} \quad \left(\frac{\partial \Delta G}{\partial P}\right)_T = \Delta V$$

Over the small range of temperature and pressure that is likely to be involved in the overstepping of a reaction, ΔS and ΔV of reaction will be almost constant; that is, the change in ΔG with change in T and P is linear. Thus the rate at which ΔG becomes increasingly negative with overstepping of a reaction should be

411

about the same in the prograde and retrograde directions. Changes in temperature and pressure are therefore likely to provide equal driving potentials for prograde and retrograde reactions.

The rates of many metamorphic reactions obey an *Arrhenius*-type relation. If the progress of a reaction is represented by ξ (Eq. 18-43), a reaction rate obeying an Arrhenius relation is given by

$$\frac{d\xi}{dt} = Ae^{-E/RT} \qquad (20\text{-}1)$$

where A is a reaction constant, E the activation energy, R the gas constant, and T the absolute temperature. The activation energy can be thought of as an energy barrier that must be crossed before the reaction can progress (Fig. 20-1). With increasing temperature, an increasing number of atoms involved in the reaction have the requisite energy to jump the barrier. Equation 20-1 can be thought of as consisting of two parts, the exponential factor and the preexponential constant. With increasing temperature the exponential term approaches unity, making the constant A equal to the reaction rate. The preexponential constant is therefore the reaction rate at infinitely high temperature. The exponential term is a fraction, and with falling temperature it becomes progressively smaller. The reaction rate at any temperature T is therefore a fraction of the rate A, this fraction decreasing exponentially with temperature (Fig. 20-1).

The free-energy change of a reaction and its kinetics are therefore affected differently by changes in temperature. As temperature rises above the equilibrium temperature of a prograde reaction, the ΔG of reaction becomes increasingly negative and the reaction rate increases. By contrast, as the temperature falls below the equilibrium temperature of a retrograde reaction the ΔG of reaction becomes increasingly negative, but the reaction rate decreases. Because of the exponential variation of reaction rates with temperature, small decreases in temperature are sufficient to slow a retrograde reaction dramatically (Fig. 20-1), and if temperature continues to fall, the reaction is almost certain to stop (Prob. 20-1). As a result, overstepping of prograde reactions is likely to be small [only a few degrees for most reactions—Walther and Wood (1984)], whereas overstepping of retrograde reactions may be large, and the larger they become, the less likely there is to be any reaction. We can conclude, therefore, that metamorphic reactions are most likely to preserve the highest temperature assemblage that the rock develops along its *P–T–t* path.

The fluid phase is also extremely important in determining what mineral assemblage is preserved in a rock. Most prograde reactions involve devolatilization. The extremely low porosity of metamorphic rocks indicates that these volatiles are almost completely expelled during reaction. Thus, when metamorphic temperatures fall, the volatiles needed for the retrograde reactions are absent. Small amounts of fluid that may remain are consumed rapidly in the formation of small amounts of retrograde minerals, such as chlorite rims on garnet crystals. A cooling metamorphic rock therefore tends to have a very low fluid content.

The fluid phase also plays an important kinetic role during metamorphism through its effect on diffusion. Prograde devolatilization reactions cause fluid pressures to equal or slightly exceed lithostatic pressures. This ensures that grains have fluid passing along dilated grain boundary channels and may even cause hydrofracturing. Diffusion rates in intergranular fluid are orders of magnitude greater than in dry rocks. Reaction rates in the presence of these fluids are therefore greater than in rocks that lack fluid.

Retrograde metamorphic rocks do occur, but they tend to be restricted to tectonic settings where H_2O and CO_2 have been able to gain access to the rock. Fault zones, for example, are commonly flanked by retrograde metamorphic rocks formed from fluids that were able to enter the rock through the fault zone. Where retrograde metamorphism has developed on a regional scale, higher-grade rocks are normally found to have been transported over lower-grade rocks, either in thrust slices or in nappes. As the lower-grade rocks are heated and undergo prograde reactions, the liberated volatiles rise into the overlying higher-grade rocks, where they cause retrograde reactions to occur.

The burial necessary to bring about regional metamorphism normally occurs during periods of plate convergence.

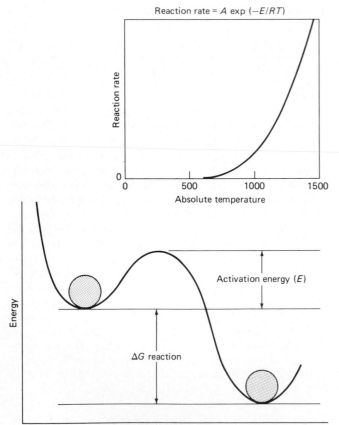

Reaction rate = A exp $(-E/RT)$

FIGURE 20-1 Relation between free-energy change of a reaction, ΔG, and activation energy, E, or energy barrier that must be overcome for the reaction to progress. Reaction rates commonly obey an Arrhenius relation (inset), where a small increase in temperature results in an exponential increase in reaction rate.

Once convergence stops, erosion and isostatic uplift brings deeply buried rocks back to the surface. Most prograde metamorphic reactions therefore occur while folding and faulting are still active. The accompanying strain provides an additional source of energy to promote reactions. Grain breakage and recrystallization also provide grain boundaries along which reactants can migrate more easily. By contrast, the small amounts of strain associated with isostatic uplift provide little extra driving force for retrograde reactions.

Because of the marked difference in the kinetics of prograde and retrograde reactions, most metamorphic rocks retain the mineral assemblages formed at peak metamorphic conditions; that is, they preserve the highest-temperature or highest-entropy assemblages formed along the $P-T-t$ paths. Only zoned porphyroblasts and retrograde rims on minerals preserve mineralogical records of conditions before and after the peak conditions. Relict textures can also indicate the former presence of diagnostic minerals.

One other source of $P-T$ information comes from fluid inclusions. During uplift, mineral grains may fracture, and if the rock is still hot enough and especially if fluids are present, they heal themselves. In doing so, they commonly trap fluid. In thin section, healed fractures are consequently visible as planar arrays of fluid inclusions. As the rock rises to the surface and cools, the fluid may split into a gas and a liquid phase, the gas forming a small bubble in the liquid. By heating such samples on a microscope hot stage until the inclusions become a single phase again, we can determine the minimum temperature of entrapment.

Figure 20-2 shows one possible path that a metasedimentary rock might follow during its history. Initially, burial and compaction leads to the lithification of the sediment. The thickening of the crust during orogenesis results in further burial and increasing pressures. At the same time the influx of heat

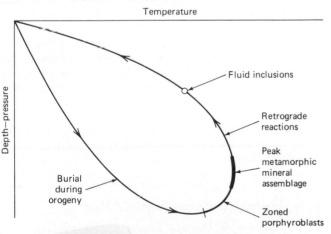

FIGURE 20-2 Possible path that a sedimentary rock might follow as a result of burial during orogenesis, followed by metamorphism, and eventually uplift due to erosion. The mineral assemblage that is likely to be preserved is developed near the maximum temperature. Zoned porphyroblasts may preserve a record of conditions just prior to the peak temperatures, and retrograde minerals may develop as rims on the high-T minerals during cooling. Fluid inclusions may be trapped in healed fractures during uplift.

from the geothermal gradient and from igneous intrusions and the generation of heat by radioactive elements in the sedimentary pile cause the temperature to rise. The rock undergoes a series of progressive metamorphic reactions during this stage, but only zoned porphyroblasts of certain minerals and relict textures preserve any record of these $P-T$ conditions. Eventually, the rock reaches its maximum temperature, but by this time it has already started to return to the surface of the Earth as a result of isostatic uplift caused by the erosion of overlying mountains. With continued rise the temperature begins to fall and reaction rates decrease exponentially, with the result that most reaction ceases. Some fluid inclusions may be trapped during uplift, but their time of entrapment is often difficult to determine precisely. Eventually, the high-temperature mineral assemblage arrives at the Earth's surface essentially unchanged.

20-3 METAMORPHIC FIELD GRADIENTS

Changes in mineral assemblage across metamorphic terranes indicate systematic changes in metamorphic grade, which can be correlated mainly with changes in temperature and to a lesser extent pressure. The record of these $P-T$ conditions has been referred to as a *metamorphic field gradient* (Spear et al. 1984), a *PT array* (Thompson and England 1984), and a *metamorphic geotherm* (England and Richardson 1977). The latter term is, unfortunately, too close to the word *geotherm*; these gradients are not geotherms, despite their early interpretation as such. We will therefore use the term "field gradient" or "*PT* array."

Barrow's (1893) mapping of isograds in the Scottish Highlands clearly established the existence of metamorphic field gradients. As more areas were studied, it became evident that the gradients were not everywhere the same, as evidenced by different sequences of metamorphic facies. In 1961, Miyashiro published an important paper in which he showed that three main types of gradient could be recognized. Because these are distinguished by the sequence of metamorphic facies, he named them *metamorphic facies series*. They are the high-temperature, low-pressure (andalusite-sillimanite) type; the moderate temperature, moderate-pressure (kyanite-sillimanite) type—this was the series mapped by Barrow; and the low-temperature, high-pressure (jadeite–glaucophane) type (Fig. 20-3). To these main facies series Miyashiro added two intermediate ones; an intermediate-temperature, low-pressure type, and an intermediate temperature, high-pressure type (Fig. 20-3). A complete continuum of $P-T$ series probably exists, but the frequency of occurrence of the three main types is so great that tectonic conditions must favor their formation.

Miyashiro (1961) also recognized that around the Pacific plate metamorphic rocks belonging to the low-T, high-P series and to the high-T, low-P series develop penecontemporaneously in *paired metamorphic belts*. The low-T, high-P series invariably occurs nearest the oceanic trench, whereas the high-T, low-P series develops in continental crust away from the oceanic plate. Although the low-T, high-P series contains metamorphosed igneous rocks from the ocean floor (ophiolite suite, Sec. 14-2), they contain almost no intrusive igneous rocks. By contrast, the high-T, low-P terranes contain an abundance of

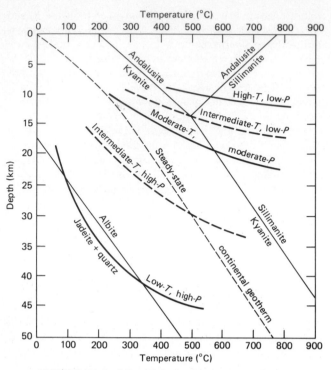

FIGURE 20-3 Mineral facies in metamorphic terranes commonly define arrays of *P–T* conditions corresponding to the three solid lines in the diagram. Miyashiro (1961) defined these as metamorphic facies series, each being named for the typical range of *P–T* conditions they cover. Less common intermediate facies series (dashed lines) are also recognized. For reference, the stability fields of the Al_2SiO_5 polymorphs, the albite = jadeite + quartz reaction, and the steady-state continental geotherm (Sec. 1-6) are included. [Modified from Miyashiro (1961).]

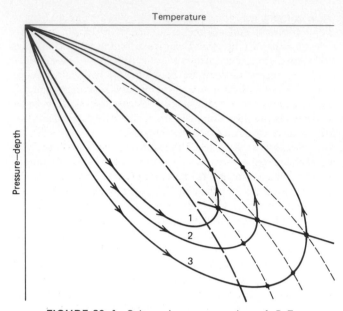

FIGURE 20-4 Schematic representation of *P–T–t* paths for three rocks, 1, 2, and 3 (solid arrowed lines). Fine dashed lines join points of equal age on the *P–T–t* paths; they also follow the geothermal gradient at any given time. The solid line joining the temperature maxima on the *P–T–t* paths is the field gradient. The heavy dashed line is the steady-state geothermal gradient. See text for discussion.

intrusive and extrusive calc-alkali rocks (Sec. 14-3). The occurrence of these paired metamorphic belts in specific tectonic settings is clear evidence that tectonic processes play an important role in determining the types of metamorphic field gradient that develop.

Figure 20-3 shows that rocks belonging to the high-*T*, low-*P* and low-*T*, high-*P* metamorphic facies series are formed, respectively, at temperatures above and below the normal steady-state continental geotherm. The elevated temperatures in the high-*T*, low-*P* series can be accounted for by large-scale intrusion of calc-alkali magma, whereas the low temperatures in the other series can be accounted for by rapid burial of cold surface sediments. In all facies series, however, the temperature rise in field gradients is normally much greater (shallower slope in Fig. 20-3) than can be accounted for by the geothermal gradients that must have produced these rocks. Although rapid advective rise of metamorphic fluids through major conduits and the proximity to igneous intrusions may account locally for steep gradients in temperature, they cannot account for the regional gradients.

This apparent paradox is resolved once it is recognized that field gradients are based on *P–T* conditions that were not developed simultaneously throughout a given metamorphic terrane. This will be evident if we consider the *P–T–t* paths of

three different sedimentary rocks that, during metamorphism, are buried to different depths. These three paths are shown in Figure 20-4 along with lines of constant time and a typical steady-state continental geotherm. When rock 1 reaches its maximum temperature and develops the mineral assemblage that will be preserved and brought to the surface, rock 3 is just reaching the deepest part of its *P–T–t* path. When rock 2 reaches its maximum temperature, rock 3 is still heating. The three rocks therefore reach their maximum temperatures at different times. The metamorphic field gradient mapped on the surface of the earth is the line joining the temperature maxima on the individual *P–T–t* paths. The slope of the field gradient is obviously very different from that of the geothermal gradient at any given time. We can conclude, therefore, that the range of *P–T* conditions recorded in the rocks of a metamorphic terrane are probably formed at different times and thus cannot be interpreted as fossilized geotherms.

20-4 PRESSURE–TEMPERATURE–TIME PATHS

Unraveling the history of most metamorphic terranes is complex and in many areas may be impossible to do in detail because pertinent parts of the record may have been eroded away or eliminated by annealing and prograde textures. Despite general similarities imposed by particular tectonic settings, each metamorphic terrane is unique. The particular *P–T–t* paths followed by rocks depend on many factors, such as the style of deformation, number of thrust faults, number of periods of folding, the amount of magma intruded, erosion rates, and du-

HQEA

ration of the orogenic episode. In view of these complexities, a clearer understanding of the principles governing P–T–t paths can be gained from a study of simple models than from a detailed analysis of any one area.

Several simple tectonic models have been investigated for which P–T–t paths can be calculated numerically (England and Richardson 1977; England and Thompson 1984; Karabinos and Ketcham 1988). In each model the crust is thickened to give the pressures recorded in metamorphic rocks. Thickening of the crust eventually leads to radiogenic heating (Prob. 1-5). At the same time, because of erosion of the overlying mountains formed during crustal thickening, rocks move toward the surface of the Earth. The actual temperature of a rock at any given time depends on the rates of these various processes.

To determine the temperature and pressure of a rock at any given time, the equation of heat transfer in a conductive, moving medium with its own sources and sinks of heat must be solved. We have already dealt separately with each aspect of this equation. First, Fourier's equation (Eq. 5-11) provides the basic relation for heat conduction:

$$\text{conductive heating:} \quad \frac{dT}{dt} = k\left(\frac{d^2T}{dz^2}\right)$$

The temperature change resulting from heat sources or sinks is some function of t and z, that is, $h(t, z)$. It might include heats of metamorphic reaction, frictional generation of heat on faults, intrusion and crystallization of magma, and radiogenic heat. We will consider only this last source, for it is normally the most important. If radiogenic elements are distributed evenly throughout the body of rock in question, then $h(t, z)$ becomes simply the rate of radiogenic heat production per unit volume of rock, or $A/C_p\rho$ (Eq. 1-6). The temperature change resulting from radiogenic heating is then:

$$\text{radiogenic heating:} \quad \frac{dT}{dt} = \frac{A}{C_p\rho}$$

In Eq. 18-58 we saw how the temperature of a metamorphic rock could be changed by the advection of fluid in which there was a temperature gradient. This equation is equally applicable when rock, rather than fluid, moves; the flux is simply the volume of rock that moves. The flux ($\text{m}^3 \, \text{m}^{-2} \, \text{s}^{-1}$) can be replaced by velocity (m s^{-1}), which is also a function, u, of t and z. The velocity depends on the rate of erosion, which we will assume to be constant. The rate of heating at any depth z due to uplift of rock is then:

$$\text{mass transport heating:} \quad \frac{dT}{dt} = -u\left(\frac{dT}{dz}\right)$$

where $u = dz/dt$; uplift will give a negative value of u and dT/dt will be positive.

The total change in temperature with time at any depth z is the sum of these three expressions; that is,

$$\frac{dT}{dt} = k\left(\frac{d^2T}{dz^2}\right) + \frac{A}{C_p\rho} - u\left(\frac{dT}{dz}\right) \tag{20-2}$$

This differential equation can be solved by the finite-difference method of Crank and Nicolson (1947) (see Sec. 5-4). Substituting

the relations in Eqs. 5-32 and 5-33 into Eq. 20-2, and making use of the same grid as in Figure 5-9, Eq. 20-2 expressed in finite-difference terms is

$$\frac{T_{m,\,n+1} - T_{m,n}}{\delta t} = \frac{k}{2(\delta z)^2}(T_{m+1,\,n} + T_{m-1,\,n} - 2T_{m,n}$$
$$+ T_{m+1,\,n+1} + T_{m-1,\,n+1} - 2T_{m,\,n+1})$$
$$+ \frac{A}{C_p\rho} - u\left(\frac{dT}{dz}\right)$$

which on rearranging gives

$$T_{m,\,n+1}\left[1 + \frac{k\,\delta t}{(\delta z)^2}\right]$$
$$= \frac{k\,\delta t}{2(\delta z)^2}(T_{m+1,\,n} + T_{m-1,\,n} + T_{m+1,\,n+1} + T_{m-1,\,n+1})$$
$$+ T_{m,n}\left[1 - k\frac{\delta t}{(\delta z)^2}\right] + \delta t\frac{A}{C_p\rho} - u\frac{\delta t(T_{m,n} - T_{m-1,\,n})}{\delta z} \tag{20-3}$$

This equation can be simplified by selecting values of δt and δz so that $k\,\delta t/(\delta z)^2 = 1$. In addition, values of $C_p\rho$ for crustal and upper mantle rocks are nearly constant at $2.5 \times 10^6 \, \text{J m}^{-3} \, \text{K}^{-1}$. Equation 20-3 thus reduces to

$$T_{m,\,n+1} = \frac{1}{4}(T_{m+1,\,n} + T_{m-1,\,n} + T_{m+1,\,n+1} + T_{m-1,\,n+1})$$
$$+ 2 \times 10^{-7}\,\delta t\,A - u\frac{\delta t(T_{m,n} - T_{m-1,\,n})}{2\delta z} \tag{20-4}$$

Values of T at any depth m and time $n + 1$ can be solved for iteratively using Eq. 20-4 if initial and boundary conditions are specified. These conditions depend on the particular model being investigated. We will consider one particular model in order to illustrate the effects of the various parameters on P–T–t paths.

Most metamorphic rocks found on the surface of the Earth have mineral assemblages that indicate crystallization at elevated pressure and, consequently, at some depth in the Earth. In ancient, eroded mountain belts these rocks are now underlain by normal thickness continental crust (about 35 km), and thus at the time of metamorphism the crust must have been thicker. Models for deriving P–T–t paths must start, therefore, with some mechanism for thickening the crust. In orogenic belts, crustal thickening is a consequence of plate convergence, and in detail is produced by folding, faulting, and simple compression and flattening. The particular mechanism affects the details of the model, but the essential consequence of each of these processes is to produce a thickened crust, which because of its content of heat-producing elements, causes temperatures to rise, and metamorphism ensues. We will consider the simplest of the crustal thickening processes, that caused by thrust faulting. Other processes could be considered, but their overall effect is similar to that caused by thrusting (England and Thompson 1984).

Consider a stable continental lithosphere with heat-producing radioactive elements concentrated in a surface layer of thickness D (Fig. 20-5a). The heat-generating capacity of this

was carried out with time intervals of 0.5 Ma, so the depth increments must be 4 km for $k \, \delta t/(\delta z)^2$ to be equal to unity. The velocity of erosion was taken to be 4×10^{-4} m a^{-1}. Note that this gives a velocity u in Eq. 20-4 of -4×10^{-4} m a^{-1}, because z is positive downward. Consequently, the last group of terms in Eq. 20-4 becomes positive. To understand fully the following discussion, Prob. 20-2 should be worked through slowly and carefully. More will be gained by solving this problem than by reading about it.

First, let us examine the change in the geotherm with time. On the left of the diagram is the sawtooth pattern of the initial faulted geotherm, with the heat-producing layer (shaded) appearing twice because of faulting. Initially, the rocks above the fault cool as heat migrates downward into the cooler rocks below the fault. By 2 Ma, however, the geothermal gradient is everywhere positive; that is, heat is transferred upward. By about 12 Ma the geotherm is back to where it was prior to thrusting. However, because of crustal thickening and the repetition of the heat-producing layer, the rocks continue heating and the geotherm rises above its initial steady-state position. As the gradient steepens, the heat flux increases (Eq. 5-3), so the rate of steepening decreases with time. In addition, erosion causes uplift and gradual thinning of the upper heat-producing layer. Eventually, as the thrust slice is eroded away, the geotherm returns to its initial steady-state form. Figure 20-6 shows geotherms only for the first 40 Ma following faulting, for this is the period during which most significant heating occurs.

The $P–T–t$ path of a rock can be followed in Figure 20-6, knowing the rate of uplift due to erosion (2 km/5 Ma). Consider, for example, the path followed by a rock that starts at a depth of 30 km. Its initial temperature is 150°C, which is too low for metamorphism to occur. By 10 Ma the rock has risen to a depth of 26 km, and its temperature is over 400°C, well within the $P–T$ range of the greenschist facies (Fig. 15-6). By 20 Ma the rock has risen to 22 km, but the $P–T–t$ path has gradually steepened, because of the steady decrease in the rate of temperature rise. Indeed, the $P–T–t$ path becomes vertical at this depth, and thus the corresponding temperature of 475°C is the maximum temperature reached by this rock. With continued uplift, the rock cools, and the metamorphic assemblage developed at the peak conditions (22 km = about 0.6 GPa and 475°C) would probably be preserved, indicating upper greenschist metamorphism. Note that despite the wide range of $P–T$ conditions at the temperature maxima on the various $P–T–t$ paths, all rocks pass through a narrow range of $P–T$ conditions during cooling.

A rock that starts at a depth of 20 km and 370°C follows a different type of $P–T–t$ path. Because it is above the thrust fault, its temperature decreases for the first 2 Ma, dropping to nearly 200°C as heat flows downward into the cooler underlying rocks. But then the temperature begins to rise, reaching a maximum after 15 Ma of 310°C, which is 60°C less than the initial temperature. During the initial cooling period, retrograde reaction would occur especially if volatiles were liberated by prograde reactions occurring beneath the fault. These retrograde minerals would then undergo prograde reactions as the rock began to heat.

In the discussion above, erosion and uplift are assumed to begin immediately following thrusting. There is evidence, however, that erosion may be delayed by as much as 20 Ma (England and Thompson 1984). The increased pressure near the base of the thickened crust may cause high-P, high-density mineral assemblages to form that would not induce isostatic uplift. The mountain chain would have low relief at this stage. Following some period of heating, rocks would revert back to the lower-density assemblages, and then isostatic uplift would initiate erosion. In this scenario, $P–T–t$ paths initially follow a 20-Ma isobaric heating period before uplift occurs. Thus a rock starting at a depth of 30 km reaches a temperature of 550°C before uplift begins (dotted path in Fig. 20-6). This rock reaches its temperature maximum at 570°C, approximately 100°C above the temperature reached by the same rock when erosion and uplift immediately follow thrusting. The effect of an erosional delay, therefore, is to raise the maximum metamorphic temperatures on $P–T–t$ paths.

The field gradient recorded in rocks originating at different depths is shown by the heavy line passing through the temperature maxima on the $P–T–t$ paths in Figure 20-6. Clearly, this line does not correspond to the geotherm at any given time. Nor are the rocks along the field gradient formed at the same time. The age of the peak metamorphic assemblage becomes progressively younger with increasing grade; that is, the maximum temperature is reached later at successively greater depths. An interesting consequence of this relation is that if rocks at depth reach the minimum melting temperature, magmas would rise into rocks that had already started to cool. This explains the general tendency of many magmatic bodies in metamorphic terranes to be postmetamorphic.

Figure 20-6 includes the phase diagram for the Al_2SiO_5 polymorphs. The field gradient produced by our simple model lies almost entirely within the kyanite stability field, except at high temperatures, where it passes into the sillimanite field. This field gradient would consequently be characterized by a metamorphic facies series similar to the Barrovian series of Scotland. Clearly, conditions must be very different if the high-T, low-P, or low-T, high-P series are to form. Let us consider how the boundary conditions for Eq. 20-4 might be modified to produce these other facies series.

The high-T, low-P facies series passes through the andalusite stability field. The geothermal gradient must therefore be increased significantly over that resulting from the boundary conditions for which Figure 20-6 was generated. First, let us consider the effect of erosion rate on the field gradient. The more rapid the erosion, the greater the advective transfer of heat due to uplift (last group of terms in Eq. 20-4). Geotherms, consequently, rise to higher temperatures at shallower depths. But the more rapid the uplift, the steeper the $P–T–t$ paths become, and the lower is the temperature maximum on these paths. These two effects therefore tend to cancel. Higher temperatures can also be generated by greater thickening of the heat-producing layer. For example, the layer could be repeated by more thrust faults. Problem 20-2 explores the consequences of thickening the layer four times by displacements on three thrust faults. Although this does bring the geotherms closer to the andalusite field, it still does not give high enough temperatures. This leaves the advective transfer of heat by ascending bodies of magma as the most likely way of raising temperatures into the andalusite field (Prob. 20-2 shows how this magmatic advection can be taken into account in Eq. 20-4). Advection of metamorphic fluids could also raise temperatures, as could tectonic processes, such as thrusting and folding, that bodily move hot metamorphic rocks to shallower depths. If advection does not occur, the only other means of raising temperatures into the andalusite field is to have initial thermal gradients that are much steeper ($60°C$ km^{-1}) than the steady-state geotherm. As far as we know, this requires that the rocks be near a divergent plate boundary, such as an oceanic ridge or continental rift, or a hot spot (Sec. 1-4). These various possibilities are discussed further in Sec. 20-5.

Rocks of the low-T, high-P facies series (blueschists and eclogites) form at temperatures below the normal steady-state geotherm. Although the early stages of many $P–T–t$ paths pass through this region, continued heating normally destroys these mineral assemblages in favor of those of the other facies series. To prevent this heating, rapid erosion and uplift are necessary. These conditions exist today where oceanic plates are subducted beneath the accretionary wedge that develops at the convergent plate boundary. In such regions, the subduction of cold oceanic lithosphere may, in addition, provide relatively low initial geothermal gradients. Finally, many of the sedimentary rocks forming the accretionary wedge that are converted to blueschists are graywackes derived from nearby island arcs and continental margins or deep-sea pelagic sediment that lack a large input of material from ancient cratons. They consequently tend to be relatively poor in heat-producing radioactive elements. These factors combine to return rocks to the surface along $P–T–t$ paths that do not rise above the normal steady-state geotherm.

20-5 OBSERVED P–T–t PATHS

The real world is, of course, far more complicated than the simple model dealt with here, and to make matters worse, only small segments of $P–T–t$ paths are likely to be preserved in rocks. The model, however, provides a framework with which to make comparisons. The records are constructed from a variety of data, including mineral assemblages of known $P–T$ range, geothermometers and geobarometers based on exchange reactions, zoned porphyroblasts, retrograde reactions, and fluid inclusions. The record commonly has large uncertainties associated with it, and at elevated temperatures high reaction rates may erase the record down to *blocking temperatures,* where slowness of diffusion prevents further reaction. In this section we survey briefly the range of $P–T–t$ paths that have been identified and discuss their tectonic significance. This is done with reference to Figure 20-7, which, in addition to showing $P–T–t$ paths, includes the $P–T$ ranges of the various metamorphic facies, the steady-state continental geotherm (Sec. 1-6), and the solidus for a muscovite granite under conditions of both excess H_2O and no additional H_2O (Huang and Wyllie 1973).

The thickening of the crust that accompanies the collision of lithospheric plates first causes pressures to rise without significant heating. During this period, many rocks pass through the blueschist facies stability field, but only where one of the plates is oceanic are these rocks likely to be preserved. Ernst (1988) recognizes two general types of $P–T–t$ path for these

rocks, depending on whether the rock remains in the blueschist facies field during uplift (path A in Fig. 20-7) or passes into the greenschist or amphibolite facies fields (path B, Fig. 20-7), in which case retrograde greenschist facies minerals rim the higher-pressure blueschist facies minerals.

Continuous subduction of cold oceanic lithosphere produces steady cooling of the rocks immediately above the subduction zone. Consequently, progressively higher P/T ratios are produced, with eclogites forming at the greatest depths. These rocks must be returned to the Earth's surface if we are to examine them. Although composed of high-pressure mineral assemblages, blueschists have lower densities than mantle rocks into which they are subducted. Blueschists may therefore decouple from the descending slab and buoyantly rise to the surface, almost retracing their path of descent. Platt (1986) has suggested that this happens when the accretionary wedge that builds at a plate boundary becomes overthickened by underplating of material onto the upper plate. To maintain equilibrium between the gravitational forces acting on the surface slope and the drag exerted by the subducting plate, the overthickened wedge develops listric normal faults that dip toward the oceanic plate and penetrate to depths where they intersect the blueschist facies rocks. Slices of these rocks are then transported upward and oceanward on the toes of these faults. The process is analogous to the way in which snow, building up in front of a snowplow, will slump forward when it becomes too deep. Regardless of the actual mechanism by which blueschists rise, their proximity to the cold descending plate in which endothermic dehydration reactions are occurring keeps the rocks "refrigerated" during ascent and preserves the low-T, high-P mineral assemblages. The resulting $P–T–t$ path, which is exemplified by the rocks of the Franciscan complex of California, is shown by path A in Figure 20-7.

If during subduction of an oceanic plate a small microcontinent or island arc with silicic rocks descends the subduction zone, the rapid, buoyant rise of such material would occur at almost constant temperature, or perhaps even initially slightly increasing temperatures if the silicic material contains significant amounts of radioactive elements. In this case, blueschists would rise into the stability fields of the greenschist or amphibolite facies and develop retrograde minerals characteristic of those facies. The resulting $P–T–t$ path (B, Fig. 20-7), for example, is found in blueschists of the western Alps.

When convergent lithospheric plates involve continental collisions, the thickened crust may initially pass through the blueschist facies field, but with the production of radiogenic heat the path soon enters the greenschist and amphibolite facies fields. A Barrovian sequence of rocks therefore develops, the greenschists forming along shallow, and the amphibolites along deeper, $P–T–t$ paths (C and C', respectively, in Fig. 20-7). Isograds plot as points along the resulting field gradient. It is interesting to note that although reactions can be written relating the mineral assemblages on either side of an isograd, the $P–T–t$ paths followed by the rocks are very different from the $P–T$ array of the field gradient, and thus the rocks may never have undergone the deduced reactions.

Perhaps the best documented $P–T–t$ path for a region involving continent-continent collision is that for the southwestern Tauern Window in the eastern Alps (Selverstone et al.

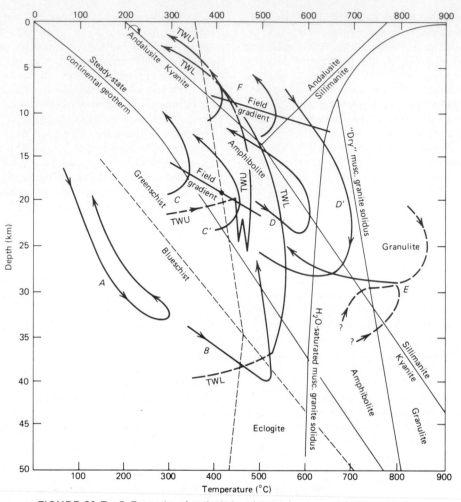

FIGURE 20-7 *P–T–t* paths of rocks belonging to the three main metamorphic facies series. Also shown is a steady-state continental geotherm (see Sec. 1-6), general *P–T* range of metamorphic facies, the Al₂SiO₅ phase diagram (Holdaway 1971), and the H₂O-saturated and "dry" solidus of muscovite granite (Huang and Wyllie 1973). Blueschists of the Franciscan complex, California (*A*), and the western Alps (*B*) (Ernst 1988); *C* and *C'*, typical Barrovian sequence; *D* and *D'* from central Massachusetts (Tracy and Robinson 1980); *E*, granulites from the Adirondacks, New York State (Bohlen et al. 1985); *F*, Trois Seigneurs Massif, Pyrenees (Wickham and Oxburgh 1985). TWL and TWU, lower and upper stratigraphic units in the Tauern Window, eastern Alps (Selverstone et al. 1984; Selverstone and Spear 1985). See text for discussion.

1984; Selverstone and Spear 1985; Blackenburg et al. 1989). Final metamorphic equilibration temperatures are indicated by the garnet–biotite geothermometer (see Eq. 17-12) (Ferry and Spear 1978; Hodges and Spear 1982), and metamorphic pressures are determined from the compositions of garnet and plagioclase coexisting with kyanite and quartz (see Eq. 17-16) (Ghent et al. 1979) or biotite and muscovite (Ghent and Stout 1981; Hodges and Crowley 1985). *P–T* conditions prior to peak metamorphic temperatures are obtained from zoned garnets according to the method of Spear and Selversone (1983; see Sec. 17-8). In addition, epidote–plagioclase–chlorite pseudomorphs after lawsonite provide textural evidence of an early stage of metamorphism in the lawsonite–albite–chlorite subfacies of the

blueschist facies. Finally, fluid inclusion data constrain the *P–T–t* path during uplift.

Two *P–T–t* paths are shown in Figure 20-7 for the Tauern Window rocks, one for the lowest stratigraphic unit exposed (TWL) and the other for the uppermost unit (TWU). Both paths begin in the blueschist facies, the lower unit being at a pressure of >1 GPa (35 to 40 km) and the upper one at a pressure of >0.7 GPa (25 to 30 km). The conditions in these two units at the peak of metamorphism were 550°C and 0.7 GPa (25 km) and 475°C and 0.55 GPa (20 km), respectively. Garnet zoning patterns in the upper unit reveal four distinct phases of tectonic activity (see sawtooth nature of *P–T–t* path in Fig. 20-7). Although the lower and upper units were initially separated by a vertical distance of 10 km, fluid inclusion data reveal that this distance had been reduced to less than 2 km by the time the temperature had decreased to 375°C. This indicates that uplift was due not only to erosion but also to tectonic thinning.

Following the collision and subduction of the European plate beneath a continental fragment to the south, pressures in the rocks increased, first as a result of imbrication of subducted units, and then in response to the development of the Austroalpine nappes. Uplift and arching of the lower unit of the Tauern Window resulted in ductile deformation and thinning of the overlying rocks at the same time the rocks were being exhumed by erosion (Selverstone 1985).

Differences in the *P–T–t* paths of rocks involved in continental collisions can be explained by differences in the types of deformation and the position of rocks relative to major structures, such as thrust faults and nappes. Tracy and Robinson (1980), for example, have shown from careful analysis of zoned garnet crystals (Tracy et al. 1976) that in central Massachusetts rocks within a few kilometers of each other followed very different *P–T–t* paths because of their structural positions. These rocks were deformed during the Devonian Acadian orogeny, first, into a series of nappes overturned to the west, followed by a period of backfolding to the east, and finally, a doming stage, which saw the formation of the Bronson Hill anticlinorium and, immediately to its east, the Merrimack synclinorium (Fig. 17-17). The early nappes are highly deformed and are found in both the anticlinorium and synclinorium. During nappe building, cold near-surface rocks were buried beneath large recumbent folds, which resulted in high pressures (kyanite field). Toward the top of these same nappes, relatively hot rocks, transported from depth, were exposed to low pressures (andalusite field). Subsequently, during the backfolding and doming stages, the rocks that had been deeply buried rose as they became hotter from radiogenic heat, producing a counterclockwise *P–T–t* path (*D*, Fig. 20-7). At the same time, the high-*T*, low-*P* rocks sank into the Merrimack synclinorium, tracing a clockwise *P–T–t* path (*D'*, Fig. 20-7). Thus, in this region the *P–T–t* paths were determined largely by the deformation paths.

In many regionally metamorphosed terranes, granulite facies rocks record the highest metamorphic conditions attained. Temperatures rise well above 600°C, which is the blocking temperature for significant diffusion in garnet. Porphyroblasts, consequently, are remarkably homogeneous, and no record of the early part of *P–T–t* paths is preserved. Retrograde reactions, however, provide evidence of conditions following the peak of metamorphism. Granulite facies terranes exposed at the Earth's surface today are underlain by crustal thicknesses of about 35 km. Mineral assemblages in these rocks indicate burial to similar depths, so at the time of metamorphism the crust must have been more than 60 km thick. Granulites thus form in thickened crust similar to that which gives rise to the Barrovian sequence, but they form along *P–T–t* paths that penetrate to greater depths where pressures reach 0.7 to 0.8 GPa and temperatures reach 700 to 800°C.

The *P–T–t* paths in many granulite terranes, such as the Grenville age Adirondacks of New York State (Bohlen et al. 1985), commonly exhibit a 200 to 300°C period of almost isobaric cooling from peak metamorphic conditions before uplift begins; this results in *P–T–t* paths that are concave toward the temperature axis (*E*, Fig. 20-7). Such paths may be spurious indicators of *P–T* conditions because of the ease with which granulite facies mineral compositions readjust themselves. If they are valid, however, how could cooling occur without simultaneous unloading? There would appear to be two possible explanations, the removal of partial melts or flushing with CO_2-rich fluids. Let us examine both of these.

P–T conditions in the granulite facies exceed the minimum melting temperatures of muscovite-bearing quartzofeldspathic rocks. Figure 20-7 includes the solidus of H_2O-saturated muscovite granite and of muscovite granite containing no additional H_2O (Huang and Wyllie 1973). Because of the low porosity of high-grade metamorphic rocks, the amounts of H_2O in granulites is not expected to be large. Thus when *P–T–t* paths cross the H_2O-saturated solidus, only small amounts of melt are formed, and not until temperatures rise above the "dry" solidus do amounts become substantial. Nonetheless, the importance of these first-formed melts lies in their strong affinity for the radioactive heat-producing elements. The buoyant rise of the melts depletes the granulite zone in these elements. Thus, following the initial thickening of the crust, temperatures would be expected to rise steadily until sufficient melt was formed to produce buoyantly rising bodies of magma that would take with them the heat-producing elements; at this point the rocks would begin to cool, but not as a result of crustal uplift.

The nearly isobaric cooling of granulite rocks could be a consequence of the influx of hot material; that is, the cooling path reflects the relaxation of a thermal pulse, which raised the temperature gradient above that which could be sustained by the production of radiogenic heat. Pulses of magma from the mantle could certainly do this, but many granulite terranes are devoid of large mafic bodies. Such bodies could, however, underplate the continental crust (Bohlen and Mezger 1989). If they were intruded early enough in the tectonic cycle, they could even result in clockwise *P–T–t* paths (Bohlen 1987). It has been suggested that granulite metamorphism results from the flushing of rocks with CO_2-rich fluids from the mantle (Newton et al. 1980; Frost and Frost 1987). The general anhydrous nature of granulites is then explained as much by fluid composition as it is by high temperature. Fluids rising from the mantle could advectively raise the temperature of granulites. Then, when the flux of fluid decreases or stops, the rocks would start cooling as they relax back to the gradient that existed prior to the flushing. This then provides another means of cooling the rocks without the necessity of unloading. Whether the cooling paths of granulites record the loss of magmas that take with them the heat-producing elements, or the relaxation of a thermal pulse from advective fluids could be answered if the conditions prior to peak metamorphism were known (Prob. 20-3).

The high-*T*, low-*P* facies series undoubtedly owes its elevated temperatures to the emplacement of magma (Miyashiro 1961) and can be thought of as contact metamorphism on a regional scale. In northern New England, for example, metamorphic rocks, which belong to this facies series, increase in metamorphic grade toward the ubiquitous granitic intrusions, and metamorphic isograds are subparallel to igneous contacts (Lux et al. 1986). In some places, isograds are overprinted by others that parallel contacts with later intrusions. Lux et al. (1986) have shown that the P–T–t paths of these rocks can be successfully modeled by thermal pulses emanating from gently dipping sill-like bodies of granitoid rocks.

Although the high temperature in these rocks can be attributed directly to the proximity of igneous intrusions, it does not account for the ultimate source of the heat causing melting. What tectonic conditions give rise to the high-T, low-P facies series? Oxburgh and Turcotte (1971) have modeled the thermal budget associated with convergent plate boundaries and shown that the paired metamorphic belts of Miyashiro (1961), that is, low-T, high-P and high-T, low P, are a natural consequence of this type of plate motion. The low-T, high-P series forms nearest the oceanic trench where temperatures are below normal because of the subduction of cold oceanic lithosphere. With continued subduction and heating, melting takes place, and ascending magma bodies transfer heat into the upper crust to produce the high-T, low-P facies series. The ascent of magma also causes crustal extension and may be responsible for the spreading that is common behind island arcs.

Oxburgh and Turcotte's model calls for crustal thickening, but in some areas, high-T, low-P metamorphic rocks are formed in very thin crust. The Chugach metamorphic complex of southern Alaska is one such region in which seismic refraction studies show the subducted oceanic plate to be no deeper than 10 km. In this area melting of the accreted sedimentary wedge may have resulted from the subduction of very young, hot oceanic crust. The upward transfer of heat was effected both by ascending magmas and fluids liberated by the subducted sediments and ocean floor rocks (Sisson and Hollister 1988).

In some regions high-T, low-P metamorphism is not closely associated with igneous rocks. In the Big Maria Moun-

tains of southeastern California, for example, massive wollastonite is developed in siliceous limestone on a regional scale (Hoisch 1987). The minimum volume ratio of fluid to rock can be calculated to have been 17:1 from Eq. 18-46, according to the method of Rice and Ferry (1982) (see Sec. 18-7). Hoisch concludes, therefore, that the high temperatures recorded by these rocks resulted from heat introduced by large fluid fluxes.

One of the largest developments of high-T, low-P rocks is in the Pyrenees. The metamorphic field gradient and schematic P–T–t paths for these rocks is shown as F in Figure 20-7. Although this metamorphic belt has been interpreted as resulting from Hercynian mountain building, it lacks many of the characteristics of mountain belts formed at convergent boundaries (Wickham and Oxburgh 1985). It lacks metamorphosed ocean floor rocks and has no large-scale nappes and thrusts. All of its rocks belong to the high-T, low-P series, that is, there is no evidence of high pressures during metamorphism. Wickham and Oxburgh interpret this metamorphism as resulting from an elevated geothermal gradient formed during continental rifting. Instead of a convergent plate boundary, they visualize an extensional basin, or en echelon series of basins, in which elevated temperatures were a result of lithospheric thinning. Large-scale melting of crustal rocks occurred at a relatively shallow depth (see intersection of field gradient with granite solidus in Fig. 20-7), and these rose to higher levels bringing with them the heat to cause metamorphism.

Perhaps the high-T, low-P facies series is not diagnostic of a single tectonic setting but indicates only regions of the Earth where advective heat transfer by magmas or fluids has played an essential role. But as in the case of other types of metamorphism, the P–T–t paths provide an important means of constraining tectonic models.

PROBLEMS

20-1. If the rate of a metamorphic reaction is determined by the rate of diffusion of Fe and Mg in garnet, calculate how much more rapid the prograde reaction would be than the retrograde one if both reactions overstep the equilibrium temperature of 800°C by 50°C, and the activation energy is 344 kJ mol^{-1} (Freer 1981).

20-2. The object of this problem is to calculate P–T–t paths for a crust that has been thickened by a series of thrust faults. Prior to faulting, radioactive heat-producing elements are concentrated in the upper 16.8 km where they give the rock a heat generating capacity of 2 μW m^{-3}. Three successive thrust slices, each of which penetrates to the base of the heat-producing layer, are emplaced on top of the original crust. The heat-producing layer is then four times as thick (67.4 km). We will assume that the thrust faults form instantaneously at time $t = 0$. Such thickening should cause the geotherm to be raised well above the initial steady state geotherm. We will calculate the position of the geotherm at 1-Ma intervals. The rate of erosion is assumed to be 0.4 m per 1000 a. All rocks have a thermal conductivity of 2.25 W m^{-1} K^{-1}, a thermal diffusivity of 10^{-6} m^2 s^{-1}, and

a heat content $(C_p \rho)$ of 2.5 × 10^6 J m^{-3} K^{-1}. The heat flow at the Earth's surface prior to faulting was 60 mW m^{-2}. Note that because the time increments are to be 1 Ma, the depth increments must be 5.616 km. We will now proceed with this problem in steps.

(a) First, calculate the steady-state geotherm prior to faulting; that is, with a single 16.8-km-thick heat-producing layer (review Prob. 1-4); take the surface temperature to be 0°C. Calculate temperatures for 5.6-km depth increments to a depth of 120 km; this will allow results to be used in the spreadsheet calculation of the numerical solution of the change in temperature with time.

(b) Next we introduce the three thrust faults. Starting at the Earth's surface where the temperature is 0°C, we proceed as in part (a), but at a depth of 16.85 we reset the temperature to zero and start the calculation over again. This is repeated at each thrust fault. Note that the heat-producing layer is terminated 16.8 km below the lowest thrust fault. Extend your calculations to a depth of 120 km. Plot a graph of the resulting geothermal gradient, which should have three cusps to it.

(c) Using the faulted geotherm for initial temperatures, calculate the geothermal gradient at 1-Ma intervals using the numerical solution of Eq. 20-4. The faulted geotherm is entered in the first column of the spreadsheet calculation with successive columns being used for the geotherms at following 1-Ma intervals. The surface temperature is always 0°C and the temperature of the deepest cells can be calculated using Eq. 20-5. Because the erosion rate is 4×10^{-4} m a^{-1}, the boundary between the lowest heat-producing layer and the underlying rocks must be moved up after an appropriate amount of time. Plot geotherms for 0, 5, 10, 15, 20, 25, 30, 35, and 40 Ma. This iterative calculation may take the computer an hour to complete.

(d) Construction of $P-T-t$ paths. Knowing the rate of erosion, follow the $P-T-t$ paths of two rocks that are initially at depths of 18 and 25 km by plotting their depths on the geotherms at successive times. This is best shown in a graph that extends to a depth of only 30 km.

(e) The metamorphic field gradient is plotted by drawing a line through the maximum temperatures on the $P-T-t$ paths.

(f) On the same diagram, plot the Al_2SiO_5 phase diagram. The triple point is at a pressure of 0.376 GPa and temperature of 501°C. The 1-atm equilibrium for the kyanite–andalusite reaction is 200°C, and that for the andalusite–sillimanite reaction is 770°C. The kyanite–sillimanite reaction passes through a point at 1 GPa and 810°C (Holdaway 1971).

(g) To which facies series of Miyashiro (1961) do these rocks belong?

(h) If, after faulting, 10 Ma elapsed before significant erosion began, how deep would a sample have to be for it just to enter the kyanite field at the maximum temperature along the $P-T-t$ path after 40 Ma since the period of faulting?

(i) Despite the great thickening of the crust containing radioactive elements by the three thrust faults and its subsequent generation of heat, rocks show very little likelihood of entering the andalusite field. What geological conditions might raise the geotherm into this field?

(j) Clearly, the diapiric rise of magma bodies from deeper in the section would elevate the geotherms, as would advection of large quantities of hot fluids. Can you think of a simple way of incorporating the rise of magma into the numerical solution?

(k) For an order-of-magnitude calculation of the effect of rising magma on the geothermal gradient of a metamorphic terrane, magma can be treated simply as an advective metamorphic fluid (Eq. 18-59). In detail, magma would form intrusive bodies that would only locally perturb thermal gradients [see Lux et al. (1986)] instead of being generally dispersed as would be metamorphic fluids. But such distinc-

tion need not concern us if we are evaluating only the overall effect of magmatic intrusion on a metamorphic terrane. Consequently, we will treat the terrane as having a permeability to magma without regard to whether flow is channelized or dispersed. The problem then is to evaluate the permeability.

In the first part of this problem the only movement of rock that occurred after thrusting was the uplift due to erosion. We must now consider, in addition, the advection of magma. First, how rapidly do calc-alkali magmas rise? Marsh and Kantha (1978) calculated that a 6-km-spherical diapir would solidify on its way to the surface if it did not rise more rapidly than 3 m a^{-1} (Sec. 3-9). We will use this figure for the ascent velocity. Of course, the percentage of a metamorphic terrane that consists of such rapidly moving bodies at any given time would be very small. We will assume that magmatic bodies constitute 5% of a metamorphic area (see Eq. 18-2 for analogy with fluid flow) and that magmatic pulses rising at 3 m a^{-1} occur every 100 years (approximate frequency of pulses in Mount St. Helens, for example). This gives a time- and space-averaged magmatic flux of 1.5×10^{-3} m a^{-1}. This can be added to the advective uplift due to erosion, but the latter is so small by comparison that it can be ignored.

Equation 20-4 can be used to determine the change in the geothermal gradient resulting from the chosen flux of magma by replacing the advective uplift term with the magmatic advective quantity of 1.5×10^{-3} m a^{-1}. Repeat part (c) of the problem using this new value.

(l) Draw $P-T-t$ paths for rocks that are initially at depths of 18 and 25 km, and indicate the resulting field gradient. To what metamorphic facies series would these rocks now belong?

20-3. Two possible explanations for the almost isobaric cooling of many granulite facies terranes involve either the extraction of a magmatic fraction that took with it the heat-producing elements or early flushing of the rocks with hot CO_2-rich fluids. Construct schematic $P-T-t$ paths for these two scenarios.

20-4. The following analyses are of a zoned garnet porphyroblast and included biotite grains. Using the method of Spear and Selverstone (1983) (Fig. 17-15) construct a $P-T-t$ path for the rock assuming that the average density of the rocks is 2.8 Mg m^{-3}. What tectonic history might this rock have experienced during metamorphism?

	Core $\longleftarrow$		$\longrightarrow$ Rim	
X_{Alm}^{G}	0.865	0.875	0.895	0.910
X_{Ann}^{Bi}	0.60	0.58	0.67	0.65

21 / Isotope Geochemistry Related to Petrology

21-1 INTRODUCTION

Some of the major petrologic advances in recent years can be attributed to isotope geochemistry. The first contributions were in the field of absolute age determinations. Not only was the long-debated question of the age of the Earth settled, but absolute dating provided a means of unraveling the chronology of that large fraction (90%) of Earth history that had previously been designated "Precambrian" because of its lack of an adequate paleontological record with which to subdivide it. With increased refinements and new techniques, absolute dating now tackles problems ranging from the cosmological, such as determining the age of heavy elements in our solar system, to the very specific, such as the time of last movement on faults at a potential nuclear power plant site.

Perhaps of still greater importance to petrology is the contribution that isotope geochemistry has made to determining the provenance of magmas. Isotopic analyses can distinguish between magmas of mantle and crustal origin. Also, degrees of contamination of mantle-derived magmas by crustal rocks can be measured through isotopic changes that could not be detected using major element chemistry. Isotopes can also indicate how much meteoric water is circulated through the rock cycle. In addition, isotopic studies reveal important information about the mantle itself, such as how its composition can change with time as a result of removal of partial melts or introduction of metasomatizing fluids.

Isotope geochemistry is such a large subject that only some of its petrologically most important aspects can be dealt with in this chapter. What is an isotope, and how does isotope geochemistry differ from what might be referred to as normal geochemistry? Atoms of an element that have different numbers of neutrons are known as *isotopes*. They consequently have different atomic masses, but their number of protons and electrons are the same (for a particular element), and thus normal chemical processes do not distinguish between most isotopes. Indeed, isotopic analyses cannot be done by normal chemical means but must be done with a mass spectrometer. At first, the

inability of chemical processes to fractionate most isotopes suggests that isotopic analyses of rocks might serve little use. This, however, is not the case. The lack of fractionation of most isotopes by chemical processes means that once a rock has developed a particular isotopic ratio, this ratio will remain unchanged except in special cases, regardless of the chemical or physical processes that take place; thus the isotopic ratio serves as an indelible fingerprint with which to identify the rock. Elements themselves cannot possibly be used in this way, because every process induces some chemical change, which can be explained only if the process is fully understood (this is usually not the case).

Isotopic ratios can change through *radioactive decay* and *mass fractionation*. Some isotopes are radioactive and decay, at known rates, into isotopes of other elements. For example, ^{87}Rb, which substitutes for K in minerals, decays to ^{87}Sr, which in turn substitutes for Ca in minerals. Rubidium and strontium are chemically so different that they will be fractionated if exposed to processes such as partial melting. If left undisturbed, however, this pair provides the basis of one of the most useful absolute age determination methods. Isotopes, with an atomic mass of less than about 20, can be fractionated to a measurable extent by some physical and chemical processes in which the mass of the isotopes makes a difference. For example, light isotopes can escape more easily than heavy ones during evaporation; consequently, ratios of the isotopes of H, He, and O in the surface layer of the oceans are depleted in the light isotopes relative to their ratios in the deep ocean.

In general, most isotopic studies of petrologic interest fall into one of three main categories: (1) absolute dating using radioactive isotopes, (2) evolution of isotopic reservoirs in the mantle and crust, and (3) stable isotopes as indicators of ancient environments. These may overlap considerably, but the divisions provide a convenient basis for a brief survey of the subject, which is all that can be done in this chapter. For more extensive coverage, the books by Faure (1986) and Jäger and Hunziker (1979) and the articles by Hart and Allègre (1980), O'Nions et al. (1980), DePaolo (1981), and O'Nions (1984) are recommended.

21-2 RADIOACTIVE DECAY SCHEMES

Some isotopes are inherently unstable and will, with time, change or "decay" to stable isotopes. The rate of decay is determined only by the instability of the radioactive nucleus and cannot be changed by external forces. This immutability is the basis of isotopic age determinations involving such element pairs as Rb–Sr, U–Pb, and K–Ar. The first element of each of these pairs has an isotope (or isotopes) that decays to an isotope of the second. Rubidium has two isotopes ^{85}Rb and ^{87}Rb; the first is stable, but ^{87}Rb decays to ^{87}Sr, which is just one of four stable isotopes of strontium, the others being ^{88}Sr, ^{86}Sr, and ^{84}Sr. Uranium has several isotopes, two of which, ^{238}U, and ^{235}U, decay to ^{206}Pb and ^{207}Pb, respectively. Potassium has three isotopes, ^{39}K, ^{40}K, and ^{41}K, but only ^{40}K is unstable. It, however, can decay in two ways, either to ^{40}Ca or to ^{40}Ar. Each of these reactions involves a change in the nucleus of the parent isotope, but in each case the way in which this takes place is different.

A nucleus can change by radioactive decay in four different ways. One is by emitting two neutrons and two protons from the nucleus. This group of four nuclear particles is equivalent to a helium nucleus and is known as an *alpha particle* (α). It has an atomic mass of 4 and a charge of $+2$. Another mode of decay involves the emission of an electron from the nucleus. This electron, which is known as a *beta particle* (β), is formed when a neutron in the nucleus changes into a proton. β particles have negligible mass. A third type of decay involves the capture of an electron by the nucleus, where it combines with a proton to form a neutron. This decay process is known as *electron capture* (cc). The fourth mode affects only uranium and thorium and involves the *spontaneous fission* of the nucleus into two nuclei of approximately equal mass.

Let us consider what changes must take place in the nucleus of $^{87}_{37}Rb$ in order for it to change to $^{87}_{38}Sr$. The superscripts here refer to the atomic mass of the *isotope*, and the subscripts refer to the atomic number of the *element*. The atomic number is the number of protons in the nucleus, which defines that element. Because the atomic masses of ^{87}Rb and ^{87}Sr are the same, the decay process cannot involve α particles. The change, instead, involves an increase by one in the number of protons. This can be achieved only by a neutron changing into a proton, which requires that a β particle be emitted from the nucleus. The reaction can be written as

$$^{87}_{37}Rb \longrightarrow ^{87}_{38}Sr + \beta$$

The decay of uranium isotopes to those of lead is more complicated than that of rubidium to strontium. Both ^{238}U and ^{235}U decay in a long chain of reactions involving intermediate radioactive isotopes of the elements Po, Th, Ra, Pa, Bi, At, Ac, Rn, Fr, and Tl. Some of these isotopes require thousands of years to decay, whereas others last only fractions of a second. Despite the complexity of the chain reactions, the overall nuclear reactions can be written easily. The decay of $^{238}_{92}U$ to $^{206}_{82}Pb$ involves a mass loss of 32, which indicates that 8 α particles must be emitted. Because each α particle has a

charge of $+2$, the loss of 8α particles would cause the charge on the nucleus to drop to 76, but it only decreases to 82, the atomic number of lead. Consequently, the loss of 8α particles must be accompanied by the emission of 6β particles. The overall reaction can then be written as

$$^{238}_{92}U \longrightarrow (chain) \longrightarrow ^{206}_{82}Pb + 8\alpha + 6\beta$$

In a similar way, the decay of $^{235}_{92}U$ can be written as

$$^{235}_{92}U \longrightarrow (chain) \longrightarrow ^{207}_{82}Pb + 7\alpha + 4\beta$$

The decay of $^{40}_{19}K$ to $^{40}_{18}Ar$ and $^{40}_{20}Ca$ involves no change in mass and so cannot involve α particles. The charge on the nucleus decreases by one when $^{40}_{19}K$ changes to $^{40}_{18}Ar$, so an electron must be captured by the nucleus, where it combines with a proton to form a neutron. The reaction can be written as

$$^{40}_{19}K + ec \longrightarrow ^{40}_{18}Ar$$

The decay of $^{40}_{19}K$ to $^{40}_{20}Ca$, on the other hand, involves an increase in the nuclear charge and must therefore involve emission of a β particle. The reaction is

$$^{40}_{19}K \longrightarrow ^{40}_{20}Ca + \beta$$

Of these three decay processes, only the emission of α particles has any significant effect on the material surrounding the decaying atom. Where this material is pleochroic, the bombardment by α particles affects the way in which light is absorbed. The result is a darkened region around the radioactive source which is known as a *pleochroic halo* (Fig. 21-1a). With time, pleochroic haloes grow more intense. Unfortunately, there are too many variables to make halo intensity a reliable means of determining absolute ages.

Having far greater effect on the surroundings than α particles are particles given off by the spontaneous fission of ^{238}U and to a very much smaller extent of ^{235}U and of ^{232}Th. These heavy nuclei split into two nuclei of approximately equal mass and liberate about 200 MeV of energy. Following fission, the two nuclei recoil in opposite directions from each other, and as they pass through the surrounding mineral, they damage its structure, leaving a *fission track* (Fig. 21-1b). Although these damaged zones are so small that they are visible only in the electron microscope, etching of polished surfaces can highlight and enlarge them to the point that they are readily visible in a normal optical microscope. The older a mineral is, the more fission tracks it will contain for a given content of ^{238}U, and on this is based the fission track method of absolute dating.

21-3 RATE OF RADIOACTIVE DECAY

All radioactive decay, whether involving α particles, β particles, or electron capture, is a statistical process; the more atoms present of the radioactive nuclide, the more chance a decay has

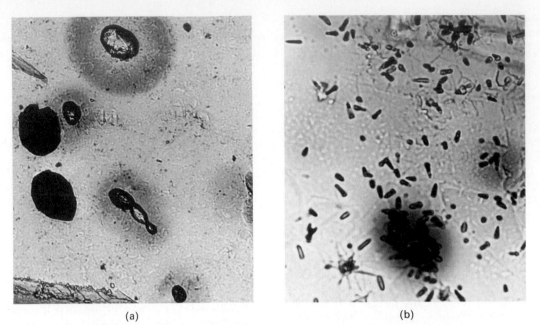

(a) (b)

FIGURE 21-1 (a) Yellow pleochroic halos surrounding zircon crystals in cordierite. Zircon typically contains small amounts of U and Th, which, on undergoing radioactive decay, affect the structure of the surrounding mineral. Plane light; the width of field is 0.3 mm. (b) Fission tracks in biotite, revealed by etching for 20 s with concentrated hydrofluoric acid. Individual tracks are formed by single uranium atoms that have spontaneously split into two atoms of approximately equal mass. Clusters of tracks near the bottom of the photograph are associated with a pleochroic halo, at the center of which must be a small radioactive mineral. The pleochroic halo near the top of the photograph is beneath the surface, so tracks within it have not been etched. Plane light; the width of field is 0.2 mm.

of occurring. This can be expressed mathematically as

$$\text{rate of decay} \propto N \qquad (21\text{-}1)$$

where N is the number of atoms present of the radioactive nuclide. According to the nomenclature of chemical reaction rates, such a process is referred to as a *first-order* reaction because the power to which N is raised in the rate expression is 1. We can write the rate of this process as the change in the number of atoms with time, inserting a minus sign to indicate that the change is negative, that is, decay:

$$\frac{dN}{dt} \propto -N \qquad \text{or} \qquad \frac{dN}{dt} = -\lambda N \qquad (21\text{-}2)$$

where the constant of proportionality, λ, is known as the *decay constant*. Typical values of λ range from 10^{-4} a^{-1} for the rapid decay of ^{14}C to ^{14}N, to 10^{-12} a^{-1} for the very slow decay of ^{147}Sm to ^{143}Nd (Table 21-1).

TABLE 21-1

Radioactive decay schemes and rate constants

Radioactive Isotope	Daughter Isotope	Type of Decay	Decay Constant (a^{-1})	Half-life $t_{1/2}(a)$
$^{14}_{6}C$	$^{14}_{7}N$	β	1.21×10^{-4}	5730
$^{40}_{19}K$	$^{40}_{20}Ca, ^{40}_{18}Ar$	β, ec	5.543×10^{-10}	1.25×10^{9}
$^{87}_{37}Rb$	$^{87}_{38}Sr$	β	1.42×10^{-11}	48.8×10^{9}
$^{147}_{62}Sm$	$^{143}_{60}Nd$	α	6.54×10^{-12}	106.0×10^{9}
$^{232}_{90}Th$	$^{208}_{82}Pb$	$6\alpha, 4\beta$	4.9475×10^{-11}	13.9×10^{9}
$^{235}_{92}U$	$^{207}_{82}Pb$	$7\alpha, 4\beta$	9.8485×10^{-10}	0.704×10^{9}
$^{238}_{92}U$	$^{206}_{82}Pb$	$8\alpha, 6\beta$	1.55125×10^{-10}	4.47×10^{9}

Equation 21-2 gives the instantaneous change in the number of atoms with time. If we wish to know the total change in N over some interval of time, the equation must be integrated:

$$\int_{N=N_o}^{N=N_p} \frac{dN}{N} = -\lambda \int_0^t dt$$

Here N_p and N_o are, respectively, the present and original numbers of atoms of the parent nuclide. Integrating yields

$$\ln \frac{N_p}{N_o} = -\lambda t$$

or

$$N_p = N_o e^{-\lambda t} \qquad (21\text{-}3)$$

This equation describes how the number of atoms of the parent nuclide decays with time.

For purposes of comparison and illustration, the concept of *half-life* is introduced; that is, the length of time necessary for the number of atoms of a nuclide to decay to one-half the initial number. In this case, $\frac{1}{2}N_o$ is substituted for N_p in Eq. 21-3, and then the half-life, $t_{1/2}$, is given by

$$t_{1/2} = \frac{0.693}{\lambda} \qquad (21\text{-}4)$$

Thus the half-life for the decay of ^{14}C is 5730 a, and for ^{147}Sm is 1.06×10^{11} a. It is instructive to compare the half-lives given in Table 21-1, keeping in mind that the age of the Earth is 4.55×10^9 a. Since the formation of the Earth, approximately half the original ^{238}U has decayed, but very little ^{147}Sm has decayed. Isotopes that have a half-life considerably greater than the age of the Earth do not make accurate radiometric clocks, except for very old rocks, because of the analytical uncertainty associated with measuring extremely small amounts of daughter isotope. Conversely, isotopes with very short half-lives can be used for dating only young rocks, because of difficulties in accurately measuring extremely small amounts of parent isotope in old rocks.

Equation 21-3 is of no practical use because we cannot measure N_o, but we do know that each time a parent nuclide decays, it produces a daughter nuclide. Therefore, $N_o = N_p + N_d$, where N_d is the present number of atoms of daughter nuclide. Substituting this into Eq. 21-3 and eliminating N_o, we obtain

$$\frac{N_p}{N_p + N_d} = e^{-\lambda t}$$

which, on rearranging, gives

$$N_d = N_p(e^{+\lambda t} - 1) \qquad (21\text{-}5)$$

For an approximate solution to this equation, the term $e^{\lambda t}$ can be expanded in a power series as $1 + \lambda t + (\lambda t)^2/2! + (\lambda t)^3/3! + \cdots + (\lambda t)^n/n!$. Because λ is so small (about 10^{-10} a^{-1}),

only the first two terms of the expansion are significant. Substituting these into Eq. 21-5 gives the approximate relation

$$N_d \approx N_p \lambda t \qquad (21\text{-}6)$$

Before using Eq. 21-5 to determine the absolute age of a rock, we must consider the possibility that although radioactive decay has produced daughter nuclides from the parent, some daughter nuclide may have been present in the environment prior to the beginning of the decay of the system under consideration. In such a case, the present number of atoms of the daughter nuclide would be the sum of the initial number present, N_d^i, and those produced by decay from the parent in the system. We can rewrite Eq. 21-5 as

$$N_d = N_d^i + N_p(e^{\lambda t} - 1) \qquad (21\text{-}7)$$

This equation, then, allows us to determine the age of a rock if the numbers of atoms of parent and daughter nuclides can be determined. Although this can be done, in practice it is more convenient and more accurate, with a mass spectrometer, to measure isotopic ratios. For example, this can be done if the daughter element has another isotope that is stable and not affected by radioactive decay. We can then divide both sides of Eq. 21-7 by the number of atoms of the stable isotope, N_s, and obtain

$$\frac{N_d}{N_s} = \frac{N_d^i}{N_s} + \frac{N_p}{N_s}(e^{\lambda t} - 1) \qquad (21\text{-}8)$$

This, then, is the working equation for determining absolute ages from measured isotopic ratios involving parent, daughter, and stable nuclides.

Equation 21-8 contains two unknowns, the initial ratio, N_d^i/N_s, and the age, t. To solve for either of these unknowns requires two equations. These could be obtained by analyzing two different minerals that are known to have crystallized at the same time or analyzing two different rocks that had a common source and time of origin. There are special cases where the initial ratio is known, in which case an age can be obtained from a single analysis. We will examine examples of both of these cases.

Equation 21-8 is also the equation of a straight line (Fig. 21-2). It has an intercept on the ordinate of N_d^i/N_s, the initial ratio, and a slope of $(e^{\lambda t} - 1)$, which is proportional to the age. If a number of different minerals crystallize together, in a magma for example, each contains the same initial isotopic ratio, that of the magma, because there can be no fractionation of isotopes between phases at high temperature, especially of the heavy isotopes. Each mineral starts life, then, with the same isotopic ratio of N_d^i/N_s; that is, they lie on the horizontal dashed line in Figure 21-2. Each mineral, however, contains different quantities of the radioactive parent (points along the dashed line). With time, various amounts of daughter product are generated, depending on the amount of parent, so the initial ratio is added to by an amount $(N_p/N_s)(e^{\lambda t} - 1)$. At any instant in time the isotopic ratio in each of the minerals lies on a sloping straight line such as that shown in Figure 21-2. With increasing time

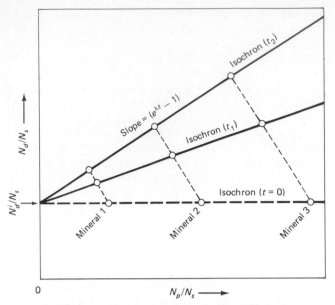

FIGURE 21-2 Plot of Eq. 21-8 where N_p is the number of atoms of the parent nuclide, N_d the number of atoms of the daughter element, and N_s the number of atoms of a nonradiogenic isotope of this same element. Three minerals formed in a rock at the same time have different concentrations of the parent nuclide but all have the same initial ratio of N_d^i/N_s. With time, the parent nuclide decays, and the amount of daughter product formed is proportional to the amount of parent present. At any time (t), the minerals plot on a straight line known as an isochron; the age can be determined from the slope of this line.

the slope of the line becomes steeper, but its intercept on the vertical axis remains the same.

In practice, the initial isotopic ratio is unknown. If, however, geological evidence indicates that a group of minerals were formed together at the same time, their isotopic ratios, when plotted in a diagram such as that of Figure 21-2, would be expected to lie on a straight line, from which could be determined both the initial isotopic ratio and the absolute age. Such a line is known as an *isochron* because all points on it have the same age. Isochrons can also be constructed using whole-rock isotopic analyses of rocks that have a range of compositions and that field evidence indicates were formed from a common source at approximately the same time—as, for example, a suite of differentiated rocks within a layered intrusion.

The *Rb–Sr method* of absolute dating is the most common technique making use of this type of analysis of the data. ^{87}Sr is formed from the decay of ^{87}Rb, but in any geological environment there is already some ^{87}Sr present, formed from earlier decay of ^{87}Rb or inherited from the formation of the solar system. Isotopic ratios of ^{87}Sr and ^{87}Rb are measured against the amount of the stable isotope ^{86}Sr, and an isochron fitted to the following equation:

$$\frac{^{87}\text{Sr}}{^{86}\text{Sr}} = \frac{^{87}\text{Sr}^i}{^{86}\text{Sr}} + \frac{^{87}\text{Rb}}{^{86}\text{Sr}}(e^{\lambda t} - 1) \qquad (21\text{-}9)$$

where ^{87}Sr^i/^{86}Sr is the initial ratio of these isotopes and λ is $1.42 \times 10^{-11}\,\text{a}^{-1}$. Figure 21-3 shows two typical isochrons fitted to data from two Monteregian intrusions in southern Quebec (Eby 1984). The slopes of the two isochrons are similar, and therefore the two intrusions have approximately the same age. The difference between the two isochrons is in their initial ^{87}Sr/^{86}Sr ratios, that for the granite from Mount Megantic having a value of 0.70518 and that for the pulaskite from Mount Shefford having a value of 0.70365. As will be shown in the next section, these differences in initial isotopic ratio can be used to distinguish mantle and crustal sources of magma, and, in the case of mantle-derived magmas, the amount of crustal contamination.

In addition to providing information on the initial ^{87}Sr/^{86}Sr ratios, the Rb–Sr method of dating is particularly useful because Rb and Sr enter common rock forming minerals, such as micas, feldspars, and amphiboles, substituting for K and Ca, respectively. Recrystallization of a rock is likely to cause radiogenic strontium to be expelled from these minerals because this strontium would have originally been present as ^{87}Rb substituting for K. These minerals, then, will have their radioactive clocks reset by recrystallization. The expelled ^{87}Sr, however, will enter nearby calcium-bearing minerals, such as apatite and plagioclase, so that the whole-rock Rb–Sr data will be unchanged. In this way, Rb–Sr dating, especially of large whole-rock samples, is able to see through metamorphic events (Fig. 21-4). Care must be taken, however, because whole-rock data

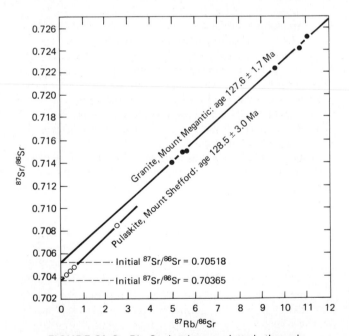

FIGURE 21-3 Rb–Sr isochrons plotted through data for rocks from two Monteregian intrusions, Quebec. The slopes of the two lines are similar; consequently, the intrusions have similar ages. Granitic rocks of Mount Megantic, however, have a higher initial ^{87}Sr/^{86}Sr ratio than do nepheline syenites (pulaskite) of Mount Shefford, indicating that Megantic rocks assimilated larger amounts of crustal rocks, which normally have a high ^{87}Sr/^{86}Sr ratio. [After Eby (1984).]

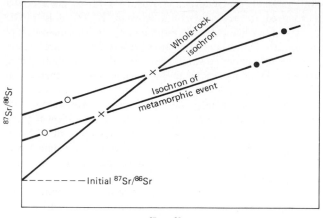

FIGURE 21-4 Large-sample, whole-rock analyses (X) for Rb and Sr isotopes can give reliable isochrons, despite metamorphic overprinting, if Rb and Sr have not moved out of the volume sampled. Local redistribution of Rb and Sr within this volume between minerals of high Rb (filled dots) and low Rb (open dots) allows isochrons to be constructed that give the age of the metamorphism.

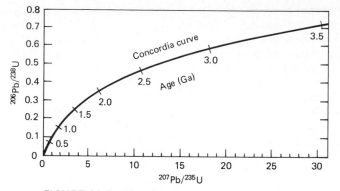

FIGURE 21-5 Plot of Eq. 21-12 showing the variation in the $^{206}Pb/^{238}U$ and $^{207}Pb/^{235}U$ ratios in a system that has remained closed to U and Pb. Ages in Ga are marked on the concordia curve.

can be affected by fluxes of fluids. Also, dating of young material is unreliable unless samples contain high concentrations of Rb, because of the long half-life ($t_{1/2} = 4.89 \times 10^{10}$ a) of the rubidium decay.

The *U–Pb method* of absolute dating is useful because the two decay schemes, ^{238}U to ^{206}Pb and ^{235}U to ^{207}Pb, which have different decay constants, provide independent measures of the age of a sample. Thus if uranium or lead has been removed from or added to a sample, the ages determined from the two methods will not agree; that is, the ages will be discordant. If, on the other hand, the sample has remained closed, the two ages will be concordant.

Using Eq. 21-5, we can write for the two decay schemes

$$^{206}Pb = {}^{238}U(e^{\lambda_{238}t} - 1) \qquad (21\text{-}10)$$

and

$$^{207}Pb = {}^{235}U(e^{\lambda_{235}t} - 1) \qquad (21\text{-}11)$$

where $\lambda_{238} = 1.55125 \times 10^{-10}$ a^{-1} and $\lambda_{235} = 9.8485 \times 10^{-10}$ a^{-1}. The slower decay rate of ^{238}U relative to that of ^{235}U means that the ratio of $^{206}Pb/^{238}U$ grows more slowly than that of $^{207}Pb/^{235}U$ during the early stages of decay. The relative changes in these isotopic ratios can be found simply by equating Eqs. 21-10 and 21-11 through a common time. This gives

$$\frac{^{206}Pb}{^{238}U} = \exp\left[\frac{\lambda_{238}}{\lambda_{235}} \ln\left(1 + \frac{^{207}Pb}{^{235}U}\right)\right] - 1 \qquad (21\text{-}12)$$

Equation 21-12, which is plotted in Figure 21-5, shows how the isotopic ratios of $^{206}Pb/^{238}U$ and $^{207}Pb/^{235}U$ change with time in a completely closed system. This line is known as

the *concordia* curve, because ages determined by either Eq. 21-10 or 21-11 on a closed system give the same (concordant) results and thus plot on the same line. The plot in Figure 21-5 is known as a concordia diagram, and is the most commonly used graph for displaying U–Pb age data.

Many geological systems have not remained closed since their formation, and because of the two different decay rates, any disturbance of the system results in different ages being determined from the two decay schemes. Because the ages do not match, they are said to be *discordant*. In the concordia diagram these discordant samples do not lie on the concordia curve, but instead, typically plot to the right of it (Fig. 21-6). Discordant samples that are genetically related commonly form a linear array in the concordia diagram (Fig. 21-6). Such a plot can contain considerable information about the history of a rock, but the interpretation is not always unambiguous and must be supported with sound geological field evidence.

The simplest and most common explanation for a linear array of discordant ages is that some initial age, given by the older intercept of the linear array with the concordia line, was

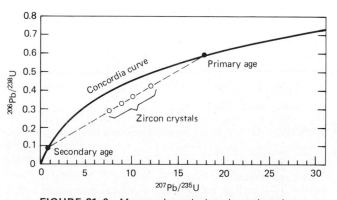

FIGURE 21-6 Many rocks and minerals, such as the zircon crystals in this diagram, do not plot on the concordia curve; that is, the ratios of $^{206}Pb/^{238}U$ and $^{207}Pb/^{235}U$ are not concordant and indicate that the system has been disturbed. One interpretation of the linear array exhibited by the zircons in this diagram is that they were originally formed 3.0 Ga ago but were later disturbed by an event 0.6 Ga ago.

disturbed by an event that occurred at the time given by the younger intercept of the linear array with the concordia line. For example, in Figure 21-6, the discordant ages shown are typical of what is obtained from a population of zircon crystals separated from a single large sample. The intercept of their array with the concordia line at 3.0 Ga could be the time of crystallization of these zircon crystals from a magma, whereas the secondary age of 0.6 Ga could be a metamorphic event which partially but not completely reset the U–Pb clocks. Numerous other, more complicated interpretations can be advanced, if supported with appropriate geological evidence. The ability of the U–Pb method to record the timing of successive events has made it a particularly powerful tool in unraveling complicated geologic histories.

The *K–Ar method* of dating differs from other common methods by involving a decay product that is an inert gas. Even at moderately low temperatures (see discussion below), this gas is a fugitive component and is typically not incorporated in minerals. Thus a newly formed mineral contains no argon to begin with, but with time, ^{40}K decays slowly to ^{40}Ar; this argon remains in place as long as the system is not disturbed. The method, in principle, then, is not affected by initial isotopic ratios, as is the Rb–Sr method.

The K–Ar method is complicated slightly by the fact that ^{40}K decays both to ^{40}Ar and ^{40}Ca. The decay constant for ^{40}K (λ_K) of 5.543×10^{-10} a^{-1} is actually the sum of the two decay constants, one from the production of ^{40}Ar ($\lambda_{Ar} = 0.581 \times 10^{-10}$ a^{-1}) and the other from the production of ^{40}Ca ($\lambda_{Ca} = 4.962 \times 10^{-10}$ a^{-1}). According to Eq. 21-5, the amount of decay product produced from ^{40}K is $^{40}K(e^{\lambda_K t} - 1)$. The decay products consist of both argon and calcium, which are present in amounts that are proportional to their respective decay rates. Thus the fraction of the total decay products consisting of ^{40}Ar at time t is

$$^{40}Ar = \frac{\lambda_{Ar}}{\lambda_K} \,^{40}K(e^{\lambda_K t} - 1) \qquad (21\text{-}13)$$

which on rearranging gives

$$t = \frac{1}{\lambda_K} \ln\left(\frac{\lambda_K \,^{40}Ar}{\lambda_{Ar} \,^{40}K} + 1\right) \qquad (21\text{-}14)$$

This equation expresses the age of a sample in terms of an easily measured isotopic ratio, $^{40}Ar/^{40}K$.

Absolute dates by the K–Ar method can be obtained on both minerals and whole rocks. These may indicate the time since the formation of a sample, but because most igneous and metamorphic rocks form at moderate to high temperatures, most dates indicate the length of time since the material dropped below some critical temperature at which diffusion of argon out of the sample became negligible. This temperature is known as the *blocking temperature*, and varies with the minerals involved, but is approximately 350°C. For large intrusive bodies or deeply buried metamorphic rocks slow cooling may produce a considerable discrepancy between the true age of a rock and its cooling age obtained from the K–Ar method. Such discrepancies can be found between Rb–Sr and K–Ar ages on the

same samples. Blocking temperatures for Rb–Sr are a couple of hundred degrees higher than those for K–Ar.

For an age determination by the K–Ar method to be accurate, the assumption that no radiogenic argon was present to begin with must be valid; also, no radiogenic argon that is produced in the mineral or rock can have escaped, nor can any radiogenic argon from an external source have been absorbed. Each of these can, in particular situations, be a source of serious error. Some environments definitely have an ambient fugacity of ^{40}Ar which can give a newly formed rock a significant initial content of radiogenic argon. For example, Mesozoic diabase dikes in Liberia, West Africa, which are associated with the early opening of the Atlantic Ocean, give K–Ar ages that are Mesozoic where the dikes cut Paleozoic rocks, but the ages are very much older where they cut Precambrian rocks, which, because of their age, contain high contents of ^{40}Ar (Dalrymple et al. 1975). Some minerals leak ^{40}Ar more easily than others by diffusion through their structure or along prominent cleavages, and thus they give erroneously young ages. The common K-bearing rock-forming minerals can be arranged according to retentiveness of ^{40}Ar, starting with the least retentive, microcline, and passing up through biotite, sanidine, pyroxene, muscovite, and ending with amphibole, the most retentive. Coarse-grained rocks tend to be less retentive than fine-grained ones. Some minerals, such as beryl, biotite, chlorite, clinopyroxene, and partially kaolinized feldspar, absorb ^{40}Ar and thus give erroneously old ages.

Typically, the outer parts of mineral grains are affected by gain or loss of argon. In the standard K–Ar dating method, there is no way of knowing precisely from where in the grains the analyzed potassium and argon comes—the analysis is simply of a bulk sample. In a modified technique that involves measuring the $^{40}Ar/^{39}Ar$ ratio, loosely bound argon on grain boundaries can be distinguished from that within the core of grains, and some of the problems of excess or lost argon can be eliminated. In the $^{40}Ar/^{39}Ar$ method, ^{40}K is not measured directly. Instead, it is first converted to ^{39}Ar by placing the sample in a flux of high-energy neutrons in a nuclear reactor. Measurement of ^{39}Ar, then, gives a measure of the ^{40}K content prior to bombardment. The sample is heated in a stepwise manner, and the $^{40}Ar/^{39}Ar$ released at each stage is determined. The daughter nuclide, ^{40}Ar, and the parent nuclide, ^{40}K, which is now in the form of ^{39}Ar, are released together as argon gas. The first argon released is the most weakly bound, probably coming from grain boundaries. At higher temperatures, more tightly bound argon from the core of grains is emitted. An age can be calculated from the $^{40}Ar/^{39}Ar$ ratio obtained at each step. If argon has been gained or lost, or the amount of potassium near grain boundaries has changed, variable $^{40}Ar/^{39}Ar$ ratios and ages would be obtained for the first few steps. With continued heating, however, the $^{40}Ar/^{39}Ar$ ratio and age should stabilize to produce a plateau on a graph of $^{40}Ar/^{39}Ar$ (or age) versus heating step (Fig. 21-7). Only when a significant plateau is obtained can any certainty be placed on the age determination.

Dating by the *Sm–Nd method* has been made possible in recent years by increased precision in mass-spectrometric techniques. ^{147}Sm decays to ^{143}Nd so slowly ($t_{1/2} = 1.06 \times 10^{11}$ a) that age determinations depend on measurements of very small

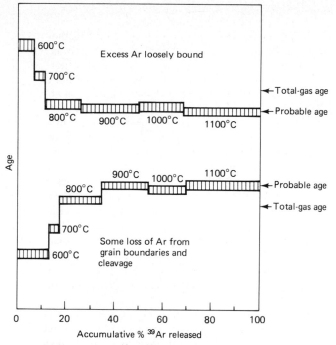

FIGURE 21-7 ^{40}Ar/^{39}Ar incremental-heating release spectra for two samples, one containing excess Ar and the other a deficiency of Ar. An age is calculated from the ^{40}Ar/^{39}Ar ratio in the gas released at each heating step. Only when the age given at each successive step remains constant (plateau) is the age likely to be reliable. Excess Ar, which might be absorbed on grain boundaries, is usually driven off at low temperature, giving erroneously old ages for the first steps. The age calculated from the total gas released (equivalent to the standard K–Ar method) is older than the plateau age if excess Ar is present. Loss of Ar from the sample gives erroneously young ages at the first heating steps, and the total-gas age is younger than the plateau age.

changes in the amount of ^{143}Nd. The method, which is only suitable for ancient rocks, is similar to that for Rb–Sr. Isotopic ratios of ^{147}Sm/^{144}Nd and ^{143}Nd/^{144}Nd are determined in a number of samples which are used to construct an isochron having the form

$$\frac{^{143}\text{Nd}}{^{144}\text{Nd}} = \frac{^{143}\text{Nd}^i}{^{144}\text{Nd}} + \frac{^{147}\text{Sm}}{^{144}\text{Nd}}(e^{\lambda t} - 1) \qquad (21\text{-}15)$$

where ^{143}Nd^i/^{144}Nd is the initial ratio of these isotopes. ^{144}Nd, the isotope against which the others are compared is radioactive, but with a decay constant of 2.89×10^{-16} a^{-1} it is effectively a stable isotope. In ancient samples (> 3.0 Ga), where the method has been used successfully, the initial ^{143}Nd^i/^{144}Nd ratio is very close to that for chondritic meteorites, which are considered to be the most representative samples we have of the primordial solar nebular material from which the planet Earth was formed. These ancient rocks, then, formed from a source that had changed little since the beginning of the Earth (O'Nions et al. 1980). One distinct advantage of the Sm–Nd

method is that both Sm and Nd, like the rest of the rare earth elements, are relatively immobile and remain in situ during weathering and even metamorphism. Rubidium and strontium, by contrast, are moved easily by solutions. The Earth's most ancient rocks from Isua, Greenland, which are 3.8 Ga old, are metamorphosed and yet yield an excellent Sm–Nd isochron.

Absolute dating by the *fission track method* differs from the previous ones in that it does not involve mass-spectrometric analysis. Its principle is simple (Fleischer and Price 1964; Fleischer et al. 1975). ^{238}U undergoes spontaneous fission at a rate of about 10^{-16} a^{-1}. The massive charged fission products damage the structure of the host mineral along a short track (5 to 20 μm). These tracks can be made visible by etching with appropriate chemicals and their number counted under an ordinary optical microscope. For a given content of ^{238}U, the age is proportional to the number of tracks per unit area. ^{235}U and ^{232}Th also undergo fission, but this happens so rarely that they do not contribute significantly to the development of fission tracks. Even the fission rate of ^{238}U is very slow, but fission tracks record every fission event, and thus their measurement is not limited by analytical precision as in mass-spectrometric analyses.

The procedure for determining an age by the fission track method first involves finding a mineral with an appropriate ^{238}U content to provide a statistically significant number of tracks. An appropriate uranium content depends on the age of the mineral. Young samples require high contents, but this same amount in older samples could produce too high a track density. The uranium should also preferably be distributed evenly through the sample. Many of the common rock-forming and accessary minerals have been used successfully. Mica, apatite, hornblende, sphene, zircon, and natural glasses are particularly good.

Once the number of fission tracks has been counted, the mineral's uranium content must be determined. This is done by first heating the sample to anneal out all of the fission tracks. Then the sample is placed, along with a piece of standard glass containing a known amount of uranium, in a nuclear reactor where both receive a dose of neutrons. These neutrons induce fission of ^{235}U, which produces new tracks that can be counted after appropriate etching.

The age of the mineral can be calculated from the three measured fission track densities: that produced by spontaneous fission in the original sample, ρ_s; that induced in the sample by the neutron flux in the reactor, ρ_i; and that produced in the "dosimeter" standard glass, ρ_D. The age is given by

$$\text{fission track age} = \frac{C\rho_s\rho_D}{\rho_i} \qquad (21\text{-}16)$$

The constant, C, includes terms for the fraction of atoms that are ^{238}U, the total number of atoms in sampled volume, length of fission tracks, fraction of tracks that are efficiently etched, and the fission decay constant.

Reliable ages up to 1 Ga have been obtained by this method. Fission tracks, however, can disappear with time, especially if the sample is heated. The age determined by the method is therefore the time since the sample was last heated above a temperature at which annealing takes place. Different minerals anneal at different temperatures, so that fission track dates on several minerals from the same rock can reveal the cooling history and erosional uplift rate of the sample. Fission track dating has been particularly useful when used in this way in conjunction with some other method of absolute dating that has a higher blocking temperature and can thus more closely record the true age of the rock.

21-4   EVOLUTION OF ISOTOPIC RESERVOIRS IN MANTLE AND CRUST

The Earth has been affected throughout geologic time by processes that have differentiated its chemical constituents into the major units of the Earth, the crust, mantle, and core. One of the most effective of these processes, and one that is still active today, is the transfer of partial melts from the mantle to the crust. The crust's composition must be largely determined in some complex way by the time-integrated effect of this process. Phase diagrams and soild–liquid partition coefficients indicate that partial melts can have very different compositions from the solid being melted; that is, partial melting can be an extremely efficient means of effecting chemical differentiation. Isotopes, at high temperature, are not fractionated during partial melting. But most radioactive isotopes have very different solid–liquid partition coefficients from their stable daughter isotopes. Consequently, partial melting does fractionate radioactive nuclides from their decay products. In this way the isotopic nature of parts of the mantle has changed throughout geologic time, and magmas extracted from the mantle today have different isotopic signatures from those extracted in the past. If we understand the evolution of the isotopic composition of the mantle, it may be possible to identify magmas derived from that source at any given time by their isotopic character.

In the preceding section we saw that the Earth's most ancient rocks, from Isua in Greenland, have initial $^{143}Nd^i/^{144}Nd$ ratios that are indistinguishable from those of chondritic meteorites. Other slightly younger Archean rocks from South Africa and Western Australia also have initial ratios which, for their particular ages, match the ratio found in chondritic meteorites. The Earth and chondritic meteorites must therefore have had not only the same initial $^{143}Nd^i/^{144}Nd$ ratio but also the same Sm/Nd ratio. Had the Earth had a higher Sm/Nd ratio, for example, the $^{143}Nd/^{144}Nd$ ratio would have grown at a faster rate than that in the meteorites. Chondritic meteorites and the Earth are therefore believed to have formed from similar primordial solar nebular material which 4.55 Ga ago had a $^{143}Nd/^{144}Nd$ ratio of 0.50682 and a Sm/Nd weight ratio of 0.309 (present Sm/Nd = 0.308).

Given the Earth's initial $^{143}Nd/^{144}Nd$ and Sm/Nd ratios, the evolution of the bulk Earth's $^{143}Nd/^{144}Nd$ ratio through geologic time can be calculated from Eq. 21-8 (see Prob. 21-5). Figure 21-8 shows how the value of this ratio has evolved from 0.50682 at the time of the Earth's accretion to 0.51265 at the present. It is important to stress that this growth line is for the bulk Earth; actual Earth material will deviate from this depending on the fractionation of the isotopes during the Earth's differentiation. The sinking of iron and nickel to form the core

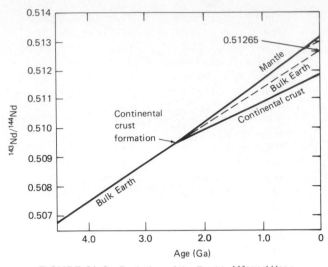

FIGURE 21-8   Evolution of the Earth's $^{143}Nd/^{144}Nd$ ratio throughout geologic time based on an initial value of 0.50682 and an initial Sm/Nd ratio of 0.309. Formation of a continental crust 2.5 Ga ago would have fractionated Nd into the crust slightly more than it would have Sm. Following crust formation, the $^{143}Nd/^{144}Nd$ ratio would therefore have grown more rapidly in the mantle than in the crust. Magmas extracted from the mantle since the time of crust formation have higher $^{143}Nd/^{144}Nd$ ratios than crustal rocks (short-dashed line). See text for discussion.

early in the Earth's history would not have affected the Nd and Sm isotopic ratios because these rare earths would simply have been excluded from the core and concentrated in overlying rocks. Differentiation of the remaining material into the crust and mantle would, however, have fractionated these elements, because some of the silicate phases involved in the fractionation incorporate different amounts of the various rare earth elements into their structures.

All of the rare earth elements (REE) have large ionic radii and preferentially enter the rather open structures of the silicates that form the crust of the Earth, rather than entering the denser structures of the minerals in the mantle. Such elements are referred to as *large ion lithophile* elements (LIL) or elements that are *incompatible* in the mantle (see Sec. 13-11). REE that exist in the mantle preferentially enter the more open structure of a melt, if present. Partial fusion and ascent of magma is one way, then, in which the REE can be fractionated into the crust. The REE are not, however, all fractionated to the same degree. With increasing atomic number, the ionic radii of the REE decreases—the so-called *lanthanide contraction*. As a consequence, Nd has a slightly larger ionic radius than Sm, which makes Nd slightly more incompatible in the mantle than is Sm. Thus, although both elements are strongly fractionated into the crust, Nd is fractionated more than Sm. At present, Nd is estimated to be about 25 times more abundant in the continental crust than in the mantle, whereas Sm is only about 16 times more abundant.

Although Sm and Nd are both fractionated into the crust, the stronger fractionation of Nd means that the Sm/Nd ratio is higher in the mantle than in the crust. Because ^{143}Nd is formed from the decay of ^{147}Sm, the higher Sm/Nd ratio in the mantle causes the $^{143}Nd/^{144}Nd$ ratio to grow more rapidly in the mantle than in the crust. Thus, once the Earth became differentiated into a mantle and crust, the growth line of the $^{143}Nd/^{144}Nd$ ratio split into two, one for the mantle at a level higher than that of the bulk Earth, and one for the continental crust at a level lower than that of the bulk Earth (Fig. 21-8).

Our next problem in understanding the evolution of the $^{143}Nd/^{144}Nd$ ratio in the Earth is to determine how and when the continental crust came into existence. Did it, for example, develop gradually throughout geologic time, or were there a number of distinct crust-building episodes, or possibly only one? These questions have, unfortunately, not yet been fully answered. We can, however, put some limits on the process. For instance, had there been any significant amount of continental crust formed prior to the development of the rocks at Isua, Greenland (3.8 Ga), the initial $^{143}Nd/^{144}Nd$ for these rocks would not match the primordial values found in meteorites. Indeed, the fact that all of the rocks older than about 3.0 Ga have initial $^{143}Nd/^{144}Nd$ ratios close to those of chondritic meteorites indicates that large amounts of continental crust could not have formed prior to that time. By the end of the Archean (2.5 Ga), however, large areas of continental crust had formed. Since then, more continental crust has undoubtedly formed, but much of it may be reworked Archean crust. For the purpose of our discussion we will assume, as a first approximation, that the continental crust formed in a single episode 2.5 Ga ago. The fact that it may have had a more prolonged development does not significantly affect our conclusions on the isotopic evolution of the crust and mantle. If it did, our initial question of how the crust developed would certainly have been answered by now.

Until the formation of the continental crust, the $^{143}Nd/^{144}Nd$ ratio evolved along the line for the bulk Earth, reaching a value of 0.5094 just prior to the first major period of crust formation (Fig. 21-8). With formation of the crust about 2.5 Ga ago, the REE were fractionated into the crust, the lighter ones more so than the heavy ones. From then on, the $^{143}Nd/^{144}Nd$ ratio in the mantle evolved at a faster rate than that in the crust because of the higher Sm/Nd ratio in the mantle. Just prior to crust formation the bulk Earth $^{147}Sm/^{144}Nd$ ratio had evolved to a value of 0.1900. Following crust formation, the $^{147}Sm/^{144}Nd$ ratio in that part of the mantle affected by differentiation probably became about 0.2245 and that of the crust became about 0.1435. With these ratios, $^{143}Nd/^{144}Nd$ ratios in the upper mantle and crust evolved along the separate lines shown in Figure 21-8 (see Prob. 21-6).

It would be interesting to know what fraction of the mantle must have been involved in forming the continental crust, and whether any pristine mantle still exists. Today the largest volumes of mantle-derived magma are erupted as basalts along the mid-ocean ridges, the so-called MORB. These come from a source that is strongly depleted in incompatible elements, and they have high Sm/Nd and consequently, high $^{143}Nd/^{144}Nd$ ratios. Because of the enormous volumes of MORB erupted, this depleted source must constitute a significant volume of the mantle. Because it is the most strongly depleted source recognized, we can take it as possibly representing the mantle formed by the extraction of the crust. Knowing the volume of the continental crust and its isotopic composition, we can calculate the volume of mantle having a MORB-like isotopic composition that would be necessary to provide the bulk Earth composition. This indicates that as little as about 600 km of the upper mantle need have been involved in the formation of the continental crust; below this it may well be pristine. If the upper mantle is not everywhere as strongly depleted as the source of MORB, a proportionately greater depth of mantle would be involved, possibly as much as 90% (Allègre et al. 1983). Considering the thickness of the mantle, it appears likely that an upper part has been depleted in incompatible elements by the process of crust formation and a separate (decoupled) lower part has maintained its primordial composition to a large extent. This conclusion is supported by other geochemical arguments and geophysical evidence (see Sec. 22-2).

Arguments similar to those used for $^{143}Nd/^{144}Nd$ can be used to trace the evolution of the $^{87}Sr/^{86}Sr$ ratio, which is tied to the Rb/Sr ratio through the decay of ^{87}Rb to ^{87}Sr. Unfortunately, initial values in ancient rocks do not provide as accurate a measure of primordial values as do those of Nd, because of the mobility of Sr and to a lesser extent Rb during metamorphism. Values obtained from meteorites can be taken as representative of primordial material similar to that which formed the Earth. It is possible to deduce the bulk Earth value for $^{87}Sr/^{86}Sr$ from the Nd data, as will be explained below.

Both Rb and Sr are LIL elements, and thus they enter melts in the mantle and are transported into the crust. Rb, however, is far more incompatible in mantle solid phases than is Sr, and therefore Rb is more strongly fractionated into the continental crust. But ^{87}Rb is the isotope that decays to ^{87}Sr. Thus the $^{87}Sr/^{86}Sr$ ratio, in contrast to that of $^{143}Nd/^{144}Nd$, is more radiogenic in the crust than in the mantle. In Figure 21-9, a continental crust, developed 2.5 Ga ago, can be seen to have the $^{87}Sr/^{86}Sr$ ratio evolve to much higher values than that of the bulk Earth or the remaining mantle, which is depleted by the formation of the crust (see Prob. 21-8).

Figure 21-8 or 21-9 can be used to show how a mantle-derived rock differs isotopically from one formed from crustal material. Consider, for example, a melt derived from a depleted mantle at the beginning of the Mesozoic era (0.245 Ga). Its $^{143}Nd/^{144}Nd$ ratio would evolve from an initial value of 0.5127 to a present value of about 0.5129, the precise value depending on the Sm/Nd ratio developed in the melt during partial fusion (Fig. 21-8). Rocks derived from crustal material, by contrast, would be expected to have values less than 0.5120. In terms of the $^{87}Sr/^{86}Sr$ ratio (Fig. 21-9), this mantle-derived melt would evolve from an initial value of 0.7026 to a present value of possibly 0.7036, again a very different value from that of a melt derived from an ancient continental crust, which would be expected to have a value near 0.7190. If instead of being derived from a depleted mantle, a melt was formed from mantle that was brought from greater depth (possibly by a mantle plume) and thus possibly isotopically closer to the composition of the bulk Earth, its isotopic signature would still be quite distinct from that of crustal-derived rocks. For magmas derived from the mantle at much earlier times, the task of distinguishing them from crustal ones obviously becomes more difficult and, in fact, impossible as their age approaches 2.5 Ga.

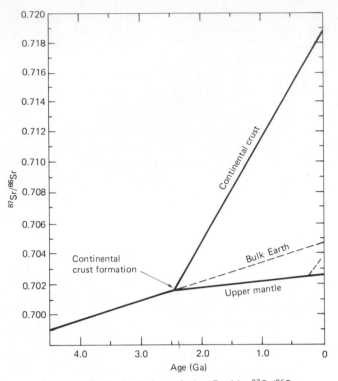

FIGURE 21-9 Evolution of the Earth's ^{87}Sr/^{86}Sr ratio throughout geologic time. The stronger fractionation of Rb relative to Sr into magmas rising from the mantle to form the crust results in the ^{87}Sr/^{86}Sr ratio growing more rapidly in continental crust than in the bulk Earth or mantle. The continental crust is shown here as forming mainly at one time 2.5 Ga ago. Magmas derived from the mantle after crust formation evolve to lower ^{87}Sr/^{86}Sr ratios than do continental crustal rocks.

Contamination of a mantle-derived magma by ancient continental crust is also clearly indicated by changes in the Nd and Sr isotopic ratios. Contamination is the simplest explanation for the difference between the initial ^{87}Sr/^{86}Sr ratios in the two Monteregian intrusive bodies plotted in Figure 21-3. Both intrusions have the same Cretaceous age; that is, their isochrons have identical slopes (about 128 Ma). The pulaskite from Mount Shefford has an initial ^{87}Sr/^{86}Sr ratio of 0.70365, which indicates a mantle source (see Fig. 21-9). The granite from Mount Megantic, on the other hand, has an initial value of 0.70518, which is well above the bulk Earth ratio for that age. The Megantic granite must therefore have incorporated some older crust. This intrusion was emplaced through a Grenville-age Precambrian basement and a thick sequence of folded lower Paleozoic rocks, any of which could have provided the isotopic contaminant. The Shefford intrusion has risen through a similar crust, but its mode of formation and emplacement must have been different from that of the Megantic granite for it to have preserved a mantle isotopic signature.

The isotopic data for Nd and Sr can be conveniently combined into a single diagram as illustrated in Figure 21-10. The front left face and base of the block diagram are, respec-

tively, Figures 21-8 and 21-9. For simplicity, only the line for the evolution of the bulk Earth has been included in this diagram along with a single point for 2.5 Ga. The evolution of the isotopic composition of the bulk Earth can be traced through the block diagram from its value at the time of accretion (4.55 Ga) to its value at the present day. The data can then be projected onto the front right face of the block diagram in terms of the ^{143}Nd/^{144}Nd and ^{87}Sr/^{86}Sr ratios. This face, which is shown separately to the right of the block diagram, is constructed for the present time, but other planes representing earlier times can be constructed. On this plane, the bulk Earth is represented by a point (+), but the evolutionary line leading to that point can be projected into this diagram, as shown by the dashed line. The isotopic ratio of the bulk Earth at any time in the past can also be read off this line, as indicated for the one age of 2.5 Ga.

Finally, some geochemists also display this diagram in terms of parameters known as the $\mathscr{E}_{Nd}$ and $\mathscr{E}_{Sr}$, which are the fractional deviations ($\times 10^4$) of the ^{143}Nd/^{144}Nd and ^{87}Sr/^{86}Sr ratios, respectively, from the values in the bulk Earth at the same time; that is,

$$\mathscr{E}_{Nd} = \left[\frac{(^{143}Nd/^{144}Nd)_{sample} - (^{143}Nd/^{144}Nd)_{bulk\ Earth}}{(^{143}Nd/^{144}Nd)_{bulk\ Earth}} \right] \times 10^4$$

(21-17)

The $\mathscr{E}_{Sr}$ is defined in the same way but using the ^{87}Sr/^{86}Sr ratios. These values are also shown in Figure 21-10. Note that the $\mathscr{E}$ values can be calculated for any desired age by adjusting the isotopic ratios for the decay of ^{147}Sm and ^{87}Rb.

A number of important rock types have been plotted in terms of their present ^{143}Nd/^{144}Nd and ^{87}Sr/^{86}Sr ratios in Figure 21-11. This diagram contains much information that needs to be worked through slowly if it is to be appreciated. To help orient yourself, first locate the present composition of the bulk Earth (+) and the evolutionary line for the bulk Earth; these are identical to those in Figure 21-10. The lines marking the present bulk Earth ratios of ^{143}Nd/^{144}Nd and ^{87}Sr/^{86}Sr divide the diagram into four quadrants. Most mantle-derived rocks plot in the upper left quadrant ($+\mathscr{E}_{Nd}$, $-\mathscr{E}_{Sr}$), and most continental crustal rocks plot in the lower right quadrant ($-\mathscr{E}_{Nd}$, $+\mathscr{E}_{Sr}$). The quadrants are, respectively, depleted and enriched in LIL elements.

Passing through the bulk Earth composition is a remarkably linear array of isotopic ratios in rock types ranging from mid-ocean ridge basalt (MORB), oceanic island intraplate basalt (OI), kimberlite (K), garnet peridotite xenoliths (GP), and continental flood basalts (Co). All of these rocks have, without doubt, formed from magma originating in the mantle. Thus they provide a sampling of mantle isotopic ratios, at least from those parts tapped by the magmas. For this reason, this range of isotopic values is commonly referred to as the *mantle array*. It indicates a strong negative correlation between Sr and Nd isotopes. Indeed, in light of the previous discussion of the relative degrees of fractionation of Sm–Nd and Rb–Sr, this variation is to be expected; that is, the continental crust was formed of material that had a lower Sm/Nd ratio and a higher Rb/Sr ratio than the bulk Earth. Upon decay of ^{147}Sm to ^{143}Nd

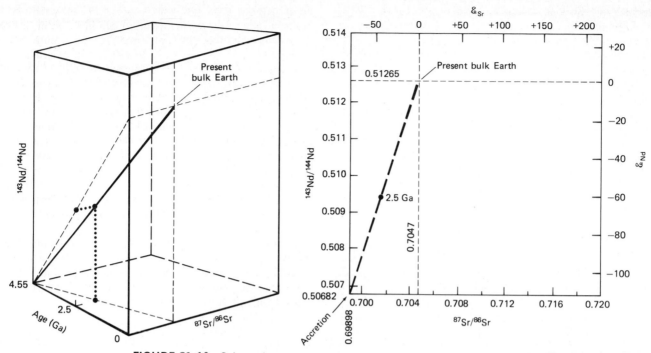

FIGURE 21-10 Schematic representation of the evolution of the $^{87}Sr/^{86}Sr$ and $^{143}Nd/^{144}Nd$ ratios in the bulk Earth (solid heavy line) throughout geologic time. This line is projected onto a plane of constant time (the present) in which $^{143}Nd/^{144}Nd$ is plotted against $^{87}Sr/^{86}Sr$. See text for definition of ε values.

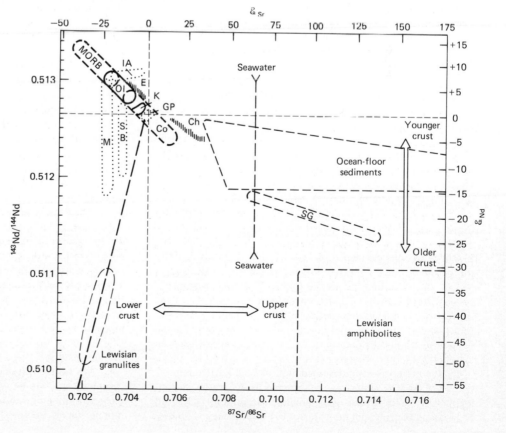

FIGURE 21-11 Plot of important rock types in terms of $^{143}Nd/^{144}Nd$ and $^{87}Sr/^{86}Sr$. Mid-ocean ridge basalts (MORB) and oceanic islands (OI) (O'Nions et al. 1977; Norry and Fitton 1983); island-arc rocks (IA) (DePaolo and Johnson 1979); kimberlite (K), garnet peridotite mantle nodules (GP), and continental flood basalts (Co) (Hawkesworth et al. 1983); central volcanic complexes of Mull (M), Skye basalts (SB), Skye granites (SG), and Lewisian granulites and amphibolites (Carter et al. 1978); andesites of Ecuador (E) and Chile (Ch) (Hawkesworth et al. 1979); and ocean floor sediments (White et al. 1985). See text for discussion.

and ^{87}Rb to ^{87}Sr, high values of ^{143}Nd/^{144}Nd correlate with low values of ^{87}Sr/^{86}Sr; that is, radiogenic Nd correlates with unradiogenic Sr, and vice versa.

The rock types that constitute the mantle array have formed from sources showing variable degrees of depletion in the LIL elements. This variation may occur laterally or vertically within the mantle. MORB have formed from the most strongly depleted source, and their restricted compositional range indicates that a rather homogeneous, depleted mantle exists beneath all ocean ridges. Basalts with similar isotopic composition do rarely occur in continental extensional regions, such as the Basin and Range province of the western United States. Oceanic islands that occur within plates (i.e., those not associated with divergent plate boundaries) are composed of rocks with a much wider range of isotopic composition along the mantle array than MORB. Most lie within the region marked OI in Figure 21-11, but they can extend to compositions which are slightly enriched over that of the bulk Earth. Their source is not nearly as depleted as that of MORB. Their range of composition may result from mixing of material derived from undepleted and depleted sources. Because oceanic islands are believed to develop over mantle plumes, their magmas probably tap a deeper part of the mantle than do MORB, and thus they have more pristine compositions. Alkaline basalts within continental plates have isotopic signatures identical to those of oceanic islands. Both must therefore have developed their signature at depths greater than the base of the thick continental lithosphere, which is so different from the oceanic lithosphere.

Continental flood basalts cover a range of composition along the mantle array, but most have compositions near that of the bulk Earth. This has had various interpretations. Their source could be in a part of the mantle that still preserves bulk Earth isotopic compositions. On the other hand, their source could be in a depleted mantle, but the magma assimilates enough continental crustal material to give a bulk Earth ratio. They could also have formed in a depleted mantle that becomes enriched by metasomatism prior to magmatism. There is, as yet, no resolution to this problem; perhaps all factors play a role in varying degrees in different regions. Kimberlites and their enclosed garnet peridotite xenoliths do have isotopic compositions that are very close to that of the bulk Earth. This implies that mantle with a bulk Earth isotopic composition does exist at depths of about 200 km beneath the continental crust in those regions where kimberlites originate.

The negative correlation between Nd and Sr isotopes in the mantle array and the mantle depletion in LIL elements, which it implies, must be the result of a long-lived phenomenon. For instance, had the depletion in ^{87}Rb occurred only recently the decay product, ^{87}Sr, would still be present in the source region. But it is not, and the amount of ^{87}Sr that is found is consistent with their having been a deficiency of ^{87}Rb in the source for a considerable period of time. All indications are that this depletion occurred during the formation of the crust, which the Nd and Sr data indicate has a mean age of between 2.46 and 2.14 Ga (Allègre et al. 1983).

It was previously mentioned that the bulk Earth composition for ^{87}Sr/^{86}Sr cannot be determined as accurately as that for ^{143}Nd/^{144}Nd because of the greater susceptibility of Rb and Sr to change during metamorphism. The strong negative correlation between these two isotopic ratios in the mantle array, however, allows the accurately known ^{143}Nd/^{144}Nd ratio to be used to determine the ^{87}Sr/^{86}Sr ratio, which at present is 0.7047.

Continental crustal rocks range more widely in isotopic composition than do mantle-derived ones, in part because of the strong elemental fractionation involved in crust-forming processes (partial melting, weathering, sedimentation), but also because of the relative volumes of the continental crust and mantle. Most crustal rocks plot in the lower right quadrant of Figure 21-11. In general, the greater the age of a segment of crust the farther to the lower right of the diagram it plots; this will be evident from inspection of the growth curves in Figures 21-8 and 21-9. Within the crust itself there appears to be fractionation of some isotopes. In Chapter 1 it was shown, from heat flow calculations for the continental crust, that U, Th, and K must be concentrated toward the Earth's surface. Rb, which behaves geochemically much like K, also appears to be concentrated upward in the crust. As a consequence, the ^{87}Sr/^{86}Sr ratio generally increases upward in the crust. No such crustal differentiation, however, affects Sm and Nd and thus the ^{143}Nd/^{144}Nd ratio is unaffected. The result in the plot of Figure 21-11 is a shift from right to left with increasing depth in the crust. This is illustrated by the ancient Lewisian rocks (2.95 Ga) of Scotland. The deepest exposed Lewisian rocks, which are in the granulite facies, have very low ^{87}Sr/^{86}Sr values (0.702 to 0.703). Shallower amphibolite facies rocks, on the other hand, have high ^{87}Sr/^{86}Sr values (0.711 to 0.720).

Because mantle-derived magmas and continental crustal rocks plot in very different positions in Figure 21-11, the effects of assimilation or magma mixing between these two should be readily evident. The isotopic variation exhibited by the mantle array is expected to result from partial melting or fractional crystallization. Isotopic variations in other directions imply mixing of different isotopic compositions. If the materials to be mixed have the same Sr/Nd ratios, mixing in Figure 21-11 results in a simple linear array between the two end-member compositions. If the Sr/Nd ratios are different in the two materials, intermediate compositions lie along curved lines (DePaolo and Wasserburg 1979). In general, continental crustal materials have lower Sr/Nd ratios than do mantle-derived rocks, and as a result mixtures of mantle and crustal rocks lie along gently curved lines that are concave toward the upper right in Figure 21-11.

The basalts of the islands of Mull and Skye on the northwest coast of Scotland formed during the early opening of the North Atlantic. Unlike lavas that have since erupted along the mid-Atlantic ridge, these early basalts ascended through continental crust, which consists of Lewisian metamorphic rocks. The almost vertical trends of the Mull and Skye basalts in Figure 21-11 are not consistent with fractional crystallization. They do, instead, form a simple linear trend from a depleted magma toward a lower crustal Lewisian granulite composition. This is a clear case of lower crustal contamination of a mantle-derived magma.

On both Mull and Skye, there are large central igneous complexes containing granites, which some geologists have interpreted as having formed from melted crustal rocks or contamination of mantle-derived magmas. Only the granites of

Skye (SG) are included in Figure 21-11, and these plot along a curved line between the mantle array and the most radiogenic Lewisian amphibolites. These granites, then, can be interpreted as a mixture of a mantle magma and crustal material. In contrast to the basalts, however, the granites assimilated higher-level crustal rocks.

Contamination of large bodies of mafic magma that have crystallized in the crust can also be demonstrated by their isotopic composition. The mid-Proterozoic (1.1 Ga) Kalka intrusion in central Australia, for example, was emplaced in quartzo-feldspathic granulite facies gneisses. Although some of the rocks of this intrusion have Nd and Sm isotopic compositions that plot very near the bulk Earth composition on the mantle array, most fall along a linear trend extending toward high $^{87}Sr/^{86}Sr$ ratios but maintaining approximately the bulk Earth $^{143}Nd/^{144}Nd$ ratio. This trend is not included in Figure 21-11, but it approximately follows the $\mathcal{E}_{Sr} = 0$ line to the right of the bulk Earth composition. This trend is directly toward the isotopic composition of the gneisses surrounding the intrusion, and is undoubtedly the product of crustal contamination of a mafic magma (Gray et al. 1981). This assimilation is believed to have been responsible for the copious precipitation of orthopyroxene in this intrusion. Other large mafic intrusions that have crystallized within the crust and which also contain abundant orthopyroxene, such as the Duluth Gabbro, the Bushveld complex, and the Palisades Sill, have isotopic trends similar to that of the Kalka rocks, plotting to the right of the mantle array. Crustal assimilation seems likely in these intrusions as well.

The amount of crustal contamination of mantle-derived magmas indicated by the isotopic ratios of Nd and Sr is commonly much greater than is permissible based on the major element chemistry. The silica content of basaltic magmas, which is typically about 50 wt %, is changed significantly when continental crustal material, which typically contains >70 wt % SiO_2, is assimilated. Many basaltic magmas whose isotopic composition indicates there has been considerable assimilation of continental crustal material (high $^{87}Sr/^{86}Sr$ values) still contain almost pristine silica contents. The most likely explanation for this paradox is that contamination involves only a small fraction of the crustal rocks, but that fraction is particularly enriched in the isotopes concerned, especially Sr. This fraction might be derived, for example, by partial fusion of the crustal rocks or by extraction of hydrous fluids.

The igneous rocks developed above a subduction zone—the calc-alkaline series—is one important group of rocks for which large-scale crustal contamination is commonly invoked to explain their composition. Five possible components may play roles in the formation of calc-alkaline magmas: the mantle wedge overlying the subduction zone, subducted rocks of the oceanic crust (principally MORB), subducted sedimentary rocks, seawater, and crustal contaminants picked up during magma ascent. Each of these components is potentially identifiable from its isotopic signature. Thus isotopic analysis of calc-alkaline rocks may shed light on the origin of these rocks. We will examine the isotopic composition of calc-alkaline rocks in two very different environments; first, where an island arc develops entirely between oceanic plates at a sufficient distance from continental areas that there can be no possible contribu-

tion from continental material, and second, where a continental plate overrides an oceanic plate.

The New Britain island arc, which has been studied isotopically by DePaolo and Johnson (1979), is sufficiently far removed from continental masses to eliminate any possible continental contribution to these rocks. They nonetheless include rocks ranging from basalt to rhyolite. Their Nd and Sr isotopic compositions fall in a very small area in Figure 21-11 (IA). They form a linear trend with a very slight positive slope extending from compositions on the mantle array at the depleted end of the oceanic island field to $\mathcal{E}_{Sr}$ values of almost zero. Other island arcs have similar isotopic compositions. The trend is not one that can be explained by fractional crystallization or varying degrees of partial melting. It must involve mixing of various components. The primary source appears to be mantle that is not as depleted as that from which most MORB are derived, but it is similar to material on anomalous ridge segments that are not as depleted as normal MORB or to some intraplate oceanic islands. The source is not simply melted average MORB, which is too depleted. The most likely major source for these rocks, then, is in the mantle wedge above the subducted plate.

The most striking feature of the isotopic composition of the island-arc rocks is their trend toward increasing $^{87}Sr/^{86}Sr$ ratios with almost constant $^{143}Nd/^{144}Nd$ ratios. Mixing of a mantle source with ocean floor sediments would not produce such a trend. Most ocean floor sediments are ultimately derived from continental areas and thus their isotopic composition places them in the lower right quadrant of Figure 21-11. Mixing of this material with mantle material would produce rocks that plot along a smooth curve with a negative slope. The most likely explanation for the isotopic trend is the incorporation of seawater in the magmas. At present seawater has an $^{87}Sr/^{86}Sr$ ratio of 0.7092, but its content of Nd is so small (<3 × 10^{-5} ppm) that addition of seawater to any rock on the mantle array results in the isotopic composition simply being shifted to the right. Seawater can be incorporated directly as pore water trapped in the subducted plate, but it can also be stored in the ocean floor rocks in the form of hydrous alteration minerals. Its transport into the overlying mantle wedge upon subduction can be either as rising solutions or as partially melted MORB. The Nd and Sr isotopes imply that most of the magma generated in oceanic island arcs must come from the mantle wedge above the subduction zone, fluxed by seawater that is released from the downgoing oceanic plate. The slight positive slope to the island arc trend in Figure 21-11 could be produced from a small contribution of MORB that has been shifted to the right as a result of seawater alteration. A small amount of melting of ocean floor basalt may therefore occur in the subducted plate. Ocean floor sediments that are subducted cannot be incorporated into these magmas in amounts of more than a few percent; otherwise, their strong isotopic signature would be detected.

Calc-alkaline rocks developed on continental crust have a very different isotopic pattern from those developed in oceanic island arcs. As an example, data for the Andes have been included in Figure 21-11 (Hawkesworth et al. 1979). In Ecuador (E in Fig. 21-11) andesites have Nd and Sr isotopic compositions that plot slightly to the right of the mantle array,

and

$$(\delta^{18}O_{H_2O} + 1000) = \frac{\delta^{18}O_{Mt} + 1000}{K_{Mt}} \qquad (21\text{-}24)$$

Combining these two equations, the del value of the fluid is eliminated, leaving

$$\frac{\delta^{18}O_{Qz}}{K_{Qz}} = \frac{\delta^{18}O_{Mt}}{K_{Mt}} \qquad (21\text{-}25)$$

Because the values of K, which are functions of temperature, are known from experiments and the del values of quartz and magnetite can be measured mass-spectrometrically, the equation can be solved for temperature. Temperatures determined in this way record the temperature when the isotopes last equilibrated, which may be different from the temperature at which the minerals first formed. Isotopic ratios may later be disturbed by exchange with circulating groundwater. Normally, unless concordant temperatures are obtained from several different mineral pairs in the same rock, little reliance can be placed on a single measurement.

Geothermometry using oxygen isotopes has proved reliable over the entire range of normal crustal temperatures, that is, from magmatic to surface temperatures. It has been used successfully on igneous, metamorphic, sedimentary, and hydrothermal rocks, and has even been used to determine paleoclimatic temperatures.

Sulfur Isotopes

The four isotopes of sulfur have masses of 32, 33, 34, and 36. Only the ratio of the two most abundant isotopes ($^{34}S/^{32}S$) is normally measured. The $\delta^{34}S$ is measured with respect to the sulfur in troilite (FeS) from the Cañon Diablo meteorite. The del values defined in this way cover a wide range of values in sulfides and sulfates from different environments. These values are of particular importance to the study of sulfide ore bodies, especially with regard to the light they may shed on the origin of these deposits.

Most $\delta^{34}S$ values of terrestrial sulfur average around the values found in meteorites, which probably closely approaches a primordial solar value. It is not surprising, therefore, that magmatic sulfides in rocks derived from the mantle have del values near zero. Sulfates in evaporites formed from seawater, on the other hand, have del values of about $+20$, which is similar to that of seawater. Even sulfates deposited in veins have much higher del values (about $+20$) than coexisting sulfides. Some magmatic sulfide deposits in mafic igneous rocks have del values as high as $+10$. The sulfide deposit at Norils'k, Russia, provides a good example. Here diabase intruded and assimilated sedimentary gypsum. This increased the del value of the primary sulfur and produced a magmatic sulfide deposit with anomalously heavy sulfur. Sulfides deposited in sediments by the action of bacteria can have del values that range widely, from -30 to $+30$.

Fractionation of sulfur isotopes between coexisting sulfides (and sulfates) is significant and can be used as the basis of a geothermometer in the same way that oxygen isotopes are used. Sphalerite and galena, pyrite and chalcopyrite, pyrite and sphalerite, and sulfide and sulfate all provide sensitive geothermometers. Considerable care must be taken, however, to ensure that the mineral pairs analyzed did indeed crystallize together. Pyrite, for example, can crystallize over a very wide temperature range.

Helium Isotopes

Helium, which has two isotopes with atomic masses 3 and 4, is an inert gas and is therefore not chemically bound in the Earth. Moreover, its mass is low enough that it is able to escape from the Earth into space. Helium in the atmosphere must, consequently, be replenished by outgassing from the Earth's interior. Within the Earth helium has two principal sources: primordial helium trapped at the time of Earth accretion, and radiogenically produced helium, which because of the concentration of radioactive elements toward the Earth's surface, is produced largely at shallow depths. Both of these sources have characteristic isotopic signatures, which can serve as a means of identifying sources.

The isotopic composition of helium produced by radioactive decay is very different from that of primordial helium. The isotopic ratio of $^3He/^4He$ (referred to as R) in the primordial solar system, which was the result initially of nuclear fusion, can be estimated from its value in meteorites, which is about 3×10^{-4}. Radioactive decay, on the other hand, produces mainly ^4He through α decay. Some ^3He is produced by fission of ^{238}U and some by the interaction of α particles with nuclei of light elements. This interaction produces a neutron flux which can combine with lithium, for example, to produce ^3He. The reaction is as follows: ^6Li combines with a neutron to produce an α particle plus ^3H; through β emission the ^3H decays to ^3He. The combined effect of these various radioactive processes is to produce an R value of 2×10^{-8}, which is very much lower than the primordial ratio. Cosmic rays are also capable of generating helium, but this is not a significant source on Earth.

Helium derived from rocks that have risen directly from deep in the mantle have R values of about 10^{-5}. In contrast, ancient granitic crustal rocks have values of about 10^{-8}, which simply reflects their high content of radiogenic helium. The ratio in the atmosphere is 1.4×10^{-6}, which must result from a mixture of primordial and radiogenic helium. In general, the higher the R value, the larger must be the contribution from a deep source. MORB typically has R values from seven to nine times that of the atmosphere, whereas oceanic island values can be as much as 30 times larger. Hot springs in areas of recent volcanism also have high R values. At Yellowstone, for example, R values are relatively constant at seven times the atmospheric value within the caldera but drop off sharply outside it. Some values within the caldera are as high as 16 times atmospheric. Similar values are obtained in Icelandic hot springs. In recent volcanic areas, helium isotopes are therefore capable of indicating a mantle component. In older rocks, however, the primordial helium component is soon diluted and swamped by the radiogenic helium derived from surrounding crustal rocks; then the mantle signature is no longer detectable.

PROBLEMS

21-1. For the following radioactive decay reactions, determine the type of decay process involved.

$$^{115}_{49}\text{In} \longrightarrow {}^{115}_{50}\text{Sn} \qquad {}^{142}_{58}\text{Ce} \longrightarrow {}^{138}_{56}\text{Ba} \qquad {}^{138}_{57}\text{La} \longrightarrow {}^{138}_{56}\text{Ba}$$

21-2. ^{147}Sm decays to ^{143}Nd with a decay constant of $6.54 \times 10^{-12} \text{ a}^{-1}$. ^{144}Nd is a stable isotope whose abundance in the bulk Earth has remained constant throughout time. If the present $^{147}\text{Sm}/^{144}\text{Nd}$ ratio in the bulk Earth is 0.1870, what would it have been at the time of Earth accretion (4.55 Ga)?

21-3. From the data presented in Figure 21-3, show that the isochron for the Megantic granite does correspond to an age of 127.6 Ma. The most Rb-rich analysis in this plot has an $^{87}\text{Rb}/^{86}\text{Sr}$ ratio of 11.04 and an $^{87}\text{Sr}/^{86}\text{Sr}$ ratio of 0.7252.

21-4. Using Eq. 21-12, construct a concordia curve in the $^{206}\text{Pb}/^{238}\text{U}$ versus $^{207}\text{Pb}/^{235}\text{U}$ diagram. Where in this diagram might zircon crystals plot that were initially formed in a 2.6-Ga-old granite that was metamorphosed at 1.0 Ga?

21-5. At the time of accretion (4.55 Ga) the Earth is believed to have had a $^{143}\text{Nd}/^{144}\text{Nd}$ ratio of 0.50682 and a $^{147}\text{Sm}/^{144}\text{Nd}$ ratio of 0.1926. Plot a graph of the value of the $^{143}\text{Nd}/^{144}\text{Nd}$ for the bulk Earth throughout geologic time and record the value of this ratio at an age of 2.5 Ga before present.

21-6. Using the information from Prob. 21-5, calculate the evolution of the $^{143}\text{Nd}/^{144}\text{Nd}$ ratios in the crust and mantle if the crust is taken as having formed 2.5 Ga before present. The $^{147}\text{Sm}/^{144}\text{Nd}$ ratios in the crust and mantle at time of formation can be taken as 0.1435 and 0.2245, respectively.

21-7. Repeat Prob. 21-5 but for $^{87}\text{Sr}/^{86}\text{Sr}$. The initial $^{87}\text{Sr}/^{86}\text{Sr}$ for the Earth was 0.69898 and the $^{87}\text{Rb}/^{86}\text{Sr}$ was 0.0857.

21-8. Repeat Prob. 21-6 but for $^{87}\text{Sr}/^{86}\text{Sr}$. The $^{87}\text{Rb}/^{86}\text{Sr}$ ratios in the crust and mantle at time of formation were 0.48423 and 0.0332, respectively.

22 / Origin of Rocks

22-1 INTRODUCTION

In previous chapters we have dealt with the specific details of the formation of rocks. In this final chapter we examine the broader question of their ultimate origin. What conditions in the Earth bring about the formation of rocks, and from where does the material come to form them? These are important questions, the answers to which are critical to interpreting the Earth's history, for only in rocks is any record of the geologic past preserved. There is certainly no unanimity among petrologists on answers to all aspects of these questions, but the theory of plate tectonics has provided a unifying paradigm that has eliminated much controversy. Difficulty in answering the questions stems from the inaccessibility of the regions in which the controlling processes operate and our limited experimental and theoretical knowledge of the behavior of material under the pressures and temperatures that exist in such regions. Because of the rapid evolution of ideas on this topic, no attempt is made in this chapter to review all aspects of these questions. Instead, some basic principles are discussed which should be of help in evaluating these ideas.

Ever since the formation of the Earth 4.55 Ga ago, heat generated by accretionary processes, radioactive decay, and gravitative differentiation—in particular of the core—has been transferred to the surface of the planet, where it has been radiated into space. Conduction, advection, and radiation have all played roles in this transfer. The opacity of rocks in the upper mantle and crust eliminates radiation as a feasible means of transferring heat through the outer half of the planet. Also, the thermal conductivity of rocks is so low that the rate of cooling of the Earth has been determined largely by advective processes, which include mantle and crustal convection, magma intrusion, and fluid transport.

All rocks, whether igneous, metamorphic, or sedimentary, owe their origins directly or indirectly to the advective cooling of the Earth. The rise of magma into the crust or its extrusion onto the surface is an obvious example of such advection. But so is the development of contact metamorphic minerals through endothermic reactions. Similarly, cold crustal rocks that are buried and undergo regional metamorphism, which is largely endothermic, and then return to the surface of the Earth provide another means of advectively removing heat from the planet's interior. The upward flux of hot fluids from metamorphic terranes is still another means of advectively cooling the Earth. Finally, sedimentary rocks are formed as a result of changes in elevation of the Earth's surface, positive changes resulting in erosion and the production of sediment and negative changes producing basins for deposition. The changes in elevation are brought about by plate tectonic motions that result from gravitational instabilities set up by the advective cooling of the planet.

Although cooling of the Earth is a continuous process, the rate of advective heat transfer may have fluctuated throughout geologic time. Evidence for this comes from the rate of production of rocks, which does not appear to have remained constant. Despite the incompleteness of the geologic record, in particular of the Precambrian, most rocks were formed during distinct episodes. According to Moorbath (1977) these occurred between 3.8 and 3.5, 2.9 and 2.6, 1.9 and 1.6, 1.2 and 0.9, and 0.6 and 0 Ga ago. Because these episodes appear to be worldwide in extent, they probably reflect major periods of advection in the mantle and possibly the core.

Certain rock types, such as tholeiitic basalt and greenschists, have formed throughout geologic time, but others are restricted to specific rock-forming episodes (see Sec. 14-8). Greenstone belts, for example, are formed in the first two episodes; massif-type anorthosites and widespread granulite facies metamorphism characterize the 1.2 to 0.9-Ga episode; and low-T, high-P metamorphic rocks are restricted to the most recent episode (the latter group may have formed in the Precambrian but are not preserved due to erosion or continued heating at depth). These changes must reflect evolution of the planet's interior as a result of progressive cooling. This evolution can be expected to continue, with the rate of production of new rocks decreasing as the Earth cools. Eventually, the rate of heat production will be insufficient to maintain advection. Then the Earth will cease to be a dynamic planet, and no new rocks will be formed.

Because the advective cooling of the Earth plays a dominant role in the production of rocks, we begin this chapter with a consideration of the mechanisms by which heat can be transferred advectively. This is followed by a discussion of how advection leads to conditions favorable to the generation of magmas and metamorphic rocks. Next, the actual processes by which partial melts coalesce and form ascending bodies of magma are treated. This, in turn, is followed by a discussion of the factors controlling the compositions of partial melts. Finally, the generation of the three main types of magma—basaltic, andesitic, and granitic—is examined in terms of their plate tectonic settings. For additional information, the book by Yoder (1976) on the *Generation of Basaltic Magma* and the review article by Wyllie (1988) are particularly informative.

22-2 ADVECTIVE HEAT TRANSFER IN THE EARTH

The Earth is believed to have formed by the accretion of particles that condensed from the original solar nebula. These included various Ca-, Al-, and Ti-oxides, metallic Fe and Ni, forsterite, enstatite, feldspar, FeS, and amphibole. As the planet grew, impact energy generated by incoming planetesimals would have increased because of the greater gravitational attraction; this, then, would have steadily raised surface temperatures. At the same time, the increasing mass of the planet would have raised internal temperatures through compression. Toward the close of the accretionary period, temperatures in the Earth ranged from about 600°C at the center to about 1600°C near the surface, with a maximum of about 2200°C at a depth of 1400 km (Ringwood 1975). These temperatures were high enough to cause partial melting only in the outer 200 km. Below this the geothermal gradient was progressively farther away from the solidus, which rises rapidly with increasing pressure.

The presence of FeS among the original condensates from which the Earth formed is of great importance to the early development of the planet. FeS and Fe form a low-temperature eutectic (about 1000°C), at which a dense iron-sulfide liquid is formed that is immiscible in silicate liquids. This liquid would have sunk rapidly through the upper molten silicate layer and then more slowly through the solid, but plastic lower part of the Earth. This sinking would have effectively mixed and homogenized any compositional layering that might have developed during the accretionary stage. The segregation of this dense liquid to form the core of the Earth was a strongly exothermic process, with enough heat being generated to raise the average temperature of the Earth by 2000°C (Ringwood 1975). Thus, although the Earth may have had a relatively cool beginning, core formation, which occurred within 10^8 years of accretion, ensured that the entire Earth was raised to moderately high temperatures early in its history.

Following core formation, temperatures were high enough to melt the Earth totally to a depth of 400 km [see Anderson (1984)]. Convection within this molten layer transferred heat rapidly to the surface, and solidification would have occurred within 10^4 years. The temperature gradient in this convecting layer would have been very nearly adiabatic (see Prob. 7-7), and as a result solidification would have taken place from the base upward, as it does in convecting bodies of magma

(Sec. 13-4 and Fig. 13-8). The core of the Earth is solidifying in this same way today, with the solid inner core being overlain by the liquid outer core. Differentiation of the outer molten layer of the Earth into a more refractory olivine-rich base and less refractory top would probably have occurred at this time [see Anderson (1982)]. We can suppose, therefore, that by some time prior to 4.0 Ga ago the Earth had differentiated into a liquid core and solid mantle.

Heat production within this differentiated Earth would have continued as a result of the decay of radioactive elements. These elements were more abundant in the early Earth, when their heat production was probably four times greater than it is today (Prob. 22-1). Temperatures within the Earth would therefore have steadily risen until a balance was achieved between the rate of heat production and the rate of heat loss. Steeper temperature gradients would have resulted in more rapid heat conduction (Eq. 5-2), but they also would have promoted convection if the viscosity of the mantle were sufficiently low (Eq. 13-10).

Because of the low thermal conductivity of rocks, temperatures within the Earth would have become extremely high and approached the melting point if conduction had been the sole means of transferring heat. As the temperature of solids approaches the melting point, however, the rate of dislocation creep increases rapidly, and the viscosity of solids decreases significantly, which then favors convection. In the extreme case if melting had occurred, convection would definitely have taken over as the dominant means of heat transfer. But even in the solid mantle, convection, albeit slow, has been the most rapid means of transferring heat. Indeed, its rate has been sufficiently rapid that convection has probably always been able to keep pace with changes in heat production. Throughout geologic time, therefore, the heat flux at the Earth's surface is thought to have been essentially equal to the rate of heat generation within the Earth (McKenzie and Weiss 1975).

Convection is easily shown to be an important mode of heat transport in the upper mantle today, and thus it must have been still more important in the past when the Earth was hotter. In Sec. 13-4 the dimensionless Rayleigh number (Eq. 13-10) was introduced to evaluate the likelihood of convection occurring in bodies of magma. In that case we considered a sheet of liquid whose upper and lower surfaces were held at different temperatures (Fig. 13-7). A Rayleigh number can also be defined for a sheet that generates its own heat internally; this would be the situation in the upper mantle, where heat is generated from radioactive decay. The Rayleigh number in this case is defined as

$$\mathrm{Ra} = \frac{\rho g\, d^5 \alpha\, A}{K k \eta} \tag{22-1}$$

where ρ is the average density, g the acceleration of gravity, d the depth of the sheet, α the coefficient of thermal expansion, A the heat productivity per unit volume, K the thermal conductivity, k the thermal diffusivity, and η the viscosity. The critical Rayleigh number for convection when the layer is bounded by rigid surfaces against which there is no slip is 2772, and when the surfaces are free to slip it is 868.

Approximate values of all terms in Eq. 22-1 can be found. The viscosity of the upper mantle is about 7×10^{20} Pa s based on postglacial uplift rates (Peltier and Andrew 1976). Deep-focus earthquakes reveal that subducted slabs can penetrate to depths of 600 to 700 km. It is reasonable to assume, therefore, that convection might affect a layer to a depth of 700 km. If we assume that the upper mantle is similar to the peridotite given in Table 1-1, its heat-generating capacity, A, is 10^{-8} W m^{-3}, and its thermal conductivity, K, is 3.35 W m^{-1} K^{-1}. The average density of the upper mantle can be taken to be 3.5 Mg m^{-3}, its thermal diffusivity, k, is 10^{-6} m^2 s^{-1}, and its coefficient of thermal expansion, α, is 3×10^{-5} K^{-1}. The Rayleigh number calculated from these values is 7×10^5, which is much greater than the critical Rayleigh number. We can conclude, therefore, that the upper mantle must be convecting today, and that in the past, when more radioactive heat was produced, the Rayleigh number would have been still greater, and convection would have been even more vigorous (Prob. 22-2).

Although it is simple to show that convection must occur in the mantle, it is not yet possible to predict the geometry or rate of this convection. In the calculation above we assumed a depth of 700 km for the convective layer. If the depth had been taken to be the entire thickness of the mantle, the Rayleigh number would have been greater, and we would have concluded that whole-mantle convection occurs. Conversely, shallower depths could have been selected that would still have given Rayleigh numbers in excess of the critical value. Convection, therefore, can occur on many different scales. Whether the mantle convects in large cells that affect the entire depth of the mantle or in tiers of shallower cells depends on many factors, such as the variation in composition and rheological properties with depth, the distribution of radioactive elements, and the amount of heat entering the mantle from the core. Unfortunately, these parameters are poorly known, so no definitive calculation can be made. Most models, however, call on convection to operate at several scales simultaneously.

The most direct evidence for mantle convection is provided by the motion of lithospheric plates. Hot, buoyant lithosphere, created at spreading oceanic ridges, cools and becomes denser as it moves away from the ridge, eventually sinking into the mantle at subduction zones. The dimensions of lithospheric plates define the largest-scale convection that we know of in the Earth. Surprisingly, it was while relating the motion of these plates to horizontal temperature gradients associated with large-scale convection that evidence was found for smaller-scale convection. The gradual deepening of oceans away from divergent plate boundaries is easily accounted for in terms of the progressive cooling and densification of newly formed lithosphere as it moves away from the spreading axis (Sclater et al. 1981). As with other processes controlled by conductive cooling (Sec. 5-3), the deepening of the ocean is proportional to the square root of time, at least for an ocean floor younger than 70 Ma (Fig. 22-1). Beyond 70 Ma, the rate of deepening decreases exponentially toward a constant depth. These observations suggest that the cooling rate of plates significantly decreases at times greater than 70 Ma. Parsons and McKenzie (1978) have suggested a model involving small-scale convection that provides a simple explanation for these observations. A description of their model follows.

New, hot lithosphere formed at a spreading axis loses heat to the overlying water as the plate moves away from the

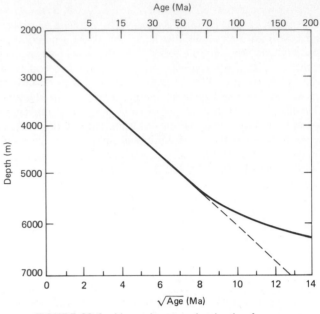

FIGURE 22-1   Linear increase in depth of ocean floor with square root of age of ocean floor during first 70 Ma is due to conductive cooling of oceanic lithosphere. [After Sclater et al. (1981).]

axis (Fig. 22-2). The water can be taken to be at a constant 0°C. The thickness of the rigid lithospheric plate at any time, t, after leaving the spreading axis depends on the depth to which the brittle-to-ductile transition has extended. This transition can be represented by an isotherm, $T_{B/D}$. Immediately below the rigid plate is a thermal boundary layer in which the viscosity decreases rapidly as temperatures increase. The bottom of this layer is actually gradational into the underlying asthenosphere in which the large-scale convection occurs, which is responsible for transporting the plate. For convenience we identify another isotherm, T_A, as the base of this thermal boundary layer.

Heat transfer through the rigid plate is largely by conduction, but advective hydrothermal fluids may play a role, especially near the spreading axis. The depth, z, to which any isotherm, T, moves in time t is given by a relation similar to Eq. 5-13, but in this case for the cooling of a half-space

$$\frac{T - T_s}{T_0 - T_s} = \operatorname{erf}\left(\frac{z}{2\sqrt{kt}}\right) \tag{22-2}$$

where T_s is the constant temperature of the cooling surface, T_0 the initial temperature, and k the thermal diffusivity. Where the ocean floor temperature is 0°C ($T_s = 0$), Eq. 22-2 becomes

$$T = T_0 \operatorname{erf}\left(\frac{z}{2\sqrt{kt}}\right) \tag{22-3}$$

Parsons and McKenzie assume T_0, the initial temperature of the rock at the spreading axis, to be 1300°C, the brittle-to-ductile transition to be at 975°C, and the base of the thermal boundary layer to be at 1260°C. Substituting these values into

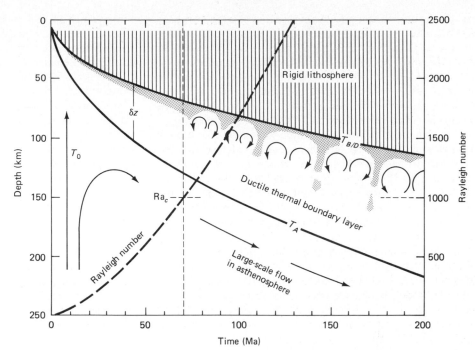

FIGURE 22-2 A thermal boundary layer at the base of the cooling and thickening oceanic lithosphere becomes gravitationally unstable after 70 Ma and begins to convect, with cooler, denser material (shaded) near the base of the lithosphere sinking into the boundary layer. This convection is proposed as the means by which ocean floor older than 70 Ma does not continue cooling as rapidly as younger ocean floor. [After Parsons and McKenzie (1978).]

Eq. 22-3 indicates that the thickness of the brittle zone is given by $z_{B/D} = 1.63\sqrt{kt}$, the depth to the base of the thermal boundary layer is given by $z_A = 3.06\sqrt{kt}$, and the difference between them—the thickness of the thermal boundary layer—is given by $\delta z = 1.43\sqrt{kt}$ (Fig. 22-2).

Rock within the thermal boundary layer cools against the base of the overlying rigid plate and becomes denser than the underlying rock. Near the spreading axis where the thermal boundary layer is thin, the Rayleigh number is below the critical value that would allow these density inversions to cause convection. Parsons and McKenzie believe that after 70 Ma, however, the thickness of this layer is great enough that the critical Rayleigh number is exceeded, and from then on convection occurs in the thermal boundary layer. As the layer thickens, convection cells become larger until a balance is achieved between the heat transported vertically by the cells and the heat conducted through the overlying rigid plate. Once convection starts, the temperature gradient in the thermal boundary layer is essentially adiabatic and heat is transferred into the base of the rigid plate much more rapidly. This then accounts for the decrease in the rate of cooling and subsidence of ocean floor older than 70 Ma (Prob. 22-3).

Small-scale convection may occur not only within the thin thermal boundary layer immediately beneath the lithosphere but in the upper mantle as a whole. Such convection would still be at a smaller scale than that defined by the plates. Its existence seems necessary to explain the heat flow through older lithosphere, but at present there is no direct evidence for such convection. The stresses imposed on the base of the rigid plate by this convection are too small to break the plate. During the early Precambrian, however, the steeper temperature gradient would have produced a thinner lithosphere that could have been broken by the small-scale flow. Indeed, the lack of long Archean mountain chains may be evidence that large lithospheric plates were unable to form because of disruptive small-scale convection. The dimensions of Archean greenstone belts may reflect the scale of this convection (McKenzie and Weiss 1975). Its pattern in plan view might have been similar to that formed in the molten wax model shown in Figure 22-3.

The change in tectonic style from greenstone belts in the Archean to long mountain chains in the Proterozoic and Phanerozoic indicates that during at least the last half of Earth

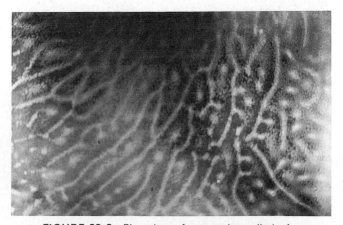

FIGURE 22-3 Plan view of convection rolls in 1-cm-thick sheet of paraffin wax caused by cooling of upper surface of wax, as seen from above. Convecting wax cools as it travels horizontally beneath a thin crust, and crystallization begins where cells descend (continuous white zones). Regions of upwelling are marked either by hot, clear wax or by small patches of white wax crystals.

history the upper mantle has been involved in large-scale convection, which must have played a dominant role in cooling the Earth. Despite its importance, there is no agreement on how this convective flow extends to depth in the present Earth, and thus its form at earlier times is completely speculative. Yet the temperature gradient within the Earth depends almost entirely on the form of this convection. In turn, crustal rocks are formed by processes that are controlled by this gradient.

The main controversy over large-scale convection centers around whether the whole mantle is involved in convective overturn, or the upper and lower mantle convect separately (Silver et al. 1988). Deep-focus earthquakes show that subducted lithospheric plates descend to depths of 600 to 700 km—below this, seismicity is lacking. Moreover, focal mechanisms indicate that although the subducted slab is in extension at shallow depths, it is in downdip compression by the time it reaches the depth of the seismicity cutoff. These facts can be interpreted as indicating that the prominent seismic discontinuity at a depth of 670 km marks a compositional boundary separating an upper and lower mantle between which matter is rarely transferred. Conversely, the discontinuity could result from a phase change, across which subducted slabs are able to cross. The lack of seimicity below the discontinuity would then indicate that the slab was simply no longer brittle. These two models are illustrated in Figure 22-4.

The existence of two long-lived isotopically distinct reservoirs in the mantle (Sec. 21-4) is strong geochemical evidence for their being separate upper and lower mantles. The reservoir from which MORB is derived is strongly depleted in LIL elements. The other is far less depleted and approaches more closely the model bulk Earth composition. The MORB source is the most commonly sampled reservoir, with the less depleted one making significant contributions only at "hot spots." These reservoirs are, consequently, interpreted to be the upper and lower mantle, respectively. Only where plumes rise into the upper mantle from the lower mantle does the less depleted reservoir make a contribution to crustal rocks. This interpretation can be further supported if the continental crust is believed to have formed from the LIL elements that were in the MORB source prior to depletion. If the MORB reservoir is uniformly depleted, its volume would have to be almost exactly that of the upper mantle if it were to account for the entire continental crust.

Although seismicity does not extend below the 670-km discontinuity, a significant body of seismic evidence indicates that subducted plates do sink below this boundary into the lower mantle (Silver et al. 1988). Because the subducted rocks are colder, they have higher seismic velocities than the lower mantle rocks into which they descend. Using seismic tomography, these regions of high velocity can be mapped as a function of depth, in the same way that x-ray tomography can be used to obtain a three-dimensional picture of the interior of the human body. For example, higher velocities are found in a belt surrounding the Pacific basin that extends from a depth of 1000 km to the core–mantle boundary (Dziewonski 1984). This belt lies beneath the circum-Pacific zone of subduction

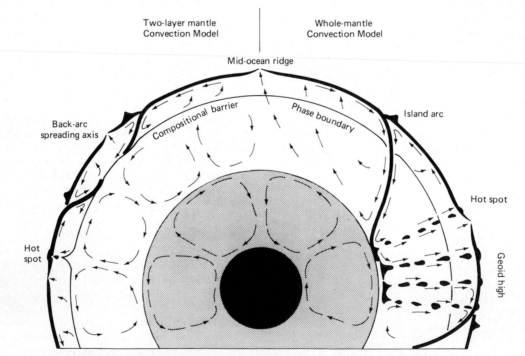

FIGURE 22-4 Possible patterns of convection in the Earth. On the left, a two-layer mantle shows subducted lithosphere descending only to the 670-km discontinuity, which is interpreted to be a compositional barrier across which heat but not matter is transferred. On the right, whole-mantle convection has subducted plates descending as far as the core–mantle boundary. In this model the 670-km discontinuity is formed by a phase transformation. (From various sources.)

and is presumably formed by the sinking of cold lithosphere into the lower mantle. These high-velocity zones may even correlate with depressions on the core–mantle boundary, in which case subducted slabs must be able to sink all the way to the bottom of the mantle.

If subducted slabs descend into the lower mantle, an equivalent mass of material must be transferred back into the upper mantle. This material would be relatively hot and be characterized by lower seismic velocities. Again, tomography reveals that two regions of low velocity occur in the lower mantle, one underlying the Pacific basin and the other underlying southern Africa. Both of these areas have the majority of the Earth's hot spots, and they form highs on the geoid (Crough and Jurdy 1980). The return flow from the lower mantle may therefore be in the form of scattered upwellings that form clusters of hot spots rather than single large convection cells. The difficulty posed by this model is that it would tend to homogenize the distinct isotopic reservoirs that we know exist in the Earth. Various models for lower mantle flow have been proposed that purport to avoid this problem. These are discussed by Silver et al. (1988).

One extremely important finding from tomography is that although the high velocities associated with subducted slabs can be traced possibly as deep as the core–mantle boundary, the low velocities associated with the hot mantle rising beneath spreading axes, while clearly detectable at depths of 150 km, are not present at depths greater than 350 km (Silver et al. 1988). Spreading axes are therefore not the expression of large ascending plumes. Instead, they appear to be passive features where hot mantle has simply risen to fill the gap left by the diverging plates. As such, they cannot be interpreted as major ascent conduits that balance the downward flow of material into the mantle at subduction zones. The return flow, instead, is more diffuse. This model of passive divergent plate boundaries is supported by the timing of events associated with rifting. For example, the Red Sea rift is now flanked by mountains that rise 2.5 km above sea level, but at the start of rifting, 30 Ma ago, this region was at sea level and did not start uplifting until 20 to 25 Ma ago. As hot mantle rose to accommodate the divergence of the plates, temperatures increased and the land subsequently was uplifted (McGuire and Bohannon 1989). Had an active mantle plume caused rifting, the uplift would have preceded the rifting.

The rate of large-scale convection associated with plate motion is sufficiently high that it plays the dominant role in cooling the Earth. At present, ocean floor is being created and subducted at a rate of 3 km^2 a^{-1}. If the average thickness of the plate is 125 km at the time of subduction, 375 km^3 of lithosphere is subducted each year. At this rate, the entire volume of the upper mantle (3×10^{11} km^3) could be processed by plate tectonic convection in 0.8 Ga or the entire volume of the mantle (9×10^{11} km^3) in 2.4 Ga.

Temperatures in the Earth depend mainly on the extent and form of convection systems. In layers, such as the lithosphere, where convection does not occur, temperatures rise steeply as a result of the slow transfer of heat by conduction. Within convecting layers, however, the temperature gradient is extremely shallow, being very nearly adiabatic (Eq. 13-14). In the mantle just beneath the lithosphere this gradient is about

0.6°C km^{-1} (Probs. 22-5 and 22-6). If the mantle is layered, steep temperature gradients exist in thermal boundary layers at the bottom and top of adjoining convecting layers. Richter and McKenzie (1981), for example, estimate that if the upper and lower mantle convect separately, a 500°C temperature increase would exist across a narrow zone at the 670-km discontinuity.

Figure 22-5 shows a possible geothermal gradient beneath old oceanic lithosphere. It has an upper 125-km-thick conductive cap in which temperatures rise to 1000°C. In younger lithosphere this temperature would be reached at shallower depths (Fig. 22-2). Beneath this, temperatures continue to rise sharply through a thermal boundary layer at the top of the convecting upper mantle. Below this, however, temperatures rise slowly until thermal boundary layers are encountered at the base of the upper mantle (670 km) where temperatures increase by 500°C. A second geotherm (dotted) shows how the temperature would continue increasing slowly if the upper and lower mantle were not compositionally layered. Below the 670-km discontinuity, temperatures again rise slowly until the thermal boundary layers at the core–mantle boundary are reached and there is again a large temperature increase of about 1000°C. Experiments on the melting of iron under high pressures sets the temperature of the core–mantle boundary at about 3500°C and at the center of the Earth at about 6600°C. Also shown in Figure 22-5 is the continental geotherm calculated in Chapter 1.

The picture of convection associated with plate motion that seems to be developing is one in which the most prominent flow is associated with the sinking of the plates at subduction zones. These rafts of cold lithosphere are metamorphosed to dense eclogite (Fig. 20-7) on sinking into the mantle, which adds substantially to the gravitational imbalance resulting from the temperature differences. These slabs clearly descend to the base of the upper mantle and probably to the core–mantle boundary. Because they affect such a large fraction of the mantle, they must be long-lived phenomena. By contrast, the upwelling at divergent plate boundaries is passive and determined simply by the position of breaks in the plate. Spreading axes are therefore able to shift at any time. The upward flow that balances the downward flow at subduction zones is distributed over broad areas, with some possibly being concentrated at hot spots. The type of convection illustrated by the wax model in Figure 22-3 is therefore similar, with the downward flow being concentrated into narrow zones while the upward flow is spread out across broad zones.

Although the mineralogical constitution of the upper mantle is discussed in Sec. 22-5, it is worth commenting here on some important phase changes that may affect convection in the mantle and thus the shape of the temperature gradient. We do this in terms of a spinel lherzolite, which is commonly thought to have a composition similar to that of much of the upper mantle. This rock type has been investigated at high pressures by numerous workers, the data in Figure 22-5 being that of Takahashi (1986).

At low pressure, lherzolite contains plagioclase, but above 1 GPa, spinel becomes the stable aluminous phase and plagioclase disappears. Above 2.5 GPa, spinel disappears and is replaced by garnet as the stable aluminous phase. Below a

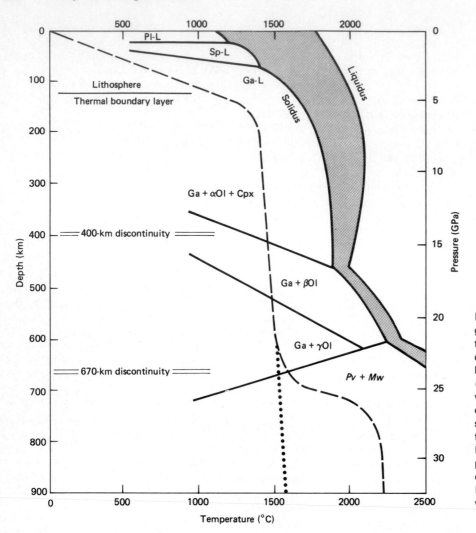

FIGURE 22-5 Possible geothermal gradient resulting from separate convecting layers above and below the 670-km discontinuity and conduction in the lithosphere. [After Richter and McKenzie (1981).] If there is whole-mantle convection, the geotherm would follow the dotted line. Also shown are the liquidus, solidus, and subsolidus phase relations for a spinel lherzolite from Kilborne Hole, New Mexico (Takahashi, 1986). Pl, plagioclase; Sp, spinel; Ga, garnet; Ol, olivine (α, β, and γ); Cpx, clinopyrozene; Pv, silicate perovskite; Mw, magnesiowüstite; L, liquid.

depth of 100 km, then, the upper mantle consists of garnet lherzolite, that is, olvine + garnet + clinopyroxene + orthopyroxene. With increasing pressure each of these minerals changes to denser phases, which may account for some of the seismic discontinuities.

The first mineral to change is olivine. At a pressure corresponding to a depth of about 400 km, α olivine transforms to a denser β phase, and if the rock also contains a spinel phase, the reaction is univariant; that is, for a given temperature, the reaction would occur at a particular depth. Because olivine typically constitutes more than 50% of the rock, this transformation significantly changes the properties of the rock and probably causes the 400-km seismic discontinuity. Because the transformation takes place rapidly, it does not inhibit the flow of rock across the discontinuity and therefore does not affect convection. Pyroxenes also undergo a transformation to the β phase plus the high-pressure silica polymorph stishovite at about the same pressure, and this may contribute further to the change in properties at the 400-km discontinuity. With increasing pressure, both olivine and pyroxene undergo further changes to a spinel γ phase. At the same time pyroxene begins to dissolve in the pyrope-rich garnet as the majorite compo-

nent [$Mg_3Al_2Si_3O_{12} + M_3(MSi)Si_3O_{12}$, where $M = Mg$, Fe, Ca] as a result of some of the silica entering octahedral coordination and combining with a divalent cation to replace Al in the garnet.

The largest change in properties in the mantle occurs at the 670-km discontinuity. At approximately this pressure, the coordination of oxygen about silicon changes from four to six with the formation of a silicate perovskite [$(Mg, Fe, Ca)(Si, Al)O_3$] from olivine and garnet; the excess Mg and Fe in the rock forms magnesiowüstite [$(Mg, Fe)O$]. These two phases presumably are the major phases in the lower mantle. Although this change provides a simple explanation for the 670-km discontinuity, the perovskite forms through a number of continuous reactions that would likely spread the transition out over a range of depth. This is not consistent with the sharpness of the seismic discontinuity, which may therefore require a compositional change for its explanation. At present, then, the experimental data do not resolve the question of whether convective currents can cross the 670-km discontinuity the same way they can the 400-km discontinuity.

Also shown in Figure 22-5 are the liquidus and solidus for this anhydrous peridotite. The geotherm, along much of its length, is within about 500°C of the solidus. If the solidus were to be lowered by fluxing with water, the geotherm would first intersect the solidus near a depth of 150 to 200 km. This could certainly account for the position of the low-velocity zone beneath the lithosphere. If the lower and upper mantle convect separately, the sharp rise in temperature through the boundary layers separating them brings the geotherm close to the solidus at this depth. Such a sharp rise might cause melting at this depth and give rise to a second low-velocity layer. The lack of such a low-velocity layer has been used as evidence for whole mantle convection. However, recent experimental work (Fig. 22-5) shows that the solidus rises steeply once silicate perovskite becomes a stable phase, reducing the possibility of the geotherm reaching the solidus at this depth.

Lherzolite melts over a 650°C temperature interval at atmospheric pressure. With increasing pressure, the liquidus and solidus approach each other and at 17 GPa are separated by less than 100°C. This indicates that the lherzolite is almost a eutectic composition at this pressure. Is this a coincidence? Probably not (Walker 1986). It seems likely that early in Earth history the upper mantle was formed as a near-minimum melt in equilibrium with solids at a depth of at least 500 km. This melting, which could have been related to core formation, would have given rise to a magma ocean (Sasaki and Nakasawa 1986). It is of interest to note that the liquidus temperature decreases as the pressure increases from 10 to 15 GPa. Olivine, which is the mineral on the liquidus under these conditions, would therefore have floated [see Herzberg (1987)]. This part of the mantle might therefore have solidified from the top down, leaving a residual liquid at its base.

22-3 CONDITIONS NECESSARY FOR ROCK GENERATION

Rocks are formed when changes in environmental conditions (T, P, X_{fluid}) cause changes in the phases constituting the Earth. This requires that material cross reaction surfaces in $T–P–X_{fluid}$–space. Changes in the environment that do not result in reaction surfaces being crossed do not produce rocks. Surface rocks on the moon, for example, undergo daily fluctuations in temperature but no new rocks are formed because no reactions take place under those conditions. Only one reaction surface, the solidus, is of importance in the generation of igneous rocks, whereas many reaction surfaces, most involving devolatilization, are crossed in the formation of metamorphic rocks. Reaction surfaces are functions of T, P, and fluid composition. Changes in any of these variables can therefore result in the formation of new rock. It is all too easy to lose sight of this important fact. For example, because magma is hot it seems reasonable to suppose that it is the product of heating. We will see, however, that most magma is formed not by heating but by decompression, which actually involves adiabatic cooling.

Reactions that form igneous rocks and most metamorphic rocks are accompanied by increases in entropy and volume because of the production of liquid or fluid. They consequently have positive slopes in terms of P and T, at least at low pressures (Eq. 8-3). At higher pressures the change in volume is smaller and the reactions become largely functions of T. At still higher pressures, the volume change may become negative, and the reaction will have a negative slope, as is seen to be the case for the liquidus of lherzolite (Fig. 22-5). Because of these changes in slope, variations in environmental conditions can have different effects at different depths in the Earth. A decrease in pressures at one depth, for example, could cause melting, whereas at another it could cause solidification.

We will now examine how changes in temperature, pressure, and fluid composition can occur. But first, it is important to keep in mind that the Earth is a dynamic planet that tends toward a state of *dynamic equilibrium*. Thus when we talk of changes in the environment, we are not referring to a change from one set of static conditions to another, but rather to a disturbance of a set of conditions that over an extended period of time have established a dynamic equilibrium . Once the conditions are perturbed fluxes of heat, stress, and matter try to reestablish the initial dynamic equilibrium. Rocks are products of the relaxation of these perturbations.

Let us start by considering how changes in temperature might occur. Nowhere is the principle of dynamic equilibrium more evident than in the development of the geothermal gradient. Heat is continuously generated in the Earth and transferred to the surface, where it is radiated into space. Because of the self-regulating nature of heat transfer, the rate of heat production in the Earth is balanced by the rate of heat loss. At any given depth, the rate at which heat is transferred depends on the mechanism of heat transfer and the thermal properties of the material at that depth. Where heat is transferred by the slow mechanism of conduction, as in the lithosphere for example, the temperature gradient must be steep in order to transfer the heat coming from below and being generated from within. In the mantle, where heat is transferred by the more rapid mechanism of convection, the temperature gradient is correspondingly less. To change the temperature at any given depth, the geothermal gradient must be perturbed.

At dynamic equilibrium the temperature at any depth in the Earth remains constant as a result of the balance between the flux of heat out, the flux of heat in, and the quantity of heat produced at that depth (see heat conservation equation 1-5). It is this balance that allowed us to calculate a steady-state geotherm in Sec. 1-6. Changes in temperature can therefore result from changes in any one of these three components. We will first examine the heat production component.

Heat production can be positive or negative, depending on whether heat is generated or consumed. But only one form of heat production is independent of all other variables, and that is radiogenic heat. Despite the slow decrease in this quantity through radioactive decay, it can be considered a constant at the time scale of normal geologic processes. It therefore cannot be responsible for changing the temperature of a rock. Certain sedimentary and igneous processes may create local concentrations of radioactive elements that may perturb the gradient, but this involves a transfer of material and is not strictly a change in the heat-producing capacity of the initial rock. Similarly, thickening of the radioactive crust results in higher temperature, but this is because of a blanketing effect and not because of a change in the heat-producing capacity of a given volume of rock.

A number of reactions take place in rocks that either generate or consume heat. These certainly can affect the temperature of a rock, but their effect is secondary because they first require changes in other environmental factors for the reactions to take place. Melting of rock and dehydration of minerals both absorb heat, whereas crystallization of magma and coarsening of grain size liberates heat. In addition, mechanical energy can be converted to heat. In magma, for example, this occurs through the dissipation of viscous forces, but the effect is small unless velocities are very high (see ash flows in Sec. 4-4). In solid rocks, shearing generates heat, but this too is usually small unless concentrated in major shear zones (Sec. 19-4).

Changes in temperature must be caused mainly by changes in heat flux. In the rigid lithosphere heat is transferred mainly by conduction, but advection can be important locally. Circulating meteoric waters can cause cooling of near-surface rocks, as for example near midocean ridges, and ascending metamorphic fluids can cause heating. Magma may also advectively introduce heat into the lithosphere. Advective flow, however, is commonly channelized, so the final distribution of heat is still by conduction. The large-scale transfer of heat in the mantle is by convection, but this flow is laminar, so that conduction must still play a role in transferring the heat in and out of any given volume of rock. The rate of heat conduction depends on the temperature gradient. Changes in the gradient will change the temperature of the rock. The gradient above a given volume of rock can be increased if the crust is thinned, which can result from erosion or extension. This would lower the temperature of the volume of rock in question. Thickening of the crust, on the other hand, would decrease the gradient and cause heating. Intrusion of magma, or increased rates of mantle convection beneath a given volume would increase the flux into the volume and cause heating. We conclude from the heat conservation equation that changes in temperature must result primarily from changes in the conduction of heat into or out of the volume of interest. But heat conduction through rocks or magma is extremely slow ($k = 10^{-6}$ m^2 s^{-1}), so that phase changes resulting from temperature changes must be equally slow.

Next we examine how changes in pressure may occur. Most pressure changes in the outer part of the Earth result from the motion of lithospheric plates. These changes can be positive or negative. At convergent plate boundaries subduction and lithospheric thickening result in increases in pressure. Where the lithosphere is thinned by extension, the pressures are decreased. Erosion and sedimentation redistribute surface loads, which also cause changes in pressure at depth. The advance and retreat of continental ice sheets can also cause small changes in pressure. Explosion craters formed by large meteorite impacts can result in rapid decreases in the load pressure beneath impact sites. This must have been an important process early in Earth history when the frequency of such impacts was far greater.

Because of the relative weakness of rocks to long-term stress (< 30 MPa), changes in pressure take place almost immediately, as evidenced by the Earth's close approach to isostatic equilibrium. Because most changes are a consequence of plate tectonics, we expect pressure changes to occur at comparable rates to plate motion, which is of the order of 0.01 m a^{-1}. This corresponds to changes in pressure of 350 Pa a^{-1} if we assume an average density of 3.5 Mg m^{-3}. Changes in pressure are consequently expected to be far more rapid than changes in temperature.

We have seen that the presence and composition of a fluid can have profound effects on the temperatures and pressures at which rocks melt and metamorphic reactions occur (Chapters 11 and 16, respectively). We have also seen in Sec. 18-8 that advective fluids are capable of transferring significant amounts of heat (Péclet number) as well as of matter at rates that are greater than those for heat conduction. Advective fluids can therefore be important in the formation of rocks.

The ultimate source of most fluids, at least during the last half of geologic time, has been the Earth's surface. Plate motion has transported this fluid in the form of hydrous and carbonate minerals into the Earth at subduction zones where rising temperature has caused its release into the overlying rocks. Some fluids may be primordial and come from outgassing of the planet. These are likely to have been more important in the early history of the Earth. The prevalence of granulite facies rocks in Precambrian terranes has been suggested to result from flushing of the lithosphere with CO_2-rich fluids derived from the mantle or deeper (Newton et al. 1980). Similarly, the isotopic evidence for metasomatism prior to the emplacement of magmas in oceanic islands (hot spots) may point to flushing with fluids enriched in incompatible elements derived from deep in the mantle.

The rates at which advection can introduce or change the composition of fluids depends on the pressure gradients on the fluids and the wetting of the minerals by the fluids. This matter was discussed at length in Sec. 18-3, where we saw that although rates may be slow, they are likely to be more rapid than the rates at which heat is transferred by conduction. Advection is therefore an important variable in the formation of rocks.

We will now consider five different environments in each of which one of the variables above plays a dominant role in generating rocks. Lithospheric plate motion provides the ultimate cause for the changes which lead to the formation of rocks, but the rate of rock production depends on which of the environmental factors changes, with pressure changes producing rocks fastest, followed by advective changes, and finally, by thermal changes.

By far the largest production of igneous rocks in the world occurs at divergent plate boundaries. This normally involves the eruption of about 20 km^3 a^{-1} of MORB at midocean ridges, but on numerous occasions throughout geologic time voluminous flood basalts have also formed when the divergent boundaries have traversed continents. In both environments magmas are formed as a result of decompression melting in zones of lithospheric extension (McKenzie and Bickle 1988).

Figure 22-6a shows the distribution of isotherms prior to lithospheric extension. The geotherm (Fig. 22-6b) rises steeply through the lithosphere because of conductive heat transfer, but then flattens in the thermal boundary layer beneath the lithosphere, reaching a temperature of about 1480°C at the top of

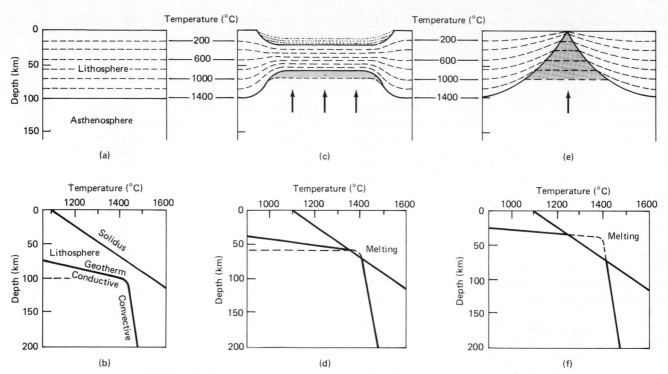

FIGURE 22-6 Development of magma by decompression melting beneath regions of lithospheric extension. (a) Position of isotherms in a conductive lithosphere and convective asthenosphere prior to extension. (b) Geotherm prior to extension and simplified solidus of lherzolite. (c) Extension of the continental lithosphere results in the rise of the asthenosphere; this brings isotherms closer to the surface, and melting takes place through decompression (shaded area). (d) Rise of the geotherm beneath the zone of extension results in temperatures exceeding the lherzolite solidus. (e) If the lithosphere is ruptured during extension, the asthenosphere rises more, but over a narrower zone; this results in more extensive melting, as seen in (f).

the convective mantle. The geotherm approaches the lherzolite solidus but does not reach it under the conditions considered here.

When the lithosphere undergoes extension it does so by thinning (Fig. 22-6c) and in the case of oceanic lithosphere this is accompanied by actual rupture at the spreading axis (Fig. 22-6e). On the lower surface of the lithosphere, mantle rises to compensate for the lost volume, whereas on its upper surface a sedimentary basin may form. The thinning, which in the continental lithosphere may occur by stretching (pure shear) or by movements on normal faults (simple shear) (White 1989), results in the conductive geotherm being steepened and the convective geotherm being brought closer to the surface. This happens because the rate of tectonic transport is faster than the rate of heat transfer. Thus each point on the preextensional geotherm moves vertically to establish the new geotherm. Eventually, this perturbed geotherm will relax and form a new steady-state geotherm, but that takes a considerable length of time (tens of Ma). In the interim, the perturbed geotherm has intersected the lherzolite solidus and produced magma.

The melting accompanying lithospheric extension is caused by the decrease in pressure as the mantle rises; it is not produced by heating of the mantle. Indeed, as the mantle rises it undergoes adiabatic cooling (Eq. 13-14), which de-

creases its temperature by about $0.6°C$ km^{-1} if the coefficient of thermal expansion is 4×10^{-5} K^{-1} and the heat capacity is 10^3 J kg^{-1} K^{-1}. The temperature is lowered still more once melting occurs because of the latent heat of fusion, which is about 420 kJ kg^{-1} (Prob. 22-7).

The amount of melting that takes place depends on how far the perturbed geotherm rises above the solidus. Extension in oceanic lithosphere is concentrated at spreading axes, and as a result the mantle rises a considerable distance with a consequent production of large volumes of melt (Fig. 22-6e and f). Extension of the continental lithosphere tends to be spread over greater distances. Consequently, the mantle does not rise as far, and correspondingly less melt is produced (Fig. 22-6c and d). The amount of melt formed in this case clearly depends on how much the continental lithosphere is stretched (Prob. 22-8).

The compositions of the first-formed melt along the solidus varies with depth, as will be seen in Sec. 22-5. Where the geotherm intersects the solidus depends on the adiabatic temperature of the convecting mantle. This probably ranges from about 1300 to 1500°C, depending on whether the mantle rises passively beneath the extending lithosphere or has a convective component rising from a deep mantle plume. In any case, temperatures, and consequently depths, cannot vary widely. This probably explains the limited range of compositions of **MORB** and continental flood basalts. Differences between MORB and continental flood basalts can probably be attributed largely to slight differences in the depth of melting and the degree of melting.

If sufficient melt is produced in the extensional zone by adiabatic decompression, magma is likely to ascend buoyantly through the lithosphere to produce lavas and intrusive bodies. This advectively introduces heat into the lithosphere, which causes contact metamorphism. In addition, the steepened gradient caused by lithospheric extension results in an increased heat flux which is independent of the intrusion of magma. Because this heat flux depends on conduction through the lithosphere, it operates on a much slower time scale than the advective heat transfer associated with the emplacement of magmas. Extension of the lithosphere consequently sets up a number of perturbations that relax at different rates. This variability produces the complexity in the geologic record. We can illustrate this by considering the type of history that might be recorded in an extensional sedimentary basins, such as the Mesozoic ones formed in eastern North America just prior to the opening of the central Atlantic.

Once extension of the lithosphere begins, sediment starts accumulating in surface depressions (Fig. 22-7). Sedimentary rocks therefore provide the earliest record of extension. This is because of the speed with which isostatic adjustments take place. The surface changes are, of course, a reflection of movement of the mantle at depth, which after rising far enough intersects the solidus and melting begins. After sufficient melt has accumulated, magma rises into the basins to form lava flows and intrusive bodies around which contact metamorphic rocks are formed. Temperatures in the sedimentary basins are initially low and are only locally raised by the intrusion of magma. Eventually, however, the thinned lithosphere with its steepened gradient begins to raise temperatures at the base of the sedimentary pile. However, pore fluids in the sediments advectively remove this heat and distribute it throughout much of the basin, causing low-grade zeolite facies metamorphism, in particular where fluid flow is channelized along faults. With time, temperatures gradually decrease until a steady-state geotherm is reestablished during which time the youngest sediments are deposited. Because the lithosphere is now thinner, radiogenic heat

production is less, and the steady-state geotherm is shallower than the preextensional geotherm.

Because of the shallow slope on the convective geotherm in the upper mantle (Fig. 22-8), material starting from anywhere on the geotherm in the upper 670 km is not likely to intersect the solidus unless there is sufficient lithospheric extension and then the point of intersection is limited to the region where the convective geotherm becomes conductive (< 100 km). If, however, convection is driven by plumes rising from deep in the mantle, as presumably happens at oceanic islands such as Hawaii, melting can occur at greater depth. If diapirs are able to rise from below 670 km or from the thermal boundary layers between the upper and lower mantle, their higher temperatures will cause them to intersect the lherzolite solidus at greater depth. Once partial melting takes place, the lowered viscosity allows the diapir to rise more rapidly. Large amounts of melting, however, do not take place, because the adiabatic cooling keeps the magma near the solidus, which at these greater depths has a shallow slope. Because of the greater depth of melting, these magmas have a different composition from the shallow MORB. Here again the cause for melting is the change in pressure on the ascending mantle and not heating.

We will now consider the case where advection of fluids plays the dominant role in forming rocks. At convergent plate boundaries, cold lithosphere is subducted into the mantle, thus lowering the average temperature in these regions. It seems paradoxical, therefore, that such regions are almost always marked by volcanoes and high-temperature metamorphic rocks. Why should regions that would be expected to be cool show such clear evidence of heat? The explanation lies in the mechanism of heat transfer. Throughout most of a lithospheric plate heat is transferred mainly by the slow process of conduction. Above subduction zones, advective fluids bring about partial melting, and bodies of magma advectively transfer heat into the overlying lithosphere. Thus, although the average temperature of the Earth at convergent plate boundaries may be low, rising bodies of magma advectively transfer heat to much higher levels in the crust than would ever be possible through conduction alone.

The ocean floor subducted at convergent plate boundaries has had a long history of cooling and hydrothermal alter-

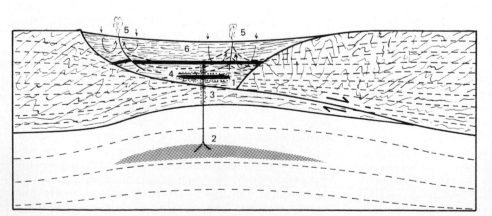

FIGURE 22-7 Chronological record of extension of continental lithosphere. (1) Sedimentary fill of extensional basins. (2) Decompression melting of ascending asthenosphere. (3) Ascent of magma after sufficient segregation of melt in source region. (4) Eruption of lavas and intrusion of dikes and sills, and development of contact metamorphic rocks. (5) Elevation of regional temperatures as a result of lithospheric thinning causes zeolite facies metamorphism, in particular near fractures where convective hydrothermal cells develop. (6) Eventually, the thinned continental crust develops a geotherm that is shallower than the preextensional one. Final sedimentation occurs at this time.

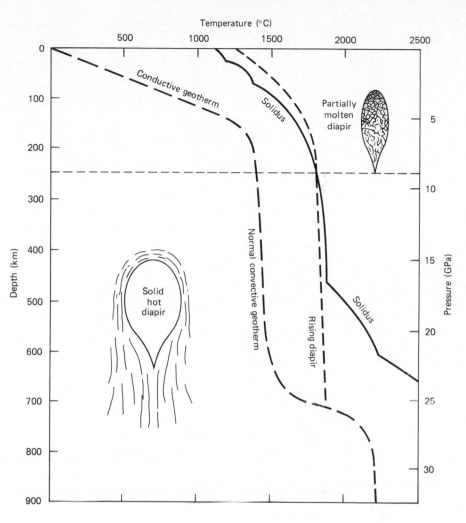

FIGURE 22-8 A hot mantle diapir rising from the top of the lower mantle intersects the lherzolite solidus at a depth of 250 km. As it continues rising, the heat of fusion causes its temperature to decrease.

ation. Following generation at an oceanic ridge, the basalt and intrusive rocks are intensely altered by convecting hydrothermal waters, which circulate through the newly formed crust. Ophiolite suites presumably provide an example of the end product of this alteration (Sec. 14-2). When this oceanic crust is subducted, increased pressures and temperatures lead to its progressive metamorphism through the blueschist to eclogite facies. In doing so, all of the water that was taken into the oceanic crust during its alteration is released and rises into the mantle above the Benioff zone, where it causes metamorphism and melting. Precisely how this takes place is not yet fully understood. For a full discussion of this topic, see the review by Wyllie (1988).

Before tackling the problem of melt generation above Benioff zones, it is important to have some understanding of the thermal structure of such regions. As already seen in Sec. 22-2, the temperature distribution in an oceanic plate can be adequately described in terms of conductive cooling of a half-space, with convective heat influx at the base of the lithosphere for ocean crust older than 70 Ma. Cooling stops, however, once the plate begins to subduct, and its temperature steadily rises as heat flows in from the mantle above and below the plate.

An important question relating to convergent plate boundaries is whether the subducting plate shears past the overriding plate or is coupled to it. In the first case, the rock in the overlying mantle wedge remains stationary and slowly loses heat to the subducting slab, a condition that is hardly conducive to the generation of magmas. On the other hand, if the overlying mantle is coupled to the subducting plate, a convective motion is set up, which continuously cycles new hot mantle into the wedge (Fig. 22-9). This mechanism has the advantage that in addition to keeping the mantle wedge hot, it provides an explanation for back-arc spreading which is present behind some but not all island arcs (Toksöz and Hsui 1978). The convection in the mantle wedge may put sufficient traction on the base of the lithosphere to cause extension and rupture even though compression is occurring simultaneously near the oceanic trench. Analogous motion has been witnessed in cooling Hawaiian lava lakes when slivers of solid crust break off and are pulled toward large slabs of crust that are sinking into the still molten lava lake (Duffield 1972).

Figure 22-9 shows the calculated distribution of isotherms in the convecting mantle wedge above the Benioff zone following 75 Ma of subduction at a rate of 8 cm yr^{-1} (Toksöz and Hsui 1978). Although the isotherms in the wedge are much

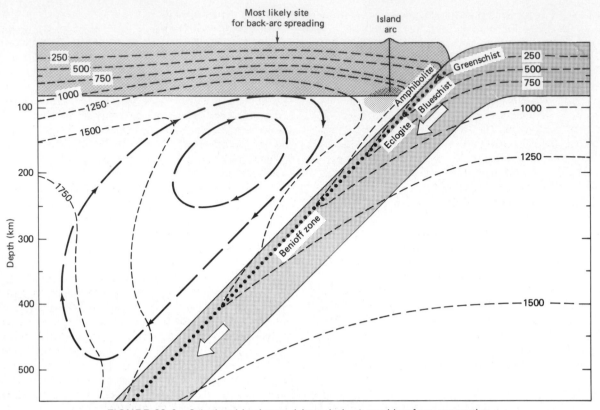

FIGURE 22-9   Calculated isotherms (short dashes) resulting from convective flow (heavy long dashes) in the mantle wedge overlying a subduction zone (Toksöz and Hsui 1978). Convection is driven by the sinking of the subducted plate. Locus of earthquakes (Benioff zone) is shown by dots. Ocean floor rocks are progressively metamorphosed as they are subducted. Melting occurs (shaded area) when Benioff zone exceeds 100 km, probably as a result of water rising into the mantle wedge from the subducted plate. This magma rises to form island-arc volcanoes.

nearer the surface than in the subducted plate, nowhere do they intersect the anhydrous solidus of lherzolite (Fig. 22-10). If convection in the wedge were strong enough to rupture the lithosphere, the upwelling mantle would intersect the lherzolite solidus and MORB-like rocks would be formed above the convecting cell at a back-arc spreading axis. But this would not produce melting beneath the island arc. Nor would the subduction of the oceanic basalts, which are much too cold to melt.

In Sec. 14-3 we saw that igneous activity above the Benioff zone does not start until the subducted slab has sunk to a depth of 100 km, with most island-arc volcanoes being located about 120 km above the Benioff zone. This relation is independent of the rate or angle of subduction, but if the angle decreases below about 25° there is no volcanism at all. These characteristics provide important evidence about the way in which magmas are generated at convergent plate boundaries. Because the position of the volcanoes is independent of the rate of subduction, the temperature of the subducted slab cannot be a critical factor in the generation of the magmas. The strong correlation with depth indicates that pressure may be an important factor. Metamorphic reactions, in particular those giving

rise to the eclogite facies (Fig. 15-7), are pressure sensitive, but they are also functions of temperature and thus should be dependent on the rate of subduction. Also, if pressure were the controlling factor, plates subducting at angles of less than 25° would eventually reach the same pressure as those sinking more steeply and magmas should be generated; but they are not.

Because the correlation between the position of volcanoes and the depth to the Benioff zone is so strong, other important factors can be overlooked. If heat is needed for the associated magmatism, it must come from convection within the mantle wedge. This convection produces a sharp bend in the isotherms above the Benioff zone where it goes below a depth of 100 km, with temperatures exceeding 1000°C between depths of 60 and 100 km (Fig. 22-9). The position of this temperature maximum is determined mainly by the depth below the Earth's surface rather than the angle of subduction. If the subducting plate plunges more steeply, the bend in the isotherms is gentler but occurs at about the same depth. As the angle of subduction decreases, the bend in the isotherms becomes sharper, and at some critical angle the convecting mantle would not be able to rise into the wedge, and temperatures would fall rapidly. This could explain why volcanoes are not present when the subduction angle is less than 25°.

Temperatures slightly in excess of 1000°C at depths of 60 to 100 km are not high enough to cause melting of dry mantle, but they are if water is present (Fig. 22-10). By the time the subducted plate has reached a depth of 120 km it has been converted to eclogite assemblages, and the water that was initially present in hydrous minerals has been released and migrated into the overlying wedge. The formation of magma in this region can therefore be attributed to the advection of water, but the elevated temperatures caused by convection also play an important role. As the Benioff zone becomes still deeper, temperatures in the overlying wedge increase, but by this time the subducted slab has been dehydrated, and thus no further melting occurs.

Although melts can form at the water-saturated solidus of lherzolite at temperatures near 1000°C, they cannot rise to the surface. A melt formed at a depth of 75 km and 1060°C, for example, adiabatically cools to a temperature of 1005°C on rising to a depth of 20 km, where it intersects the solidus and crystallizes (Fig. 22-10). This has led some petrologists to invoke still higher temperatures in the mantle wedge, so that melts that are undersaturated in water can be formed (Marsh 1979). It should be pointed out, however, that the calculated isotherms in Figure 22-9 are based on convection in a solid,

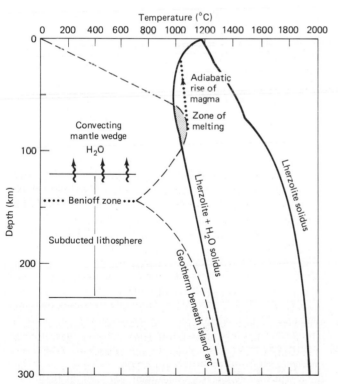

FIGURE 22-10 Geotherm beneath the island-arc volcanoes in the model of Figure 22-9. The water-saturated solidus of lherzolite (Kushiro et al. 1968) intersects the geotherm between depths of 60 to 100 km. The water necessary for melting is liberated by dehydration reactions in the subducted oceanic lithosphere.

viscous mantle with no account taken of the effects of melting. Once melting occurs, the decreased viscosity modifies the position of the isotherms and promotes still more melting. Magma rising from 75 to 20 km, for example, elevates the isotherms in its path and raises the temperature of the source region. If the lithosphere is ruptured and back-arc spreading starts, isotherms in the lithosphere would again be raised. Also, if the lithosphere contains thickened continental crust containing radioactive elements, there would be additional heat. Although it is not certain how all these factors affect the production of magmas at convergent plate boundaries, there is little doubt that water plays a dominant role.

Advective transfer of heat by ascending bodies of magma is the most important means of raising lithosphere temperatures above the values that normally result from conduction. Most large bodies of silicic magma in the continental crust are now recognized to derive their heat from basaltic magmas that rise from the mantle (Secs. 13-10 and 14-3, and Fig. 14-8). Because of density, many basaltic magmas probably spread laterally at the base of the crust to form large sills. Heat liberated by the cooling and crystallization of these bodies raises the temperature of overlying rocks and forms melts of granitic composition, which being less dense than the basaltic magma, form gravitationally stable layers on top of the basalt with minimal mixing (Huppert and Sparks 1988). The granitic melt, however, is unstable with respect to the overlying crust and may rise into it, advectively transferring the mantle-derived heat into the upper crust, where it brings about high-T, low-P metamorphism.

The rate at which the basaltic magma transfers heat to the granitic melt is geologically rapid, being in the tens to hundreds of years (Huppert and Sparks 1988). This is because all but the thinnest basaltic sills (<10 m) convect, thus keeping their roof rocks at high temperatures. Initially, the roof rocks are simply heated, but eventually melting of a granitic fraction takes place. The actual temperature at which this occurs depends on the availability of H_2O. As the zone of melted rock thickens, its Rayleigh number increases to the point where convection occurs in the granite as well (Eq. 13-10). This then allows the granitic layer to transfer heat more rapidly to its upper surface and cause more melting.

The heat required for melting the roof of the granitic layer is supplied initially by the underlying convecting basaltic magma. This, however, cools the basalt, which, on becoming more than 60% crystallized (Eq. 13-24), becomes too viscous to convect. From then on the granitic layer must cool, but it is sufficiently superheated that convection continues for some time. Even melting of the roof continues as heat is liberated by the crystallizing granitic magma. Convection of this crystal mush continues until it is 60% solidified. After this, heat can only be transferred by conduction.

If the sheets of granitic melt formed above basaltic sills are thick enough, they become unstable relative to the denser overlying crust, and diapiric bodies rise from them. This advectively transfers heat still higher in the crust. The height to which the diapirs rise depends on many factors (Sec. 3-9), but eventually they are halted by the cooler, more viscous crust, and then heat transfer into the surroundings is largely by the slow process of conduction (advective meteoric waters may play a role near the surface). The rate of the heat transfer from the mantle by advection through the basalt and granite (order about 10^2 to 10^3 years) is so much faster than the rate of heat loss by conduction (order about 10^4 to 10^5 years) that the thermal gradient in the vicinity of the diapirs is greatly steepened and high-T, low-P metamorphic rocks are formed.

Finally, let us consider the case where conditions in the Earth are changed by crustal thickening. This has already been considered in Chapter 20 in connection with $P-T-t$ paths of metamorphic rocks. Initially, crustal thickening lowers the geotherm, because tectonic processes are faster than the rate at which heat conducts through rock. Eventually, however, the geotherm is elevated by the thickened crust, which behaves like a layer of insulation on the Earth. Furthermore, thickening of the layer containing abundant radioactive elements provides additional heat. The elevated geotherm eventually returns to its former steady state once erosion or tectonic processes has thinned the crust to its original thickness.

Calculated $P-T-t$ paths followed by rocks during crustal thickening by a single thrust fault are shown in Figure 20-6. At depth, these paths increase in temperature while pressure decreases due to erosion of the thickened crust. They cross numerous metamorphic reactions, each of which involves a heat of reaction and possibly the evolution of fluid that could advectively remove heat from the rock. Neither effect was considered in calculating the paths, in part because of complexity, but also because of uncertainties in some of the quantities. The effects on solid metamorphic rocks are probably small, but the same cannot be said if melting occurs. First, the enthalpy change on melting is large—about 300 kJ kg^{-1}, and second, once the rock is melted, it may rise and advectively transfer heat. $P-T-t$ paths will therefore be strongly affected if significant amounts of melt are formed.

The lowest temperature at which melt can form is defined by the water-saturated granite solidus (Fig. 20-7). The amount of melt formed at this temperature is unlikely to be great, simply because large quantities of water are not available. Whatever water is present, however, enters this minimum melt. Because of the slope on the water-saturated solidus, these melts are unable to rise in the crust without solidifying. Water-saturated granitic melts are therefore not likely to be volumetrically or thermally important.

If $P-T-t$ paths reach higher temperatures, less water-saturated melts can form. If the amount of water available in the rock at grain junctions is negligible, the next important melting curve traversed by $P-T-t$ paths is that of "dry" muscovite granite (Huang and Wyllie 1973). The only water present in this rock is bound up in muscovite (0.6 wt % of rock), which is released on melting. The solidus is consequently for a granitic liquid containing 0.6 wt % H_2O (review Sec. 11-5 and Fig. 11-7). Unlike the water-saturated solidus, this one slopes in the opposite direction, so that the fraction of liquid in partially melted rock steadily increases with decreasing pressure, at least to a high level in the crust (about 5 km).

The melting process buffers temperatures near the solidus. $P-T-t$ paths cannot go far beyond this until all rock that is capable of melting has done so. The attainment of temperatures near 800°C, but no higher, in many granulite facies rocks may be explained by this buffering effect. Magmas that rise from

this zone of melting adiabatically cool by about 0.4°C km^{-1}; in addition, further cooling accompanies the decompression melting due to the latent heat of fusion (about 300 kJ kg^{-1}). Consequently, magma remains close to the solidus during ascent. On reaching a depth between 10 and 5 km it becomes water-saturated and must crystallize, releasing the heat that was imparted to the magma by fusion during ascent. This results in an advective influx of heat into the upper crust, where it causes high-T, low-P metamorphism.

22-4 GENERATION AND ACCUMULATION OF MELTS

Because seismic evidence reveals that the Earth's mantle and crust are almost everywhere solid, the formation of bodies of magma requires a deviation from the norm. This may result from changes in pressure, advective heat transfer, fluid fluxing, or thermal insulation—their importance probably decreasing in that order. Given sufficient change in any of these factors, a rock crosses the solidus, and a series of steps ensue that may lead to the development of a body of magma. First, melt is formed where the phases involved in the solidus reaction come together. The enthalpy of fusion is sufficiently large that only small fractions of the rock melt—usually much less than 30%. For a partial melt to form a body of magma, it must first form a continuous phase throughout the rock so that the melt can flow. Finally, the melt must be able to separate from the source rock in a reasonable period of time, which requires that permeabilities be sufficiently high. Only then are bodies of magma formed that are large enough to have sufficient buoyancy to exceed the yield stress of the overlying lithosphere. In this section we examine these magma-forming processes.

The first melt in a rock must form where minerals involved in the solidus reaction come together. The reaction will involve eutectic, peritectic, or cotectic assemblages. In each case it involves the maximum number of phases in the rock, which typically come together at grain corners (normally four) or grain edges (three); grain faces juxtapose only two minerals. The first melt appears as small globules at grain corners or along grain edges. As the amount of melt increases, especially if it flows, the site of melting is less specific. At first, however, melt is restricted to multigrain corners.

Once melt is formed between the grains of appropriate composition, equilibrium requires that the globules of melt have shapes that minimize the interfacial free energies with the juxtaposed grains. This is achieved by the grains adjusting their area and angle of contact with the liquid by dissolving and reprecipitating from the melt. At equilibrium, a characteristic dihedral angle develops, which is determined simply by the ratio of the surface energies on each of the different interfaces (Eq. 12-32). If the dihedral angle is greater than 120°, the melt globules remain as globules at grain corners with no penetration along grain edges (Fig. 22-11). If the dihedral angle is less than 120° but greater than 60°, the melt partially extends along grain edges; separate globules connect only after they have grown large enough as a result of higher degrees of melting. If the dihedral angle is less than 60°, melt extends all the way along grain edges, thus connecting corners with other grain corners to form a continuous network. Which of these configurations exists when rocks partially fuse is of obvious importance in determining how much melting is necessary before melt can be extracted.

A number of experimental studies have been carried out to determine the dihedral angle between silicate melts and common minerals, and in almost all cases the angle is less than 60°. The fact that these experiments were successful is, in itself, important because it indicates that if equilibrium textures can be achieved in the laboratory in hundreds of hours, nature most certainly would have achieved them in a time scale of years. Melts as varied as olivine basalt, ultramafic liquid, and granite have been investigated at pressures up to 2 GPa [see review by Waff (1986); see also Jurewicz and Watson (1985) and Cooper and Kohlstedt (1984)]. Only orthopyroxene–orthopyroxene junctions with basaltic melt have a large dihedral angle (70°), but this decreases to 52° when the melt is water-saturated (Fujii and Osamura 1986). Quartz–quartz junctions with granitic melt (<0.2 wt % H_2O) were found to have the relatively large angle of 59° (Jurewicz and Watson 1985).

Despite the preponderance of experimental evidence that most dihedral angles are less than 60°, a number of natural rocks contain evidence to the contrary. Spinel lherzolite nodules in nephelinite from Dreiser Weiher, Germany, contain small isolated glassy or microcrystalline blebs that are almost spherical, suggesting that dihedral angles in this case must exceed 120° (Maaløe and Printzlau 1979). Nicolas (1986) describes similar-shaped aggregates of orthopyroxene, clinopyroxene, and spinel in many lherzolite nodules from basalt. There is no obvious explanation for this discrepancy, other than field observations are always open to various interpretations. We will continue our discussion of melt extraction assuming the experimental results to be valid. It is important to keep in mind, however, that the dihedral angles may be larger than 60°, and this would have a profound effect on the amount of fusion that is necessary before melt can be extracted from a rock.

Based on the experimental data that indicate dihedral angles to be less than 60°, we assume that even the smallest degree of fusion results in melt being distributed throughout the rock along a network of channels along most grain edges, except, perhaps around orthopyroxene grains. Melt, however, is not able to penetrate along grain faces, and thus grains remain welded together while melt is able to flow through the channels. Only after the fraction of melt reaches 0.2 to 0.4 (depending on dihedral angle) can melt penetrate along grain faces and disaggregate the rock. Note that if the dihedral angles are less than 60°, the melt forms a continuous phase throughout the rock, regardless of how small the fraction of melt present. If this melt experiences a pressure gradient, it can theoretically be extracted from the rock, but the permeability would be very low if the fraction of melt were small. Also, if dihedral angles are less than 60°, melt will be dispersed along all grain edges regardless of where the melt is formed (Watson 1982). In this respect the rock behaves in an analogous manner to a piece of blotting paper or sponge. This also means that a certain amount of melt is needed along grain edges to minimize surface free energies, and this melt cannot be extracted from the rock. This fraction is on the order of 3%.

A body of magma forms by the separation and collection of melt from a partially fused source rock. The rate of this accumulation depends on permeability, which in turn is strongly dependent on dihedral angles. If these angles are larger than 60°, the permeability will be zero up to a critical fraction of fusion where melt globules in neighboring grain corners first touch (Fig. 22-11). When the dihedral angle is less than 60°, however, the channels along grain edges always form a continuous network. As the dihedral angle decreases or the amount of melt increases, the cross-sectional area of the channels increases and so does the permeability (Fig. 22-11). Just as the flow of fluid through a cylindrical conduit increases with the square of its radius (Eq. 3-10), so does the velocity through the trigonal prisms between grains increase with its cross-sectional area. Small increases in degree of melting result in large increases in permeability. For example, if grain diameters are 1 mm and the dihedral angle is 50°, the bulk permeability when only 0.01%

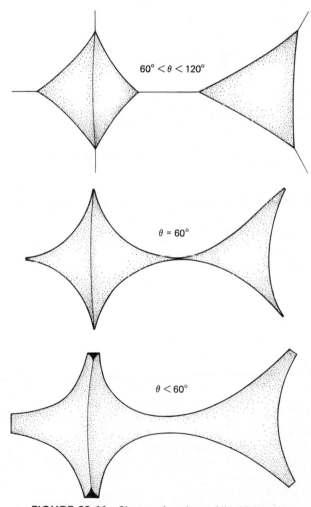

FIGURE 22-11 Shapes of pockets of liquid formed between grains during partial fusion as a function of dihedral angle, θ, between grains and liquid. Silicate melts form dihedral angles with most minerals of less than 60°. As a result, even small fractions of melt form an interconnected network of channels through the rock along grain edges.

melt is present is 10^{-16} m^2, whereas it is 5×10^{-15} m^2 when 0.05% melt is present, and 2.5×10^{-14} m^2 when 1% melt is present (von Bargen and Waff 1986).

Because of the small dihedral angles, the permeability of partially molten rock is moderately high, even at small degrees of melting. For example, partially molten rock is far more permeable than metamorphic rock is to fluids (about 10^{-18} m^2; review Sec. 18-3). Knowing the viscosity of the melt and the driving pressures (buoyancy or volume expansion on melting), flow velocities can be calculated using Darcy's law (Eq. 18-2). These velocities are greater than those at which plates move, and therefore significant bodies of magma can form in geologically reasonable times from small amounts of partial melt extracted from the source region (Prob. 22-11).

Although the grains in a partially molten rock remain in contact until the degree of melting exceeds 20 to 40%, they are not rigid particles but can deform both by creep and dissolution and redeposition through the melt phase (see pressure solution, Sec. 12-5). Thus because the average density of the solids is normally greater than that of the melt, the solids tend to compact and displace the melt upward. The rate at which the solid matrix deforms in large part determines the rate at which the melt is expelled. Quantitative treatment of this process (McKenzie 1984) indicates that in regions of upwelling mantle where there is more than a few percent of fusion, melt separates from matrix rapidly. Indeed, the process is so rapid that the percentage of melt in the source region at any given time is never likely to exceed more than a few percent, although up to 3% may have to remain in the source to satisfy surface energy requirements (the blotting paper effect).

The ease with which melts separate from a partially molten source plays an important role in determining the major compositional range of magmas and their trace element contents. Because the melt separates from the source after only a few percent melt has formed, equilibrium melting does not take place. Instead, the process approaches very closely that of extreme fractional melting (Sec. 10-4), and the trace element behavior obeys Rayleigh fractionation (Sec. 13-11 and Eq. 13-37).

We conclude that melts separate from their source rapidly after only small degrees of fractional melting to form bodies of magma that rise toward the surface of the Earth. Near the source, rocks are ductile, so magma probably moves as diapirs, whereas at higher levels where the lithosphere is brittle dikes will be more important, especially if the lithosphere is undergoing extension (see Chapter 3). It has been suggested that in the source region, magma may even move as waves of liquid passing through a porous medium. Analogous behavior can be created in fluidized sand at high pore–fluid flow rates. At some critical flow rate of water through sand, particles become fluidized and separate from one another (the quick condition). If the flow rate is increased still more, large bodies of relatively sand-free water rise through the sand. Particles fall rapidly through these regions, allowing waves to propagate upward without lifting the sand particles. This motion is different from that of a diapir because there is no net movement of sand particles during the passage of a wave. The supposed analogous waves of magma, which are named *magmons*, have wavelengths of kilometers and velocities of centimeters per year (Scott and Stevenson 1986). The episodic activity in many volcanic regions may be due to the rise of magmons through the mantle.

22-5 COMPOSITION OF THE SOURCE OF MAGMAS

The compositions of magmas and the temperatures and pressures at which they form is strongly dependent on the chemistry and mineralogy of the rocks in the source region. For granitic magmas generated in the continental crust these source rocks are well known and form the metamorphic core of any deeply dissected mountain range. Magmas of basaltic composition, however, have deeper sources that are not accessible for direct sampling. Our knowledge of these rocks is therefore deduced largely from petrological, geochemical, and geophysical evidence. This section examines the limits that can be placed on the nature of the source of basaltic magmas.

Estimates of the composition of source regions can be made in three ways. First, rare samples of deep-seated rocks are brought to the surface as xenoliths in certain volcanic rocks. These indicate the types of rock penetrated during ascent of the magma and may actually provide samples of the source region. Second, experimental studies on basaltic magmas at high pressures indicate the phases that must be stable in the source region. Third, model compositions can be constructed that are capable of generating basaltic magmas while satisfying the general physical and chemical properties that the mantle is thought to have based on geophysical evidence.

Certain volcanic rocks of alkaline affinity, in particular those emplaced rapidly by explosive activity, may contain xenoliths of rocks through which the magma passed en route to the surface. While some fragments are clearly of local crustal rocks, others are of rocks that are completely foreign to the upper crust and have much deeper sources. Pressure-sensitive mineral assemblages indicate that these have come from depths ranging from the lower crust to as much as 200 km in the upper mantle.

These xenoliths provide the most direct evidence of the rocks from which magmas may be derived, but care must be taken in interpreting this evidence. Xenoliths may give a biased sample of the mantle. For example, they do not occur in all types of basalt, nor are all xenoliths likely to be preserved to the same extent. Mantle-derived xenoliths are restricted to alkaline rocks, especially to the ultra-alkaline silica-poor varieties, such as kimberlite, alnoite, and melilite nephelinites; they never occur in tholeiitic rocks. The ultra-alkaline rocks also tend to be restricted to continental regions, where there is rifting. All rock types passed through by these magmas may not be sampled equally, because of differences in physical properties. Moreover, some rocks may survive the trip to the surface more easily than others. Finally, although the xenoliths indicate the rock types picked up by a magma during ascent, they may not include samples from the source region itself.

Mantle xenoliths fall into two general types, the extremely abundant peridotites (about 95%) and the less abundant eclogites (about 5%) (Carter 1970). The peridotites in kimberlites typically range from garnet lherzolite (ol + cpx + opx + gar) to harzburgite (ol + opx), whereas those in the alkali basalt–

nephelinite series range from spinel lherzolite to harzburgite. The garnet and spinel lherzolites overlap in composition and can be shown to be equivalent rocks equilibrated at different pressures—the garnet lherzolite being the higher pressure rock (see discussion below and Fig. 22-12). The eclogite xenoliths occur mainly in kimberlites and consist essentially of two minerals, pyrope-rich garnet and omphacite (diopside–jadeite).

In some areas sufficiently large numbers of xenoliths have been derived from a range of depths that it is possible, through the use of geothermometers and geobarometers, to construct geothermal gradients. These allow us to examine the stratigraphy of the upper mantle. Because the xenoliths are brought to the surface so rapidly, their minerals are literally quenched, leaving no time for reequilibration.

Figure 22-12 shows geotherms beneath southeastern Australia, Saudi Arabia, and southern Africa. The lherzolite xenoliths from southeastern Australia indicate a steep geothermal gradient, especially at shallow depths where the indicated temperatures would correspond to a surface heat flux of 90 mW m^{-2} (Griffin et al. 1984). Temperatures indicated by the deeper garnet lherzolite xenoliths correspond to a more normal surface heat flux of 60 mW m^{-2}. The high temperatures of the shallow xenoliths are interpreted to result from the rise of diapirs from mantle plumes. The xenoliths from Saudi Arabia also indicate higher than normal temperatures. The mantle here has risen passively from depth in response to the opening of the Red Sea rift (McGuire and Bohannon 1989). By contrast the xenoliths in the kimberlites from southern Africa (Boyd and Gurney 1986) indicate a much shallower geotherm, but one that is consistent with ancient shield areas (Pollack and Chapman 1977).

Mantle xenoliths vary considerably in composition, and most are not suitable source rocks for basaltic magmas. Indeed, many appear to be refractory residues formed by the removal of basaltic fractions; this is particularly true of those of harzburgitic composition. It will be recalled from Sec. 14-2 that harzburgites that underlie ophiolite suites are also interpreted to be the residual mantle from which MORB have been extracted. As the contents of clinopyroxene and garnet or spinel increase, the more likely is the rock to be able to generate a basaltic fraction. Eclogite xenoliths would therefore appear to be ideal candidates for basaltic source rocks. Experiments, however, indicate that eclogites have near-eutectic compositions and are simply formed from basaltic magma that has crystallized at high pressure. They can also be formed from basaltic rock that has been transported to depth, where it has recrystallized into an eclogite assemblage. Eclogites are consequently not the source of basaltic magmas (Yoder and Tilley 1962). The spinel and garnet lherzolites are far more likely candidates. An analysis of a particularly *fertile* spinel lherzolite from Kilborne Hole, New Mexico, is given in Table 22-1 (Takahashi 1986).

If a basaltic magma is derived from its source without undergoing any compositional change (primary magma), and the solidus phases can be experimentally determined as a function of pressure, the minerals in the source region can be determined if the depth of origin can be independently estimated. Note that this does not mean that we would know the bulk composition of the rock, only the minerals that must be present in the source. Although the technique is simple in principle, most basalts, even those with high magnesium numbers, are found to have undergone some fractionation en route to the surface, and thus their high-pressure solidus phases may not be

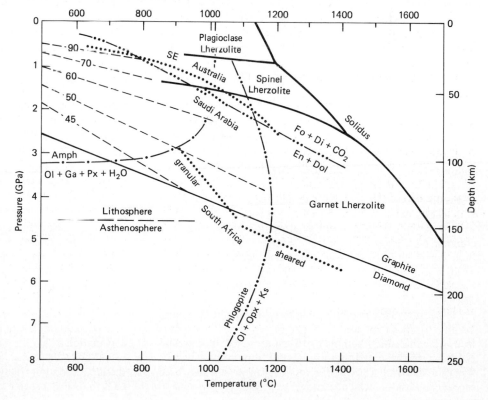

FIGURE 22-12 Pressure–temperature conditions in the upper mantle based on xenoliths brought up in alkaline magmas (dotted lines). Data are from southeastern Australia (Griffin et al. 1984), Saudi Arabia (McGuire and Bohannon 1989), and southern Africa (Boyd and Gurney 1986). Conductive geotherms in the lithosphere are based on surface heat fluxes ranging from 90 to 45 mW m^{-2}. The phase boundaries are from the following sources: solidus (Takahashi 1986); plagioclase to spinel lherzolite (Kushiro and Yoder 1966); spinel to garnet lherzolite (Danckwerth and Newton 1978; O'Neill 1981); forsterite (Fo) + diopside (Di) + CO$_2$ to enstatite (En) + dolomite (Dol) (Eggler 1976); graphite to diamond (Kennedy and Kennedy 1976); amphibole and phlogopite (Kushiro 1970).

those found in the source. For example, consider an olivine tholeiite (no normative quartz) that is fractionated into the quartz-tholeiite field by early separation of olivine. High-pressure experiments on the quartz tholeiitic would indicate erroneously that the source contains coesite instead of olivine.

Experiments reveal that the typical mineral assemblages of basaltic rocks change to different assemblages with only moderate increases in pressure. The most common constituent of crustal rocks—plagioclase feldspar—disappears by reacting with olivine to produce pyroxene and spinel (Kushiro and Yoder 1966). The precise reaction depends on the proportions and compositions of the minerals (Prob. 22-13). For example,

(plagioclase)   (olivine)   (clinopyroxene)   (enstatite)   (spinel)

$$CaAl_2Si_2O_8 + 2Mg_2SiO_4 = CaMgSi_2O_6 + 2MgSiO_3 + MgAl_2O_4 \tag{22-4}$$

$$CaAl_2Si_2O_8 + Mg_2SiO_4 = CaAl_2SiO_6 + 2MgSiO_3 \tag{22-5}$$

$$NaAlSi_3O_8 + Mg_2SiO_4 = NaAlSi_2O_6 + 2MgSiO_3 \tag{22-6}$$

The clinopyroxene in Eq. 22-4 is diopside, whereas that in Eq. 22-5 is the Ca-Tschermak's molecule, and that in Eq. 22-6 is jadeite. These form a continuous solid solution known as *omphacite*. Because all of the minerals involved in these reactions belong to solid solution series the conversion of a plagioclase-bearing rock to a pyroxene–spinel-bearing one takes place over a range of pressures and temperatures. However, the range is small and is not particularly sensitive to temperature. For natural rock compositions, the transformation takes place at approximately 1 GPa at 1200°C (Fig. 22-12).

At still higher pressures, the assemblage spinel + pyroxene becomes unstable and is replaced by the assemblage garnet + olivine (Danckwerth and Newton 1978; O'Neill 1981). The reaction can be written in terms of the magnesian end members as

(spinel)   (enstatite)   (pyrope)   (forsterite)

$$MgAl_2O_4 + 4MgSiO_3 = Mg_3Al_2Si_3O_{12} + Mg_2SiO_4 \tag{22-7}$$

The reaction in a natural spinel lherzolite (Takahashi 1986) is found to take place at 2.5 GPa at 1400°C (Fig. 22-12).

What can we conclude about the mineralogy of the source region of basalts from these experiments? At low pressures, olivine is normally the primary mineral on the liquidus of basalts, and above 2.5 GPa it should again be present along with garnet. Between 1 and 2.5 GPa, however, olivine could be absent, depending on the bulk composition of the rock. The common occurrence of xenoliths of garnet lherzolite in kimberlites and of its lower-pressure equivalent spinel lherzolite in alkali basalts, consequently, makes this type of peridotite, the most likely parental material from which basaltic magmas are derived.

Because many mantle xenoliths, especially those of harzburgitic composition, were thought to be refractory residues from which basaltic fractions had been extracted, model compositions were invented that are capable of producing basaltic liquids and leaving a peridotitic residue. The most widely used model composition, which is known as *pyrolite* (=pyroxene +

olivine ± pyrope garnet), was introduced by Ringwood (1975). It consists of three parts alpine peridotite (harzburgite—79% ol, 20% opx, 1% sp—see Sec. 14-2) and one part Hawaiian tholeiite (see analysis 1 in Table 22-1).

Although fertile mantle xenoliths of pyrolite composition are rare, some do occur. Analysis 2 in Table 22-1 is of a spinel lherzolite xenolith from Kilborne Hole, New Mexico, which almost exactly matches pyrolite. This xenolith has been the subject of careful melting experiments up to pressures of 14 GPa (Takahashi 1986). The results give a good idea of the composition of melts that can be expected from a pyrolite-like mantle. They are discussed in the following section.

Finally, we must consider the composition of possible fluid phases in the source regions of basaltic magmas. In the upper mantle, porosities and permeabilities are not likely to be high, and thus only extremely small quantities of a free fluid phase could exist. Larger quantities of fluid can be bound in hydrous and carbonate minerals, and should these be involved in devolatilization reactions, substantial volumes of fluid could be evolved. As was seen in Chapter 11, even small quantites of fluid can dramatically affect the solidus temperature of rocks and the composition of melts.

Surprisingly, the olivine crystals in many mantle xenoliths contain fluid inclusions which are composed largely of CO_2. Care must be taken in assigning a mantle origin to these inclusions. Fractures in xenoliths formed during the decompression accompanying ascent from the mantle can be filled with volatiles derived from the surrounding magma at shallow depths. The fractures may then heal, leaving isolated fluid inclusions. These, however, form planar arrays that can be distinguish from inclusions trapped on the surface of crystals as they grow. There is no doubt that some of the CO_2 in xenoliths originated as a separate fluid phase in those parts of the mantle sampled by alkali basalts. However, the presence of CO_2 in the upper mantle is limited by its ability to react and form carbonates, and in the deeper parts of the upper mantle where conditions are more reducing CO_2 may combine with H_2O to form CH_4 and C, the carbon existing as diamond.

Both amphibole and phlogopite occur in some mantle xenoliths, and carbonatites are certainly derived from the mantle. Both H_2O and CO_2 can therefore be derived from mantle minerals by appropriate reactions (review Sec. 11-7). Amphibole is stable to depths of about 100 km at high activities of water and low geothermal gradients, whereas phlogopite is stable to depths of about 180 km under similar conditions (Fig. 22-12). Phlogopite is of interest not only because it provides a source of water but also of potassium, which is otherwise not present in any significant amount in other minerals of lherzolite.

In the presence of a CO_2-rich fluid the common mineral assemblage of basaltic rocks—olivine + augite—reacts to form enstatite + dolomite (Eq. 11-33) at pressures just inside the garnet lherzolite field (Fig. 22-12). Because this reaction lies at pressures below most geothermal gradients, CO_2 should react in mantle rocks to form carbonate. Why then should mantle-derived xenoliths contain CO_2 fluid inclusions? One possible explanation is that the magmatic event responsible for bringing xenoliths to the surface may also elevate the geotherm above the decarbonation reaction. This would not only produce free CO_2 but also provide a flux of gas capable of driving the explosive volcanism with which mantle xenoliths are so commonly associated. As will be seen in the following section, volatile-bearing magmas can form from mantle containing hydrous or carbonate minerals at temperatures well below those of the anhydrous lherzolite solidus (see also Sec. 11-7).

22-6 PARTIAL MELTING IN THE SOURCE REGION

Now that we have some idea of the materials present in the source regions of magmas, we are in a position to consider the actual melting process and the compositions of magmas formed. To do this we must know the phase relations between the minerals at the pressures of interest and the effects a fluid phase has on the composition and temperature of the magma. We discuss basaltic magmas first because they are the most abundant and because they may bear a parental relation to other magmas. This is followed by a treatment of granitic magmas, and finally of andesitic ones.

Melting in the mantle probably results from decompression rather than from heating. Decompression can result from the passive rise of the mantle into zones of lithospheric extension or from the rise of diapirs from deeper in the mantle, possibly emanating from mantle hot spots or plumes. In either case the temperature of the source region of basaltic magmas must be nearly that of the convective geotherm in the upper mantle, which on rising adiabatically to the base of the lithosphere has a temperature of 1400 to 1500°C. If the temperature in the upper mantle cannot vary greatly, the actual depths at which basaltic magmas are generated depends largely on the

position of the solidus, which, in turn, is strongly dependent on the presence and composition of fluids.

Figure 22-13 shows schematically the solidus for a lherzolitic mantle under fluid-absent, H_2O-saturated, and CO_2-saturated conditions. The highest solidus temperatures are obtained when no fluid is present (Takahashi, 1986). With addition of water, the solidus decreases dramatically and intersects the stability fields of amphibole and phlogopite (Kushiro et al. 1968). The solubility of CO_2 in basaltic melts is not great at low pressures, and consequently, the solidus of lherzolite is not lowered much by the presence of CO_2 up to pressures of 2 GPa. However, the solubility increases as the pressure approaches 2.9 GPa (90 km) and the solidus drops rapidly, reaching a low temperature at the reaction of forsterite and diopside with CO_2 to produce enstatite and dolomite (Eggler 1976; Wyllie and Huang 1976). At depths below this reaction the solidus temperature again increases. For fluid compositions containing both H_2O and CO_2, the solidus is at intermediate temperatures between the curves for the H_2O- and CO_2-saturated solidus.

The solidus in a $P-T$ diagram can never be a straight line if the assemblage of solid phases changes with pressure. This is evident from a consideration of Schreinemaker's rules (Sec. 8-4). The solidus in a $P-T$ diagram is a univariant line marking the first appearance of liquid. If it is intersected by a univariant reaction between the solid phases, the field occupied by the liquid must subtend an angle of less than 180° at the invariant point. Thus we expect to see cusps on the solidus where plagioclase lherzolite changes to spinel lherzolite, and spinel lherzolite changes to garnet lherzolite, and forsterite, diopside, and CO_2 react to form enstatite and dolomite (Fig. 22-5). These cusps are of importance because it is here that a perturbed

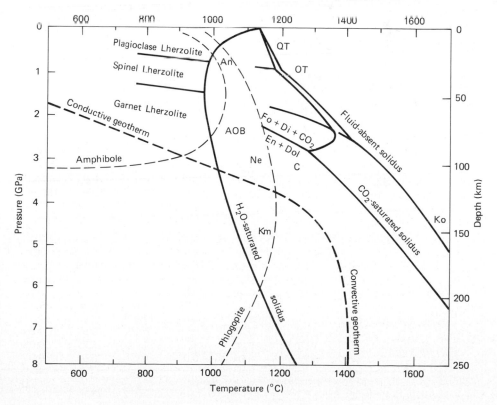

FIGURE 22-13 Approximate phase relations for a lherzolite mantle under fluid-absent, H_2O-saturated, and CO_2-saturated conditions. Sources of information same as in Figure 22-13, with H_2O-saturated solidus from Kushiro et al. (1968), and CO_2-saturated solidus from Eggler (1976) and Wyllie and Huang (1976). Probable source regions are indicated for the following rocks: quartz tholeiite (QT), olivine tholeiite (OT), komatiite (Ko), andesite (An), alkali olivine basalt (AOB), nephelinite (Ne), carbonatite (C), and kimberlite (Km).

geotherm which is raised toward the surface is most likely to intersect the solidus. The decreased variance at the cusps also restricts the compositional range that melts can have.

We have seen that the mantle beneath spreading oceanic ridges probably rises passively in response to lithospheric extension. If the geotherm in Figure 22-13 is raised toward the surface, and the mantle lacks significant quantities of fluid, the perturbed geotherm is likely to intersect the fluid-absent solidus at the cusp formed by the reaction between plagioclase and spinel lherzolite. The restricted range of composition of MORB may be due to the decreased variance at this cusp (Presnall et al. 1979).

As discussed in Sec. 11-7, the marked cusp in the solidus of the CO_2-saturated solidus may play a role in the development of the ultra-alkaline, strongly silica-undersaturated igneous rocks and their explosive emplacement. If a perturbed geotherm rises in a mantle that contains a CO_2 or CO_2–H_2O fluid phase, melt first forms in the vicinity of the prominent cusp at the carbonation reaction. Experiments reveal that these melts are indeed strongly silica undersaturated. As the magma begins to rise buoyantly, the solubility of CO_2 decreases rapidly and the ensuing exsolution of gas could bring about explosive activity and the development of diatremes. If sufficient H_2O is present in the fluid, the cusp intersects the stability fields of phlogopite and amphibole. These hydrous minerals would then appear as magmatic phases, as for example in kimberlites and alnoites.

The composition of melts formed at, or near, the solidus varies with pressure. This is best illustrated with the use of phase diagrams, but the effect is readily apparent from electron microprobe analyses of first-formed liquids. Table 22-1 includes analyses and CIPW norms of glasses formed in melting experiments at different pressures on the fertile spinel lherzolite xenolith from Kilborne Hole (Takahashi 1986). The first-formed melt at atmospheric pressure has the composition of a quartz tholeiite, but above 1 GPa it is an olivine tholeiite. As pressure increases so does the normative olivine content, and between 5 and 8 GPa near-solidus melts containing from 30 to 35% MgO are identical to komatiitic magmas. Such high MgO contents can be formed at lower pressures only by large degrees of partial melting ($> 50\%$), and as we have seen in Sec. 22-4, melts separate and segregate long before they attain such high fractions. We can conclude, therefore, that komatiites are most likely formed by small degrees of partial melting at depths of 150 to 250 km (Ko in Fig. 22-13). The convective geotherm would have to be at a much higher temperature than it is at present in order to intersect the lherzolite solidus in this depth interval. This probably explains the restriction of high-magnesian komatiites to the Archean.

TABLE 22-1
Composition of mantle materials[a]

| | Pyrolite | Kilborne Hole Lherzolite | Composition of First-formed Melts in Kilborne Hole Lherzolite at: | | | | | Peridotitic Komatiite |
			1 atm	1 GPa	3 GPa	8 GPa	14 GPa	
SiO_2	45.20	44.48	54.2	49.2	46.9	46.6	45.2	44.9
TiO_2	0.71	0.16	0.7	0.6	0.9	0.2	0.2	0.2
Al_2O_3	3.54	3.59	15.1	17.7	11.0	4.6	4.2	5.3
FeO	8.47	8.10	5.6	6.7	7.8	8.8	7.9	10.4
MnO	0.14	0.12	0.1	0.1	0.2	0.2	0.2	0.2
MgO	37.48	39.22	8.4	9.5	19.2	34.9	37.8	33.6
CaO	3.08	3.44	12.3	11.4	12.2	3.9	3.7	5.0
Na_2O	0.57	0.30	2.1	2.9	1.2	0.3	0.3	0.4
K_2O	0.13	0.02	0.0	0.0	0.0	0.0	0.0	0.0
P_2O_5	0.06	0.03	0.0	0.0	0.0	0.0	0.0	0.0
Cr_2O_3	0.43	0.31	0.1	0.1	0.4	0.4	0.5	0.3
NiO	0.20	0.25	0.0	0.0	0.0	0.0	0.0	0.0
Total	100.01	100.02	98.6	98.2	99.8	99.9	100.0	100.3

CIPW norms

Q	0.0	0.0	5.22	0.0	0.0	0.0	0.0	0.0
Or	0.77	0.12	0.0	0.0	0.0	0.0	0.0	0.0
Ab	4.82	2.54	17.77	24.54	10.15	2.54	2.54	3.38
An	6.72	8.39	31.78	35.28	24.63	11.20	10.11	12.67
Cpx	6.46	6.70	23.60	17.21	28.63	6.45	6.51	9.65
Opx	15.81	12.57	18.76	1.89	1.46	26.92	17.36	13.39
Ol	63.31	68.86	0.0	17.99	32.63	51.82	62.37	60.39
Chr	0.63	0.46	0.15	0.15	0.59	0.59	0.74	0.44
Il	1.35	0.30	1.33	1.14	1.71	0.38	0.38	0.38
Ap	0.14	0.07	0.0	0.0	0.0	0.0	0.0	0.0

[a] Model mantle composition: pyrolite after Ringwood (1975); spinel lherzolite, Kilborne Hole, New Mexico, and composition of first-formed melts, after Takahashi (1986); peridotitic komatiite after Arndt et al. (1977).

Another consequence of the increasing MgO content of the first-formed melts with increasing pressure is that by 14 GPa the melts have compositions that are nearly the same as that of the lherzolite as a whole. This indicates that the composition of the lherzolite must be near a eutectic or minimum at this pressure, which agrees with the observation made in Sec. 22-2 that the liquidus and solidus of lherzolite are separated by less than 100°C toward the base of the upper mantle.

The composition of first-formed melts in fluid-saturated lherzolite also changes with pressure. The addition of water to lherzolite not only lowers the solidus dramatically, but it increases the silica content of the melt into the andesitic range (An in Fig. 22-13; see also Fig. 11-20). As pressure increases above 2 GPa, especially when the fluid contains CO_2 as well as H_2O, melts become nepheline normative and resemble alkali olivine basalts (AOB in Fig. 22-13). At still higher pressures and mole fraction of CO_2, the first-formed melts become increasingly undersaturated in silica, passing through nephelinitic to carbonatitic compositions (Ne and C in Fig. 22-13), as was discussed in Sec. 11-7. Finally, at very high pressures and with mixed CO_2–H_2O fluids, kimberlitic magmas form in equilibrium with phlogopite (Km in Fig. 22-13).

The compositional variation of melts with pressure and fluid composition can only be understood clearly through a study of multicomponent phase diagrams [see, for example, Wyllie (1988)]. These tend to be complex and are beyond the scope of this book. We will, however, consider a relatively simple system, that of silica–forsterite–nepheline, which contains a number of important minerals and invariant points that play roles in the generation of major rock types.

At low pressure in the system silica–forsterite–nepheline (Fig. 22-14), the join forsterite–albite is a thermal divide with tholeiitic rocks plotting on the quartz side of the join and alkaline ones on the nepheline side. For clarity, the only low-pressure phase relations shown are those of the boundary curve between forsterite and enstatite and the ternary peritectic involving forsterite, enstatite, and albite (Schairer and Yoder 1961). A plagioclase lherzolite that plots at the point marked X would produce its first melt at the ternary peritectic, which because it lies in the triangle enstatite–albite–silica would give rise to a quartz tholeiite magma.

With increasing pressure, the peritectic involving forsterite, enstatite, and albite shifts toward nepheline, and by 1.0 GPa it lies to the nepheline side of the forsterite–albite join. At these higher pressures, the peritectic reaction involves enstatite plus liquid going to forsterite plus albite. Above 2.5 GPa the mineral jadeite is stable, and at 3 GPa, the peritectic, which now involves reaction of enstatite with liquid to form forsterite and jadeite, plots in the triangle forsterite–nepheline–jadeite, and the join enstatite–jadeite is now the thermal divide in the system. The lherzolite of composition X consists of olivine, enstatite, and jadeite at 3 GPa and on first melting would produce a liquid at the ternary peritectic involving these three minerals. Intrusion of this liquid to a region of low pressure would result in the formation of a strongly undersaturated nepheline normative magma.

Clearly, melting of a simple lherzolitic composition in this ternary system produces melts that range from quartz tholeiite to strongly undersaturated liquids depending on the pressure.

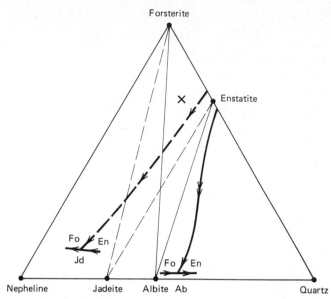

FIGURE 22-14 System forsterite–nepheline–quartz showing position of boundary between liquidus fields of forsterite and enstatite at atmospheric pressure (Schairer and Yoder 1961) and at 3 GPa (Kushiro 1968). Solid tie lines between phases refer to low pressure and dashed ones to high pressure. At low pressure, forsterite reacts with peritectic liquid to form enstatite and albite; at high pressures this reaction is reversed, with enstatite reacting with peritectic liquid to form forsterite and jadeite.

The higher the pressure, the more undersaturated the melt is likely to be. Also, at constant pressure, introduction of a fluid phase rich in CO_2 has the same effect as increasing the pressure; that is, melts become more undersaturated in silica. Introduction of a hydrous fluid, on the other hand, has the same effect as decreasing the pressure; that is, the melts become richer in silica.

These changes in composition of the first liquid formed in this simple system from a lherzolitic starting material appear to explain the general compositions of a number of major rock types. MORB, for example, are formed at relatively shallow depths beneath spreading ocean ridges and have olivine tholeiite compositions. Basaltic magmas that come from greater depth, as evidence by mantle xenoliths with high-pressure mineral assemblages, are alkaline, and those that are strongly undersaturated in silica contain high carbonate contents and presumably are formed in equilibrium with a CO_2-rich fluid. Magmas developed above Benioff zones are probably formed at depths near 100 km (about 3 GPa), and might therefore be expected to be undersaturated in silica if formed from a lherzolitic mantle. However, water released from the subducted slab would shift the first-formed melts into the quartz-saturated field and produce andesitic magmas. The water would also promote crystallization of amphibole in these magmas.

We turn now to the question of how the magma evolves as melting takes place. Does melting proceed in an equilibrium manner, or does it take place fractionally, or perhaps through batch melting? The composition of the segregated magma that eventually rises into the lithosphere depends critically upon this stage. We will illustrate this process by considering melting in the simple pseudoternary system forsterite–diopside–pyrope at 4 GPa (Davis and Schairer 1965).

Let us consider first the equilibrium melting of a garnet lherzolite of composition X (Fig. 22-15). On reaching a temperature of 1670°C the first melt forms with the eutectic composition, E. With continued heating, the temperature remains constant while the fraction of eutectic liquid increases. Simultaneously, the composition of the remaining solid changes from X toward s_1. When the solids reach s_1, the rock is 33% melted (s_1X/s_1E) and no garnet remains. Consequently, the eutectic liquid begins to change its composition along the cotectic toward F, while the solids change their composition from s_1 toward olivine. For example, when the liquid has composition l_2, the solid have composition s_2. When the liquid reaches l_3, the solid, s_3, reaches olivine and thus no diopside remains. Consequently, the melt leaves the cotectic and moves toward olivine. As it does so, the temperature rises rapidly, and melts are unlikely to progress far in this direction because of the large amounts of heat required. For example, when the melt reaches l_4, the original rock is 66% melted (s_3X/s_3l_4) and the temperature is already 1800°C. The results of equilibrium melting are a range of possible melt compositions along the line El_3X, with the actual composition depending on the temperature attained.

Consider next the fractional melting of the same garnet lherzolite. Melting again begins at 1670°C with the formation of a eutectic liquid at E, but this time the melt is removed as soon as it is formed. This results in the bulk composition of the rock changing from X toward s_1. One-third of the original rock can be extracted as eutectic liquid before the bulk composition of the rock reaches s_1. At this stage the remaining rock consists only of forsterite and diopside; no garnet remains. But olivine and diopside can melt only at the binary eutectic, F, which has a temperature of 1745°C. Melting therefore ceases after the removal of the last drop of ternary eutecic liquid and does not begin again until the temperature reaches 1745°C. The liquid that forms at this temperature is very different from the first one and has the composition of the binary eutectic, F. Melting continues at this eutectic until all of the diopside has been consumed and then it ceases. Melting would not occur again until the temperature reached the melting point of pure forsterite, which is most unlikely. The results of fractional melting are a series of magmas with distinct compositions corresponding to invariant points. No magmas of intermediate composition are formed.

Melting in the mantle is likely to proceed under conditions that are intermediate between these two extremes. However, it will be recalled from the discussion of compaction (Sec. 22-4) that fractions of melt as small as a few percent are likely to separate from the source rock. We can conclude therefore that fractional melting provides a better model of natural melting than does equilibrium melting. Primary magmas, that is, ones derived directly from the source with no change in composition, are expected to have relatively fixed major element compositions, although several different compositions may be present due to melting at successive invariant points. Magmas that show a gradational range of compositions are likely to be the product of differentiation of the magmas after they leave the source region.

Although the major element composition of successive melts extracted from a source rock undergoing fractional melting may remain relatively constant, the same is not true of trace elements with small distribution coefficients (Sec. 13-11). Incompatible elements, such as P, K, and Zr, enter the first melt and are removed from the system; that is, they undergo Rayleigh fractionation (Sec. 13-11). The abundance of such elements could therefore be used to monitor the extraction of magma from a source region whereas the major elements could not (Prob. 22-15).

Fractional melting of the mantle provides a means of producing magmas that are rich in volatiles. Although a separate fluid phase may exist in the solid mantle, it can constitute only an extremely small fraction of the rock. Equilibrium melting could therefore not produce a significant amount of fluid-saturated magma. On the other hand, if segregation of liquid takes place after only a few percent of melting, the fluid-saturated melt, which is bound to form first because of its lower melting point, will be able to collect and form bodies of magma.

Granitic melts can form directly from the partial fusion of a variety of continental crustal rocks, including pelites, arkosic sandstones, and granites (i.e., any rock containing quartz + plagioclase + K-feldspar or muscovite). This partial melting process (*anatexis*) is relatively well understood, in part because of extensive experimental studies on relevant synthetic systems (Sec. 10-19), but also because of field evidence. Most high-grade metamorphic terranes contain rocks known as *migmatites*, which are intimate mixtures of igneous granite and refractory metamorphic rocks (Fig. 22-16). Most migmatites are thought to form by partial fusion, but some may form through injection of granite into metamorphic rocks. Migmatites provide a

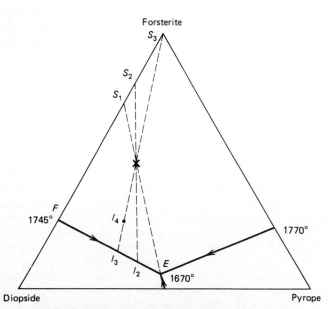

FIGURE 22-15 Pseudoternary liquidus diagram for the system forsterite–diopside–pyrope at 4 GPa. The ternary eutectic, *E*, is actually a piercing point. See text for discussion of equilibrium and fractional melting of composition *X*. [after Davis and Schairer (1965).]

FIGURE 22-16 Migmatite showing dark refractory amphibolite that was partially disrupted by the flow of the surrounding granite. Contrast this deformation with the brittle behavior of the rocks at a later time when a diabase dike was intruded (background). Coast of Labrador, near Hopedale.

the metamorphic foliation. Migration of the melt is therefore channelized rather than dispersed. The distribution of the metamorphic rocks within the migmatite normally provides clear evidence that migmatites flow as coherent masses, with the solid fraction being stretched and pulled apart during flow. This movement may, in fact, be important to the segregation process, with liquid migrating into fractures or low-pressure regions where competent layers are necked down by stretching.

During progressive metamorphism, migmatites first appear toward the top of the amphibolite facies where temperatures exceed 650°C (Fig. 15-6). Granite must contain about 4 wt % H_2O to melt at these temperatures (Whitney, 1988). Intergranular pore spaces can contribute no more than 0.3 wt % H_2O; the fluid must therefore be derived largely from other sources. Dehydration reactions provide the largest amount. Water is released locally by progressive metamorphic reactions, such as those accompanying the transition from amphibolite to granulite facies. The abundance of migmatites at this boundary is probably related to the availability of this water. In Benioff zones, water is released into the overlying mantle wedge by dehydration reactions taking place in the subducted plate. Because the fluid does not wet mineral grains (Sec. 18-3), it is probably forced upward in channels by pressure generated by the dehydration reactions. Also, water can be released during crystallization of hydrous basaltic and andesitic magmas that rise from the Benioff zone into the lower crust.

Because the amount of intergranular water in metamorphic rocks is very small, no significant amount of water-saturated melt can form from it. Instead, melting is delayed until dehydration reactions take place or water infiltrates the rock from an external source. When melting accompanies a dehydration reaction, the melt is undersaturated in water. The degree of undersaturation depends on the amount of water released by the reaction.

Three reactions are capable of releasing large quantities of water at temperatures where melting of a granitic fraction can occur in the crust (Whitney 1988). The lowest temperature of these involves the reaction of muscovite and quartz to form sillimanite and K-feldspar (Eq. 15-1). In Figure 15-4a this reaction is seen to intersect the water-saturated beginning of melting curve of granite at about 0.3 GPa and about 680°C. Above this pressure the reaction proceeds with the formation of an undersaturated melt ("Dry" Musc. Granite Solidus in Fig. 20-7). If the rock also contains albitic feldspar, the melting and breakdown of muscovite will occur at a slightly lower temperature according to the reaction

muscovite + quartz + Na-feldspar

$$= sillimanite + K\text{-feldspar} + liquid \quad (22\text{-}8)$$

This melting curve, which is shown in Figure 22-17, gives rise to peraluminous granites with K/Na ratios greater than 1 at temperatures below 750°C.

The next reaction is similar but involves the breakdown of biotite according to the reaction

biotite + quartz + Na-feldspar

$$= K\text{-feldspar} + pyroxene + liquid \quad (22\text{-}9)$$

glimpse of the partial fusion process frozen in progress, and though the melts are viscous and siliceous, basaltic melts may segregate from mantle peridotites in a similar manner.

Migmatites are striking rocks with light-colored granite veins anastomosing through dark-colored refractory metamorphic rocks. The contrast in color and composition indicates that partial fusion results in almost total segregation of the melt from the refractory residue. Granitic material is not dispersed along grain boundaries throughout the rock. Instead, it is segregated into veins and sheets, most of which parallel

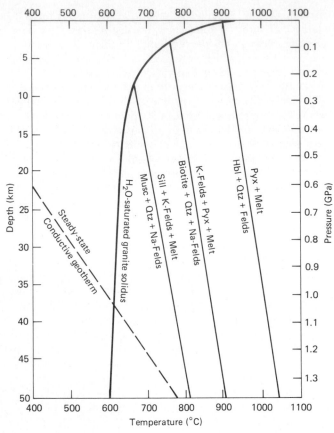

FIGURE 22-17 Beginning of melting curves for granitic rocks. The lowest temperature melting is for water-saturated conditions. Little melting is likely to occur at this temperature because of the low porosity of the rocks. Instead, melting is more likely to occur under water-undersaturated conditions at temperatures marking the breakdown of muscovite, biotite, or hornblende. [After Whitney (1988).]

followed by metamorphic rocks, temperatures are not likely to go much above the temperature of the solidus associated with the breakdown of muscovite unless there is an additional heat source (Fig. 20-7). Most melting in the crust must therefore be associated with the advection of basaltic magmas from the mantle.

Ascending basaltic magma on reaching the base of the continental crust is likely to spread laterally because of the lower density of the overlying rocks. The temperature of the crust at this depth is about 500°C (Fig. 22-8). As the basaltic magma cools and crystallizes, the temperature of the overlying rocks rises and partial melting of a granitic fraction occurs (Huppert and Sparks 1988). The temperature and amount of this melt depend on how much water is released by the crystallizing basaltic magma and how much is generated in the rocks from dehydration reactions. Formation of a cap of granitic magma over a basaltic magma prevents any subsequent intrusions of basalt penetrating to higher levels in the crust. Later intrusions of basalt pond at the base of the granite, possibly forming pillowlike structures (Fig. 13-29). If the basaltic magma mixes with the granitic one (Sec. 13-10), intermediate magmas of granodioritic to tonalitic composition may form (intermediate magmas may also form directly by partial fusion of crustal rocks). The low-density granitic cap to these bodies explains why in regions of granitic magmatism, basaltic dikes are either emplaced before the granitic magmas are formed, or they intrude only in peripheral regions, presumably beyond the extent of the granitic cap. Once the cap becomes large enough, bodies of siliceous magma rise into the crust advectively transferring heat to higher levels and thus steepening the regional geothermal gradient.

Finally, we come to the question of the origin of calc-alkaline andesites. No rock type has a more predictable mode of occurrence—it forms volcanoes near convergent plate boundaries where the depth to the Benioff zone exceeds 100 km, provided that the subduction angle is greater than 25°. Despite this certainty, there is no agreement on the source of andesites. Are they formed in the subducting slab, in the overlying mantle wedge, or even in the crust? Are they formed by partial melting of subducted MORB, ocean floor sediments, lherzolitic mantle above the Benioff zone, or do they require assimilation of crustal material? Geochemical evidence relating to these questions, which is given in Sec. 21-4, suggests, but does not prove, that andesites are derived from a mantle that is less depleted than the type from which MORB are derived, and that a certain fraction of oceanic water must be involved. This evidence favors the mantle wedge as the source of this magma. We will now consider whether this interpretation is supported by any other lines of evidence.

As discussed in Sec. 22-3, convergent plate boundaries must, in general, be the coolest parts of the Earth, because of the subduction of cool lithospheric plates. To develop magmas in the overriding plate, the subducting one must induce convection in the overlying mantle wedge. The calculated results of such convection are given in Figure 22-9 (Toksöz and Hsui 1978), but the region in which magmas may form is given in more detail in Figure 22-18. First, note that temperatures in the subducted ocean floor rocks beneath the island arc are less than 750°C and are consequently at least 200°C below the water-saturated solidus at this depth. We can conclude, therefore, that

Because this reaction does not involve muscovite, the melts are less aluminous and typically have K/Na ratios of 1 or less. The melts also range from granitic to granodioritic in composition and have temperatures between 750 and 850°C.

The final reaction involves the breakdown of hornblende to pyroxene and a meta-aluminous to peralkaline melt with a low K/Na ratio at the amphibolite-granulite facies boundary. These melts have temperatures of 900 to 1000°C, which probably require an influx of heat from a basaltic magma. With increasing pressure, the composition of the granite minimum (or eutectic) shifts to less quartz-rich compositions (Fig. 10-27). As a result, partial melting of crustal rocks at depths of about 50 km gives rise to syenitic magmas (Wyllie 1977). Fractional melting of an amphibole-bearing source can produce both quartz and nepheline normative melts (Presnall and Bateman 1973). Extraction of the first-formed melt, which is silica oversaturated, leaves a critically undersaturated residue, which in the next stage of melting produces nepheline normative magma.

The temperatures necessary to bring about the dehydration reactions might result from crustal thickening during orogenesis. However, as was seen in the discussion of *P–T–t* paths the ocean floor basalts (eclogitic at depth) and their underlying sheeted dikes, layered gabbroic intrusions, and residual harzburgitic mantle rocks could not melt. Nor are the ocean floor rocks likely to rise diapirically into the overlying wedge where temperatures are higher, because once they are converted to eclogite their average density of 3.6 Mg m^{-3} is significantly greater than that of garnet peridotite, which is only 3.4 Mg m^{-3}. If sediment is subducted, a granitic component could melt, but this would not account for basalts and andesites. Moreover, most sediment accumulates as an accretionary wedge in front of the arc rather than being subducted.

Temperatures in the mantle wedge above the Benioff zone are not high compared with normal steady-state geotherms, nor are they low, despite their proximity to the subducting plate. They are prevented from falling by the convection induced by the subducting plate (Fig. 22-9). The temperatures in the wedge are not high enough to generate melts if no water is present. The fact that melts do exist in this region is therefore evidence that water must be present, and the subducted ocean floor rocks are the most obvious source of this fluid. It is significant that volcanism occurs where the Benioff zone goes below the depth to which amphibole is stable (dotted line in Fig. 22-18). Below this, essentially all water that had been bound in minerals is released as the rocks are converted to eclogite

assemblages. The pressures generated by the dehydration reactions would cause the fluids to rise into the overlying wedge despite the poor wetting ability of the fluids.

Fluids rising from the Benioff zone enter progressively hotter rocks, and at approximately the 1000°C isotherm they cross the water-saturated solidus of garnet lherzolite, and melting begins (region 1 in Fig. 22-18). The up-dip extent of this melting region is limited by the position of both the isotherms and the stability field of amphibole (note the sharp bend in the dotted line in Fig. 22-18). The down-dip extent, however, is limited only by the availability of water. The melting region is bounded on its upper side by the solidus involving the breakdown of amphibole to a water-undersaturated melt. Thus the melts range from water-saturated in the lower part of this region to water-undersaturated at its top. Note that the isotherms in Figure 22-18 take into account only the convection induced by the subducting slab. Once significant volumes of melt are formed in the mantle wedge, the buoyant rise of this magma will distort the isotherms and extend the zone of melting vertically.

If the overriding plate has a continental crust, another region of possible melting occurs just above the Moho where water-saturated granite melts can form (region 2 in Fig. 22-18). This region is bounded on its lower side by the compositional Moho and on its upper, by the water-saturated granite solidus.

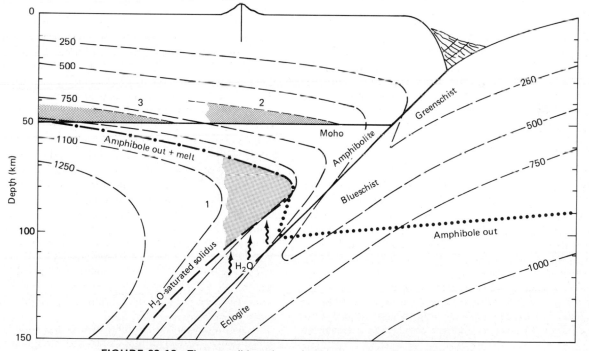

FIGURE 22-18 Three possible regions of melting near convergent plate boundary based on positions of isotherms in convection model of Toksöz and Hsui (1978) (see Fig. 22-9). Region 1, water liberated from the subducted plate allows garnet lherzolite to melt under water-saturated conditions. Region 2, granitic melt could form in continental lithosphere only under water-saturated conditions (not likely; see text). Melting in this region is more likely to occur at higher temperatures as a result of the breakdown of mica or amphibole (see Fig. 22-17); this occurs when magma rises from region 1. Region 3, granitic melt formed during breakdown of muscovite can form in the lower continental crust at the top of convection cell; this might be associated with continental rifting.

Free water must be available for melts to form in this region. Water is not likely to enter directly from the Benioff zone because melting in the underlying mantle wedge consumes all available H_2O. However, these mantle-derived melts, on rising into the base of the crust and crystallizing, liberate water and heat, which promotes crustal melting. The lateral extent of this melting thus depends on the flux of magmas from the mantle.

A third possible region of melting occurs in the lower crust above the top of the convecting cell in the mantle wedge (region 3 in Fig 22-18). This is the same place that crustal rifting and back-arc spreading may occur. The melting region is again bounded on its lower side by the Moho, but its upper surface is the solidus marked by the breakdown of muscovite to water-undersaturated melt. Fusion here is, therefore, not dependent on an influx of H_2O. Again, the positions of these regions of granite melting change as diapiric bodies rise into the overlying crust and distort the isotherms.

Estimated magmatic temperatures for andesites based on geothermometers and experimental studies range from a low of 850°C for hornblende-bearing varieties to 1100°C for two-pyroxene varieties (Gill 1981). Allowing for adiabatic cooling during ascent, these temperatures indicate a likely magma source in the mantle wedge. This must also be the source of the commonly associated basaltic rocks.

Basalts were initially thought to be parental to andesites. This idea was based mainly on the almost continuous chemical variation of calc-alkali rocks from basalt, through andesite, and dacite, to rhyolite. Experiments on andesites, however, have failed to provide any simple means of differentiating basaltic liquids into andesitic ones. If andesite is derived from basalt by fractional crystallization, minerals crystallizing in the basalt must be present as primary liquidus phases in the andesite. Magnetite is one mineral that must fractionate if basalt is to differentiate to andesite; magnetite crystallization reduces the iron content while increasing the silica content of the residual melt. Experiments indicate that under normal oxygen fugacities magnetite is a late-crystallizing mineral in andesite [see, for example, Eggler and Burnham (1973)], and many andesites do not contain magnetite phenocrysts at all. These facts are clearly unreconcilable with andesite being derived from basalt.

In many island arcs, andesites contain lower crustal xenoliths of a variety of ultramafic to mafic rocks, including hornblende-bearing olivine clinopyroxenite, hornblende gabbro, and horblende-free gabbro (Conrad and Kay 1984). These cumulates record the differentiation of basaltic magma in the lower crust or upper mantle. The fact that andesite so commonly brings up these xenoliths shows that andesite itself has a deeper source than the site of this basaltic differentiation. Again this would argue in favor of andesite having an independent origin from basalt.

Experiments indicate that unless andesites contain small but signifcant amounts of H_2O ($>2\%$), crystallization sequences do not match those found in the natural rocks. This volatile content would certainly account for the large volumes of andesitic magma that are erupted explosively (see Sec. 14-3). Could the volatile content determine whether andesitic or basaltic melts are formed by partial fusion of the mantle? Melting experiments on garnet lherzolite indicate that it might (Kushiro 1972). With increasing pressure the composition of the first-

formed melt becomes progressively less siliceous, with nepheline normative compositions forming above 1 GPa (Fig. 22-14). However, if water is present, the melts are richer in silica, and at 2 GPa the first-formed water-saturated melt is andesitic in composition (Fig. 11-20). Thus basalt and andesite could be derived from a common source but under different activities of water.

The amount of water available for melting must decrease as the Benioff zone deepens. The prevalence of andesitic volcanoes along the volcanic front of the arc could thus be explained if andesites require high activities of water for their generation. In deeper zones, where there is less water, magmas would be basaltic, with compositions becoming progressively more undersaturated in silica with increasing depth. This would explain the common distribution of basaltic rocks in island arcs with tholeiitic ones being closest to the convergent plate boundary and alkaline ones farthest from it, as is found in going from east to west across Japan (Kuno 1960).

The origin of andesites remains uncertain, but fluids undoubtedly play a critical role in their formation. Fluids may also be important in the genesis of other igenous and metamorphic rocks. Isotopic data (Sec. 21-4) indicate that hot-spot magmatism is preceded by mantle metasomatism, which introduces incompatible elements into otherwise relatively depleted source regions (Menzies and Hawkesworth 1987). If these elements are transported by fluid, as seems most likely given the low rates of solid diffusion, the fluid may at the same time act as a flux and cause melting. There is evidence that granulite facies metamorphism, especially in early Precambrian terranes, may be the product of large-scale flushing by CO_2-rich fluids which are probably derived from the mantle (Newton et al. 1980). In both examples rocks may be formed in response to mantle outgassing. If the upper mantle were completely melted during formation of the core, most of its fluids would have been expelled. Fluids in the upper mantle at subsequent times would therefore have to come from either subducted sediments or the lower mantle.

22-7   SUMMARY AND CONCLUSIONS

The formation of a rock, whether it be igneous, metamorphic, or sedimentary, is a rare geological event that requires extraordinary conditions. All rocks are formed directly or indirectly in response to perturbations in the steady-state geotherm, and they represent just one of the ways the Earth has of dissipating its internal heat.

Throughout most of the Earth the geotherm is determined mainly by convection, except in the lithosphere where conduction is the principal means of heat transfer. But even here the oceanic part of the lithosphere is involved in convection through ocean-floor spreading and subduction. Perturbations in the geotherm can have shallow or deep origins. Continents, for example, preserve a complex record of changing lithospheric plate configurations, each of which would have had its own shallow thermal pattern. Seismic evidence indicates that subducted lithospheric slabs may descend as far as the core–mantle boundary, which must cause a counterflow from this same depth, perhaps in the form of mantle plumes. The ascent of this material not only transfers heat from the core to the upper mantle but may also transfer elements that would otherwise remain in an untapped lower mantle reservoir.

The upper mantle, although now largely solid, behaves plastically and is able to convect. Temperature gradients in the upper mantle are consequently nearly adiabatic and extremely shallow ($<1°C$ km^{-1}). This means that no matter from where material rises in the upper mantle its temperature, on reaching the base of the lithosphere, is approximately the same, that is, about 1400°C. Throughout the upper mantle the adiabatic gradient is about 400°C below the fluid-absent solidus of garnet lherzolite, the rock type most likely comprising this part of the Earth. Because of the limited range of temperatures in this part of the mantle, melting is not likely to occur as a result of temperatures rising to the solidus. Instead, the solidus is more likely to be lowered to the geotherm by changes in composition. These changes could result from the rise of mantle diapirs of a different composition or by infiltration metasomatism of CO_2 or H_2O fluids.

In the conductive lithosphere and thermal boundary layer at the top of the convecting mantle the temperature gradient is steep, and consequently rocks that are displaced upward find themselves at temperatures that may be far enough above the geotherm to intersect the fluid-absent solidus of lherzolite and cause melting by decompression. This type of melting takes place in zones of lithospheric extension as a result of passive upwelling of the mantle.

Basaltic magmas are formed by partial fusion of peridotite in the upper mantle. Melting first takes place at grain corners where low-melting mineral assemblages are in close proximity. The dihedral angle between the melt and crystals is sufficiently small that the melt flows by capillary action along all grain edges. Once it forms a continuous network it is able to rise buoyantly through porous flow while the residual solids undergo compaction. Channels or veins of melt probably develop, as happens with granitic melts in migmatites. Once large enough bodies of melt have coalesced, they may rise diapirically through the mantle if their density is less than that of the surroundings. Their ascent is normally halted at the base of the rigid lithosphere, and only when fractures can form easily is basaltic magma likely to rise higher. This most commonly occurs in regions of lithospheric extension, such as mid-ocean ridges and continental rift valleys.

The rise of magma into the base of the lithosphere elevates the conductive thermal gradient in the overlying rocks and causes metamorphism. Fluids released by metamorphic reactions rise into the upper part of the lithosphere, advectively transferring heat as they do so. The heat from the basaltic magma at the base of the lithosphere also causes melting of granitic fractions in regions of continental crust. These siliceous melts form diapiric bodies that rise and advectively introduce heat into the upper crust, where it brings about high-temperature, low-pressure metamorphism.

The main mechanism by which the Earth has cooled throughout most of geologic time has been through creation of new, hot ocean floor and the subduction of old, cold, hydrated ocean floor. Few of these rocks are preserved in the geologic record—obducted ophiolites being one of the rare exceptions. The water released during the metamorphism of subducted ocean floor, however, has led to the development of magmas above Benioff zones at convergent plate boundaries that have, throughout geologic time, slowly added to the rocks of the continental crust.

Heat production in the Earth was greater in the early Precambrian than it is now. Convection would therefore have been more rapid, and the lithosphere would have been thinner. Despite these changes, certain rocks, such as tholeiitic basalt and granite, have formed with remarkably constant composition throughout geologic time. Other rocks, however, may owe their origins to changes in the nature of convection—komatiite in the Archean, widespread granulites in the Archean and Proterozoic, alkaline magmas in the early Proterozoic and Cretaceous, massif-type anorthosite in the Proterozoic, and blueschists in the Phanerozoic. Some of these rocks that are prevalent at particular times may also involve changes occurring in the lower mantle or core. As our understanding of the physics of the deeper parts of the Earth increases, it may be possible to use the petrologic record to interpret a truly global history of the Earth.

PROBLEMS

22-1. If the present average mantle of the Earth contains 0.025 ppm ^{238}U, 0.0002 ppm ^{235}U, 0.103 ppm ^{232}Th, and 0.033 ppm ^{40}K, plot graphs of the heat production from each of these isotopes and their total heat production per kilogram of rock throughout the past 4.5 Ga. Compare the present heat production of this composition with that of 4.0 Ga ago. Which were the most important heat-producing isotopes during the early history of the Earth? The decay energies in mW kg^{-1} are 0.0937 for ^{238}U, 0.569 for ^{235}U, 0.0269 for ^{232}Th, and 0.0279 for ^{40}K. Use the decay constants in Table 21-1.

22-2. Using the rates of heat production from the mantle at the present and 4.0 Ga ago, as calculated in Prob. 22- (i.e., 0.61 × 10^{-11} and 2.20 × 10^{-11} W kg^{-1}, respectively), and assuming a mantle density of 3.5 Mg m^{-3}, calculate the Rayleigh number bers at these two times for a 700-m-deep layer. The physical properties of this layer are as follows: $\alpha = 3 \times 10^{-5}$ K^{-1}, $K = 3.35$ W m^{-1} K^{-1}, $k = 10^{-6}$ m^2 s^{-1}, and $\eta = 7 \times 10^{20}$ Pa s. Would you expect the mantle to convect, and if so, how would the convection 4.0 Ga ago have compared with that of today?

22-3. If early in Earth history, magma was emplaced at spreading oceanic ridges with a temperature of 1400°C, how long would you expect the ocean to deepen at a rate proportional to $t^{1/2}$ according to the Parsons and McKenzie (1978) model for small-scale convection in a thermal boundary layer? All conditions, other than the initial intrusion temperature, are assumed to be the same as the present day; that is, the thermal boundary layer has a density of 3.33 Mg m^{-3}, α of 3 × 10^{-5} °C^{-1}, k of 8 × 10^{-7} m^2 s^{-1}, and viscosity of 7.76 × 10^{19} Pa s, the temperatures at the top and bottom of the boundary layer are 975 and 1260°C, respectively, and the temperature of the ocean floor is 0°C. Use a critical Rayleigh number (Eq. 13-10) of 10^3.

22-4. Given that the average radius of the Earth is 6.371×10^6 m, that of the core is 3.486×10^6 m, and the depth of the upper mantle–lower mantle boundary is 650 km, calculate the volumes of the Earth, core, lower mantle, and upper mantle. How many times greater is the volume of the lower mantle than the volume of the upper mantle?

22-5. If the upper mantle has a coefficient of thermal expansion of 4×10^{-5} °C^{-1}, a heat capacity of 10^3 J kg^{-1} °C^{-1}, and a temperature of 1500°C, what would the adiabatic temperature gradient be in °C km^{-1}?

22-6. If freely convecting mantle at a depth of 200 km has a temperature of 1400°C, construct an adiabatic geothermal gradient to a depth of 600 km. Use the same physical properties for the mantle as in Prob. 22-5. (*Hint:* Use Eq. 13-14 to express T as a function of z.)

22-7. If adiabatic decompression during lithospheric extension caused 10 wt % of the mantle to melt, how much lowering of temperature would this cause if the latent heat of fusion of the mantle were 420 kJ kg^{-1} and the heat capacity were 10^3 J kg^{-1} K^{-1}?

22-8. The conductive geotherm through a 100-km-thick lithosphere is described by the relation $T_z = 14z$, where z is depth in kilometers. Below this the temperature rises adiabatically in the convecting mantle. For simplicity the thermal boundary layer has been ignored.

  (a) Calculate expressions for the temperature gradient immediately after the lithosphere has been stretched by factors of 1.5 and 2.0.

  (b) The solidus of the mantle is described by the relation $T_s = 1100 + 4.4z$, where z is depth in kilometers. At what depth and temperature would the geotherm intersect the solidus if the lithosphere is stretched by factors of 1.5 and 2.0?

  (c) If the geotherm is assumed to be at its maximum temperature above the solidus at the point where it switches from being conductive to convective, calculate by how much the geotherm exceeds the solidus temperature for lithospheric stretching by factors of 1.5 and 2.0. How do the degrees of melting compare in these two cases?

22-9. A 500-m-thick basaltic sill at its liquidus temperature of 1200°C is emplaced into country rocks of granitic composition at a temperature of 500°C. If during convection all of the heat lost in becoming 60% solidified goes into making an overlying molten layer of convecting granitic magma that is 50% crystalline, how thick is the layer of granite? The basaltic liquid has a latent heat of crystallization of 400 kJ kg^{-1} and a density of 2.7 Mg m^{-3}. The granitic liquid has a latent heat of crystallization of 300 kJ kg^{-1} and a density of 2.3 Mg m^{-3}. The heat capacity of all materials is 1.34 kJ kg^{-1} K^{-1}. The fraction of

basalt crystallized as a function of temperature in degrees Celsius is $X_b = 7200T^{-1} - 6$ and that of the granite is $X_g = 0.65(1000 - T)/150$ (Huppert and Sparks 1988).

22-10. If in Prob. 22-9 the base of the granitic layer is kept at a temperature of 1175°C by the convecting basaltic magma, and its upper surface is kept at the temperature at which the rock contains 65% crystals (acts as a solid), that is, 850°C, how thick must the layer be in order to convect? The coefficient of expansion of the granitic magma is 5×10^{-5} K^{-1}, its thermal diffusivity is 10^{-6} m^2 s^{-1}, and its viscosity is 2.5×10^5 Pa s.

22-11. If following 1% partial melting of the mantle the melt is able to buoyantly rise through the solid matrix along the interconnecting channels along grain edges, calculate the velocity of rise if the permeability is 2.5×10^{-14} m^2 (this assumes dihedral angles for the melt phase of 50°), the viscosity of the melt is 1 Pa s, and the density contrast between the melt and solid is 600 kg m^{-3}. Would this velocity be rapid enough to supply magma at a divergent plate boundary if the rate of plate motion is taken to be on the order of 10 mm yr^{-1}? (Review Sec. 18-3 and Eq. 18-2.)

22-12. If a dike of alnoite contains spherical xenoliths of eclogite, what must the minimum rate of intrusion have been for the xenoliths to rise from the mantle? The densities of the xenolith and magma were 3.6 and 2.6 Mg m^{-3}, respectively, the viscosity of the magma was 10 Pa s, and the diameter of the largest xenolith is 10 cm.

22-13. Do reactions 22-4 and 22-5 account for all possible reactions that can eliminate anorthite from assemblages involving the six minerals in these two reactions if we specify that forsterite must be present? Consider Ca-Tschermak's molecule as a separate mineral. [*Hint:* You must determine if there is degeneracy. To do this, review your answer to Prob. 17-3; ignore garnet from this problem, and add Ca-Tschermak's molecule. The plotted position of this molecule in the projection from forsterite (Fo is always present) will provide you with the answer.]

22-14. What composition magmas will be formed by **(a)** the equilibrium partial fusion and **(b)** the fractional partial fusion of a garnet peridotite composed of 60% forsterite, 30% pyrope, and 10% diopside at 4 GPa? (Use Fig. 22-15.) The total fraction of the rock that melts in both cases is 30 wt %. **(c)** What compositions do the residual solids have in the fractional melting case? **(d)** Which mechanism requires the higher temperature?

22-15. If a source rock contains 10 ppm Eu and the bulk distribution coefficient (rock/liquid) is 0.1, what concentration of Eu would be found in melts formed by 2, 5, 15, and 20% fusion of this material under complete equilibrium and complete fractional melting? (Review Eqs. 12-20, 13-36, and 13-39. Be careful of what the fraction F refers to and how the distribution coefficient is defined.) How might the fraction of melting be determined in rocks?

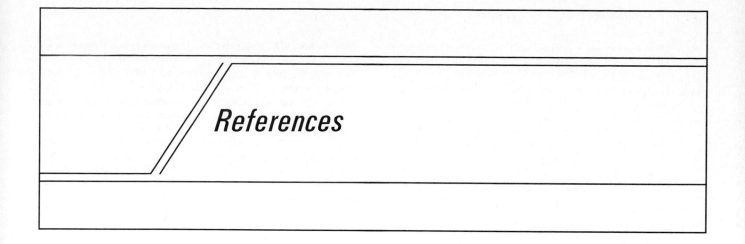

References

ABBOTT, R. N., JR., 1978, Peritectic reactions in the system An-Ab-Or-Qz-H₂O: *Canad. Mineral.*, v. 16, p. 245–256.

ALBEE, A. L., 1965, A petrogenetic grid for the Fe–Mg silicates of pelitic schists: *Amer. J. Sci.*, v. 263, p. 512–536.

ALLÈGRE, C. J., HART, S. R., AND MINSTER, J.-F., 1983, Chemical structure and evolution of the mantle and continents determined by inversion of Nd and Sr isotopic data: II. Numerical experiments and discussion: *Earth Planet. Sci. Lett.*, v. 66, p. 191–213.

AMSTUTZ, G. C., 1974, *Spilites and Spilitic Rocks*, Springer-Verlag, Berlin.

ANDERSEN, O., 1915, The system anorthite–forsterite–silica: *Amer. J. Sci.*, v. 39, p. 407–454.

ANDERSON, A. T., 1975, Some basaltic and andesitic gases: *Rev. Geophys. Space Phys.*, v. 13, p. 37–55.

ANDERSON, A. T., JR., AND MORIN, M., 1968, Two types of massif anorthosites and their implications regarding the thermal history of the crust: in Isachsen, Y. W., ed., *Origin of Anorthosites and Related Rocks*, N. Y. State Mus. Sci. Serv., Mem. 18, p. 57–69.

ANDERSON, D. L., 1982, The chemical composition and evolution of the mantle: Advances in Earth and planetary sciences: in Akimato, S., and Manghnani, M. H., eds., *High-Pressure Research in Geophysics*, D. Reidel Publishing Co., Hingham, Mass., p. 301–318.

———, 1984, The earth as a planet: paradigms and paradoxes: *Science*, v. 223, p. 347–355.

ANDERSON, E. M., 1936, The dynamics of the formation of cone-sheets, ring-dikes, and caldron-subsidences: *Proc. Roy. Soc. Edinburgh*, v. 56, p. 128–163.

ANDERSON, G. M., AND BURNHAM, C. W., 1965, The solubility of quartz in supercritical water: *Amer. J. Sci.*, v. 263, p. 494–511.

ARNDT, N. T., NALDRETT, A. J., and PYKE, D. R., 1977, Komatiite and iron-rich tholeiite lava of Munro Township, northeast Ontario: *J. Petrol.*, v. 18, p. 319–369.

ARNDT, N.T., AND NISBET, E.G., 1982, What is a komatiite? in Arndt, N. T., and Nisbet, E. G., eds., *Komatiites*, George Allen & Unwin (Publisher) Ltd., London, p. 19–27.

BAILEY, D. K., 1974, Continental rifting and alkaline magmatism: in Sørensen H., ed., *The Alkaline Rocks*, John Wiley & Sons Ltd., Chichester, West Sussex, U. K., p. 148–159.

BAKER, B. H., AND WOHLENBERG, J., 1971, Structure and evolution of the Kenya Rift Valley: *Nature*, v. 229, p. 538–542.

BAKER, B. H., CROSSLEY, R., AND GOLES, G. G., 1978, Tectonic and magmatic evolution of the southern part of the Kenya Rift Valley: in Neumann, E.R., and Ramberg, I. B., eds., *Petrology and Geochemistry of Continental Rifts*, D. Reidel Publishing Co., Dordrecht, The Netherlands, p. 29–50.

BAKER, P. E., 1968, Comparative volcanology and petrology of the Atlantic island arcs: *Bull. Volcanol.*, v. 32, p. 186–206.

BARLEY, M. E., 1986, Incompatible-element enrichment in Archean basalts: a consequence of contamination by older sialic crust rather than mantle heterogeneity: *Geology*, v. 14, p. 947–950.

BARRON, L. M., 1972, Thermodynamic multicomponent silicate equilibrium phase calculations: *Amer. Mineral.*, v. 57, p. 809–823.

BARROW, G., 1893, On an intrusion of muscovite biotite gneiss in the southeast Highlands of Scotland, and its accompanying metamorphism: *Geol. Soc. London Quart. J.*, v. 49, p. 330–358.

BASALTIC VOLCANISM STUDY PROJECT (BVSP), 1981, *Basaltic Volcanism on the Terrestrial Planets*, Pergamon Press, Inc., Elmsford, N. Y., 1286 p.

BELL, T. H. RUBENACH, M. J., AND FLEMING, P. D., 1986, Porphyroblast nucleation, growth, and dissolution in regional metamorphic rocks as a function of deformation partitioning during foliation development: *J. Metamorphic Geol.*, v. 4, p. 37–67.

BICKLE, M. J., AND MCKENZIE, D., 1987, The transport of heat and matter by fluids during metamorphism: *Contrib. Mineral. Petrol.*, v. 95, p. 384–392.

BIRCH, F., 1966, Compressibility: elastic constants: in Clark, S. P., Jr., ed., *Handbook of Physical Constants*, Geol. Soc. Amer., Mem. 97, p. 97–173.

BLACKENBURG, F. v., VILLA, I. M., BAUR, H., MORTEANI, G., and STEIGER, R. H., 1989, Time calibration of a *PT*-path from the Western Tauern Window, eastern Alps: the problem of closure temperatures: *Contrib. Mineral. Petrol.*, v. 101, p. 1–11.

BOHLEN, S. R., 1987, Pressure–temperature–time paths and a tectonic model for the evolution of granulites: *J. Geol.*, v. 95, p. 617–632.

BOHLEN, S. R., BOETTCHER, A. L., AND WALL, V. J., 1982, The system albite-H₂O-CO₂: a model for melting and activities of water at high pressures: *Amer. Mineral.*, v. 67, p. 451–462.

BOHLEN, S. R., AND MEZGER, K., 1989, Origin of granulite terranes and the formation of the lowermost continental crust: *Science*, v. 244, p. 326–329.

Bohlen, S. R. Wall, V. J., and Boettcher, A. L., 1983, Experimental investigations and geological applications of equilibria in the system FeO–TiO₂–Al₂O₃–SiO₂–H₂O: *Amer. Mineral.*, v. 68, p. 1049–1058.

Bottinga, Y., and Weill, D. F., 1972, The viscosity of magmatic silicate liquids: *Amer. J. Sci.*, v. 272, p. 438–475.

Bottinga, Y., Weill, D. F., and Richet, P., 1982, Density calculations for silicate liquids—I. Revised method for aluminosilicate compositions: *Geochim. Cosmochim. Acta*, v. 46, p. 909–919.

Bowden, P., and Turner, D. C., 1974, Peralkaline and associated ring complexes in the Nigeria–Niger province, West Africa: in Sørensen, H., ed., *The Alkaline Rocks*; John Wiley & Sons Ltd., Chichester, West Sussex, U. K., p. 330–351.

Bowen, N. L., 1913, The melting phenomena of the plagioclase feldspars: *Amer. J. Sci.*, v. 34, p. 577–599.

———, 1915a, The crystallization of haplobasaltic, haplodioritic, and related magmas: *Amer. J. Sci.*, v. 40, p. 161–185.

———, 1915b, Crystallization-differentiation in silicate liquids: *Amer. J. Sci.*, v. 39, p. 175–191.

———, 1928, *The Evolution of the Igneous Rocks*, Princeton University Press, Princeton, N. J., 334 p.

———, 1940, Progressive metamorphism of siliceous limestone and dolomite: *J. Geol.*, v. 48, p. 225–274.

Bowen, N. L., and Andersen, O., 1914, The binary system MgO–SiO₂: *Amer. J. Sci.*, v. 37, p. 487–500.

Bowen, N. L., and Schairer, J. F., 1932, The system FeO–SiO₂: *Amer. J. Sci.*, v. 24, p. 177–213.

———, 1935, The system MgO–FeO–SiO₂: *Amer. J. Sci.*, v. 29, p. 151–217.

Boyd, F. R., 1961, Welded tuffs and flows in the rhyolite plateau of Yellowstone Park, Wyoming: *Geol. Soc. Amer. Bull.*, v. 72, p. 387–426.

Boyd, F. R., and Gurney, J. J., 1986, Diamonds and the African lithosphere: *Science*, v. 232, p. 472–477.

Boyd, F. R., and Nixon, P. H., 1975, Origins of the ultramafic nodules from some kimberlites of northern Lesotho and the Monastery Mine, South Africa: in Ahrens, L. H., Dawson, J. B., Duncan, A. R., and Erlank, A. J., eds., *Physics and Chemistry of the Earth*, Vol. 9, Pergamon Press Ltd., Oxford, p. 431–454.

Brady, J. B., 1983, Intergranular diffusion in metamorphic rocks: *Amer. J. Sci.*, v. 283A, p. 181–200.

Brady, J. B., and McCallister, R. H., 1983, Diffusion data for clinopyroxenes from homogenization and self-diffusion experiments: *Amer. Mineral.*, v. 68, p. 95–105.

Brady, J. B., and Yund, R. A., 1983, Interdiffusion of K and Na in alkali feldspars: homogenization experiments: *Amer. Mineral.*, v. 68, p. 106–111.

Brandeis, G., and Jaupart, C., 1987, Crystal sizes in intrusions of different dimensions: constraints on the cooling regime and the crystallization kinetics: in Mysen, B. O., ed., *Magmatic Processes: Physicochemical Principles*, Geochem. Soc. Spec. Publ. No. 1, p. 307–318.

Brandeis, G., Jaupart, C., and Allègre, C. J., 1984, Nucleation, crystal growth and the thermal regime of cooling magmas: *J. Geophys. Res.*, v. 89, p. 10161–10177.

Bridgwater, D., and Coe, K., 1970, The role of stoping in the emplacement of the giant dikes of Isortoq, South Greenland; in Newall, G., and Rast, H., eds., *Mechanism of Igneous Intrusion*, Liverpool Geol. Soc., Geol. J. Spec. Issue No. 2, p. 67–78.

Bryan, W. B., 1972, Morphology of quench crystals in submarine basalts: *J. Geophys. Res.*, v. 77, p. 5812–5819.

Buddington, A. F., 1959, Granite emplacement with special reference to North America: *Geol. Soc. Amer. Bull.*, v. 70, p. 671–748.

Buddington, A. F., and Lindsley, D. H., 1964, Iron-titanium oxide minerals and synthetic equivalents: *J. Petrol.*, v. 5, p. 310–357.

Bunch, T. E., Dence, M. R., and Cohen, A. J., 1967, Natural terrestrial maskelynite: *Amer. Mineral.*, v. 52, p. 244–253.

Bundy, F. P., Bovenkerk, H. P., Strong, H. M., and Wentorf, R. H., Jr., 1961, Diamond-graphite equilibrium line from growth and graphitization of diamond: *J. Chem. Phys.*, v. 35, p. 383–391.

Burnham, C. W., 1979, The importance of volatite constituents: in Yoder, H. S., Jr., ed., *The Evolution of the Igneous Rocks: Fiftieth Anniversary Perspectives*, Princeton University Press, Princeton, N.J., p. 439–482.

Burnham, C. W., and Davis, N. F., 1974, The role of H₂O in silicate melts: II. Thermodynamic and phase relations in the system NaAlSiO₃O₈–H₂O to 10 kilobars, 700° to 1100°C: *Amer. J. Sci.*, v. 274, p. 902–940.

Burnham, C. W., Holloway, J. R., and Davis, N. F., 1969, Thermodynamic properties of water to 1000°C and 10,000 bars: *Geol. Soc. Amer., Spec. Pap.* 132, p. 1–96.

Cameron, E. N., Jahns, R. H., McNair, A. H., and Page, L. R., 1949, *The Internal Structure of Granitic Pegmatites*, Econ. Geol., Monogr. 2, 115 p.

Campbell, I. H., 1985, The difference between oceanic and continental tholeiites: a fluid dynamic explanation: *Contrib. Mineral. Petrol.*, v. 91, p. 37–43.

Campbell, I. H., and Turner, J. S., 1986, The influence of viscosity on fountains in magma chambers: *J. Petrol.*, v. 27, p. 1–30.

Carmichael, D. M., 1969, On the mechanism of prograde metamorphic reactions in quartz-bearing pelitic rocks: *Contrib. Mineral. Petrol.*, v. 20, p. 244–267.

———, 1978, Metamorphic bathozones and bathograds: a measure of post-metamorphic uplift and erosion on a regional scale: *Amer. J. Sci.*, v. 278, p. 769–797.

Carmichael, I. S. E., and Ghiorso, M. S., 1986, Oxidation–reduction relations in basic magma: a case for homogeneous equilibria: *Earth Planet. Sci. Lett.*, v. 78, p. 200–210.

Carmichael, I. S. E., Nicholls, J., and Smith, A. L., 1970, Silica activity in igneous rocks: *Amer. Mineral.*, v. 55, p. 246–263.

Carslaw, H. S., and Jaeger, J. C., 1959, *Conduction of Heat in Solids*, 2nd ed., Oxford University Press, Inc., New York, 510 p.

Carter, J. L., 1970, Mineralogy and chemistry of the earth's upper mantle based on the partial fusion–partial crystallization model: *Geol. Soc. Amer. Bull.*, v. 81, p. 2021–2034.

Carter, S. R., Evenson, N. M., Hamilton, P. J., and O'Nions, R. K., 1978, Neodymium and strontium isotope evidence for crustal contamination of continental volcanics: *Science*, v. 202, p. 743–747.

Castellan, G. W., 1971, *Physical Chemistry*, Addison-Wesley Publishing Company, Reading, Mass., 866 p.

Chapin, C. E., and Elston, W. E., eds., 1979, Ash-flow tuffs: *Geol. Soc. Amer., Spec. Pap.* 180, 211 p.

Chinner, G. A., 1961, The origin of sillimanite in Glen Clova, Angus: *J. Petrol.*, v. 2, p. 312–323.

Christensen, J. N., Rosenfeld, J. L., and DePaolo, D. J., 1988, Rates of tectonometamorphic processes: direct measurement using Sr isotopic profiles of garnet porphyroblasts: *Eos Trans. AGU*, v. 69, p. 508.

Christiansen, R. L., 1979, Cooling units and composite sheets in relation to caldera structure: in Chapin, C. E., and Elston, W. E., eds., Ash-flow tuffs: *Geol. Soc. Amer., Spec. Pap.* 180, p. 29–42.

Christiansen, R. L., and Lipman, P. W., 1966, Emplacement and thermal history of a rhyolite lava flow near Fortymile Canyon, Southern Nevada: *Geol. Soc. Amer. Bull.*, v. 77, p. 671–684.

CLARK, S. P., JR., 1966, Handbook of physical constants: *Geol. Soc. Amer.*, Mem. 97, 587 p.

CLEMENS J. D., AND WALL, V. J., 1981, Origin and crystallization of some peraluminous (S-type) granitic magmas: *Canad. Mineral.*, v. 19, p. 111–131.

CLOUGH, C. T., MAUFE, H. B., AND BAILEY, E. B., 1909, The cauldron-subsidence of Glen-Coe, and the associated igneous phenomena: Geol. Soc. London Quart, J., v. 65, p. 611–678.

COE, K., 1966, Intrusive tuff of West Cork, Ireland: Geol. Soc. London, Quart. J., v. 122, p. 1–28.

COLEMAN, R. G., 1977, *Ophiolites*, Springer-Verlag, New York, 229 p.

COOPER, R. F., AND KOHLSTEDT, D. L., 1984, Solution-precipitation enhanced diffusional creep of partially molten olivine–basalt aggregates during hot-pressing: *Tectonophysics*, v. 107, p. 207–233.

CONRAD, W. K., AND KAY, R. W., 1984, Ultramafic and mafic inclusions from Adak Island: crystallization history, and implications for the nature of primary magmas and crustal evolution in the Aleutian arc: *J. Petrol.*, v. 25, p. 88–125.

COX, K. G., 1978, Kimberlite pipes: *Sci. Amer.*, April 1978, p. 120–130.

———, 1980, A model for flood basalt vulcanism: *J. Petrol.*, v. 21, p. 629–650.

CRANK, J., AND NICOLSON, P., 1947, A practical method for numerical evaluation of solutions of partial differential equations of the heat-conduction type: *Proc. Cambridge Philos. Soc.*, v. 43, p. 50–67.

CROUGH, S. T., AND JURDY, D. M., 1980, Subducted lithosphere, hotspots, and the geoid: *Earth Planet. Sci. Lett.*, v. 48, p. 15–22.

CURIE, P., 1885, Sur la formation des cristaux et sur les constantes capillaires de leurs différentes faces: *Soc. Minéral. France Bull.*, v. 8, p. 145–150.

DALRYMPLE, G. B., GROMMÉ, C. S., AND WHITE, R. W., 1975, Potassium–argon age and paleomagnetism of diabase dikes in Liberia: initiation of central Atlantic rifting: *Geol. Soc. Amer. Bull.*, v. 86, p. 399–411.

DALY, R. A., 1933, *Igneous Rocks and the Depths of the Earth*, Hafner Press, New York, 598 p. (reprinted, 1968).

DANCKWERTH, P. A., AND NEWTON, R. C., 1978, Experimental determination of the spinel peridotite to garnet peridotite reaction in the system $MgO-Al_2O_3-SiO_2$ in the range 900°–1100°C and Al_2O_3 isopleths of enstatite in the spinel field: *Contrib. Mineral. Petrol.*, v. 66, p. 189–201.

DANES, Z. F., 1972, Dynamics of lava flows: *J. Geophys. Res.*, v. 77, p. 1430–1432.

DARKEN, L. S., AND GURRY, R. W., 1945, The system iron–oxygen: I. The wüstite field and related equilibria: *J. Amer. Chem. Soc.*, v. 67, p. 1398–1412.

DAVIS, B. T. C., AND SCHAIRER, J. F., 1965, Melting relations in the join diopside–forsterite–pyrope at 40 kilobars and at one atmosphere: *Carnegie Instit. Washington, Yearbook 64*, p. 123–126.

DAWSON, J. B., 1962, The geology of Oldoinyo Lengai: *Bull. Volc.*, v. 24, p. 349–387.

DAWSON, J. B., AND HAWTHORNE J. B., 1973, Magmatic sedimentation and carbonatitic differentiation in kimberlite sills at Benfontein, South Africa: *Geol. Soc. London Quart. J.*, v. 129, p. 61–85.

DECKER, R., AND DECKER, B., 1981, *Volcanoes*, W.H. Freeman and Company, Publishers, San Francisco, 244 p.

DENBIGH, K., 1955, *The Principles of Chemical Equilibrium*, Cambridge University Press, Cambridge, 491 p.

DENSE, M. R., 1971, Impact melts: *J. Geophys. Res.*, v. 76, p. 5552–5565.

DEPAOLO, D. J., 1981, Nd isotopic studies: some new perspectives on Earth structure and evolution: *Eos*, v. 62, p. 137–140.

———, 1985, Isotopic studies of processes in mafic magma chambers: I. The Kiglapait Intrusive, Labrador: *J. Petrol.*, v. 26, p. 925–951.

DEPAOLO, D. J., AND JOHNSON, R. W., 1979, Magma genesis in the New Britain island-arc: constraints from Nd and Sr isotopes and trace-element patterns: *Contrib. Mineral. Petrol.*, v. 70, p. 367–379.

DEPAOLO, D. J., AND WASSERBURG, G. J., 1979, Petrogenetic mixing models in Nd-Sr isotopic patterns: *Geochim. Cosmochim. Acta*, v. 43, p. 615–627.

DIETZ, R. S., 1961, Vredefort ring structure: meteorite impact scar: *J. Geol.*, v. 69, p. 499–516.

———, 1964, Sudbury structure as an astrobleme: *J. Geol.*, v. 72, p. 412–434.

DIXON, S., AND RUTHERFORD, M. J., 1979, Plagiogranites as late stage immiscible liquids in ophiolite and mid-ocean ridge suites; an experimental study: *Earth Planet. Sci. Lett.*, v. 45, p. 45–60.

DOWTY, E., 1980, Crystal growth and nucleation theory and the numerical simulation of igneous crystallization: in Hargraves, R. B., ed., *Physics of Magmatic Processes*, Princeton University Press, Princeton, N.J., p. 420–485.

DOYLE, C. D., 1987, The relationship between activities of divalent cation oxides and the solution of sulfide in silicate and aluminosilicate liquids: in Kushiro, I., and Perchuk, I. L., eds., *Physical Chemistry of Magma*, Advances in Physical Geochemistry Series, Vol. 7, Springer-Verlag, New York, chap. 9.

DUCHESNE, J.-C., 1984, Massif anorthosites: another partisan review: in Brown, W. L., ed., *Feldspars and Feldspathoids*, D. Reidel Publishing Co., Dordrecht, The Netherlands, p. 411–433.

DUFFIELD, W. A., 1972, A naturally occurring model of global plate tectonics: *J. Geophys. Res.*, v. 73, p. 619–634.

DZIEWONSKI, A. M., 1984, Mapping the lower mantle: determination of lateral heterogeneity in P velocity up to degree and order 6: *J. Geophys. Res.*, v. 89, p. 5929–5952.

EBY, G. N., 1980, Minor and trace element partitioning between immiscible ocelli-matrix pairs from lamprophyric dikes and sills, Monteregian Hills petrographic province, Quebec: *Contrib. Mineral. Petrol.*, v. 75, p. 269–278.

———, 1984, Geochronology of the Monteregian Hills alkaline igneous province, Quebec: *Geology*, v. 12, p. 468–470.

EGGLER, D. H., 1973, Role of CO_2 in melting processes in the mantle: *Carnegie Inst. Washington, Yearbook 72*, p. 457–467.

———, 1974, Effect of CO_2 on the melting of peridotite: *Carnegie Inst. Washington, Yearbook 73*, p. 215–224.

———, 1976, Composition of the partial melt of carbonated peridotite in the system $CaO-MgO-SiO_2-CO_2$: *Carnegie Inst. Washington, Yearbook 75*, p. 623–626.

EGGLER, D. H., AND BURNHAM, C. W., 1973, Crystallization and fractionation trends in the system andesite-$H_2O-CO_2-O_2$ at pressures to 10 kb: *Geol. Soc. Amer. Bull.*, v. 84, p. 2517–2532.

EMELEUS, C. H., 1987, The Rhum layered complex, Inner Hebrides, Scotland: in Parsons, I., ed., *Origins of Igneous Layering*, D. Reidel Publishing Co., Dordrecht, The Netherlands, p. 263–286.

EMSLIE, R. F., 1985, Proterozoic anorthosite massifs: in Tobi, A. C., and Touret, J. L. R., eds., *The Deep Proterozoic Crust in the North Atlantic Provinces*, D. Reidel Publishing Co., Dordrecht, The Netherlands, p. 39–60.

ENGEL, A. E. J., ITSON, S. P., ENGEL, C. G., STICKNEY, D. M., AND CRAY, E. J., JR., 1974, Crustal evolution and global tectonics: A petrogenetic view: *Geol. Soc. Amer. Bull.*, v. 85, p. 843–858.

ENGLAND, P. C., AND RICHARDSON, S. W., 1977, The influence of erosion upon the mineral facies of rocks from different metamorphic environments: *Geol. Soc. London Quart. J.*, v. 134, p. 201–213.

ENGLAND, P. C., AND THOMPSON, A. B., 1984, Pressure–temperature-time paths of regional metamorphism: I. Heat transfer during the evolution of regions of thickened continental crust: *J. Petrol.*, v. 25, p. 894–928.

EPP, D., 1984, Possible perturbations to hotspot traces and implications for the origin and structure of the Line Islands: *J. Geophys. Res.*, v. 89, p. 11273–11286.

ERNST, W. G., 1988, Tectonic history of subduction zones inferred from retrograde blueschist *P–T* paths: *Geology*, v. 16, p. 1081–1084.

ESKOLA, P., 1920, The mineral facies of rocks: *Norsk. Geol. Tidsskr.*, v. 6, p. 143–194.

ESSENE, E. J., 1982, Geologic thermometry and barometry: in Ferry, J. M., ed., *Characterization of Metamorphism through Mineral Equilibria*, Mineralogical Society of America, Reviews in Mineralogy, Vol. 10, Washington, D.C., p. 153–206.

ETHERIDGE, M. A., WALL, V. J., COX, S. F., AND VERNON, R. H., 1984, High fluid pressures during regional metamorphism and deformation: implications for mass transport and deformation mechanisms: *J. Geophys. Res.*, v. 89, p. 4344–4358.

ETHERIDGE, M. A., WALL, V. J., AND VERNON, R. H., 1983, The role of the fluid phase during regional metamorphism and deformation: *J. Metamorphic Petrol.*, v. 1, p. 205–226.

EWART, A., 1976, Mineralogy and chemistry of modern orogenic lavas—some statistics and implications: *Earth Planet. Sci. Lett.*, v. 31, p. 417–432.

FAURE, G., 1986, *Principles of Isotope Geology*, 2nd ed., John Wiley & Sons, Inc., New York, 589 p.

FEDOTOV, S. A., 1978, Ascent of basic magmas in the crust and the mechanism of basaltic fissure eruptions: *Int. Geol. Rev.*, v. 20, p. 33–48.

FERRY, J. M., 1980, A comparative study of geothermometers and geobarometers in pelitic schists from south-central Maine: *Amer. Mineral.*, v. 65, p. 720–732.

———, 1983, On the control of temperature, fluid composition, and reaction progress during metamorphism: *Amer. J. Sci.*, v. 283A, p. 201–232.

FERRY, J. M., AND SPEAR, F. S., 1978, Experimental calibration of the partitioning of Fe and Mg between biotite and garnet: *Contrib. Mineral. Petrol.*, v. 66, p. 113–117.

FINCHAM, C. J. B., AND RICHARDSON, F. D., 1954, The behaviour of sulfur in silicate and aluminate melts: *Philos. Trans. Roy. Soc. London*, v. A233, p. 40–62.

FISHER, G. W., 1973, Nonequilibrium thermodynamics as a model for diffusion-controlled metamorphic processes: *Amer. J. Sci.*, v. 273, p. 897–924.

FISHER, R. V., AND SCHMINKE, H. V., 1984, *Pyroclastic Rocks*, Springer-Verlag, New York, 339 p.

FLEISCHER, R. L., AND PRICE, R. B., 1964, Techniques for geological dating of minerals by chemical etching of fission fragment tracks: *Geochim. Cosmochim. Acta*, v. 28, 1705–1714.

FLEISCHER, R. L., PRICE, R. B., AND WALKER, R. M., 1975, *Nuclear Tracks in Solids: Principles and Applications*, University of California Press, Berkeley, Calif. 605 p.

FOLAND, K. A., AND FAUL, H., 1977, Ages of the White Mountain intrusives—New Hampshire, Vermont, and Maine: *Amer. J. Sci.*, v. 277, p. 888–904.

FRASER, D. G., 1977, *Thermodynamics in Geology: Proceedings of NATO Advanced Study Institute*, Oxford, 1976, D. Reidel Publishing Co., Dordrecht, The Netherlands, 410 p.

FREER, R., 1981, Diffusion in silicate minerals and glasses: a data digest and guide to the literature: *Contrib. Mineral. Petrol.*, v. 76, p. 440–454.

FREESTONE, I. C., 1978, Liquid immiscibility in alkali-rich magmas: *Chem. Geol.*, v. 23, p. 115–123.

FRENCH, B. M., 1972, Shock-metamorphism features in the Sudbury structure: a review: in Guy-Bray, J., ed., *New Developments in Sudbury Geology*, Geol. Assoc. Canada, Spec. Pap. 10, p. 19–28.

FROST, B. R., AND FROST, C. D., 1987, CO_2, melts and granulite metamorphism: *Nature*, v. 327, p. 503–506.

FUJII, N., AND OSAMURA, K., 1986, Effect of water saturation on the distribution of partial melt in the olivine–pyroxene–plagioclase system: *J. Geophys. Res.*, v. 91, p. 9253–9259.

FYFE, W. S., PRICE, N. J., AND THOMPSON, A. B., 1978, *Fluids in the Earth's Crust*, Elsevier/North-Holland, Amsterdam, 583 p.

GASS, I. G., 1970, Tectonic and magmatic evolution of the Afro-Arabian dome: in Clifford, T. N., and Gass, I. G., eds., *African Magmatism and Tectonics*, Oliver & Boyd, Edinburgh, p. 285–300.

GEISER, P. A., AND SANSONE, S., 1981, Joints, microfractures, and the formation of solution cleavage in limestone: *Geology*, v. 9, p. 280–285.

GHENT, E. D., 1976, Plagioclase–garnet–Al_2SiO_5–quartz: a potential geobarometer/geothermometer: *Amer. Mineral.*, v. 61, p. 710–714.

GHENT, E. D., ROBINS, D. B., AND STOUT, M. Z., 1979, Geothermometry, geobarometry, and fluid compositions of metamorphosed calcsilicates and pelites, Mica Creek, British Columbia: *Amer. Mineral.*, v. 64, p. 874–885.

GHENT, E. D., AND STOUT, M. Z., 1981, Geobarometry and geothermometry of plagioclase–biotite–garnet–muscovite assemblages: *Contrib. Mineral. Petrol.*, v. 76, p. 92–97.

GIBBS, J. W., 1875, On the equilibrium of heterogeneous substances: *Conn. Acad. Arts Sci. Trans.*, v. 3, p. 108–248.

GILL, J. B., 1981, *Orogenic Andesites and Plate Tectonics*, Springer-Verlag, New York.

GOLDSCHMIDT, V. M., 1911, Die Kontaktmetamorphose im Kristiania-gebiet: *Oslo Vidensk. Skr., I, Math.-Nat. Kl.*, no. 11.

GRAHAM, C. M., AND ENGLAND, P. C., 1976, Thermal regimes and regional metamorphism in the vicinity of overthrust faults: an example of shear heating and inverted metamorphic zonation from southern California: *Earth Planet. Sci. Lett.*, v. 31, p. 142–152.

GRAY, C. M., CLIFF, R. A., AND GOODE, A. D. T., 1981, Neodymium-strontium isotopic evidence for extreme contamination in a layered basic intrusion: *Earth Planet. Sci. Lett.*, v. 56, p. 189–198.

GRAY, N. H., 1971, A parabolic hourglass structure in titanaugite: *Amer. Mineral.*, v. 56, p. 952–958.

———, 1973, Estimation of parameters in petrologic materials balance equations: *Math. Geol.*, v. 5, p. 225–236.

———, 1978, Crystal growth and nucleation in flash-injected diabase dikes: *Canad. J. Earth Sci.*, v. 15, p. 1904–1923.

GREEN, D. H., 1972, Archean greenstone belts may include terrestrial equivalents of lunar maria? *Earth Planet. Sci. Lett.*, v. 15, p. 263–270.

———, 1975, Genesis of Archean peridotitic magmas and constraints on Archean geothermal gradients and tectonics: *Geology*, v. 3, p. 15–18.

GREENWOOD, H. J., 1967a, Mineral equilibria in the system MgO–SiO_2–H_2O–CO_2: in Abelson, P. H., ed., *Researches in Geochemistry*, Vol. 2, John Wiley & Sons, Inc., New York, p. 542–567.

———, 1967b, Wollastonite: stability in H_2O–CO_2 mixtures and occurrence in a contact metamorphic aureole near Salmo, British Columbia, Canada: *Amer. Mineral.*, v. 52, p. 1669–1680.

———, 1975, Buffering of pore fluids by metamorphic reactions: *Amer. J. Sci.*, v. 275, p. 573–593.

———, 1976, Metamorphism at moderate temperatures and pressures: in Bailey, D. K., and MacDonald, R., eds., *The Evolution of the Crystalline Rocks*, Academic Press, Inc. (London) Ltd., London, p. 187–259.

GREIG, J. W., AND BARTH, T. F. W., 1938, The system $Na_2O \cdot Al_2O_3 \cdot$ $5SiO_2$ (nephelite, carnegieite)-$Na_2O \cdot Al_2O_3 \cdot 6SiO_2$ (albite): *Amer. J. Sci.*, v. 35A, p. 93–112.

GRIFFIN, W. L., WASS, S. Y., AND HOLLIS, J. D., 1984, Ultramafic xenoliths from Bullenmerri and Gnotuk maars, Victoria, Australia: petrology of a subcontinental crust–mantle transition: *J. Petrol.*, v. 25, p. 53–87.

GROVER, J. E., 1980, Thermodynamics of pyroxenes: in Prewitt C. T., ed., *Reviews in Mineralogy, Vol. 7, Pyroxenes*, Mineralogical Society of America, Washington, D.C., p. 341–418.

GUY-BRAY, J., AND GEOLOGICAL STAFF, INTERNATIONAL NICKEL CO., 1966, Shatter cones at Sudbury: *J. Geol.*, v. 74, p. 243–245.

HÄKLI, T. A., AND WRIGHT, T. L., 1967, The fractionation of nickel between olivine and augite as a geothermometer: *Geochim. Cosmochim. Acta*, v. 31, p. 877–884.

HARGRAVES, R. B., ed., 1980, *Physics of Magmatic Processes*, Princeton University Press, Princeton, N.J., 585 p.

HARKER, R. I., AND TUTTLE, O. F., 1956, Experimental data on the P_{CO_2}–T curve for the reaction calcite + quartz = wollastonite + CO_2: *Amer. Mineral.*, v. 265, p. 239–256.

HART, S. R., AND ALLÈGRE, C. J., 1980, Trace-element constraints on magma genesis: in Hargraves, R. B., ed., *Physics of Magmatic Processes*, Princeton University Press, Princeton, N.J., p. 121–159.

HARTE, B., AND HUDSON, N. F. C., 1979, Pelite facies series and temperatures and pressures of Dalradian metamorphism in eastern Scotland: in Harris, A. L., Holland, C. H., and Leake, B. E., eds., *The Caledonides of the British Isles—Reviewed, Geol. Soc. London, Spec. Publ. 8*, Scottish Academic Press Ltd., Edinburgh, p. 323–337.

HATHERTON, T., AND DICKINSON, W. R., 1969, The relationship between andesitic volcanism and seismicity in Indonesia, the Lesser Antilles and other island arcs: *J. Geophys. Res.*, v. 74, p. 5301–5310.

HAUGHTON, D. R., ROEDER, P. L., AND SKINNER, B. J., 1974, Solubility of sulfur in mafic magmas: *Econ. Geol.*, v. 69, p. 451–467.

HAWKESWORTH, C. J., ERLANK, A. J., MARSH, J. S., MENZIES, M. A., AND VAN CALSTEREN, P., 1983, Evolution of the continental lithosphere: evidence from volcanics and xenoliths in southern Africa: in Hawkesworth, C. J. and Norry, M. J., eds., *Continental Basalts and Mantle Xenoliths*, Shiva Publishing Ltd., Nantwich, Cheshire, U.K., p. 111–138.

HAWKESWORTH, C. J., NORRY, M. J., RODDICK, J. C., BAKER, P. E., FRANCIS, P. W., AND THORPE, R. S., 1979, $^{143}Nd/^{144}Nd$, $^{87}Sr/^{86}Sr$, and incompatible element variations in calc-alkaline andesites and plateau lavas from South America: *Earth Planet. Sci. Lett.*, v. 42, p. 45–57.

HAY, R. S., AND EVANS, B., 1988, Intergranular distribution of pore fluid and the nature of high-angle grain boundaries in limestone and marble: *J. Geophys. Res.*, v. 93, B8, p. 8959–8974.

HEALD, E. F., NAUGHTON, J. J., AND BARNES, I. L., JR., 1963, The chemistry of volcanic gases: 2. Use of equilibrium calculations in the interpretation of volcanic gas samples: *J. Geophys. Res.*, v. 68, p. 545–557.

HEINRICH, E. W., 1966, *The Geology of Carbonatites*, Rand-McNally & Co., Chicago.

HEKINIAN, R., 1982, *Petrology of the Ocean Floor*, Elsevier Science Publishing Co., Inc., Amsterdam.

HELGESON, H. C., 1967, Solution chemistry and metamorphism: in Abelson, P. H., ed., *Researches in Geochemistry*, John Wiley & Sons, Inc., New York, p. 362–404.

HELGESON, H. C., DELANY, J. M., NESBITT, H. W., AND BIRD, D. K., 1978, Summary and critique of the thermodynamic properties of rock-forming minerals: *Amer. J. Sci.*, v. 278A, p. 1–229.

HENDERSON, P., 1982, *Inorganic Geochemistry*, Pergamon Press Ltd., Oxford, 353 p.

HERZBERG, C. T., 1987, Magma density at high pressure: Part 2. A test of the olivine flotation hypothesis: in Mysen, B. O., ed., *Magmatic Processes: Physicochemical Principles*, Geochem. Soc. Spec. Publ. 1, p. 47–58.

HESS, H. H., 1960, Stillwater Igneous Complex, Montana: a quantitative mineralogical study: *Geol. Soc. Amer.*, Mem. 80, 230 p.

HESS, P. C., 1969, The metamorphic paragenesis of cordierite in pelitic rocks: *Contrib. Mineral. Petrol.*, v. 24, p. 191–207.

———, 1980, Polymerization model for silicate melts: in Hargraves, R. B., ed., *Physics of Magmatic Processes*, Princeton University Press, Princeton, N. J., p. 3–48.

HIGGINS, M. W., 1971, Cataclastic rocks: *U.S. Geol. Surv., Prof. Pap. 687*, 97. p.

HILDRETH, W., 1981, Gradients in silicic magma chambers: implications for lithospheric magmatism: *J. Geophys. Res.*, v. 86, p. 10153–10192.

HOBBS, B. E., MEANS, W. D., AND WILLIAMS, P.F., 1976, *An Outline of Structural Geology*; John Wiley & Sons, Inc., New York, 571 p.

HODGES, K. V., AND CROWLEY, P., 1985, Error estimation and empirical geothermobarometry for pelitic systems: *Amer. Mineral.*, v. 70, p. 702–709.

HODGES, K. V., AND SPEAR, F. S., 1982, Geothermometry, geobarometry, and the Al_2SiO_5 triple point at Mt Moosilauke, New Hampshire: *Amer. Mineral.*, v. 67, p. 1118–1134.

HOFMANN, A. W., 1980, Diffusion in natural silicate melts: a critical review: in Hargraves, R. B., ed., *Physics of Magmatic Processes*, Princeton University Press, Princeton, N. J., p. 385–417.

HOFMANN, A. W., GILETTI, B. J., YODER, H. S. JR., AND YUND, R. A., eds., 1974, *Geochemical Transport and Kinetics*, Academic Press, Inc. (London) Ltd., London, 353 p.

HOISCH, T. D., 1987, Heat transport by fluids during Late Cretaceous regional metamorphism in the Big Maria Mountains, southeastern California: *Geol. Soc. Amer. Bull.*, v. 98, p. 549–553.

HOLDAWAY, M. J., 1971, Stability of andalusite and the aluminum silicate phase diagram: *Amer. J. Sci.*, v. 271, p. 97–131.

HOLLAND, T. J. B., 1980, The reaction albite = jadeite + quartz determined experimentally in the range $600 - 1200°C$: *Amer. Mineral.*, v. 65, p. 129–134.

HOLLISTER, L. S., 1966, Garnet zoning: an interpretation based on the Rayleigh fractionation model: *Science*, v. 154, p. 1647–1651.

———, 1970, Origin, mechanism and consequences of compositional sector zoning in staurolite: *Amer. Mineral.*, v. 55, p. 742–766.

HUANG, W. L., AND WYLLIE, P. J., 1973, Melting of muscovite-granite to 35 kbar as a model for fusion of metamorphosed subducted oceanic sediments: *Contrib. Mineral. Petrol.*, v. 42, p. 1–14.

HUPPERT, H. E., AND SPARKS, R. S. J., 1980, The fluid dynamics of a basaltic magma chamber replenished by influx of hot, dense ultrabasic magma: *Contrib. Mineral. Petrol.*, v. 75, p. 279–289.

———, 1985, Cooling and contamination of mafic and ultramafic magmas during ascent through continental crust: *Earth Planet. Sci. Lett.*, v. 74, p. 371–386.

———, 1988, The generation of granitic magmas by intrusion of basalt into continental crust: *J. Petrol.*, v. 29, p. 599–624.

HUPPERT, H. E., SPARKS, R. S. J., TURNER, J. S., AND ARNDT, N. T., 1984, Emplacement and cooling of komatiite lavas: *Nature*, v. 309, p. 19–22.

ILDEFONSE, J.-P., AND GABIS, V., 1976, Experimental study of silica diffusion during metasomatic reactions in the presence of water at 550°C and 1000 bars: *Geochim. Cosmochim. Acta*, v. 40, p. 297–303.

INGERSOLL, L. R., ZOBEL, O. J., AND INGERSOLL, A. C., 1954, *Heat Conduction*, University of Wisconsin Press, Madison, Wis., 325 p.

IRVINE, T. N., 1974, Petrology of the Duke Island Ultramafic Complex, southeastern Alaska: *Geol. Soc. Amer.*, Mem. 138.

———, 1977, Chromite crystallization in the join Mg_2SiO_4–$CaMgSi_2O_6$–$CaAl_2Si_2O_8$–$MgCr_2O_4$–SiO_2: *Carnegie Inst. Washington, Yearbook 76*, p. 465–472.

———, 1979, Rocks whose composition is determined by crystal accumulation and sorting: in Yoder, H. S., Jr., ed., *The Evolution of the Igneous Rocks: Fiftieth Anniversary Perspectives*, Princeton University Press, Princeton, N. J., p. 245–306.

———, 1980, Magmatic infiltration metasomatism, double-diffusive fractional crystallization, and adcumulus growth in the Muskox Intrusion and other layered intrusions: in Hargraves, R. B., ed., *Physics of Magmatic Processes*, Princeton University Press, Princeton, N. J., p. 325–385.

———, 1982, Terminology for layered intrusions: *J. Petrol.*, v. 23, p. 127–162.

IRVINE, T. N., AND BARAGAR, W. R. A., 1971, A guide to the chemical classification of the common volcanic rocks: *Canad. J. Earth Sci.*, v. 8, p. 523–548.

IRVINE, T. N., KEITH, D. W., AND TODD, S. G., 1983, The J-M Platinum–Palladium reef of the Stillwater Complex, Montana: II. Origin by double diffusive convective magma mixing and implications for the Bushveld Complex: *Econ. Geol.*, v. 78, p. 1287–1334.

IRVINE, T. N., AND SMITH, C. H., 1967, The ultramafic rocks of the Muskox Intrusion, Northwest Territories, Canada: in Wyllie, P. J., ed., *Ultramafic and Related Rocks*, John Wiley & Sons, Inc., New York, p. 38–49.

ISACHESEN, Y. W., ed., 1968, Origin of anorthosites and related rocks: *N. Y. State Mus. Sci. Serv.*, Mem. 18.

JACKSON, E. D., 1967, Ultramafic cumulates in the Stillwater, Great Dyke, and Bushveld Intrusion: in Wyllie, P. J., ed., *Ultramafic and Related Rocks*, John Wiley & Sons, Inc., New York, p. 20–38.

JAEGER, J. C., 1968, Cooling and solidification of igneous rocks: in Hess, H. H., and Poldervaart, A., eds., *Basalts*, Vol. 2, John Wiley & Sons, Inc., New York, p. 503–536.

JÄGER, E., AND HUNZIKER, J. C., eds., 1979, *Lectures in Isotope Geology*, Springer-Verlag, Berlin, 329 p.

JAHNS, R. H., AND BURNHAM, C. W., 1969, Experimental studies of pegmatite genesis: I. A model for the derivation and crystallization of granitic pegmatites: *Econ. Geol.*, v. 64, p. 843–864.

JOESTEN, R. L., 1974, Local equilibrium and metasomatic growth of zoned calcsilicate nodules in a contact aureole, Christmas Mountains, Big Bend region, Texas: *Amer. J. Sci.*, v. 274, p. 876–901.

———, 1976, High-temperature contact metamorphism in a shallow crustal environment, Christmas Mountains, Big Bend region, Texas: *Amer. Mineral.*, v. 61, p. 776–781.

———, 1977, Evolution of mineral assemblage zoning in diffusion metasomatism: *Geochim. Cosmochim. Acta*, v. 41, p. 649–670.

———, 1983, Grain growth and grain-boundary diffusion in quartz from the Christmas Mountains (Texas) contact aureole: *Amer. J. Sci.*, v. 283A, p. 233–254.

JOESTEN, R., AND FISHER, G., 1988, Kinetics of diffusion-controlled mineral growth in the Christmas Mountains (Texas) contact aureole: *Geol. Soc. Amer.* Bull., v. 100, p. 714–732.

JOHANNSEN, A., 1931, *A Descriptive Petrography of the Igneous Rocks*, Vol. I, *Introduction, Textures, Classification and Glossary*, University of Chicago Press, Chicago, 267 p.

JOHNSON, A. M., 1970, *Physical Processes in Geology*, Freeman, Cooper & Company, San Francisco, 577 p.

JOHNSON, A. M., AND POLLARD, D. D., 1973, Mechanics of growth of some laccolithic intrusions in the Henry Mountains, Utah, I: *Tectonophysics*, v. 18, p. 261–309.

JUREWICZ, S. R., AND WATSON, E. B., 1985, The distribution of partial melt in a granitic system: the application of liquid phase sintering theory: *Geochim. Cosmochim. Acta*, v. 49, p. 1109–1121.

KALSBEEK, F., AND JEPSEN, H. F., 1984, The late Proterozoic Zig-Zag Dal basalt formation of eastern North Greenland: *J. Petrol.*, v. 25, p. 644–664.

KARABINOS, P., AND KETCHAM, R., 1988, Thermal structure of active thrust belts: *J. Metamorphic Geol.*, v. 6, p. 559–570.

KEAREY, P., 1978, An interpretation of the gravity field of the Morin anorthosite complex, southwest Quebec: *Geol. Soc. Amer. Bull.*, v. 89, p. 467–475.

KENNEDY, C. S., AND KENNEDY, G. C., 1976, The equilibrium boundary between graphite and diamond: *J. Geophys. Res.*, v. 81, p. 2467–2470.

KEPEZHINSKAS, K. B., AND KHLESTOV, V. V., 1977, The petrogenetic grid and subfacies for middle-temperature metapelites: *J. Petrol.*, v. 18, p. 114–143.

KERN, R., AND WEISBROD, A., 1967, *Thermodynamics for Geologists*, Freeman, Cooper & Company, San Francisco, 304 p.

KERRICK, D. M., 1972, Experimental determination of the muscovite + quartz stability with $P_{H_2O} < P_{total}$: *Amer. J. Sci.*, v. 272, p. 946–958.

KERRICK, D. M., AND JACOBS, G. K., 1981, A modified Redlich–Kwong equation for H_2O, CO_2, and H_2O–CO_2 mixtures at elevated pressures and temperatures: *Amer. J. Sci.*, v. 281, p. 735–767.

KINGERY, W. D., 1960, *Introduction to Ceramics*, John Wiley & Sons, Inc., New York, 781 p.

KINGSLEY, L., 1931, Cauldron subsidence of the Ossipee Mountains: *Amer. J. Sci.*, v. 222, p. 139–168.

KIRKPATRICK, R. J., 1974, Kinetics of crystal growth in the system $CaMgSi_2O_6$–$CaAl_2SiO_6$: *Amer. J. Sci.*, v. 274, p. 215–242.

———, 1975, Crystal growth from the melt: a review: *Amer. Mineral.*, v. 60, p. 798–814.

KJARSGAARD, B. A., AND HAMILTON, D. L., 1988, Liquid immiscibility and the origin of alkali-poor carbonatites: *Mineral. Mag.*, v. 52, p. 43–55.

KLEIN, F. W., 1984, Eruption forecasting at Kilauea volcano, Hawaii: *J. Geophys. Res.*, v. 89, p. 3059–3073.

KNOPF, A., 1936, Igneous geology of the Spanish Peaks region, Colorado: *Geol. Soc. Amer. Bull.*, v. 47, p. 1727–1784.

KOMAR, P. D., 1972, Flow differentiation in igneous dikes and sills: profiles of velocity and phenocryst concentration: *Geol. Soc. Amer. Bull.*, v. 83, p. 3443–3448.

KORZHINSKII, D. S., 1970, *Theory of Metasomatic Zoning*, Clarendon Press, Oxford, 162 p.

KOSTER VAN GROSS, A. F., AND WYLLIE, P. J., 1966, Liquid immiscibility in the system Na_2O–Al_2O_3–SiO_2–CO_2 at pressures to 1 kilobar: *Amer. J. Sci*, v. 264, p. 234–255.

KRAUSKOPF, K. B., 1968, A tale of ten plutons: *Geol. Soc. Amer. Bull.*, v. 79, p. 1–18.

KRETZ, R., 1966, Interpretation of the shape of mineral grains in metamorphic rocks: *J. Petrol.*, v. 7, p. 68–94.

———, 1973, Kinetics of the crystallization of garnet at two localities near Yellowknife: *Canad. Mineral.*, v. 12, p. 1–20.

KUNO, H., 1960, High alumina basalt: *J. Petrol.*, v. 1, p. 125–145.

KUSHIRO, I., 1968, Compositions of magmas formed by partial zone melting of the earth's upper mantle: *J. Geophys. Res.*, v. 73, p. 619–634.

———, 1969, The system forsterite–diopside–silica with and without water at high pressures: *Amer. J. Sci.*, v. 267A, p. 269–294.

————, 1970, Stability of amphibole and phlogopite in the upper mantle: *Carnegie Inst. Washington, Yearbook 68*, p. 245–247.

————, 1972, Effect of water on the composition of magmas formed at high pressures: *J. Petrol.*, v. 13, p. 311–334.

————, 1980, Viscosity, density, and structure of silicate melts at high pressures, and their petrological applications: in Hargraves, R. B. ed., *Physics of Magmatic Processes*, Princeton University Press, Princeton, N.J., p. 93–120.

KUSHIRO, I., SYONO, Y., AND AKIMOTO, S., 1968, Melting of a peridotite nodule at high pressures and high water pressures: *J. Geophys. Res.*, v. 73, p. 6023–6029.

KUSHIRO, I., AND YODER, H. S., JR., 1966, Anorthite–forsterite and anorthite–enstatite reactions and their bearing on the basalt–eclogite transformation: *J. Petrol.*, v. 7, p. 337–362.

LACHENBRUCH, A. H., AND SASS, J. H., 1977, Heat flow in the United States and the thermal regime of the crust: in Heacock, J. G., ed., *The Earth's Crust*, Amer. Geophys. Union, Monogr. 20, p. 626–675.

LE BAS, M. J., LE MAITRE, R. W., STRECKEISEN, A., AND ZANETTIN, B., 1986, A chemical classification of volcanic rocks based on the total alkali-silica diagram: *J. Petrol.*, v. 27, p. 745–750.

LINDSLEY, D. H., 1983, Pyroxene thermometry: *Amer. Mineral.*, v. 68, p. 477–493.

LOFGREN, G., 1974, An experimental study of plagioclase crystal morphology: isothermal crystallization: *Amer. J. Sci.*, v. 274, p. 243–273.

————, 1980, Experimental studies on the dynamic crystallization of silicate melts: in Hargraves, R. B., ed., *Physics of Magmatic Processes*, Princeton University Press, Princeton, N.J., p. 487–551.

LONDON, D., 1987, Internal differentiation of rare-element pegmatites: effects of boron, phosphorous, and fluorine: *Geochim. Cosmochim. Acta*, v. 51, p. 403–420.

LONSDALE, P., 1985, Nontransform offsets of the Pacific-Cocos plate boundary and their traces on the rise flank: *Geol. Soc. Amer. Bull.*, v. 96, p. 313–327.

LORENZ, V., 1975, Formation of phreatomagmatic maar-diatreme volcanoes and its relevance to kimberlite diatremes: *Phys. Chem. Earth*, v. 9, p. 17–27.

LUTH, W. C., JAHNS, R. H., AND TUTTLE, O.F., 1964, The granite system at pressures of 4 to 10 kilobars: *J. Geophys. Res.*, v. 69, p. 759–773.

LUX, D. R., DeYOREO, J. J., GULDOTTI, C. V., AND DECKER, E. R., 1986, Role of plutonism in low-pressure metamorphic belt formation: *Nature*, v. 323, p. 794–797.

LYNN, H. B., HALE, L. D., AND THOMPSON, G. A. 1981, Seismic reflections from the basal contacts of batholiths: *J. Geophys. Res.*, v. 86, p. 10633–10638.

MAALØE, S., AND PRINTZLAU, I., 1979, Natural partial melting of spinel lherzolite: *J. Petrol.*, v. 20, p. 727–741.

MacDONALD, G. A., 1972, *Volcanoes*, Prentice-Hall, Inc., Englewood Cliffs, N.J., 510 p.

MACGREGOR, A. M., 1951, Some milestones in the Precambrian of southern Rhodesia: *Trans. Proc. Geol. Soc. S. Afr.*,v. 54, p. 27–71.

MacLEAN, W. H., 1969, Liquidus phase relations in the FeS–FeO–Fe_3O_4–SiO_2 system, and their application in geology: *Econ. Geol.*, v. 64, p. 865–884.

MANNING, D. A. C., AND PICHAVANT, M., 1983, The role of fluorine and boron in the generation of granitic melts: in Atherton, M. P., and Gribble, C. D., eds., *Migmatites, Melting, and Metamorphism*, Birkhäuser Boston, Inc., Cambridge, Mass., p. 94–109.

MARSH, B. D., 1978, On the cooling of ascending andesitic magma: *Philos. Trans. Roy. Soc. London*, v. A288, p. 611–625.

————, 1979, Island arc volcanism: *Amer. Sci.*, v. 67, p. 161–172.

————, 1982, On the mechanism of igneous diapirism, stoping, and zone melting: *Amer. J. Sci.*, v. 282, p. 808–855.

MARSH, B. D., AND CARMICHAEL, I. S. E., 1974, Benioff zone magmatism: *J. Geophys. Res.*, v. 79, p. 1196–1206.

MARSH, B. D., AND KANTHA, L. H., 1978, On the heat and mass transfer from an ascending magma: *Earth Planet. Sci. Lett.*, v. 39, p. 435–443.

MARSH, B. D., AND MAXEY, M. R., 1985, On the distribution and separation of crystals in convecting magma: *J. Volcanol. Geotherm. Res.*, v. 24, p. 95–150.

MARTIN, D., GRIFFITHS, W., AND CAMPBELL, I. H., 1987, Compositional and thermal convection in magma chambers: *Contrib. Mineral. Petrol.*, v. 96, p. 465–475.

MARTIN, H., 1987, Petrogenesis of Archaean trondhjemites, tonalites, and granodiorites from eastern Finland: major and trace element geochemistry: *J. Petrol.*, v. 28, p. 921–953.

MATHIAS, M., 1974, Alkaline rocks of southern Africa: in Sørensen, H., ed., *The Alkaline Rocks*, Wiley-Interscience, New York, p. 189–202.

McBIRNEY, A. R., 1975, Differentiation of the Skaergaard intrusion: *Nature*, v. 253, p. 691–694.

McBIRNEY, A. R., AND NOYES, R. M., 1979, Crystallization and layering of the Skaergaard intrusion: *J. Petrol.*, v. 20, p. 487–554.

McBIRNEY, A. R., TAYLOR, H. P., AND ARMSTRONG, R. L., 1987, Paricutin re-examined: a classic example of crustal assimilation in calc-alkaline magma: *Contrib. Mineral. Petrol.*, v. 95, p. 4–20.

McGETCHIN, T. R., AND ULLRICH, G. W., 1973, Xenoliths and maars and diatremes with inferences for the Moon, Mars, and Venus: *J. Geophys. Res.*, v. 78, p. 1833–1853.

McGUIRE, A. V., AND BOHANNON, R. G., 1989, Timing of mantle upwelling: evidence for a passive origin for the Red Sea rift: *J. Geophys. Res.*, v. 94, p. 1677–1682.

McKENZIE, D., 1984, The generation and compaction of partially molten rock: *J. Petrol.*, v. 25, p. 713–765.

McKENZIE, D., AND BICKLE, M. J., 1988, The volume and composition of melt generated by extension of the lithosphere: *J. Petrol.*, v. 29, p. 625–679.

McKENZIE, D., AND BRUNE, J. N., 1972, Melting on fault planes during large earthquakes: *Geophys. J. Roy. Astron. Soc.*, v. 29, p. 65–78.

McKENZIE, D., AND WEISS, N., 1975, Speculations on the thermal and tectonic history of the Earth: *Geophys. J. Roy. Astron. Soc.*, v. 42, p. 131–174.

McTAGGART, K. C., 1960, The mobility of nuées ardentes: *Amer. J. Sci.*, v. 258, p. 369–382.

MEL'NIK, Y. P., 1972, Thermodynamic parameters of compressed gases and metamorphic reactions involving water and carbon dioxide: *Geochem. Int.*, v. 9, p. 419–426.

MENZIES, M. A., AND HAWKESWORTH, C. J., eds., 1987, *Mantle Metasomatism*, Academic Press, Inc. (London) Ltd., London, 477 p.

MEYER, H. O. A., 1979, Kimberlites and the mantle: *Rev. Geophys. Space Phys.*, v. 17, p. 776–788.

MITCHELL, R. H., 1986, *Kimberlites: Mineralogy, Geochemistry and Petrology*, Plenum Press, New York, 460 p.

MIYASHIRO, A., 1961, Evolution of metamorphic belts: *J. Petrol.*, v. 2, p. 227–311.

————, 1973, *Metamorphism and Metamorphic Belts*, John Wiley & Sons, Inc., New York, 492 p.

MOORBATH, S. 1977, Ages, isotopes and evolution of Precambrian continental crust: *Chem. Geol.*, v. 20, p. 151–187.

MORSE, S. A., 1969, The Kiglapait layered intrusion, Labrador: *Geol. Soc. Amer.*, Mem. 112.

———, 1970, Alkali feldspars with water at 5 kb pressure: *J. Petrol.*, v. 11, p. 221–253.

———, 1980, *Basalts and Phase Diagrams*, Springer-Verlag, New York.

———, 1982, A partisan review of Proterozoic anorthosites: *Amer. Mineral.*, v. 67, p. 1087–1100.

———, 1986, Convection in aid of adcumulus growth: *J. Petrol.*, v. 27, p. 1183–1214.

MUAN, A., 1955, Phase equilibria in the system $FeO–Fe_2O_3–SiO_2$: *Trans. AIME*, v. 203, p. 965–976.

———, 1958, Phase equilibria at high temperatures in oxide systems involving changes in oxidation states: *Amer. J. Sci.*, v. 256, p. 171–207.

MYERS, J. S., 1975, Cauldron subsidence and fluidization: mechanisms of intrusion of the Coastal Batholith of Peru into its own volcanic ejecta: *Geol. Soc. Amer. Bull.*, v. 86, p. 1209–1220.

NALDRETT, A. J., 1969, A portion of the system Fe–S–O between 900 and 1080°C and its application to sulfide ore magmas: *J. Petrol.*, v. 10, p. 171–201.

NAVROTSKY, A., HON, R., WEILL, D. F., AND HENRY, D. J., 1980, Thermochemistry of glasses and liquids in the systems $CaMgSi_2O_6–CaAl_2Si_2O_8–NaAlSi_3O_8$, $SiO_2–CaAl_2Si_2O_8–NaAlSi_3O_8$ and $SiO_2–Al_2O_3–CaO–Na_2O$: *Geochim. Cosmochim. Acta*, v. 44, p. 1409–1423.

NELSON, K. D., 1981, A simple thermal mechanical model for mid-ocean ridge topographic variation: *Geophys. J. Roy. Astron. Soc.*, v. 65, p. 19–30.

NESBITT, B. E., 1988, Gold deposit continuum: a genetic model for lode Au mineralization in the continental crust: *Geology*, v. 16, p. 1044–1048.

NEWALL, G., AND RAST, N., eds., 1970, *Mechanism of Igneous Intrusion*, Liverpool Geol. Soc., Geol. J. Spec. Issue No. 2, 380 p.

NEWTON, R. C., SMITH, J. V., AND WINDLEY, B., 1980, Carbonic metamorphism, granulites and crustal growth: *Nature*, v. 288, p. 45–50.

NICOLAS, A., 1986, A melt extraction model based on structural studies in mantle peridotites: *J. Petrol.*, v. 27, p. 999–1022.

NICOLAS, A., AND PRINZHOFER, A., 1983, Cumulative or residual origin for the transition zone in ophiolites: structural evidence: *J. Petrol.*, v. 24, p. 188–206.

NORDLIE, B. E., 1971, The composition of the magmatic gas of Kilauea and its behavior in the near-surface environment: *Amer. J. Sci.*, v. 271, p. 417–463.

NORRY, M. J., AND FITTON, J. G., 1983, Compositional differences between oceanic and continental basic lavas and their significance: in Hawkesworth, C. J., and Norry, M. J., eds., *Continental Basalts and Mantle Xenoliths*, Shiva Publishing Ltd., Nantwich, Cheshire, U.K., p. 5–19.

NORTON, D., AND TAYLOR, H. P., 1979, Quantitative simulation of the hydrothermal systems of crystallizing magmas on the basis of transport theory and oxygen isotope data: an analysis of the Skaergaard intrusion: *J. Petrol.*, v. 20, p. 421–486.

OFTEDAHL, C., 1960, Permian rocks and structures of the Oslo region: in Holtedahl, O., ed., *Geology of Norway*, Nor. Geol. Unders., v. 208, p. 298–343.

OLIVER, H. W., 1977, Gravity and magnetic investigations of the Sierra Nevada batholith, California: *Geol. Soc. Amer. Bull.*, v. 88, p. 445–461.

O'NEILL, H. ST. C., 1981, The transition between spinel lherzolite and garnet lherzolite, and its use as a geobarometer: *Contrib. Mineral. Petrol.*, v. 77, p. 185–194.

O'NIONS, R. K., 1984, Isotopic abundances relevant to the identification of magma sources: *Philos. Trans. Roy. Soc. London*, v. A310, p. 591–603.

O'NIONS, R. K., HAMILTON, P. J., AND EVENSEN, N. M., 1977, Variations in $^{143}Nd/^{144}Nd$ and $^{87}Sr/^{86}Sr$ ratios in oceanic basalts: *Earth Planet. Sci. Lett.*, v. 39, p. 13–22.

———, 1980, The chemical evolution of the Earth's mantle: *Sci. Amer.*, v. 242, p. 120–133.

OSBERG, P. H., 1971, An equilibrium model for Buchan-type metamorphic rocks, south-central Maine: *Amer. Mineral.*, v. 56, p. 570—586.

OSBORN, E. F., AND SCHAIRER, J. F., 1941, The ternary system pseudowollastonite–akermanite–gehlenite: *Amer. J. Sci.*, v. 239, p. 715–763.

OXBURGH, E. R., 1980, Heat flow and magma genesis: in Hargraves, R. B., ed., *Physics of Magmatic Processes*, Princeton University Press, Princeton, N.J., p. 161–200.

OXBURGH, E. R., AND MCRAE, T., 1984, Physical constraints on magma contamination in the continental crust: an example, the Adamello complex: *Philos. Trans. Roy. Soc. London*, v. A310, p. 457–472.

OXBURGH, E. R., AND TURCOTTE, D. L., 1971, Origin of paired metamorphic belts and crustal dilation in island arc regions: *J. Geophys. Res.*, v. 76, p. 1315–1327.

PARSONS, B., AND MCKENZIE, D., 1978, Mantle convection and the thermal structure of the plates: *J. Geophys. Res.*, v. 83, p. 4485–4496.

PARSONS, B., AND SCLATER, J. G., 1977, An analysis of the variation of ocean floor bathymetry and heat flow with age: *J. Geophys. Res.*, v. 82, p. 803–827.

PATTERSON, E. M., 1951, A petrochemical study of the Tertiary lavas of northeast Ireland: *Geochim. Cosmochim. Acta*, v. 2, p. 283–299.

PEACOCK, S. M., 1987, Thermal effects of metamorphic fluids in subduction zones: *Geology*, v. 15, p. 1057–1060.

PECK, D. L., MOORE, J. G., AND KOJIMA, G., 1964, Temperatures in the crust and melt of the Alae lava lake, Hawaii, after the August 1963 eruption of Kilauea volcano—a preliminary report: *U.S. Geol. Surv. Prof. Pap. 501-D*, p. 1–7.

PELTIER, W. R., AND ANDREWS, J. T., 1976, Glacial isostatic adjustment: I. The forward problem: *Geophys. J. Roy. Astron. Soc.*, v. 46, p. 605–646.

PHILPOTTS, A. R., 1964, Origin of pseudotachylites: *Amer. J. Sci.*, v. 262, p. 1008–1035.

———, 1968, Igneous structures and mechanism of emplacement of Mount Johnson, a Monteregian intrusion, Quebec: *Canad. J. Earth Sci.*, v. 5, p. 1131–1137.

———, 1970, Mechanism of emplacement of Monteregian intrusions: *Canad. Mineral.*, v. 10, p. 395–410.

———, 1972, Density, surface tension and viscosity of the immiscible phase in a basic, alkaline magma: *Lithos*, v. 5, p. 1–18.

———, 1974, The Monteregian province: in Sørensen, H., ed., *The Alkaline Rocks*, John Wiley & Sons Ltd., Chichester, West Sussex, U.K., p. 293–310.

———, 1976, Silicate liquid immiscibility: its probable extent and petrogenetic significance: *Amer. J. Sci.*, v. 276, p. 1147–1177.

———, 1981, A model for the generation of massif-type anorthosites: *Canad. Mineral.*, v. 19, p. 233–253.

———, 1982, Compositions of immiscible liquids in volcanic rocks: *Contrib. Mineral. Petrol.*, v. 80, p. 201–218.

———, 1989, *Petrography of Igneous and Metamorphic Rocks*, Prentice-Hall, Inc., Englewood Cliffs, N.J., 178 p.

PHILPOTTS, A. R., AND DOYLE, C. D., 1983, Effect of magma oxidation state on the extent of silicate liquid immiscibility in a tholeiitic basalt: *Amer. J. Sci.*, v. 283, p. 967–986.

PHILPOTTS, A. R., AND MARTELLO, A., 1986, Diabase feeder dikes for the Mesozoic basalts in southern New England: *Amer. J. Sci.*, v. 286, p. 105–126.

PLATT, J. P., 1986, Dynamics of orogenic wedges and the uplift of high-pressure metamorphic rocks: *Geol. Soc. Amer. Bull.*, v. 97, p. 1037–1053.

POLLACK, H. N., AND CHAPMAN, D. S., 1977, On the regional variation of heat flow, geotherms, and lithospheric thickness: *Tectonophysics*, v. 38, p. 279–296.

POLLARD, D., 1973, Derivation and evaluation of a mechanical model for sheet intrusions: *Tectonophysics*, v. 19, p. 233–269.

POLLARD, D. D., AND JOHNSON, A. M., 1973, Mechanics of growth of some laccolithic intrusions in the Henry Mountains, Utah: II: *Tectonophysics*, v. 18, p. 311–354.

POLLARD, D. D., MULLER, O. H., AND DOCKSTADER, D. R., 1975, The form and growth of fingered sheet intrusions: *Geol. Soc. Amer. Bull.*, v. 86, p. 351–363.

PRESNALL, D. C., AND BATEMAN, P. C., 1973, Fusion relations in the system $NaAlSi_3O_8$–$CaAl_2Si_2O_8$–$KAlSi_3O_8$–SiO_2–H_2O and generation of granitic magmas in the Sierra Nevada batholith: *Geol. Soc. Amer. Bull*, v. 84, p. 3181–3202.

PRESNALL, D. C., DIXON, J. R., O'DONNELL, T. H., AND DIXON, S. A., 1979, Generation of mid-ocean ridge tholeiites: *J. Petrol.*, v. 20, p. 3–35.

PRESNALL, D. C., DIXON, S. A., DIXON, J. R., O'DONNELL, T. H., BRENNER, N. L., SCHROCK, R. L., AND DYCUS, D. W., 1978, Liquidus phase relations on the join diopside–forsterite–anorthite from 1 atm to 20 kbar: their bearing on the generation and crystallization of basaltic magmas: *Contrib. Mineral. Petrol.*, v. 66, p. 203–220.

PRESS, F., AND SIEVER, R., 1982, *Earth*, W.H. Freeman and Company, Publishers, San Francisco, 613 p.

PUTNIS, A., AND HOLLAND, T. J. B., 1986, Sector trilling in cordierite and equilibrium overstepping in metamorphism: *Contrib. Mineral. Petrol.*, v. 93, p. 265–272.

PYKE, D. R., NALDRETT, A. J., AND ECKSTRAND, O. R., 1973, Archean ultramafic flows in Munro Township, Ontario: *Geol. Soc. Amer. Bull.*, v. 84, p. 955–978.

RAMBERG, H., 1981, *Gravity, Deformation and the Earth's Crust*, 2nd ed., Academic Press, Inc. (London) Ltd., London, 452 p.

RAMBERG, I. B., 1976, Gravity interpretation of the Oslo Graben and associated igneous rocks: *Nor. Geol. Unders.*, v. 325, 194 p.

RANKIN, A. H., AND LE BAS, M. J., 1974, Liquid immiscibility between silicate and carbonate melts in naturally occurring ijolite magma: *Nature*, v. 250, p. 206–209.

READ, H. H., 1923, Petrology of the Arnage district: *J. Geol. Soc. London*, v. 79, p. 447–486.

———, 1957, *The Granite Controversy*, Thomas Murby & Co., London, 430 p.

REYNOLDS, D. L., 1954, Fluidization as a geological process and its bearing on the problem of intrusive granites: *Amer. J. Sci.*, v. 252, p. 577–613.

RHODES, J. M., DUNGAN, M. A., BLANCHARD, D. P., AND LONG, P. E., 1979, Magma mixing at mid-ocean ridges: evidence from basalts drilled near 22°N on the Mid-Atlantic Ridge: *Tectonophysics*, v. 55, p. 35–62.

RHODES, R. C., 1975, New evidence for impact origin of the Bushveld Complex, South Africa: *Geology*, v. 3, p. 549–554.

RICE, J. M., AND FERRY, J. M., 1982, Buffering, infiltration, and the control of intensive variables during metamorphism: in Ferry, J. M., ed., *Characterization of Metamorphism through Mineral Equilibria*, Reviews in Mineralogy, Vol. 10, Mineralogical Society of America, chap. 7, Washington, D.C., p. 263–326.

RICHARDSON, S. W., GILBERT, M. C., AND BELL, P. M., 1969, Experimental determination of kyanite–andalusite and andalusite–sillimanite equilibria: the aluminum silicate triple point: *Amer. J. Sci.*, v. 267, p. 259–272.

RICHTER, D. H., EATON, J. P., MURATA, K. J., AULT, W. U., AND KRIVOY, H. L., 1970, Chronological narrative of the 1959–60 eruption of Kilauea Volcano, Hawaii: *U.S. Geol. Surv., Prof. Pap.* 537-E, 73 p.

RICHTER, F. M., AND McKENZIE, D., 1981, On some consequences and possible causes of layered mantle convection: *J. Geophys. Res.*, v. 86, p. 6133–6142.

RINGWOOD, A. E., 1975, *Composition and Petrology of the Earth's Mantle*, McGraw-Hill Book Company, New York, 618 p.

ROBERTS, J. L., 1970, The intrusion of magma into brittle rocks: in Newall, G., and Rast, H., eds., *Mechanism of Igneous Intrusion*, Liverpool Geol. Soc., Geol. J. Spec. Issue No. 2, p. 287–338.

ROBERTSON, P. B., DENCE, M. R., AND VOS, M. A., 1968, Deformation in rock-forming minerals from Canadian craters: in French, B. M., and Short, N. M., eds., *Shock Metamorphism of Natural Materials*, Mono Book Corp., Baltimore, p. 433–452.

ROBIE, R. A., HEMINGWAY, B. S., AND FISHER, J. R., 1978, Thermodynamic properties of minerals and related substances at 298.15°K and 1 bar (10^5 Pascals) pressure and at high temperatures: *U.S. Geol. Surv. Bull.*, v. 1452, 456 p.

ROBINSON, P., HOLLOCHER, K. T., TRACY, R. J., AND DIETSCH, C. W., 1982, High grade Acadian regional metamorphism in south-central Massachusetts: in Joesten, R. L., and Quarrier, S. S., eds., *Guidebook for Fieldtrips in Connecticut and South Central Massachusetts*, New England Intercollegiate Geological Conference, 74th Annual Meeting, State Geological and Natural History Survey of Connecticut, Guidebook No. 5, p. 289–339.

ROBINSON, P., JAFFE, W., ROSS, M., AND KLEIN, C., JR., 1971, Orientation of exsolution lamellae in clinopyroxenes and clinoamphiboles: consideration of optimal phase boundaries: *Amer. Mineral.*, v. 56, p. 68–94.

ROEDDER, E., 1951, Low temperature liquid immiscibility in the system K_2O–FeO–Al_2O_3–SiO_2: *Amer. Mineral.*, v. 36, p. 282–286.

———, 1979, Silicate liquid immiscibility in magmas: in Yoder, H. S., Jr., ed., *The Evolution of the Igneous Rocks: Fiftieth Anniversary Perspectives*, Princeton University Press, Princeton, N.J., p. 15–57.

ROEDER, P. L., AND EMSLIE, R. F., 1970, Olivine liquid equilibrium: *Contrib. Mineral. Petrol.*, v. 29, 275–289.

ROSENFELD, J. L., 1968, Garnet rotations due to the major Paleozoic deformations in southeast Vermont: in Zen, E-an, White, W. S., Hadley, J. B., and Thompson, J. B., Jr., eds., *Studies in Appalachian Geology, Northern and Maritime*, Wiley-Interscience, New York, p. 185–202.

———, 1970, Rotated garnets in metamorphic rocks: *Geol. Soc. Amer.*, Spec. Pap. 129, 105 p.

ROSS, M. E., 1986, Flow differentiation, phenocryst alignment, and compositional trends within a dolerite dike at Rockport, Massachusetts: *Geol. Soc. Amer. Bull.*, v. 97, p. 232–240.

ROY, R. F., BLACKWELL, D. D., AND BIRCH, F., 1968, Heat generation of plutonic rocks and continental heat-flow provinces: *Earth Planet. Sci. Lett.*, v. 5, p. 1–12.

RUMBLE, D., FERRY, J. M., HOERING, T. C., AND BOUCOT, A. J., 1982, Fluid flow during metamorphism at the Beaver Brook fossil locality, New Hampshire: *Amer. J. Sci.*, v. 282, p. 886–919.

RYAN, M. P., KOYANAGI, R. Y., AND FISKE, R. S., 1981, Modeling the three-dimensional structure of macroscopic magma transport systems: application to Kilauea Volcano, Hawaii: *J. Geophys. Res.*, v. 86, p. 7111–7129.

RYERSON, F. J., AND HESS, P. C., 1978, Implication of liquid–liquid distribution coefficients to mineral–liquid partitioning: *Geochim. Cosmochim. Acta*, v. 42, p. 921–932.

SASAKI, S., AND NAKAZAWA, K., 1986, Metal silicate fractionation in the growing Earth: energy source for the terrestrial magma ocean: *J. Geophys. Res.*, v. 91, p. 9231–9238.

SAUTTER, V., JAOUL, O., AND ABEL, F., 1988, Aluminum diffusion in diopside using the $^{27}Al(p,\gamma)^{28}Si$ nuclear reaction: preliminary results: *Earth Planet. Sci. Lett.*, v. 89, p. 109–114.

SAXENA, S. K., 1973, *Thermodynamics of Rock-Forming Crystalline Solutions*, Springer-Verlag, New York, 188 p.

SCARFE, C. M., LUTH, W. C., AND TUTTLE, O. F., 1966, An experimental study bearing on the absence of leucite in plutonic rocks: *Amer. Mineral.*, v. 51, p. 726–735.

SCHAIRER, J. F., AND BOWEN, N. L., 1955, The system $K_2O-Al_2O_3-SiO_2$: *Amer. J. Sci.*, v. 253, p. 681–746.

———, 1956, The system $Na_2O-Al_2O_3-SiO_2$: *Amer. J. Sci.*, v. 254, p. 129–195.

SCHAIRER, J. F., AND YODER, H. S., JR., 1960, The nature of residual liquids from crystallization, with data on the system nepheline–diopside–silica: *Amer. J. Sci.*, v. 258A, p. 273–283.

———, 1961, Crystallization in the system nepheline–forsterite–silica at one atmosphere pressure: *Carnegie Inst. Washington*, Yearbook 60, p. 141–144.

———, 1964, Crystal and liquid trends in simplified alkali basalts: *Carnegie Inst. Washington*, Yearbook 63, p. 65–74.

SCLATER, J. G., JAUPART, C., AND GALSON, D., 1980, The heat flow through oceanic and continental crust and the heat loss of the Earth: *Rev. Geophys. Space Phys.*, v. 18, p. 269–311.

SCLATER, J. G., LAWVER, L. A., AND PARSONS, B., 1975, Comparison of long-wavelength residual elevation and free air gravity anomalies in the North Atlantic and possible implications for the thickness of the lithospheric plate: *J. Geophys. Res.*, v. 80, p. 1031–1052.

SCLATER, J. G., PARSONS, B., AND JAUPART, C., 1981, Oceans and continents: similarities and differences in the mechanisms of heat loss: *J. Geophys. Res.*, v. 86, p. 11535–11552.

SCOTT, D. R., AND STEVENSON, D. J., 1986, Magma ascent by porous flow: *J. Geophys. Res.*, v. 91, p. 9283–9296.

SELIG, F., 1965. A theoretical prediction of salt-dome patterns: *Geophysics*, v. 30, p. 633–643.

SELVERSTONE, J., 1985, Petrologic constraints on imbrication, metamorphism, and uplift in the SW Tauern Window, eastern Alps: *Tectonics*, v. 4, p. 687–704.

SELVERSTONE, J., AND SPEAR, F. S., 1985, Metamorphic $P-T$ paths from pelitic schists and greenstones from the south-west Tauern Window, eastern Alps.: *J. Metamorphic Geol.*, v. 3, p. 439–465.

SELVERSTONE, J., SPEAR, F. S., FRANZ, G., AND MORTEANI, G., 1984, High-pressure metamorphism in the SW Tauern Window, Austria: $P-T$ paths from hornblende–kyanite–staurolite schists: *J. Petrol.*, v. 25, p. 501–531.

SHAND, S. J., 1917, The pseudotachylyte of Parijs (Orange Free State), and its relation to "trap-shotten-gneiss" and "flinty-crush-rock": *Geol. Soc. London Quart. J.*, v. 72, p. 198–221.

SHAW, H. R., 1965, Comments on viscosity, crystal settling, and convection in granitic magmas: *Amer. J. Sci.*, v. 263, p. 120–152.

———, 1969, Rheology of basalt in the melting range: *J. Petrol.*, v. 10, p. 510–535.

———, 1980, The fracture mechanism of magma transport from the mantle to the surface: in Hargraves, R. B., ed., *Physics of Magmatic Processes*, Princeton University Press, Princeton, N.J., p. 201–264.

SHAW, H. R., HAMILTON, M. S., AND PECK, D. L., 1977, Numerical analysis of lava lake cooling models: Part I. Description of the method: *Amer. J. Sci.*, v. 277, p. 384–414.

SHAW, H. R., AND SWANSON, D. A., 1970, Eruption and flow rates of flood basalts: in Gilmour, E. H., and Stradling, D., eds., *Proceedings of the Second Columbia River Basalt Symposium*, Eastern Washington State College Press, Cheney, Wash., p. 271–299.

SHERIDAN, M. F., 1979, Emplacement of pyroclastic flows: a review: in Chapin, C. E., and Elston, W. E., eds., *Ash-Flow Tuffs*, Geol. Soc. Amer. Spec. Pap. 180, p. 125–138.

SHERIDAN, M. F., AND WOHLETZ, K. H., 1983, Hydrovolcanism: basic considerations and review: *J. Volcanol. Geotherm. Res.*, v. 17, p. 1–29.

SHIRLEY, D. N., 1987, Differentiation and compaction in the Palisades Sill, New Jersey: *J. Petrol.*, v. 28, p. 835–865.

SHOEMAKER, E. M., GAULT, D. E., MOORE, H. J., AND LUGN, R. V., 1963, Hypervelocity impact of steel into Coconino sandstone: *Amer. J. Sci.*, v. 261, p. 668–682.

SHORT, N. M., 1966, Effects of shock pressures from a nuclear explosion on mechanical and optical properties of granodiorite: *J. Geophys. Res.*, v. 71, p. 1195–1215.

SILVER, P. G., CARLSON, W., AND OLSON, P., 1988, Deep slabs, geochemical heterogeneity, and the large-scale structure of mantle convection: investigation of an enduring paradox: *Annu. Rev. Earth Planet. Sci.*, v. 16, p. 477–541.

SIMMONS, G., 1964, Gravity survey and geological interpretation, northern New York: *Geol. Soc. Amer. Bull.*, v. 75, p. 81–98.

SIMPSON, C., 1985, Deformation of granitic rocks across the brittle-ductile transition: *J. Struct. Geol.*, v. 7, p. 503–511.

SISSON, V. B., AND HOLLISTER, L. S., 1988, Low-pressure facies series metamorphism in an accretionary sedimentary prism, southern Alaska: *Geology*, v. 16, p. 358–361.

SKIPPEN, G. B., 1974, An experimental model for low pressure metamorphism of siliceous dolomitic marble: *Amer. J. Sci.*, v. 274, p. 487–509.

SMITH, R. B., AND CHRISTIANSEN, R. L., 1980, Yellowstone Park as a window on the Earth's interior: *Sci. Amer.*, v. 242, p. 84–95.

SMITH, R. L., 1979, Ash-flow magmatism: in Chapin, C. E., and Elston, W. E., eds., *Ash-Flow Tuffs*, Geol. Soc. Amer., Spec. Pap. 180, p. 5–27.

SMITH, R. L., AND BAILEY, R. A., 1968, Resurgent cauldrons: *Geol. Soc. Amer.*, Mem. 116, p. 613–663.

SØRENSEN, H., ed., 1974a, *The Alkaline Rocks*, John Wiley & Sons Ltd., Chichester, West Sussex, U.K., 622 p.

———, 1974b, Alkali syenites, feldspathoidal syenites and related lavas: in Sørensen, H., ed., *The Alkaline Rocks*, John Wiley & Sons Ltd., Chichester, West Sussex, U.K., p. 22–52.

SPARKS, R. S. J., 1978, The dynamics of bubble formation and growth in magmas: a review and analysis: *J. Volcanol. Geothermal Res.*, v. 3, p. 1–37.

———, 1986, The role of crustal contamination in magma evolution through geological time: *Earth Planet. Sci. Lett.*, v. 78, p. 211–223.

SPARKS, R. S. J., HUPPERT, H. E., KERR, R. C., MCKENZIE, D. P., AND TAIT, S. R., 1985, Postcumulus processes in layered intrusions: *Geol. Mag.*, v. 122, p. 555–568.

SPARKS, R. S. J., HUPPERT, H. E., AND TURNER, J. S., 1984, The fluid dynamics of evolving magma chambers: *Philos. Trans. Roy. Soc. London*, v. A310, p. 511–534.

SPARKS, R. S. J., AND MARSHALL, L., 1986, Thermal and mechanical constraints on mixing between mafic and silicic magmas: *J. Volcanol. Geotherm. Res.*, v. 29, p. 99–124.

SPEAR, F. S., AND SELVERSTONE, J., 1983, Quantitative $P-T$ paths from zoned minerals: theory and tectonic applications: *Contrib. Mineral. Petrol.*, v. 83, p. 348–357.

SPEAR, F. S., SELVERSTONE, J., HICKMOTT, D., CROWLEY, P., AND HODGES, K. V., 1984, $P-T$ paths from garnet zoning: a new technique for deciphering tectonic processes in crystalline terranes: *Geology*, v. 12, p. 87–90.

SPERA, F. J., 1980, Aspects of magma transport: in Hargraves R. B., ed., *Physics of Magmatic Processes*, Princeton University Press, Princeton, N.J., p. 265–323.

SPERA, F. J., YUEN, D. A., AND KIRSCHVINK, S. J., 1982, Thermal boundary layer convection in silicic magma chambers: effects of temperature-dependent rheology and implications for thermogravitational chemical fractionation: *J. Geophys. Res.*, v. 87, p. 8755–8767.

SPRY, A., 1969, *Metamorphic Textures*, Pergamon Press Ltd., Oxford, 350 p.

STEIGER, R., AND JÄGER, E., 1977, Subcommission on geochronology: convention on uniform decay constants in geo- and cosmochronology: *Earth Planet. Sci. Lett.*, v. 36, p. 359–362.

STOLPER, E., AND WALKER, D., 1980, Melt density and the average composition of basalt: *Contrib. Mineral. Petrol.*, v. 74, p. 7–12.

STRECKEISEN, A., 1976, To each plutonic rock its proper name: *Earth-Sci. Rev.*, v. 12, p. 1–33.

———, 1979, Classification and nomenclature of volcanic rocks, lamrophyres, carbonatites, and melilitic rocks: recommendations and suggestions of the IUGS Subcommission on the Systematics of Igneous Rocks: *Geology*, v. 7, p. 331–335.

STRENS, R. G. J., 1968, Stability of the Al_2SiO_5 solid solutions: *Mineral. Mag.*, v. 36, p. 839–849.

SUPPE, J., 1985, *Principles of Structural Geology*, Prentice-Hall, Inc., Englewood Cliffs, N.J., 537 p.

SWANSON, S. E., 1977, Relation of nucleation and crystal-growth rate to the development of granitic textures: *Amer. Mineral.*, v. 62, p. 966–978.

TAKAHASHI, E., 1986, Melting of a dry peridotite KLB-1 up to 14 GPa: implications on the origin of peridotitic upper mantle: *J. Geophys. Res.*, v. 91, p. 9367–9382.

TAKAHASHI, E., AND SCARFE, C. M., 1985, Melting of peridotite to 14 GPa and the genesis of komatiite: *Nature*, v. 315, p. 566–568.

TARLING, D. H., 1980, Lithosphere evolution and changing tectonic regimes: *Geol. Soc. London Quart. J.*, v. 137, p. 459–465.

THAYER, T. P., 1964, Principal features and origin of podiform chromite deposits and some observations on the Guleman-Soridag district, Turkey: *Econ. Geol.*, v. 59, p. 1497–1524.

THOMPSON, A. B., 1976a, Mineral reactions in pelitic rocks: I. Predictions of $P–T–X$(Fe–Mg) phase relations: *Amer. J. Sci.*, v. 276, p. 401–424.

———, 1976b, Mineral reactions in pelitic rocks: II. Calculation of some $P–T–X$(Fe–Mg) phase relations: *Amer. J. Sci.*, v. 276, p. 425–454.

THOMPSON, A. B., AND ENGLAND, P. C., 1984, Pressure temperature-time paths of regional metamorphism: II. Their inference and interpretation using mineral assemblages in metamorphic rocks: *J. Petrol.*, v. 25, p. 929–955.

THOMPSON, J. B., JR., 1957, The graphical analysis of mineral assemblages in pelitic schists: *Amer. Mineral.*, v. 42, p. 842–858.

———, 1959, Local equilibrium in metasomatic process: in Abelson, P. H., ed., *Researches in Geochemistry*, Vol. 1, John Wiley & Sons, Inc., New York, p. 427–457.

———, 1967, Thermodynamic properties of simple solutions: in Abelson, P. H., ed., *Researches in Geochemistry*, Vol. 2, John Wiley & Sons, Inc., New York, p. 340–361.

THOMPSON, R. N., MORRISON, M. A., DICKIN, A. P., AND HENDRY, G. L., 1983, Continental flood basalts . . . arachnids rule OK?: in Hawkesworth, C. J., and Norry, M. J., eds., *Continental Basalts and Mantle Xenoliths*, Birkhäuser Boston, Inc., Cambridge, Mass. (Shiva Publishing Ltd., Nantwich Cheshire, U.K.).

THORARINSSON, S., 1968, On the rate of lava- and tephra-production and the upward migration of magma in four Icelandic eruptions: *Geol. Rundsch.*, v. 57, p. 705–718.

———, 1981, Tephra studies and tephrochronology: a historical review with special reference to Iceland: in Self, S., and Sparks, R. S. T., eds., *Tephra Studies*, D. Reidel Publishing Co., Dordrecht, The Netherlands, p. 1–12.

THORNTON, C. P., AND TUTTLE, O. F., 1960, Chemistry of igneous rocks: I. Differentiation index: *Amer. J. Sci.*, v. 258, p. 664–668.

THORPE, R. S., ed., 1982, *Andesites: Orogenic Andesites and Related Rocks*; John Wiley & Sons Ltd., Chichester, West Sussex, U.K.,

TILLEY, C. E., 1948, Earlier stages in the metamorphism of siliceous dolomite: *Mineral. Mag.*, v. 28, p. 272–276.

———, 1952, Some trends of basaltic magma in limestone syntexis: *Amer. J. Sci.*, Bowen Volume, p. 529–545.

TOKSÖZ, M. N., AND HSUI, A. T., 1978, Numerical studies of back-arc convection and the formation of marginal basins: *Tectonophysics*, v. 50, p. 177–196.

TRACY, R. J., 1982, Compositional zoning and inclusions in metamorphic minerals: in Ferry, J. M., ed., *Characterization of Metamorphism through Mineral Equilibria*, Reviews in Mineralogy, Vol. 2, Mineralogical Society of America, Washington, D.C., p. 355–397.

TRACY, R. J., AND ROBINSON, P., 1980, Evolution of metamorphic belts: information from detailed petrologic studies: in Wones, D. J., ed., *The Caledonides in the U.S.A.*, Va. Polytech. Inst. State Univ. Mem. 2, p. 189–195.

TRACY, R. J., ROBINSON, P., AND THOMPSON, A. B., 1976, Garnet composition and zoning in the determination of temperature and pressure of metamorphism: central Massachusetts: *Amer. Mineral*, v. 61, p. 762–775.

TULLIS, J., AND YUND, R. A., 1987, The brittle–ductile transition in feldspathic rocks: Eos *Trans. AGU*, v. 68, p. 1464.

TURCOTTE, D. L., AND SCHUBERT, G., 1982a, *Geodynamics Applications of Continuum Physics to Geological Problems*, John Wiley & Sons, Inc., New York, 450 p.

———, 1982b, Heat transfer: in *Geodynamics, Applications of Continuum Physics to Geological Problems*, John Wiley & Sons, Inc., New York, chap. 4, p. 134–197.

TURNER, D. C., 1963, Ring-structures in the Sara-Fier complex, northern Nigeria: *Geol. Soc. London Quart. J.*, v. 119, p. 345–366.

TURNER, F. J., AND WEISS, L. E., 1963, *Structural Analysis of Metamorphic Tectonites*, McGraw-Hill Book Company, New York, 545 p.

TURNER, J. S., 1973, *Buoyancy Effects in Fluids*, Cambridge University Press, Cambridge, 368 p.

TUTTLE, O. F., AND BOWEN, N. L., 1958, Origin of granite in the light of experimental studies in the system $NaAlSi_3O_8$–$KAlSi_3O_8$–SiO_2–H_2O: *Geol. Soc. Amer.*, Mem. 74, 153 p.

TUTTLE, O. F., AND GITTINS, J., eds., 1966, *Carbonatites*, John Wiley & Sons, New York, 591 p.

VERHOOGEN, J., TURNER, F. J., WEISS, L. E., WAHRHAFTIG, C., AND FYFE, W. S., 1970, Heat sources and thermal evolution of the Earth: in *The Earth*, Holt, Rinehart and Winston, New York, chap. 12, p. 637–661.

VIDALE, R. J., 1974, Vein assemblages and metamorphism in Dutchess County, New York: *Geol. Soc. Amer. Bull.*, v. 85, p. 303–306.

VILJOEN, M. J., AND VILJOEN, R. P., 1969, Evidence for the existence of a mobile extrusive peridotitic magma from the Komati Formation of the Onverwacht group: Upper Mantle Project: *Geol. Soc. S. Africa*, Spec. Publ. 2, p. 87–112.

VOGEL, T. A., AND WILBAND, J. T., 1978, Coexisting acidic and basic melts: geochemistry of a composite dike: *J. Geol.*, v. 86, p. 353–371.

VON BARGEN, N., AND WAFF, H. S., 1986, Permeabilites, interfacial areas and curvatures of partially molten systems: results of numerical computations of equilibrium microstructures: *J. Geophys. Res.*, v. 91, p. 9261–9276.

WAFF, H. S., 1986, Introduction to special section on partial melting phenomena in Earth and planetary evolution: *J. Geophys. Res.*, v. 91, p. 9217–9221.

WAGER, L. R., AND BROWN, G. M., 1967, *Layered Igneous Complexes*; Oliver & Boyd, Edinburgh, 588 p.

WAGER, L. R., BROWN, G. M., AND WADSWORTH, W. J., 1960, Types of igneous cumulates: *J. Petrol.*, v. 1, p. 73–85.

WAGER, L. R., AND DEER, W. A., 1939, Geological investigations in East Greenland: Part III. The petrology of the Skaergaard Intrusion, Kangerdluqssuaq, East Greenland: *Meddr. Grønland*, v. 105, p. 1–352.

WALDBAUM, D. R., AND THOMPSON, J. B., JR., 1969, Mixing properties of sanidine crystalline solutions: *Amer. Mineral.*, v. 54, p. 1274–1298.

WALKER, D., 1986, Melting equilibria in multicomponent systems and liquidus/solidus convergence in mantle peridotite: *Contrib. Mineral. Petrol.*, v. 92, p. 303–307.

WALKER, D., AND DELONG, S. E., 1982, Soret separation of mid-ocean ridge basalt magma: *Contrib. Mineral. Petrol.*, v. 79, p. 231–240.

WALTHER, J. V., AND ORVILLE, P. M., 1982, Volatile production and transport in regional metamorphism: *Contrib. Mineral. Petrol.*, v. 79, p. 252–257.

WALTHER, J. V., AND WOOD, B. J., 1984, Rate and mechanism in prograde metamorphism: *Contrib. Mineral. Petrol.*, v. 88, p. 246–259.

WATERS, A. C., 1955, Volcanic rocks and the tectonic cycle: *Geol. Soc. Amer.*, Spec. Pap. 62, p. 703–722.

WATSON, E. B., 1976, Two-liquid partition coefficients: experimental data and geochemical implications: *Contrib. Mineral. Petrol.*, v. 56, p. 119–134.

———, 1982, Melt infiltration and magma evolution: *Geology*, v. 10, p. 236–240.

WATSON, E. B., AND BRENAN, J. M., 1987, Fluids in the lithosphere: 1. Experimentally-determined wetting characteristics of CO_2–H_2O fluids and their implications for fluid transport, host-rock physical properties, and fluid inclusion formation: *Earth Planet. Sci. Lett.*, v. 85, p. 497–515.

WEIGAND, P. W., AND RAGLAND, P. C., 1970, Geochemistry of Mesozoic dolerite dikes from eastern North America: *Contrib. Mineral. Petrol.*, v. 29, p. 195–214.

WEILL, D. F., HON, R., AND NAVROTSKY, A., 1980, The igneous system $CaMgSi_2O_6$–$CaAl_2Si_2O_8$–$NaAlSi_3O_8$: variations on a classic theme by Bowen: in Hargraves, R. B., ed., *Physics of Magmatic Processes*, Princeton University Press, Princeton, N. J., p. 49–92.

WHITE, A. J. R., AND CHAPELL, B. W., 1983, Granitoid types and their distribution in the Lachlan Fold Belt, southeastern Australia: *Geol. Soc. Amer. Mem. 159*, p. 21–34.

WHITE, N., 1989, Nature of lithospheric extension in the North Sea: *Geology*, v. 17, p. 111–114.

WHITE, W. M., DUPRÉ, B., AND VIDAL, P., 1985, Isotope and trace element geochemistry of sediments from the Barbados Ridge–Demerara Plain region, Atlantic Ocean: *Geochim. Cosmochim. Acta*, v. 49, p. 1875–1886.

WHITNEY, J. A., 1988, The origin of granite: the role and source of water in the evolution of granitic magmas: *Geol. Soc. Amer. Bull.*, v. 100, p. 1886–1897.

WICKHAM, S. M., AND OXBURGH, E. R., 1985, Continental rifts as a setting for regional metamorphism: *Nature*, v. 318, p. 330–333.

WILCOX, R. E., 1954, Petrology of Paricutin volcano, Mexico: *U.S. Geol. Surv. Bull.*, v. 965-C, p. 281–353.

WILLIAMS, A. F., 1932, *The Genesis of Diamond*, Ernest Benn Ltd., London, 2 vols., 636 p.

WILLIAMS, H., AND MCBIRNEY, A. R., 1979, *Volcanology*, Freeman, Cooper & Company, San Francisco, 397 p.

WILSON, C. J. N., 1984, The role of fluidization in the emplacement of pyroclastic flows: 2. Experimental results and their interpretation: *J. Volcanol. Geotherm. Res.*, v. 20, p. 55–84.

WILSON, J. R., AND LARSEN, S. B., 1985, Two-dimensional study of a layered intrusion—The Hyllingen Series, Norway: *Geol. Mag.*, v. 122, p. 97–124.

WILSON, J. T., 1973, Mantle plumes and plate motions: *Tectonophysics*, v. 19, p. 149–164.

WILSON, L., AND HEAD, J. W., III, 1981, Ascent and eruption of basaltic magma on the Earth and Moon: *J. Geophys. Res.*, v. 86, p. 2971–3001.

WINKLER, H. G. F., 1949, Crystallization of basaltic magma as recorded by variations of crystal size in dikes: *Mineral. Mag.*, v. 28, p. 557–574.

WINTSCH, R. P., 1975, Feldspathization as a result of deformation: *Geol. Soc. Amer. Bull.*, v. 86, p. 35–38.

WOOD, B. J., AND FRASER, D. G., 1976, *Elementary Thermodynamics for Geologists*, Oxford University Press, Oxford, 303 p.

WOOLSEY, T. S., MCCALLUM, M. E., AND SCHUMM, S. A., 1975, Modeling of diatreme emplacement by fluidization: *Phys. Chem. Earth*, v. 9, p. 29–42.

WRIGHT, T. L., AND DOHERTY, P. C., 1970, A linear programming and least squares computer method for solving petrologic mixing problems: *Geol. Soc. Amer. Bull.*, v. 81, p. 1995–2008.

WRIGHT, T. L., AND FISKE, R. S., 1971, Origin of the differentiated and hybrid lavas of Kilauea volcano, Hawaii: *J. Petrol.*, v. 12, p. 1–65.

WYLLIE, P. J., 1977, Crustal anatexis: an experimental review: *Tectonophysics*, v. 13, p. 41–71.

———, 1979, Petrogenesis and physics of the Earth: in Yoder, H. S., Jr., ed., *The Evolution of the Igneous Rocks: Fiftieth Anniversary Perspectives*, Princeton University Press, Princeton, N. J., p. 483–520.

———, 1980, The origin of kimberlites, *J. Geophys. Res.*, v. 85, p. 6702–6910.

———, 1988, Magma genesis, plate tectonics, and chemical differentiation of the Earth: *Rev. Geophys.*, v. 26, p. 370–404.

WYLLIE, P. J., AND HUANG, W. L., 1976, Carbonation and melting reactions in the system CaO–MgO–SiO_2–CO_2 at mantle pressures with geophysical and petrological applications: *Contrib. Mineral. Petrol.*, v. 54, p. 79–107.

WYLLIE, P. J., AND TUTTLE, O. F., 1960, The system CaO–CO_2–H_2O and the origin of carbonatites: *J. Petrol.*, v. 1, p. 1–46.

YOCHELSON, E. L., ed., 1980, The scientific ideas of G. K. Gilbert: *Geol. Soc. Amer.*, Spec. Pap. 183, 148 p.

YODER, H. S., JR., 1965, Diopside–anorthite–water at five and ten kilobars and its bearing on explosive volcanism: *Carnegie Inst. Washington Yearbook 64*, p. 82–89.

———, 1976, *Generation of Basaltic Magma*, National Academy of Sciences, Washington, D. C., 265 p.

YODER, H. S., JR., STEWART, D. B., AND SMITH, J. R., 1957, Ternary feldspars: *Carnegie Institution of Washington Yearbook 55*, p. 206–214.

YODER, H. S., JR., AND TILLEY, C. E., 1962, Origin of basalt magmas: an experimental study of natural and synthetic rock systems: *J. Petrol.*, v. 3, p. 342–532.

ZEN, E-AN, 1966, Construction of pressure temperature diagrams for multicomponent systems after the method of Schreinemakers—a geometric approach: *U.S. Geol. Surv. Bull.*, v. 1225, 56 p.

Chapter 1

1-2 1.174 km   **1-3** 0.557 μW m^{-3} **1-7** 68.12 mW m^{-2}

Chapter 2

2-1 $\bar{V}_{SiO_2} = 26.11 \times 10^{-6}$ m^3 mol^{-1}, $\bar{V}_{CaO} = 18.68 \times 10^{-6}$ m^3 mol^{-1}
2-3 $\rho_{basalt} = 2.7$ Mg m^{-3}, $\rho_{granite} = 2.3$ Mg m^{-3}
2-7 Max. V at top = 6.27 m s^{-1} **2-9** Max. V at top = 1.52 m s^{-1}

Chapter 3

3-1 49.75 km   **3-5a** 47 m s^{-1} **b** 1.65 m   **3-8** 3.72 m

Chapter 4

4-1 0.055 m s^{-1}, laminar   **4-5a** 27.25   **b** 4.5 $\times$ 10^{-3} m s^{-1}
c Intrusion 73.8 m s^{-1}; subsidence 2.7 m s^{-1} **4-7a** 17.6 m

Chapter 5

5-1 Continental 30°C km^{-1}, ocean ridge 160°C km^{-1}
5-3 5.85 years   **5-5** 390 years   **5-7** 125.2 days
5-12 7.3 s

Chapter 7

7-1 1 $\times$ 10^4 J   **7-8** 1.22 $\times$ 10^{-8} °C Pa^{-1} 0.287°C km^{-1}
7-10 0 J mol^{-1}, equilibrium

Chapter 8

8-2 1894.9 K   **8-4a** +9276 J mol^{-1} **8-8a** T_{eq} = 570 K

Chapter 9

9-4 15.319 kJ mol^{-1} **9-5a** $X_{Or}^{Ab_{ss}} = 0.113$ $X_{Or}^{Or_{ss}} = 0.887$

Chapter 10

10-1b 1337°C, 35.5 wt % An   **10-2b** $\gamma_{Di} = 0.71$
10-4c 84% K-feldspar   **10-6** 122.57 kJ mol^{-1}
10-12e 49% Ab, 51% Ne   **10-19** 1100°C

Chapter 11

11-1 0.61   **11-4a** 0.423   **b** 0.69   **11-7a** 1225 K
11-9 87.5%   **11-12a** 40% Fe$_2$O$_3$, 60% FeO
11-14 1000°C 10$^{-10.5}$ bar

Chapter 12

12-1 $r_c = 0.25$ μm   **12-4** 0.86   **12-5c** -34.3°C

Chapter 13

13-3 Fo$_{77}$ **13-5a** 311.7 Pa s   **13-8** 2.7 Mg m^{-3}, 3.32 Pa
13.9b 1.9 km day^{-1} **13-10b** 0.436 mm day^{-1}
13-14 1034°C   **13-16c** -0.316 mm
13-20e 60 wt % plagioclase

Chapter 14

14-2 Alkali basalt 3.6%, tholeiite 21.5%   **14-7** 72.5
14-8 $\log_{10} a_{SiO_2} = -0.634$ **14-12** 5.2 cm
14-13b 1.96 cm s^{-1} **14-14b** 8.16 km

Chapter 15

15-1 $V_{fluid} = \frac{1}{5} V_{rock}$

Chapter 16

16-4a 0.178 GPa, $\gamma = 1.69$ **16-5a** 0.22 GPa, $\gamma = 1.98$
16-10 0.048 mol

Chapter 17

17-7   0.40 GPa, 570°C

Chapter 18

18-1   10%   **18-2**   19%   **18-4** 7.67×10^{-16} m^2
18-6   11 Ma   **18-9a** $\Delta\mu_{SiO_2} = -4057$ J   **18-10c**   1108 cm^3

Chapter 19

19-1 5.6×10^{-3} **19-3**   31 MJ

Chapter 20

20-1   36.66   **20-2h**   30 km

Chapter 21

21-2   0.193   **21-5**   At 2.5 Ga, 0.5094   **21-7**   At 2.5 Ga, 0.7015

Chapter 22

22-3   191 Ma   **22-5**   0.6°C km^{-1} **22-7** -42°C
22-9   339 m   **22-11**   0.5 m a^{-1} **22-12**   0.54 m s^{-1}

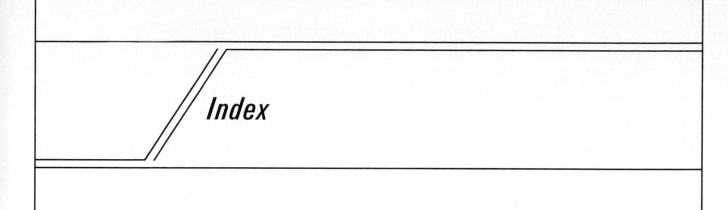

Index

A